SECTION 2:
Tree Use and Management

Preface

The phrase "Plant the right tree in the right place" has been around for a number of years. The process of putting this seemingly simple idea into practice may not be as well understood as we would like. Only a few crude tools have been developed to help select and plant proper trees in urban and suburban landscapes. Few, if any, present the topic in a detailed, complete package. The goal of this book is to provide this detail and to guide you through the process of evaluating a planting site, choosing trees that are best suited for the site, and planting and managing them after installation.

The details presented are based on many years of accumulated research conducted by a variety of people from around the world, and on many interviews and discussions with practitioners. A glance at the bibliography will help you appreciate the breadth of knowledge represented in this work. This research has been synthesized and combined with personal work experience across the United States and Canada.

A properly chosen tree that is correctly planted can provide shade and beauty for a long time. A poorly suited tree or one that is improperly planted may die soon after planting. One that is not suited to a site may need regular pruning or pest control, or it could damage curbs, sidewalks, patios, and other structures. Fruit may create a slippery goo on the sidewalk that could become a hazard if the improper tree is chosen for the site.

Most of the commonly planted trees across North America, Puerto Rico, and the Caribbean Islands are presented in this work. Other, lesser known trees are also discussed in hopes that some will become more frequently used. Those who choose, care for, or plant trees for any type of site will find useful information in this book. Many of the principles discussed here can be applied to sites in Europe, South America, and other places in the world.

ACKNOWLEDGMENTS

This was a big project. Although many people helped me accumulate the information within this book, I take all responsibility for its synthesis and interpretation. Your experiences may differ from the information presented within these pages. If so, I encourage you to let me know so that we can all learn from your experience.

I developed a portion of this material during a six-month stay at The Morton Arboretum in Lisle, Illinois. I cannot express enough thanks to the staff at the Arboretum, especially Gary Watson.

The following people provided me with constructive comments by reviewing portions of this manuscript. Their input is much appreciated and made this project more relevant: Dr. B. Appleton, Virginia Polytechnical Institute; Dr. L. Costello, University of California; E. B. Gilman, my wife; Dr. R. Harris (retired), University of California; P. Kelsey, Morton Arboretum; Dr. T. O. Perry (retired), North Carolina State University; Dr. D. Struve, Ohio State University; Dr. G. Ware, Morton Arboretum; Dr. Gary W. Watson, Morton Arboretum; D. Weigle, University of Florida.

Thanks to the following people for reviewing portions of the tree text: Dr. B. L. Appleton, horticulturist; Sharon Dolliver, urban forester; Larry Morris; Dr. W. Fountain, horticulturist; Dr. V. D. Ammon, plant pathologist; Dr. T. E. Nebeker, entomologist; D. Slater, urban forester; Dr. T. G. Ranney, horticulturist; Clark Beavans, urban forester; Steve Scot, urban forester; D. Price, urban forester; G. Russel, urban forester; N. Letson, urban forester; C. Weber, urban forester; Dr. E. L. Barnard, plant pathologist; Dr. W. Dixon, entomologist; Dr. C. H. Gilliam, horticulturist; D. Fare, horticulturist. Many other people also contributed to the succesful completion of this project. Thanks to all of you!

Thanks to the following people for taking time out of their schedules to help build the data base that eventually became this book: Dr. Kim Coder, University of Georgia, Athens, Georgia; Dr. Donna Fare, Tennessee Tech University, Cookeville, Tennessee; Dr. Don Gardener, Savannah, Georgia; Howard Garrett, Dallas, Texas; Don Hodel, Los Angeles, California; Way Hoyt, Ft. Lauderdale, Florida; Roger Huff, Norfolk, Virginia; Sam Knapp, Riverside, California; Tom Larson, Laguna Hills, California; David Morgan, Lubbock, Texas; Joanne Phillips, Tallahassee, Florida; J. C. Raulston, North Carolina State University, Raleigh, North Carolina; Benny Simpson, Texas A&M University, Dal-

TREES

for Urban and Suburban Landscapes

Online Services

Delmar Online
To access a wide variety of Delmar products and services on the World Wide Web, point your browser to:
http://www.delmar.com
or email: info@delmar.com

thomson.com
To access International Thomson Publishing's home site for information on more than 34 publishers and 20,000 products, point your browser to:
http://www.thomson.com
or email: findit@kiosk.thomson.com

Internet

for Urban and Suburban Landscapes

Edward F. Gilman

University of Florida, Gainesville
Environmental Horticulture Department

an International Thomson Publishing company

Albany • Bonn • Boston • Cincinnati • Detroit • London • Madrid
Melbourne • Mexico City • New York • Pacific Grove • Paris • San Francisco
Singapore • Tokyo • Toronto • Washington

NOTICE TO THE READER

Publisher does not warrant or guarantee any of the products described herein or perform any independent analysis in connection with any of the product information contained herein. Publisher does not assume, and expressly disclaims, any obligation to obtain and include information other than that provided to it by the manufacturer.

The reader is expressly warned to consider and adopt all safety precautions that might be indicated by the activities herein and to avoid all potential hazards. By following the instructions contained herein, the reader willingly assumes all risks in connection with such instructions.

The publisher makes no representation or warranties of any kind, including but limited to, the warranties of fitness for particular purpose or merchantability, nor are any such representations implied with respect to the material set forth herein, and the publisher takes no responsibility with respect to such material. The publisher shall not be liable for any special, consequential, or exemplary damages resulting, in whole or in part, from the readers' use of, or reliance upon, this material.

Cover art courtesy of Don Almquist
Cover Design: Christina Almquist

Delmar Staff
Publisher: Tim O'Leary
Acquisitions Editor: Cathy L. Esperti
Senior Project Editor: Andrea Edwards Myers
Production Manager: Wendy A. Troeger
Marketing Manager: Maura Theriault

Printed in the United States of America

For more information, contact:

Delmar Publishers
3 Columbia Circle, Box 15015
Albany, New York 12212-5015

Intemational Thomson Publishing Europe
Berkshire House 168-173
High Holbum
London WC I V7AA
England

Thomas Nelson Australia
102 Dodds Street
South Melbourne, 3205
Victoria, Australia

Nelson Canada
1120 Birchmount Road
Scarborough, Ontario
Canada M1K 5G4

Intemational Thomson Editores
Campos Eliseos 385, Piso 7
Col Polanco
11560 Mexico D F Mexico

International Thomson Publishing Gmbh
Königswinterer Strasse 418
53227 Bonn
Germany

Intemational Thomson Publishing Asia
221 Henderson Road #05-10
Henderson Building
Singapore 0315

Intemational Thompson Publishing-Japan
Hirakawacho Kyowa Building, 3F
2-2-1 Hirakawacho
Chiyoda-ku, 102 Tokyo
Japan

1 2 3 4 5 6 7 8 9 10 XXX 02 01 00 99 98 97

Library of Congress Cataloging-in-Publication Data
Gilman, Edward F.
Trees for urban and suburban landscapes / Edward F. Gilman.
p. cm.
Includes indexes.
ISBN 0-8273-7053-9 (textbook)
1. Ornamental trees. 2. Trees in cities. 3. Ornamental trees—United States. 4. Trees in cities—United States. I. Title.
SB435.5.G55 1996
635.9'77—dc20

96-17000
CIP

Table of Contents

SECTION 1: Introduction

las, Texas; Tom Smiley, Bartlett Tree Research Laboratory, Charlotte, North Carolina; Ted Stamen, Riverside, California; Michael Walterscheidt, Texas A&M University, Dallas, Texas; Dr. Gary Watson, Morton Arboretum, Lisle, Illinois; Chuck Weber, Huntsville, Alabama; Dr. Will Witty, University of Tennessee, Knoxville, Tennessee; and many others.

Special thanks to Robin Morgan, USDA Forest Service, Atlanta, Georgia, for planting the seed for this project.

The help and guidance of Phyllis Stambaugh and Diane Weigle, Environmental Horticulture Department, University of Florida, through the early parts of this project are much appreciated.

Thanks to the following people for providing several of the photographs:
Dr. Benny Simpson, Texas A&M University: *Acacia wrightii, Arbutus texana, Leucana retusa,* and *Prunus mexicana;* Way Hoyt, arborist, Ft. Lauderdale, Florida: *Tabebuia heptophylla;* Gene Joyner: *Bulnesia arbora, Cresentia cujete, Bischofia javanica.*

I hope you enjoy reading and using this book.

DEDICATION

To all the dedicated teachers in my life, including and especially my wife, Betsy. And to my children, Samantha and Megan, and Franklin B. Flower, Spencer H. Davis, Ida A. Leone, and Pat and Cliff Gilman.

SECTION 1

Introduction

Chapter 1

Selecting the Right Tree for the Right Place

INTRODUCTION

Selecting the right tree for the right place can save time, money, and later disappointments. As with any major investment, it is crucial to take the time to investigate and invest wisely. Choosing a tree for a particular site is no exception.

One method of choosing a tree for a particular site is to drive around town to find out which species grow well in landscapes with similar site attributes. A problem with this approach is that most people do not do it; when they overdo it, the result is urban landscapes with little species diversity. Another problem with this approach is that the soil conditions at your planting site may be different from visually similar sites around town. Many professionals who specify trees for urban and suburban landscapes visit arboreta and botanic gardens. This is good because it potentially brings new plants to our urban landscapes. Others rely on books and computer software to choose trees. This is reasonable, but the specific planting site must first be evaluated to determine the cultural and physical attributes required of trees at the site.

Trees grow when selected according to plant requirements and site conditions. To maintain the desired habit or form to fulfill a design function, cultural requirements must be met. The site has to be evaluated to understand the cultural requirements. To accomplish a complete site evaluation, one must visit the site and get dirty. There is no substitute. Trees that could grow on the site can then be listed using various references, including this book. Inclusion of ornamental attributes, such as shape, color, texture, and the like, then narrows the list to a reasonable number of trees to select for planting. Easy! Right? Well, let's see.

There are five components to choosing trees for a planting site. Begin by evaluating site attributes, potential site modifications, and tree maintenance (management) capabilities. Then choose desirable tree attributes and select appropriate trees for the site. It is a process that many people do not undertake. Appendix 1 contains a site evaluation form to help guide you through the evaluation process. At the very minimum, always evaluate the above-ground and below-ground attributes discussed in the following pages. Review the other sections of this book when appropriate for the planting site.

EVALUATE CRUCIAL SITE ATTRIBUTES

Hardiness

Tree adaptations to regions of the country are designated by their hardiness zones (see Section II introduction). It is not a perfect system of determining whether a tree will grow in a particular site, but it is the best available at this time. The hardiness zone map, developed by the United States Department of Agriculture, specifies the average lowest winter temperature expected for most regions of North America. It is a fairly reliable guide in this respect.

This book includes a range of hardiness values for each tree. The smallest number in the range indicates the coldest hardiness zone tolerated by the tree. If the tree is planted in a colder zone, it is likely to be damaged or killed by the cold temperatures. The southern, or warmest, zone represented by the highest number in this range is less reliable because plants have not been extensively tested for this attribute. In other words, some trees will be able to tolerate the climate in a higher (warmer) zone number than indicated in the range. High day and night temperatures combined with high relative humidity are thought to limit the southern ranges of some trees.

When choosing trees for a planting site, first note the hardiness zone number of the planting site on the hardiness map. Trees with a hardiness zone range that includes this number are best suited for the site. You might plant one or two trees beyond the recommended planting range as a test now and then, but do not plant large numbers of trees out of the recommended range. This type of species testing is one of the functions of botanic gardens and arboreta. Some universities also conduct such testing. Visit one near you to evaluate the hardiness of a plant in question.

A number of trees grow poorly in hardiness zones 9 and 10 in Florida and other subtropical areas in the Caribbean region, but do nicely in these same zones in California. This may be due to cooler summer temperatures in California's coastal areas and lower relative humidity. The warmest zone indicated in this book for these trees is usually 8 or 9, indicating that they are not suited for planting in parts of the deep southeastern United States. A note in parentheses is included for these trees if they grow well in southern California and Arizona (see, for instance, *Cedrus deodara*).

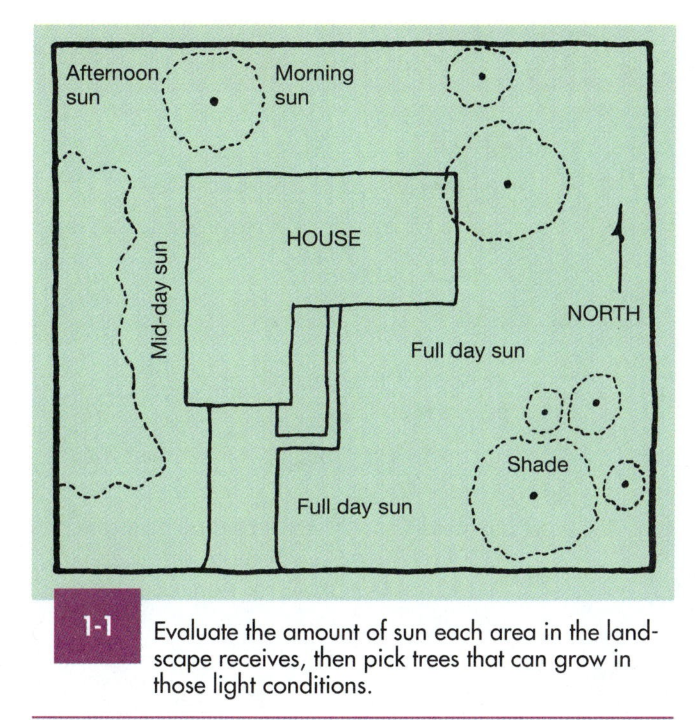

1-1 Evaluate the amount of sun each area in the landscape receives, then pick trees that can grow in those light conditions.

Above-ground site analysis

Light exposure Note how many hours of direct sun the planting site receives in the summer (Figure 1-1). Remember to account for the seasonal change in the sun angle when evaluating sites in other seasons. Trees requiring full sun need at least six hours of direct sun, but all-day sun often produces the best form and growth. Trees suited for full sun to partial sun/partial shade will adapt to a site receiving three to six hours of direct sun. Shade-loving trees are adapted to sites with filtered sun or filtered shade, or those receiving less than three hours of direct sun. Most large trees grow best in full sun. Some small trees grow best in sites that receive shade for part of the day.

Some planting sites in highly urbanized areas are close to tall buildings with reflective surfaces such as glass. Sunlight reflected from glass or a white wall increases the heat load on a tree planted near a building. Even though the site might receive less than full sun because of shadows cast by buildings, the plant must tolerate a high heat load during the sunny part of the day. Therefore, drought-tolerant trees that grow in full sun *and* partial shade are best suited for such a site. In addition, providing a large area of soil for roots to explore often helps trees withstand reflected light by providing more soil from which to absorb water. Leaves on trees receiving reflected light in a restricted soil area often turn brown along the edges in summer (called marginal leaf scorch) unless provided with ample water.

Slope exposure Trees with thin bark, such as cherries, plums, maples, mountain ash, and others, transplant poorly on southern and western slopes, especially at higher elevations. Bark often splits from direct sun exposure and desiccation. Trunks are sometimes coated with white latex paint to reflect light and help prevent this injury. There are also trunk wraps, polypropylene tubes, and other protective devices designed to reduce this damage. Transpiration and evaporation from the soil are also enhanced on south and west aspects, making it more difficult to maintain adequate soil moisture. Because of this, plan on providing more irrigation to southern and western exposures to help prevent desiccation. Northern slopes are more protected from direct sun exposure, so the soil there stays moist longer.

Wind Wind increases the amount of water lost from a tree to the atmosphere. This may not be a big problem for drought-tolerant trees if roots can explore surrounding soil uninhibited by urban structures such as compacted soil, curbs, buildings, and streets. However, trees growing in restricted soil spaces are especially susceptible to wind desiccation in summer, because the root system is too confined. Good examples of this type of site are street trees planted in a sidewalk cutout (planting wells in a sidewalk) with tall buildings nearby. Unless the site is properly designed (see Chapter 2), or the soil is sandy, roots frequently concentrate in the small area of the cutout because they are unable to extend beneath the pavement due to soil compaction or waterlogged conditions. Because the roots are concentrated in a small area, the soil in the root zone dries quickly. Such trees could also become unstable and blow over. Well-managed irrigation can overcome some of this water deficit, but irrigation is difficult to deliver to trees in highly urbanized sites, unless the irrigation system was properly designed and installed and has been well maintained.

The best method for managing water loss in a windy site is with proper site design (see Chapter 2) and proper species selection. Species tolerant of drought (see Appendices 7, 8, and 9) usually grow best in windy areas unless soil is poorly drained. Species that tolerate both wet soil and drought are best suited for poorly drained, windy sites (see Appendix 2). In wet soil, ill-suited trees often appear drought stressed because oxygen is unavailable for root growth. The underdeveloped root system cannot absorb enough water to keep pace with transpiration demands.

Salt Airborne salt can affect trees through twigs and foliage, or through roots after it is deposited on the ground and leaches into the soil. Trees planted within one-eighth mile of salt-water coastlines should possess some degree of tolerance to aerosol salt spray. Those exposed to direct spray along the dunes should be highly salt-tolerant. Salt-tolerant trees are often deformed by direct exposure to salty air, but they survive and grow. Foliage on salt-sensitive trees burns, so such trees become deformed and grow poorly when exposed to salty air.

Trees within several hundred feet of major roadways that receive de-icing salt regularly should also be tolerant of aerosol salt spray. Those planted adjacent to salted roadways or parking lots should be tolerant of soil salt because salty water splashes or runs off from pavement into the root zone. Appendix 3 lists salt-tolerant trees.

Overhead wires Trees are often planted too close to power lines. When branches reach wires, the utility company must prune or trim them to ensure uninterrupted utility service. Unfortunately, this costs utility companies

TABLE 1-1 PLANTING TREES WITHIN 40 FEET OF WIRES OR STREET LIGHTS

IF: planting hole is this distance from wire/light	**AND IF:** you would like to train a tall tree over the wire/light*	**AND IF:** you would like to keep branches away from wires/lights
	THEN: Pick trees with these characteristics	
0–6 feet	not recommended	mature height about 10 feet less than wire/light height
6–18 feet	expected height at least twice wire/light height	mature height about 10 feet less than wire/light height, or
18–40 feet	not recommended	mature height about 10 feet less than wire/light height or mature canopy spread about 10 feet less than twice the distance to the wire/light
more than 40 feet	——— any tree can be planted ———	

* If this is the intention, be sure all interested parties are made aware of the plans, including the utility company, urban forester, tree pruning contractor, property owner, and so on. This is a maintenance-intensive strategy, and may not be suited for cities that receive ice storms.

(and ultimately the customers) more than $1 billion (1995 dollars) each year in the United States. Costs could be lowered by planting only properly sized trees near wires (Table 1-1 and Figure 1-2). It is best to plant trees as far away from wires as possible. Planting farther from the wire will allow you to choose from a greater variety of trees.

1-2 Only small trees should be planted beneath wires; trees with narrow canopies can be planted within 40 feet of the wire.

Some communities that do not receive ice and heavy snow storms allow trees to be trained to grow well over wires. This allows planting of large-maturing trees close to the wires. Be sure to notify the utility companies in your area before embarking on this practice. They will have to coordinate and train the pruning crews on the specialized techniques required for this kind of planting.

Street and security lights Planting a tree that requires regular pruning to keep it away from a street or security light is a costly malpractice. If the tree management budget does not allow for training large trees away from or over lights, plant them at least 30 feet away to minimize interference with the light. Large trees planted too close will require regular pruning to prevent foliage from blocking the light. Many people forget the option of moving a light so that a tree can be planted or maintained in a particular spot without regular pruning. Moving the light can often provide a permanent solution to a potential conflict between lights and trees. Follow the recommendations for the situation that most closely matches your planting site.

Lights less than 15 feet away.
Lights shorter than 20 feet. Train large-maturing trees (Appendix 9) planted close to a light so that they grow over it, allowing the light to shine beneath the canopy (Table 1-1). The people responsible for maintaining the tree after planting should be alerted to the early pruning needed to implement this plan. If not, the tree might be improperly pruned to keep it under the light. This creates recurring pruning costs and provides no shade to the area. Medium-sized trees might be poorly suited for areas near low lights because they block the light and branches may not grow high enough to train over the light. Very small trees (10 to 15

feet tall) might be suited if the canopies will not reach the light fixture. Trees with a narrow canopy are well suited for planting here provided they will not grow into the light.

Lights 20 to 40 feet tall. Large and medium-sized trees may not be suited for planting close to a fixture 20 to 40 feet tall because branches will grow into and block the light. Keeping the tree pruned away from the light will be a regular chore. Small trees are better suited for planting close to lights 20 to 40 feet tall.

Lights more than 40 feet tall. Small and medium-sized trees are well suited for planting near tall light fixtures. Their canopies will not grow into or block the light. Large-maturing trees can be planted if root space is limited, or other site conditions will prevent trees from reaching their mature size.

Lights 15 to 30 feet away.

Choose trees that are narrow enough so they will not grow into the light. Use the guidelines in Table 1-1. If you choose to disregard this advice and plant trees that will grow into lights, select those with an open canopy so light can shine through. Trees such as callery pears and oaks with dense canopies are often pruned destructively near lights to provide adequate illumination and safety.

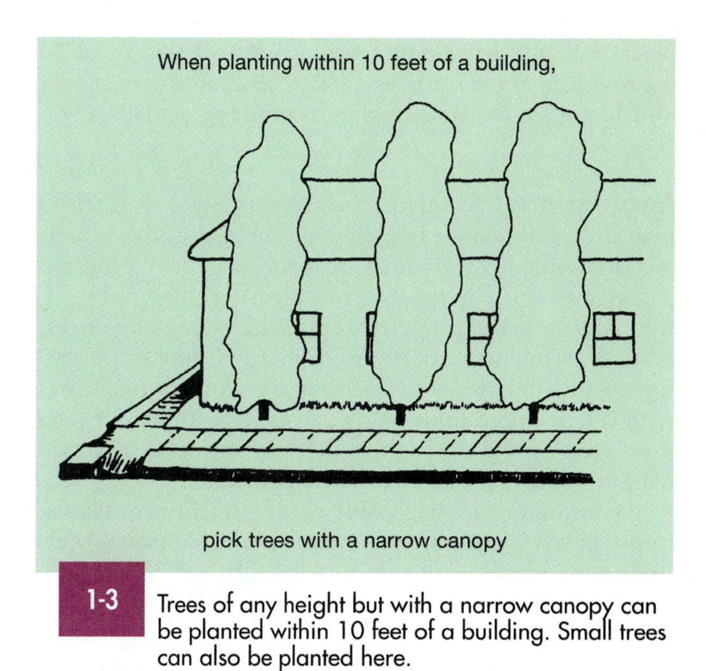

1-3 Trees of any height but with a narrow canopy can be planted within 10 feet of a building. Small trees can also be planted here.

Buildings Trees are most stable in the ground when they develop a uniform root system distributed more or less evenly around the tree. When roots meet a building wall, they are deflected laterally; in well-drained soils, some may grow downward. If a tree is close to a building, the root system can become one-sided and unbalanced. Large-maturing trees planted within 10 feet of a one- or two-story building could become hazardous and blow over because of an unbalanced root system, especially if wind blows from the building side of the tree. Trees growing in compacted and poorly drained soil are most susceptible to this because there are few (if any) deep roots helping to support the tree in an upright position.

A tree with an open or narrow canopy is probably the most appropriate choice within 10 feet of a building (Appendix 10). Plant a grove of them to create a closed canopy of shade in this area, if you wish (Figure 1-3). Small or medium-sized trees are also suitable for planting this close to a building.

Signs Signs and trees sometimes conflict with each other. To help prevent this, plant small trees near tall signs and large trees near low signs. Large-maturing trees could be in the way of a low sign for several years after planting, but lower limbs can be thinned and eventually removed so the sign remains visible. A tall canopy over a sign directs people's eyes to the sign.

Trees can block building facades, interesting architectural features, and signs in downtown areas. There are several solutions to this potential problem. Large-sized nursery stock can be purchased which has most branches taller than low signs. Trees with a thin canopy can be selected to allow people to see. Trees with an upright or columnar habit will not spread to cover adjacent buildings and signs. Trees can be regularly thinned to allow people to see through them. The signs on the buildings can be lowered to make them visible under the tree canopy. The island of Hilton Head, South Carolina has enacted ordinances which require signs to be low. This is one reason that the tree canopy of this island is so extensive.

Other trees Young trees with a decurrent growth habit (oaks, maples, and other trees with a broad canopy) that require full or partial sun, if planted under or near the canopy of established trees, often bend toward the sunlight and develop a one-sided canopy. This may be the intention in some cases and could help create a special effect. In other situations, shade-tolerant trees are better suited for planting in the shade of established trees because they often develop a more uniform canopy.

Chemicals produced in the leaves, trunk, roots and fruit of some trees slow or prevent growth of other plants. This is called allelopathy. For example, walnut trees produce a substance called juglone that inhibits trees, shrubs, and herbaceous plants from growing near the trees. It might be difficult to establish other trees nearby. Sycamore, eucalyptus, and hackberry inhibit growth of grasses and herbaceous plants. Sumacs inhibit Douglas fir. Redwood bark mulch can slow root growth on some trees. Some grasses may also inhibit growth of trees. Much of this is not well understood and we have a great deal to learn about this phenomenon.

Vandalism People sometimes intentionally destroy or injure trees after planting. Trees may be unintentionally injured in downtown planting sites such as sidewalk cutouts, where many people regularly walk close to the trees. When planting in an area where trees may be vandalized or injured, consider eliminating trees that have thin bark (e.g., red maple). Trees with thin bark are identified in Section II. In addition, it is best to purchase trees with a trunk at least 1.5 to 2 inches in diameter, as trees of this size appear to perform better than smaller ones in areas where vandalism is a problem. Some landscape architects choose trees at least 4 inches in trunk diameter in areas prone to vandalism.

Regulations Municipal, county, or state regulations may affect how close a tree can be planted to a street, intersection, traffic light, or other structure or piece of property. These laws typically regulate tree planting along rights-of-way and on other property controlled by a government. State departments of transportation might prohibit planting trees that mature over a certain size along thoroughfares with certain speed limits and curb configurations. Regulations vary. Always check them before planting along a road maintained by a department of transportation.

Communities are beginning to set guidelines or regulations against planting certain tree species. Many of these rules are a response to the invasive nature of some non-native trees. Unfortunately, a few communities have developed or are considering approved tree lists, meaning that only trees on the list can be planted. This is not a wise course of action because it limits choice and discourages learning.

Below-ground site analysis

Evaluation procedures Important soil attributes that affect tree selection are pH, drainage, depth, salinity, distance to the water table, and rooting space restrictions. Many plantings fail because these factors are improperly evaluated or even ignored. Each factor is discussed in detail later in this chapter.

For newly developing sites, the best time to begin planning for good tree and landscape plant growth is before construction begins. This early evaluation allows you to identify good quality soil and make provisions to remove and stockpile it. Do not allow it to be buried or mixed with other debris and poor soil, because this will degrade its quality. Good quality soil is precious and should not be wasted. It can be spread over the site once the job is complete to promote good tree growth.

Preconstruction planning also gives you the opportunity to work with the contractors to prevent excessive soil compaction in areas where trees will be planted. These areas can be isolated from heavy equipment and other vehicles with sturdy fences, and fines can be levied on contractors for violations. It is crucial that soil designed to support tree root growth not become compacted, because growing trees in compacted soil is challenging (to say the least).

Unfortunately, you are often left to choose trees for a site after it has been modified and the grading and construction of buildings, walks, and other hard surfaces has been completed. Even if you performed a preconstruction site analysis, a thorough site evaluation should be conducted after construction because equipment operations may have negatively affected important soil attributes. For example, equipment often compacts a clay soil during construction, turning a moderately well-drained soil that would have supported tree growth into a collection of poorly drained pools of mud.

Many landscapes fail in this respect. One species or cultivar might be planted regardless of soil conditions, in hopes that it will grow as seen in a catalogue or book. The job looks great on paper and immediately after installation. All the trees are the same size and shape because they were growing in a uniform nursery environment. But trees often grow at different rates after planting, and some may not even survive on certain spots in the landscape due to high soil pH, compacted soil, or drainage problems. In this case, lack of proper site evaluation will spoil the intended design objective of uniform canopies. The key to maintaining uniform canopies is to create uniform soil conditions. Species selection alone cannot accomplish this.

More site evaluation and soil tests will be needed in a landscape that has been highly urbanized or disturbed than in an undisturbed site. Without a soil management plan, poor quality subsoil often ends up as topsoil on these disturbed sites. Rubble and other debris can become mixed with soil, making the mix ill suited for tree growth. Separate soil tests should be conducted wherever you have reason to believe the soil is different.

You may collect soil data about the site using soil surveys available at the Soil Conservation Office. Soil surveys describe the percolation rate, hardpan characteristics, slope, soil type, and, in some cases, the trees that naturally grow on the soil type. These are most useful for residential developments carved out of woodlands or agricultural fields or other relatively undisturbed sites. More focused, on-site soil evaluation is needed for most urban and suburban landscapes where soil has been moved with heavy equipment. For example, soil in *each* sidewalk cutout along a street may have to be evaluated separately because conditions often vary greatly. You may test soil in three cutouts and find it well drained. The other 15 or 20 cutouts at the site that you did not test may have poor drainage. If you were to base your tree selection decisions on the three cutouts you tested, you might choose a tree requiring good drainage for survival. The planting is likely to fail because most trees would drown.

Holes for testing soil attributes can be dug with a backhoe or a soil auger. A backhoe is preferred because you can see a large portion of the soil when the hole is dug. Specific guidelines on the proper number of soil tests are impractical because each site is unique. Uniqueness requires judgment on your part. For example, many residential sites in clay soil become compacted during construction, whereas those in sandy soil may not. Therefore, a more detailed evaluation of soil attributes is often required in clay soil. The professional judgment of a local soil specialist, horticulturist, or urban forester can provide unique clues to the nature of the soil, so you can begin to list the trees likely to grow well at the site.

Soil pH Soil pH governs the availability of nutrients to plants and also affects the activity of soil microorganisms. Parts of the Northeast, Midwest, Rocky Mountains, Texas, Great Plains, Southwest, coastal Florida, and many urban areas throughout the country have alkaline (high pH) soil. Trees such as pin oak, sweet gum, red maple, queen palm, many pines, and some others develop chlorosis and grow poorly in these soils. Some, like junipers and goldenchaintree, prefer alkaline soil. Choose trees that are adapted to alkaline soils if the pH is above 7 (Appendix 4). Do not guess the pH. Have it tested! It is a simple, quick, and inexpensive test. Consider using two different labs to test the soil. If results do not agree, use a third lab.

A pH test should be conducted in several areas of the site (Figure 1-4), wherever soil color or texture appears distinctly different. Site pH may vary too much to plant the same species or cultivar across the entire job site. To collect samples for testing from an open area such as a lawn, where soil may be fairly uniform, dig about 10 small holes 5 to 10 feet apart with a trowel or shovel. Remove a slice of soil from the side of each hole from the surface down to 12 inches deep. You might choose to use a soil coring device to collect the samples if one is available. Mix soil together in a clean plastic bag or clean bucket or jar and take a subsample (about a pint) to a lab to be tested. The test report you obtain from the lab will include corrective measures for adjusting pH if needed.

If soil at the site is not uniform because of natural soil variability or significant soil disturbance, several samples are needed. For example, urban sites can be so diverse and variable that one soil test may be required for each island in a parking lot in order to select the right tree for the project. One tree may not be suited for the entire project due to soil variability. To plant the same tree species over an entire site with variable soil pH, choose one that tolerates both the highest and lowest pH at the site.

Most trees can grow in soils with a pH between 4.8 and 7.2. If soil pH is less than 4.8, check with local specialists regarding proper species selection. Trees planted in a soil with a pH of between 7.2 and 7.8 should be tolerant of slightly alkaline conditions. Only trees tolerant of alkaline soil should be planted in soil with a pH above 7.8 (Appendix 4). Few trees grow well in soils with a pH above 9.0.

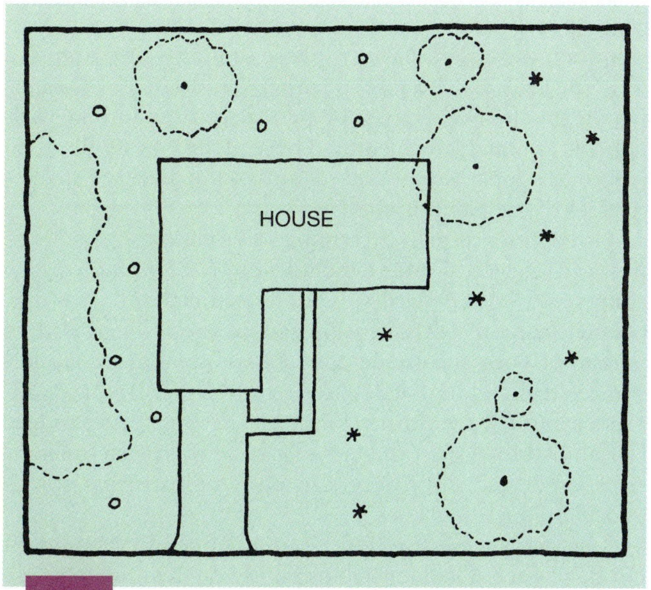

As you dig soil cores, take note of soil color and texture. Indicate on your landscape sketch which areas contain loose soil and which contain dense clay or other types of soil. If you are unable to do this, hire someone who can, because this procedure provides important information to the manager of the completed landscape. For example, trees planted in loose, well-drained areas may need more frequent irrigation until they are well established. Those in poorly drained, clayey soil may need less water and could even drown before they become established.

HOUSE

1-4 Combine soil from similarly marked areas into one composite sample.

Soil texture Soil texture is more an indicator of other soil attributes that influence tree growth than a growth-limiting factor itself. For example, clayey soils often drain poorly, especially if the terrain is flat and the soil was disturbed. When planting in a clayey soil, pay particular attention to evaluating soil drainage.

Sandy soils drain faster than clayey soils. If irrigation cannot be provided in a sandy soil during hot dry weather after trees are established, drought-resistant trees should be chosen for the site. This is especially important if roots will be confined to a small area such as a parking lot island. Sandy soils also leach faster than clayey soils, carrying nitrogen, potassium, and other essential elements below the root zone. At first this might appear to have little impact on species selection, but it will affect fertilization management in the established landscape. If the designer or other person specifying trees has reason to believe that the landscape management budget is not sufficient to maintain an adequate fertilization program, consider choosing species that are native to the sandy soil type at the planting site. These trees may be more adapted to soils with lower nutrient content.

Areas with compacted soil, poor drainage, and low oxygen Compacted and poorly drained soils contain little oxygen—a gas that tree roots need to survive and grow. Many trees die or grow poorly and succumb to an insect or disease problem because they are planted in soil that is too wet during certain times of the year. Only species and cultivars tolerant of wet sites can survive in these difficult soils (Appendix 5).

To check for compaction and drainage, dig several holes at least 18 inches deep (preferably 24 to 30 inches) in each section of the site (Figure 1-5). If soil is very difficult to dig with a shovel, it may be compacted. Another sign of compaction is standing water that persists for a day or more after a period of rain. If a pickax appears to be the best tool for digging a planting hole, then soil is probably too compacted. If soil is fairly easy to dig into with a shovel, it is probably not compacted.

Soil bulk density is one reliable measure of compaction. It is used by professional landscape architects, urban foresters, and others. A quick method (Lichter and Costello 1994) of measuring density is to dig a 12-inch-deep hole with a shovel, placing *all* excavated soil in a bucket or other container. Line the hole with a plastic bag and fill the bag

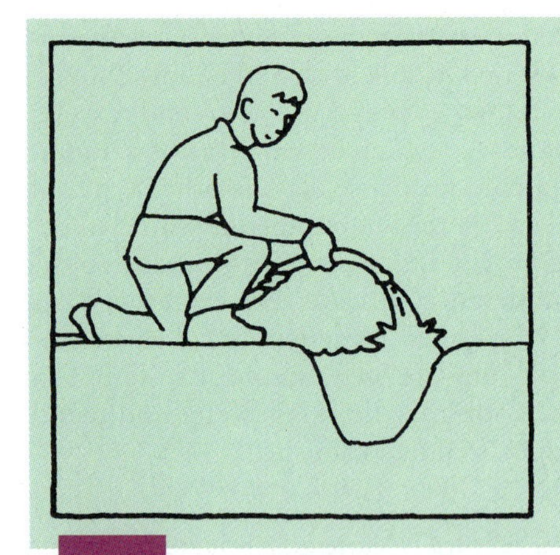

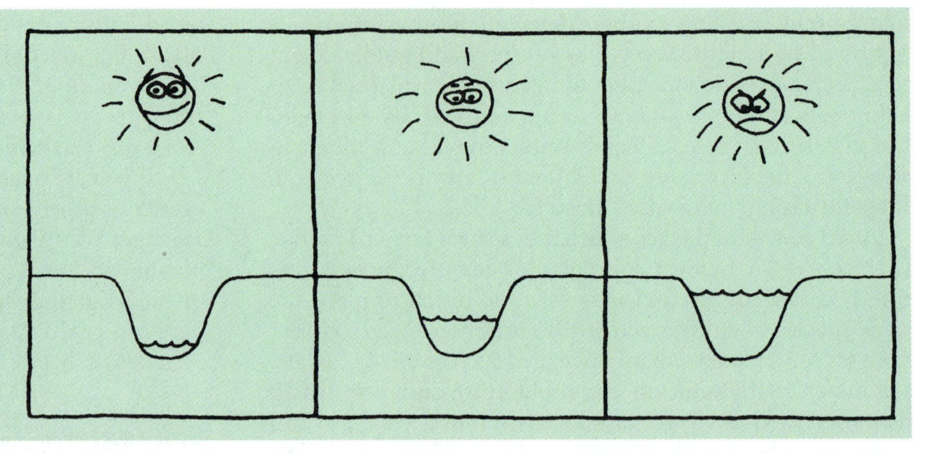

1-5 Checking soil drainage. Dig and fill several 18-inch-deep holes with water (extreme left). If water drains away within one hour, drainage is good (left center). If water takes several hours to a day to drain from the hole, drainage is fair (right center). There is a high water table or poor drainage if water stands in the hole for more than a day (extreme right).

with water. Measure the volume of water in the bag. Place soil in an oven for several days at 180 to 200°F until it is completely dry. Weigh the dried soil. Bulk density is calculated by dividing the weight of dried soil by the volume of water. This technique can underestimate bulk density by 3 to 9 percent compared to the more traditional (and more accurate) undisturbed soil core technique. If the soil is too compacted, root and tree growth will be poor (Table 1-2). For those who have access to a penetrometer, 2.3MPa can be considered a critical soil resistance value because root growth is slow above this value (Day and Bassuk 1994).

If soil is very compacted and hard all the way down to the bottom of the planting hole, then wet-site-tolerant trees are most appropriate. Occasionally soil is loose underneath and compacted only on the surface. If you can break up the surface compacted layer for 15 feet or more around the tree before planting, drainage and tree growth may improve. In this situation, trees can be chosen regardless of their wet-site tolerance.

One method for determining if soil is poorly drained is to smell it. Sour-smelling soil may be a gray color, indicating that it contains little oxygen. Occasionally, the smell may be strong enough to detect while standing close to the hole. More often, a soil aggregate or clump must be broken open close to your nose to detect the smell.

If the soil does not smell sour when you dig, it may still be anaerobic (without oxygen) during certain times of the year. This could limit tree growth unless the species to be planted are tolerant of wet sites. To complete a thorough soil evaluation, check the soil several times during the year (especially during cool, wet weather) prior to planting. Many poorly drained soils are wettest in the dormant season when there is little if any foliage to draw water from the soil. You may also be able to gather soil drainage information from the Soil Survey or persons more familiar with that particular planting site. These techniques will allow you to develop a year-round picture of the drainage characteristics of the planting site.

Other clues to help determine soil drainage characteristics and oxygen content include presence of subsurface compacted layers (hardpans) and artificial soil layering. Oxygen content and diffusion rate in the soil can also be measured with special devices. These machines may become useful and find wider acceptance as professionals gain more experience with them and they become easier to use. On undisturbed sites, the presence of only wet-site-tolerant trees, such as baldcypress, can also indicate that soil is poorly drained.

Trees with aggressive root systems (Appendix 11) should not be planted in compacted or poorly drained soil, because surface roots often form. These can disrupt lawn mowing operations and can damage curbs, sidewalks, pavement, and other nearby structures.

TABLE 1-2 CRITICAL BULK DENSITY VALUES FOR DIFFERENT SOIL TEXTURES

SOIL TEXTURE	CRITICAL BULK DENSITY[1] RANGE g/cc[2]
Clay, silt loam	1.4–1.55
Silty clay, silty clay loam, silt	1.4–1.45
Clay loam	1.45–1.55
Loam	1.45–1.6
Sandy clay	1.55–1.65
Sandy clay loam	1.55–1.75
Sandy loam	1.55–1.75
Loamy sand, sand	> 1.75

[1]Bulk densities greater than these values could restrict root growth
[2]grams per cubic centimeter
Adapted from Morris and Lowery *(1988) and* Harris *(1990).*

Subsurface compacted layers Soil loosely spread over a compacted subsoil creates special challenges. Roots often grow only in the loose soil and will not penetrate the compacted subsoil. Large-maturing trees could become unstable and hazardous as they grow due to a shallow root system. Therefore, only small and medium-sized trees are recommended (Appendices 7 and 8) if less than 2 feet of loose soil will be spread over a compacted subsoil. The lowest areas in this landscape are likely to be very wet during certain times of the year. Evaluate this carefully and, if needed, choose trees tolerant of wet sites for the low areas (Appendix 5). The best time to perform this evaluation is within a day or two after a significant rainfall. If you conduct site evaluation during a drier time of the year, you could mistakenly conclude that drainage is fine.

Artificial soil horizons Most soils in urban areas and many suburban landscapes are disturbed with heavy machinery before trees are planted. Poor quality subsoil with a fine texture or high clay content is often brought to the surface or used as fill soil. Construction debris and other soils are often layered on top of one another, creating an artificial soil profile (horizon). This layering of dissimilar soil types keeps the ground unusually wet by disrupting the natural percolation of water. Suspect a water percolation problem if there are sharp boundaries between different soil types. This is often indicated by distinct differences in soil color (Figure 1-6). It is often best to mix different soil layers together with a rototill or other device (see "Potential site modifications" later in this chapter).

Soil salinity Some soils in coastal areas, or in regions of the country receiving less than 30 inches of rainfall annually, have a high salt content. A site that receives regular applications of de-icing salts may also have elevated levels of soil salt. Salts dry out roots, making it difficult or impossible for trees to establish and grow. If you are unfamiliar with the area, or if you suspect that salts could be a problem, have the soil tested for this. If soil conductivity (E.C.) is over 1 millimhos per centimeter (mmhos per cm), then salt content may be too high for many trees. Choose trees that have good tolerance to soil salts (Appendix 3), leach the salts with water (if possible), or replace the soil with good quality material.

Even if soil conductivity is low, you may still have a salt problem, especially if de-icing salts are used nearby. Have the soil tested and interpreted by a soil scientist for excess sodium and sodium absorption ratios. Sodium absorption ratio should be less than 5. These are the most reliable indicators of a salt-contaminated soil. Note, though, that reliable data on salt tolerance of trees are scarce, and reports vary on the tolerance of certain species. Use Appendix 3 only as a guide.

Irrigation water may also be salty. Have it tested if you use well water along the coast or if you live in an area with rainfall of less than 30 inches per year. If conductivity of irrigation water is above 1 mmho per cm, it may cause poor growth on some trees and other plants from salt buildup. If good water is not available, choose from among the more salt-tolerant trees. Find out what has been growing well in the area with the same irrigation water.

Determine if de-icing salts (especially those with a high sodium content) will be regularly used on the adjacent roadway. If so, only salt-tolerant trees should be selected. However, even some of these trees cannot survive close to a highway or other high-speed thoroughfare if salt is regularly used. Salty runoff water from the road can also wash lethal amounts of salt into the root zone. If possible, plant on the uphill side of the road to avoid salty runoff water, or construct a berm to divert runoff away from the potential rooting zone. Some communities install trees in elevated planters along downtown streets to reduce the amount of salty water washing into the root zone (Figure 1-7). When designing parking lots and other pavement areas receiving de-icing salts, make provisions for storing salt-laden snow away from tree root zones.

1-6 Sharp boundaries in the soil may indicate a water drainage problem. Note how roots often grow best in the top layer.

1-7 Raised planters open to soil at the bottom increase the amount of good soil available for root growth, and they can be used as seats by pedestrians.

Other soil contaminants Bring a soil scientist to the site if you have reason to believe that the planting site contains contaminants such as petroleum waste products, heavy metals, or other potentially hazardous residues. Not only could these cause poor plant growth, but they could also become potentially dangerous to people. If soil is mixed with construction-type materials, such as bricks, concrete, or other debris, consider replacing it, or sift out the debris. Concrete-based debris can increase soil pH and it takes up valuable space in the soil. Roots cannot penetrate debris; they need soil to grow in.

Soil depth If bedrock comes close to the surface, leaving only a thin layer of soil on top, or if there is little soil, consider planting only small or medium-maturing trees to help prevent large surface roots from forming. If you choose to plant large-maturing trees on a site with little soil, select only those with noninvasive roots (Appendix 11) and relatively open canopies. Large-maturing trees in soil less than two feet thick could topple over in storms as they grow older because they lack deep roots.

Distance to water table Dig several holes with a shovel, four-inch auger, or backhoe two to three feet deep and wait two to four hours. Any tree can be planted if no water appears in the hole. If water appears in the hole, select trees that tolerate wet sites. Measure the distance between the soil surface and the top of the water. If this distance is less than 18 inches, only small or medium-sized wet-site trees are recommended (Appendices 7 and 8). Large trees may topple in wind storms as they grow older because of shallow root systems (Figure 7-17). Possible exceptions to this rule are baldcypress, black gum, and some other trees that can grow with root systems submerged in water (Appendix 5).

Distance to the water table often varies during the year. It might be several inches below the surface in the cooler season but drop several feet in the growing season because transpiration pulls it from the soil. This type of site should be considered poorly drained. To help avoid making erroneous conclusions about depth to the water table, determine it during the coolest and wettest season.

Underground utilities Do not plant a tree before determining where underground utilities are located. You might damage the utilities, or you could get hurt digging the planting hole (Figure 1-8). Consult your cable company, water and sewer departments, electric utility, and telephone and gas companies before digging. There may also be fiber optic cable nearby. Repairing this type of cable is very expensive. Roots of large maturing trees planted within 10 feet of underground municipal utility lines could be damaged when the lines have to be serviced. For this reason, some communities restrict planting near these utilities. Roots usually will not penetrate well-designed, properly installed utilities that do not leak water.

Rooting space restrictions Match ultimate tree size to the size of the planting space to keep trees healthy and prevent damage to surrounding sidewalks, curbs, and pavement. Some horticulturists suggest that at least two cubic feet of soil is required to support each square foot of canopy area (ground surface area outlined by the edge of the branch canopy) under nearly ideal conditions (Lindsey and Bassuk 1992). This means that by knowing the ultimate canopy size desired for a particular planting site, you could calculate the required soil volume for adequate root growth. The site could then be designed to provide for this amount of soil.

This concept has significant implications for urban spaces that are designed to support trees. If we expect trees to provide shade to pavement in urban landscapes, we must create favorable conditions for root growth under pavement or deeper in the soil. These conditions must be created during the planning and design phase of a project. It is usually very difficult to engineer these conditions into the site once it is built.

It is not always apparent if soil can support root growth, especially in highly disturbed sites beneath pavement. Soil under pavement is typically poorly aerated and compacted, a condition that is inhospitable for roots unless it is sandy and well drained. Roots will mostly be confined to the soil space not covered by pavement. This could stunt growth if the tree is completely surrounded by pavement (Figure 1-9). However, some wet-site-tolerant trees (Appendix 5) such as Honeylocust are adapted to produce roots under pavement; many other trees are not adapted. Roots often grow directly under pavement along cracks. Trees with roots under pavement are likely to be less stunted than those with roots confined to a small island. The next chapter will discuss alternate site designs that might enhance root growth under pavement.

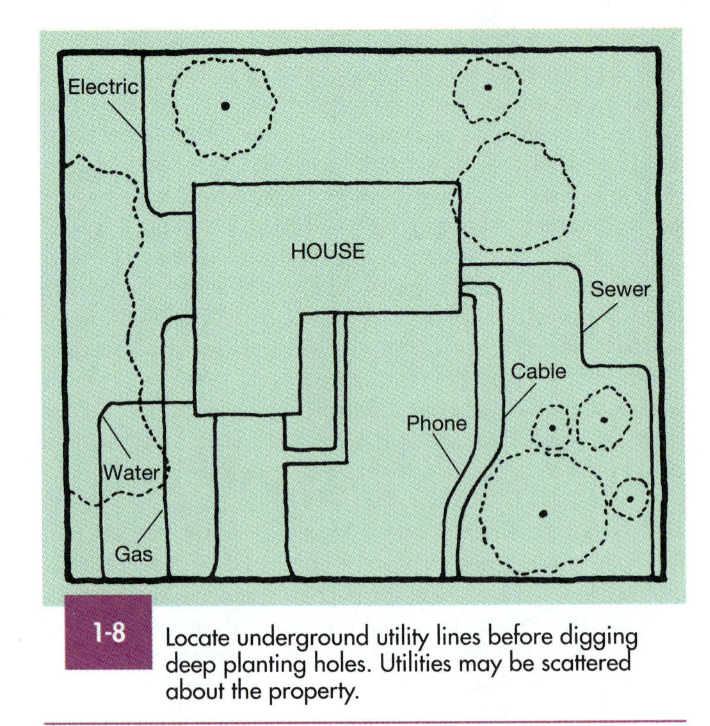

1-8 Locate underground utility lines before digging deep planting holes. Utilities may be scattered about the property.

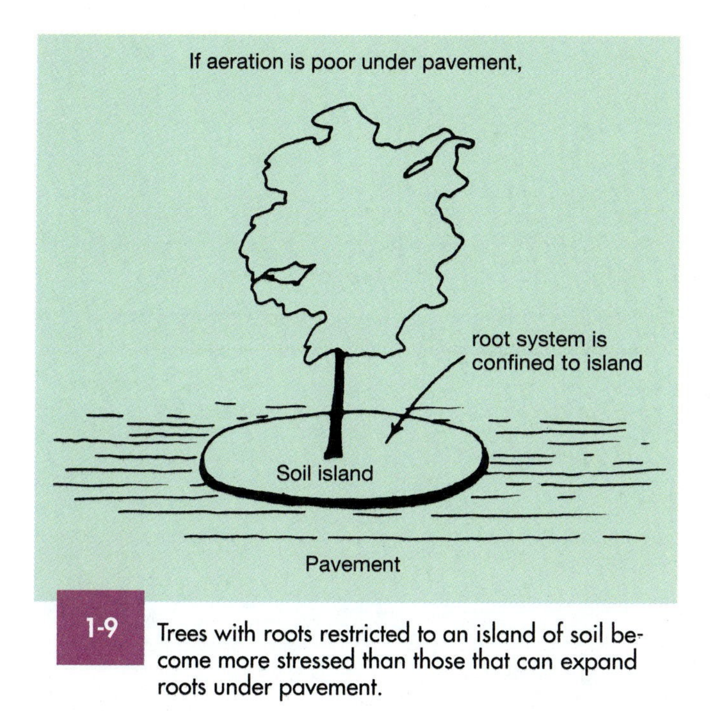

1-9 Trees with roots restricted to an island of soil become more stressed than those that can expand roots under pavement.

POTENTIAL SITE MODIFICATIONS

Modifications made to the site can help accommodate a wider variety of tree species. One or more may be appropriate for your landscape. Decide how the site will be modified before choosing plant material. Site modifications such as moving street lights or overhead wires, using specialized design techniques, grading, improving drainage, making soil amendments, and so forth, when made before planting, can have an impact on soil conditions that affect tree growth. If a site is modified after the initial evaluation, above- and below-ground attributes may have to be reevaluated before appropriate plant material can be chosen.

Moving lights and wires

Street lights and overhead power lines can be moved to make room for trees. Though this is not commonplace, it is done surprisingly often once the suggestion is made. In many instances, it is a more permanent solution to a design problem, allowing trees to be planted along a street in an area where they could not be planted without moving fixtures.

Some communities provide utility corridors. Utilities must be contained within the corridor, allowing trees to be planted away from the corridor, perhaps on the other side of the street. This eliminates the conflict between wires and trees.

Changing soil pH

It is best to plant trees adapted to the existing soil pH than to attempt to change soil pH. Applications of sulfur or limestone to the soil usually provide only a temporary change in pH. They have to be applied regularly to maintain the adjusted pH level.

Low, acidic pH can be temporarily raised with regular applications of ground dolomite or limestone. Mix it into the soil as best you can before planting for the quickest effect. Most alkaline soils (high pH) cannot be corrected permanently by any practical means. Unless the high pH was caused by overliming, applications of sulfur will usually provide only a temporary drop in soil pH. This could help prevent nutrient deficiency symptoms for a short while by allowing uptake during this short period of time. A better solution is to use plants that tolerate high pH (Appendix 4). Soil pH can be lowered (made more acid) with ammonium sulfate (least effective, least likely to damage plants), sulfur, aluminum sulfate, or sulfuric acid (most effective, greatest risk of damaging plants).

Improving drainage

Water running off a site can carry soil, pesticides, and fertilizers that contribute to environmental degradation. Soil is often graded (shaped) to keep as much water on the site as possible. Water can be channeled by creating subtle swales and berms, providing better drainage over much of the site, but wetter conditions will exist where water collects. This might improve drainage enough to expand the range of trees suited for the site. In most instances, redirected water should be kept on site, not channeled to streets or streams. This reduces runoff and sedimentation in streams by allowing it to percolate through the soil or evaporate.

Grade changes are common on large-scale landscape jobs, but often are not practical for the homeowner. Consider installing gutters on a home or office building and directing water to an area that drains well. Or collect it in a cistern (large container) and use it for irrigation. French drains (a sloping trench filled with large stones) can also be constructed to channel water to a lower area. Be sure to divert runoff away from existing trees, because flooding could kill them, especially if soil drains poorly.

Drainage in compacted soil can be improved by breaking through the compacted layer and into loose subsoil beneath. This allows water to flow into the underlying soil. This is normally done with a heavy piece of equipment.

Placing gravel by itself at the bottom of a planting area does not improve drainage or plant growth in wet soil. Excess water collected in the gravel layer must be piped to a lower area such as the street or sewer. Perforated pipe surrounded by geotextile fabric to keep soil from entering the pipe is recommended for this task (see Chapter 2). These systems are sometimes used to draw standing water out of the soil.

One or two vertical pipes are sometimes installed in the backfill soil in compacted soil in the hope of improving aeration and root growth in deeper soil depths. Studies show that they slightly improve aeration very close to the pipe but have little influence away from the pipe (Day and Bassuk 1994). To be effective, many vertical vent pipes may have to be installed very close together making this impractical. In addition, some practitioners report that these devices clog with debris, making them ineffective.

If the site is located in a low spot where surface water collects from surrounding areas, tilling or otherwise loosening the soil will not help. Correct poor drainage with underground or French drains, diversion ditches to prevent water from draining into the area, or addition of soil to raise the elevation.

Improve compacted soils

Because roots grow poorly in compacted soil, it should be tilled or broken up with specialized heavy equipment prior to planting. Avoid doing this beneath the canopy of existing trees. Significant root damage could occur and the trees might die.

A coring machine can be used to enhance growth in compacted soil. It pulls soil cores out of the soil, creating holes for water and air penetration. The soil cores are left on the surface of the soil. This is sometimes referred to as vertical mulching. A different technique pulls tines over the site to create holes in the soil, but the tines compact the sides of the holes and may not improve drainage and plant growth appreciably. A more sophisticated machine, manufactured by Verti-Drain of Holland, drives 16-inch-long tines into the soil. When the tines are fully inserted into the soil, they are deflected laterally several inches. This breaks up the soil in a manner similar to inserting a garden pitchfork into the soil and displacing it several inches. These techniques can increase water percolation and reduce water runoff. A pitchfork can also be used to loosen and aerate compacted soil on a small site.

Several two- to three-foot-deep trenches can be dug from the planting hole like spokes in a wheel (Figure 1-10). A backhoe or trenching machine can dig trenches quickly. Some people also refer to this technique as vertical mulching. Amended or original soil can be placed back into the trench, although there is no evidence that amended soil increases root growth more than backfilling with original soil. Although this may not provide all the benefits of loosening the soil around the entire planting hole, it may be less expensive, and roots should be able to grow well in the loose, aerated soil in the trenches. This technique can also increase root growth on existing trees in some circumstances.

Gypsum applications can reduce the effects of compaction only if the compaction was caused by sodium-saturated exchange complexes in the soil. Gypsum will not reduce compaction if compaction is caused by other factors. A specialized soil test performed by a soil lab can reveal if sodium has saturated the exchange complexes.

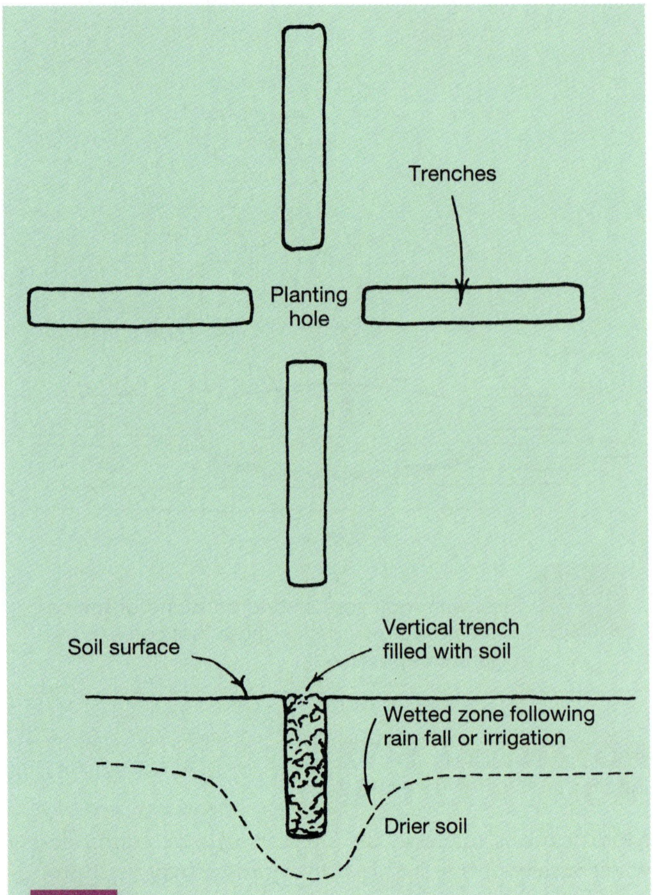

Adding fill soil

Ideally, top (fill) soils brought onto the site should be mixed together before spreading. This helps prevent drainage problems caused by layering of different soil types. It is costly but will reduce landscape maintenance by providing a more uniform soil for root expansion. If mixing is not practical, purchase fill soil that is as uniform as possible. The texture of fill soil should be about the same as or slightly coarser than the texture of existing soil. At the very least, blend the interface between the two soils by tilling several inches of the fill soil into the existing soil. These techniques help move water through the soil profile and reduce the likelihood of creating a perched water table. If different types of fill soil are used, the landscape will probably require more labor to maintain due to their different moisture- and nutrient-holding capacities and poor internal drainage.

1-10 A trencher can be used to dig four or more trenches out from the planting hole. Loosely fill the trenches with the same soil that came out of the trench, or amend it with organic matter if you wish. This provides loosened soil channels for root growth in a compacted site. Vertical trenches can be cut along a slope to slow runoff water.

If loose, well-aerated soil is spread over compacted soil, tree roots will grow mostly in the fill soil, not in the compacted subsoil (see Chapter 6). If the fill soil layer is less than two feet thick, a shallow root system will develop, with few deep roots to provide anchorage for large-maturing trees. Water is likely to pond in the lowest areas on the site. Plant wet-site-tolerant trees there. If fill soil is more than two feet thick, large-maturing trees can probably be planted with little danger of blowover.

Soil replacement

Recently, a small group of arborists began experimenting with replacing soil from beneath existing trees with fresh topsoil. Moistened, existing soil is removed with a strong stream of water and a vacuum. The technique leaves most roots intact. The theory is that with time, existing soil in sites where roots are confined to a small space becomes less able to support adequate root growth. Fresh soil is carefully replaced around existing roots and brought up to grade. The tree reportedly responds by regenerating additional roots in the fresh soil, especially if existing soil was compacted or chemically contaminated. This new technique needs much more evaluation before it becomes common practice.

A modification of this method is to remove all soil from vertical trenches dug as in Figure 1-10 and replace it with the loosened soil that came from the trench. If the tree is established and growing in an open lawn area, a similar increase in root growth could be achieved by killing the grass and simply adding several inches of mulch. This certainly would be less costly than replacing soil.

The same techniques can be used at new planting sites. If soil is considered too compacted, contaminated, or poor to support root growth, consider replacing it or digging long trenches backfilled with good soil.

Mitigating soil salt contamination

Sodium-affected (alkali) soils can be modified by adding gypsum or sulfur. This improves soil by displacing sodium from the soil particles. However, gypsum will provide little, if any, benefit if the soil has a high calcium content. It is a poor amendment in clay soils because of its small particle size. Gypsum will not help mitigate soils impacted by other salts. Follow the recommendations on the label for application rates. Check with a soil specialist first to be sure your soil is really alkali—this treatment does not work and is a waste of time in other soil types.

If drainage is adequate, salts can be leached from the soil with excess water. This is a specialized technique that is best done in conjunction with a soil scientist. Berming can occasionally be used to divert salty water from the root zone and prevent a problem from occurring.

Other soil improvements

Other than temporarily changing soil pH, improving drainage, or loosening compacted soil, it is difficult to permanently improve most existing soils for trees. Most other so-called improvements are temporary and have not been shown to impact species selection.

Colloidal phosphate, a byproduct of phosphate mining, can be used as an inorganic amendment for sandy soils. It is expensive and only locally available, but its improving effects (enhanced water-holding and exchange capacity) can last 10 years or more. Synthetic, organic, and other soil amendments are being developed and marketed and may be useful in the future. Research shows little, if any, benefit from products currently on the market. If soil amendments are used, they should be mixed into at least the top 12 inches of *large* areas of soil before planting. Amending only backfill soil around an individual plant provides no benefit to trees unless roots will be confined to this small area by adverse conditions in surrounding soil. For example, it may pay to amend the soil in a sidewalk cutout if you believe the compacted soil beneath the sidewalk will prevent root growth there (see Chapter 2).

Expanded slate or sintered fly ash incorporated into a compactable soil can help prevent future compaction. At least 20 percent of the soil volume should be comprised of this soil amendment to be effective. Long-term testing in Washington, D.C., indicates that this strategy helps prevent compaction for at least 22 years, even in heavily used turf areas (Patterson and Bates 1994).

Organic matter incorporated into sandy soil can temporarily improve water- and nutrient-holding capacities. Organic matter can help aerate clay soil by separating aggregates or clumps. If clumps remain separated, air and water should penetrate better. If people and heavy equipment are kept off the soil after organic matter is mixed in, clay soil clumps may stay apart and some of the new airways may remain intact even after organic matter has decayed and disappeared. This might improve tree growth.

Reducing runoff

Trenches dug into the soil profile can also improve drainage (Figure 1-10). Water collects in the loosened soil in the trench and infiltrates into the soil profile instead of running off. Creating terraces with retaining walls helps to level a site, allowing water to percolate through soil instead of running off. Trees planted in soil retained by a wall should be appropriately sized. Large trees, especially those with aggressive roots, should not be planted here unless the tree is at least 10 feet from the wall. Small- and medium-maturing trees can be planted closer.

EVALUATE MAINTENANCE PRACTICES

Irrigation

Irrigation capabilities at the planting site should be considered before selecting trees for it. Irrigation affects not only species selection, but also maximum recommended size of the nursery stock and the tree production method best suited for the site.

Determining maximum tree size at planting

If trees will be irrigated infrequently (just several times) after planting, only seeds, seedlings, or small saplings of drought-tolerant trees should be planted (Appendices 7, 8, and 9). Small-sized nursery stock have small tops, so roots come into balance with the tree soon after planting. When roots come into balance with the top of the tree, they usually survive and grow well with rainfall alone in humid climates, or with just the turf and landscape irrigation system in dry desert climates. With infrequent irrigation, larger trees will grow poorly, die back, or be killed from lack of water, because they take much longer to establish. If trees will be irrigated regularly only until they are established (see Chapter 5), drought-tolerant trees should be chosen, and nursery stock of any size can be planted. If trees receive irrigation during establishment and then regularly during the life of the tree, or if you are planting in the plant's native range and soil type, any tree regardless of drought tolerance can be planted.

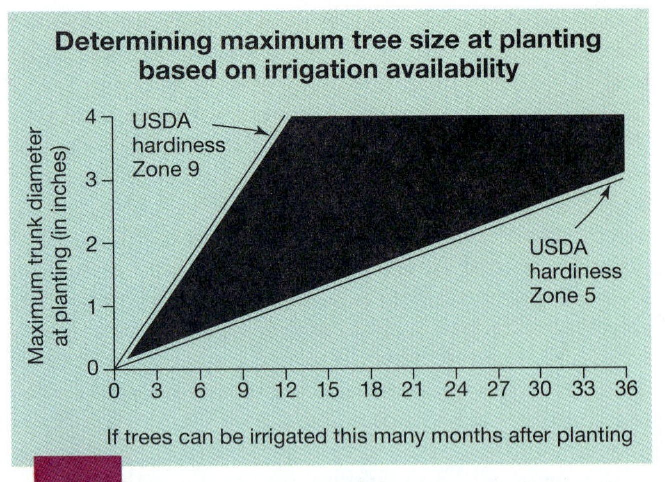

1-11 Maximum recommended tree size at planting depends on climate and how long trees can be irrigated after planting. For example, if irrigation can be provided for six months after planting in USDA hardiness zone 9, maximum trunk diameter at planting is two inches. With six months' irrigation, maximum trunk diameter for planting in zone 5 would be about one-half inch. Irrigation should be provided during dry weather in the growing season for 24 months after planting a two-inch tree in USDA hardiness zone 5.

Maximum tree size at planting can be determined by knowing the number of months trees will receive irrigation after planting (Figure 1-11). For example, if irrigation can be provided for only three months after planting, trees no larger than one inch in caliper should be planted in USDA hardiness zone 9. If trees can be watered for six months after planting, larger trees can be installed. Because root growth is slower in colder climates, irrigation during the growing season in the first 12 months is required to establish a one-inch tree in hardiness zone 5. Consult Chapter 5 for specifics on the amount and frequency of irrigation. Consult Chapter 3 for further comparisons of small- versus large-sized nursery stock.

Another factor affecting tree size at planting is weed control and the mulch replacement program. If weed growth is not controlled chemically or with mulch application around the tree during the establishment period, small nursery stock may be the best choice. While large trees are trying to establish, weeds often usurp the limited water supplied. Small trees establish quickly and therefore are better able to compete with weeds.

Determining appropriate tree production method Irrigation practices following planting influence the establishment rate of field-grown trees differently from those raised in containers. If trees will receive regular and frequent irrigation after planting, then trees grown in a nursery by any production method appear to perform equally well after planting to the landscape. If postplanting irrigation will be infrequent, then hardened-off field-grown trees may perform better than those from containers. Container trees take longer to establish their roots in the landscape than field-grown trees. This is discussed in more detail in Chapter 3.

Planting in irrigated turf In western climates, some trees such as oaks and mandrone grow poorly in lawns and other areas receiving frequent irrigation. Place these trees in spots that receive less frequent irrigation.

Pruning

Trees must be pruned regularly to maintain good health and longevity. Unfortunately, tree pruning budgets are often too low to allow pruning every five years following planting. If this is the case, consider planting species that require only a moderate amount of pruning to develop and maintain good structure (see Appendix 14). With limited pruning budgets, it is also crucial that high-quality nursery stock be planted, because poor-quality trees require more pruning to develop strong branch structure in the first several years after planting (see Chapter 3). Small nursery stock also requires more pruning to develop good structure than does good-quality, large-sized stock. Some permanent scaffold branches may already be formed on the large stock, thus requiring less training.

For one reason or another, some property managers routinely clip and shape into balls large-maturing trees in parking lot islands and buffer strips around parking lots. This also occurs near signs and other urban structures. This practice keeps them from becoming large trees. If early interviews determine that a client intends to maintain trees in this odd manner, plant small-maturing trees instead of those that will require regular clipping.

Fertilization and soil chemical modification

If there are no funds to conduct a monitoring and treatment program in alkaline soil for micronutrient deficiencies, plant only trees tolerant of alkaline soil. If a tree (for example, pin oak or queen palm) that is not tolerant of alkaline soil has to be planted in a soil with alkaline pH for historic or other special reasons, then be prepared to conduct a regular monitoring and treatment program designed to prevent micronutrient deficiencies. There is nothing too terribly wrong with this approach provided the landscape designer is prepared to conduct an educational session with the client to make it clear that this is a special, exceptionally high-maintenance planting. If the client is unable to conduct this sort of program for the life of the tree, then a more appropriate species should be chosen. It is the responsibility of the person specifying plant material to conduct this training with the client and to make appropriate choices. This is a recognized, established principle and anything less could be considered malpractice.

Alternate de-icers

Sodium and chloride in de-icing salts have a large negative impact on trees and other plants. If de-icing salts have to be applied regularly near a planting area, consider using a deicer that does not contain sodium or chloride.

Pest control

Trees rated as pest-sensitive (Appendix 12) can be planted in urban and suburban landscapes. They often provide benefits for a long time and may never succumb to major problems. The particular problem that afflicts the tree may not occur in your area. For example, some crabapple cultivars are very prone to scab disease, but scab may not be a big problem in your region of the country, so the tree may not be affected. However, trees rated as pest-sensitive are known to have serious problems in some parts of their planting range. Be prepared to address these problems when planting trees rated as pest-sensitive. Before planting, check with local tree specialists as to the severity of these problems on pest-sensitive trees in your area.

Cleanup

Trees with large fruit (royal poinciana or honeylocust), hard fruit (hickory), or very fleshy fruit (fig, persimmon, cherry, or queen palm) can create a mess or hazard on a sidewalk or pavement beneath the canopy. Pedestrians can slip and fall on the fruit and it can be unsightly. If cleanup budgets are low, consider planting trees, such as the fruitless 'Rotundiloba' cultivar of sweetgum, without this type of messy fruit in areas with high pedestrian traffic (Appendices 15 and 16). Ethephon sprays can be used on some species in some states to halt fruit production (Elam and Baker 1996), but proper timing is crucial.

CHOOSE DESIRABLE TREE ATTRIBUTES

Up to this point in the evaluation process, trees have been chosen primarily for their ability to grow at the site. *This is the most crucial criterion for tree selection at a planting site:* not how big they are, not what color foliage or flowers they have, nor any other functional or aesthetic quality. This is all secondary to first determining which trees are adapted to the cultural and physical conditions at the planting site! Now it is time to consider tree function and shape, aesthetics, and other desirable attributes.

Function

Trees provide us with many benefits, from the obvious, such as shade and production of oxygen, to the no less important but perhaps less noticeable, such as erosion control, protection of water resources, support of wildlife, reduced run off, and stream bank stabilization. The function(s) we would like our trees to perform may dictate tree size, shape (form), life span, canopy density, color, growth rate, fruit characteristics, location of the tree and other attributes. For example, trees that provide shade to east and west walls or provide shade to an air conditioning unit can provide a financial benefit by reducing air conditioning costs.

Mature Size

Earlier discussion pointed out that although small, medium, or large trees belong in large spaces, small-maturing trees are usually best suited to small soil spaces. Narrow trees are well suited for narrow overhead spaces, but wide-spreading trees also adapt well to narrow spaces. Small trees are often best suited for planting near power lines, and so forth. Beyond the basic guidelines discussed previously, the choice of tree size is purely functional and aesthetic.

Large trees are the obvious choice for providing shade to large open spaces and for planting along streets. Medium or large trees will cast the most shade onto a building. Small trees are often suggested for planting in downtown areas, but they provide little shade. Small or medium-sized trees may be best for planting near a deck or patio. Small trees with showy trunks, fruits, foliage, or flowers attract attention to an area and make nice specimens in the yard. These are only a few examples of what could be part of a long discussion. You will undoubtedly add to this discussion.

Form

Match form to function Tree form varies from tall and thin (Lombardy poplar or Italian cypress) to short and wide (live oak, bur oak). After considering the physical attributes of and limitations at the planting site, the most appropriate form for a particular planting site depends on the function that tree will provide. For example, you would not plant a Lombardy poplar or Italian cypress for a shade tree or for erosion control. A live oak or sugar maple is more appropriate. A pyramidal tree with drooping branches, such as pin oak, may not be a good choice for planting near a street intersection, because low branches would block visibility. One with an upright or vase shape, such as zelkova, would be more appropriate.

Form affects maintenance requirements Selecting the proper tree form can have a big impact on tree maintenance requirements after planting. For example, many urban landscape situations call for trees near pavement. Small, spreading trees, such as multitrunked versions of amur maple, crabapples, kousa dogwood, or Jerusalem thorn, planted near a walk or pavement require regular pruning if planted too close, whereas a small upright tree or a larger tree can be trained to grow over the walk or street (Figure 1-12a and b). Trees with a pyramidal form usually require less pruning to develop strong branch structure than those with other forms. Trees with rounded, oval, or spreading canopies often need periodic pruning in the first 20 years after planting to ensure good structure (see Chapter 2). Remember that many trees that normally produce several trunks can be trained in the nursery to a more upright form (Appendix 17). These can make nice street trees. Look for these at your nursery.

Maintaining uniform canopies Many commercial landscape designs call for a number of uniform trees planted in rows, or in other formal arrangements. Cultivated varieties (cultivars) of popular species are often planted because cultivars have the potential to grow with a uniform canopy habit. If soil conditions at the planting site are uniform, then canopies have a good chance of maintaining a fairly uniform habit and size in the landscape. However, more often than not, trees grow at different rates following planting because soil conditions vary. This might

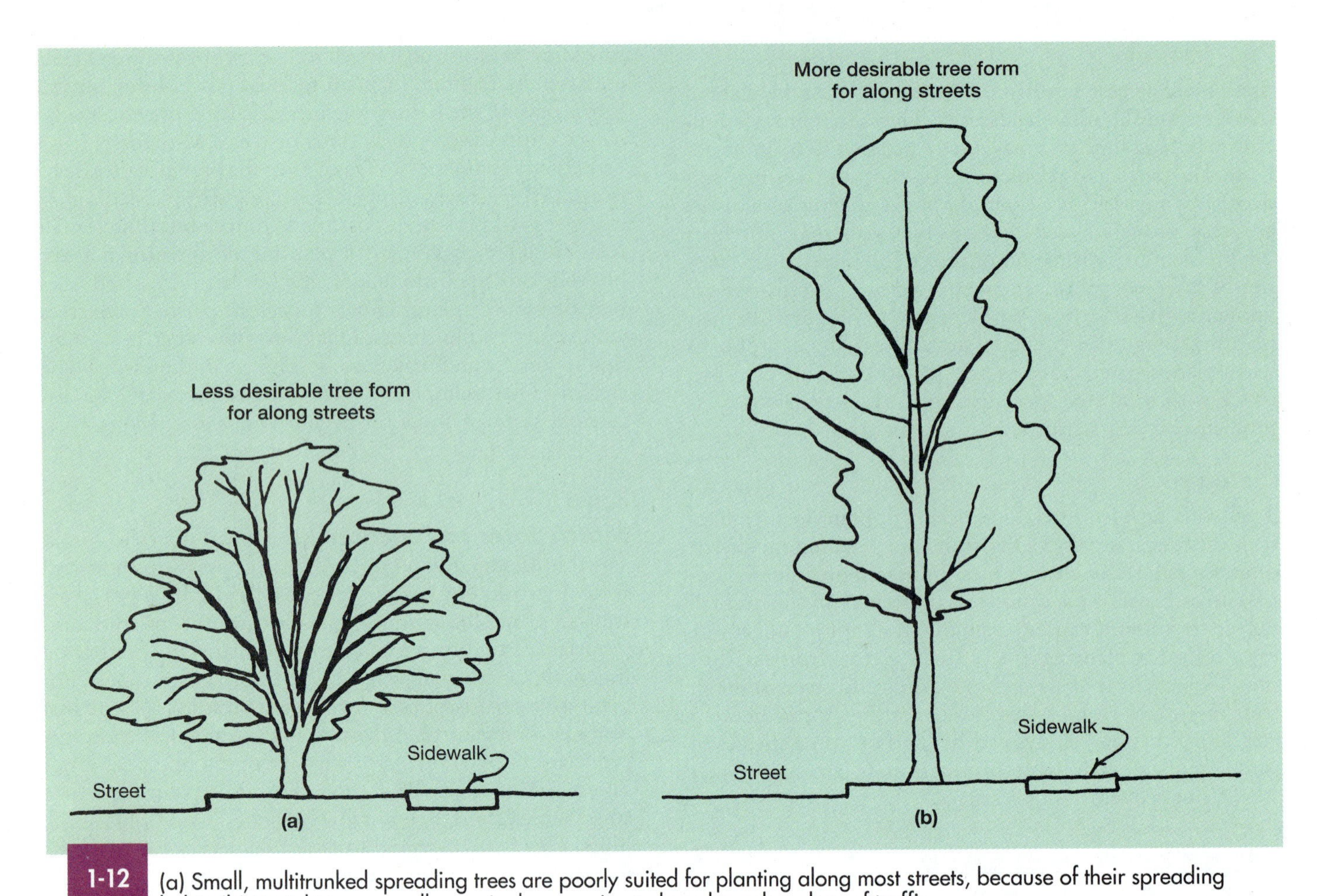

1-12 (a) Small, multitrunked spreading trees are poorly suited for planting along most streets, because of their spreading habit. (b) Upright trees usually require less pruning to keep branches clear of traffic.

spoil an otherwise well-planned landscape design because not all trees will have identical canopies.

A good method of maintaining uniform canopies in a landscape design is to create uniform soil conditions and then pick a species that will grow in that soil. Another (more expensive) method is to choose the species or cultivar first and then create the soil conditions to support that tree.

Longevity

Large-maturing trees often live longer than small trees. It would appear that large, long-lived trees might be the logical choice for planting in most landscape situations because they would provide a lasting effect. However, with reasonable placement and care, long-lived trees will probably outlast many of today's streets, homes, and buildings. Many structures are renovated or expanded 30 to 50 years after construction. The renovation is often so extensive that it becomes difficult to protect a tree's extensive root system. For this reason, concern about tree longevity may be inappropriate in highly urbanized landscapes, unless special provisions are undertaken to protect older trees.

Life span is of least concern in the most stressful urban sites. Trees in parking lot islands, small nonengineered sidewalk cutouts, and other small soil spaces have short lives almost regardless of species. Life span may be an important tree selection criterion in large open areas, such as parks and residential or some commercial landscapes, where little disturbance is expected beneath the tree and there is plenty of soil space for root expansion. At these sites, long-lived trees may be a better investment than those with a shorter life because they will have to be replaced less often.

Canopy density

Trees with dense canopies cast more shade and make the best screens. Those with thin canopies allow enough light penetration for grass growth near the tree. There is no evidence that denser trees buffer noise better than trees with an open canopy. The apparent noise buffering effects are probably a result of the visual screen provided by plants.

Deciduous versus evergreen

Deciduous trees are often planted in downtown urban sites in northern landscapes (zones 2 through 8) so that the sun can shine through in winter to warm the sidewalk, pavement, or building. This helps keep pedestrians more comfortable and minimizes heating costs. The same type of site in warm climates often is planted with evergreen trees to provide year-round shade. Many citizens in Florida, California, and other warm climates prefer or even demand a year-round canopy.

Growth rate

Fast-growing trees provide their benefits quickly, but their wood is often (but not always) more brittle than that of slow-growing trees. Information in Section II includes an indication of the growth rate and in some cases wood strength of each tree.

Wood strength

All trees can break apart in very strong winds. No tree, with the possible exception of cabbage palm, is immune. However, some are more prone to breakage than others in ice, snow, and wind storms. Consider using these trees in limited quantities and in out-of-the-way places. Surprisingly, wood strength per se is not related to a tree's ability to withstand an ice storm. Trees with a few coarse branches, such as walnut, escape damage from ice storms more often than those with a finer branching habit, regardless of wood strength (Hauer, Wang, and Dawson 1993). Trees with bark included in crotches of major branches are also susceptible. Information on each tree in this book indicates if it is susceptible to breakage (see Section II).

Ornamental traits

Some trees develop outstanding bark (Appendix 18), some grow nicely with several trunks (Appendix 17), and some have showy fruit, flowers (Appendix 19), foliage, or fall color (Appendix 20). After developing a tree list by considering all of the previously mentioned factors first, trees can be selected for these traits.

DIVERSITY OF SPECIES

It has been a part of urban forestry teaching for so long that it seems redundant to state that communities and others should plant a variety of species. Most professionals do not recommend arbitrary selection of several species on the same site or along the same street for diversity's sake. This usually looks bad. One street or a portion of a commercial landscape often looks best when planted with one species, especially if trees will be placed in a row. This ties the design together. What many people suggest is a mix of species throughout the community.

Perhaps it began in the 1800s in this country with planting of American elm, certainly one of the most elegant street and shade trees to grace our cities. So many were planted that when disease came, disaster struck. Many communities now have only fragments of the original plantings due to devastation from Dutch elm disease. Those with significant numbers of American elm along their streets, such as Washington, D.C., spend considerable sums of money on regular monitoring and treatment programs.

The London plane and sycamore trees were very popular and overplanted in many communities until the last decade or two, when large numbers began dying from disease in certain communities. Austrian pine can be seen planted almost exclusively in many midwestern landscapes. They too grew fine until a tip blight disease began killing them in some regions. In the 1960s, honeylocust began to emerge as the perennial favorite street tree in many communities in the northern United States. Now several problems are beginning to catch up with this tree. Japanese black pine was almost exclusively planted along the New Jersey and Long Island shorelines in the 1970s and 1980s, until an insect began its devastating march. Bradford callery pear was considered a standard tree in the 1970s and 1980s in many areas of the country, until it was determined that the structure was weak and they break apart.

The stories are all similar. Only the tree species and the insect, disease, or other causal agent changes. If we listen, these experiences teach lessons. Plant a variety of trees in the community. This might require more work and creativity on your part. Read this book, read other books, and talk to nursery operators and teachers. Do not rely on the "magic pill" tree—though it might be easier in the short term to rely on the same five types of trees that have been used for years in your community. Reach out, touch, and plant some new trees for a more lasting urban forestry program.

It is impossible to predict where the next major devastating insect or disease will strike. One state, Florida, has almost exclusively planted live oaks along streets, in parking lots, in commercial landscapes, and in every other imaginable place. There are cities in the southeastern United States that are overplanted with willow oaks. Red oak is currently one of the most popular trees in the northern United States, and baldcypress is finally being recognized as a tree well adapted to many urban and suburban landscape sites. Will these too be overplanted? The reasons for this are complex, but changes must be made if we are to develop an urban forest that can be sustained.

SPACING BETWEEN TREES

Trees are often spaced apart according to their mature canopy spread. Tree spacings of 50 to 60 feet are commonplace. This allows the open grown form of the tree to develop. There are also nice examples of trees in urban and suburban landscapes spaced much closer. A more natural upright form of the tree usually develops with close spacing (Figure 1-13 a and b). There are advantages and disadvantages associated with both strategies (Table 1-3).

MAKE THE FINAL TREE SELECTION

Trees often live for decades (sometimes longer), so spend the time conducting a proper analysis as outlined in this chapter. Time spent during this process does not cost anything. It pays you back in lower air conditioning bills, reduced tree and sidewalk maintenance costs, healthy trees, and compliments from friends, neighbors, and citizens in the community. Everyone wins!

(a)

(b)

1-13 (a) Trees planted more than 30 feet apart develop an open, spreading canopy. (b) Those planted close together form a more upright canopy akin to that in the forest.

TABLE 1-3 TREE SPACING IMPACT ON URBAN FORESTRY PROGRAM

Advantages	Disadvantages
CLOSE SPACING (20 TO 30 FEET):	
1. quick shade	1. outer trees bend away from others and are one-sided
2. cathedral-like canopy	2. if one or several die, adjacent trees may be susceptible to wind damage
3. less pruning of drooping branches required due to the upright canopy	3. more costly than farther spacing
4. more natural, upright form develops	
FARTHER SPACING (40 TO 60 FEET):	
1. less costly than close spacing	1. except for trees with an upright habit, drooping branches need regular removal
	2. large lower branches often develop, requiring removal and creating a large wound that can initiate decay
	3. longer time needed to form a closed canopy

Chapter 2

Special Planting Situations

INTRODUCTION

This chapter discusses some of the most commonly encountered urbanized landscape situations. Various design suggestions are presented as guidelines from which specifications can be developed. Some of the specifications and design suggestions have been in practice for a number of years; others are new and have not been thoroughly tested. Discuss these ideas with engineer and architect colleagues before specifying them for landscape sites. Many of them will have to be modified and fine-tuned to the conditions at your site.

STREET TREES

Form and habit

Consider your design objectives, pruning budget, equipment, and crew capabilities before selecting a form of tree best suited for your streets. Although all trees need some pruning in the early years, the amount, type, timing, and technician skill level required often depend on tree habit and form (Table 2-1). The pruning requirements of three major tree forms—upright, pyramidal, and rounded—are compared here (see Chapter 1 for more details on street tree selection).

TABLE 2-1 ADVANTAGES AND DISADVANTAGES OF PLANTING TREES WITH DIFFERENT CANOPY HABITS OR FORMS

TREE HABIT	ADVANTAGES	DISADVANTAGES
Upright	• Little pruning required to remove drooping branches • Trunk usually stays small or medium-sized	• Trees cast only a small amount of shade • Major branches on some trees can develop weak included bark and split from tree • Pruning occasionally needed to develop good branch structure
Pyramidal	• Strong trunk and branch structure often develops with little or no pruning • Lower branch removal can be performed by relatively unskilled labor • Lower branches are small and create only a small wound when removed • Trees cast a moderate amount of shade	• Lower branches often droop and need regular removal
Rounded	• On young trees, fewer branches droop compared to pyramidal trees • Trees cast abundant shade	• Highly skilled pruning crews needed • Regular pruning needed to develop strong structure • Branches that eventually droop, requiring removal, often are large and leave a large wound

Upright Many trees along streets, in parking lot islands, and in other high-traffic areas require periodic lower branch removal after planting to allow pedestrians and vehicles to pass safely. This maintenance requirement can be reduced by planting trees with an upright or narrow canopy (see Appendix 10 for a list of trees with an upright, narrow habit; note, though, that some trees in Appendix 10 may not be suited for planting along streets due to other characteristics). Branches on these trees grow mostly upright and do not droop into the path of pedestrians and vehicles (Figure 2-1 a-d). Compared to rounded or spreading trees, upright trees provide limited shade, due to the narrow canopy. The abundant shade cast by trees with a rounded canopy usually comes with a bigger price tag. That price tag is a higher pruning budget required to create a well-structured, rounded tree. The advantages and disadvantages of these two sides of the story must be balanced in your decision-making process.

Do not become complacent after planting trees with an upright growth habit. Some develop aggressive upright branches, with weak, included bark in the crotch (Figure 2-2 a-c), if corrective pruning is not done when the tree is young. This pruning should be directed at preventing development of codominant branches on the lower trunk.

2-1 (a) Upright trees cast shade primarily in the early morning and late afternoon but require less pruning. (b) Spreading trees cast significant shade all day long. Branches eventually droop and require removal. (c) Pyramidal-shaped trees require removal of drooping branches even when they are young but branches remain small in diameter (d), creating only a small wound when removed.

(a)

(b)

(c)

(d)

2-2 (a & b) Weak crotches have bark included between the trunks. (c) Strong crotches are wider, without included bark.

Without this, branches could split from the tree later. However, this characteristic may be typical of the tree and it may not be practical to try to correct it with pruning.

Despite the risk of developing included bark, many upright trees stay together and will not break apart easily, even without pruning. If the life expectancy of the tree at the site is less than 30 years, because of site constraints, you may not need to be concerned about trees with included bark. Many trees along streets planted in sidewalk cutouts and other small soil spaces, such as parking lot islands, live less than 30 years.

Rounded Because fewer branches droop on young oval, spreading, or round-shaped trees (decurrent growth habit) than on pyramidal trees, this type of tree might not be pruned in the first 10 years after planting, because there may be few complaints from citizens about drooping branches (Figure 2-1). However, without some pruning, tree structure may deteriorate due to weak branch attachments (Figures 2-2 and 2-3). During this 10-year period, trees can develop several to many aggressive, upright trunks, some with weak, included bark in the crotches. This could lead to breakage in storms as trees grow (Figure 2-4 a and b). In addition, without regular pruning, fast-growing branches that initially grew upright on the lower trunk can begin to droop, and may have to be removed. This often leaves a large wound in the trunk because low, fast-growing branches can develop large diameters (Figure 2-5 a and b). This large wound can weaken the trunk and initiate decay. Therefore, preventive pruning to develop good structure when trees are young is recommended on all trees, especially those with a rounded canopy shape. Pruning at two, five, and ten years after planting should be considered a minimum requirement. Pruning at one, three, five, and ten years would be even better.

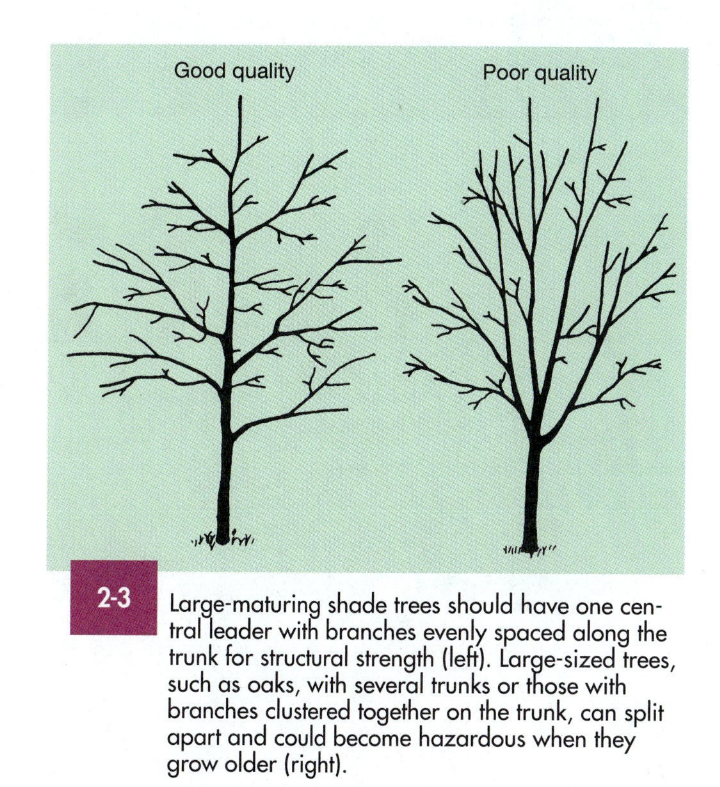

2-3 Large-maturing shade trees should have one central leader with branches evenly spaced along the trunk for structural strength (left). Large-sized trees, such as oaks, with several trunks or those with branches clustered together on the trunk, can split apart and could become hazardous when they grow older (right).

Pyramidal Trees with a pyramidal form (excurrent growth habit), such as deodar cedar, southern magnolia, and pin oak, frequently have a dominant central trunk far up into the canopy. They eventually cast abundant shade. Trees with a dominant trunk are considered to be strong and durable in urban landscapes. Little pruning is required to create this strong tree structure. However, lower branches on pyramidal trees often droop toward the ground. These have

2-4 (a) Round-shaped (decurrent) trees may form poor structures and can break apart if not pruned regularly when young. (b) This often creates a large wound in the trunk. Trees with this extensive damage usually require removal.

(a)

(b)

(a) (b)

to be removed regularly to allow pedestrian and vehicular passage beneath the canopy (Figure 2-1).

Several prunings may be required, but most of the work can be performed by crews on the ground. Little tree climbing is required. Crews that remove only drooping, lower branches need fewer skills and less specialized training than those pruning higher in the tree. The diameter of drooping branches on pyramidal-shaped trees is usually relatively small, resulting in only a small wound when removed. If double leaders or multiple trunks form in the canopy, they can be easily corrected by removal when lower branches are pruned.

Initial size

Trees often grow slowly along streets in urban sites with limited soil space for root growth. Even though a large-sized tree may grow slowly after planting, it provides an initial impact at the site. Trees less than four inches in trunk diameter provide limited impact at planting and are more easily damaged by mechanical injury from vandals and

2-5 (a) Round-shaped trees often develop large low branches without regular pruning. As the tree grows, these droop into the way of traffic and must be removed to allow for vehicle passage. (b) The large wound created by branch removal can initiate decay in the trunk.

others than larger trees. Bark on large trees is thicker, which makes the tree more resistant to mechanical impact. For these reasons, larger trees often are preferred over small nursery stock for planting along streets, in parking lots, and in other highly urbanized sites where people will be in close contact with the tree. Of course, larger trees take a long time to establish, are expensive to purchase, and are difficult to handle (see Chapter 3 for a complete comparison of large versus small trees).

Fruit and foliage characteristics

Some trees produce large fruit and foliage that falls to the ground and clogs sewers. When large hard fruits from horsechestnut, hickories, pitch apples, and other trees are kicked up by car tires, they could hit windshields and startle drivers. Large, rounded fruits could also become projectiles. Other trees drop soft, fleshy fruits that become slippery when they squash on the sidewalk or pavement, creating a potential slip hazard for pedestrians. Some produce fruit that rolls and causes slipping problems for pedestrians. The fruit issue is less important if the nearby sidewalk is infrequently used or if there is no walk. It is also less important for a highway median strip, as pedestrians are usually a good distance from these trees. Trees without these characteristics are often considered better suited for planting along streets and near parking lots, patios, and decks (Appendices 15 and 16).

Some trees, such as cottonwood, tuliptree, and sycamore, drop foliage over an extended period of time, beginning in the summer. The dropping needles and cones of pine trees cause concern for some individuals. Semi-hard fruits (such as acorns and many palm fruits) litter the ground, but they break easily under foot traffic and are washed down sewers or into lawns with rain. These characteristics normally do not limit the trees' usefulness as street trees. Remember that no trees are completely "clean." All trees drop something during certain times of the year. Some are simply messier than others.

Street or highway medians

Trees in strips of soil between opposing lanes of multilane boulevards (highway medians) should be chosen according to the width of the median (Table 2-2). Trees that have a mature size considered too big for the median often grow slowly due to confinement of the root system. A poorly developed root system could also cause the tree to fall over as it becomes larger. Roots on large trees quickly fill the soil space, causing a pot-bound effect and more stress than for trees that are appropriately sized. Increased irrigation, fertilization, and pest control measures raise the maintenance costs of this planting scheme. Roots on trees that are too big for the space could eventually break curbing if roots grow under the pavement. One exception to this rule-of-thumb appears to be palm trees, which have small-diameter roots that rarely disrupt curbing. They are often successfully planted in smaller spaces than outlined in Table 2-2.

Trees in medians should be tolerant of drought to account for increased air and soil temperature near pavement, even if they receive irrigation (Appendices 7, 8, and 9). Trees should be tolerant of salt (Appendix 3) if de-icing salt is regularly applied near the root zone. Upright trees may be more appropriate than low-branching, multitrunked trees with drooping branches in instances where drivers need visibility (refer back to Figure 1-12).

Trees near sidewalks

The space between a street curb and the sidewalk is called a treelawn, streetscape, or parkway, depending on the region of the country. Small trees are best suited for planting in narrow treelawns because their roots do not grow large. Save large-maturing trees like oaks for spaces greater than seven or eight feet wide (Table 2-2), because their roots can

TABLE 2-2 PLANTING AREA GUIDELINES

IF YOUR SITUATION MATCHES ONE OF THE FOLLOWING:			CHOOSE THIS SIZE TREE TO FIT:
A **Total planting area (lawn, island, or soil strip)**	B **Planting strip width**	C **Distance from pavement or wall**	**Maximum tree size at maturity[1]**
Less than 100 square feet	3 to 4 feet	2 feet	Small (less than 30 feet tall)
100 to 200 square feet	4 to 7 feet	4 feet	Medium (less than 50 feet tall)[2]
More than 200 square feet	More than 7 feet	More than 6 feet	Large (taller than 50 feet)[2]

[1]Refer to Appendices 7, 8, and 9 for lists of small, medium, and large trees.

[2]In compacted or poorly drained soil, increase recommended distances and planting area size to compensate for aggressive surface root formation.

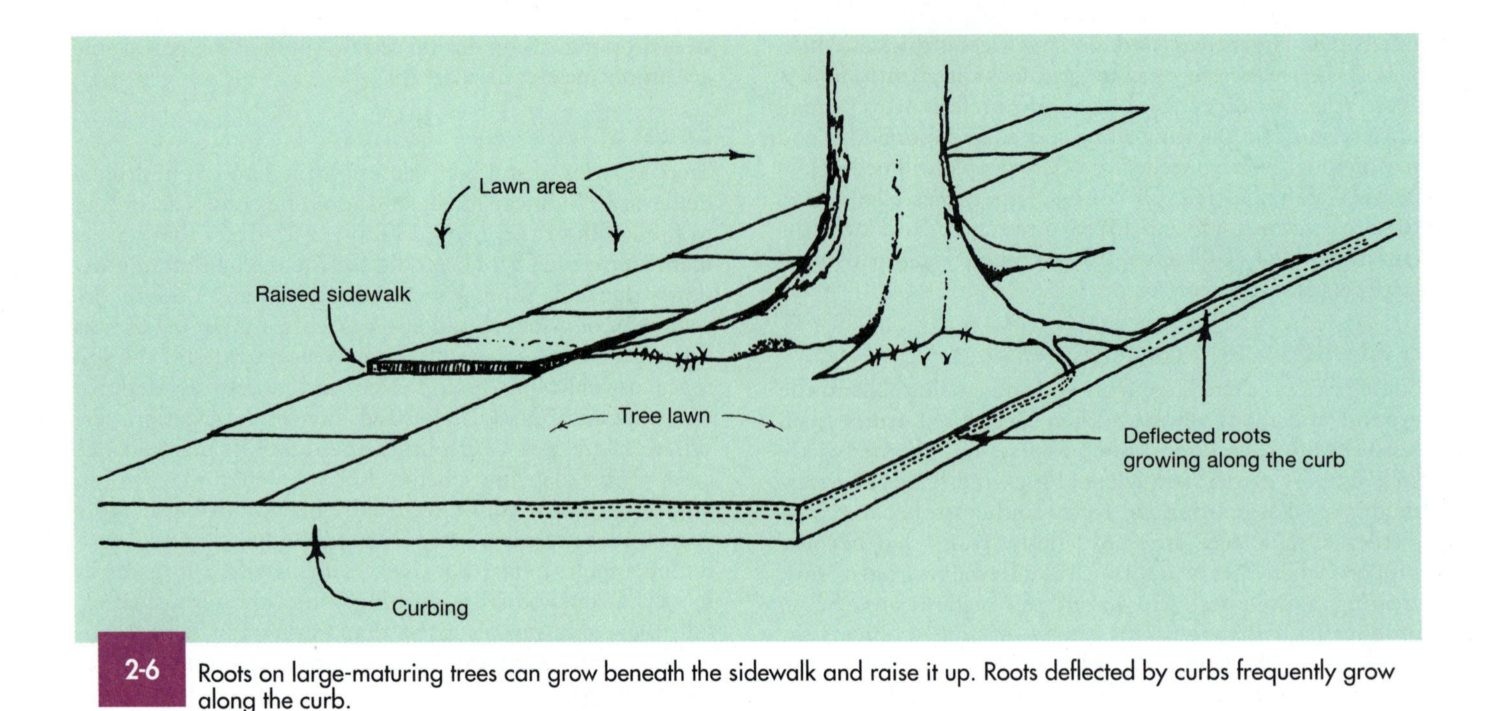

2-6 Roots on large-maturing trees can grow beneath the sidewalk and raise it up. Roots deflected by curbs frequently grow along the curb.

raise the sidewalk and pavement if planted too close. Do not plant trees with wide, buttressed trunks or invasive roots in a treelawn (see Appendix 11). Even when these guidelines are followed, roots often grow just under the sidewalk slab and lift it up (Figure 2-6). Root barriers or specially designed sidewalk subbases may help prevent this in certain circumstances (see "Preventing root damage to sidewalks, pavement, and curbs" later in this chapter).

Trees planted along streets without a sidewalk or with a sidewalk attached to a curb should be set back from the curb or walk according to their mature size. Though prevention of walk damage cannot be assured, the guidelines in Table 2-2 will help keep curbing and walks intact and will help trees produce a well-formed root system. Trees planted closer can cause significant walk or curb damage soon after planting. Roots at the base of trees often become deformed or injured on trees planted too close, causing poor tree growth or trunk rot. In compacted or poorly drained soil, increase the distances given in Table 2-2 to account for aggressive surface root formation, especially on large-maturing trees.

Shade will be provided to a street, pavement, sidewalk, or other hard surface quickest if trees are planted close by (refer back to Figure 1-13). Setting the tree far back from the walk or street (e.g., 10 feet) reduces the chances of roots damaging the hard surface, but prolongs the time needed for the tree to provide shade to the hard surface. These two objectives must be balanced to create the best planting strategy in your community.

Street-side islands

Trees are often planted in islands of soil surrounded by pavement in downtown locations. These can be treated as parking lot islands (see "Parking lots" later in this chapter).

Street-side buffer strips

Street-side buffer strips can be treated as parking lot buffer strips (see "Parking lots" later in this chapter).

Highways and turnpikes

Trees planted along turnpikes and highways need no special attributes other than drought tolerance (Appendices 7, 8, and 9). Large nursery stock often transplants poorly on these sites because trees are watered infrequently. Small saplings, seedlings, and even seeds are more appropriate. Trees with thin bark might be damaged more than thick-barked trees when mowers and wildlife brush against them. Some people object to planting nonnative trees such as Amur and Norway maples, Chinese tallow, English oak, and others that seed themselves into the surrounding landscape. Although this might be considered desirable in some respects (i.e., trees will begin reproducing themselves, thus adding to the total tree population), their invasive habit may displace native trees from the habitat under certain circumstances.

SIDEWALK CUTOUTS

If possible, plant in another area. A sidewalk cutout (also called treepit or tree well) is a very difficult place to grow a tree. Sidewalks around cutouts and along streets with large-maturing trees nearby often need replacement about every 10 years; sidewalks often last 25 years or longer without trees nearby. Yet we continue to imagine the big tree in the small cutout in the walk. This is mostly fantasy because tree life averages about 10 years in this type of site, although there are certain instances (for example, in sandy soil) in which trees live longer than the average. Tree life can be even less than 10 years in the most stressful sites.

When choosing trees for such a site, be sure that soil drainage and soil compaction tests were conducted carefully. Unless soil is sandy or loamy, drainage and aeration are often poor and soil beneath the sidewalk is compacted. Roots of most trees grow poorly in these conditions and may remain in the loosened soil in the cutout. Those that manage to escape the soil in the cutout often grow directly under the sidewalk slab or along the curb, raising them up or displacing them. Trees that tolerate both wet sites and drought often grow better in the unpredictable conditions found in cutouts than other trees (Appendix 2). If soil under the walk and in the cutout is sandy and well drained, choose drought-tolerant trees without concern for flood tolerance.

Minimum cutout size is often considered four feet by four feet for small trees, but should be increased to six feet by six feet for larger-maturing trees. Surface roots close to the trunk on healthy trees in sidewalk cutouts often raise sidewalks. If given the opportunity, plant in an area with more open soil, or design the site for adequate root growth (see "Preventing root damage to sidewalks, pavement, and curbs" later in this chapter). Trees with small-diameter main roots heave sidewalks less than other trees. Palms and small-maturing trees usually fit this criteria.

Trees in cutouts should be free of long thorns on trunks and main branches. Short thorns or spines might be tolerated if the maintenance budget allows for regular trimming of low branches. Spines on leaves, such as those on some hollies, are usually acceptable but leaves on palms with spiny petioles should be at least seven feet from the ground. Trees should be free of fleshy fruit that would become slippery on walks. Small hard fruits like acorns are probably acceptable provided it breaks under the pressure of foot traffic. Trees with fruit that is large or hard enough to resist breakage under foot traffic should probably be planted elsewhere (see Appendix 16).

Trees planted in cutouts with a tree grate need regular attention. Grates require periodic enlarging to accommodate trunk and root growth (Figure 2-7). If this is not performed, trees can die from strangulation of the trunk. Designate the agency that will maintain the tree grates *before* installing trees in grates, or do not install them.

2-7 Tree grates must be widened periodically to accommodate trunk growth, or the trunk could become girdled.

Three basic designs

It is very expensive to plant trees in sidewalk and pavement cutouts. Money is often wasted on poorly conceived plans. There are three basic methods of designing sidewalk cutouts. Each has associated costs and benefits to the community.

For communities with a short-term view Design the site so roots essentially remain in the small soil area of the cutout (this is most common).

Result: Tree grows slowly, dies and is replaced 5 to 10 years after planting; the sidewalk stays mostly intact.

Costs: Inexpensive to install; recurring but known cost of replacing tree every 5 to 10 years.

Community benefit: Little. Trees are rarely healthy and never become large; they provide little shade. This is a poor investment because money is spent with little benefit to anyone.

For communities with a medium-term view Design the site so roots grow out of the soil in the cutout and into the soil under the walk.

Result: Tree grows well and is fairly healthy, but sidewalk often heaves and the tree is blamed.

Costs: Larger initial cost; recurring costs and root damage from fixing pavement.

Community benefit: Significant. Tree grows well and becomes large, providing significant shade, but is often damaged during pavement repair.

For communities with a long-term view Design the site so roots grow out of the soil in the cutout and under the walk in an engineered fashion.

Result: Tree grows well and is healthy; sidewalk remains intact.

Costs: Largest initial cost but no recurring costs.

Community benefit: Significant. Tree grows well and becomes large, providing significant shade.

No studies have done this cost-benefit comparison, but it would not be surprising if the long-term costs of each of these three strategies were similar. If this is the case, it makes little sense to pursuit the now-common short-term strategy.

Promoting root growth

One of the biggest problems with encouraging healthy trees in poorly designed cutouts is that roots near the trunk often raise the surrounding sidewalk. Techniques designed to encourage root growth beneath the pavement around individual sidewalk or pavement cutouts without affecting pavement integrity are in practice today (Figures 2-8 and 2-9 a-d). Remember that all these designs include compromises and that they are not suited for all sites. Many are

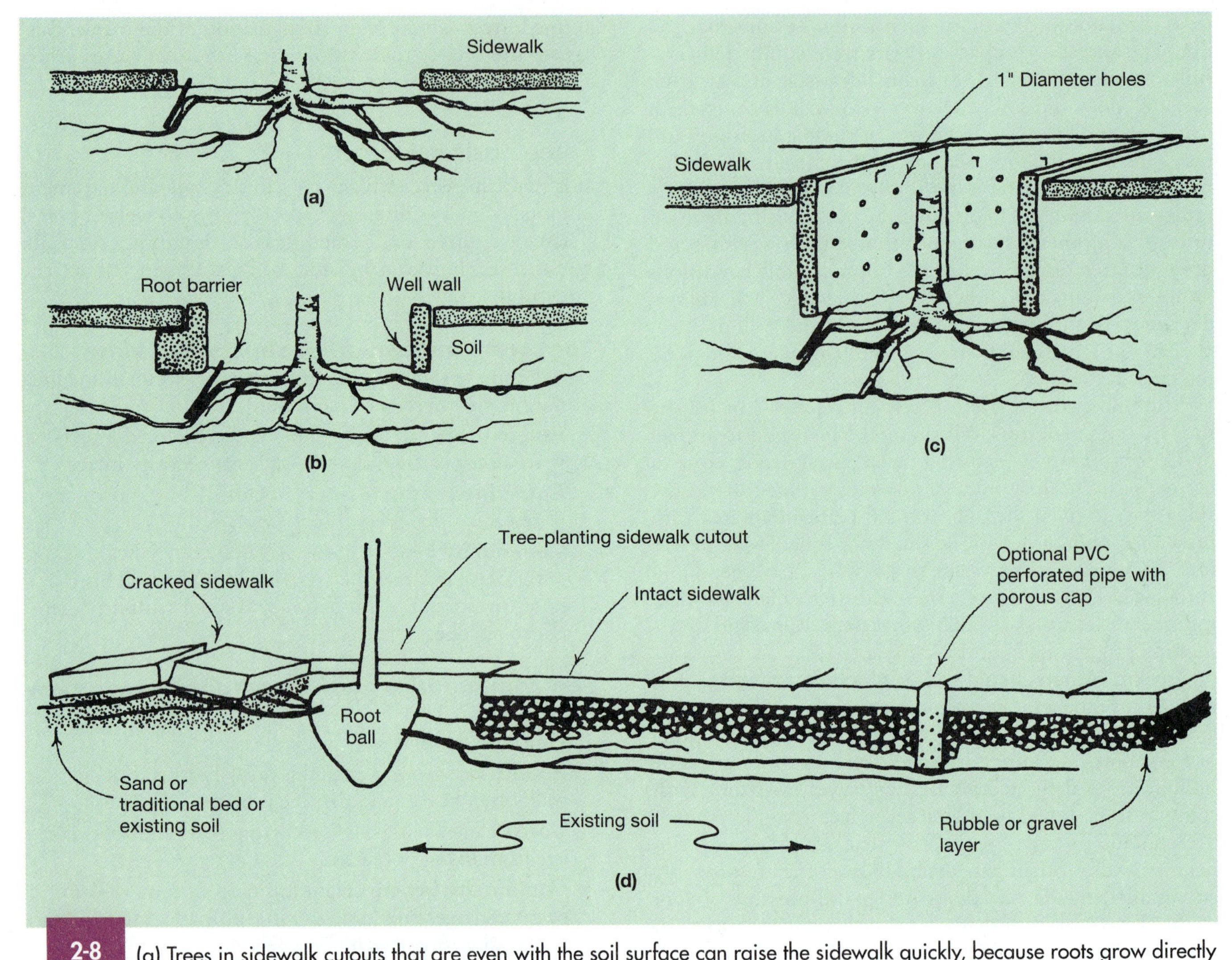

2-8 (a) Trees in sidewalk cutouts that are even with the soil surface can raise the sidewalk quickly, because roots grow directly under the walk. A root barrier might help direct roots under the sidewalk and could prolong the life of the walk. However, deflected roots often come back to the surface farther from the tree. (b) Roots can be directed well under the sidewalk by planting in a shallow well covered with a tree grate (grate not shown). If the end of the sidewalk rests directly on soil, roots growing directly beneath the well wall may raise the wall but not the sidewalk next to the tree. If the sidewalk rests on the well wall, it will be raised as roots grow beneath. A root barrier may help prolong sidewalk life. (c) Deep wells have been used successfully in well-drained soils in parts of California. They are not suited for poorly drained soil. (d) Coarse gravel (at least two-inch particle size) used as a sidewalk base helps keep roots well below the walk.

quite expensive. You may choose to modify certain parts of a design by substituting a concept from another.

Perhaps the most sophisticated design is to suspend the entire sidewalk over the soil on pilings or angle irons. The concrete sidewalk slabs do not touch the soil (a modification of this is shown in Figure 2-9b). This provides air and water to the trees, and roots can grow in the loose soil under the walk. This is rarely done but has been used very successfully along Pennsylvania Avenue in Washington, D.C.

Another method places stone pavers over a structural soil mix (see "Engineered soil" later in this chapter) or, less preferably, over loose soil connecting cutouts; the sidewalk can also be cantilevered over loose soil. Pavers tend to heave from freezing and thawing of the soil, especially in clay soil. Another alternative to conventional design is to provide a long strip of loose soil connecting a row of cutouts (Figure 2-10 a and b). This creates a corridor of soil suitable for root penetration. These techniques allow pedestrians to walk freely over the area and provides air and water to the root system.

Pavers can also be eliminated if the row of trees is elevated slightly above the surrounding walk. This can be accomplished by building an 18-inch-high wall around the planting strip and backfilling with good soil (refer back to Figure 1-7). Trees are installed in this raised planter, which is open to the soil on the bottom. The walls of such planters can also act as seats for pedestrians. Short of building a wall, the long strip of soil without pavers can be surrounded by wooden or brick benches to discourage pedestrians from walking on and compacting the soil. Trees will be healthier and live longer in these systems than in individual, single-tree cutouts.

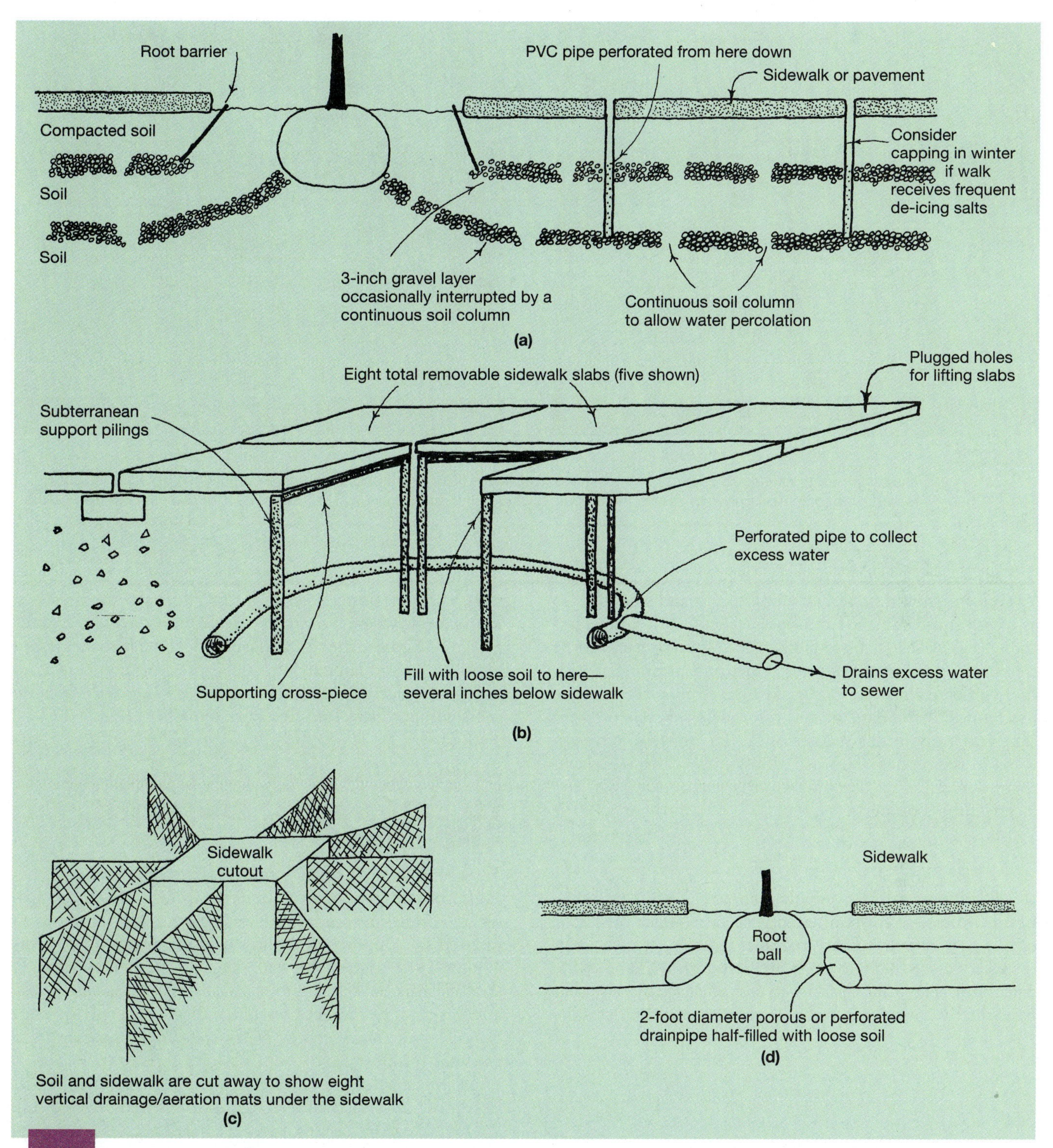

2-9 Four potential designs for sidewalk cutouts. (a) Gravel layers help keep soil aerated and could promote better root growth. Continuous columns of soil uninterrupted by gravel provide for drainage of water that would accumulate above gravel layers. Perforated pipe provides additional aeration to the gravel layer. The gravel layer may have to be covered with a soil-separator (geotextile) fabric to keep sandy soil from washing into gravel. Soil has to be kept out of the gravel for best air movement through the gravel. (b) The eight slabs surrounding the cutout are suspended on support pilings; they do not touch the soil. They are removable to provide access to lights, irrigation, drainage and other utilities. Soil under the slabs is loosely placed into the enlarged cutout to provide for improved root growth. Be sure to provide backflow protection so sewer water does not move into the perforated pipe. (c) Vertical aeration and drainage mat is installed before the sidewalk to provide for air movement under the sidewalk. (d) Porous or perforated drainage pipe half-filled with soil is installed beneath the walk. Two pipes are shown, but more could be used. A slow-release fertilizer could be incorporated into the loose soil in the pipe. Also, an irrigation system could be installed in the pipe. Roots may grow well in the loose soil in the pipe.

(a)

(b)

2-10 (a) The large sidewalk cutouts shown here are more hospitable than traditional single-tree cutouts. (b) Bricks or cobblestone laid onto loose soil between connected cutouts allows better soil aeration. This improves root growth and tree health.

Cutouts should be placed along streets and sidewalks so as to minimize damage from buses, trucks, cars, and pedestrians (Figure 2-11). Improperly placed cutouts can put the tree in the path of pedestrians. Opening car doors can hit the trunks of improperly placed trees. This often leads to mechanical trunk damage and can shorten the life of a tree that is already under stress from other unsuitable conditions in the cutout. Coordination among various government departments and contractors is needed to design and locate cutouts properly.

Soil replacement

If soil contains debris, is highly compacted, has a very high pH, is otherwise contaminated with salts, waste products, or oils, or is of poor quality, consider replacing it with a better soil. This is likely to promote better growth and could extend life span. Due to the high cost of this procedure, this is normally considered only for small areas, such as sidewalk cutouts along streets, parking lot islands, or along foundations where building waste products may have accumulated. Replace the soil as deep as possible if roots are not likely to expand laterally outside of the cutout. There is no point in replacing soil deeper than the water table because roots will not grow into the water table.

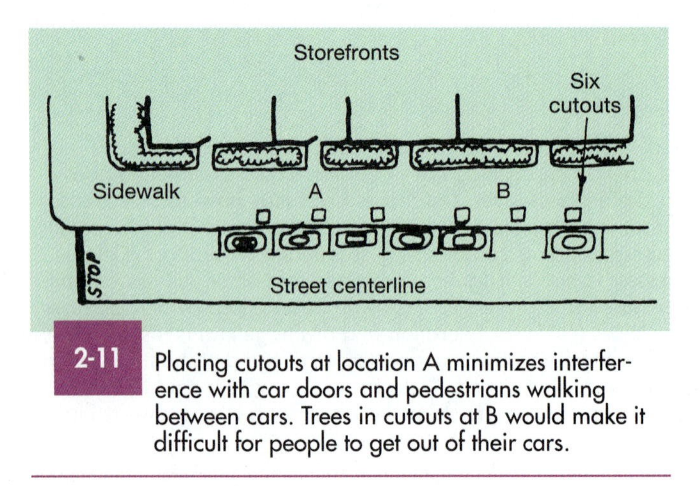

2-11 Placing cutouts at location A minimizes interference with car doors and pedestrians walking between cars. Trees in cutouts at B would make it difficult for people to get out of their cars.

PREVENTING ROOT DAMAGE TO SIDEWALKS, PAVEMENT, AND CURBS

Trees lift sidewalks because they are planted too close or because roots grow along the underside of the pavement slab where conditions are favorable for growth. When trees are planted too close to the walk, the trunk and roots eventually meet the edge of the walk. This cracks or lifts the walk and deforms the trunk and roots, often causing tree injury and internal decay. As previously discussed, this damage can be largely prevented by planting a proper distance from the walk (Table 2-2).

While this does not always prevent damage, it is a good beginning for a successful strategy. Properly engineered soil, altered sidewalk designs and root barriers could also provide solutions in certain instances.

Engineered soil

The soil under sidewalks, adjacent to sidewalks, and in sidewalk cutouts is usually compacted to prevent the walk from settling. This prevents many tree roots from growing in soil under the walk. Growth can be severely restricted, creating unhealthy trees, especially in sidewalk cutouts. The problem arises from the fact that loose soil is required for root growth, whereas compacted soil is needed to stabilize the walk. If the soil is not compacted, the walk will settle.

Perhaps the best way to create root space without compromising soil strength for sidewalk support is to use an engineered soil mix (sometimes referred to as structural soil mix) beneath the sidewalk. This strategy appears to be especially beneficial for clayey and loamy soil where compaction beneath walks can become quite severe. Research in the United States is only beginning, but scientists in the Netherlands have developed a soil mix that supports the weight of pavers placed over the soil, settles very little, meets compaction requirements of engineers and architects (95 percent of compactibility), and provides good growing conditions for roots (Figure 2-12). This mix is used to replace existing soil before pavers are installed. It is spread in lifts 12 inches thick and compacted. Settlement is minimal if it is compacted with 187 to 250 pounds per square inch (p.s.i.) during installation. A layer of engineered soil at least two feet thick (preferably three) is needed for proper root growth. Tests without the root barrier on the left in Figure 2-12 show that roots do not lift the pavers. The barrier would help direct roots away from the edge of the pavers if they must be installed close to the trunk. The design on the right of Figure 2-12 is proposed for poured concrete or asphalt sidewalks. However, the engineered soil mix presented in Figure 2-12 is not suited for use beneath this surface and it will not support vehicle traffic.

Another method being studied in the United States also relies on a structural soil mix. It is a skip-graded (sometimes called gap-graded) mix, meaning that certain particle sizes are completely missing. The soil mix is built with an aggregate material ¾ to 1½ inches in diameter and a filler soil such as a sandy loam. Aggregates must be used in sufficient volume so that when the soil mix is compacted to meet specifications, all aggregate particles are touching. When the mix is compacted, all the weight is transferred from particle to particle. Because the filler soil in the voids between aggregate particles does not become compacted, it remains suitable for root growth. This technology is developing rapidly and specifications are likely to change as research results become available.

It is crucial to determine the void space between compacted aggregate particles. This can only be determined by testing the specific aggregate you intend to use. Void space has varied from 35 to 40 percent for a number of aggregates. Fill no more than 95 percent of the void space with filler soil. If too much is used, the aggregate particles will not touch each other and the mix will not function properly. Reported ratios of aggregate to filler soil vary from 2.6:1 to 2.8:1 on a volume basis, although the proper ratio for your mix will depend on the materials used in the mix (Lindsey 1994a and 1994b).

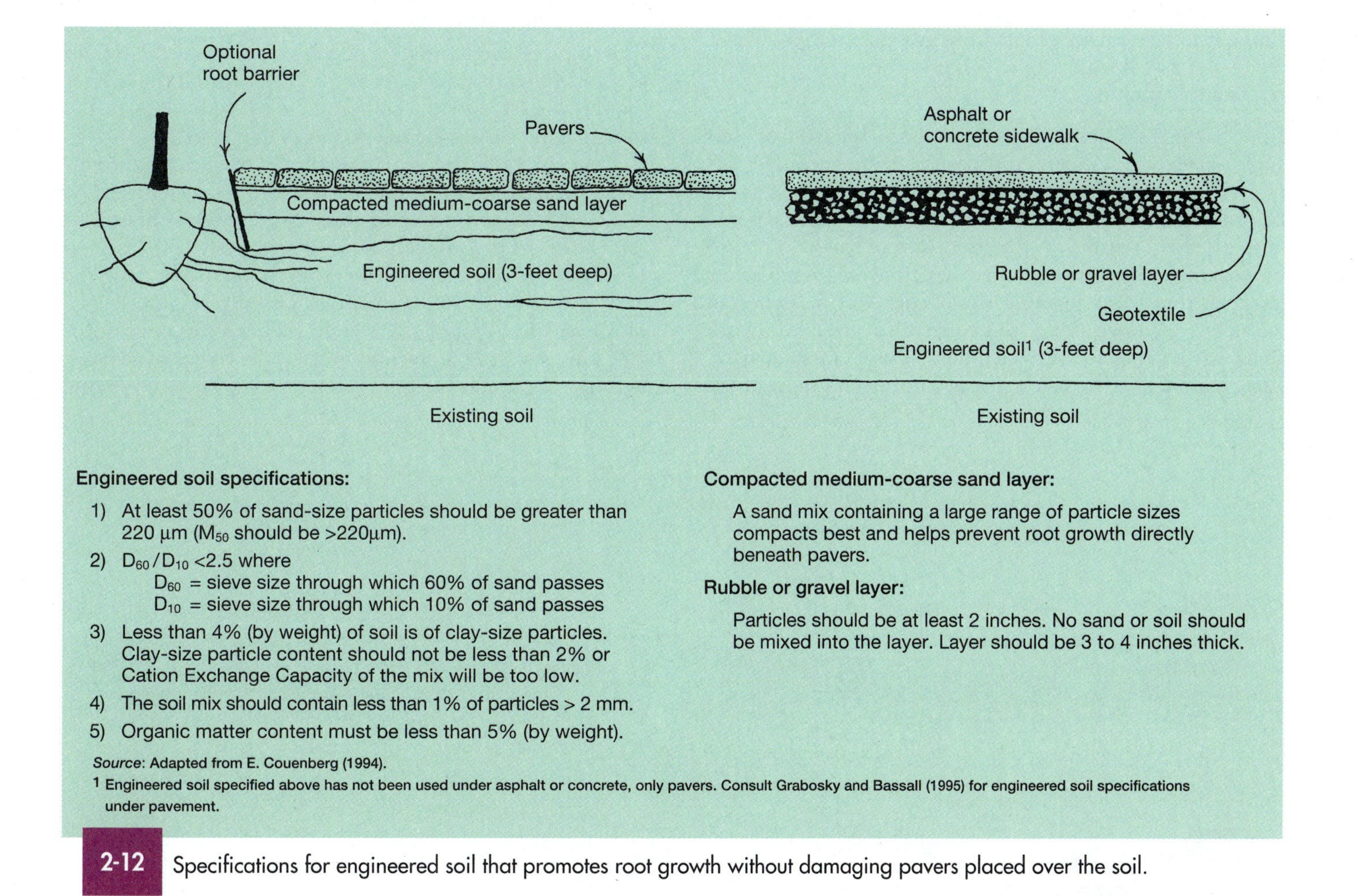

2-12 Specifications for engineered soil that promotes root growth without damaging pavers placed over the soil.

The newest, and perhaps most promising, procedure for supporting both root growth and heavy pavement loads is being developed at Cornell University (Grabosky and Bassuk 1995). The mix was placed under a street on campus, and at other locations, in 1994 and trees were planted at street-side. The test is designed to demonstrate that a special mixture of stones and soil can support both the weight of automobiles and trucks on the pavement above and also allow for adequate root growth.

Alternate sidewalk designs

A promising new concept in sidewalk construction may help prevent roots from damaging walks. Roots may be kept well beneath the sidewalk slab by installing a coarse rubble or gravel instead of sand as a base for the walk (Figure 2-13). Crushed glass has also been suggested for this. Instead of proliferating in the traditional sandy sidewalk subbase, roots grow in the soil below the rubble. This is likely to prolong the life of the sidewalk (Kopinga 1993). Exact specifications for proper gravel particle size have not been tested, but gravel greater than two inches in diameter has been effectively used in Amsterdam. Because the objective is to maintain large air spaces between particles, the particles should all be the same size, not a mixture of different sizes. *No sand or other soil or organic material should be mixed with the gravel.* If it is mixed in, roots are likely to grow in the mix. The air spaces between large particles are thought to help prevent roots from growing in the subbase.

Root barriers

Root barriers are designed to prolong the life of structures such as sidewalks, curbs, and pavement by keeping roots away. They are typically installed parallel to the structure to be protected. Unfortunately, there are no published reports, and little field testing, on the effect of root barriers on curb or pavement integrity. Related research on root barriers suggests that roots meeting the barrier will be deflected laterally and down under the barrier, but roots can work their way back to the surface under certain circumstances. Research on root barriers is just beginning and most of the following discussion is based on the few studies that have been conducted and published to date (Gilman 1995b; Wagar 1985).

Root barriers are made from plastic, geotextile fabric, herbicide-impregnated fabric, metal, and wood. Unfortunately, there is precious little research comparing the effectiveness of root barriers. Recent studies on sycamore and live oak show that fabric barriers impregnated with herbicide (Biobarrier™) are no more effective than those without herbicide (geotextile fabric). Both deflect roots laterally and under the barrier equally well (Gilman 1995b). Other work shows that poplar roots grow through geotextile and Biobarrier, whereas roots of other trees tested did not (Wagar and Franklin 1994). Further research is needed to test this on a larger variety of species in different soil types.

Root barriers that completely encircle the tree are occasionally used around the edge of sidewalk cutouts containing well-drained, coarse-textured soils to help prevent sidewalk lifting. This technique appears to direct roots beneath the edge of the cutout and into the soil under the walk. Some practitioners report extended sidewalk life due to this treatment and they recommend it highly. Others are not convinced of its effectiveness (Bernhard and Swiecki 1993). Unfortunately, this technique has not been thoroughly tested in the compacted and poorly drained soils typical of many urban landscapes. Two studies show that once under the barrier, some roots quickly grow to the soil surface (Gilman 1995b; Wagar 1985). In any situation, be sure to place the barrier far enough away from the tree to account for future trunk growth. Never install the barrier around the edge of the root ball, as this will deform the root system causing the tree to fall over or lean.

There is also little research on proper barrier depth or installation techniques. We do know that the barrier must be installed with about two or three inches above the soil or mulch. Roots will quickly grow over the barrier if the top of it is in the soil. Popular literature suggests abutting the barrier to the walk, pavement or curbing (Figure 2-14). However, installing it six to eight inches away might provide longer-lasting protection because the center of the root will be that much further away from the structure being

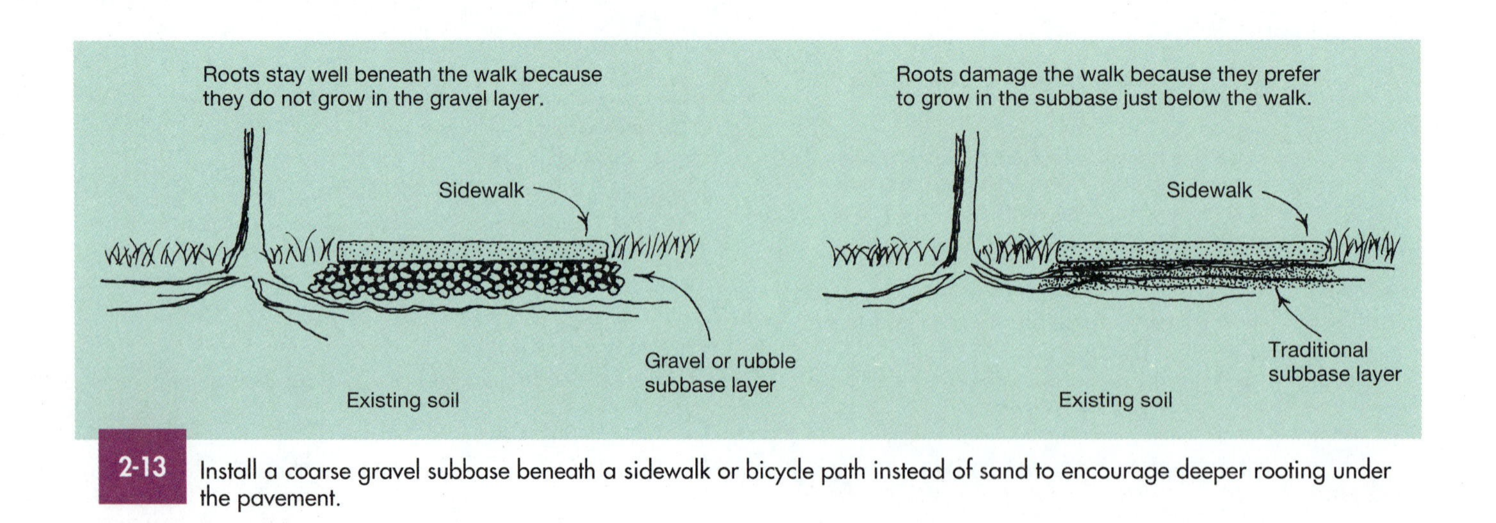

2-13 Install a coarse gravel subbase beneath a sidewalk or bicycle path instead of sand to encourage deeper rooting under the pavement.

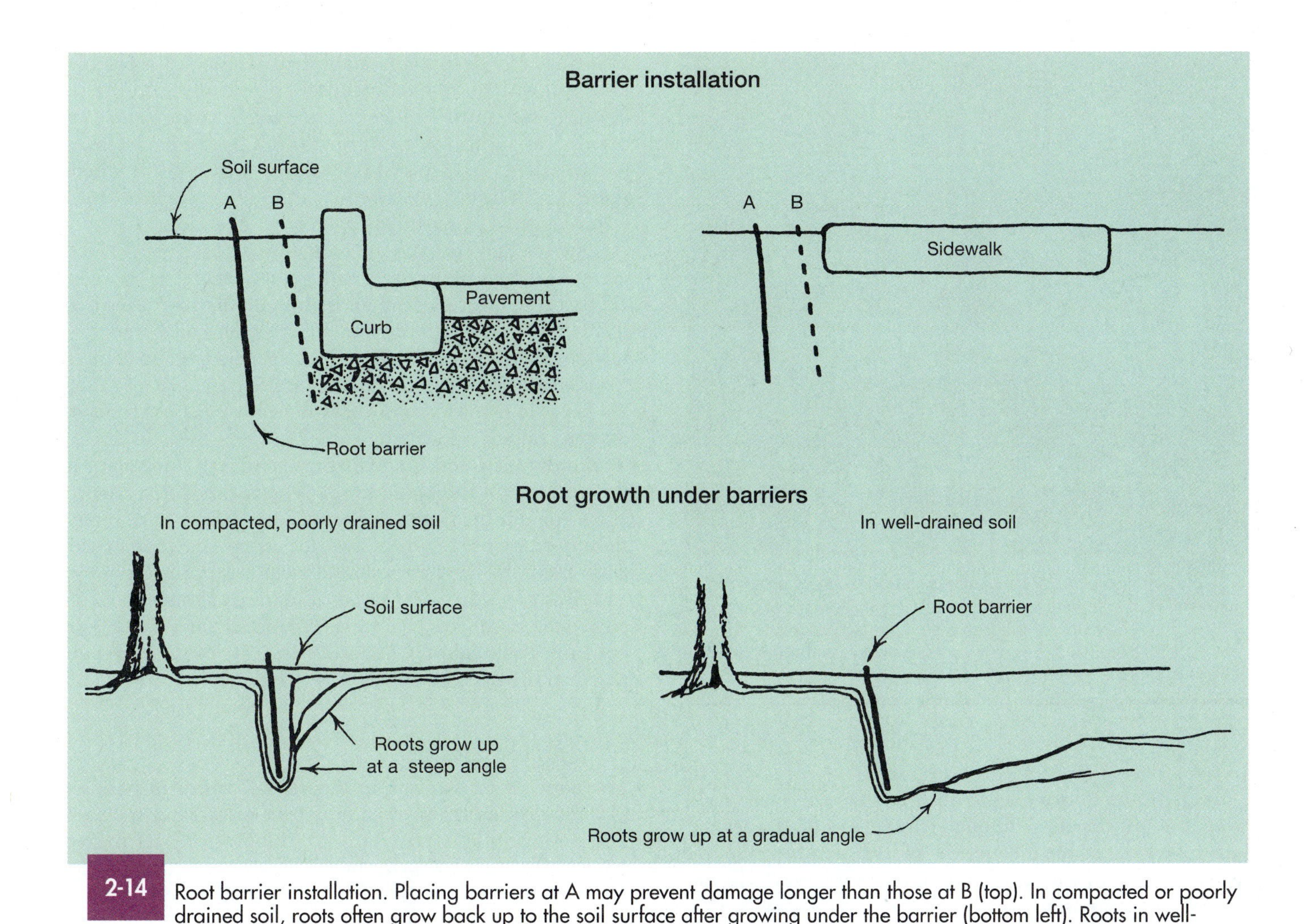

2-14 Root barrier installation. Placing barriers at A may prevent damage longer than those at B (top). In compacted or poorly drained soil, roots often grow back up to the soil surface after growing under the barrier (bottom left). Roots in well-drained soil may grow back to the surface at a shallower angle (bottom right).

protected. This will allow for significant root diameter growth before the root touches and disturbs the structure. There is probably little reason to install a barrier deeper than about six to eight inches beneath the bottom of the structure you are trying to protect, such as a curb or sidewalk.

Although the following has received only limited field testing, consideration could also be given to extending the barrier completely under the sidewalk before the sidewalk is poured. The standard vertical barrier deflects roots under the walk. The horizontal barrier under the walk would prevent roots from growing up. This would be a costly, but potentially effective, procedure. The vertical and horizontal barrier has to be made from one piece of material. If two pieces are used without sealing them together, roots will grow between them.

Roots that meet barriers are deflected laterally; they can also be deflected and grow down along the barrier if soil aeration and drainage are adequate. The process of digging the soil to install a barrier enhances aeration next to the barrier and probably allows better downward root growth on the tree side of the barrier. Roots grow freely along the barrier parallel to the soil surface because the soil is loosened here during installation. Barriers with vertical "ribs" might help direct some roots down so they do not grow along the barrier. Research is needed in this area.

One study in loose, well-drained soil showed that roots stayed deep and grew more or less even with the soil surface or grew up at a shallow angle once they grew under the barrier (Barker 1993). Unfortunately, many urban sites do not have loose, well-drained soils.

In the compacted or poorly drained soils typical of many urban sites, some deflected roots quickly return to the soil surface after they grow under the barrier (Figure 2-15). They grow straight up, sometimes in the loosened soil filled in around the barrier, once they reach the other side. As recommended by some manufacturers, this might be partially prevented by placing the barrier against the wall of the installation hole that is away from the tree. Backfill only on the tree side of the barrier and be sure to pack the soil tightly against the barrier.

Some roots reach the soil surface within six inches of the barrier, where they can potentially disrupt the walk. Many deflected roots grow up at a gradual angle, reaching the top six inches of soil several feet from the barrier. Although this helps protect the side of the walk closest to the barrier, it

2-15 Roots in poorly drained soil often grow up toward the soil surface after growing under a root barrier. The trunk of the tree from where these roots originated is 2.5 feet from the barrier on the upper left (not shown in this photograph).

puts the roots in position to disrupt the side of the walk away from the barrier. Although barriers in a poorly aerated site appear to reduce the number of roots reaching the soil surface, those that do reach the surface are large in diameter and can potentially cause significant damage to walks, pavement, or curbing.

These experiences cast doubt on the usefulness of root barriers in compacted or poorly drained soils. Ironically, this is just where barriers are most needed. To properly protect sidewalks, curbs, and pavement in this type of site, it might be best to consider alternate sidewalk designs. In contrast, roots deflected beneath the barrier in well-drained soil grow more or less parallel to the soil surface for at least several feet once they reach the bottom of the barrier (Barker 1993). In well-drained soil, fewer roots reach the soil surface if a barrier is installed. This may prolong the life of an adjacent sidewalk or nearby curb or pavement.

PARKING LOTS

Suitable tree attributes

Trees close to pavement should have a spreading or rounded habit to cast the most shade. Many of these require regular pruning of lower branches once they begin to droop into the way of traffic. This may be a small price to pay for abundant shade. Upright trees will not require as much pruning, but will not cast abundant shade (refer to Table 2-1).

Soil drainage must also be considered carefully. If water stands in a parking lot island after it rains, or if soil is compacted, then soil in the island may not support long-term tree growth. Even wet-site-tolerant trees may have trouble adapting and becoming established. Consider replacing poor soil with good-quality topsoil, providing a subsurface drainage system to remove excess water, or both. This can dramatically improve tree growth. If a drainage system is not installed, replace soil only down to the point where there is standing water, as roots are unlikely to grow into water-logged soil, regardless of its other qualities.

Due to high soil and air temperatures in parking lots, even if soil stays moist, trees with good tolerance to drought are best suited for parking lot islands and narrow soil strips (buffer strips). In clayey and many loam soils, wet-site-tolerant trees adapted to drought are generally good choices for islands, and they can become quite large because they are capable of producing roots beneath pavement where oxygen content is low (Appendix 2).

If only one tree is to be planted in each island, plant it in the center of the island or soil strip at the widest part to allow for uniform root and trunk expansion. Small trees should be planted at least two feet from the edge of the curb; medium-sized trees should be at least four feet away; and trees more than 40 feet tall at maturity should be six to eight feet from the edge of the island, if possible. These guidelines will help maintain the integrity of the curb and will help the tree anchor into the island.

Parking lot islands

The concept of planting in individual islands in parking lots should be mostly abandoned. In most cases they are not designed large enough to support long-term tree growth. If trees grow well, they often raise the surrounding curbs and pavement. If they grow poorly, the curb and pavement stay intact but the tree provides little (if any) benefit to the site. Instead, trees should be planted in linear soil strips at least eight feet wide (Figure 2-16). This provides for better tree growth by allowing roots from several trees to use the same soil space. Canopies of round and spreading trees provide significant shade to pavement in this type of design.

Alternatively, you could plant small trees in a large number of small islands. Their combined canopies could cast significant shade. Islands can be connected by strips of soil covered with open paver blocks to allow penetration of air and water. Exposing more soil to the air generally enhances tree growth, especially if you can design the site to prevent pedestrians from walking directly on the soil.

If you decide to use islands, they should be at least eight feet wide, to allow adequate trunk and root growth and to prevent car bumpers and doors from injuring tree trunks. In most circumstances, a narrower strip will not allow adequate trunk and root growth. Shade trees grow best in islands greater than 400 square feet. There are places where the natural gravelly or sandy soil at the site allows root penetration deep into the soil beneath the pavement. In these sites, you may see a large-sized tree growing in an island much smaller than 400 square feet. This is unusual and should not be used as a model for most other parts of the country; normally, roots grow poorly beneath pavement or they grow directly under the slab and damage it. Use Table 2-2 as a guideline for matching tree size with island size.

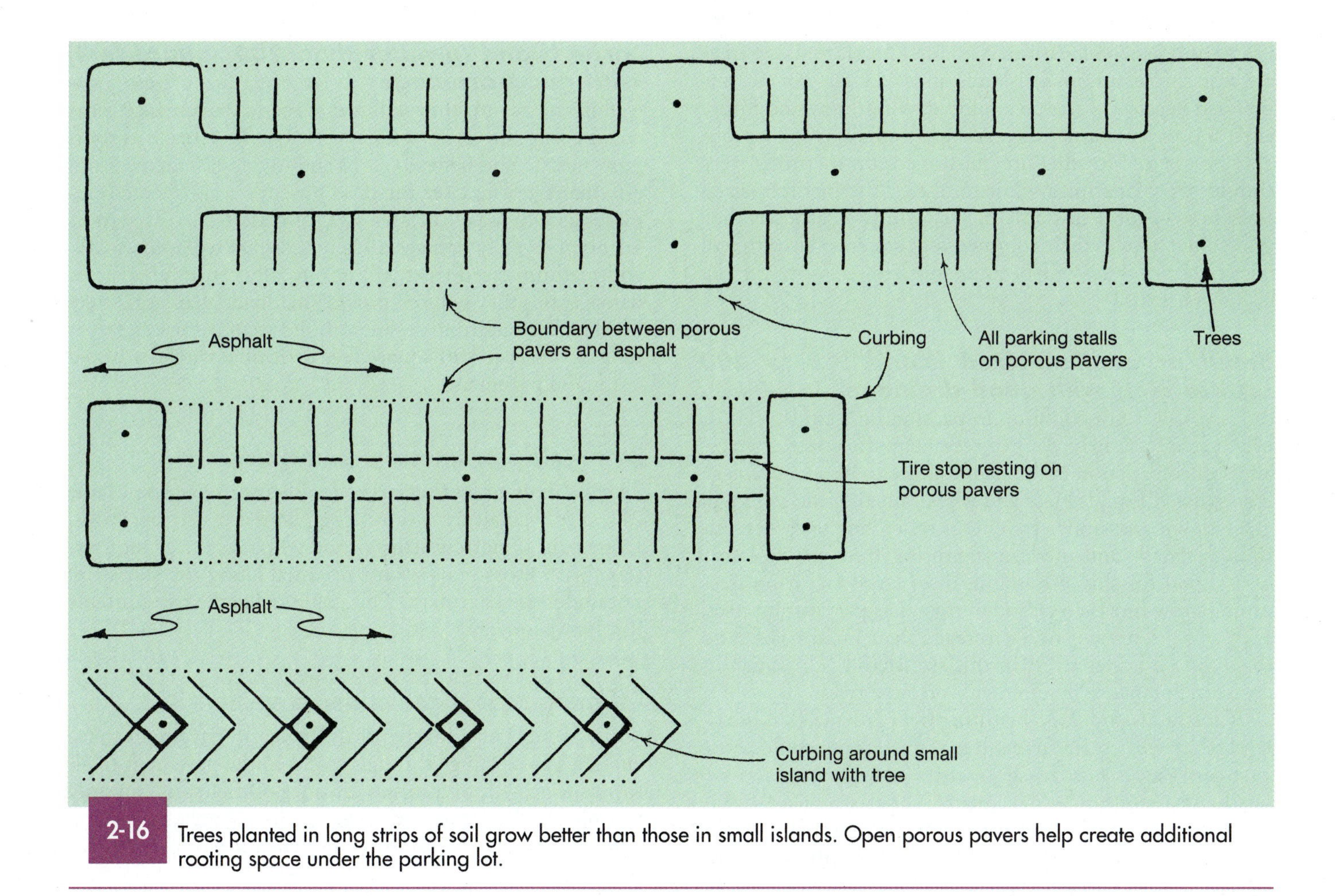

2-16 Trees planted in long strips of soil grow better than those in small islands. Open porous pavers help create additional rooting space under the parking lot.

To prevent soil compaction, parking lots should be designed so that people walk on pavement, not on the soil where tree roots will be growing. Soil compaction from pedestrian traffic may be one of the leading causes of poor growth of trees in parking lot islands and other small soil areas where tree roots are growing. Installing porous pavement or open paver blocks over soil that has not been overly compacted (refer back to Table 1-2) may be a good way to maintain both quality soil structure for root growth and solid footing for pedestrians (Figure 2-17). A structural soil mix might also be useful beneath porous pavement, but it has not been shown to hold up under the weight of automobiles. Use it only for pedestrian traffic areas.

The size of the open soil area should be considered (Table 2-2). Find the island size or buffer strip width and soil drainage in the following sections that most closely match your planting site. Each section contains site-specific tree specifications.

2-17 Open paver blocks in parking lot stalls allow water and air to reach the root systems of trees planted close by.

Tiny islands less than 45 square feet This tiny island is really too small for most trees. If you must plant, small-maturing trees are most appropriate. Trees that mature at any size can be planted, but they will be dwarfed or bonsaied by the small soil space. If roots can grow freely under the pavement in well-drained soil, the tree may get reasonably large, but pavement is likely to be damaged by roots unless special provisions are made (see "Creating root space under pavement" later in this chapter).

Small island (45 to 100 square feet) with poor drainage Trees planted in poorly drained or compacted soil will be shallow rooted. Except for some wet-site-tolerant trees, roots of most trees are not likely to escape the island and grow beneath pavement. Therefore, large trees are not suited for small islands and they are not recommended because of the possibility of windthrow and future curb and pavement damage. Although small trees are probably best suited for this site, most medium-maturing trees should grow fine for a while. Some of the medium-sized trees may become unstable in very windy weather as they grow older, due to their confined root systems. Growth will be slow and trees will become stunted as roots fill the island space.

Small or medium-sized island (45 to 200 square feet) with good drainage Loose, well-drained soil is not common in parking lot islands or most urban sites. You may want to reevaluate the water drainage and soil compaction at the planting site to be sure drainage is adequate! If soil is loose and well-drained, some roots will grow down and under pavement to anchor the trees. Although small- and medium-maturing trees are probably best suited for this size island, most large-maturing trees should grow fine for a period of time. Large-maturing trees, such as oaks, are not likely to reach their mature size, and they will be under stress in this size island. This may increase maintenance requirements on large trees.

If medium- and large-maturing trees grow well in the island, they will create shade, but surface roots could cause curb and pavement damage as roots grow next to and beneath curbs and under pavement. Property managers may see the need for curb and pavement replacement to make the area safe for pedestrians. Tree roots are often severely damaged when curbs are replaced. Small trees will not usually cause this type of damage, but they will not provide the shade that larger trees do. The best alternatives are to increase the size of the island or to live with disrupted curbs and pavement.

Medium-sized island (100 to 200 square feet) with poor drainage Small- and medium-maturing trees are probably best suited for this site, but some large-maturing trees, such as oaks, may grow fine for a period of time. Again, trees planted in poorly drained, compacted soil will be shallow rooted. The root system is often confined to the parking lot island and roots are not likely to grow beneath pavement in this type of soil. Large-maturing trees are not well suited to this size island because of the possibility of windthrow and future curb damage from roots growing along the curb. Large-maturing trees will not reach their mature size, and they will be under more stress than smaller trees. This will increase maintenance requirements on large-maturing trees.

Large island (greater than 200 square feet) with poor drainage Poor drainage at this site is likely to confine roots to the soil within the island. Many wet-site-tolerant trees will grow fine here, but large-maturing trees could become top heavy and fall over in very strong winds as they grow older, due to their confined, shallow root systems.

Large island (greater than 200 square feet) with good drainage If large-maturing trees grow well in the island, they will create shade, but surface roots could eventually cause curb and pavement damage as roots grow next to and beneath curbs and under pavement. Property managers may see the need for curb and pavement replacement to make the area safe for pedestrians. Tree roots are often severely damaged when curbs are replaced. Small- and medium-sized trees will not usually cause this type of damage, but they will not provide the shade that larger trees do. The best alternatives are to build islands for large trees that are at least 400 square feet or to live with disrupted curbs and pavement.

Buffer strip or linear planting strip

The width of a planting strip should govern tree size (Table 2-2). Only small trees should be planted in narrow strips. Large trees in narrow strips of soil will have linear root systems with most major roots oriented along the soil strip, not under the pavement. This could make the tree unstable if it grows too large.

Creating root space under pavement

Some sandy or gravelly soils beneath pavement have enough gas exchange through cracks in the pavement to allow root growth of a variety of trees. Other soils can only be colonized by roots of trees, such as honeylocust and baldcypress, capable of growing beneath pavement when the oxygen supply is limited (Appendix 5).

Conditions suitable for root growth of a variety of trees can be created by providing air to the soil beneath pavement. This is easier said than done, though several methods might be tried. For an existing site, vertical holes can be drilled into pavement and soil cores removed to make room for a perforated pipe (Figure 2-18). A cap that allows air and water movement into the pipe is placed over the hole to keep large debris and feet out. This is intended to enhance aeration, although research shows that it increases aeration only in a very limited zone of several inches around the pipe. The pipe must be U-shaped to be effective. This is difficult to install in an existing site. Experience also shows that these pipes tend to clog with sediment and become ineffective within several years. To help prevent clogging, the ends of the pipes are best located in soil above grade so that runoff water and debris are kept out.

The most innovation is possible when a parking lot is designed. The objective is to provide for root and air channels without compromising the strength of the soil supporting pavement or sidewalks. Porous pavement or open pavers can be installed for 20 feet or more around an island (Figures 2-16 and 2-17). Provided the soil has not been compacted beyond the critical bulk density (refer back to Table 1-2), this will increase rooting space by allowing greater aeration and water infiltration into the soil. It can also reduce runoff from the parking lot.

Because research in this area is only beginning, these are presented as possibilities, not tried-and-true techniques. It

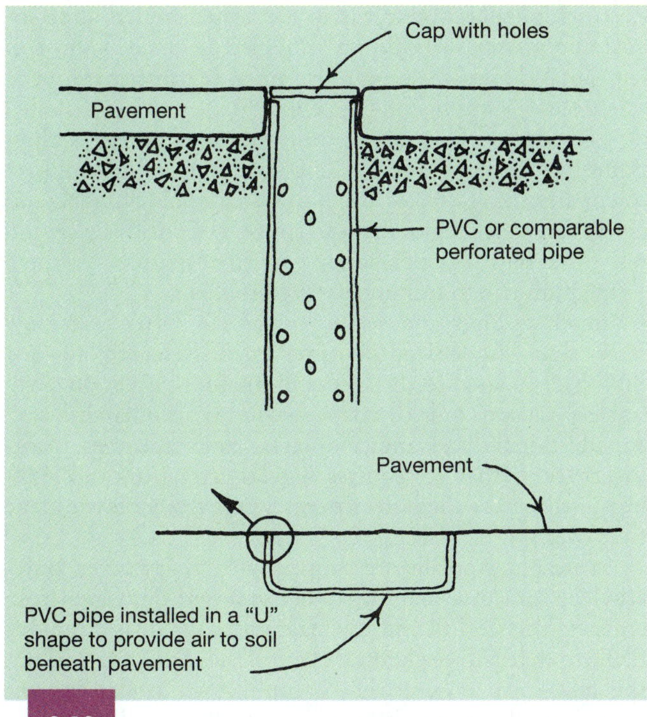

2-18 A perforated pipe can be installed vertically to help enhance air exchange beneath pavement. It should be connected to a horizontal pipe and to another vertical pipe nearby to form a U for best aeration. The U shape is essential for supplying good airflow to the subsoil. Vertical pipes alone are much less effective. Pipes should be no larger than 3 inches in diameter with no less than 6 percent perforation.

will be interesting to follow how these and other techniques actually help in promoting root growth beneath pavement.

To curb or not to curb

Instead of installing curbing around a parking lot island or median strip in the traditional manner, consider placing the curb on top of the pavement edge (Figure 2-19a and b). Curbing can be constructed from treated wood, which is lighter than concrete, to prevent pavement sag. Also consider building islands without curbs or with only shallow curbs. Deep curbs often deflect roots, preventing them from growing beneath pavement, which also restricts root growth and stresses the trees. Building parking lots without curbs around islands could increase tree health by allowing more roots to escape the island and grow beneath the pavement.

A possible disadvantage of this is that the property owner might perceive the raised pavement near aggressive surface roots as a hazard. Cracks created by roots also allow water to enter the soil beneath pavement; freezing and thawing in cracks then causes further pavement damage. But remember, most pavement cracks after it is installed, often without help from trees! There is a tradeoff to having healthy trees in this very stressful, urban planting site. Often, with current technology, if trees are growing well, then pavement is damaged. If pavement remains intact, trees are growing poorly.

RESIDENTIAL AND COMMERCIAL LOTS

Swimming pools and septic tanks

Be sure to plant all trees at least 10 feet away from a pool wall. Medium-maturing trees should be located at least 20 feet away. Large ones should be at least 30 feet away from a pool. This helps prevent leaves and fruit from falling into the pool and is far enough away so that roots will not unduly interfere with the pool wall or plumbing. Palms can be planted closer to pools because the roots are not aggressive and canopies do not spread very far.

Conifers appear to be especially sensitive to chlorine volatilizing from pool water on a hot summer day. Foliage can be permanently damaged. Some communities are beginning to reduce chlorine applications to pool water by using ozone (O_3) for disinfection.

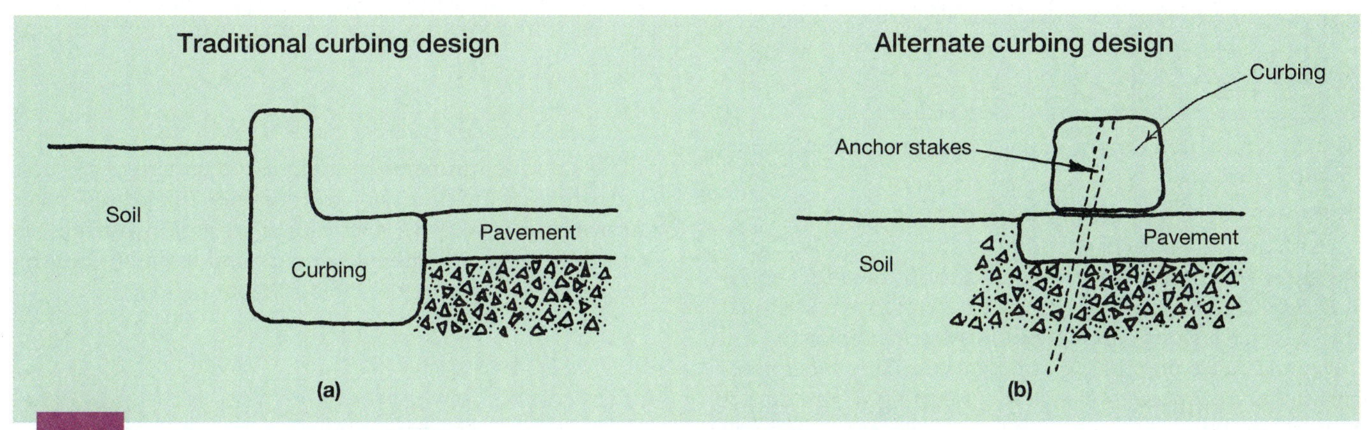

2-19 (a) Curbs designed in the traditional fashion often deflect roots laterally and under the curb. This can result in broken curbs as roots enlarge. (b) Curbs placed on top of the pavement may allow more roots to grow under the pavement and could be less expensive to install. They can be constructed from concrete, treated wood, or synthetic materials.

There may be local regulations specifying minimum tree setback distances from septic tanks and drain fields. In general, roots on most trees grown in an open soil area extend about two to three times beyond the edge of the canopy (see Chapter 7). To judge how close to a septic tank or drain field to plant, estimate ultimate canopy diameter of the proposed tree. Plant at least this far from the area. For example, a tree expected to grow a 40-foot-wide canopy should be planted at least 40 feet away. Most aggressive, large roots should stay clear of the drain field or septic tank if planted according to this guideline. Trees with aggressive roots (Appendix 11) and those tolerant of wet soils (Appendix 5) often cause the most serious problems.

Driveways, walks, patios, and decks

Use the guidelines in Table 2-2 for planting trees near asphalt or concrete driveways. Keep trees with aggressive roots (Appendix 11) at least 10 to 15 feet from the edge of pavement, patios, and walks, especially if soil is poorly drained or compacted. Trees with soft, fleshed fruit could create a mess on the deck or patio during some periods of the year (Appendices 15 and 16). You might want to consider planting these trees far enough away so that the canopy will not grow over the deck or patio.

Retaining and ornamental walls

Roots on trees planted too close to a wall can crack the wall as they grow beneath it. Some roots that are deflected laterally by the wall can get large and crack the wall as they slowly push it. This type of damage can be reduced by following the guidelines in Table 2-2.

LANDFILL SITES

The most limiting factors for tree growth on former municipal landfills are dry, shallow, poor-quality soil and displacement of soil oxygen by methane and carbon dioxide from decomposing refuse. Any landfill containing organic matter will decompose and produce methane and carbon dioxide for many decades. Oxygen displacement is not a concern, however, on former fly ash, well-incinerated, or other landfills that contain only nonbiodegradable garbage.

Three questions must be answered before you select trees for landfill sites:

1. Is gas actively pumped from the landfill?
2. Has refuse been capped with a clay layer or other impervious material?
3. How deep is the final soil cover over the cap or refuse?

As long as gas is actively pumped from the refuse, landfill gases (primarily methane and carbon dioxide) should not enter the soil cover. If the gas recovery system is turned off, even for a short period, landfill gas could enter the soil cover and cause poor growth or plant mortality. Vent pipes that are not connected to an active pumping system do not reduce landfill gas contamination of the cover soil and will not aid in plant establishment. If gas is not pumped from the landfill, trees with shallow root systems will perform best. In one test, shallow-rooted Norway spruce and Japanese black pine grew better than the deeper-rooted honeylocust and green ash (Gilman 1989c). Tolerance to low soil oxygen and wet soil might also provide some measure of tolerance to landfill sites (Appendix 5).

The top of a landfill or mine site may be the highest point in the area. Damage caused by strong winds and frequent lightning strikes can limit tree establishment and growth in some regions of the country. Consider planting only trees with an open canopy and that mature at a small or medium size to minimize these problems.

Small saplings and seedlings survive better and grow faster than larger-sized planting stock. Research suggests that this is related to the ability of the root system on small nursery stock to quickly adapt to conditions in landfill soil. An additional disadvantage of large, landscape-sized planting stock is that it requires regular irrigation and staking until established to prevent wind from blowing the trees over.

Some old landfills are capped with a clay layer, polyethylene, or other material to help prevent rainwater from entering the landfill and to entrap methane. It is often difficult to determine whether a landfill has been capped, but the municipal engineering or public works department should be able to answer this question for you. It is important to determine this because sites with a cap present fewer problems for trees. Gas contamination of the cover soil tends to be localized, not widespread as it is in a landfill without a cap. Although many caps eventually crack as refuse settles, allowing landfill gas to migrate into the cover soil in these localized areas, tree growth on the rest of the landfill should be better than on a landfill without a cap. All recently closed landfills have caps.

Final cover soil less than two feet thick

Consider adding soil so that at least two feet of final cover soil exist over the cap, if one is present, or over the refuse if there is no cap. In most cases the law requires at least two feet of cover. Trees, shrubs, and grasses grow poorly in cover soil less than two feet thick. Without two feet of cover, this will be a very draughty site for which regular irrigation is recommended. Small-maturing trees that tolerate drought are best suited for sites with less than two feet of soil (Appendix 7).

Final cover soil two to four feet thick

Small- and medium-sized, drought-tolerant trees are best suited for sites with a two- to four-foot-thick final cover (Appendices 7 and 8). Occasional irrigation during extended drought will enhance growth and maintain healthy trees, especially during the establishment period.

Final cover soil more than four feet thick

A final cover of four feet or more of soil is the most desirable planting environment, but is very unusual for a landfill site. Many trees should thrive on this site. There are few restrictions on plant selection other than those mentioned in Chapter 1.

Chapter 3

Selecting Trees from the Nursery

INTRODUCTION

Tree selection does not end with choosing the appropriate species or cultivar for the planting site. Suitable nursery stock must be chosen based on planting site conditions and intended after-care. Site conditions and after-care capabilities should dictate maximum tree size at planting, root ball characteristics, appropriate tree production method, and tree structure.

Nursery stock must be inspected carefully to pick the best quality tree. Trees of poor quality may be inexpensive in some cases but might perform poorly in the landscape. Quality factors to evaluate include root ball size, shape, and structure; nursery planting depth; presence of included bark; trunk form and branch arrangement; old pruning cuts; presence of pests and disease; leaf color; top die-back; trunk strength; and canopy uniformity.

DETERMINING MAXIMUM TREE SIZE AT PLANTING

The maximum tree size at planting should be governed not by the budget for the job, but by the irrigation capabilities after planting and climate (refer back to Figure 1-11). Also be sure you are able to handle the tree at the job site without causing damage to it. If irrigation cannot be provided for the recommended period after planting, choose smaller nursery stock to ensure survival. This technique will help avoid the now-common practice of planting nursery stock that is too large for the irrigation capabilities at the site.

Site drainage also should govern maximum tree size at planting. Large-sized nursery trees (more than two inches in trunk diameter) may be poorly suited for wet sites because roots in the bottom portion of the root ball could become submerged in water. This can stress the tree by killing the deeper roots in the root ball. Smaller trees have shallower root balls and so are better suited for planting on wet sites. If large-sized nursery stock is absolutely necessary for a poorly drained site, purchase trees with shallow root balls, such as those grown in low-profile containers, or plant on a mound to keep roots out of the water (see Chapter 4).

Table 3-1 outlines the important criteria that should become part of the process of selecting nursery stock for a planting site. Follow these guidelines to simplify the process of selecting the correct nursery stock size.

ROOT BALL SHAPE AND DEPTH

Root balls of any shape can be planted in well-drained loamy or sandy soil. However, a relatively tall root ball might help keep deeper roots moist, as deeper soil layers dry more slowly than the soil near the surface. Shallow root balls may dry quicker on well-drained sites than deeper root balls. This is only speculation and has not been thoroughly tested. Low-profile root balls are probably better suited than traditional-shaped or standard root balls for planting in poorly drained and compacted sites (Figure 3-1). Low-profile root balls are produced in low-profile containers (Milbocker 1994), or come from field nurseries with compacted subsoil or high water table.

The root ball shape from a low-profile root ball resembles the natural shape of many tree and shrub root systems, that is, shallow and wide. (This is covered in more depth in Chapters 6 and 7.) Roots in the deeper part of a traditional-shaped root ball could die if they are submerged in water or

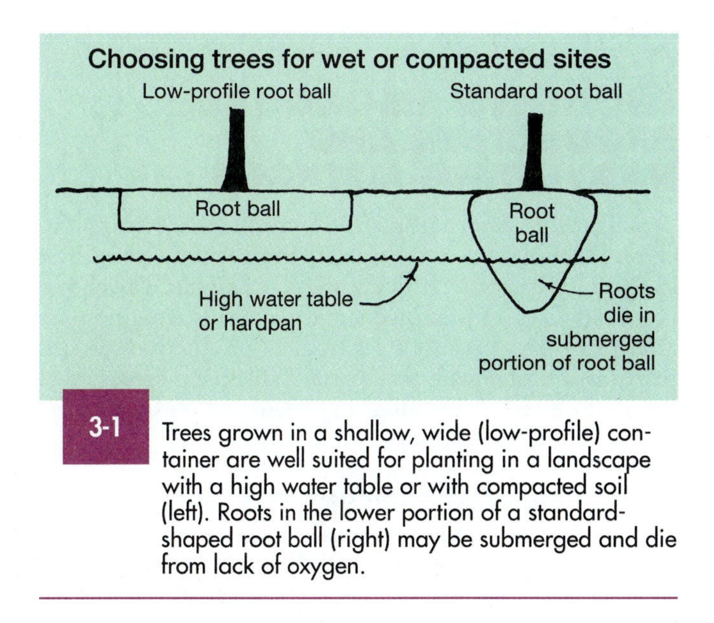

3-1 Trees grown in a shallow, wide (low-profile) container are well suited for planting in a landscape with a high water table or with compacted soil (left). Roots in the lower portion of a standard-shaped root ball (right) may be submerged and die from lack of oxygen.

TABLE 3-1 COMPARISON OF SMALL-SIZED WITH LARGE-SIZED NURSERY TREES[1]

CRITERION	SMALL-SIZED NURSERY STOCK	LARGE-SIZED NURSERY STOCK (MORE THAN 2-INCH TRUNK DIAMETER)
Landscape establishment period	quick	slow
Irrigation period after planting	brief	extended
Susceptibility to drought or flooding	for a brief period after planting	sensitive for a long time
Cost of nursery stock	inexpensive	expensive
Number of trees planted per dollars spent	large	small
Pruning needs[2]	high	moderate
Suitability for compacted or poorly drained sites	well-suited	could be poorly suited

[1]This table applies regardless of the ultimate (or mature) height of the trees.

[2]Assuming good-quality trees were purchased. If trees with poor structure are installed, large trees could also require a great deal of pruning to improve structure.

placed in compacted soil with poor aeration. This stress could also make the trees sensitive to drought injury for a longer period of time by slowing the establishment rate.

ROOT BALL TEXTURE

Some horticulturists working in clayey soil prefer planting trees with a clayey root ball. They feel that a sandy or loamy root ball dries out quicker because water is pulled from the ball into the landscape soil by capillary action. Although this idea may have merit and makes sense in theory, there is little scientific data to support it. Sandy or loamy root balls will dry out quicker when planted in any landscape soil, clayey or sandy, simply because sandy soil holds less water than clayey soil.

CHOOSING AMONG TREE PRODUCTION AND HARVESTING METHODS

To ensure greater transplant survival, it is essential to choose trees grown in the nursery production system best suited for the characteristics of the planting site. The choice is simple for a well-drained site if root balls remain moist after planting—because trees from all production systems perform almost equally well there (Table 3-2). Professional judgment based on the following discussion must guide the selection process in less hospitable sites.

Comparing production methods

Root balls of field-grown trees are similar to those grown in fabric containers except that fabric-container-grown root balls are about half the volume (Gilman, Beeson, and Black 1992). This makes them easier to handle. Contrary to popular belief, research now shows that about the same percentage of roots are harvested from both production methods (Gilman, Beeson, and Black 1992; Gilman and Beeson 1996; Harris and Gilman 1991) (Figure 3-2a and b). Because the root ball is smaller, there is less water storage capacity in the fabric container root ball than in the larger-sized root ball of the field-grown tree. Combined with a dense root system, this lesser reserve makes trees produced in fabric containers more sensitive to desiccation immediately after digging than trees grown directly in field soil. Quality nursery operators make provisions for delivering the irrigation needed to prevent desiccation immediately after harvesting.

Trees produced in fabric containers are as viable in the landscape as those grown directly in the field if they are handled carefully and irrigated properly after digging. However, be sure to purchase field- and fabric-container-grown trees that have been dug at least several weeks earlier to ensure that the tree will survive the shock of digging. Until they are established in the landscape, trees harvested from fabric containers will require more frequent irrigation than those from a field nursery (Harris and Gilman 1993). (This is covered in more depth in Chapter 5.)

A container root ball has many times more fine roots than that of a similar-sized tree harvested from a fabric container or balled in burlap (B&B) from a field (Gilman and Beeson 1996). These roots dry the container media quickly which makes most trees produced in containers very sensitive to drought injury after planting. They are more sensitive to lack of water after planting than hardened-off field-grown or bare-root trees, which have only a fraction of the fine roots compared to containers (Gilman 1994; Gilman and Black 1997; Marler and Davies 1987) (see "Hardened-off field-grown trees" later in this chapter). Hardened-off field-grown trees are better able to tolerate

TABLE 3-2 CHOOSING TREES BY PRODUCTION METHOD

PRODUCTION METHOD	PLANTING TIME	ROOT BALL WEIGHT	NEED STAKING	IF: IRRIGATION AFTER PLANTING IS:	THEN: ROOT GROWTH AND TRUNK GROWTH WILL BE:	AND: SURVIVAL WILL BE:
Above-ground container	Anytime soil is not frozen	light	frequently	frequent	good to excellent	excellent
				infrequent	fair to good	fair
In-ground fabric container	Anytime soil is not frozen	light to moderate	usually	frequent	excellent	excellent
				infrequent	good[1]	good[1]
B&B	Anytime soil is not frozen	heavy	sometimes	frequent	excellent	excellent
				infrequent	good[1]	good[1]
Bare-root	Spring in north	very light	usually	frequent	excellent	excellent
	Spring and fall in mid-south			infrequent	good[1]	good[1]

[1]Survival and growth are good for trees hardened off in the nursery by pre-digging or root pruning several weeks to several months before planting. Survival and growth are only fair for fabric container and balled-in-burlap (B&B) trees dug and immediately transplanted to the landscape without regular irrigation.

Note: Table is based on Beeson and Gilman 1992a and 1992b; Dana and Blessing 1994; Gilman and Black 1997; Gilman and Beeson 1996; Harris and Gilman 1991 and 1993; and Hensley 1994.

(a)

(b)

3-2 (a) The same amount of roots are found in a root ball from a fabric container as from a (b) balled-in-burlap tree. Because the root ball is smaller on a fabric-container-grown tree, the roots are concentrated in a smaller area.

infrequent irrigation because roots regenerate and grow quickly into the landscape soil. Root growth from containers is slower, especially when the landscape and root ball stay fairly dry after planting.

Despite differences in root growth and establishment rates after transplanting (Beeson and Gilman 1992a), shoot growth on container-grown trees appears to be similar to that on hardened-off field-grown trees, provided irrigation is supplied until trees are established. However, trees planted from containers appear to be more susceptible to desiccation and death, and they grow slower if the root ball is not kept moist during the early establishment period. For example, citrus trees from containers grew less than bare-root trees in a sandy soil (Marler and Davies 1987). In contrast, recently dug trees that have not been hardened-off from a field or fabric-container nursery are *more* susceptible to desiccation than trees from plastic containers (Beeson and Gilman 1992b; Gilman and Beeson 1996; Harris and Gilman 1993). Because freshly dug trees require careful irrigation management, they should not be planted into the landscape unless frequent irrigation can be provided. Nurseries are better equipped to handle trees' special irrigation needs immediately after digging.

In a recent study, live oak trees grown in containers treated with copper on the inside surface were more stressed than those grown in conventional containers in the first three weeks after planting to the landscape without irrigation. However, with daily irrigation there was no difference in stress levels (Gilman 1994a). The slight increase in water stress may be due to the lack of roots on the outside of root balls planted from copper-coated containers (Arnold and Struve 1993; Gilman and Beeson 1995). Those from conventional containers have roots on the outer edge of the root ball to provide intimate contact with moist landscape soil. This may provide for more water uptake than for trees from copper-coated containers. When regular irrigation was discontinued four to six weeks after planting, stress levels were about equal and root growth into landscape soil was similar for trees planted from both types of containers. In contrast, Arnold (1996) showed no increase in water stress on trees planted from one-gallon copper-treated containers.

Further study showed that scarlet and red oak grown in copper-coated containers had greater regrowth after planting to the landscape than trees grown in conventional containers (Struve 1994). The central leader was left intact and did not die back as often in trees from copper-coated containers. However, there was no difference in regrowth between container types for red maple or sweetgum. Trees from all container types survived equally well. One advantage of planting from copper-coated containers is that root balls have few, if any, circling roots, making them superior to trees grown in conventional, smooth-sided containers.

Bare root

Bare-root deciduous trees are usually available only when dormant (early spring in the north, early spring and fall in the mid-south), and in a limited size range (typically two-inch trunk diameter or less). It is the least expensive tree to purchase. Unlike all other harvesting methods, the root system can be thoroughly inspected for defects because there is no soil or media covering the roots. A larger root ball can be harvested than on B&B trees because bare-root trees are light weight.

Bare-root trees are very sensitive to drying if not properly stored and shipped. If provisions are made to keep roots in the shade and moist during storage, transport, and at the planting site, and they are regularly irrigated after planting, they can perform as well as trees from other production methods (Hensley 1994). However, there have been reports of tree spade dug trees surviving (Cool 1976) and growing (Magley and Struve 1983) better than bare-root trees. There have been few other in-depth studies comparing transplantability of bare-root trees with those from other production methods.

Balled-in-burlap (B&B)

Root development within the root ball of a field-grown tree can best be described as variable. There is wide variation among species and among individuals within a species. There appears to be more variation among root balls of oak trees than among individuals of other species, such as crape myrtle, baldcypress and southern magnolia. This may account for some of the difficulty in transplanting live, white, chinkapin, and bur oaks. The root system on some species can be influenced by root pruning, irrigation and fertilization management, and other cultural practices.

Tree production practices in the field nursery impact transplantability. Some tree species receiving drip irrigation at the base of the trunk for the duration of the production cycle often have a concentration of fine roots in the root ball (Gilman, Beeson, and Black 1992; Ponder and Kenworthy 1976) (Figure 3-3). Fine root growth on some species can also be stimulated close to the trunk by making fertilizer applications only within the area of the future root ball, not outside the root ball (Beeson and Gilman 1995). Many nurseries fertilize trees in this manner. However, one study showed that after trees received overhead irrigation and broadcast fertilization for two years, concentrating the irrigation or fertilizer close to the trunk did not influence fine root growth (Beeson and Gilman 1995). Apparently, to increase fine roots near the trunk, irrigation and fertilizer application to the root ball area must begin soon after the tree is planted in the nursery field.

Some field nurseries routinely root-prune certain species (such as oaks) once each year to help them survive the transplanting process. Others prune only once, several weeks to several months before harvesting. Some nurseries do not root-prune trees during production. Certain trees like crape myrtle and red maple are not routinely root-pruned because their root systems are dense and fibrous. Although root-pruning can affect tree survival after transplanting, it has little impact on growth after transplanting (Gilman 1992). In warm climates, without frequent irrigation after transplanting, trees pre-dug from a field nursery or root-pruned at least several weeks or months prior to planting to a landscape are less stressed after planting than freshly dug trees. (Research in cooler climates is mostly

3-3 The three nursery trees on the right received irrigation inside and outside the root ball; those in the center received irrigation only in the root ball; those on the left received no irrigation in the nursery. Notice the increased amount of small fine roots in trees in the center group.

lacking.) Nevertheless, if irrigation can be provided regularly until trees are established, root-pruning prior to transplanting has little impact on survival of trees with a trunk diameter less than five inches. Research on larger trees has not been done. Nurseries are usually best prepared to provide the intensive irrigation management required for freshly dug trees. Freshly dug trees usually cannot be irrigated properly in a landscape setting. Some deciduous trees in cooler regions (hardiness zones 2 through 6) are not irrigated regularly after transplanting when they are dug or planted at the end of the dormant season. Although the cool weather and lack of foliage allows the tree to survive, the onset of warm summer weather can cause dieback. Arborists often root-prune very large trees prior to moving them with the intention of improving transplant survival. Nursery operators usually like to harvest only root-pruned trees in summer. Trees not previously root-pruned may not survive summer digging without daily irrigation.

Trees grown in the field and not harvested as bare-root stock are dug with a soil ball that is immediately wrapped in burlap. The burlap is secured around the root ball with nails, string, or wire. Treated or untreated natural fiber burlap usually decomposes in the soil and does not have to be completely removed at planting. However, after the plant is set at the planting site, professionals remove all burlap, wire, and string from the trunk and the top of the ball. Some also slice or partially remove natural burlap on the side of the ball to ensure good contact between backfill soil and root ball soil.

Synthetic burlap is occasionally used on root balls so that the nursery operator can pre-dig the tree several months before it is planted in the landscape. This pre-digging helps the tree survive transplanting. However, synthetic burlap should be removed at least from the upper portion of the root ball before backfilling, and it is probably best to remove all synthetic burlap. Although there are few published reports of synthetic burlap preventing root regeneration or penetration into backfill soil, there are anecdotal accounts from landscapers and researchers of synthetic burlap preventing root growth out of the root ball. In these instances, it was not clear whether the synthetic burlap inhibited growth or whether some other site characteristic (such as too much or too little water) was responsible. In many instances, roots grow freely through the holes in synthetic burlap. However, many of these roots are restricted from growing to a large diameter because the fabric does not decompose or stretch.

Synthetic burlap cannot be removed from a root ball tightly secured in a wire basket. To ensure continued health of the tree, remove some or all of the wire basket and cut away the synthetic burlap. Some horticulturists recommend removing the entire basket before planting, despite the fact that roots appear to engulf and grow around the wire (Lumis 1990; Watson 1994). Many nurseries prefer to use treated instead of synthetic burlap to help hold the ball together after digging, because under most circumstances the burlap decomposes after planting. Synthetic burlap should probably not be used on tree root balls.

A recent innovation should eliminate the need for synthetic burlap in a wire basket. It allows the tree to be held in the nursery for many months after digging without producing large-diameter roots out of the root ball. Trees are dug and placed in natural burlap and then secured with wire or string to hold the root ball tight. The intact root ball is lowered into a pre-dug hole in the nursery that is lined with black nursery ground cloth, a synthetic burlap,

3-4 B&B root balls lowered into a hole in the nursery lined with fabric can be held in the nursery for several months. The fabric holds the root ball together when the tree is sold even if the natural burlap inside the wire decomposes. The fabric *must* be removed before the tree is planted in the landscape.

or fabric sleeve sewn to fit nicely around the root ball (Figure 3-4). In several weeks to several months, depending on the season of the year and species, regenerated roots will grow through the fabric and hold it onto the outside of the root ball. This will allow the tree to be held in the nursery for many months and then lifted without losing soil from the ball. The natural burlap may have decomposed, but the synthetic or fabric on the outside of the ball will be intact and will hold the soil in the root ball. Of course, the synthetic fabric on the outside of the ball must be stripped away from the root ball and removed at the planting site.

Tree spade

Following transplanting, trees moved with a tree spade generally act like B&B trees, with the following exception. If the planting hole was dug with a tree spade in clayey soil and the sides of the hole are glazed like a smooth clay pot, then root growth into the landscape soil could be limited in some circumstances (Birdel, Whitcomb, and Appleton 1983). Enlarge the hole before planting so that roots can penetrate loosened backfill soil.

Trees moved with a spade into loamy or sandy soil can usually be planted into a hole dug with the same size spade. The hole does not have to be enlarged. The tree often ends up a little higher than it was in the nursery. This is much better than planting too deep. Simply create a gently sloping mound of soil to cover the sides of the root ball and mulch as usual.

Hardened-off field-grown trees

Field-grown trees dug several weeks or months prior to shipping to the landscape site are said to be "hardened off." Freshly dug trees are not hardened off. During the hardening-off period, roots begin to regenerate within the root ball, and the tree may drop some leaves. The tree may also make other adjustments that are not now well understood. Roots often grow through the burlap wrapped around the root ball of a hardened-off tree. Some nurseries provide overhead irrigation to the foliage during the hardening-off period, especially for summer digging. Recent research shows that there appears to be no benefit to conducting this for more than three weeks on evergreen oaks, and it often is not necessary at all if the root system is well developed. There is little research on other trees from which to make further recommendations.

In-ground containers

Plastic (pot-in-pot) Some nursery operators grow trees in plastic containers sunk into the ground. The container holding the root ball is slipped into an empty container of the same size permanently installed in the ground. This prevents the tree from blowing over and buffers root ball temperature. Some tree species benefit from growing in this system by producing more root dry weight and a more uniform root system than those in above-ground containers. This increases the root-to-shoot ratio (Ruter 1993a and 1993b). Roots are often lacking on the south and west sides of the root ball on trees grown above ground, due to high temperatures in the root ball; roots are more evenly distributed in a pot-in-pot root ball because temperatures can be up to 10°F cooler. This leads to a more uniformly distributed root system in the landscape after planting which could help to ensure tree stability as the tree grows in the landscape. There are only a few published comparative tests, but they show that trees grown in this fashion transplant similar to those grown above ground in traditional, smooth-sided plastic containers (Ruter 1993a and 1993b).

Perforated fabric (mesh-bag) A recently developed in-ground container was designed as a modification of the original fabric container (the manufacturer is Lacebark Inc., of Stillwater, OK). It is made from a knitted polyester fabric with 3/32-inch-diameter holes. Like the original fabric container, it is sunk into the ground and filled with field soil, not media. Roots are allowed to grow through numerous small holes in the container but are not supposed to enlarge as they did in earlier models of fabric containers. Although there are no published comparative tests, trees grown in this fashion should transplant similar to trees grown in fabric containers. If roots are truly girdled by the fabric and no large roots develop through the fabric, then trees may transplant more consistently than those from earlier models of fabric container. The author's observations have been that fewer large-diameter roots develop through this fabric than through others.

Fabric Trees grown in fabric containers are easier to lift than the same size balled-in-burlap trees, because root balls are smaller. However, they must be handled very carefully because, unlike a rigid plastic container, there is little structure or rigidity to the root ball. Roots are easily broken inside the root ball. Fabric-container-grown trees also require more frequent irrigation than balled-in-burlap trees until they are established in the landscape (Harris and Gilman 1993). They require staking to hold them up in the landscape. With these extra precautions, fabric-container-grown trees transplant similar to traditional balled-in-burlap trees (Table 3-2).

Containerized field-grown

Some nursery operators grow trees directly in the ground or in fabric containers in the ground, harvest them with soil root balls, and place them into containers. This containerization process produces a hybrid between a field-grown and a container tree. In many respects, a tree grown in this fashion is similar to a well-hardened-off field-grown tree. Although there is little research data on transplant success of containerized trees, once roots grow out into the media in the container, such trees should perform as well as hardened-off field-grown trees.

Above-ground containers

Trees are grown in a variety of above-ground containers in the nursery. The variety is likely to increase in the future as growers search for increased operations efficiency. Containers are usually filled with well-drained artificial (referred to as soilless by many nursery operators) media in moist, humid climates like those in the eastern United States. Because root balls are designed to drain quickly to prevent root rot in the nursery, daily irrigation is required in the summer. In drier climates, like those in the western United States, container media are designed to hold moisture, and containers may not need daily irrigation in the nursery.

Roots may be missing on the south and west sides of above-ground containers because high media temperature killed them. Some speculate that this encourages development of a one-sided root system in the landscape. This could cause the tree to become unstable as it grows.

The physical properties of containers create a perched water table that forms a wet zone at the bottom of the container. The water table disappears when the root ball is removed from the container and planted in the landscape. Moisture is also drawn out of the container media into the finer-textured landscape soil. Consequently, the root ball dries quicker in the landscape than it did in the container, making container trees very sensitive to dry soil conditions following planting (Costello and Paul 1975). To maintain optimum growth after planting into well-drained soil, container trees should be watered at least as often as they were in the nursery (Gilman and Beeson 1996). In the summer in the southern United States, daily irrigation may be best in well-drained soil for a number of weeks or months after planting, especially for trees with trunk diameters over two inches. Irrigation can be less frequent for cooler climates and in poorly drained, compacted soil. Irrigation can be tapered off as roots begin to grow out into the landscape soil (see Chapter 5 for more details).

Nursery operators choose among the various container types based on a variety of production concerns. Choosing among them is of lesser concern for persons writing tree specifications or purchasing trees for a landscape. Until research shows differently, we can assume that in many situations trees transplant about equally well from most above-ground container production systems. However, some container designs help prevent formation of potentially lethal circling roots better than others. Consideration should be given to planting trees from these new containers.

Plastic container types

Smooth-sided plastic containers. Most trees and shrubs offered in containers today are grown in these containers. They are manufactured in a variety of shapes and sizes. Circling roots on the outer edge of the root ball can cause problems for the tree after planting. To prevent this, roots circling the outside of the container root ball should be separated and straightened, or cut to prevent them from girdling the tree later.

Tall and narrow containers. Trees from containers that are much taller than they are wide might survive better than those from other container shapes in well-drained, draughty sites with little irrigation. Roots at the bottom of the container are less susceptible to desiccation because the deeper soil dries slower than soil at the surface. Although roots in the top portion of the root ball may dry out, those at the bottom have a better chance of elongating into the landscape soil before drying up and dying. In compacted or poorly drained sites, though, roots at the bottom of a tall root ball may not help the tree become established, because they often die from lack of oxygen (Gilman, Leone, and Flower 1987). A short and wide container is better suited for sites with poor drainage.

Short and wide (low-profile) containers. Circling roots are less of a problem in a wide container because, if they do develop, they are located far from the trunk, which lessens the likelihood of the roots girdling the trunk. However, circling roots may be present close to the trunk if the tree was in a smaller container when it was younger. Although trees from low-profile containers can be planted in many soils, they are especially suited for poorly drained and compacted sites (refer to Figure 3-1). Research indicates that if irrigation is limited after planting, more roots are regenerated from this root ball, and top growth is better, than on trees with a traditional-shaped root ball (Milbocker 1994). However, there is no difference in growth between container types if irrigation can be provided regularly after planting.

Air-pruned containers. Some containers are designed with many holes in the sides and bottom. Some are actually bottomless. Roots at the extreme outside edge of the root ball are pruned and branch because their tips are killed by air entering the holes (Privett and Hummel 1992). Few roots are evident on the outside of the root ball but the ball stays together nicely. This is normal and is a response to the air-pruning effect. This almost eliminates circling roots and, according to manufacturers, creates a denser root ball on some species (Whitcomb and Williams 1985). Other studies show no increase in root density (Newman and Follett 1987). The lack of circling roots reduces the likelihood of roots strangling the trunk as the tree grows older. There is little reason to believe that trees from these containers will transplant any differently than those from traditional, smooth-sided containers.

Ridged container. Vertical ridges in a variety of configurations along the insides of containers are designed to prevent roots from circling the outside of the root ball. Roots are often deflected down vertical ridges to the bottom of the container. Some of these containers also have holes along the ridges. Trees from ridged containers should transplant similar to those from traditional, smooth-sided containers, although detailed research is lacking. The advantage of purchasing trees grown in ridged containers is that there may be fewer roots circling the outside of the root ball, especially near the top.

Copper-coated containers. A recently introduced practice in container tree production is to coat the inside of the container with a copper compound (Spin Out™, produced by Griffin Corporation of Valdosta, GA, is one). Although root-to-shoot ratio is usually unaffected by the copper, circling roots do not form in these containers on most species (Struve et al. 1994). Because few, if any, roots circle the root ball, no potentially girdling roots form on these plants. Roots are often more evenly distributed along the sides of a copper-treated container, with less root density at the bottom compared to an untreated container. When the ball is slipped from the container, few roots will be evident on the outside of the root ball but the ball will stay together nicely. This is normal and is a desirable response to the copper. Early tests indicate that following planting into the landscape, root elongation and growth of the shoots and trunk are similar to or slightly greater than growth on trees from conventional, smooth-sided containers (Struve 1994). Be sure the copper compound was applied to the entire inside surface of the container. Stripping the inside with copper does not usually reduce root circling.

Wooden boxes or metal containers There are no data to suggest that trees grown in wooden boxes or metal containers transplant any differently than those grown in other containers.

Peat pots The bottom portion of a peat pot can be left on the root ball at planting. The dried top portion should be removed to prevent moisture from wicking out of the root ball.

Plastic poly-bag Trees are occasionally grown in media placed inside a plastic bag (called a poly-bag). The poly-bag is a nonridged or flexible container sometimes with a gusseted bottom that prevents root circling. Trees from poly-bags should transplant as do traditional container-grown trees, provided the root ball remains intact. It is rather fragile in the plastic bag.

Collected native trees

Native trees, especially oaks, are collected from natural stands and transplanted to landscape sites in certain regions. If held in a nursery for two years after digging, the nursery industry considers the collected trees to be nursery grown (American National Standards Institute 1990). Research on laurel oak indicates that root balls on collected trees are similar to those on nursery-grown trees that are not root-pruned prior to transplanting (Gilman, Beeson, and Black 1992). The only real difference is that collected trees have slightly less fine root mass than trees grown in the nursery. In addition, they often have a thinner canopy until they become established, but ironically may have a better form because they were not topped as are some nursery-grown trees. Collected trees survive, but may grow slower after transplanting than trees from other production systems. No research has been conducted on other tree species or in other parts of the country.

ROOT BALL SIZE

The American National Standards Institute (1990) recommends minimum root ball sizes for field-grown and fabric container-grown trees based on trunk diameter or tree height (Table 3-3). In addition, Florida grades and standards (Florida Department of Agriculture 1996) make minimum container size recommendations for trees grown in above-ground containers and fabric containers (Table 3-3). Adhering to these standards helps trees to establish successfully in the landscape.

ROOT BALL DEFECTS

Root ball defects can occur on all trees, regardless of the method of production. Once formed, severe defects on the main roots close to the trunk are difficult to correct, and they can have a significant impact on the ability of landscape plants to survive and grow. Tree roots deformed within the first several months of propagation in the nursery can doom a tree to early death (Harris 1992). Unfortunately, unless a number of root systems are examined prior to purchase, the buyer will not see these defects. The consequences of root deformations may not become evident until the trees are older and well established. Trees with intact circling roots should not be planted.

Circling roots can be found on trees in containers or those grown in field soil if they were in a container or propagation bed when young. Roots that are deflected by container walls may circle the outside of the root ball unless the container is specially designed to prevent this. Kinking can occur if roots are folded into a propagation bed at the seedling stage. In addition, girdling roots on transplanted Norway maple (Watson, Clark, and Johnson 1990), and perhaps other trees, can form when regenerated roots grow perpendicular to a cut root. As the tree grows, these roots may meet the trunk and begin to strangle it. These defects stay with the tree until correction. Even if the defect is corrected by cutting the girdling roots, new roots can regenerate to form more girdling roots at the same spot. It is best not to purchase trees with these defects.

TABLE 3-3 RECOMMENDED MINIMUM ROOT BALL SIZES AND MINIMUM AND MAXIMUM TREE HEIGHTS FOR NURSERY TREES

	MINIMUM ROOT BALL DIAMETER (INCHES)				MINIMUM TREE HEIGHT (FEET)[2]		
TRUNK CALIPER[1] (INCHES)	FIELD-GROWN SHADE TREES[2]	FIELD-GROWN SMALL TREES[2]	FABRIC CONTAINER-GROWN TREES[3]	MINIMUM CONTAINER SIZE (GALLONS)[3]	STANDARD TREES	SLOWER-GROWING TREE SPECIES OR CULTIVARS	MAXIMUM TREE HEIGHT (FEET)[2]
0.5	12	—	—	1	5	3.5	6
0.75	14	16	—	3	5	4.5	8
1.0	16	18	12	5	6	5.0	10
1.25	18	—	14	7	7	6.0	11
1.5	20	20	16	15	8	7.0	12
1.75	22	22	—	—	9	7.5	13
2.0	24	24	18	20	10	8.0	14
2.5	28	28	18	25	11	9.0	15
3.0	32	32	20	45	12	9.5	16
3.5	38	38	24	65	13	10.0	17
4.0	42	42	30	95	14	10.5	18
4.5	48	48	36	95	14.5	11.0	18.5
5.0	54	54	36	95	—	—	—

[1]Trunk diameter (caliper) is measured 6 inches from the ground unless the trunk is more than 4 inches in diameter. If so, measure trunk diameter 12 inches from the ground.
[2]American National Standards Institute (ANSI Z60.1, 1990) used by American Association of Nurserymen. Note: Small trees 2, 3, 4, or 5 feet tall need root balls 10, 12, 14, or 16 inches.
[3]Florida Grades and Standards, 1996.

The root ball might have obvious abnormalities, such as kinked or circling roots at the surface (Figure 3-5a–c). If these are visible and less than about one-third the diameter of the trunk, they can be severed where they begin to circle, to correct the defect, with little harm to the tree. This may temporarily slow growth but should have a positive impact on future tree survival and growth. More often, defects are hidden inside the root ball. To find these, some soil or media may have to be removed from the root ball. Bare-root trees are easily examined.

Another serious defect can be diagnosed without removing soil or media from the root ball. After removing any stakes, simply push the trunk back and forth once or twice while securing the base of the root ball so it does not move. The trunk on a good-quality tree will bend along its length and will not move in the soil or medium (Figure 3-6a). The trunk on a tree with a defective root system will often pivot at its base and move in the root ball before it bends (Figure 3-6b). The trunk on the defective tree will appear to be loose in the root ball. This is often caused by a large circling or kinked root close to the trunk. Few branch roots form on the outside curve of a circling root, causing the tree to become unstable (Figure 3-6). This can also occur if the tree was planted too deep in the nursery.

Cultivating equipment used along nursery rows can throw soil around tree trunks and on top of the roots. This raises the soil level over the root system in the nursery and essentially buries the roots too deep (Figure 3-7a and b). Trees can also be planted too deep in the field or in containers, creating the same unwanted effect. When the nursery crew harvests the tree, this soil or medium without roots can make up the top third of the root ball or more. This is not a good practice, because the additional soil cuts off oxygen to the roots and can cause trunk rot where moist soil contacts bark. If you were unfortunate enough to purchase such trees, take them back to the nursery. Do not plant them! If, for some reason, you find yourself compelled to plant this inferior nursery stock, remove the excess soil to find where the first large diameter root meets the trunk. Place this topmost root just under the soil surface

3-5 (a) Roots circling close to the trunk can eventually slow growth and girdle the trunk. (b) Roots circling the top of the root ball are especially troublesome. Do not plant trees with this defect. Notice that few roots grow from the outside portion of a circling root. There are more roots on the ball without circling roots. (c) Lack of roots often leads to tree instability. There are few circling roots on trees grown in air-pruning or copper-coated containers.

3-6 (a) Root defects can cause trunk instability. A good-quality tree bends along the trunk when it is pushed to one side, and the trunk does not move in the soil. (b) A poor-quality tree pivots at or below the soil line, and the trunk bends little as it is pushed.

(Figure 3-7). Notice in Figure 3-7 that even though the root ball is sized appropriately, a smaller root system is harvested on a tree with soil on top of the root ball. The undersized root system places additional stresses on the tree.

Because the most serious defects inside the root ball are close to the trunk, they can be found simply by removing soil or media from three inches around the trunk four to six inches deep. This can be done with a garden hose or your fingers if the medium or soil is loose. This procedure will not injure the tree. Although it would be best to check all trees for these defects, this is not always practical. Most professionals perform this check on a certain percentage of trees (e.g., 5 percent). If they find no problems on this small sample, they accept all the trees and check each one as it is planted. If problems are found on the sample trees, more trees should be checked in the nursery before purchasing. This test can be performed quickly on trees in containers because there is no string or burlap to remove. On B&B trees, untie string or strapping from the trunk of a field-grown plant and loosen the burlap so you can work your fingers down along the trunk into the root ball. The point where the first root is attached to the trunk should be within two inches of the soil or medium surface. Circling roots less than about one-third the trunk diameter can be cut at the point where they begin to circle. If this procedure is conducted in the nursery, resecure the burlap and string

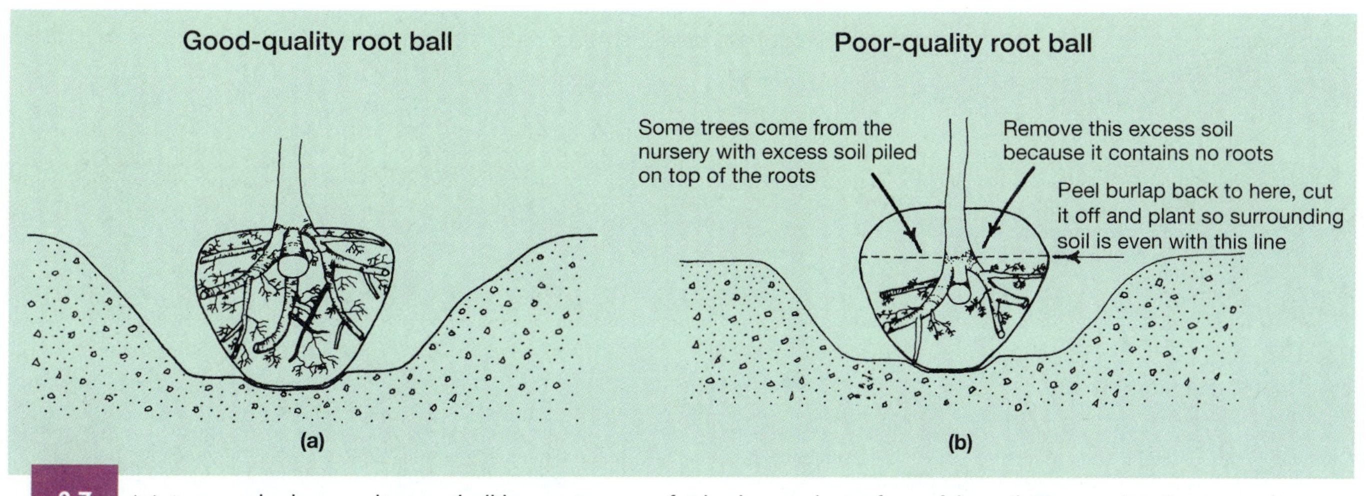

3-7 (a) A properly dug, quality root ball has major roots fairly close to the surface of the soil. You can brush some soil away from the base of the trunk and see these roots. This will not harm the tree. (b) Trees sometimes come from a nursery with excess soil over the major roots. There are not enough roots in this type of root ball to ensure survival and roots are too deep. Do not plant these trees. If for some unfortunate reason you are compelled to plant them, remove the excess soil so that major roots are located close to the soil surface, not deep in the soil.

before shipping. If performed in the landscape, remove it after planting.

There should be no evidence that roots larger than about one-fifth the trunk diameter were growing out the bottom of the container. Cutting larger escape roots to remove the root ball from the container could cause some stress on the tree and leaf drop.

Slip the root ball from the container. The root ball should stay together but be somewhat pliable. While the root ball is out of the container, check to see if the tree is pot-bound. A pot-bound (or root-bound) plant has many roots circling around the outside of the root ball. Also, the root ball may be very hard (Figure 3-8). In addition to potentially girdling the tree, a mass of circling roots on the outside of a root ball can act as a physical barrier to root penetration into the landscape soil following planting. Do not purchase pot-bound plants.

3-8 Too many roots on the outside of the root ball indicate a pot-bound tree. Do not plant trees with this defect.

The root ball on field-grown trees should be secured tightly with pins, twine, or wire and the trunk should be sturdy in the root ball. A loose or droopy root ball indicates that the tree was not properly cared for; roots may be broken and in poor contact with soil. As a result, it might perform poorly in the landscape after planting. If the trunk needs a stake to prevent it from falling over, the root system may be inferior and the tree should not be planted.

MULTITRUNK VERSUS SINGLE-TRUNK TREES

Street trees

Trees trained to one trunk are usually more appropriate for planting along streets and near walks than those with several trunks (Figure 3-9). Trees with one trunk are easier to train so that branches grow over the tops of vehicles and pedestrians, and they are more durable in the landscape.

Small-maturing trees, such as crape myrtle, Amur maple, some acacia species, and many others that usually grow with several trunks (Appendix 17), can be trained in a nursery to grow with one trunk. Although the multitrunked forms of these trees may not be suited for planting close to most streets or near walks, the same tree with a single trunk might be. If you cannot find the tree trained in the manner you like, contract with a nursery to grow it according to your specifications. Nursery operators have also developed upright cultivars of some multitrunked species, such as tree lilac and European hornbeam, that are more appropriate for along streets.

Lower branches and entire trunks on multitrunked trees often have to be removed several years after planting because they obstruct pedestrians and traffic. Planting a properly trained single-trunked version of the tree would help prevent this type of destructive pruning from becoming necessary (Figure 3-9).

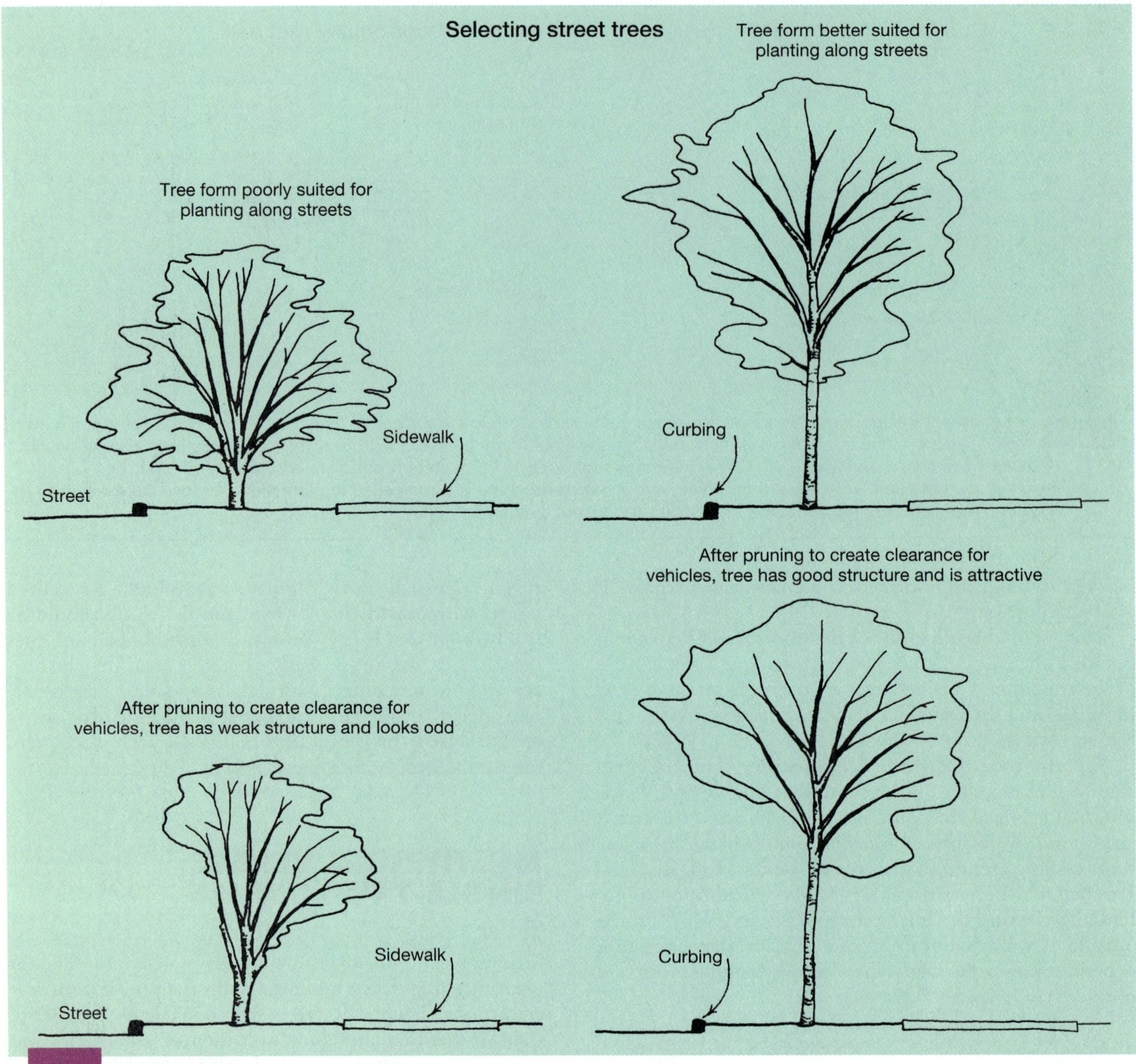

3-9 Multitrunked trees such as crape myrtle and Amur maple are usually not suited for planting along streets and near walks (left). However, nursery operators often offer these same tree species, as well as many other trees, in a single-trunked version (Appendix 17). These can be planted near walks and streets due to the more upright growth habit developed in the nursery (right).

Landscape trees

Trees with one trunk are usually considered structurally stronger than their multitrunk counterparts. However, small-maturing trees are often planted with several trunks originating close to the ground. This makes a nice specimen, especially on trees that have attractive, showy bark or trunk structure. Although trunks may eventually split apart on older specimens of small trees with multiple trunks, little if any personal or property damage usually results. Multitrunked small-maturing trees have a definite place in the landscape.

TRUNK FORM AND BRANCH STRUCTURE

Trunk form

Strong trunks taper and are thickest near the ground. They do not require a stake to support them. Trees that were staked in the nursery may not develop proper trunk taper and could fall over after the stake is removed (Figure 3-10a and b). Their trunks are often the same diameter at the ground as they are several feet up the tree. This could be a sign of a weak trunk.

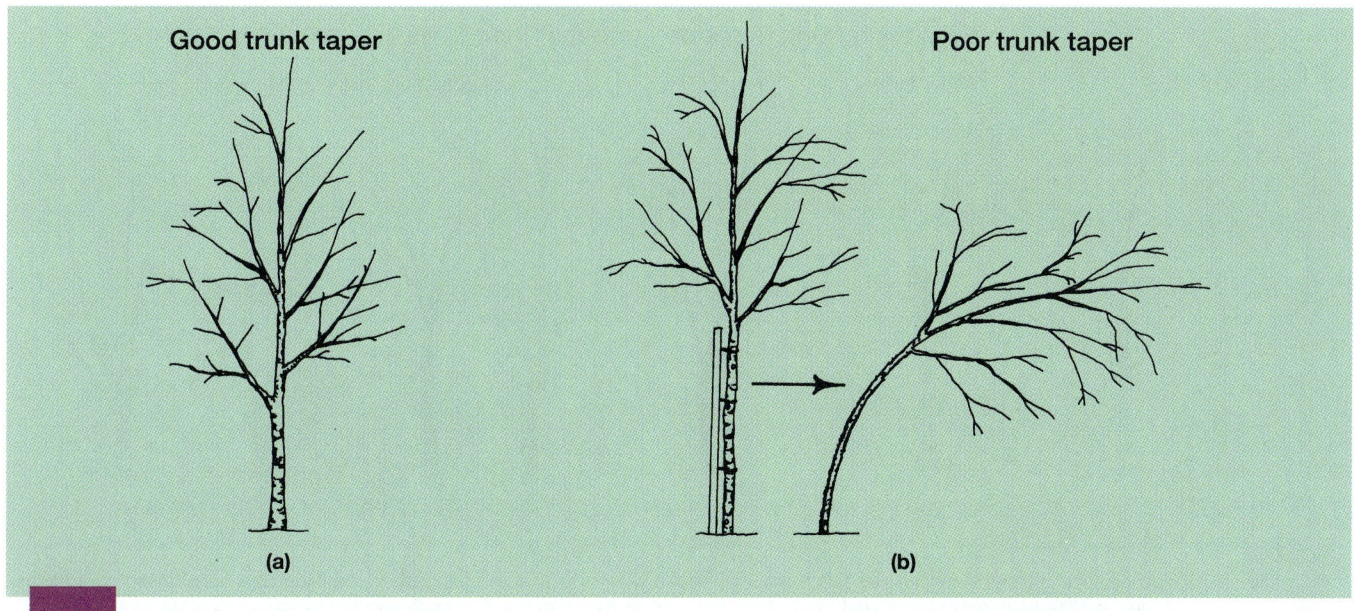

3-10 (a) Trees that were not staked too long in the nursery, and those with lower branches on the trunk, develop good trunk taper. They are able to stand up by themselves after planting. (b) Trees that were staked too long, or those with few lower branches, develop poor trunk taper and sometimes cannot stand erect when the stake is removed.

To check for trunk strength, separate the tree from all surrounding trees and remove all stakes to see if it will stand erect. If the tree is in leaf and it stands erect, it is probably strong enough. Wet the foliage with a hose to check for certain. This will simulate a rainfall and weigh the branches down.

Trunk strength is more difficult to evaluate on dormant trees. Buy the tree if it stands erect without a stake in the nursery. Be prepared to thin the canopy after leaves emerge in the spring if the trunk or branches droop over. Landscape staking will be needed on trees without sufficient caliper and strength to stand erect.

Small-maturing trees Trees with several trunks often develop included bark in the crotches. This condition can cause one of the trunks to split from the rest of the tree resulting in severe damage to the tree. (Refer back to Figure 2-2.) This can occur during a snow, ice, or rain storm, or it might simply snap on a windy day. If the broken stem is small, the damage may be negligible, but if a major trunk splits, then the character and health of the tree may be altered significantly. The tree may even have to be replaced. A well-formed, stronger, multitrunk tree has branches less than two-thirds the trunk diameter, wide branch angles, and no included bark. (Refer back to Figure 2-2.)

Large shade trees Trees that grow to be more than about 40 feet tall (Appendices 8 and 9) should have a single trunk, but that trunk need not be arrow-straight. A slight bend, subtle dogleg, or small zig-zag is acceptable (Figure 3-11a and b). If the trunk is unusually bent, forks in the bottom half of the tree, or has a severe dogleg, it should not be planted in a landscape (Figure 3-11c). Large, multitrunk specimens can break apart, particularly if they develop codominant stems with included bark (Figure 3-12). It may pay in the long run to spend a little more money on large-maturing trees with one dominant trunk instead of planting poor-quality multitrunked trees that require follow-up pruning soon after planting. Purchasing single-leadered trees will reduce the amount of pruning needed later.

Weeping trees A number of weeping trees, such as weeping Norway spruce or weeping cherry, form sprawling shrubs without training. Therefore, nurseries often train or graft them so branches weep from atop a single trunk. The trunk rarely continues to grow in a straight upright fashion. This is normal.

Clear trunk Branches should be distributed along the trunk (Figure 3-13a), not clumped toward the top (Figure 3-13b). At least half of the foliage should originate from branches on the lower two-thirds of the tree. Branches in the lower half of the tree help distribute the stress on a trunk when the wind blows.

Branch arrangement

Branch arrangement and spacing is less crucial on trees that will be small at maturity. Simply choose those with a pleasing branch arrangement, one that will fit the needs of the planting site. Major branches and trunks should not touch. Also, branches that remain less than two-thirds the diameter of the trunk and those with a U-shaped crotch (Figure 3-14b) are stronger than those that grow larger and those with a V-shaped crotch (Figure 3-14a).

On large-maturing shade trees with a trunk diameter of less than two inches, main (the largest diameter) branches should be about six inches apart (Figure 3-15a). Smaller-sized

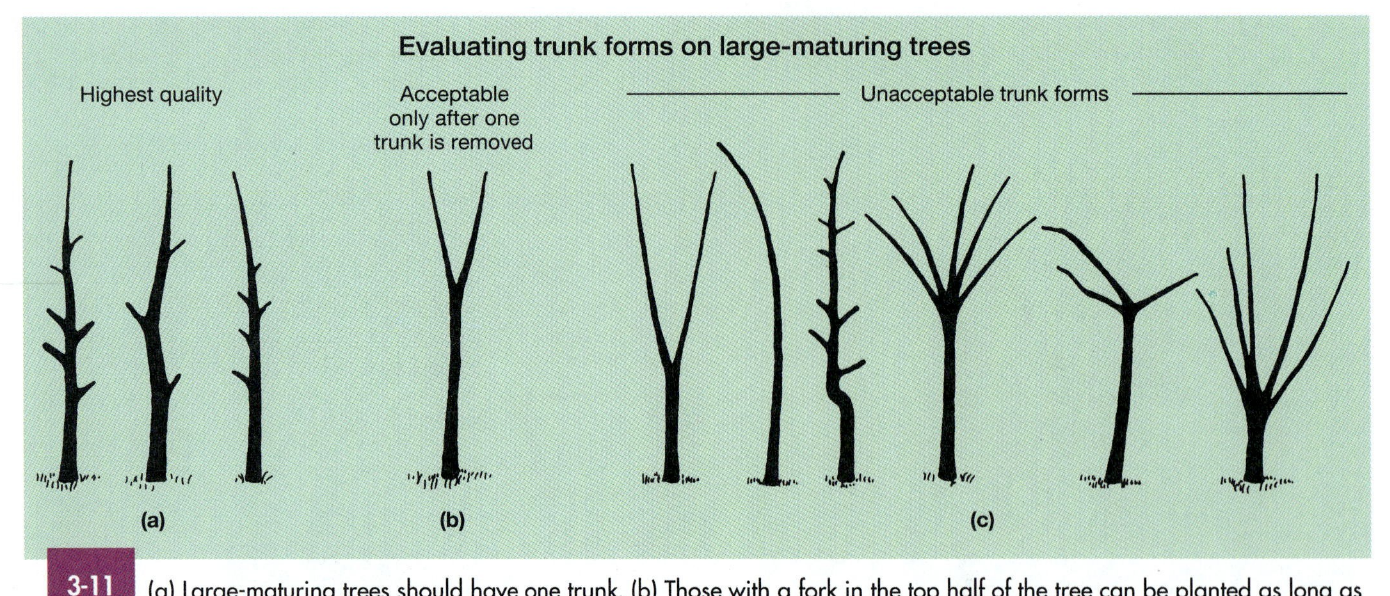

3-11 (a) Large-maturing trees should have one trunk. (b) Those with a fork in the top half of the tree can be planted as long as one of the two trunks is removed or cut back. (c) Other trunk forms are not acceptable for planting in most landscapes.

3-12 When included bark (roundish dark area at top of split crotch) develops in a branch crotch, the branch is not well secured to the trunk. A tree split like this normally is cut down.

3-13 (a) Branches on most trees should be distributed along the upper two-thirds of the trunk. Half the branches should originate from the lower two-thirds of the trunk. (b) Poorer-quality trees have no branches along the lower half of the trunk.

branches can be closer than this. Although most branches on these sapling-sized trees are not permanent and will be removed as the tree grows, it is best to avoid trees with branch arrangements like those in Figure 3-15b. Trees with this inferior branch arrangement that are not properly pruned could split apart when they get older.

Trees with a trunk diameter between two and four inches might have one or two branches on the tree that could become permanent; that is, they will remain on the tree for a long time and will not have to be removed. Permanent branches should be spaced at least 18 inches apart and should not have bark included in the branch crotches (Figure 3-14). Trees larger than 4 inches in diameter are likely to have permanent branches, and these too should be at least 18 inches apart. Choose trees with a branch arrangement similar to that illustrated on the left of Figure 3-15.

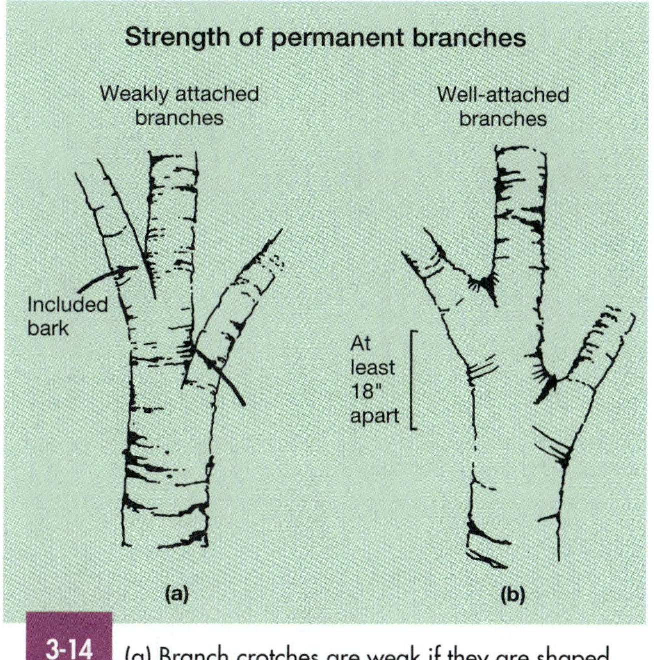

3-14 (a) Branch crotches are weak if they are shaped like the letter V with bark included or pinched in the crotch. (b) Crotches shaped like the letter U are usually stronger. Permanent branches should be at least 18 inches apart on trees maturing at more than 40 feet tall.

OTHER IMPORTANT FACTORS

Tree height

The American National Standards Institute Nursery Standards (ANSI Z60.1, 1990), used by many nursery operators, recommend minimum and maximum tree heights for nursery-grown trees based on the diameter of the trunk (Table 3-3). Trees taller than the recommended height may have a weak trunk. Trees shorter than the recommended height that are not normally slow-growing or dwarf-type cultivars may have root defects or other problems. It may be best to avoid these unless the cause of the slow growth can be determined.

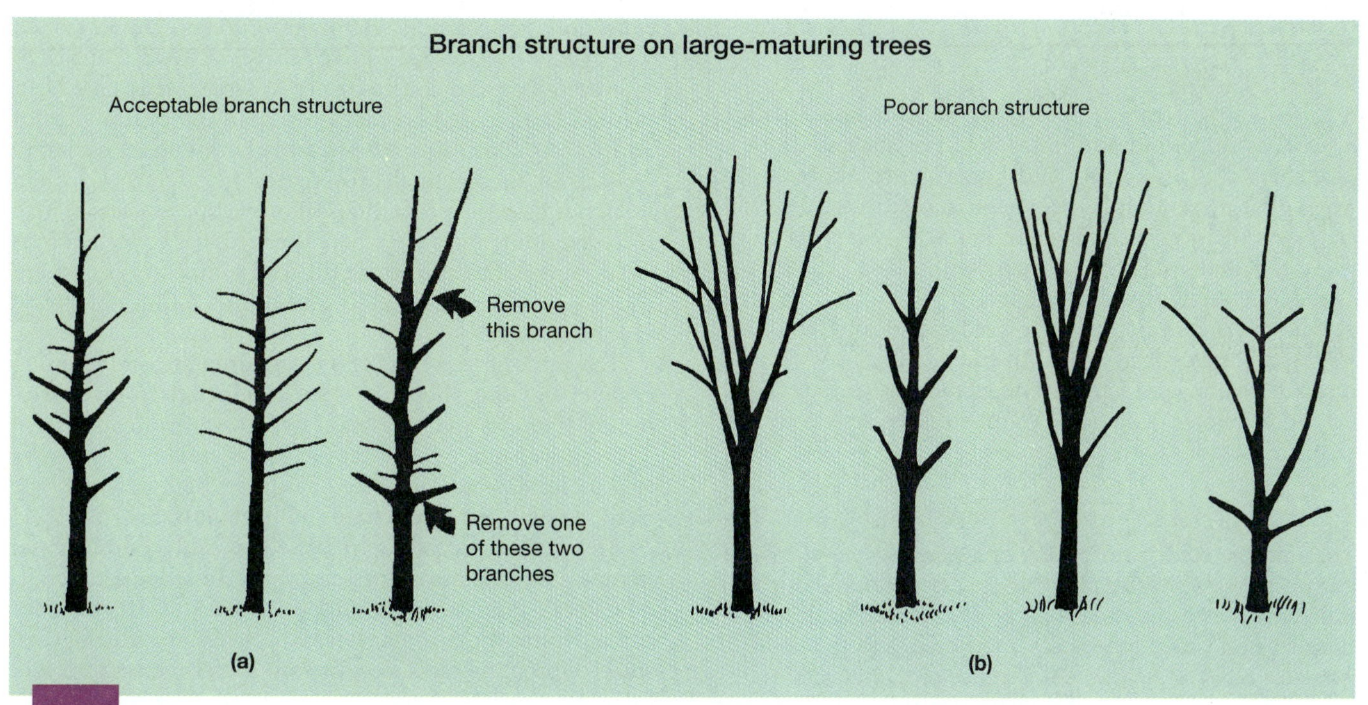

3-15 (a) The biggest branches on large-maturing trees (such as oaks) should be spaced along the trunk, and some should be larger than others (left and center). If a pair of large branches are opposite one another (right, lower arrow), one of the branches should be removed at or soon after planting. Large branches oriented nearly vertically should be removed (right, upper arrow). (b) Trees with most branches clustered together, those with few branches, or those with all major branches opposite each other may not be durable. (Note: Only the largest diameter branches are shown. The many smaller branches and twigs that would occur on some trees are not shown in this illustration.)

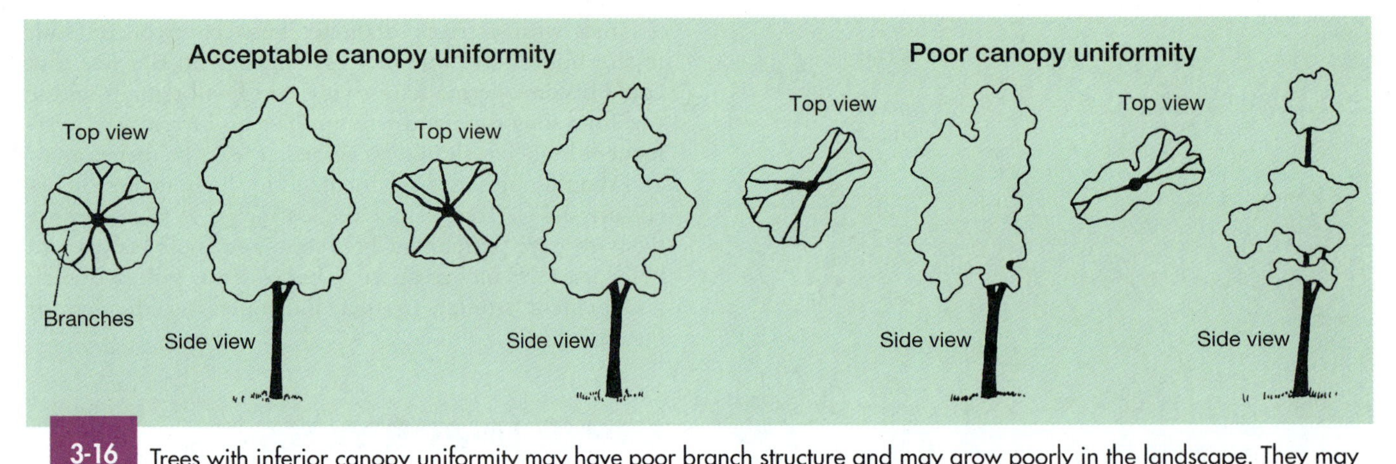

3-16 Trees with inferior canopy uniformity may have poor branch structure and may grow poorly in the landscape. They may be more difficult to train into well-formed trees than those with a more uniform canopy.

Canopy uniformity

Trees with uniform canopies look more attractive to most people than those that are unbalanced. However, uniformity of the canopy on nursery-sized trees is far less important than trunk form and branch arrangement. For example, a tree with a somewhat irregular canopy, one dominant trunk, and good branch arrangement is superior to a tree with a uniform canopy and a double trunk with bark included in the crotch. The canopy on the superior tree will usually fill in as the tree grows. If the canopy is flattened, one-sided, or very unbalanced, consider purchasing a different tree (Figure 3-16).

Canopy fullness

A thin canopy does not necessarily indicate poor quality, disease, or insect infestation. Many trees, such as Chinese pistache, royal poinciana, and honeylocust, appear open when young and fill in as they grow in the landscape. Others, like callery pears and many southern magnolia cultivars, are full of foliage even when they are small. Trees such as evergreen oaks usually lose most or all leaves for a brief period in the spring. Pines lose needles and southern magnolias lose leaves as new growth emerges in spring. This is normal. *You* have to judge whether the thin canopy is due to poor growing conditions in the nursery or is a natural characteristic of the tree.

Seed or propagule source

Trees propagated from plants originating from the same hardiness range as the planting site, or within about 200 miles, are likely to be adapted to the climatic conditions at the planting site. Those from another area of the country may not be as well adapted, because they may lack tolerance to the local climate. However, tree cultivars have been developed and varieties chosen for tolerance to cold temperatures, high soil pH, wet soil, drought, or pests; these may be better adapted than trees originating from within the same hardiness range as the planting site. Although trees from within the same hardiness zone may be well adapted to the heat and cold normally experienced at the planting site, a cultivar with tolerance attributes that match the specific conditions at the site might perform even better. For example, pin oak is adapted to hardiness zone 5, but—regardless of where the seed originated from—it would be a poor choice for a site with high soil pH. A honeylocust or redbud cultivar, even one developed outside the region in another hardiness zone, is probably a better choice.

Do not confuse seed source with provenance (Arnold 1995). Provenance refers to the geographic origin of the genotype, whereas seed source is the geographic site of seed collection. For example, you might purchase nursery stock propagated from a seed source (a tree) in lower Louisiana. However, that seed-source tree in Louisiana may have been planted from a seed borne by a tree in the coastal area of South Carolina. Your tree probably maintains the characteristics of the original provenance (the South Carolina tree), not the seed source (lower Louisiana). The lesson here is to determine not only where the seed was collected from (seed source), but also where the seed source tree originated (provenance). Unfortunately, this information is not always easy to come by.

Tolerance of trees to alkaline soil, drought, and flooding can depend on where the seed originated from (provenance). For example, Shumard oak grown from an eastern Texas provenance, where soil pH is 6.8, might not perform well on alkaline soil; one from a caliche (high pH) soil type in the central part of the state could adapt nicely.

Trees could be better matched to soil conditions at planting sites if nursery operators recorded these details when collecting seeds. If more buyers ask where the seeds came from, more operators will keep records. Urban forestry programs will benefit from this because trees will be better matched to sites.

Old pruning cuts

Evaluate old pruning cuts to determine the quality of the nursery stock. Properly made pruning cuts indicate that the nursery has high pruning standards and is capable of

3-17 (a) A proper pruning cut is made along the ink line drawn next to the word "yes" just outside the branch collar. It is usually round in cross-section and callus closes over the wound in a circular pattern. (b) An improper cut made through the collar (next to the word "no" on left) results in an elongated wound that begins to callus from the sides only.

growing high-quality trees (Figure 3-17a). Improper cuts indicate a poor understanding of tree care and biology (Figure 3-17b).

Insects and disease

Here are a few tips on how to avoid trees with major insect and disease problems.

Speckled or spotted foliage could indicate the presence of a leaf spot disease, but spots are usually harmless. They can also indicate that foliage is infested with spider mites, lacebugs, or some other pest that sucks plant juices from the foliage. It is best to purchase trees without these pests. They might also indicate sunburn or chemical injury.

Trees should be free of scale insects, which can be a troublesome pest in the landscape. These bugs can be hard to detect because their color is often similar to that of twigs and branches. Look for raised ridges or bumps on the twigs. These may be part of the tree, or they could be scales. Pick several off with your fingernail. Green or white tree tissue will be evident under the bark if the bump is a normal part of the tree. The bark of the twig will be left more or less intact if the bump was a scale insect. Recently planted trees have enough stress placed on them without having to defend against a scale infestation. Do not plant trees with a scale infestation.

Open cankers and other wounds on the trunk or branches could indicate poor growing techniques, cold injury from previous years, or disease. Trees with these problems should usually be avoided.

Trunk injury and broken branches

Avoid trees with scars and other open wounds along the trunk. Open pruning wounds are acceptable if they are small. The presence of large, open pruning wounds could indicate an unplanned or poor pruning program at the nursery. Small broken branches should be pruned back to healthy tissue; trees with large broken branches should usually be left in the nursery. Do not purchase trees with bark stripped down the trunk from an improper pruning cut. Check for injury to the trunk caused by stakes rubbing against it, and be sure the trunk was not girdled in the nursery from stake ties that were left on the tree too long.

Trunk wrap

Trunk wrap or other material covering the trunk should be removed so you can inspect the trunk. The trunk can be rewrapped after inspection to help prevent minor trunk damage during shipping (for further information, refer to Chapter 5).

Leaf color and size

Leaves should be colored as on other trees of the same type. If leaves are smaller, lighter colored, or yellower than others, the tree may lack vigor and grow poorly in the landscape.

Die-back

Nursery trees should have foliage to the ends of all branches. Dead tips indicate problems. Twigs can be checked in the dormant season by scraping several small twigs with your fingernail. If tissue is greenish or white beneath the bark, then the twig is probably alive. If the wood under the bark on small twigs is brown and dry, that part of the branch out to the tip is probably dead.

Included bark

Major branches should not have bark included in the crotch. Included bark is an indication that the branch is not well attached to the trunk. It could separate from the tree as it grows older (Figure 3-12).

Stakes

All stakes should be removed from the tree before purchase to check if the trunk stands on its own. If not, do not buy it.

Weeds in the root ball

Weeds in the root ball will be planted in your landscape along with the tree. This can slow the establishment rate of the tree, and the weeds can spread in the landscape. Buy trees without weeds or remove any weeds before planting. If the removed weeds have already produced seeds, the weeds are likely to germinate next year.

PALMS

Palms without visible trunk development should not be transplanted from a field nursery, because recovery is dismal. Instead, plant young palms from containers (Meerow 1992). With the exception of Bismark palm and very high value palms, field-grown palms usually are not root-pruned prior to transplanting. If pruning is conducted, it is best performed two to three months prior to moving the palm. This allows for root regeneration from the root initiation zone at the base of the trunk. Of the palms tested, coconut and queen palms initiated few roots during this time period (Broschot and Donselman 1984a and 1984b).

Table 3-4 summarizes desirable root ball sizes for different-sized palms (Florida Department of Agriculture 1996). Coconut and sabal palms are usually successfully transplanted with a smaller root ball than indicated in Table 3-4 because most cut roots die back to the trunk. Essentially all roots on these palms growing into landscape soil after planting will be new ones from the root initiation zone located at the base of the trunk.

Palms planted in hardiness zones 7 through 10A are best planted in the spring and summer when soil temperatures are most conducive to root growth. In zones 10B and 11, soil temperatures are usually suitable for adequate growth year round.

TABLE 3-4 MINIMUM ROOT BALL SIZE FOR PALMS

PALM SIZE	RADIUS OF BALL[1] (INCHES)	DEPTH OF BALL[2] (INCHES)
less than 15 feet	8	12
15 to 25 feet	10	18
more than 25 feet	10 + 1 for each additional 5' of height	24

[1]Measured from the base of the turnk on single-trunk palms. Measured from the base of the trunk furthest from the center of the cluster on multitrunk palms.

[2]Measured from the soil surface to the bottom of the root ball. Root ball should have straight sides.

Source: Table adapted from Meeroa 1992.

Chapter 4

Planting Techniques

HANDLING TREES PRIOR TO PLANTING

Transporting

Root balls are fragile and should be handled carefully. Those in hard plastic containers or boxes are most resistant to abusive handling; those in soft, fabric containers and balled-in-burlap are most sensitive. Never pick a tree up by the trunk—always carry or lift it by the root ball. Never drop the tree. This will disrupt contact between fine roots and soil. *Roots must be in intimate contact with soil to absorb water.* Because only a slight movement of soil in the root ball can crack the ball and dislodge roots, be sure the burlap is secured tightly around the soil. Roots break and the root ball dries quickly when cracks appear in the ball.

Tie trees securely to the truck so they do not roll around during transport. Rolling or other movement during shipping can crack the root ball and break roots. Trees transported on open trucks lose more water than those shipped in a closed truck and can arrive at the planting site in poor condition. Be sure trees are irrigated just prior to shipping to help minimize desiccation. Do not allow closed trucks to remain standing in the sun in hot weather unless they are air-conditioned. Trees could be injured if the temperature inside the truck is maintained at more than about 100°F.

Some nurseries shrink-wrap black plastic around the outside of the root ball prior to transporting. This practice appears to reduce water loss from the tree and root ball during transport to the planting site. Trees come to the job site with less water stress than those without plastic. Once they reach the planting site, do not let plastic-covered root balls remain in the sun. Keep them in the shade or remove the plastic to prevent temperatures in the root ball from reaching lethal levels. Plastic must be completely removed prior to planting.

Prior to shipping, some nursery operators routinely spray trees with an antidesiccant (sometimes referred to as an antitranspirant). Some of these reduce water loss during shipping and increase survival (Englert et al. 1993), but could hinder photosynthesis for several weeks after planting. This could lengthen the period of establishment by slowing root growth. Others find that antidesiccants provide no benefit to survival except for certain species during certain times of the year (Harris and Bassuk 1995). With careful handling and proper after-care, antidesiccants probably are not routinely needed; however, more research in this area would be welcome. Antidesiccant wax applied to the stem and branches at transplanting to desiccation-sensitive trees, such as Washington hawthorn, can increase survival rates (Bates and Niemiera 1996).

Mechanical or chemical defoliation of deciduous trees prior to shipping has been practiced occasionally by some nursery operators, especially on deciduous trees shipped in summer. It is a costly and time-consuming technique. There is not much research in this area, but one recent study shows that this can be detrimental to tree survival (Harris and Bassuk 1995). With proper shipping techniques, including transport in a closed area and use of cover to protect trees from direct sun and wind exposure, good holding practices at the planting site, and good irrigation management after planting, defoliation is probably not needed.

Branches normally are tied together close to the trunk to prevent breakage during shipping (Figure 4-1). They can be secured with string, plastic straps, fabric, or other material.

4-1 Trees are damaged less during shipping if branches are wrapped close to the trunk.

Holding area at the planting site

It is best to plant trees the day they arrive at the planting site. If trees cannot be planted for an hour or two, irrigate them as soon as they are inspected and unloaded from the truck. Trees left out and not planted can deteriorate in quality quickly when not properly cared for. If trees cannot be planted the day they arrive at the planting site, establish a holding area at the site. The holding area should be as shaded as possible and away from the wind. It should also have provisions for irrigation. This area should be set up before trees are brought to the planting site.

As soon as balled-in-burlap (B&B) trees arrive in the holding area, cover the sides of the root balls with soil, compost, mulch, sawdust, or other organic matter to help prevent root desiccation. This also helps irrigation water penetrate the root ball. Do not cover the top, as this could restrict the flow of irrigation into the root ball. Trees in containers should remain upright so irrigation water can seep into the root ball. Group them close together to provide mutual shading of the root balls; direct sun hitting the sides of the containers often increases temperatures inside the root balls to lethal levels. The exposed container wall on root balls on the outside of the grouping should be covered with organic matter or soil to prevent root death from sun and overheating. Roots can die in a matter of hours, so prompt action is essential. Root systems on bare-root trees should be covered with moist sphagnum or other moisture-holding material and kept out of the wind and sun to help keep them alive. Unless trees will be stored in complete shade, do not cover root balls or trees with any type of plastic, as this could increase temperature to lethal levels.

Irrigation

Plastic containers and boxes Trees grown in containers in the eastern United States are irrigated daily in the nursery during the summer because they are grown in porous media to allow adequate drainage. In warm southern climates, some trees receive water twice each day in the summer. In the deep south, trees may be watered daily in the spring and fall as well. In the northwest, irrigation is applied less frequently. Irrigation in most areas is cut back during the cooler months. When trees leave the nursery, they should continue to receive the same amount of water at the same frequency, unless they are placed in a shaded location. The best way to determine the irrigation amount and frequency trees received just prior to your purchase is to ask the nursery operator. Even one day without water can cause significant root death. This can cause quick decline in tree vigor and can extend the stressful establishment period. Trees not watered for several days in warm weather could die.

Trees grown in containers in the dry portions of the western United States often receive irrigation less frequently because root balls contain a denser medium that holds moisture. Regardless of where trees are raised or when they are purchased, supply trees with the amount and frequency of irrigation they received in the nursery just prior to purchase. The nursery operator probably knows more about the water needs of these trees than anyone else.

Balled-in-burlap and fabric containers Nursery operators sometimes protect roots that grow through the burlap after trees are dug by wrapping (e.g., shrink-wrapping) the sides and bottom of the root ball in black plastic or fabric, or by otherwise keeping them from drying out. Be sure there is a drainage hole through the plastic at the bottom of the root ball. Do not remove roots growing through the burlap unless the burlap is synthetic and must be removed anyway. Removing regenerated roots wastes valuable carbohydrates used to produce those roots. With regenerated roots intact, the energy level of the tree will remain as high as possible.

B&B trees must receive regular irrigation after leaving the nursery and prior to planting. In sunny, warm weather, daily irrigation is suggested for trees dug during the growing season. If trees were pre-dug at least several weeks prior to your purchase, ask the nursery operator how often the trees were irrigated during the two or three weeks before purchase. Apply irrigation at the same interval. You can be sure that the nursery was supplying the tree with enough water to keep the tree alive.

Water rolls off the surface of the root ball if it is applied too fast. This wastes water and keeps roots too dry. A drip emitter, spray emitter, or other low-volume irrigation head can usually deliver water slowly enough to prevent excessive runoff. Large root balls may need two or more emitters for thorough coverage. If this is not available, slowly apply water with a hose or sprinkler long enough to allow thorough wetting inside the ball. Be sure the entire root ball gets wet, but do not allow it to sit in a pool of standing water.

Bare root Bare-root trees, especially the roots, should be kept in complete shade prior to planting to prevent root desiccation. Fine roots can become dry and die if exposed to direct sun for just a few minutes. Cover the roots with a moist towel or moist straw, saw dust, or wood shavings. Roots should be sprayed with water frequently enough to keep them moist. Frequency will depend on the weather and the nature of the storage facility or holding area. In cool or wet weather, two or three sprays during the day may be enough to keep roots from drying. Hourly applications may be needed in dry, hot, or windy weather. Also consider soaking the entire root system in a water bath for two to six hours just prior to planting. You might also heel them into sawdust, bark, or soil in the holding area if they will be planted later in the day. Cut off damaged or broken roots with a sharp hand pruner. Remove potentially girdling or badly kinked roots.

PLANTING THE TREE

Three of the most common causes of poor plant establishment are planting too deep, and under- or overwatering. If appropriate trees are planted at the right depth and they are irrigated properly, the planting will be successful. As simple as this appears to be, problems often arise that lead to poor establishment or plant failure.

Well-drained soil

Do not dig the planting hole deeper than the height of the root ball (Figure 4-2a), because this buries the roots too deep. Never plant so the top of the root ball is below the soil surface (Figures 4-2a–c). If anything, err on the high side by planting slightly higher than the surrounding soil. Soil placed over the root ball can inhibit or prevent water from reaching the roots, especially on trees planted from containers. This is due to the large difference in texture between soil and container media. Water is bound tighter in the finer-textured substrate (soil) over the root ball, so it does not readily percolate down into the coarser substrate (container medium). Always set the bottom of the root ball on solid soil that has not been disturbed or loosened. If bottom soil has been loosened, repack it so the root ball will not settle. Setting the root ball on loosened soil will cause the tree to settle and the tree often sinks too deep into the soil. This slows establishment and can kill the tree by suffocating the roots.

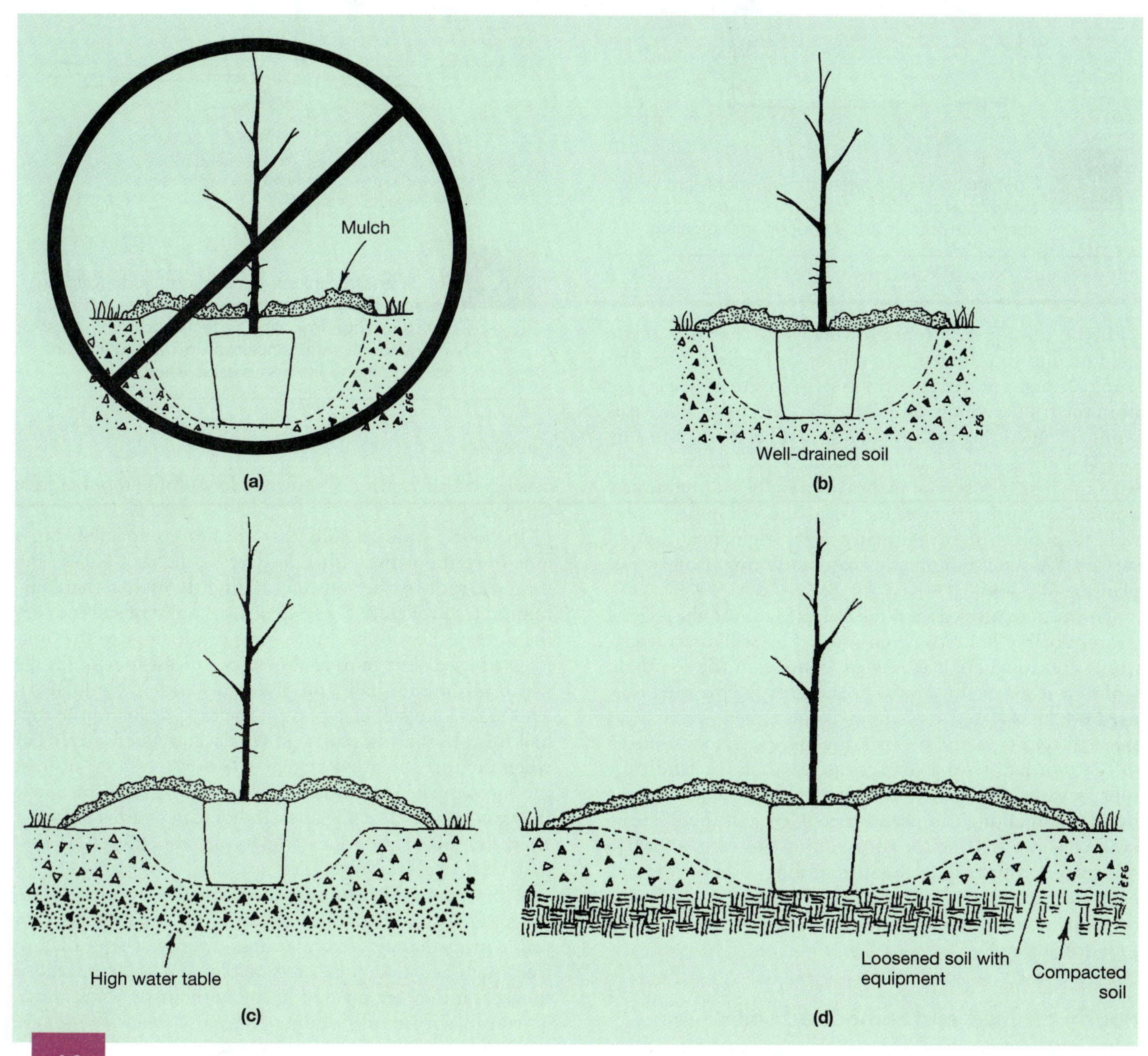

4-2 (a) Never plant trees or shrubs too deep and never place soil over the root ball. (b) In most well-drained soils, locate the top of the root ball even with the soil line. (c) In poorly drained, loose soil, install the root ball slightly higher than the surrounding soil. (d) In compacted sites, loosen the hard soil around the planting hole as far away from the tree as possible (preferably 15 to 20 feet) and install the root ball slightly above grade. Remove the berm of soil around the root ball if the root ball stays too wet.

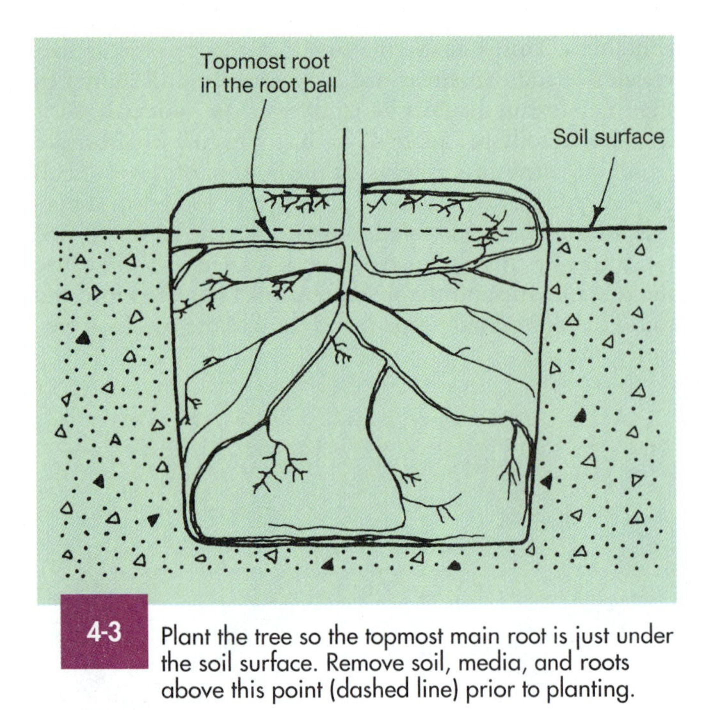

4-3 Plant the tree so the topmost main root is just under the soil surface. Remove soil, media, and roots above this point (dashed line) prior to planting.

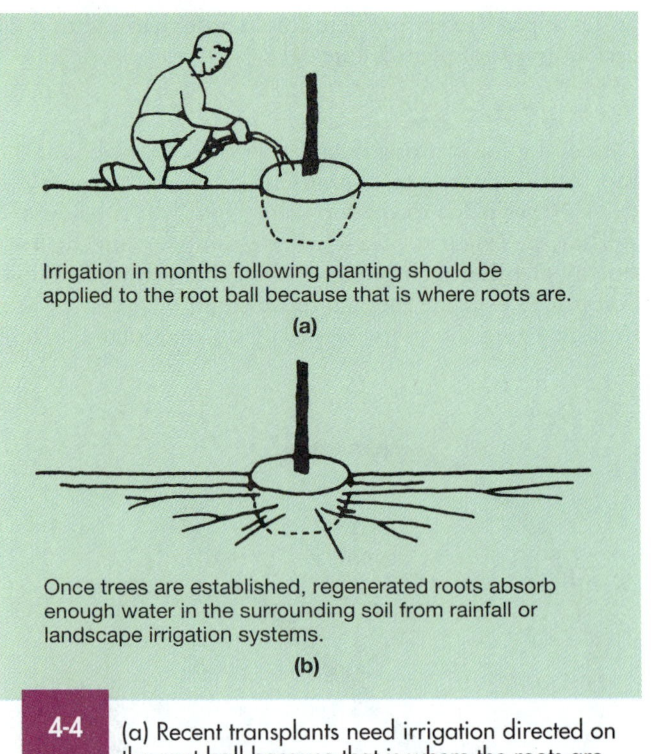

4-4 (a) Recent transplants need irrigation directed on the root ball because that is where the roots are. (b) Landscape irrigation systems and rainfall provide enough water once roots grow into the landscape soil and the tree is established.

If you cannot see a root growing from the trunk at the soil line, slip your fingers into the soil or medium along the trunk. When you feel the topmost root growing directly from the trunk, remove all soil, media, and roots above this point. Plant so the topmost root is just below the soil surface (Figure 4-3; refer also to Figure 3-7). Planting a tree with the first root several inches or more below the surface is just like planting the tree too deep. You will notice as you walk through a forest that many roots originate at the soil surface. We should mimic this natural rooting habit by not planting too deep.

Form a one- to two-inch-tall soil ring around the edge of the root ball to hold irrigation water. This will direct irrigation water to where it is needed, namely, in the root ball. Unless it is unusually dry, soil outside of the planting hole need not be irrigated at planting, because there are no roots there (Figure 4-4a and b). In many instances, soil or media in the root ball have a coarser texture than the landscape soil, so water will drain away from the root ball and wet the backfill soil enough to promote root growth there. However, consider wetting the surrounding soil if the root ball soil texture is finer than the surrounding soil or if the surrounding soil is very dry. In drier climates, or during drought, wetting the surrounding soil is likely to encourage root growth.

Poorly drained and compacted soil

Follow all the guidelines outlined in the preceding section on well-drained soil, but note the following precautions. One of the biggest killers of trees planted in poorly drained, clayey sites is saturated soil. Too much water in the root ball suffocates roots. This can be prevented by planting about a third of the root ball higher than the surrounding soil (Figure 4-2c and d), and by managing irrigation properly. Another good alternative is to plant a low-profile root ball (see Chapter 3).

In poorly drained soils that are not compacted, dig a hole only about two-thirds as deep as the root ball so the topmost root in the root ball is slightly higher than surrounding soil (Figure 4-2c). Create a mound of soil to cover the sides of the root ball, forming a gentle slope to the original soil level. The mound should be at least four times the width of the root ball, and preferably wider (Figure 4-5). This technique will help prevent the top portion of the root ball from becoming saturated with water when it rains or when the turf irrigation system is left on too long. At least the roots in the top portion of the root ball will be above the standing water at all times. In hot, dry weather on some sites, the top of the root ball may dry quickly, so be prepared to irrigate accordingly.

Most plants (except aquatics) will not become established, or they will establish slowly, if they are inundated by water immediately after planting. Even wet-site-tolerant trees such as baldcypress, red maple, and alders establish quickest if they are planted in soil without standing water. They can withstand flooding later, after their roots have grown into surrounding landscape soil.

When planting trees in compacted soil, rototill or deep-rip the area prior to planting 15 to 20 feet or more in diameter around the tree (Figure 4-2d). Soil can be ripped 18 to 36 inches deep with a ripping tool attached to the back of a large tractor or bulldozer. This type of large equipment is suited for loosening large areas of compacted soil. Disk to

4-5 Several years after planting slightly above the existing, poorly drained soil, there is a gentle slope away from the trunk.

smooth the soil after it is ripped. This technique will reduce the time it takes to establish a tree by dramatically increasing the rate of root growth and root penetration into the landscape soil (Gilman, Leone, and Flower 1987; Smalley and Wood 1995). If this is not practical, dig the planting hole at least three times wider than the root ball, making sure the sides are sloped. As roots reach the sloped sides, they are deflected to the soil surface where they proliferate (Watson, Kupkowski, and von der Heide-Spravke 1992). This results in a shallow root system.

Root growth will be slow in compacted soil that has not been loosened, and the tree will establish slowly. Many trees are more susceptible to drought, insect, and disease attack during the establishment period, so it is most desirable to minimize the time it takes to establish a tree. The best way to do this is to loosen compacted soil. Some increase in root growth can be obtained by digging five vertical trenches 18 inches deep extending several feet from the planting hole (Bassuk 1994). Roots grow easily in the loosened soil in the trenches (refer back to Figure 1-10).

Some horticulturists suggest that the one- to two-inch-tall soil berm routinely constructed around the root ball in well-drained soil to retain water can be omitted in poorly drained soil. They argue that too much water can enter the root ball if the berm is constructed. However, this cannot happen unless the tree is irrigated too frequently. Rain hitting the top of the root ball is just as likely to enter the root ball whether a berm is in place or not. All of it will run into the root ball with or without a berm if the backfill soil is loose with a high percolation rate. In fact, berms could prevent some runoff water from the surrounding landscape from entering the root ball, which could help prevent roots from drowning. A saturated root ball after planting usually indicates that the root ball was too deep and inappropriately shaped, the tree was too large, or it was installed too deep for the conditions at the site. Consider selecting low-profile root balls, planting smaller nursery stock, or planting trees that tolerate wet soil for such a site (refer back to Chapter 3). The small soil berm around the root ball is usually not the cause of the problem; it is likely to do more good than harm. Berms more than two inches tall can cause problems because the soil in the berm often works its way toward the tree covering the root ball and burying the lower trunk.

Balled-in-burlap

Synthetic burlap does not decompose in soil. Cut away synthetic and plastic burlap as far down the root ball as possible so that soil along the side of the root ball is in direct contact with backfill soil. It is best to remove all synthetic or plastic burlap and throw it away, because roots that grow through it often become girdled (Figure 4-6a and b). This could cause the tree to topple over, or the tree might slowly decline as roots become completely strangled by the fabric. In moist climates, biodegradable or natural burlap can be left along the sides of the root ball because it will usually rot in the soil. However, there are reports of natural, untreated burlap not decomposing in drier parts of the country (regions receiving less than about 20 inches of annual rainfall). Therefore, it may be best to somehow completely remove the burlap in these climates. It should always be removed from the top of the ball so the sun will not dry it out (Figure 4-6c). If allowed to dry out, the burlap may repel water, making it difficult to rewet the root ball. The concept of water wicking out of the soil through burlap exposed to the sun is unsubstantiated by research.

It may be difficult to distinguish between natural and synthetic burlap. If so, burn a small portion with a match. Synthetic burlap often melts and does not produce a flame; natural burlap usually burns with a flame.

Bare-root

Spread out roots of bare-root trees before backfilling. Some people lay the roots over a mound of firmly packed soil in the center of the planting hole and carefully place soil between groups of roots; others wash soil between roots. In either case, position the topmost root where it meets the trunk just under the soil surface. The shallowest roots usually are positioned about parallel with the soil surface, or angled down slightly. Do not break or bend roots a great deal. The planting hole should be dug wide and deep enough to accommodate roots without bending them into the hole. On grafted trees, be sure to position the graft union above the soil line.

Do not cut roots to fit into the planting hole. Dig a long hole or trench if necessary to hold long roots. There is no need to trim or cut any roots unless a portion is obviously dead, broken, or dried. Cut broken roots cleanly with a sharp pruning tool. Pruning roots indiscriminately at planting will not stimulate root regeneration and is not recommended. Although some horticulturists recommend removing the dried ends (approximately ½ inch) of the biggest roots, there is no scientific evidence that this improves growth or survival.

(a)

(b)

(c)

4-6 (a) Roots often grow through synthetic burlap, but many become strangled because they cannot enlarge in diameter. (b) Note the swollen root tissue from a root growing through the fabric (right). (c) Therefore, it is best to remove burlap.

Plastic containers and boxes

Pot-bound root balls from containers have large or many roots on the outer edge of the ball (refer back to Figure 3-8). It is best not to plant trees in this condition, but if you must, these roots should be cut with a knife or pruning tool to prevent them from girdling the tree later, especially if they are near the top of the root ball. Make three or four slices an inch deep from the top of the root ball to the bottom. If in doubt about whether a root is large enough to cut, go ahead and cut it. Recent research shows that if there is any shoot growth reduction from root-pruning container-grown trees at planting, the effect is negligible (Dana and Blessing 1994; Gilman et al. 1996).

Recent studies show that slicing the root ball from top to bottom in several locations does not consistently increase root growth after planting (Dana and Blessing 1994). However, one study showed that slicing pot-bound root balls enhanced distribution of regenerated roots in the backfill soil profile. Instead of growing almost exclusively from the bottom of the root ball, slicing encouraged root regeneration along the slice from the top to the bottom of the ball (Figure 4-7).

Wire baskets and wire-wrapped root balls

Wire baskets made from heavy-gauge wire are often used to help hold a B&B root ball intact during shipping and handling. This is a very effective means of ensuring that roots inside the root ball remain in contact with soil. Wire baskets degrade very slowly in soil and can be intact 20 years after planting. There is little research documenting the detrimental effects of wire baskets on trees. In fact, some research shows that roots engulf the wire with little apparent harm to the tree (Lumis 1990; Watson 1994). More time is needed to evaluate whether the wire will girdle the trunk as it expands, because wire baskets have been used for only several decades.

Despite two research studies suggesting that wire has little effect on trees, many horticulturists recommend removing at least the top 12 to 18 inches (two or three levels) of wire from the root ball on large-maturing trees (Figure 4-8). This will allow the major roots and trunk to grow without any possibility of becoming girdled by the wire. One to several minutes are required to remove the wire basket. Many people would consider this a good investment in time. The bottom half of baskets more than about 40 inches in diameter can probably be left intact if planting in poorly drained or compacted soil, as most of the root system will develop in the surface soil layers. Few are likely to develop through the lower portion of the basket. Baskets do not have to be removed from small-maturing trees and large shrubs.

To remove the top portion of the basket from a tree placed in a hole dug with a tree spade, the top of the planting hole will have to be dug wider than the root ball, a procedure that is not often practiced on trees planted with a tree spade (Figure 4-9). This extra hand-digging creates the clearance needed to cut and remove the wire all around the root ball after it is installed in the planting hole. Trees with

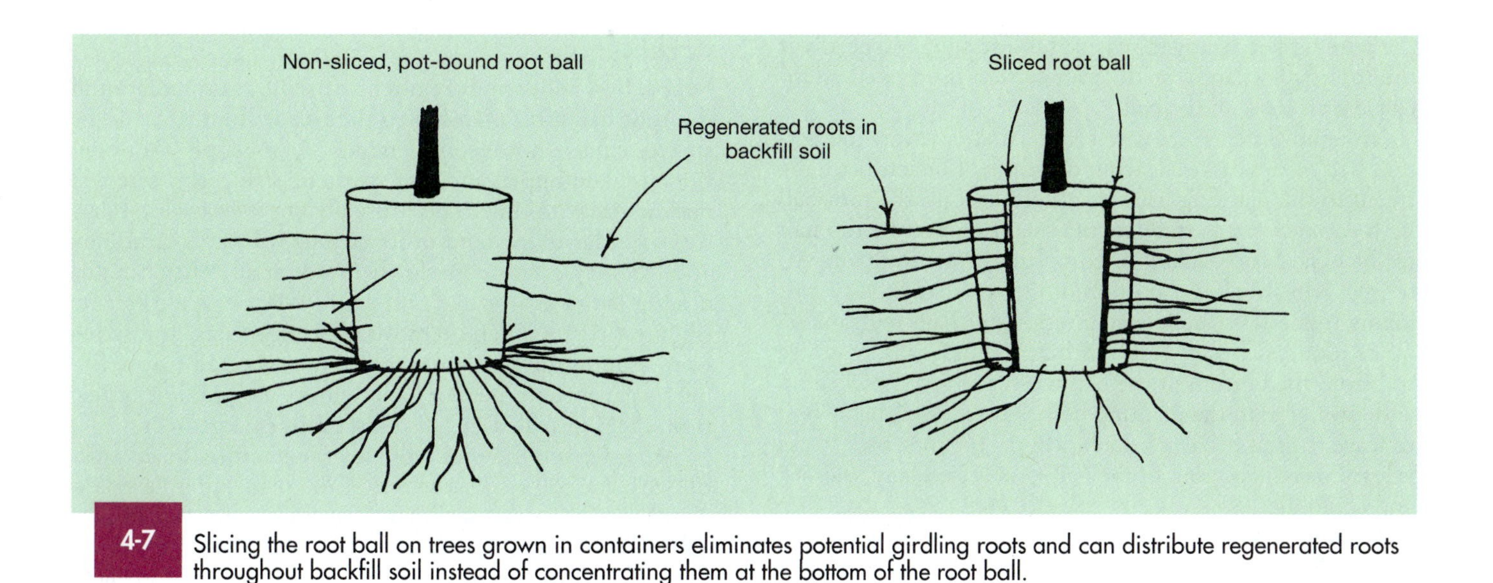

4-7 Slicing the root ball on trees grown in containers eliminates potential girdling roots and can distribute regenerated roots throughout backfill soil instead of concentrating them at the bottom of the root ball.

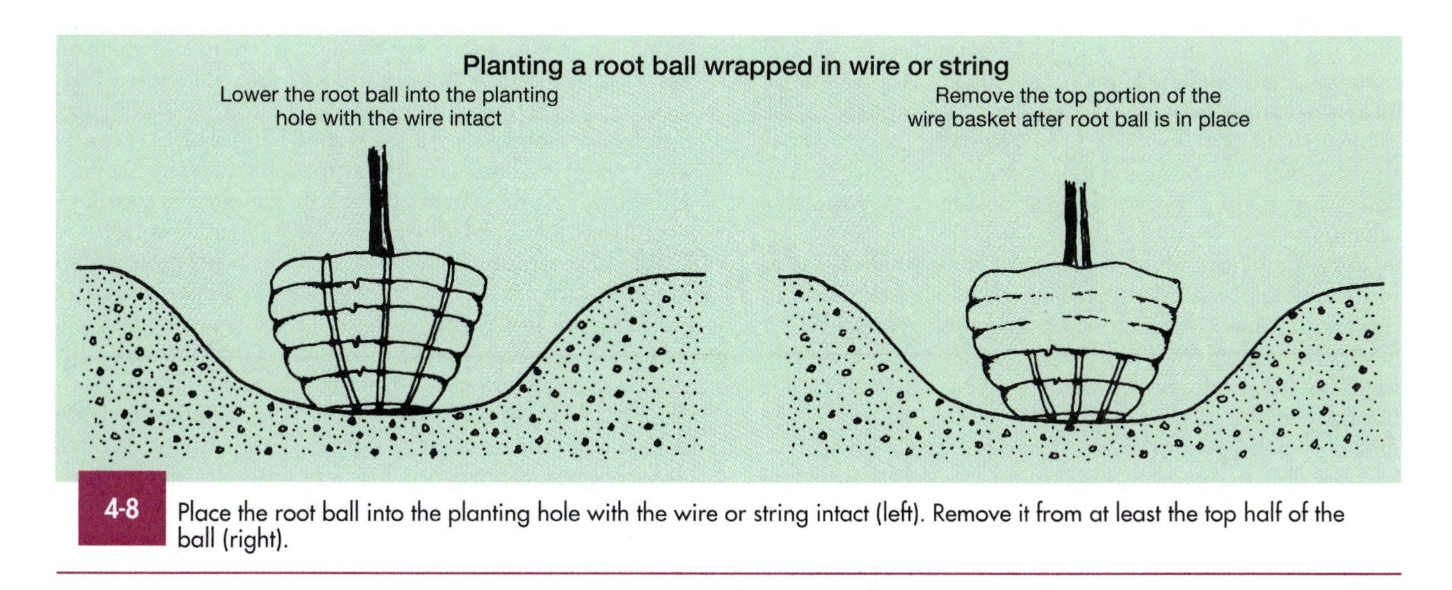

4-8 Place the root ball into the planting hole with the wire or string intact (left). Remove it from at least the top half of the ball (right).

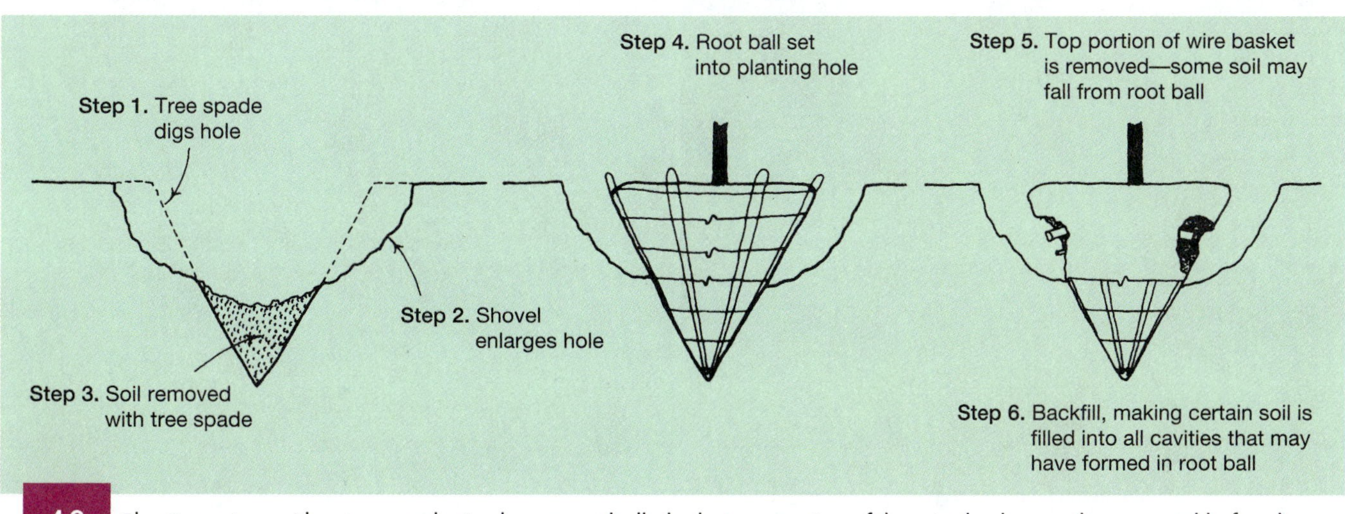

4-9 Planting a tree with a tree spade. In clayey root ball, the bottom portion of the wire basket can be removed before lowering the tree into the planting hole.

the wire basket removed may require staking, especially if the root ball is sandy or the root system is not well established or is loose in the ball.

To remove the entire wire basket, first cut away the bottom half of the basket, leaving the top half intact. Place the tree into the planting hole and fill in around the root ball with enough soil to stabilize the tree. Remove the top half of the basket and backfill to complete the planting job. After removing the lower portion of the wire basket, clayey or loamy root balls usually stay intact. Root balls with inadequate root systems, or those with very sandy soils, may lose soil from the bottom of the ball when the tree is picked up and lowered into the planting hole. If so, it is probably best to leave the lower half of the basket on the root ball. If soil is lost from the bottom of the ball, the tree is likely to shock and could die.

Fabric containers

Root balls of field-grown trees in fabric containers are very fragile and must be handled carefully. Soil inside the ball can become loose from just a moderate disturbance. Never drop the ball, because roots will lose contact with soil and trees will shock and are likely to die quickly. *Always* remove all fabric from the ball before carefully sliding it into the planting hole. Make a slit in the fabric from the bottom of the root ball to the top and gently pull the fabric from the ball. Most people lay the tree on its side to perform this operation.

Some fabric container designs allow only small-diameter roots to develop outside the fabric; fabric on these trees will be easy to remove without disturbing the root ball. Other fabrics allow large roots to develop through the fabric, making fabric removal more challenging. A hand pruner can be used to cut large-diameter roots flush with the inside of the fabric to make removal easier. Some nursery operators use special tools designed to quickly remove fabric from the root ball.

Backfilling

Straps, ties, and string are often left tightly secured around the root ball after planting so the tree will not have to be staked. This is not recommended. Some people are under the mistaken impression that synthetic string and wire will decompose with time. This is usually not the case. Even hay rope, binder twine, and other strings with a natural look and feel may decompose slowly or not at all. They can girdle the trunk as it grows and could kill the tree if left on (Figure 4-10). Cut and remove *all* straps, wires, and string from around the trunk and top of the root ball before filling soil in around the root ball. Remove all synthetic burlap from the root ball, as it can restrict root enlargement.

After placing the root ball into the planting hole, backfill with the soil that came out of the planting hole unless your site analysis revealed that the soil was unsuited for root growth (refer back to Chapter 1). According to the best information now available, amendments of any kind (gels, polymers, liquids, dusts, powders, fertilizers, bark, organic matter, humic acid, etc.) incorporated only into the planting hole around trees and shrubs generally provide little or no benefit to root growth or drought tolerance (Henderson and Hensley 1992; Ingram, Black, and Johnson 1981; Paine et al. 1992; Schulte and Whitcomb 1975; Smalley and Wood 1995). Day, Bassuk, and Van Es (1995) found an increase in shoot growth with peat-amended backfill, whereas pine bark-amended backfill inhibited growth unless nitrogen was added to the backfill (Smalley and Wood 1995). In most instances, once they start roots grow so fast (12 inches to 10 feet in the first year; this is covered in greater depth in Chapter 7) regardless of amendments that they soon leave the backfill soil and grow into the surrounding undisturbed soil. The best backfill in most landscape situations appears to be the loosened original soil dug from the planting hole. The best soil amendment is water. However, some horticulturists use amendments because their price is minute compared to the cost of the trees and

4-10 String left on the trunk at planting often strangles the tree (left), resulting in poor growth and then death. The tree at right is declining from the string left at the base of the trunk (right).

planting. With the exception of those that encourage water retention in poorly drained soil, amendments in backfill soil probably do little harm.

The exception to this rule is where existing soil is so terrible or contaminated, such as in a parking lot island or in a small cutout in a sidewalk, that all soil has to be replaced with good-quality soil (Craul 1992). This is a good practice and is usually recommended.

Loosen and break up large clods of soil before backfilling. Clods of soil used as backfill create air pockets around the ball and could hinder root growth and establishment. After filling the bottom half of the space around the root ball with backfill soil and tamping *lightly* with your foot, push a hose with running water in and out of the backfill soil all around the hole. This will settle the soil around the root ball and eliminate air pockets. This is usually not needed if the planting hole is at least twice the width of the root ball because the use of a shovel can work the soil eliminating air pockets. Add soil so it is even with the top of the root ball, but do not cover the root ball with soil (Figure 4-11). Only mulch should be placed on top of the root ball. Construct a one- or two-inch-tall soil berm at the edge of the root ball to entrap irrigation water (refer to Figure 4-2). If it is installed farther from the tree, water could perco-late only into backfill soil, missing the root ball entirely and keeping roots dry. This is especially important when B&B root ball soil texture is considerably finer than backfill soil texture.

4-11 Placing soil over the root ball buries roots too deep. This cuts off oxygen and can keep roots and trunk too wet.

TIME OF YEAR

In cooler climates, a number of B&B and bare-root trees recover best from transplanting in late winter or early spring when roots on some trees begin active growth (Table 4-1).

TABLE 4-1 TREES THAT TRANSPLANT BEST IN SPRING IN THE NORTHERN UNITED STATES (after *Selecting and planting trees,* The Morton Arboretum)

Apricot, Manchurian, *Prunus armeniaca*
Baldcypress, *Taxodium distichum*
Beech, *Fagus spp.*
Birch (except River), *Betula spp.*
Cherry, *Prunus spp.*
Corktree, *Phellodendron amurense*
Dogwood, *Cornus*
Goldenraintree, *Koelreuteria paniculata*
Hawthorn, *Crataegus spp.*
Hemlock, *Tsuga spp.*
Hickory, *Carya spp.*
Hornbeam, *Carpinus spp.*
Ironwood, *Ostrya virginiana*
Katsuatree, *Cercidiphyllum japonicum*
Larch, *Larix spp.*
Linden, Silver, *Tilia tomentosa*
Magnolia, *Magnolia spp.*
Maple, Freeman, *Acer x freemanii*
Maple, Paperbark, *Acer griseum*
Maple, Red, *Acer rubrum*
Oaks (except English, Pin), *Quercus spp.*
Pawpaw, *Asimina triloba*
Pear, Callery, *Pyrus calleryana*, and other pears
Pecan, *Carya illinoensis*
Persimmon, Common, *Diospyros virginiana*
Planetree, London, *Platanus x acerifolia*
Plum, American, *Prunus americana*
Plum, Purple-leaf, *Prunus cerifera*
Poplar, *Populus spp.*
Sassafras, Common, *Sassafras albidum*
Silverbell, Carolina, *Halesia carolina*
Sourgum (Black Tupelo), *Nyssa sylvatica*
Sourwood, *Oxydendrom arboreum*
Sweetgum, *Liquidambar styraciflua*
Tuliptree (Yellow poplar), *Liriodendron tulipifera*
Walnut, Black, *Juglans nigra*
Willow, *Salix spp.*
Yellowwood, *Cladrastis kentukea*
Zelkova, Japanese, *Zelkova serrata*

Many nursery operators have trouble transplanting bare-root and B&B trees, including oaks and filberts, in the fall because roots do not regenerate until bud swell in the spring (Harris and Bassuk 1994). Swanson (1977) recommends transplanting in spring rather than fall in open, windy, cold climates with a relatively low humidity. Good and Corell (1982) found that fall planting can be successful on some species if conducted at least four weeks before soil temperatures drop below 40°F. This probably allowed for some root regeneration before winter. In zone 9, roots on live oak regenerate best during the months of July and August when shoots are not growing (Beeson 1995). A large summer flush of root growth is common on a great number of other tree species. Some nurseries now dig live oak, magnolias, and elms and certain other species in summer to take advantage of this. The native fringe tree recovers best when transplanted in fall, not spring (Harris, Knight, and Fanelli 1997).

Many nurseries and landscape contractors prefer not to dig trees while shoots are actively elongating because young, succulent shoot tissue is sensitive to drying out. Roots on many species grow poorly at this time, so diggers wait until leaves have fully expanded and turn dark green. However, nurseries continue to report that, when pushed for time, they dig certain species during active shoot elongation with great success. Survival during this time of year appears to be greatest for trees that have previously been root-pruned. It should be clear that we have a lot to learn about the proper time for transplanting field-grown trees. For now, rely on local wholesale nursery operators for guidance. They probably best understand the optimum digging times for your area because they do it year in and year out.

Trees in containers in warm climates where soil does not freeze for extended periods (hardiness zones 7 through 11) are routinely planted year-round. Soil temperatures are usually warm enough for some root growth in winter, especially in zones 8 through 11. Trees planted in the fall can establish roots into the landscape soil before warm summer temperatures draw moisture from and stress the trees (Whitcomb 1986b). This could give them an edge over trees planted in spring, which will have few roots out into the landscape soil when warm temperatures arrive soon after planting. Despite this reasoning, one study reports slightly better growth from planting southern magnolia in spring than in fall (Hensley et al. 1988). Containers are a fairly new commodity in the cooler climates, so there is less research and experience to rely on there. Such trees are likely to transplant fine during the seven to nine warmest months of the year.

Chapter 5

Establishing Trees

ESTABLISHMENT PERIOD

In the eastern United States, by the end of the establishment period a tree has regenerated enough roots to keep it alive without supplemental irrigation. In the drier parts of the central and western United States, the turf and landscape irrigation system, combined with rainfall, can provide enough water for survival after establishment. Trees in unirrigated landscapes in dry climates may need supplemental irrigation well beyond the end of the establishment period. This is especially important if the trees are not adapted to a dry climate.

When the tree is established, many roots will have grown a distance equal to approximately three times the distance from the trunk to the branch tips (Gilman 1988a and 1988c; Watson and Himelick 1982). During the establishment period, shoots and trunk grow slower than they did before transplanting (Figure 5-1). When their growth rates become fairly consistent from one year to the next, the tree is considered established.

The establishment period lasts 3 to 12 months (or more) per inch of trunk diameter, depending on climate, irrigation management, and method by which the tree was produced in the nursery (Beeson and Gilman 1992a; Watson, Himelick, and Smiley 1986) (Figure 5-2). For example, a 2-inch diameter tree kept well irrigated requires 6 to 24 months to fully establish. Establishment is quicker in warm climates (3 months per inch trunk diameter in hardiness zone 9) and slower in cooler climates (12 months per inch trunk diameter in zone 5). There is little research in zones 2 through 4, but establishment probably takes longer than 12 months per inch diameter.

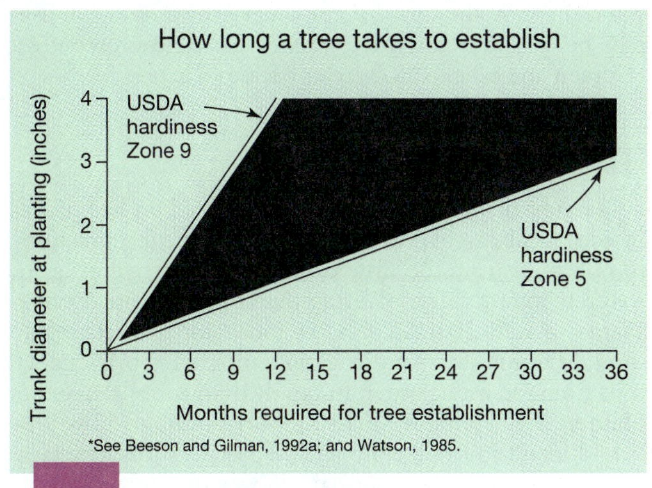

5-2 Tree establishment time depends on tree size and climate. It is important to know when trees are established because established trees do not usually need supplemental irrigation.

Other than small seedlings, most trees need regular irrigation during the establishment period to supplement rainfall. Adequate irrigation promotes quick establishment by encouraging rapid root growth into the landscape soil (Harris and Gilman 1993). Trees irrigated infrequently after planting often establish slowly because roots grow

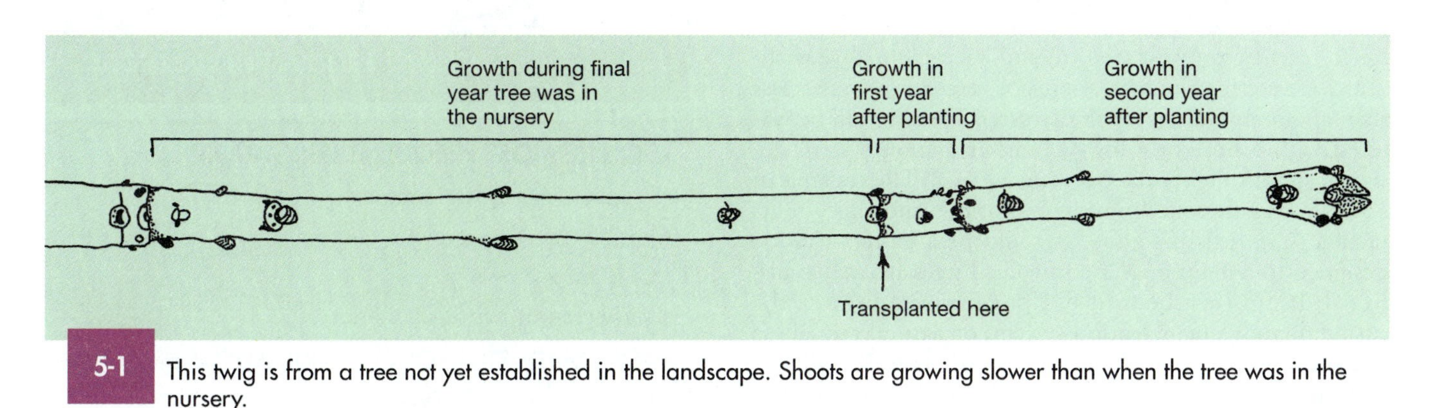

5-1 This twig is from a tree not yet established in the landscape. Shoots are growing slower than when the tree was in the nursery.

slowly. However, if they are overwatered they could die. This is common in poorly drained, clayey soils.

Trees transplanted from containers take longer to establish than field-grown trees even if root ball soil moisture is maintained at optimum levels until the tree is established (Beeson and Gilman 1992a; Blessing and Dana 1988). This appears to be due to slower root growth from container trees (Gilman and Beeson 1996a). This means that container trees will require supplemental irrigation for a longer period following planting than field-grown trees. Recent research indicates that they could take one or two months longer per inch trunk diameter to become established. For example, a container-grown tree with a three-inch trunk diameter could take three to six months longer to establish than a tree from a field nursery. If irrigation is cut off too soon, the mortality rate for container-grown trees will usually be higher than for hardened-off field-grown trees (Gilman and Black 1997) (refer back to Chapter 3).

5-3 Newly planted trees are best irrigated with a low-volume system such as the black, flexible tubing shown here, designed to deliver water directly to the root ball.

IRRIGATION

When new plants die, blame is often placed on bad plants, insects, or disease. However, more plants die from planting too deep, or from too little or too much water during the first few months after planting than from any other cause. Plants in well-drained soils are more apt to succumb to drought from infrequent irrigation; those in poorly drained soils from too much water. In poorly drained soils, trees not planted slightly higher than the surrounding soil will be very difficult to irrigate properly, especially if trees are large with deep root balls. Root balls in wet sites often become too wet and trees die or grow poorly from too much water.

Specific recommendations for watering are impractical because of the enormous variety of soil and environmental conditions, but guidelines are presented here from which to work. Apply less irrigation volume than suggested if the root ball becomes saturated; apply more if it is too dry. Some research indicates that in well-drained soils, applying a little water frequently is much better than applying a large volume only once each week. Most trees planted in well-drained soil struggle because they receive irrigation too infrequently.

An irrigation system for trees

In well-drained soil, an irrigation system designed to water turf and landscape beds usually cannot supply enough water to recently transplanted trees to keep them alive without overwatering the turf. A possible exception is trees less than about an inch in trunk diameter. Trees should be watered with a hose or a low-volume irrigation system designed specifically for the trees (Figure 5-3). The system can be on a separate zone from the rest of the landscape, or attached to an existing zone. It could be a temporary one designed to remain in place only until trees are well established. It can then be turned off or removed because the turf and landscape irrigation system, or natural rainfall, is adequate once trees are established and roots are spread out.

In poorly drained sites, trees can easily be killed by too much water in the weeks and months immediately following planting. Trees planted in or near turf that is regularly irrigated are especially susceptible to overirrigation. Irrigation water and rainfall run over the soil surface, collect in the loose soil in the planting hole, and can drown the tree roots. There are three ways to prevent this. One: Locate trees away from the turf irrigation system, that is, do not plant them in the turf. Place them in beds with a separate irrigation system designed to deliver water only to the newly installed trees. Two: Adjust the turf irrigation system so root balls of recently planted trees do not become saturated. This may not be practical if the turf becomes too dry with less irrigation. Three: Plant trees slightly higher than the surrounding soil (refer to Figure 4-2). This keeps a portion of the root ball out of the standing water in the planting hole. These techniques are highly recommended for planting and establishing trees on sites with poor drainage.

Where, how often, and how much

Container-grown trees need more frequent irrigation than hardened-off field-grown trees (Table 5-1). However, freshly dug field-grown trees are more sensitive to drying

TABLE 5-1 IRRIGATION REQUIREMENTS FOR TREES TRANSPLANTED FROM DIFFERENT PRODUCTION METHODS

REQUIRE MOST FREQUENT IRRIGATION

- Freshly dug fabric container
- Freshly dug field-grown
- Container-grown
- Hardened-off fabric container

REQUIRE LEAST FREQUENT IRRIGATION

- Hardened-off field-grown

out than trees planted from containers for at least several weeks after digging (Harris and Gilman 1993). Container-grown trees dry quicker because the textural difference between the soil root ball of a field-grown plant and the surrounding backfill soil is not as great as between a container-medium root ball and the surrounding soil. In addition, container-grown trees are planted with all roots intact in a small root ball, which dries quickly due to transpiration, whereas root density in a field-grown root ball is much less.

Many trees planted in warm weather in well-drained soil are usually not irrigated frequently enough after planting to maintain optimum growth rate, especially in warm climates. Budgets often do not allow for frequent visits over an extended period of time. Therefore trees grow and establish slower than they could if irrigation were applied often. Work done at the University of Florida (Gilman and Black 1997) in hardiness zones 8 and 9 indicates that two- to four-inch-diameter trees receiving frequent irrigation after planting establish their roots in the landscape soil quicker and become tolerant of drought sooner than those receiving only periodic water (about once every seven to ten days). This is not surprising because, after planting, container root balls dry several hours to one day after irrigation (Nelms and Spomer 1983; Costello and Paul 1975). The root ball has to be irrigated when it dries out, even when surrounding soil is moist, because water will not move from the surrounding soil into the root ball unless the soil is saturated. There is no way roots on recently transplanted trees can use the moisture in backfill soil until roots grow into the backfill.

Research in California in hardiness zone 9B on xerophytic shrubs planted from five-gallon containers indicates that, when provided with about two-thirds of the potential evapotranspiration, such shrubs establish similarly whether irrigated once a day or once a week (Paine et al. 1992).

There is little published work on irrigating trees from other areas of the country, which makes generalized recommendations difficult. However, root growth on recently planted trees appears to be most rapid in soil kept continually moist—more so than in soil that is allowed to dry regularly. The challenge for the landscape manager is to find the irrigation amount and frequency that keeps soil *in the root ball* moist. We have a long way to go before there is a true understanding of the proper water amount and frequency needed on recently transplanted trees.

Placement Apply irrigation *directly to the root ball*, because that is where all the roots are located at planting. Although roots grow quickly after planting in hardiness zone 8, more than 50 percent of total root length remains in the root ball one year after planting containers (Gilman 1990a). In the cooler climates, roots may not grow for 5 or 6 months after planting in the fall, and when they do, many only grow 12 to 18 inches the first growing season (Harris, Knight, and Fanelli 1997). This emphasizes the need for concentrating the irrigation on the root ball in the months following planting.

At times, the soil in the root ball is a much finer texture (for example, clay) than the landscape soil (for example, sand). Since irrigation and rainwater will tend to percolate into the sand quicker than into the clay, the root ball can become bone dry and the tree often dies. To prevent this, be sure that the one- to two-inch-high berm of soil is at the edge of the root ball, not beyond. This forces water to percolate into the root ball. If it is constructed farther from the tree, much of the applied irrigation will miss the root ball because it will percolate into the sandy backfill soil.

You may wet the surrounding soil with additional water if it becomes dry, but this is usually not necessary in regions receiving more than 30 inches of annual rainfall unless soil is extremely dry. Normal rainfall in humid climates, or turf and landscape irrigation systems in drier climates, should provide enough soil moisture in most years to allow adequate tree root growth into the soil surrounding the root ball. In dry climates or dry weather, where no landscape irrigation will be provided, wet the area outside the root ball to keep pace with root growth, especially if there is competing vegetation such as grass and weeds. Roots on well irrigated, vigorous trees grow about one (hardiness zone 5) to three (hardiness zone 9) inches per week during the growing season in the first few years after planting. Growth is slower in colder climates and more rapid in warmer climates.

Frequency Determine when to water by familiarizing yourself with the characteristics of your planting site. Strive to maintain a moist root ball and backfill soil, but avoid saturation conditions. The proper frequency of irrigation is rarely identical for two different planting sites, because climate, soil drainage, root ball soil texture, season of planting, rainfall after planting, amount of sunlight, wind speed, and size of trees differ among planting sites. Table 5-2 summarizes the effects of these factors on irrigation management following planting.

In the warmest parts of the country, to maintain the same rapid growth after planting that occurred in the nursery, trees planted from containers into well-drained soil should receive irrigation as frequently as they received it in the nursery (Table 5-3). Hardened-off field-grown trees also benefit from frequent irrigation after planting, although they can survive with less frequent irrigation than trees from containers. Hardened-off trees might survive with one or two applications each week, but root growth and establishment are likely to be quickest with nearly daily irrigation. In cooler regions, daily irrigation is not necessary and could drown roots.

In poorly drained soil, scheduling irrigation is not as simple. Without a trained person monitoring the site, scheduling is more difficult because the moisture status of the soil can change quickly following a rainfall or change in the weather. For example, several cloudy days can greatly reduce the irrigation needs, even in summer. Water from a turf or landscape irrigation system can also quickly influence soil moisture status in a root ball planted in poorly drained soil. Apply irrigation before the root ball becomes dry, but do not keep it saturated. Saturated soil contains little oxygen, an element needed to keep roots alive.

A root ball from a container dries quickly during the warm season, even when the surrounding landscape soil

TABLE 5-2 FACTORS AFFECTING IRRIGATION REQUIREMENTS AFTER PLANTING

INCREASES IRRIGATION FREQUENCY REQUIREMENT AFTER PLANTING	REDUCES IRRIGATION FREQUENCY REQUIREMENT AFTER PLANTING
well-drained soil	poorly drained soil
sandy-textured root ball	clayey-textured root ball
planting in warm season	planting in cool or dormant season
dry weather	rainy weather
sunny days	cloudy days
windy days	calm days
container-grown and freshly dug field-grown nursery stock	hardened-off field-grown nursery stock
sloping ground	flat ground
warm climate	cool climate
southern or western exposure	northern or eastern exposure

TABLE 5-3 IRRIGATION GUIDELINES FOR QUICKLY ESTABLISHING TREES IN WELL-DRAINED SITES DURING THE GROWING SEASON

SIZE OF NURSERY STOCK	HARDINESS ZONES		
	2–6[1]	7–8[1]	9–11[2]
<2-inch caliper	Daily for 1 week; every other day for 1–2 months; weekly until established	Daily for 1–2 weeks; every other day for 2 months; weekly until established	Daily for 2–4 weeks; every other day for 2 months; weekly until established
<2–4-inch caliper	Daily for 1–2 weeks; every other day for 2 months; weekly until established	Daily for 2 weeks; every other day for 3 months; weekly until established	Daily for 1–2 months; every other day for 4 months; weekly until established
4-inch caliper	Daily for 2 weeks; every other day for 3 months; weekly until established	Daily for 2–4 weeks; every other day for 4 months; weekly until established	Daily for 2 months; every other day for 5 months; weekly until established

[1]Delete daily irrigation when planting in fall and early spring. Little irrigation is needed when planting in winter in zones 2–6.
[2]Delete daily irrigation when planting in winter.

NOTES:

Irrigation frequency may be slightly reduced for hardened-off field-grown trees. Reduce frequency in cool, cloudy, wet weather if soil is poorly drained (refer back to Figure 1-5). Eliminate daily irrigation in poorly drained soil. Following a rainfall, wait until all free moisture drains out of the soil (see Figure 1-5). Establishment takes 3 months per inch trunk caliper in hardiness zones 9 through 11, 6 months in zones 7 and 8, and 12 months in cooler regions.

Recommended minimum irrigation to keep trees alive eliminates daily irrigation. Start by watering every other day immediately after planting. Minimum frequency for survival could be once each week in the coolest parts of zones 2–6.

Irrigation can cease once deciduous trees drop foliage in fall.

At each irrigation, apply 1.5 gallons (cool climates) to 3 gallons (warm climates) for each inch of trunk diameter to the root ball.

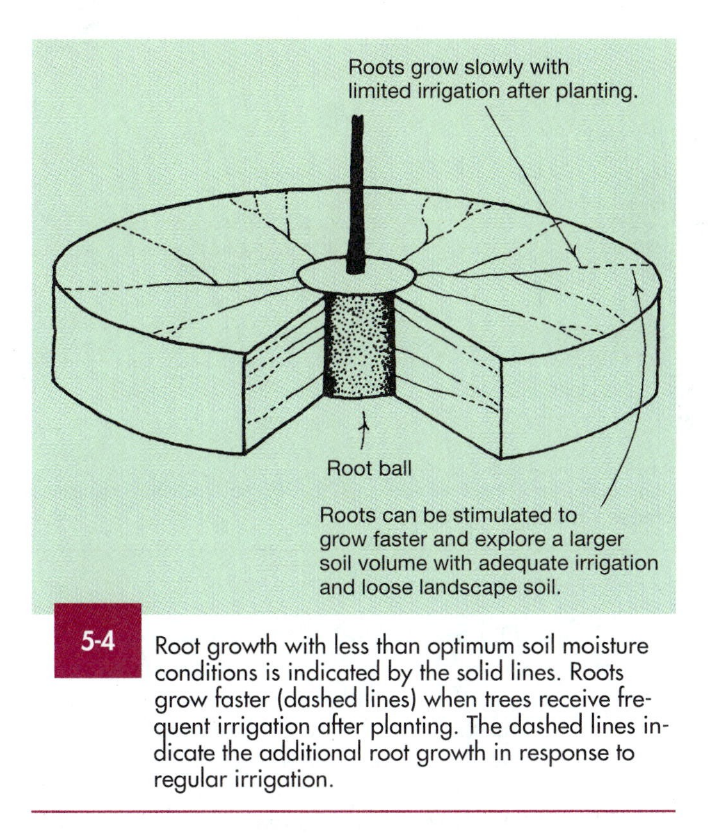

5-4 Root growth with less than optimum soil moisture conditions is indicated by the solid lines. Roots grow faster (dashed lines) when trees receive frequent irrigation after planting. The dashed lines indicate the additional root growth in response to regular irrigation.

remains moist. It may dry out nearly as quickly as a tree planted in well-drained soil, because water cannot move from backfill soil into the container medium unless soil is saturated beyond field capacity. This is hard to believe until you see it happen. If the root ball and soil are saturated, do not water the tree. Raise it up out of the water, replant it slightly higher than the surrounding soil, and try to improve drainage.

Maintaining adequate moisture in the root ball and surrounding soil allows faster and more extensive root growth (Figure 5-4). Root growth is also much more rapid in loose soil than in compacted soil. A tree with a rapidly growing root system becomes established quicker than one receiving only occasional water after planting. Without irrigation, trees may manage to stay alive if rainfall is frequent enough, but they often grow very slowly. The tree could die back during a dry period. The same tree species, irrigated regularly and growing in loose soil, will be more resistant to drought sooner because it has developed a more extensive root system.

Amount The best method of determining how much irrigation to apply to container-grown trees in the first several weeks after planting in well-drained soil is to ask the nursery operator how much was applied in the nursery just prior to purchase. Apply this amount directly to the root ball each time you irrigate. Container-grown trees planted in poorly drained soil may need less irrigation volume than they received in the nursery, especially if the lower part of the root ball remains saturated.

As a rule-of-thumb based on research in well-drained sand with container-grown and field-grown trees in hardiness zones 8 through 11, 1.5 to 3.0 gallons per inch trunk diameter applied to the root ball each time the tree is irrigated during the growing season should be enough to maintain adequate root growth. For example, a tree with a two-inch trunk diameter needs about three to six gallons each time it is irrigated. Guidelines for cooler climates are not as clear because less research has been conducted there. Extrapolating from warm climate studies, no more than one to two gallons per inch trunk diameter should be needed at each irrigation. Plants installed during the cooler months need less volume, depending on the weather and soil drainage (Table 5-2). Those planted in compacted or poorly drained soil with a portion of the root ball above the surrounding soil may need more water because of the drying effect of the mounded planting.

Within the first week after planting, gently dig a small hole in the loosened backfill soil about as deep as, and just outside, the root ball to check soil moisture. Squeeze some soil in the palm of your hand from the top and the bottom of the hole. If water drips out between your fingers, reduce the volume of water applied at each irrigation (Figure 5-5a). If soil crumbles and falls out of your hand as you open your fingers, increase the volume of irrigation (Figure 5-5c). If soil stays together as you open your fingers, moisture in the backfill soil is probably just right (Figure 5-5b), but the root ball might be bone dry.

To check for this, insert your finger into the root ball. If it is dry, increase your irrigation amount by one gallon per inch trunk diameter, or increase irrigation frequency. Too much water is in the root ball if water squeezes out between your fingers as you press them together in the soil, or if the root ball smells sour. With some practice, this technique becomes a quick and easy way to evaluate soil moisture.

Some people use a root feeder or other soil probes to judge soil moisture in clayey and finer-textured loamy root balls (not container root balls). If it is easy to push into the root ball, soil moisture may be fine. If it is hard to penetrate, then you may need to increase the amount of irrigation. This technique takes some practice to develop and it varies greatly with soil type. Once mastered, it can be very useful. There is no evidence that probes with gauges of various types, marketed as soil moisture evaluation devices, help judge when newly transplanted trees need water, although current designs may help schedule irrigation for established landscapes. Recent work in Nevada suggests that a stem flow gauge might be useful in scheduling irrigation for established woody plants in arid climates. Little work has been done on transplanted trees.

There is no evidence in the eastern United States that increasing the amount (volume) of irrigation as the tree becomes established enhances the rate of establishment. Research shows that maintaining a regular frequency after planting is more important than increasing the volume (Gilman and Black 1997), provided that turf, ground cover, and weeds are not competing for moisture under the canopy. However, one may speculate that increasing the volume applied and area covered by the irrigation as the tree

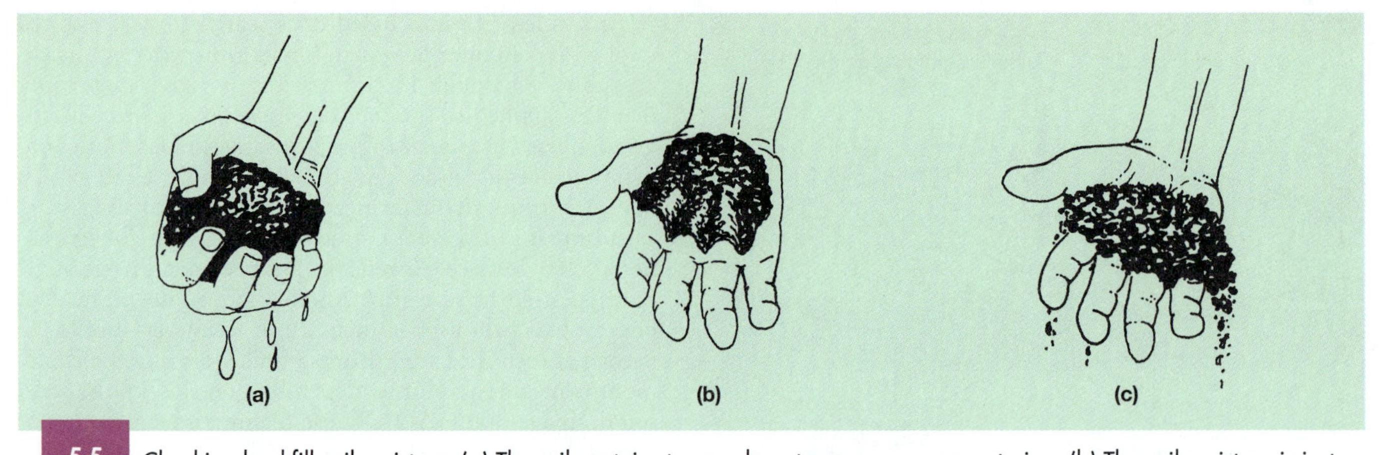

5-5 Checking backfill soil moisture. (a) The soil contains too much water; you are overwatering. (b) The soil moisture is just right; check for root ball dryness. (c) The soil is too dry; increase watering.

grows may enhance growth in drier parts of the country (areas receiving less than 20 inches annual rainfall).

Tapering off

Two (for trees with less than two inches trunk diameter) to eight (for trees with two- to four-inch trunk diameters) weeks after planting in hardiness zones 9 through 11, begin to cut back irrigation frequency (Table 5-3). Apply the same 1.5 to 3 gallons per inch trunk diameter to the root ball each time you irrigate, but skip a day between irrigations for several months. After this, gradually decrease the frequency of irrigation, but irrigate at least once each week until trees are established (Figure 5-2). In well-drained soils, rainwater should not be counted on to provide much water to the root ball in the first month or two after transplanting during the growing season.

One to four weeks after planting in cooler climates, cut back to every-other-day irrigation for one to four months. Then maintain weekly irrigation until trees are established. You may be able to skip an irrigation if the weather has been cloudy, rainy, or cool since the last irrigation. Check the soil moisture status in the root ball to determine this. Be sure not to apply water if the root ball is saturated.

Managing with less

Trees often survive and eventually establish with less irrigation than outlined in this text (see Table 5-3 for minimum irrigation recommendations). Nevertheless, you are risking tree die-back and possible death. If hot weather or drought were to occur before trees are established, the trees could die back. For example, in the warmest parts of the country, trees planted in the fall or winter and maintained with a minimum of irrigation may survive and appear to be established by the end of spring. Sometimes these turn brown and die when the summer heat arrives, because they do not have enough roots in the landscape soil to keep up with the transpirational demands of the new foliage.

Trees without adequate irrigation during the establishment period may also develop a weak, multitrunk habit (Figure 5-6). This may be initiated in response to tip dieback on the main trunk and branches in dry weather. Following the return of wet weather or irrigation, several new shoots often emerge from the living tissue back from the dead tips. These developing shoots often become equally dominant (Struve 1994). The result is several trunks instead of the one that was present at planting. This produces an inferior, poorly structured tree that could split apart as it grows older (refer to Figure 3-12). If proper irrigation cannot be provided, consider planting smaller trees that establish quicker (refer to Figure 5-2).

Comparing irrigation strategies

The practice of intense irrigation management in the months immediately following transplanting could save resources in the long term. Irrigation cost is concentrated up front in a short period of time, not spread over a long period. The same amount of money may be spent on both strategies, but trees in intensively managed areas will establish quicker, due to faster root growth, and be resistant to drought and other stresses such as borers sooner. This may be the best buy for your limited irrigation budget. The concentrated costs of intensive irrigation may be more than offset by reduced mortality and healthier trees.

STAKING

Is it needed?

Good-quality trees often do not require staking following planting. Read this section to determine if staking is needed. If not, do not use them. They are expensive to install and remove, and if they stay on the tree too long, they can injure or kill it.

Trees planted in a location protected from the wind, such as in a courtyard, among other trees, or near a home with existing shrubs and trees, may not need to be staked for anchorage. Trees may also not need to be anchored if the canopy is relatively thin. Trees transplanted with a tree spade often can be installed without stakes, due to the

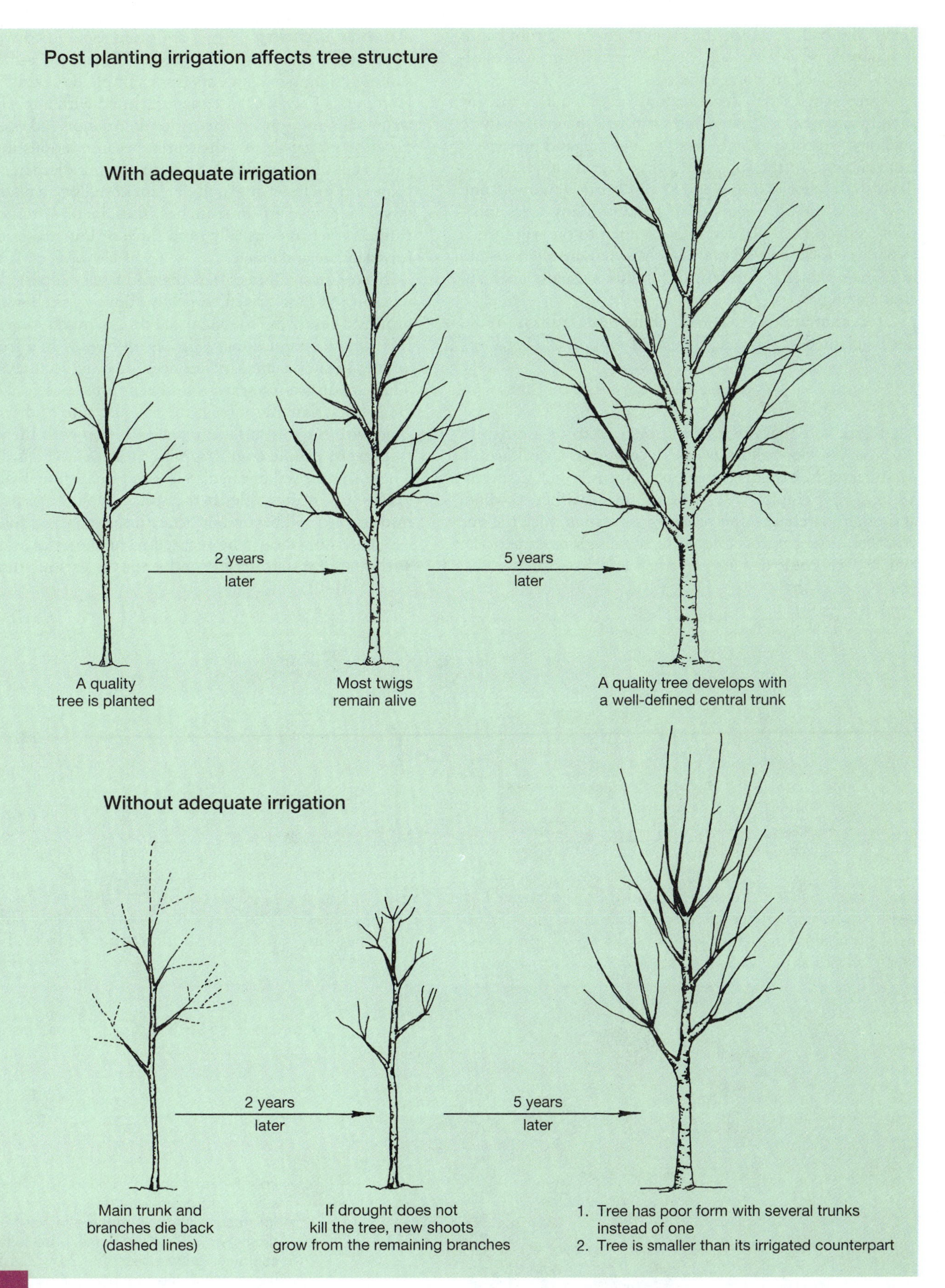

5-6 Good-quality nursery stock receiving adequate irrigation during establishment develops into a well-structured tree (top). A poorly formed tree with multiple trunks often develops without adequate irrigation (bottom).

heavy root ball, provided the root system in the root ball is well developed and can support the tree. Bare-root trees often do not need anchorage stakes.

A tree with a large, dense canopy often requires anchorage until roots can grow to sufficient length and density to anchor it. Anchorage stabilizes the root ball and prevents it from moving in the soil or blowing over on a windy day. Even slight root ball movement can break new roots and slow plant establishment dramatically. Many trees more than eight feet tall planted in areas open to the wind, such as parking lots and parks, are likely to require anchor staking. Trees planted from fabric containers usually need anchor staking.

Trees that are too tall, with thin, weak trunks, should not be planted. If they are, they need support staking.

Types of staking

There are two types of staking, each used for a different purpose: for root ball *anchorage* or to *support* a weak trunk. Manufacturers have developed many types of staking mechanisms. They employ wire, metal, rubber, synthetics, wood, and other materials. After reading this section, pick the one you feel most comfortable with. Although most perform equally well, some damage the trunk less than others.

Anchor staking Trees are often secured to wood or metal stakes driven into the ground (Figure 5-7a–d). Some staking and guying systems (for example, Duckbill®, from Commerce City, CO) are manufactured with a mechanism that locks the guy in the ground. Anchor stakes should never be placed against the trunk, because serious injury to the trunk often results. Trees with a trunk diameter of less than two inches can usually be anchored by two stakes or guys. Trees two to three inches in diameter often require three. Larger trees usually need three or four stakes or guys for adequate anchorage.

Stakes on small trees (less than 3 inches caliper) can be 5 feet long and are driven vertically about 24 inches into the ground (Figure 5-7b). Many stakes are much longer than this, which is unnecessary. Stakes on trees with a trunk diameter greater than 2 inches need be only 30 inches long. They are driven into the ground at approximately a 20° angle so they slant away from the tree. Be sure to drive them even with the soil surface (or at least even with the mulch) to prevent people from tripping over them (Figure 5-7a). These shorter stakes can also be used to anchor small trees. Stakes are most stable in the soil if they are driven into undisturbed soil beyond or beneath the planting hole.

The trunk should be secured to anchor stakes or guys with material that is wide and smooth. Various rubber or

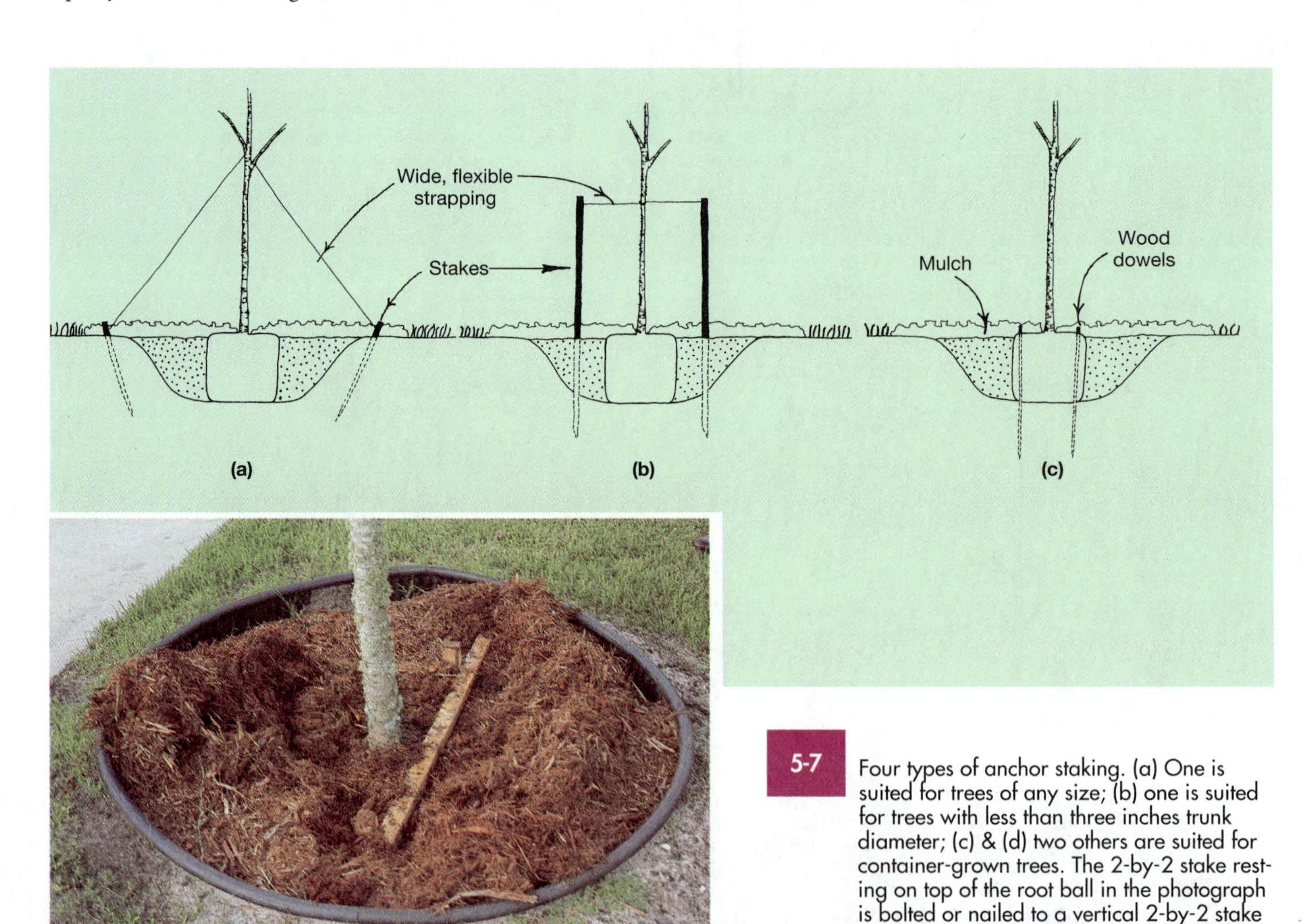

5-7 Four types of anchor staking. (a) One is suited for trees of any size; (b) one is suited for trees with less than three inches trunk diameter; (c) & (d) two others are suited for container-grown trees. The 2-by-2 stake resting on top of the root ball in the photograph is bolted or nailed to a vertical 2-by-2 stake driven vertically on either side of the ball.

(a)

(b)

5-8 (a) Rubber straps are nicely suited for attaching stakes to trunks. (b) Wire threaded through hose can girdle the tree if left on too long.

rubber-like products are specially designed for this purpose (Figure 5-8a). Elastic webbing, belting, rope, and polyethylene tape are also acceptable. They hold the tree firm and damage the trunk less than any other type of securing mechanism. They can be installed to hold the root ball firm in the soil, yet allow for some trunk movement in the wind. This is desirable because trunk movement encourages thickening of the trunk (Harris 1992). A trunk held rigid can develop poorly. Wire slipped through garden hose is often used to secure the trunk to a stake. This system of attachment can damage the trunk as it moves in the wind and rubs against the hose and wire. Research shows that the hose does little to protect the trunk from the wire. It can also begin to girdle the tree if it is not removed 6 to 12 months after planting (Figure 5-8b).

Do not allow slack in the material used to hold stakes to the trunk, but be careful not to overtighten the material. Tighten it only enough to eliminate slack—no more. Some slack in wire usually develops in the weeks following planting. This causes rubbing on the trunk and can injure it. Injury can be prevented by installing turnbuckles in the wire at the time of installation. The turnbuckles can be adjusted to eliminate slack at any time. Rubber-like material, properly installed with some stretch, is superior because it is less likely to develop slack and thus need not be adjusted.

In most instances, anchor stakes should be removed within one year after planting. Unfortunately, this does not always happen, and trees have died due to the trunk-girdling effects of attached wires and other supports. If anchorage is required for more than a year, there is a problem with the tree. Do not use anchor stakes if you have not developed a schedule to remove them within one year after planting.

Two alternative staking techniques, used to anchor root balls from containers, are especially suited for places where pedestrians walk close to the trees. Each can be used in other landscapes as well. One technique uses a one-half to three-quarter-inch diameter, *untreated*, wood dowel driven through the root ball well into the soil beneath. The top of the dowel should be driven flush with the top of the root ball. One or two are enough for small, 3-gallon trees; two or three are needed for trees up to 15 gallons. This technique has not been used extensively on, and may be poorly suited for, larger trees. The dowels should be driven straight down through the root ball several inches inside the edge of the ball. If the dowel is driven slowly, the sharpened point should help prevent it from penetrating or damaging major roots.

Another nearly invisible staking technique employs three *untreated* two-inch-by-two-inch stakes. Two are sharpened and driven vertically into the ground along the outside of the root ball on opposite sides of the tree; the third rests on top of the root ball and is secured to the two vertical stakes with nails or bolts (Figure 5-7d). The vertical stakes must be driven about two feet into undisturbed soil under the ball. Cover the stakes with mulch to prevent tripping.

Tall, heavy-trunked transplanted palms usually require anchorage. They can be anchored with guy wires, or more commonly with wood supports (Figure 5-9a and b). Wood supports should not be nailed to the trunk. Instead, wrap three or four boards (two-by-fours) two feet long in several layers of burlap and fasten these to the trunk of the palm with metal strap. Enough pieces of wood should be used to keep the metal strapping from touching the trunk. The trunk is often wrapped in burlap instead of the wood, but this could keep the trunk too moist and encourage decay. Then nail support posts to the padded boards, being careful that the nails do not penetrate into the trunk. Remove the stakes 6 to 12 months later. Palms without leaves may not need staking unless they are exceptionally tall. Smaller sized palms can sometimes be sufficiently anchored with three steel bars twice the length of the root ball depth driven through the root ball. Do not plant the palms too deep instead of staking. The palms will decline.

Support staking Support staking is used to hold a weak trunk straight, in the upright position (Figure 5-10). Trees grown under poor nursery practices often require this type of staking. The best alternative to support staking is not to plant a tree that has a thin, weak trunk (refer back to

5-9 (a) Palm trees are properly anchored with sturdy lumber nailed to short, padded two-by-fours strapped to the trunk. (b) Supporting lumber attached to the side of the trunk is not as sturdy.

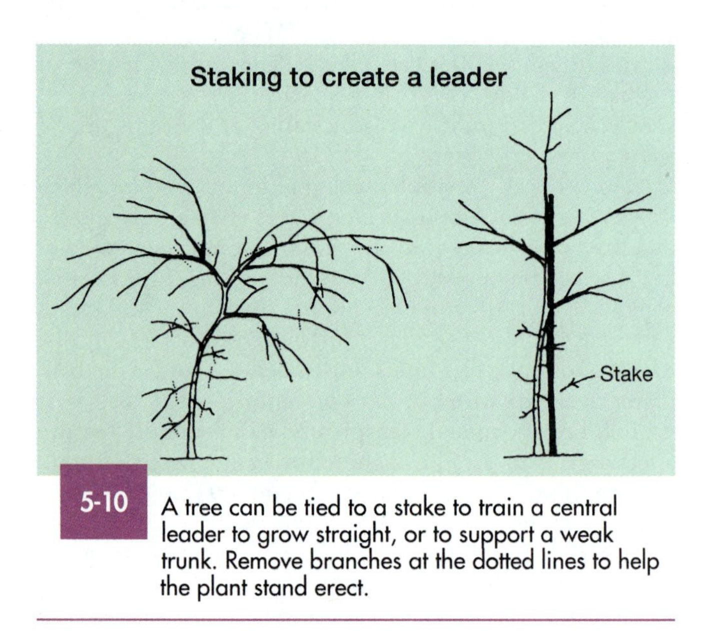

5-10 A tree can be tied to a stake to train a central leader to grow straight, or to support a weak trunk. Remove branches at the dotted lines to help the plant stand erect.

Chapter 3). This will usually eliminate the need for this type of staking.

Support stakes should be secured to the trunk at the lowest position that will hold the tree erect. You can find this point by holding the top of the trunk with your hand, so the tree stands upright, and then sliding it down the trunk to the point where the top of the tree begins to bend over. Move up six inches and attach the stake. Cut the stake above the point of attachment to help prevent trunk damage.

It is important to remove the support stake from the tree as soon as possible, because supportive trunk tissue develops slowly on staked trees. Unfasten the stake from the trunk 6 (in warm climates) to 12 (cool climates) months after staking. Do this immediately following a rain shower, because the weight of the water will cause the trunk to bend over if the tree is still too weak. If the plant stands erect, remove the stake. If not, repeat the process approximately every two months until sufficient trunk strength develops. If the tree requires staking for more than one year, it may never develop the strength needed to support itself. Leaving small branches along the lower trunk will also help the trunk increase in diameter and strength. These branches can be removed once the tree can support itself usually by the time the trunk has a diameter of about 2 inches. The canopy can also be thinned to reduce weight. This helps hold the tree erect.

MULCHING

Weed suppression during the tree establishment period is crucial. Weeds absorb water in the soil that is meant to be available for tree roots. Tree growth can come to a halt if weeds are allowed to grow in and over the root ball. Mulch cuts off light to the soil, thus reducing weed germination. Mulching combined with regular herbicide application or hand-weeding enhances growth after planting by reducing weed growth. Second to watering following planting, maintaining an adequate mulched area under the canopy will give you the biggest return in tree growth and establishment rate. Adding fertilizer or irrigating more frequently does little to overcome the detrimental effects of weeds.

Benefits of properly applied mulch include enhanced root growth (for organic mulches), prevention of soil compaction, erosion control, prevention of allelopathic effects from grasses (Fales and Wakefield 1981), and reduced fertilizer runoff. Aeration in the mulch layer is better than in soil, which enhances root growth directly below and in the mulch. Mulch prevents raindrops from impacting directly on the soil, which reduces erosion and helps keep soil pore spaces open. This allows better water percolation and air movement through the soil below the mulch. A mulched soil is also less likely to become compacted from foot and vehicle traffic (Lichter and Lindsey 1993). Fertilizer granules lodge in between mulch particles, which potentially reduces runoff. Mulch also reduces evaporation from the soil, which is usually considered a benefit, although it could keep poorly drained soil too wet. These benefits combine to improve plant growth in most circumstances (Green and Watson 1989; Litzow and Pellett 1983a). Apply one pound of nitrogen per 1,000 feet^2 over the mulch to prevent microorganisms from competing for nitrogen. Without nitrogen, there could be a slight inhibition of growth the first year after planting (Smalley and Wood 1995).

Citrus appears to be the only tree that should not be mulched. Horticulturists in Florida find that roots often rot when mulch is applied. Apply it only to the portion of the landscape outside the edge of the root ball. Arid land species may not benefit in the first three months after planting from an application of mulch if the landscape is irrigated (Hild and Morgan 1993).

Before applying mulch, remove any soil covering the root ball. Soil placed over a root ball can prevent water from entering the root ball and could kill the tree. If the top of the root ball is deeper than the surrounding soil, raise it up to the appropriate depth before applying mulch.

Apply mulch in a three- to five-inch-deep layer (Figure 5-11). Create a circle of mulch at least two feet in diameter for each inch of trunk diameter (Table 5-4). Increase the size of the mulched area as the tree grows until the tree is fully established (refer to Figure 5-2). Pull mulch at least six inches away (12-inch diameter) from tree trunks and plant stems. Mulch placed on the trunk and piled too high (Figure 5-12a) can cause bark decay and can suffocate roots, leading to poor plant performance (Figure 5-12b). It also

TABLE 5-4 MULCH APPLICATION FOR ESTABLISHING TREES

TRUNK DIAMETER (INCHES)	MULCH DIAMETER (FEET)
1	4
2	6
3	8
4	10

5-11 A large mulched area around a newly installed tree helps the tree become established by reducing weed competition.

(a) (b)

5-12 Mulch that is too deep can lead to problems. (a) Never allow mulch to touch the trunk or be spread this thick. (b) Trees can decline from a wet trunk and poor aeration in the root ball.

TABLE 5-5 SOIL REACTION FROM MULCH APPLICATION

ALKALINE (BASIC) FORMING MULCHES	ACID-FORMING MULCHES
• grass clippings	• shredded cypress
• fresh hardwood chips (except oak)	• conifer wood chips
• gravel	• oak wood chips
• volcanic pumice	• oak leaf mold
	• crushed granite
	• pine bark

Source: Kelsey 1995.

makes checking the root ball for moisture difficult or impossible. Mulching is best done at the end of a project so that it will not become compacted from foot traffic.

Many different types of organic mulch are used in landscapes, depending on availability (Table 5-5). Most common are tree bark, wood shavings from lumber mills, wood chips, pine needles, and composted yard waste. Byproducts from the food industry are also used occasionally, depending on the region of the country. For example, macadamia nut hulls are used in parts of Hawaii and peanut shells in Alabama. Each is effective at reducing weed growth. Pine bark outlasts many other organic mulches (Skroch et al. 1992). Pumice, gravel, and other inorganic products are occasionally used for mulching. The decomposition of freshly ground wood chips over the root ball consumes nitrogen, and this process could result in deficiency symptoms and slow growth on trees (Hensley, McNeil, and Sundheim 1988; Smalley and Wood 1995). Application of supplemental nitrogen over the mulch is recommended to help prevent these symptoms.

Freshly chipped wood can increase soil conductivity for a week or two, and this can damage roots. Composted mulch does not have this potential problem. Fresh hardwood chips can increase pH in the root zone and induce chlorosis in susceptible plants (Kelsey 1995).

Polyethylene or polypropylene under mulch increases its service life. Except for sedges and some other perennial weeds, these synthetic mulches reportedly resist weed growth more effectively than organic mulches (Derr and Appleton 1989). These are usually overlaid with organic mulch because they are unattractive.

PROTECTING THE TRUNK

Trunk wraps

Trunk wraps are of little benefit to transplanted trees, and can even be harmful, but they can provide some protection during shipping (Appleton 1993). Their use should not be mandatory at planting. If trees are shipped to the job site with trunks wrapped and they have not been previously inspected for damage, unwrap all of them for inspection. Be sure to remove all string and twine from the trunk at this time, as they can girdle and kill the tree.

The most commonly used material, paper wrap, does not buffer temperatures as presumed (Litzow and Pellett 1983b). Burlap and plastic devices are also occasionally used. Parts of the guard or wrap could fall off the trunk and this might encourage large temperature differences in the trunk tissue. The exposed bark could heat to a higher temperature than the covered portions, possibly causing cracks to develop (Appleton 1993). Alternatives to trunk wraps and guards occasionally used in the winter to prevent sunscald include whitewash, slaked lime, or white latex paint, which can be sprayed, brushed, or rolled onto the trunk. Although used in some regions, there is a report that all these can cause injury to the trunk. Some nurserypersons and landscapers mark the north side of the tree by hanging a piece of ribbon from a branch. Thus, when planting in the new location with the ribbon facing north, the trunk is exposed to sun rays just as it was before, which minimizes danger of sunscald.

In some cooler regions of the country, horticulturists wrap tree trunks in the winter on recently planted young, thin-barked trees such as birch, ash, maple, and linden. This is thought to protect them from injury initiated by direct sun on the trunk (referred to as sunscald, sun scorch, or frost cracks) and to protect from rodent injury (Figure 5-13). Unfortunately, there is little research that has validated this practice for purposes of reducing sun-induced injury. Trunk wrap probably does little harm to the tree provided it remains intact, is light colored, and is removed the following summer, along with any tape and string used to tie it on the trunk. If it is left on too long (more than one year), some trunk insect and disease problems could be encouraged. For instance, dogwood borer infestation is higher

5-13 If trunk wraps are used, wrap from the bottom of the tree toward the top, and overlap the material. Use electrical tape or some other flexible or degradable adhesive instead of the string shown here. String left on the trunk could girdle it.

in trees with plastic trunk guards than in those without guards. The frequency of infestation was higher in trees with tight-fitting guards than in trees with loose-fitting guards (Owen, Sadof, and Raupp 1991). Fire ants, earwigs, and spiders can also be found under wraps.

As an alternative to wrapping the trunk, some protection from direct sun can be provided by short, temporary branches left along the lower trunk of transplanted trees. Encourage nursery operators to leave these intact. These can remain on the tree until the upper branches shade the trunk.

The decision to use a wrap should be based on site conditions. For example, trunk wrap may be more appropriate in a windy, sunny parking lot than in a cool, shaded location near a home. If a trunk wrap is used, choose materials that are light colored and will either biodegrade or photodegrade. One such polypropylene fabric photodegrades in six to nine months in Virginia (Appleton 1994). The material is constructed so that no ties or string are needed to secure the material to the trunk. Begin wrapping at the ground and spiral the material around the trunk up to the first major branches. Overlap each layer by half a width and tie it with a knot at the base of the branch. Do not use nylon string to secure trunk wrap, because it can strangle the trunk if it is left on.

5-14 (a) Corrugated plastic drainpipe (or similar product) is sometimes used to protect trunks from mechanical, sun, deer, or rodent damage.
(b) Also available are specially designed protection devices that are easier to install. Notice how the lower branches left on the tree also help protect the trunk from direct sun exposure; they help keep equipment away, too.

Protection from rodents and deer

Keep mulch six inches from the base of the trunk to discourage rodents. Corrugated plastic protection tubes (sometimes called tree shelters) protect newly planted, small tree seedlings from rodents, deer, and sunscald. They also speed shoot growth of small seedling trees. However, some tree species develop weak trunks inside the shelters and are unable to hold themselves erect (Kjelgren 1994). Others develop smaller root systems than trees planted without shelters (Burger, Svihra, and Harris 1992). Trees can be enclosed in wire or mesh screening. Some horticulturists slice a 6- to 12-inch-long section of black plastic drainpipe so that it can be slipped around the lower trunk (Figure 5-14a). Be careful not to injure the trunk or yourself when installing this pipe—use thick gloves. Others use plastic trunk guards that can be spiraled around the trunk and lower branches or other specially designed devices for trunk protection (Figure 5-14b). Be sure not to leave gaps between the plastic spirals or bark could be injured due to the difference in temperature between protected and unprotected areas.

Protection from equipment

Ideally, mowers and string trimmers should not be operated close enough to trees to warrant the following protection devices. Maintaining a mulch layer around trees helps keep equipment away (Table 5-4). Unfortunately, landscapes are not always designed in this ideal fashion. Wood or metal stakes can protect the tree from accidental injury from lawn maintenance equipment. Three or more stakes are usually driven through the edge of the mulch layer into the ground, but they are not attached to the tree. Mowers can bump into the stake but not the tree. These stakes can remain indefinitely. Thick, black plastic drainpipe slipped around the base of the trunk, or other plastic devices, are also used to provide some protection against string trimmers and mowers (Figure 5-14). But fire ants in the southeastern part of the country often nest in them and can kill bark. Some tree managers wrap trunks to help protect them from vandals and equipment damage, but there is little evidence documenting the effectiveness of this practice.

FERTILIZATION

Research on landscape-sized trees transplanted from field nurseries indicates that if there is a benefit to fertilizing the root ball at planting, it is small or undetectable (Shoup, Reavis, and Whitcomb 1981; van de Werken 1981). Healthy nursery trees are well fertilized in the nursery, have plenty of stored carbohydrates, and usually need no additional elements at planting. However, fertilizing several months after planting has provided a significant (but small) growth response (Wright and Hale 1983). In no case has research on landscape plants indicated that fertilizer application at planting increases survival. In fact, salts from soluble fertilizers can reach levels that kill roots if too much is applied, especially if trees are not irrigated regularly. Postplanting management efforts in the months following planting are better spent on irrigation, mulching, and weed control than on fertilization.

There is also research on fertilizing trees planted from containers. One study showed that fertilization at planting, or even one year after planting, did not increase growth of oaks planted into a soil with moderate fertility (Perry and Hickman 1992). Others showed little significant benefit during the first year from fertilizing one- to two-gallon-sized trees at planting (Corley, Goodroad, and Robacker 1988; Conover and Joiner 1974). However, many fertilized plants were larger three years after planting than trees not receiving fertilizer. One study on arid land shrubs planted from five-gallon containers showed no benefit from nitrogen application at planting (Paine et al. 1992). However, fertilizing several months after planting increased tree growth (Hamilton and Drosdoff 1946; Kirby and Potter 1956). In addition, magnolia and Japanese zelkova fertilized with nitrogen at planting grew more than nonfertilized trees (Harris, Paul, and Leiser 1977; Hensley, McNeil, and Sundheim 1988). In every case where fertilized trees grew more than non-fertilized trees, the non-fertilized trees still grew—they just grew a little slower.

If trees planted from containers can be irrigated regularly after planting, fertilizer spread on and around the root ball at planting might enhance tree growth, especially in soil that has low fertility. Because regularly irrigated container trees suffer little transplant shock, they might respond sooner to an early application of fertilizer than trees transplanted from the field. This is only speculation and has not been extensively tested.

Weighing this evidence, it appears that if trees are planted into a landscape that receives regular (once or twice a year) fertilizer applications for turf and shrubs, in most instances enough will reach the tree roots to maintain adequate growth. In contrast, in a low-maintenance situation, such as along a highway where fertilizer is not regularly applied, fertilizing at planting may be a good idea. It may be the only fertilizer the tree receives for a long time.

As there is little evidence that slow-release fertilizer applied at planting discourages growth after planting, many horticulturists add some. However, do not add too much. Recent research indicates that root growth can be suppressed by adding more than one pound of nitrogen per thousand square feet (Warren 1993). This is not much fertilizer! Applying more than the manufacturer's recommendations could also burn and kill roots and stress the tree.

To fertilize at planting, apply a small amount of slow-release fertilizer (fertilizing is covered in more depth in Chapter 8) to the top of the backfill soil and root ball at planting or any time after planting. Although a few horticulturists mix a small amount in with backfill soil, thoroughly mixing fertilizer with soil is time-consuming. In addition, there is little if any benefit from mixing or injecting fertilizer in with soil, compared to surface application (Gilman 1989a; Harris, Paul, and Leiser 1977; Hensley, McNeil, and Sundheim 1988; van de Werken 1981; Yeager, Ingram, and Larson 1989), unless the fertilizer will run off during the rain or when the landscape is irrigated. Placing fertilizer at the bottom of the planting hole makes little sense, because few roots grow down under the root ball unless soil is well drained. Most grow out the side of the root ball more or less parallel to the soil surface.

It may be a good idea to broadcast nitrogen on top of the mulch layer if one is applied. Two studies showed an inhibition of growth when an organic mulch was applied over the root ball without adding nitrogen (Hensley 1988; Smalley and Wood 1994). The inhibition of growth is due primarily to microorganisms utilizing the nitrogen in the mulch. One pound of nitrogen per 1,000 feet^2 is adequate to overcome this.

Except for palms, nitrogen is usually the only macroelement that provides a growth benefit to trees. Potassium and phosphorus do not unless the soil is lacking in these elements or they are rendered unavailable. Results from the soil sample you collected during your site analysis will indicate if any elements are lacking in the soil. Do not waste energy by applying potassium and phosphorus if they are not needed. Neither has been shown to increase root growth on recently planted trees.

Although a number of products are sold as transplant aids, there is no scientific evidence that other types of fertilizers, gels, vitamins, powders, emulsions, organic acids, kelp extracts, or the like, whether natural or synthetic, help trees become established. Once again, it is best to focus postplanting efforts on proper irrigation management and weed suppression in and around the root ball.

PRUNING AT PLANTING

Little (if any) pruning should be necessary at transplanting if quality trees were purchased. Do not prune the plant to compensate for root loss. The latest research indicates that in most instances pruning does not help the plant overcome transplant shock (Shoup, Reavis, and Whitcomb 1981). Carbohydrates and hormones produced in leaves are needed for root regeneration, so the more leaves on the tree, the better the root growth (Abod and Webster 1990). *If you feel that the top may die back from lack of irrigation after planting, and you have to remove branches (leaves on palms) by pruning before they die anyway, then you purchased nursery stock too big for your irrigation capabilities.* Refer back to Figure 1-11 to determine maximum tree size at planting. If you still feel compelled to prune field-grown trees at planting to compensate for root loss, thin the canopy; do not randomly top the tree or round it over.

Some people suggest pruning broken, dead, diseased, badly injured, or insect-infested branches at this time. This is fine, but if the tree requires much of this type of pruning, you bought a poor-quality tree. Take it back and get a healthy one.

Although pruning at planting appears to have little, if any, positive impact on transplant survival and growth after planting, structural defects should be corrected if pruning is not planned for the next few years (Figure 5-15). The main objective when pruning young, medium- and large-maturing shade trees is to develop a dominant leader or trunk by removing clustered and competing branches (Figure 5-16). Clustered and fast-growing branches can form included bark in the crotch, which creates a weak point on the tree.

When removing a branch, cut back to the branch collar if one is present (Figure 5-17a–g). Estimate the proper cutting line if the collar is not apparent (Figure 5-18).

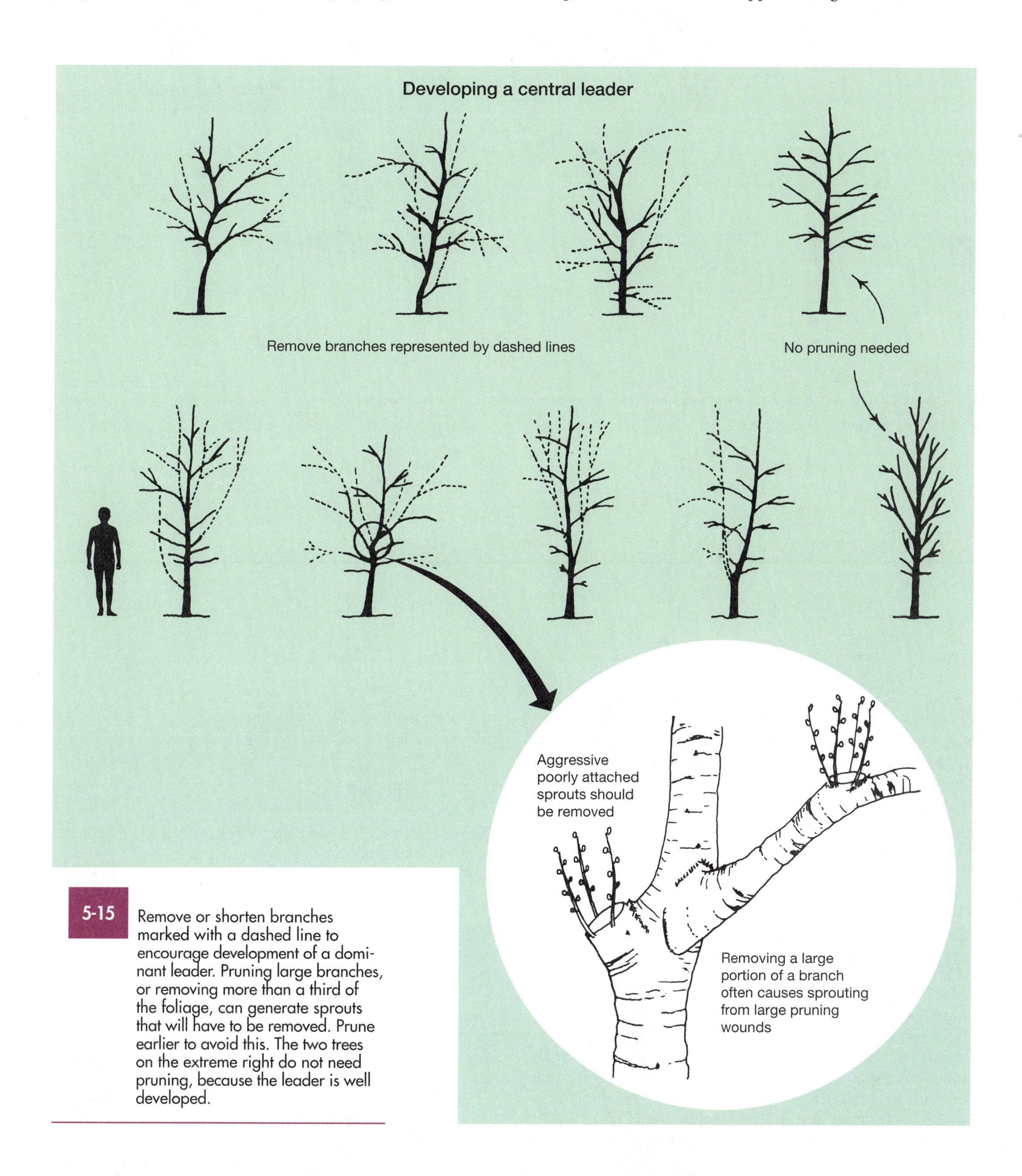

5-15 Remove or shorten branches marked with a dashed line to encourage development of a dominant leader. Pruning large branches, or removing more than a third of the foliage, can generate sprouts that will have to be removed. Prune earlier to avoid this. The two trees on the extreme right do not need pruning, because the leader is well developed.

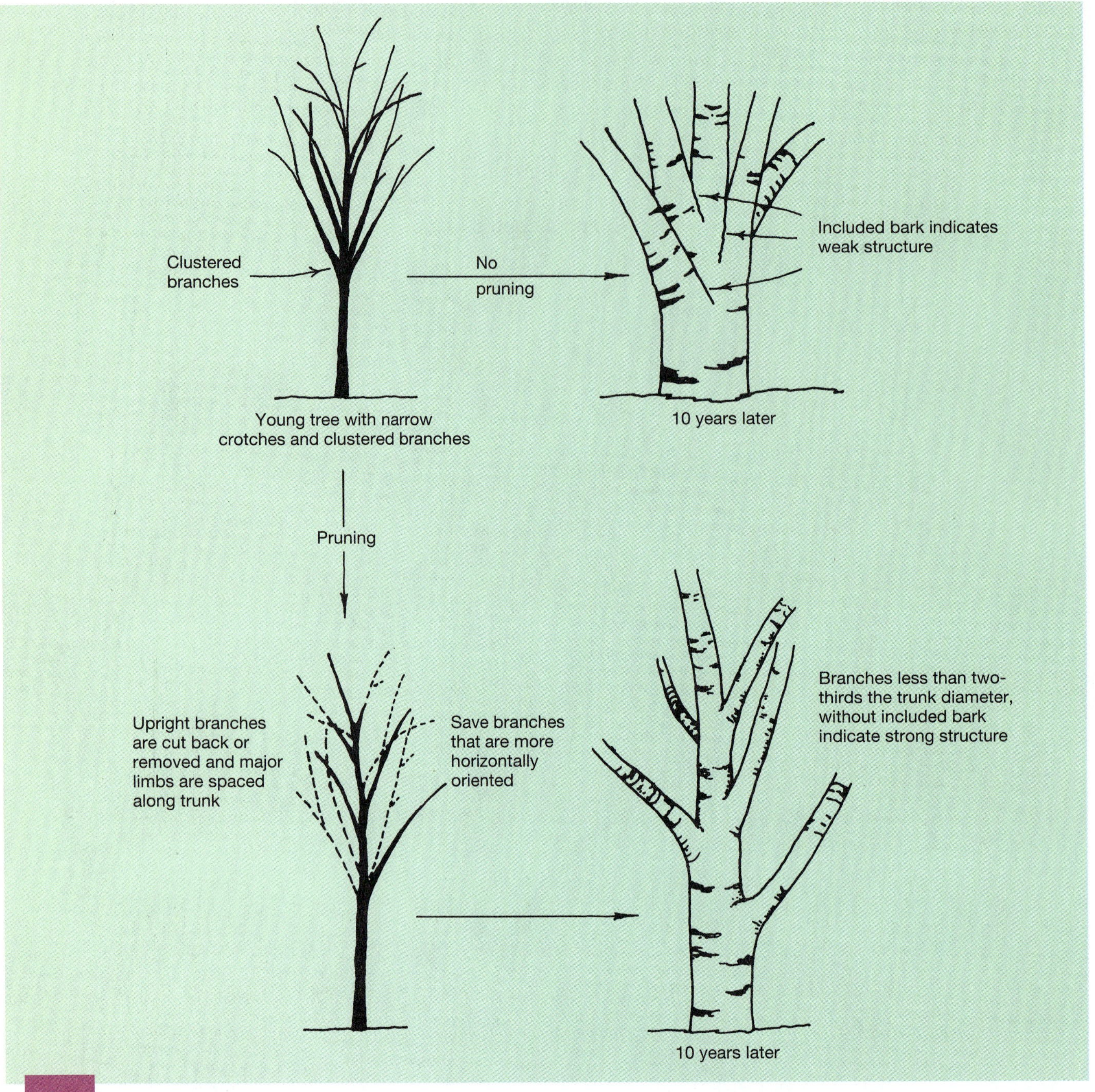

5-16 Rapid-growing limbs clustered together on the trunk often form included bark and are not well attached to the tree (top). Early pruning can remove limbs with included bark when they are small (bottom).

PRUNING DURING ESTABLISHMENT

Most trees with a decurrent growth habit (rounded or oval canopy form) that mature greater than about 40 feet tall need regular pruning in the first 25 years after planting to establish and maintain strong structure. Do not allow poor form to continue to develop, because it is often difficult to correct later (Figure 5-19a–g). One of the most useful techniques is the drop-crotch cut (Figure 5-20a and b). This type of cut reduces the growth rate on the cut branch, allowing the trunk to dominate the tree. Without pruning, structural defects often develop that could lead to tree failure (refer to Figure 2-4).

If good-quality trees with well-developed trunk and branch structure were planted (this is often not the case), pruning should not be needed for several years after planting. Prune the tree every three to five years until the permanent scaffold branch structure has developed (Figure 5-21). If a community, property manager, or homeowner is not able to provide this important care in the first 25 years after planting, perhaps trees should not be planted at this site. If

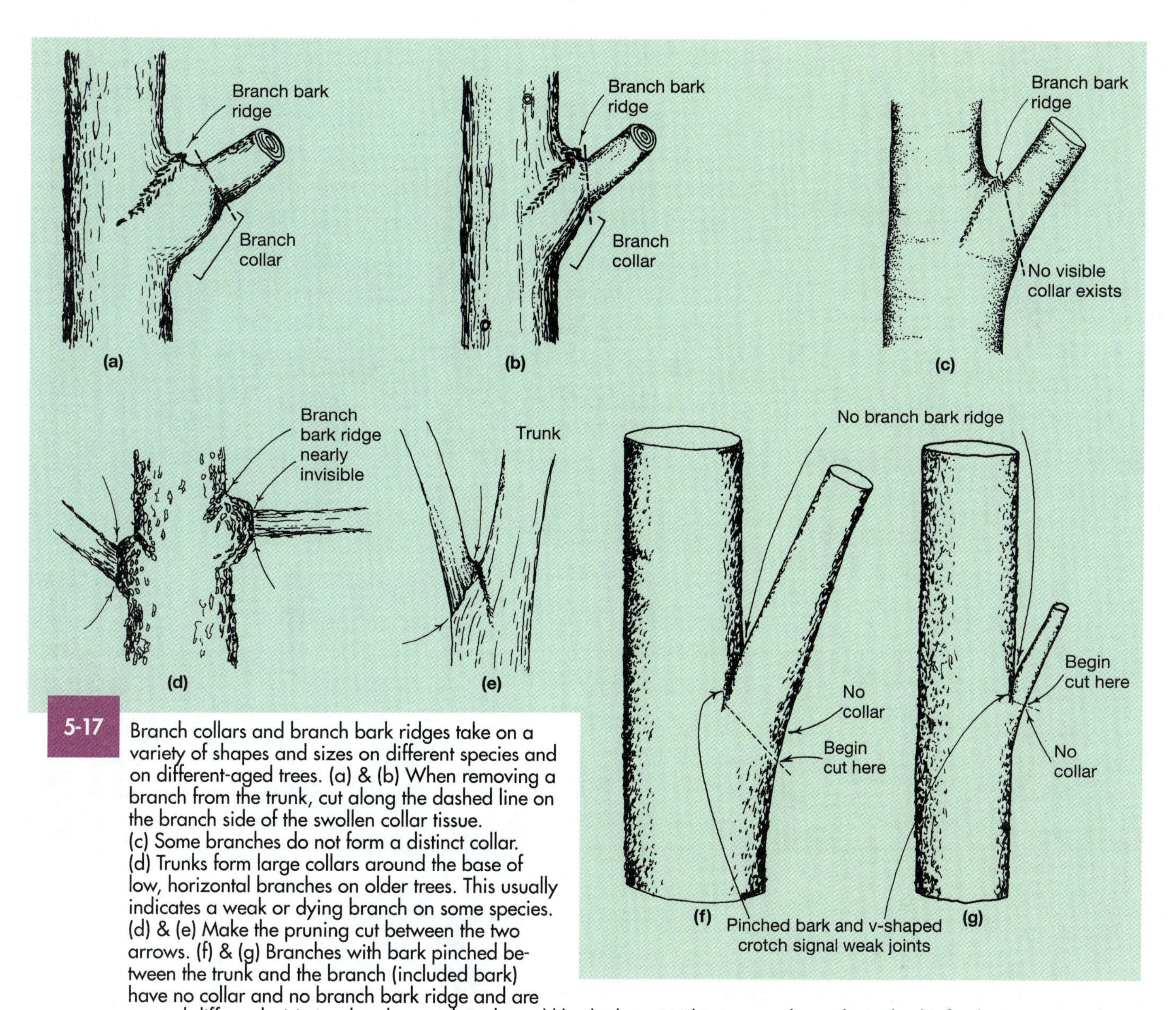

5-17 Branch collars and branch bark ridges take on a variety of shapes and sizes on different species and on different-aged trees. (a) & (b) When removing a branch from the trunk, cut along the dashed line on the branch side of the swollen collar tissue. (c) Some branches do not form a distinct collar. (d) Trunks form large collars around the base of low, horizontal branches on older trees. This usually indicates a weak or dying branch on some species. (d) & (e) Make the pruning cut between the two arrows. (f) & (g) Branches with bark pinched between the trunk and the branch (included bark) have no collar and no branch bark ridge and are pruned differently. Notice that the crotch is shaped like the letter V. This is a weak crotch. Make the final pruning cut along the dashed line. The cut should end in the crotch at the point where the branch tissue actually meets the trunk tissue. This may be several inches (young trees) or feet (older trees) down into the crotch. The cut may have to be finished with a chisel to prevent injuring trunk tissue.

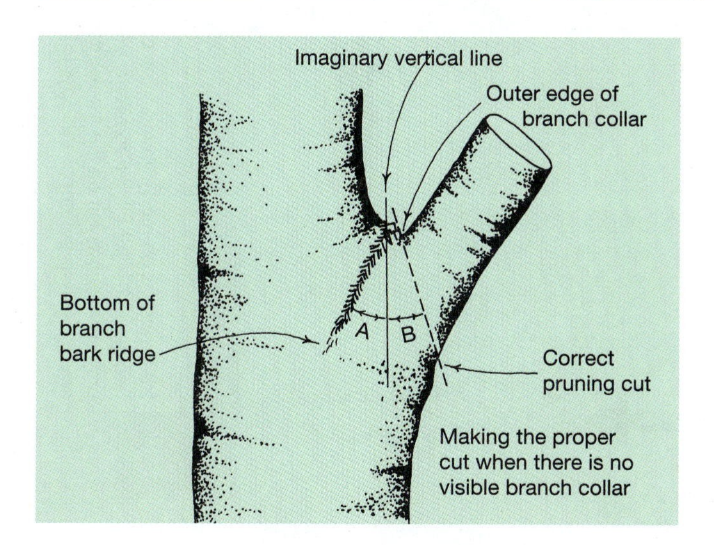

5-18 When the bottom of the branch collar is hard to see, estimate angle A by drawing an imaginary vertical line as shown here. This line is parallel with the trunk. Beginning on top of the branch at the outer edge of the branch bark ridge, make a pruning cut so angle B is the same as angle A. The cut often ends even with the bottom of the branch bark ridge.

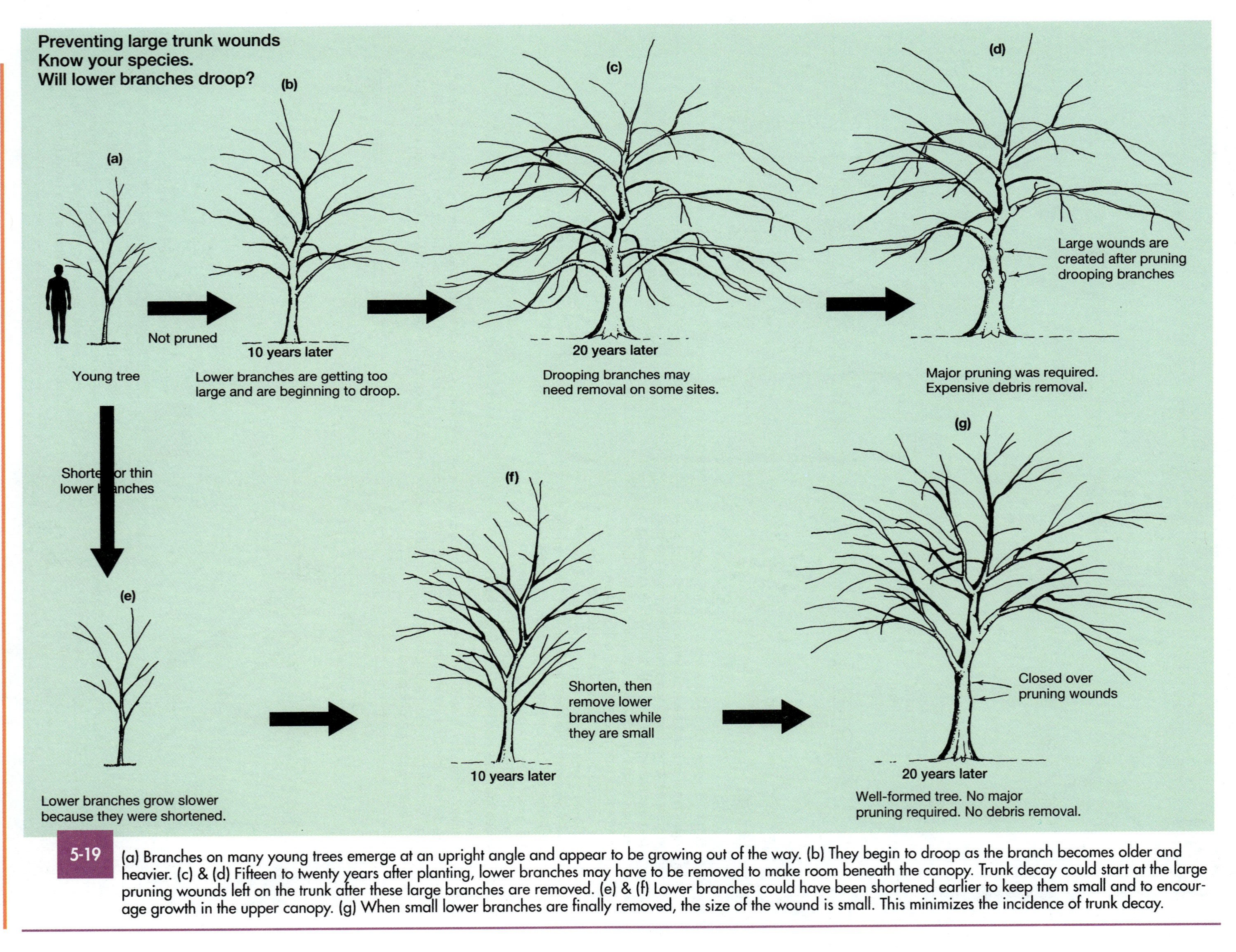

5-19 (a) Branches on many young trees emerge at an upright angle and appear to be growing out of the way. (b) They begin to droop as the branch becomes older and heavier. (c) & (d) Fifteen to twenty years after planting, lower branches may have to be removed to make room beneath the canopy. Trunk decay could start at the large pruning wounds left on the trunk after these large branches are removed. (e) & (f) Lower branches could have been shortened earlier to keep them small and to encourage growth in the upper canopy. (g) When small lower branches are finally removed, the size of the wound is small. This minimizes the incidence of trunk decay.

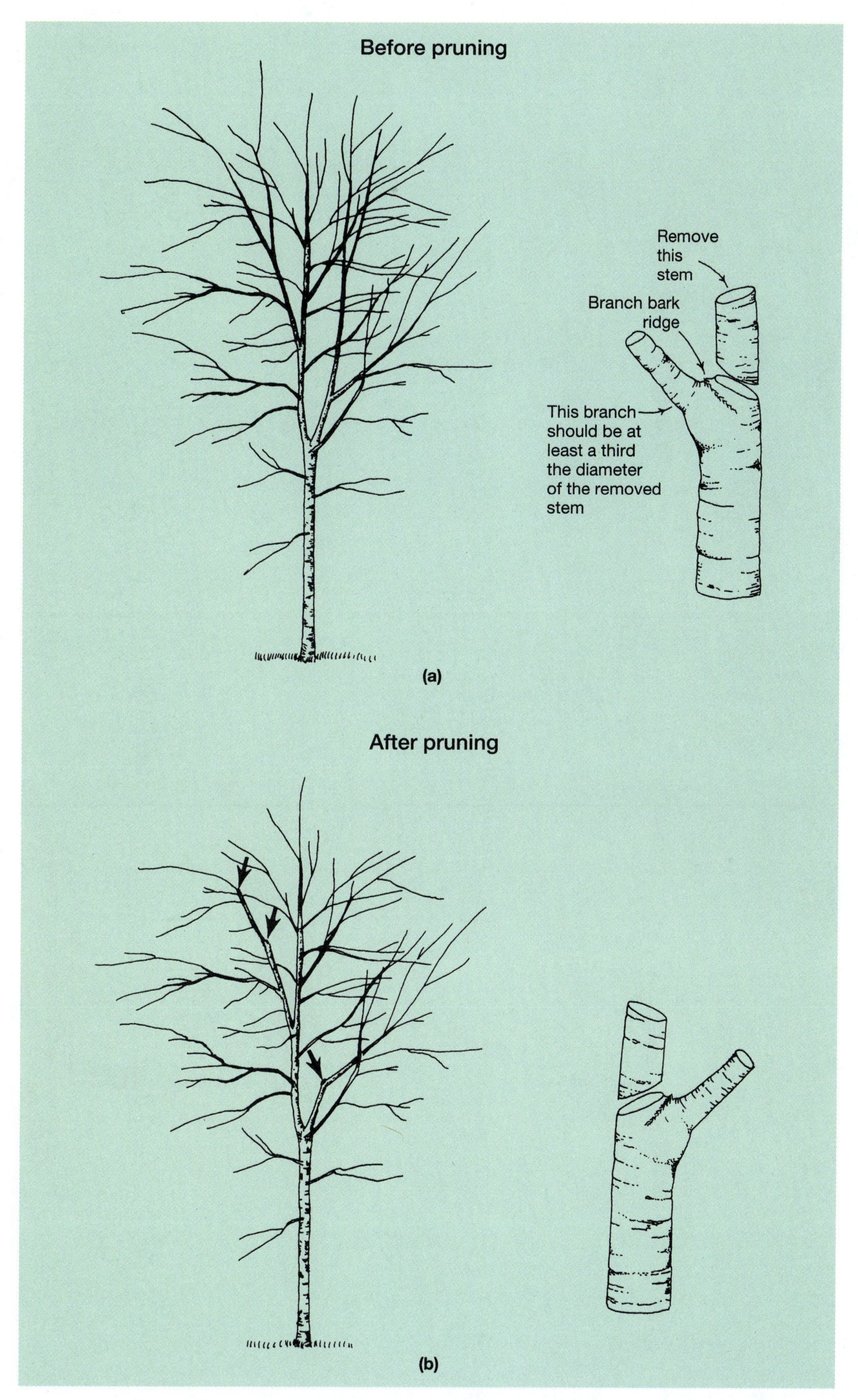

5-20 (a) & (b) Drop-crotching removes an upright portion of a branch or leader back to a more horizontal one (see arrows), encouraging less aggressive, horizontal growth on the cut branch. It is also used to reduce the height of the branch by cutting to a lateral large (vigorous) enough to assume the terminal role. Do not injure the branch bark ridge when making the cut.

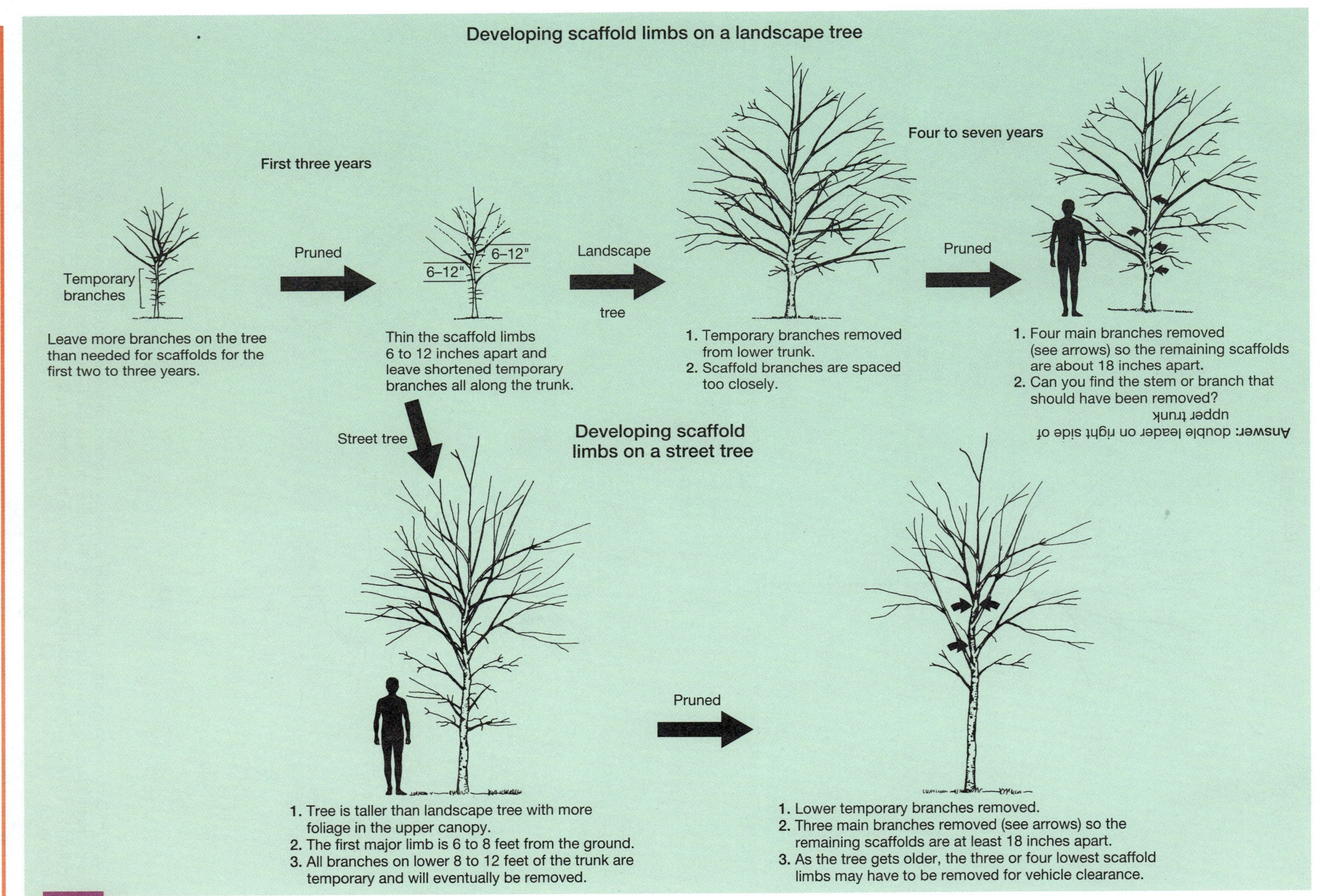

5-21 Training large-maturing trees to grow with one dominant leader can minimize the possibility of branch failures. Scaffold branch arrangement depends on tree location and function.

you choose not to prune, at least plant trees with an excurrent growth habit (conical shape), such as spruces, firs, pines, or planetree, that require less regular pruning than others to create a strong-structured tree (Appendix 14).

If poor-quality trees were planted, the costs associated with pruning to develop good structure in the first several years after planting should be considered part of the planting costs. You might find that when you add the costs associated with pruning poor-quality trees after planting to the initial low cost of the trees, you could have spent less money in the long run purchasing and planting good-quality trees.

SUMMARY

Planting at the appropriate depth and maintaining rapid root growth by controlling weeds and managing soil moisture and aeration with irrigation and mulch is crucial for successfully establishing trees. Many products sold to enhance root growth and plant establishment have not been scientifically tested in the field and are of questionable value (Table 5-6). Avoid the pitfalls of planting to establish trees quickly (Table 5-7).

TABLE 5-6 FACTORS INFLUENCING ROOT GROWTH AND ESTABLISHMENT RATE OF TRANSPLANTED TREES

ENHANCES GROWTH	LIMITS GROWTH	LITTLE OR NO EFFECT
planting in loose soil	pruning to compensate for root loss	fertilizing at planting**
proper irrigation management	grass and weeds close to trunk	root stimulant products
mulch over root balls and beyond	little or no irrigation	peat or other organic matter in planting hole
leave as much foliage on trees as possible	compacted soil	water-absorbing polymers or gels
proper planting depth	planting too deep	mycorrhizae*

* Research on seedlings indicates growth enhancement in certain circumstances including acid or neutral soils.
** Fertilizing during the establishment period can increase growth rate in some circumstances.

TABLE 5-7 IMPROPER PLANTING AND ESTABLISHMENT PRACTICES

ABOVE-GROUND	BELOW-GROUND
touching branches not pruned	tree planted too deep
leader has been topped	planting hole less than twice root ball width
heavily damaged or infested branch not removed	synthetic burlap left on root ball
branch larger than ⅔ trunk diameter not pruned or removed	straps left over root ball
branch with included bark not pruned	highly amended backfill soil
wires or string left on trunk or branches	roots wrapped around inside of planting hole
root ball tying ropes left on trunk	circling roots left intact
wire used for guying	top tier of wire on wire basket not removed*
staking required to hold a weak trunk upright	no irrigation after planting
tree staked more than one year	soil placed over root ball

* Wire baskets greater than 40 inches in diameter may be left intact.

Chapter 6

Root Response to Culture and Planting

INTRODUCTION

Roots on trees and shrubs planted in nurseries and landscapes are influenced by cultural and management practices. Competition with turf grass reduces tree and shrub root density near the surface of the soil. Drip irrigation causes a localized increase in root growth. Nutrient applications can increase or decrease root density, depending on application techniques and the amount and concentration applied. The structure and density of roots inside the root ball of a nursery-grown tree can be manipulated by tree production techniques in the nursery, such as root-pruning.

Root growth and extension after planting are somewhat predictable within a wide range of values, although they vary with species, climate, production method, competition from other plants, size, and health. Tree and shrub roots on established plants commonly extend from two to three times the distance from the trunk to the edge of the branches (Gilman 1988a; Watson and Himelick 1982). This relationship is established within three years following planting of two- to three-inch caliper trees.

CULTURAL FACTORS AFFECTING ROOT GROWTH

Grass and weed management

Soil is cultivated, treated with herbicides, or mulched to control competing vegetation such as weeds and grass. Tree root density is greatest on sites without competing plants. Many landscapes receive applications of mulch to discourage turf and weed competition, buffer soil temperatures, and for other reasons (see Chapter 5). This practice increases tree growth after planting by encouraging root growth just below and in the mulch (Watson 1988). The effects of these soil management practices are most pronounced on roots close to the surface of the soil.

Grass is most competitive with tree roots if mowing is conducted infrequently. Turf roots are deeper and root density greater when turf is not mowed or is infrequently mowed than when turf is regularly mowed (Beard and Danial 1965; Biswell and Weaver 1933). Competition is most keen in the top foot of soil because roots on both turf and trees proliferate in this well-aerated, nutrient-rich zone.

Reduced surface-root density on trees in cultivated or grassed soil may be compensated for by increased root growth deeper in the soil, provided the site is well drained and not compacted. However, extensive or deep cultivation on wet or compacted sites is damaging to tree roots and is not recommended, because low soil oxygen content prevents regeneration of deeper roots to replace the surface roots damaged by cultivation.

Trees have a wider-spreading root system under a grass sod than under cultivation, due to the root-pruning effect of cultivation (Coker 1959; Mitchell and Black 1968). Consequently, root balls dug from a nursery with grass close to the trunk may contain fewer roots than those dug from cultivated or mulched plots or plots receiving herbicide to maintain bare soil near the trunk.

Trees in field nurseries often are grown in weed-free strips of bare soil separated by grassed alleyways. Root growth and density are greater under the bare-ground herbicide strip than in the grassed alley. Mineral uptake by tree roots in the grassed area is small compared with uptake from the herbicide strip, even for established trees (Atkinson et al. 1979). Nurseries often apply fertilizer only on the bare soil strip to take advantage of the increased root density there.

Irrigation

The effects of irrigation on tree root growth are variable. Most research has been conducted in orchards with apple trees, and it shows that irrigation increases root density only in the top 6 inches and reduces density at the 6- to 12-inch soil depth. In contrast, one study found no effect of irrigation on apple root distribution. Variation in response among studies probably resulted from fluctuations in other uncontrollable factors that also influence root growth, such as soil type and climatic conditions.

The method of irrigation influences root distribution. Compared to a nonirrigated control, a more vertically uniform root distribution is often created with low-volume irrigation (Huguet 1976). In other words, there are more deeper roots. Low-volume irrigation delivers water to the soil at a slow rate, which allows it to penetrate deeper into the soil. Flood irrigation can limit root growth to the surface layers of the soil. Furrow and overhead irrigation produced equally dense root systems in apple trees. However, the main horizontal roots were deeper with furrow irrigation, perhaps due to deeper water penetration.

In arid climates, low-volume drip irrigation increases root density within a 12- to 24-inch radius of the drip head on peach and apple trees. There is usually no effect on the portions of the root system not wetted by the drip head, and there may be few roots in this drier area (Goode, Higgs, and Hyrycz 1978). In one study, root distribution on citrus trees under drip irrigation was no different from that on trees under sprinkler irrigation. It is important to provide for adequate aeration in soil wetted by a drip system by monitoring soil moisture beneath the dripper. If soil stays too wet, root density can be locally reduced or forced to grow only close to the soil surface.

In temperate climates, drip irrigation placed six inches from the base of the trunk had no effect on root system depth in sugar maple, honeylocust, and pin oak. Compared to nonirrigated trees, drip irrigation during a three-year study increased fine-root weight within the root ball in pin oak and sugar maple, but not in honeylocust (Ponder and Kenworthy 1976). Chinese elms receiving low-volume irrigation within nine inches of the trunk had four times the amount of fine roots near the trunk than trees receiving irrigation over a larger area (Figure 6-1a and b). However, live oak roots did not respond to irrigation treatments, and there appeared to be no concentration of roots beneath the drip emitter (Beeson and Gilman 1995). Live oak, red maple, and southern magnolia receiving low-volume drip irrigation from one drip emitter at the base of the trunk had roots extending to well beyond the branch dripline. This was similar to root distribution on other species receiving occasional overhead irrigation (Gilman 1988c). Although root growth is enhanced near the drip emitter on a number of species, there appears to be sufficient soil moisture for root growth well beyond the drip emitter in temperate climates, because enough rainfall occurs to encourage growth in this soil.

Provided that soil compaction, high water table, or other factors are not limiting the penetration of roots, depth of wetting exerts a powerful influence on depth of rooting. Roots penetrate deeper into soil as the wetting depth increases (Cullen, Turner, and Wilson 1972). Because deep roots grow at the expense of roots in the middle soil profiles, there is no reduction in root density near the surface of the soil as wetting depth increases. However, a larger percentage of the root system is close to the surface in soils kept continually moist than in those with periodic drying cycles (Doss, Ashley, and Bennett 1962). Periodic drying cycles may encourage deeper rooting by killing or slowing growth on roots in the drier soil close to the surface.

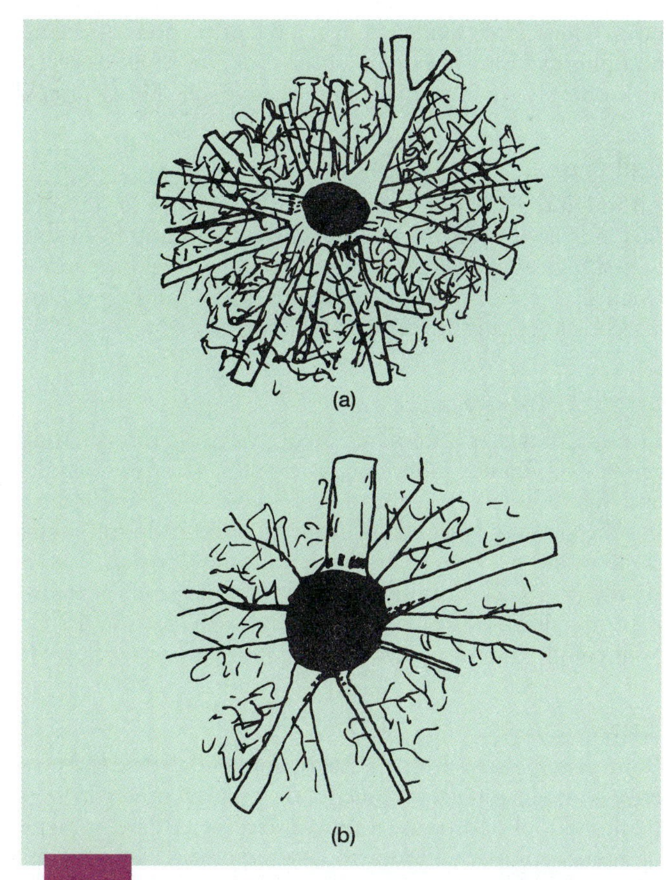

6-1 (a) Fine roots proliferated near the soil surface and close to the trunk on Chinese elms receiving all irrigation within nine inches of the trunk. (b) Fine roots were not concentrated in this area on trees receiving irrigation over a larger area.

Fertilization and nutrient levels

Localized nitrogen applications increase root density and lateral root branching only in the immediate area of application. If half of the root system is fertilized, root growth will be enhanced only in the fertilized part (Coutts and Philipson 1976; Smith 1965). On some species, addition of fertilizer close to the trunk increases fine-root density there, but at the expense of growth on other portions of the root system. The increase in density is due to closer spacing between lateral roots at higher nutrient concentrations (May, Chapman, and Aspinall 1964).

Root density can be reduced in the area around a drip emitter when too high a concentration of nitrogen (400 ppm from ammonium nitrate) is applied through the drip irrigation system. This can be caused by a temporary but dramatic reduction in soil pH (e.g., from 6.2 to 3.7) due to the acidifying effect of ammonium nitrate application (Edwards et al. 1982). Because plants generally utilize no more than about 80 ppm nitrogen in the soil solution, excess application is probably wasteful and slows root growth (Niemiera and Wright 1982).

Overfertilization with nitrogen can retard development of fine roots in citrus and perhaps in other tree species. As applied nitrogen increases from 50 to 300 ppm, shoot growth is encouraged at the expense of root growth (Yeager and Wright 1981). This may lead to greater water stress in landscape trees after transplanting, as a reduced root system struggles to support a large top. Phosphorus at levels greater than 85 ppm had no effect on root growth or shoot-to-root ratio. There is no evidence in the scientific literature that phosphorus stimulates tree or shrub root growth, unless very little phosphorus is available in the native soil. This is rare.

Soil type

Although there is not much scientific evidence to support this, nursery operators report that root balls dug from soil with a high clay or organic matter content often have more small-diameter, fine roots than those from sandy soil. Research is needed in this area to help sort this out.

Shoot-pruning

Pruning branches stimulates shoot growth and slows root growth (Gilliam, Cobb, and Fare 1986; Ranney, Bassuk, and Whitlow 1989). Light to moderate shoot-pruning during the growing season slows root growth for up to two or three months, as the plant replaces removed shoot tissue (Fordham 1972). Pruning branches on young trees at transplanting slows postplanting shoot growth (Larson 1975; Whitcomb 1986b), probably because root growth is slowed.

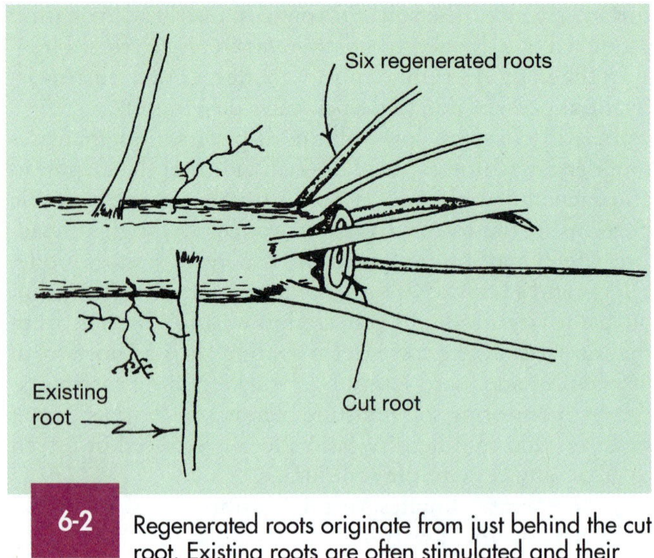

6-2 Regenerated roots originate from just behind the cut root. Existing roots are often stimulated and their growth rate increases in response to root-pruning.

Root-pruning

Root-pruning trees in fruit, forest, and landscape tree nurseries is an old and varied practice. It has been used as a horticultural tool to produce a sturdier tree; force development of a more compact, fibrous root system; retard top growth; and increase transplant survival and posttransplant growth. The timing, frequency, severity, and location of root-pruning are governed more by practical experience and tradition than by scientific studies. Only recently have the effects of root-pruning been studied.

A recognized plant response to root-pruning is reduction of shoot growth. Spring root-pruning reduced shoot growth of three-year-old apple trees by 25 percent and of four-year-old trees by 40 percent. Most literature shows that early-season root-pruning reduces trunk expansion and shoot growth more than later-season pruning. However, the greatest reduction in white spruce height was caused by root-pruning in mid-September, at the end of the growing season (Mullen 1966). A single root-pruning of five-year-old blue spruce reduced top growth so that eight-year-old unpruned plants were the same size as ten-year-old pruned plants (Watson and Sydnor 1987). Root-pruning southern magnolia reduced leaf number, tree height, trunk caliper, and total-tree leaf area and weight compared to unpruned trees (Gilman and Kane 1990). Reduced top growth may result from induced water stress in the tree, limited mineral absorption by a smaller root system, or reduced hormone synthesis.

Each species appears to have a characteristic shoot-to-root ratio, although this ratio can vary during the year (Kramer 1983). When the ratio is disturbed, plants respond by redirecting energy to replace the removed parts. Root-pruning, while reducing shoot growth, stimulates root growth as the plant attempts to restore the prepruning shoot-to-root ratio. Roots regenerated in response to root-pruning originate primarily at or just behind the cut (Figure 6-2). However, a portion of regenerated roots can originate from at least four inches behind the cut, depending on species (Gilman 1988c).

Sycamore can regenerate more than 30 new roots on each cut root, whereas live oak, Chinese elm, crabapple, and linden regenerate 10 to 20 (Gilman 1988b). One of these roots frequently dominates and outgrows the others within two to four years following planting (Watson 1985).

Existing roots behind the cut end of the root are stimulated to grow at a faster rate. This likely accounts for the increase in fibrous roots within the root ball, in response to root-pruning, reported for a number of species. Stimulation of existing roots can also lead to trunk girdling on Norway maple and perhaps on other trees. Existing roots are often oriented perpendicular to the pruning cut. As the trunk and roots grow larger and meet, this root can grow part way around the trunk and girdle it (Watson, Clark, and Johnson 1990).

Smaller shoot-to-root ratios are induced by root-pruning in the tree nursery, and are associated with improved posttransplant tree seedling performance (Benson and Shephard 1977). Some researchers report no benefit for survival and posttransplant growth from pretransplant root-pruning of seedling-sized forest species. Root-pruned 10-foot-tall southern magnolia trees grew at a slightly faster rate following transplanting than unpruned trees during the first year after transplanting, but there was no difference in growth the second and third year after planting. Survival was not affected by root-pruning, probably because all trees were irrigated regularly after transplanting (Gilman 1992).

In general, there is no difference in the shoot-to-root ratios between trees pruned once and those pruned twice prior to transplanting. In one study, increasing the number of root-prunings enhanced posttransplant growth (Bacon and Bachelard 1978); in another, growth was reduced by multiple pretransplant root-prunings (Benson and Shephard 1977). There does not appear to be a clear advantage to multiple root-pruning of trees prior to transplanting.

Root-pruning live oak two inches inside the edge of the proposed root ball one year prior to harvest, and then again at the edge of the root ball six months before digging, increased dry weight of fine roots inside the root ball sixfold compared to unpruned plants (Gilman and Yeager 1987). Root-pruning Colorado blue spruce eight inches inside the edge of the harvested root ball five years before digging resulted in a four-fold increase in root surface area within the root ball (Watson and Sydnor 1987). It appears safe to conclude that root density within the root ball can be increased by harvesting the ball beyond the point of root-pruning.

Root-pruning technique varies from one nursery to the next. Most prune only part of the root system at one time. This imposes water stress on the tree and may help prepare it for transplanting, but leaves enough roots intact to supply the foliage with an uninterrupted water supply. Intact roots also help prevent trees from blowing over. Many nursery operators cut roots, for example, on the north and south sides of the tree, leaving the other 50 percent intact. They return several months to a year later to prune the other roots. Roots are cut with a sharp shovel or mechanical implement. The root ball is harvested several inches outside of where the roots were pruned.

Whereas this technique works well for trees root-pruned two or more times before digging, here is one for pruning a tree only once about six to ten weeks before transplanting in the spring or summer. Cut all surface roots 100 percent around the tree down to a specified depth. Be sure to leave the deepest roots intact. You will have to perform several root excavations (preferably on each species you root-prune) to determine the proper root-pruning depth. Strive to cut about 75 percent of the root system, but do not cut too deep or too many roots will be severed. If roots are not pruned all the way around the trunk, one or two large surface roots may not be cut until the tree is dug several months later, causing it to shock or die.

Bismarck palm is one palm that most growers root-prune before digging from a field nursery. A number of growers routinely root-prune other palms in the nursery. They say this helps create a dense root ball and eliminates the need to remove leaves when the palm is transplanted. There is little research to support this practice, but it works for many growers.

ROOT GROWTH AFTER PLANTING

Root development following transplanting varies with climate, method of tree production used by the nursery, tree size, species, and cultural practices. Many other factors affect root form, depth, and spread.

Root system form

Roots proliferate in the area of the landscape that is most conducive to root growth. For example, if one side of a tree receives adequate water and nutrition, and the soil is well aerated, roots will grow well on that side and slowly on the other. Root growth on one side of the tree may be poor if it is frequently flooded, or is very dry or compacted, or if there is a thick stand of trees on that side with established root systems. This could lead to an asymmetrical root system that could affect the management and stability of a large tree.

Container grown plants

Mean root spread diameter on Chinese junipers planted from three-gallon containers in hardiness zone 8 was five feet one year after planting. Root spread diameter averaged 12 feet, and maximum spread was 16 feet, 3 years after planting (Figure 6-3). Poplar, honeylocust, southern magnolia, live oak, and red maple in hardiness zones 6 through 8 had root spread diameters of about 15 feet 3 years after planting. Live oak in zone 9 had an 18-foot root spread one year after planting.

Many trees produced for the landscape trade are grown in containers for a period of time during the production cycle. Roots deflected by container walls can cause root deformations that may lead to long-term tree growth problems. Although kinks in the roots caused by container production are associated with restricted flow in the xylem or phloem, the long-term significance of kinks in lateral roots is uncertain. It is known that circling roots can contributed to tree instability after planting.

Researchers have described the fate of circling roots caused by container tree production compared to other production methods. One showed that root morphology on trees in the forest appears similar to that on trees planted from containers (Carlson, Preisig, and Promnitz 1980). Most often, studies show that root morphology is altered by container production practices. The resulting defects include kinked, circling, or girdling roots, all of which can restrict growth of landscape plants.

The results of one study showed that root deformation within the container may not have a long-term effect on the growth of red maple and perhaps other species. The majority of the root system emerged from adventitious roots initiated after planting, above and presumably removed from the potential girdling effects of the circling roots in the bottom of the container (Figure 6-4). Perhaps some species adapted to wet sites, such as red maple, can avoid the potential problems of container-induced root deformations by developing adventitious roots close to the soil surface after planting. Long-term growth may be most affected by these container-induced root deformations on species producing a less adventitious root system originating from deformed roots produced in the container (Gilman and Kane 1990).

Long-term growth can be affected by circling roots near the top of the container (Figure 6-5). As the trunk enlarges, it makes contact with the circling root. If these roots are not cut, they could choke the tree by restricting trunk growth. Growth may be slowed, or the tree could become unstable, or it could die.

Container-grown trees planted in a nursery or landscape sometimes develop lateral roots on only two or three sides of the plant. High temperatures inside traditional black containers in the nursery cause root death on the south

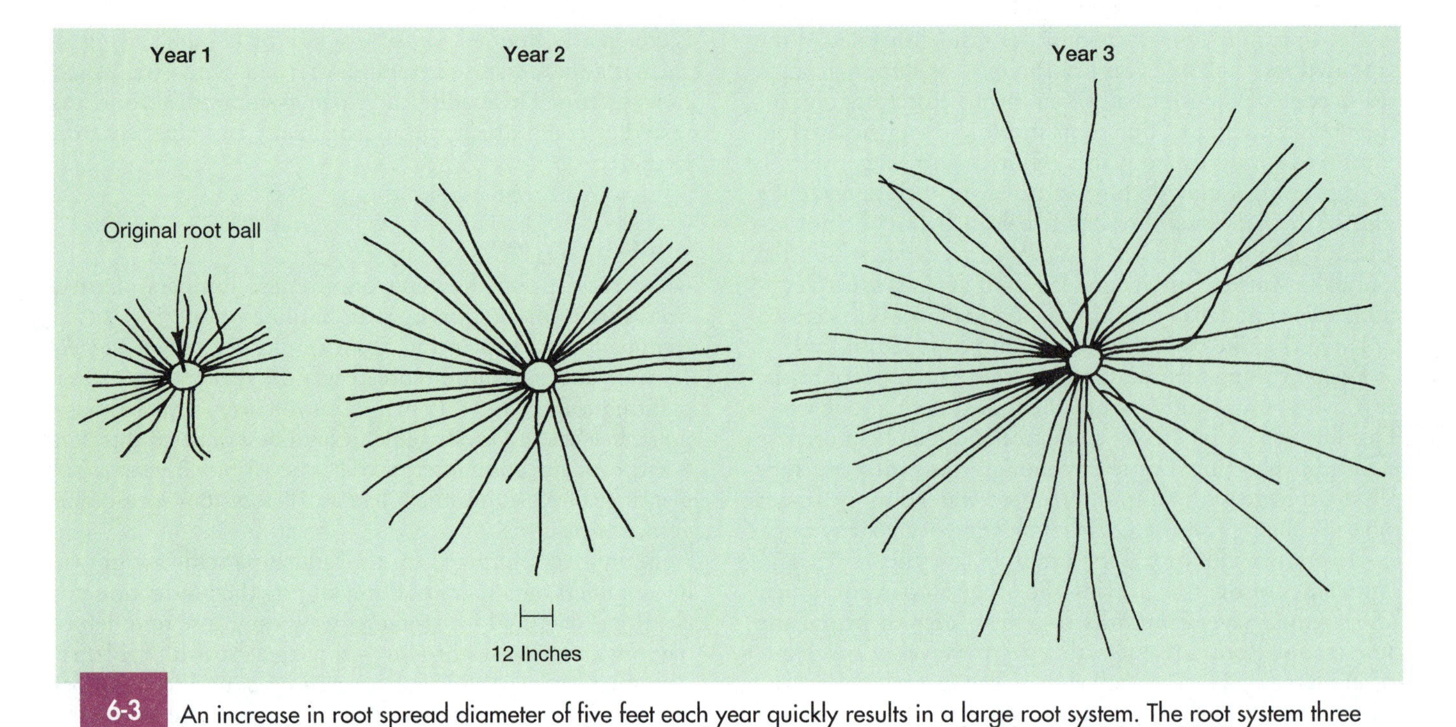

6-3 An increase in root spread diameter of five feet each year quickly results in a large root system. The root system three years after planting covers an area nearly ten times larger than at one year after planting (Gilman 1990a).

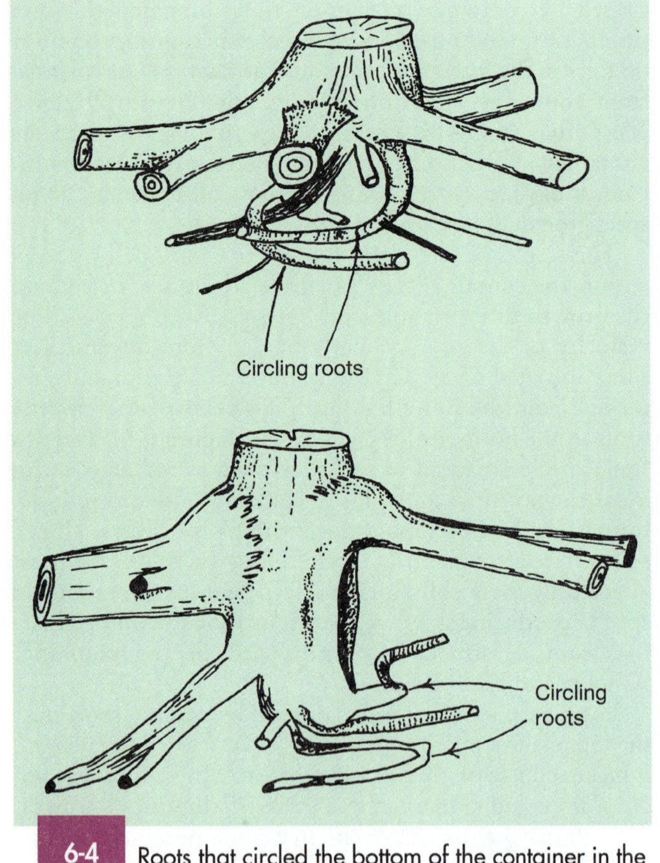

6-4 Roots that circled the bottom of the container in the nursery can still be seen three years after planting into the landscape. However, the major roots on red maple grew above these circling roots and should continue to develop normally.

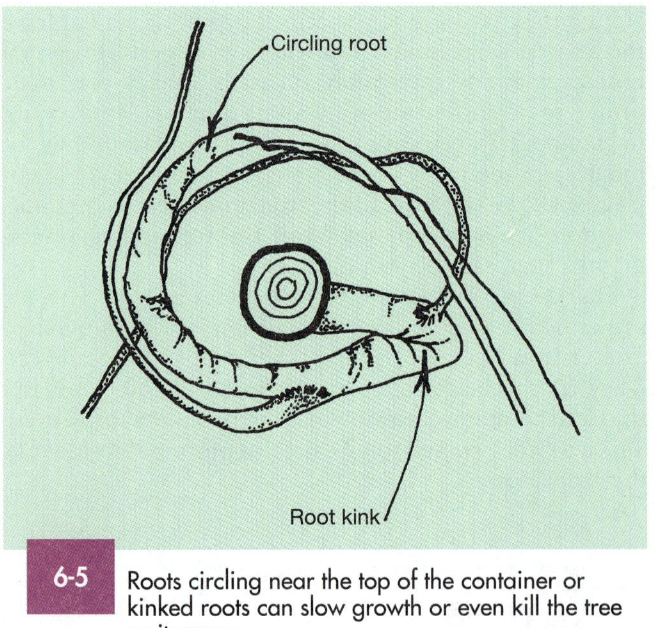

6-5 Roots circling near the top of the container or kinked roots can slow growth or even kill the tree as it grows.

and west sides of root balls exposed to direct sun (Ingram 1981). This eliminates roots from this portion of the root ball and could lead to uneven root distribution in landscape soil (Ruter 1993a and 1993b). Plants in containers that reflect light, or that are sunk into the ground, have more uniform root systems due to lower root ball temperatures. Root circling and kinks that develop on container-grown plants may also be responsible for uneven root development following planting (Marler and Davies 1987).

Field-grown plants

Unlike container-grown plants, only a small portion (2 to 20 percent) of the root system length is harvested with the tree on field-grown plants that are not root-pruned (Gilman 1988c; Gilman and Beeson 1996; Watson and Himelick 1982). A considerably larger portion of the root system length is harvested on trees that are root-pruned prior to digging (Figure 6-6). Despite the small fraction of root length harvested in the root ball of a field-grown tree, more than 75 percent of the original root weight is inside the root ball (Gilman and Beeson 1996). This probably accounts for the tremendous survival rate of trees transplanted from field nurseries. Although roots on healthy trees begin to regenerate within a week or two after severance during the warm season, water demand of the top requires that roots stay moist at all times. Trees will be under some water stress until the root system is restored to the original pretransplant size, which can take from one to as long as ten years or more for larger trees.

Root spread diameter increases about three feet per year for trees in the northern United States (hardiness zone 5) (Watson 1985). Root spread diameter of eight-year-old Sitka spruce increased at a rate of four feet per year. Root spread diameter was 20 feet 2 years after transplanting 5-inch-caliper live oak from fabric containers (hardiness zone 8), amounting to a 7.5-foot-per-year increase in root spread diameter. Trees planted bare-root have growth rates similar to those of trees planted with a soil ball (Hensley 1994).

In summary, expect root system diameter of recently planted trees to expand at a rate of 3 to 7.5 feet per year, depending on climate, competition from turf and other plants, species, and postplanting care. Preliminary studies in subtropical regions (hardiness zones 9 through 11) of the country indicate a much faster rate of root growth after planting, especially if irrigation is provided regularly. Live oak, Chinese elm, and poplars gave a 15- to 20-foot increase in root spread diameter in the first year after planting into friable soil.

Root system

Root ball

6-6 Only a small fraction of the root system length is moved with a tree transplanted from a field nursery if the tree is not root-pruned prior to transplanting. However, more than 75 percent of the root system weight is transplanted with the tree.

GROWTH IN URBAN SITES

Some practitioners think that tree roots extend beneath streets. Penetration beneath pavement is not likely to be extensive except in the most friable, moist, well-drained soil, because soils beneath streets are traditionally compacted to densities higher than those conducive to root growth. This reduces the oxygen concentration and can increase carbon dioxide to 20 percent under roadways. These soil air conditions usually limit root growth. Despite this, roots of trees tolerant of wet sites (e.g., honeylocust, magnolia, sycamore) and other species, including eucalyptus, can be found growing directly under pavement in some situations. This can even be seen in clayey soils (Figure 6-7).

Some roots may grow in well-drained soil beneath parking lots. The soil beneath parking lots is not compacted as much as beneath some streets. Because roots draw moisture from the soil, some soils with high clay content shrink, causing pavement to settle and crack (Mathene and Clark 1995). Some roots may continue to live after pavement is placed over an existing root system; however, roots of planted trees in many sites are often deflected by curbing and do not extend under the street. Rather, they often grow parallel to the curb and beneath the sidewalk, because this area is highly favorable to growth of tree roots. Roots often extend into the lawn area beyond the sidewalk. As lateral roots enlarge in diameter, sidewalk slabs can be lifted, making them hazardous (refer to Figure 2-6). Soil temperature beneath roads may also reach lethal levels, preventing roots from exploring soil directly beneath asphalt.

6-7 Roots of this oak grew beneath this pavement on a sandy soil. Notice the cracked pavement over each root.

Chapter 7

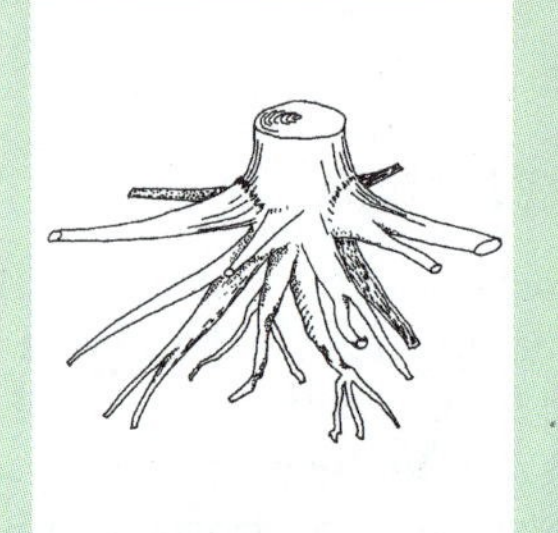

Root Form, Depth, Spread, and Season of Growth

INTRODUCTION

Root form, depth, and spread vary considerably from one site to another. In short, roots grow where oxygen, water, and the elements essential for growth are located. Root depth is somewhat predictable within a broad range, but managers of landscapes and nurseries could more precisely manage trees and shrubs under their care by digging in soil to determine the location of roots. This could be performed easily in a number of exemplary sites within the area of operation. This would help employees gain a better understanding of the resources they manage and would go a long way to educate consumers in the basic principles of tree care.

Roots on open-grown trees spread generally to about three times the edge of the branches, and perhaps farther for trees in the forest (Figure 7-1). Most are fairly close to the soil surface, not deep in the soil (Stout 1956).

Root system morphology of trees can influence survival, stability, growth, and transplantability. The root system that develops on trees seeded in place (trees in the forest) is a function of seedling genetics; soil characteristics such as texture, compaction, fertility, depth to the water table, and moisture content; soil insect activity; and other factors. Moreover, root form on planted trees is also influenced by preplanting nursery production practices, tree age at planting, planting method, and postplanting cultural practices. Cultural practices and soil conditions have a bigger influence on root system form and distribution than do tree species and seedling genetics.

ROOT DEVELOPMENT ON YOUNG TREES

Tap root

The first root to emerge from a tree seed is called the primary or tap root, but it does not continue to develop on every tree (Figure 7-2a and b). It branches if injured or cut. Tap roots are most obvious on seedlings and are often found on pines, gums, and trees with large seeds (e.g., oaks and walnuts) (Preisig, Carlson, and Promnitz 1979). Soil conditions permitting, a tap root occurs most frequently on

7-1 Tree roots spread way beyond the branches to about three times the canopy spread. More than half of the root system is located beyond the edge of the canopy. Few roots grow deep into the soil on most sites.

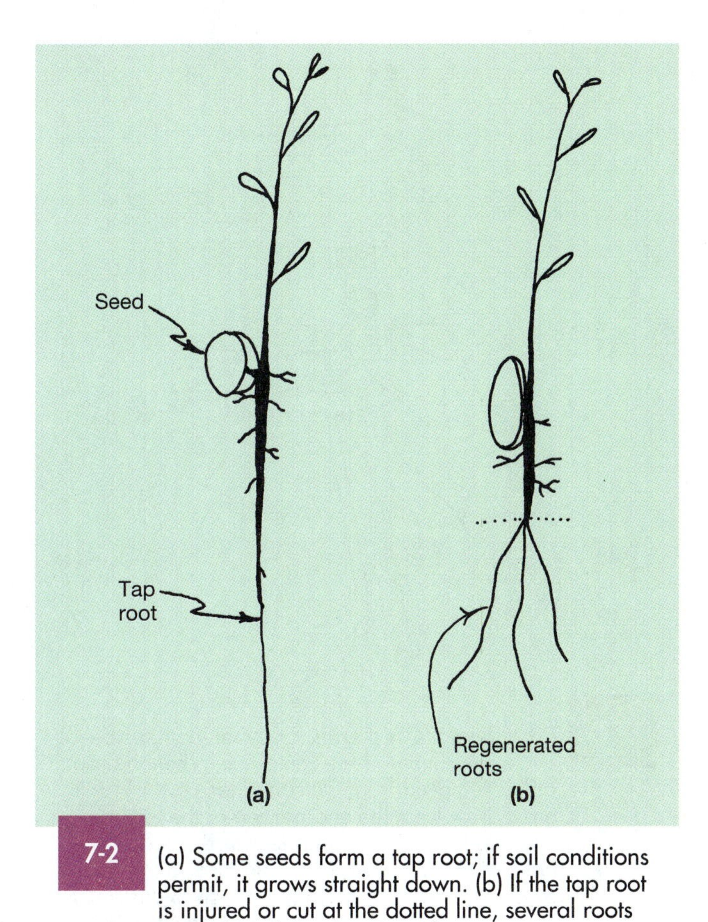

7-2 (a) Some seeds form a tap root; if soil conditions permit, it grows straight down. (b) If the tap root is injured or cut at the dotted line, several roots regenerate from the cut end.

trees in a naturally regenerated forest growing where the seed germinated in place (i.e., the tree was not transplanted). Many trees, such as maples, do not develop tap roots under most circumstances (Figure 7-3).

7-3 Some trees, including those propagated from cuttings, form a highly branched root system immediately after germination, and there is no dominant tap root.

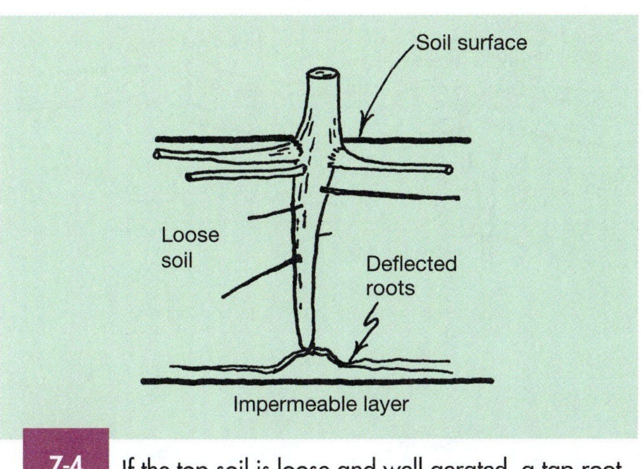

7-4 If the top soil is loose and well aerated, a tap root may grow down, but eventually it will be deflected by an impermeable layer such as rock, a hardpan, a water table, or a low-oxygen layer. These types of layers can be several inches or many feet below the soil surface.

Tap roots sometimes occur on trees in nurseries, but they are much less common in landscape sites. Occasionally, a tap root persists and grows straight down into the soil to depths of three to eight feet or more, until it meets a water table, impermeable soil layer, compacted soil, or other low-oxygen environment. It branches at this point or is deflected horizontally (Figure 7-4). If present, the tap root is usually largest just beneath the trunk and decreases rapidly in diameter deeper in the soil. Occasionally, the portion of the tap root just below the ground on oaks, pines, and sweet gum is larger in diameter than the trunk.

In the shallow, poor, or disturbed soils typical of urban areas, if a tap root is present it branches into several roots close to the soil surface and becomes indistinguishable from the rest of the root system (Figure 7-5). Tap roots do not develop on trees growing in compacted soils or soils with poor drainage or a high water table. These soils are common in many areas of the country (Patterson 1976). Tap roots are thought to contribute to water adsorption and tree stability.

The tap root on a tree that was not seeded in place is almost never intact as a single, straight, descending root and is frequently absent altogether (Figure 7-6a and b) (Carlson, Preisig, and Promnitz 1980). In other words, most trees

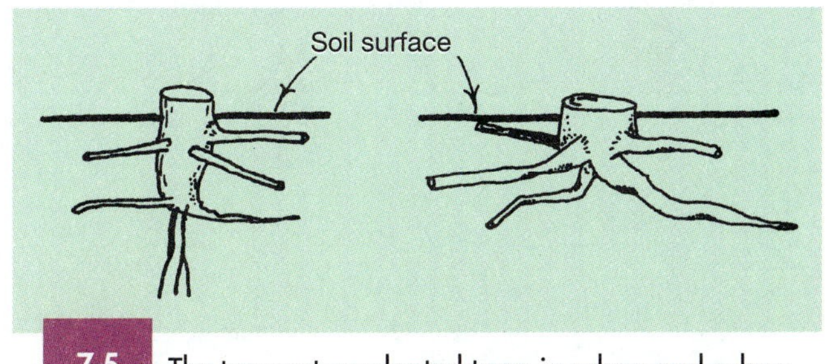

7-5 The tap root on planted trees in urban and suburban landscapes is often aborted (left), or completely missing (right). This is often due to unsuitable conditions for root growth in the deeper soil layers.

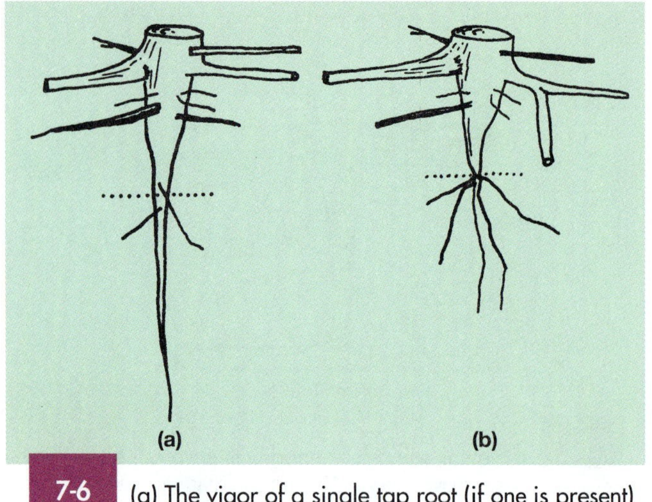

7-6 (a) The vigor of a single tap root (if one is present) is destroyed when the tree is moved from the nursery and transplanted to the landscape. (b) The tap root was cut at the dotted line and five roots regenerated in its place.

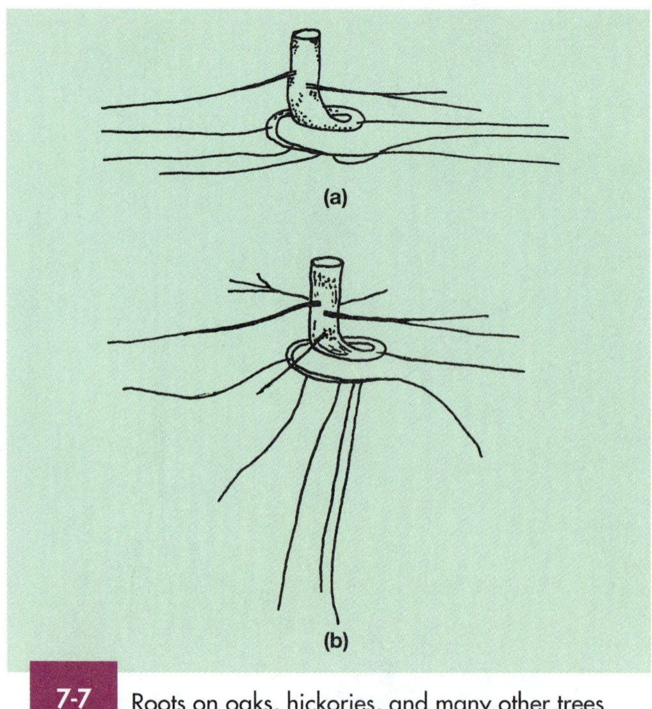

7-7 Roots on oaks, hickories, and many other trees circle the bottom of the container. When the root ball is removed from the container, several roots often grow from the end portion of the circling root. (a) In many landscapes, these roots grow parallel to the soil surface. (b) Occasionally, soil oxygen content will be high enough to allow these roots to grow vertically into the soil.

planted in urban and suburban landscapes do not have tap roots. Because roots are pruned when trees are transplanted, root branches develop and the dominance of a single tap root is destroyed. Several roots may emerge from the pruned or intact roots circling the bottom of a container-grown plant (Figure 7-7a and b). Typically, these are located beneath the trunk and are smaller in diameter than a single tap root.

Oblique roots

Many species are capable of producing oblique roots that grow down at a steep angle (Figure 7-8a and b). There may be 10, 20, or more of these roots on a single tree. They originate from the primary root or shallow lateral roots close to the trunk during the first one to four years after seed germination, and are usually smaller in diameter than the tap root, if one is present. Oblique roots can penetrate several to many feet into the soil if soil conditions permit

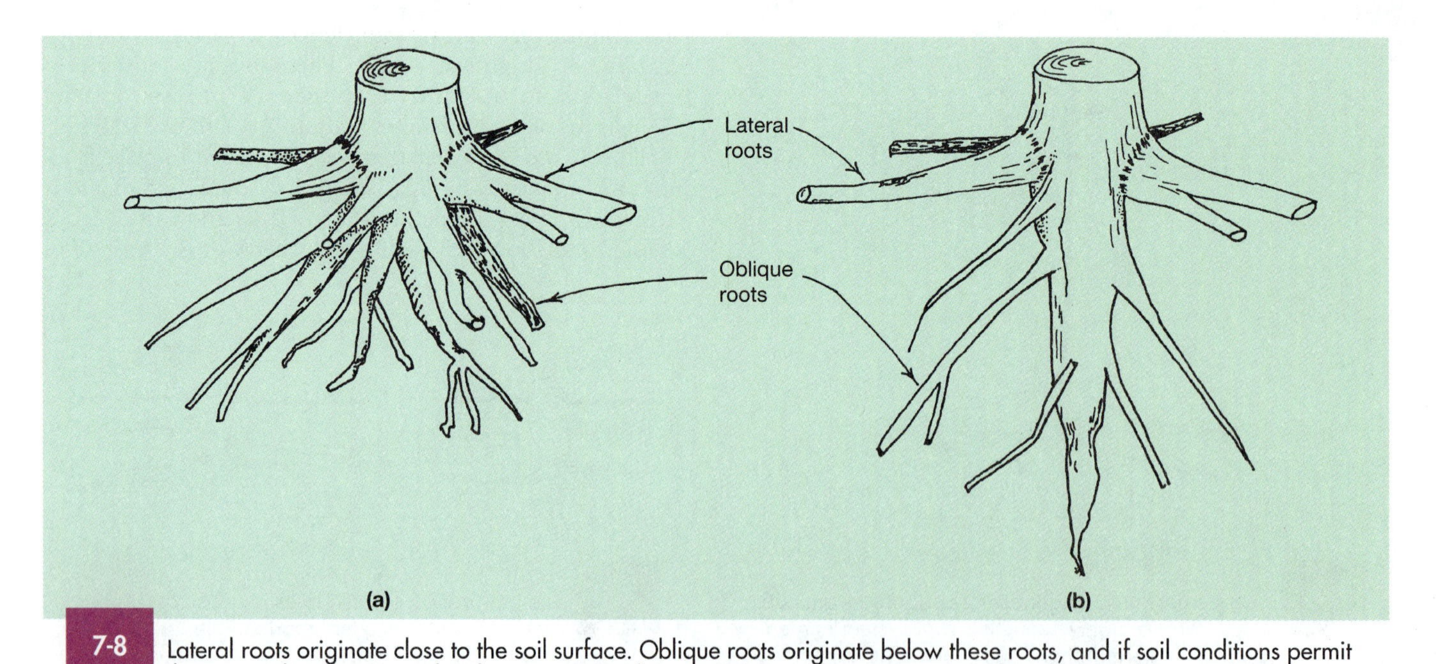

7-8 Lateral roots originate close to the soil surface. Oblique roots originate below these roots, and if soil conditions permit they grow down at an angle. The root system shown in (a) is much more common than the one shown in (b).

(McMinn 1963). They may branch several times and generally bear only a few small-diameter fine roots. Their function is probably to stabilize the tree in the soil and provide for some water uptake. Horizontal lateral roots typically do not emerge from these roots.

Lateral roots

Lateral roots originate either from the primary root or as adventitious roots growing directly from the base of the trunk. One of their main functions is to hold the tree erect. Between 4 and 11 major lateral roots are generated during the first 3 to 7 years after seed germination and grow horizontally through the soil (Figure 7-9). These will generally remain as the largest-diameter, most dominant lateral roots for at least 30 years (Coutts 1983). *The largest five lateral roots comprise about three-quarters of the total root system.* Their points of attachment to the trunk are usually at or near groundline on young trees and are associated with a slight swelling of the trunk commonly called root flare.

These major roots branch within several feet of the trunk to form a network of long, untapered, rope-like roots one-half to several inches in diameter (Figure 7-10). They advance parallel to the soil surface at a depth of 12 to about 36 inches, depending on species, soil profile, soil texture, soil density, soil moisture content, competition from other plants, and maintenance practices. In many soils, most are in the top one foot of soil. A large portion of the total root system length and surface area is represented by these and the fine roots emerging from the lateral roots. Other, less dominant, small-diameter lateral roots emerge from the trunk at points between the major roots.

Lateral root growth is largely determined by local growing conditions. If by chance a root tip on one lateral root grows into an area of superior nutrient, oxygen, or water content, growth on this root will be stimulated. This lateral root will grow larger than other major lateral roots (Coutts and Philipson 1976; Eissenstat and Caldwell 1988).

Roots cannot "sense" that an area of superior nutrient or water content is nearby; instead, they grow into it by random chance. This process can lead to an unevenly distributed root system with only a few large, well-developed laterals. Providing a uniform soil environment on a planting site may help encourage a more uniformly distributed root system. It is not known for certain if this will increase tree health and vigor or enhance transplant survivability or posttransplant growth, but a well-distributed root system surely cannot harm the tree.

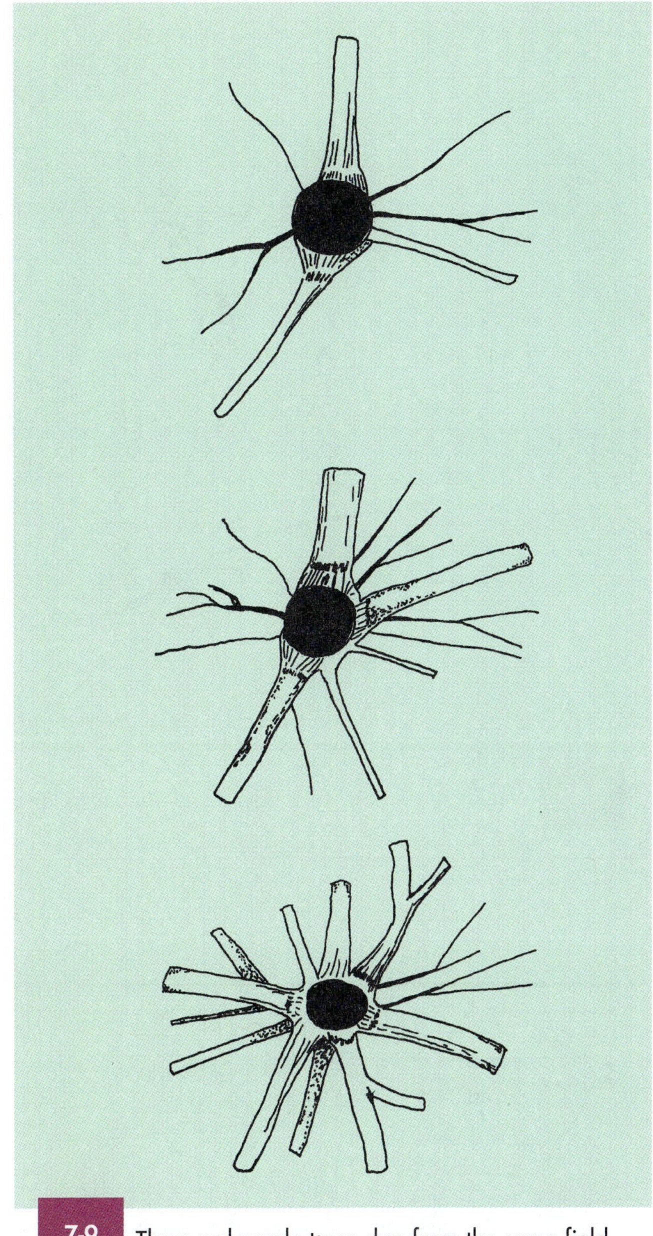

7-9 Three red maple trees dug from the same field show that root systems vary widely, even within the same species. The top one has two or three major lateral roots; the center one has three or four lateral roots; and the bottom one has eight or nine major lateral roots.

Fine roots

Fine roots originate from all along lateral roots (Figure 7-11). They advance outward, down, and most frequently up toward the soil surface, branching three or four times to form fans or mats terminating in hundreds to thousands of fine, short tips. They are not concentrated at the branch dripline (at the edge of the canopy) but are distributed throughout the total area covered by the root system (Hermann 1977; Reynolds 1974). They often die back to main lateral roots in drought, but new ones emerge within several days of the next rainfall. Fungi often infect the roots forming mycorrhizae, which essentially replace the smallest diameter roots called *root hairs*. Mycorrhizae are essential to the survival of many trees.

At least half the fine roots on a tree or shrub are beneath the branch canopy. Fine root density on young trees decreases with distance from the trunk (Gilman 1990b). Because fine roots originate largely from lateral roots, they are generally located within the top eight inches of soil. Some fine roots emerge from deeper roots but their development is often limited. Most absorption of water and minerals takes place through the fine roots or through mycorrhizae,

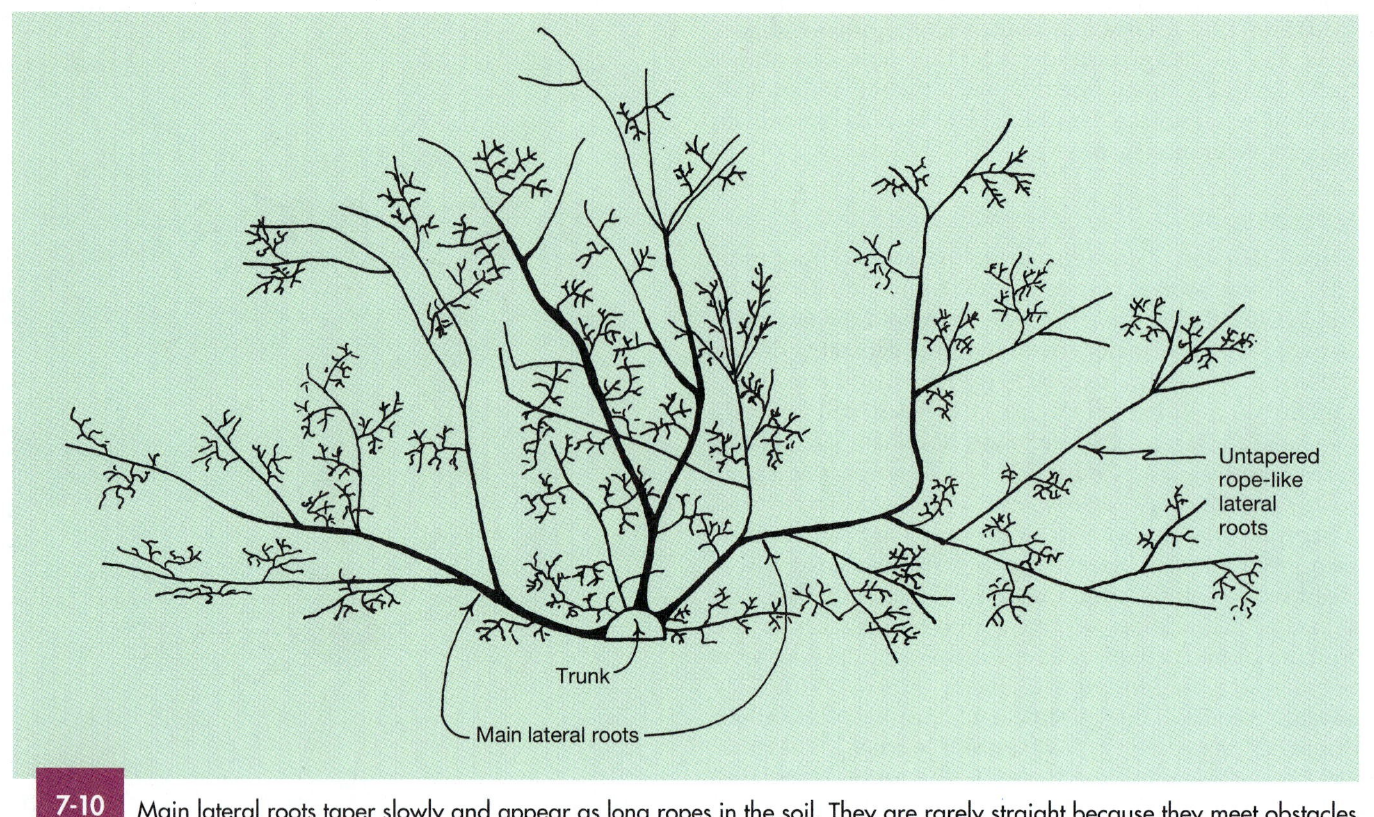

7-10 Main lateral roots taper slowly and appear as long ropes in the soil. They are rarely straight because they meet obstacles in the soil.

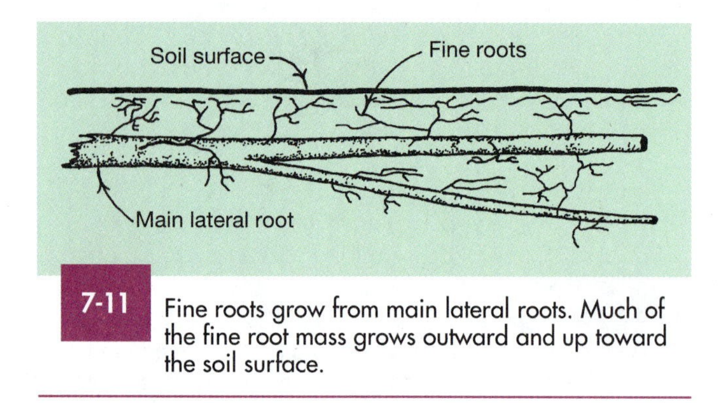

7-11 Fine roots grow from main lateral roots. Much of the fine root mass grows outward and up toward the soil surface.

because they represent a large portion of total root surface area (Jensen and Petterson 1977).

ROOT DEVELOPMENT ON OLDER TREES

Deep roots

As trees grow older, deep roots constitute a decreasing portion of the total root system, even if a tap root was dominant in the seedling and sapling stages (Figure 7-12) (McMinn 1963). Deep roots still function in support and water absorption. They are usually located within the

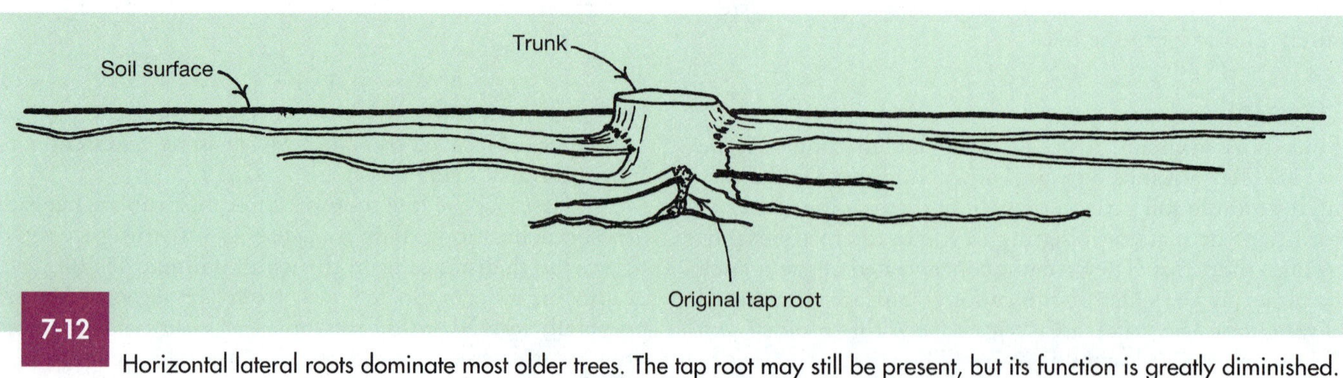

7-12 Horizontal lateral roots dominate most older trees. The tap root may still be present, but its function is greatly diminished.

branch dripline (under the canopy), with a concentration directly beneath or close to the trunk. Tap roots are frequently unrecognizable or very small in relation to the rest of the root system.

Lateral roots

Lateral roots make up a major portion of the root system. They are more or less evenly distributed around the trunk in sandy, well-drained soil, although there are many exceptions. Because lateral roots emerge from the trunk and root system close to the soil surface, an easily recognizable root flare develops as a pronounced swelling at the base of the trunk. This is thought to form in response to wind movement of the trunk. Lateral roots play an important role in keeping large trees erect.

The 4 to 11 major lateral roots are oval in cross-section close to the trunk because they enlarge vertically more than horizontally (Wilson 1964). They taper rapidly and branch into numerous one-half- to three-inch-diameter round, long, rope-like roots with many side branches. Side branches also grow to be long and untapered and branch several times into a network of fine roots (Figure 7-10). Lateral roots grow away from the trunk, generally far out beyond the edge of the branches. Roots frequently change direction, apparently in response to meeting obstacles and roots of adjacent trees and shrubs.

Sinker roots

Sinker roots originate from the lower side of lateral roots. Soil conditions permitting, they grow straight down for three to six feet or more (Figure 7-13). Most are located within the branch dripline. They are smaller in diameter than lateral roots, branch infrequently, and initiate only a small number of finer roots. They may function in water uptake and help to anchor the plant in the soil. In compacted or other soil with a low oxygen content, sinker roots emerge and grow parallel to the soil surface on top of the adverse soil layer (Perry 1982). In trees not tolerant of flooding, vertical roots are killed in water-logged soil, making them susceptible to windthrow.

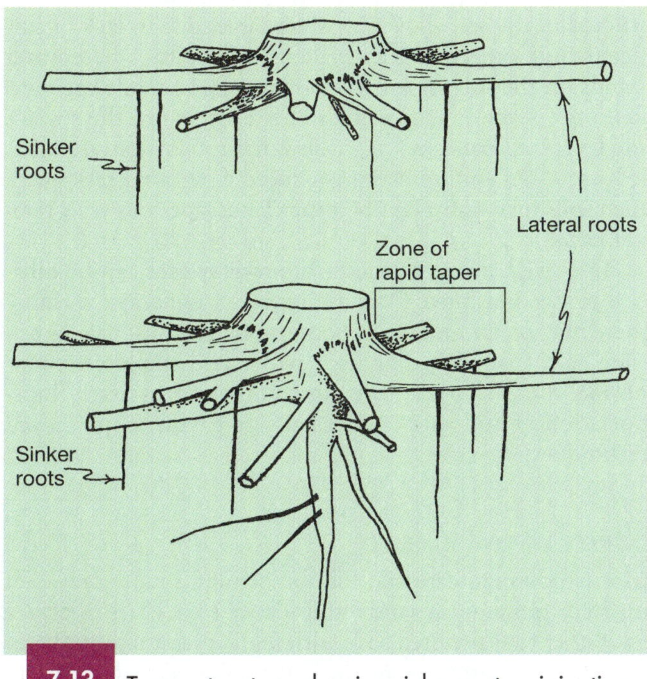

7-13 Two root systems showing sinker roots originating from the lower side of lateral roots. More than 75 percent of trees have this type of root system.

Fine roots

Fine root density on large trees is greater than that on sapling-aged trees (Gholz, Hendry, and Cropper 1985; McMinn 1963). Root density may be greater near the surface in the older tree than on the sapling if the natural organic layer or mulch on the soil surface is left relatively undisturbed and there is little competition from nearby plants. Root density is usually greater beneath the canopy than beyond (Gilman 1990a).

ROOT SPREAD

Numerous studies in forests, orchards, and tree nurseries clearly indicate that roots spread well beyond the dripline; they are not concentrated at the dripline (refer back to Figure 7-1). Studies on oak, hickory, maple, ash, magnolia, pear, apple, juniper, and other trees show that the ratio of root spread to branch spread on nursery-grown trees depends on the tree. Despite variability, the average root spread is about three times the branch spread (Gilman 1988b and 1988d; Stout 1956; Watson and Himelick 1982). Maximum root spread ranges from about 1.5 times the dripline for open-grown green ash to 3.7 for southern magnolia. Roots extending farthest from the trunk are usually found near the soil surface. The horizontal extent of roots is reached at a certain tree age. This implies that the root spread-to-branch spread ratio may decrease as trees age.

The ratio of root spread to branch spread appears to be related to the shape of the crown. For example, roots of the narrow, columnar-shaped 'Torulosa' juniper (*J. chinensis*) spread to about three times the branch spread, whereas root tips of the spreading 'Hetzii' juniper were only one-and-one-half times as far from the trunk as were the branch tips (Figure 7-14). Root spread was identical on both trees (Gilman 1990a). Frequently, tree roots extend to encompass a roughly circular area four to seven times the area beneath the branches. The diameter of this area may be one, two, or more times the height of the tree or shrub. Less than half of root system length is located beneath the branches (Figure 7-15) (Gilman 1988b and 1988d).

ROOT DEPTH

Ninety-nine percent of the tree root system is located in the top three feet of soil (Figures 7-1, 7-16) (Coile 1937; Gale and Grigal 1987), although there are a few examples of much deeper penetration, particularly in the southwestern United States and other areas where soils are loose and

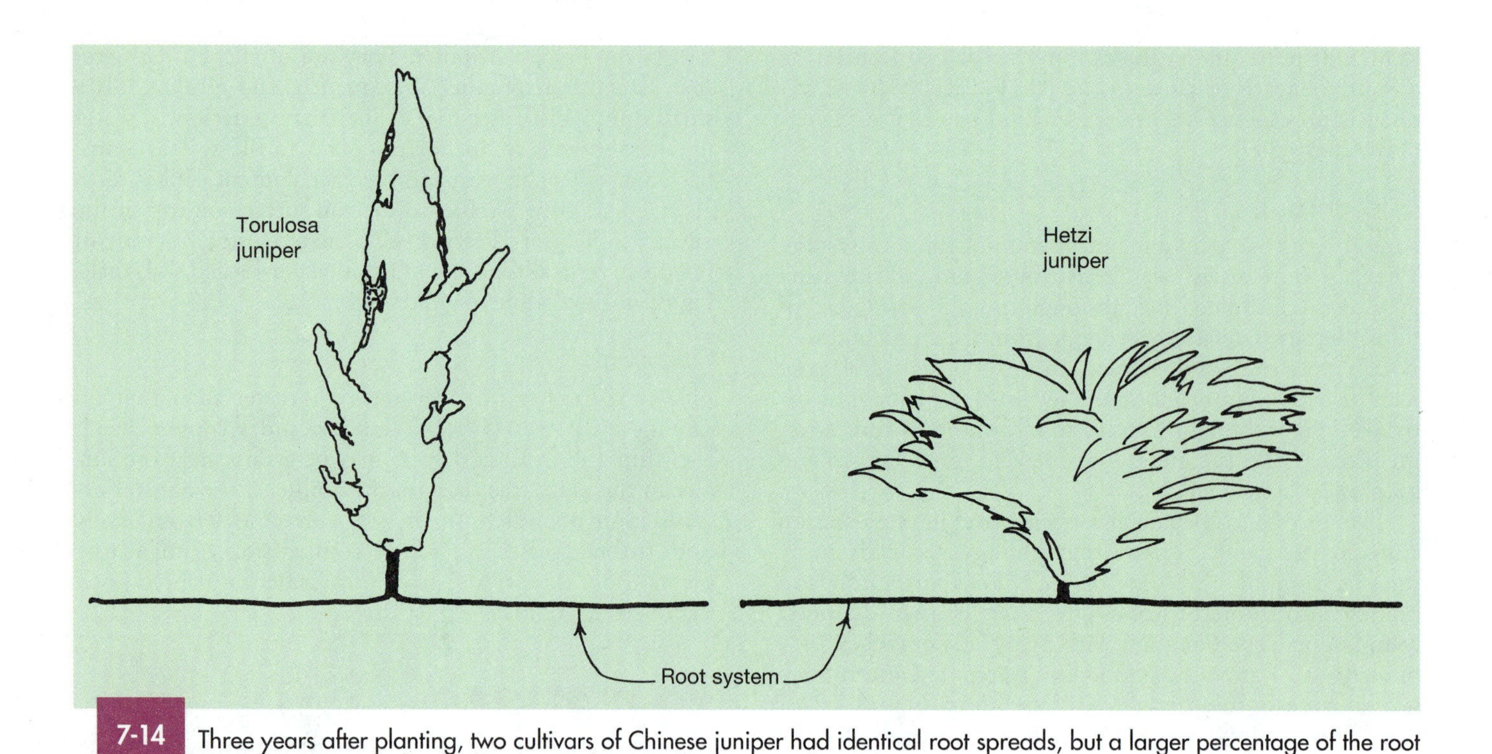

7-14 Three years after planting, two cultivars of Chinese juniper had identical root spreads, but a larger percentage of the root system was outside the dripline on Torulosa juniper because the canopy is very narrow.

7-15 If soil were removed from around the root system, roots could be seen growing well beyond the edge of the canopy. More than half of the root system is beyond the branch tips. The dashed line represents the edge of the canopy.

deep. Maximum rooting depth is usually established during the seedling or early sapling stage for trees in the forest (Lyr and Hofman 1967), and probably within the first several years after planting for transplanted trees (Carlson, Preisig, and Promnitz 1980). Most fine roots of nursery-grown trees are in the top 18 inches of soil, even on well-drained, sandy soils. The root system is usually shallower at greater distances from the trunk (Figure 7-1). Fine roots are typically found growing in the leaf litter in the forest, among turf roots, and in and directly beneath organic mulch in the landscape.

Effects of water table

Except for baldcypress and some other wet-site-tolerant trees, root development in soils with a high water table is restricted to the soil above the water table, due to lack of sufficient soil oxygen at greater depths. On sites where water stands on the surface for several days after a moderate rainfall, root development will also be restricted to the surface soil layers. A "pancake" or shallow, flat root system develops (Figure 7-17a and b). Plants produced in field nurseries with such soils may transplant best into landscape soils with poor drainage.

Trees with a flat, pancake-type root system can be difficult to dig and move from a nursery or landscape, because there may be no roots in the bottom of a traditional-shaped root ball. Consider digging a wider and shallower-than-normal root ball to capture more of the root system. Transportation of odd-shaped root balls may require specialized techniques.

Effects of soil texture

Most studies indicate that root branching is more prevalent on trees growing in sandy and loamy soils than in clayey and other fine-textured soils, although at least one study reports more branching in the finer-textured soil (Pan and Bassuk 1985). The root system in sandy soil has shallow major lateral roots with vertically descending sinker roots close to the trunk. A tap root may be present if the tree has the capability to generate one.

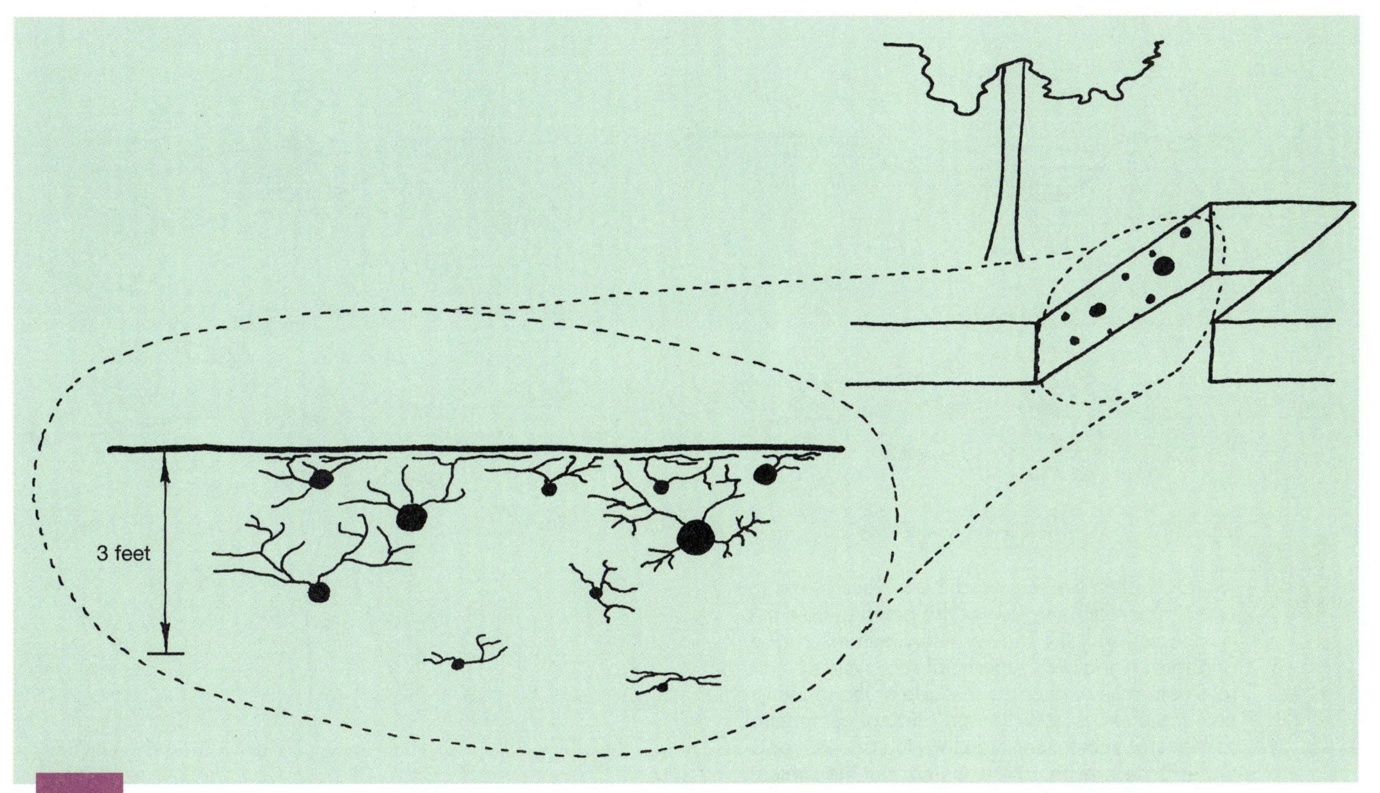

7-16 A trench cut in the soil at the dripline shows that most of the root system is located within the top three feet of soil. Fine roots can be seen growing from main lateral roots, which are represented by the dark circles.

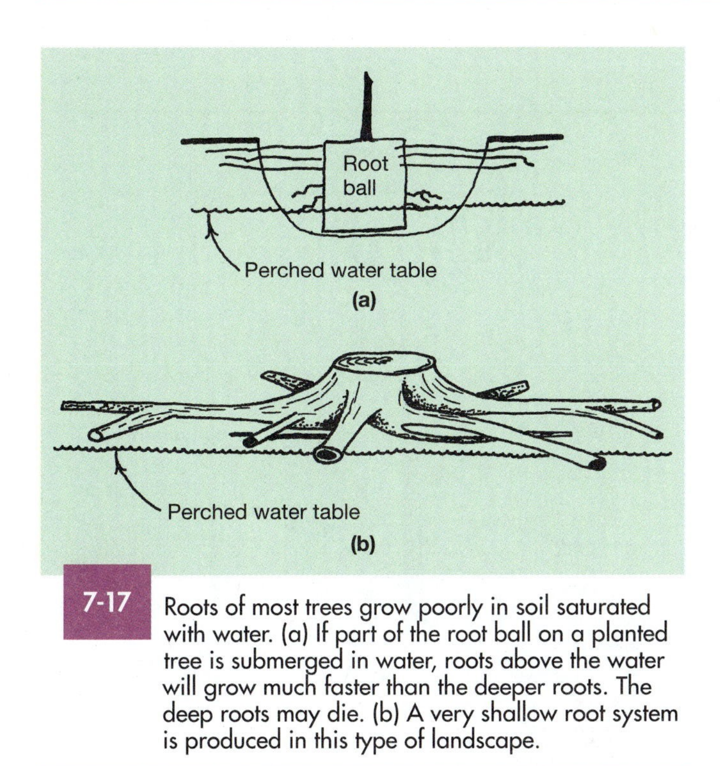

7-17 Roots of most trees grow poorly in soil saturated with water. (a) If part of the root ball on a planted tree is submerged in water, roots above the water will grow much faster than the deeper roots. The deep roots may die. (b) A very shallow root system is produced in this type of landscape.

In some clayey soils, several major lateral roots slope down and grow into the subsoil until they reach a zone where oxygen is limiting. The main lateral root system is deeper in well-aerated, heavier soils than in lighter, sandy soils (Rogers and Vyvyan 1934). In a heavy clay soil with low oxygen content or poor drainage, root systems can resemble those produced in a soil with a high water table (Figure 7-17). Many clayey soils in urban areas drain poorly and are compacted, so root systems often resemble those in Figure 7-17b. If cracks penetrate deep into heavy, clayey soil, oxygen can reach deeper profiles, so some roots on species capable of producing deep roots may grow down through or near the cracks (Figure 7-18). Roots often appear contorted and twisted because soil cracks are rarely straight.

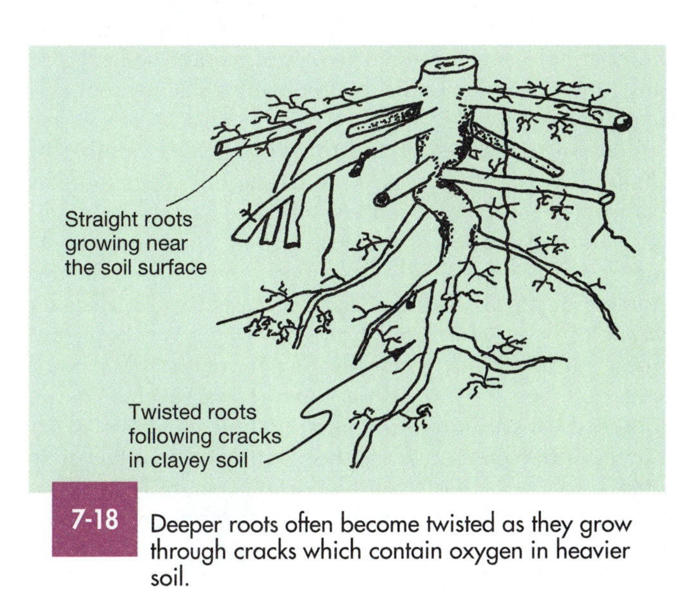

7-18 Deeper roots often become twisted as they grow through cracks which contain oxygen in heavier soil.

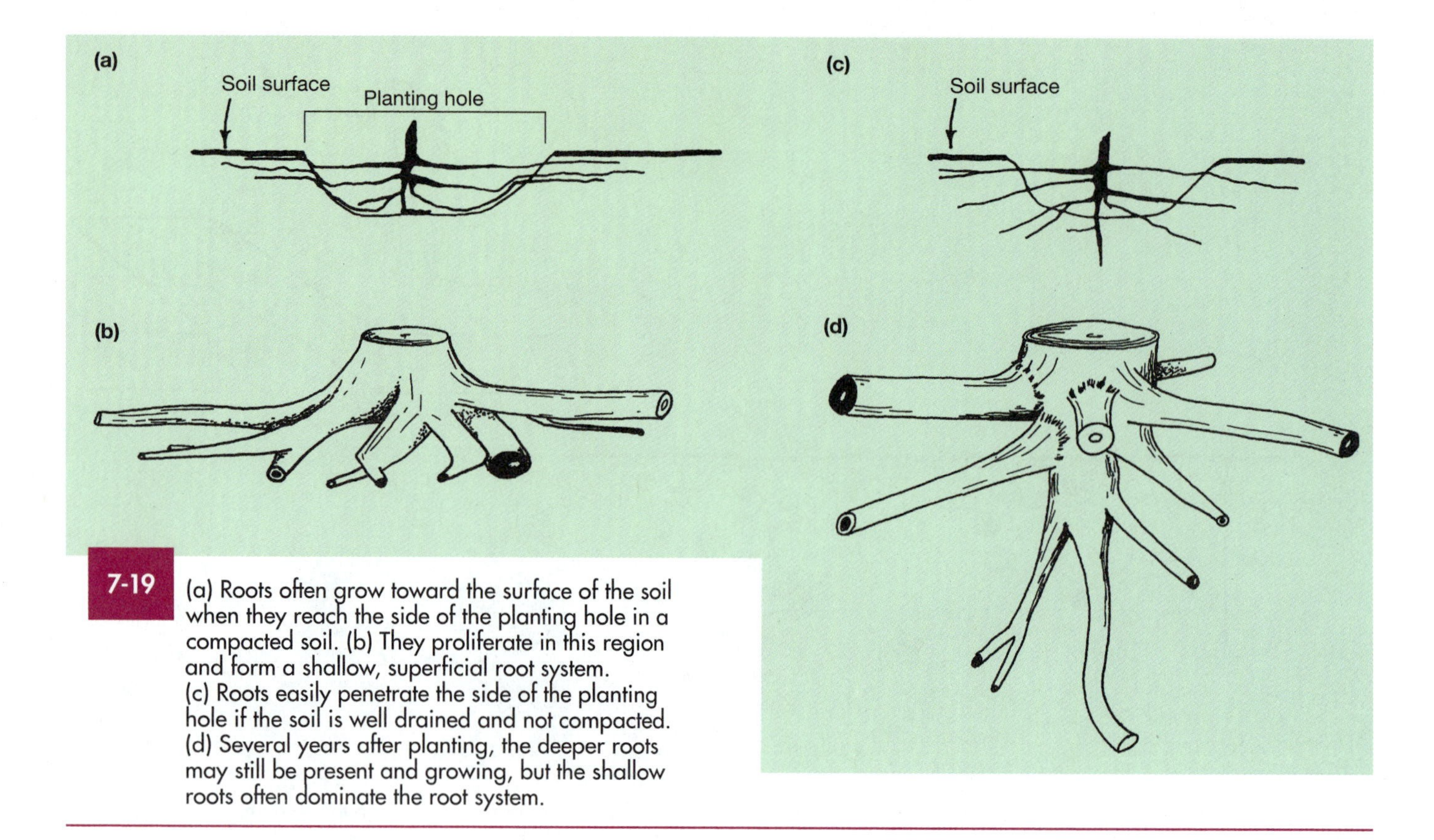

7-19 (a) Roots often grow toward the surface of the soil when they reach the side of the planting hole in a compacted soil. (b) They proliferate in this region and form a shallow, superficial root system. (c) Roots easily penetrate the side of the planting hole if the soil is well drained and not compacted. (d) Several years after planting, the deeper roots may still be present and growing, but the shallow roots often dominate the root system.

Effects of soil compaction

Compacted soil is common in many new landscapes. In compacted soil, root growth is restricted by low soil oxygen supply and physical impedance to root penetration. It is physically difficult for roots to push through compacted soil. Trees and shrubs respond to compaction by producing a shallow root system (Gilman, Leone, and Flower 1987; Ziza, Halverson, and Stout 1980). Roots on recently transplanted trees in compacted soil are redirected up toward the soil surface from deeper layers when they meet the side of the original planting hole (Figure 7-19a–d). Whereas a number of trees can survive periods of reduced soil oxygen, only those capable of quickly producing a shallow root system grow well.

No tap root develops and there is a greater number of shallow roots and increased branching of deeper roots in compacted soils. These shallow roots are longer, straighter, and larger in diameter than deeper roots, indicating vigorous growth (Figure 7-18). Only roots growing from the top portion of root balls or those reaching the soil surface by virtue of being redirected from the deeper soil layers are likely to develop as a significant part of the permanent structural root system. Roots often spiral around in the loosened backfill soil in the planting hole until they reach the soil surface and can leave the planting hole, or they may grow into cracks in the compacted soil (Figure 7-20). This could cause trunk girdling later when the trunk meets these circling roots. A tree with a spiraling root system could become unstable as it grows.

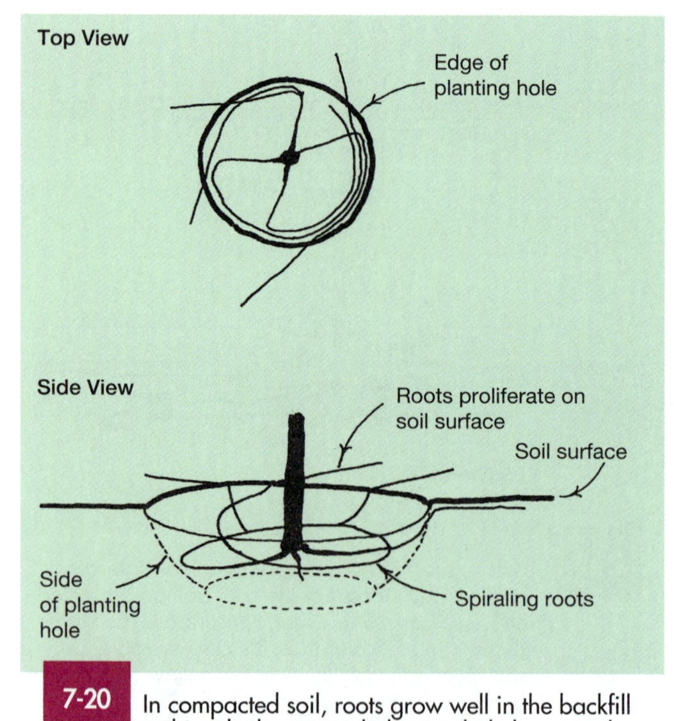

7-20 In compacted soil, roots grow well in the backfill soil inside the original planting hole because the soil is loose. Main roots reaching the sides of the holes are often deflected and can spiral around the planting hole. These circling roots could eventually slow growth or girdle the tree as the trunk grows and meets the root.

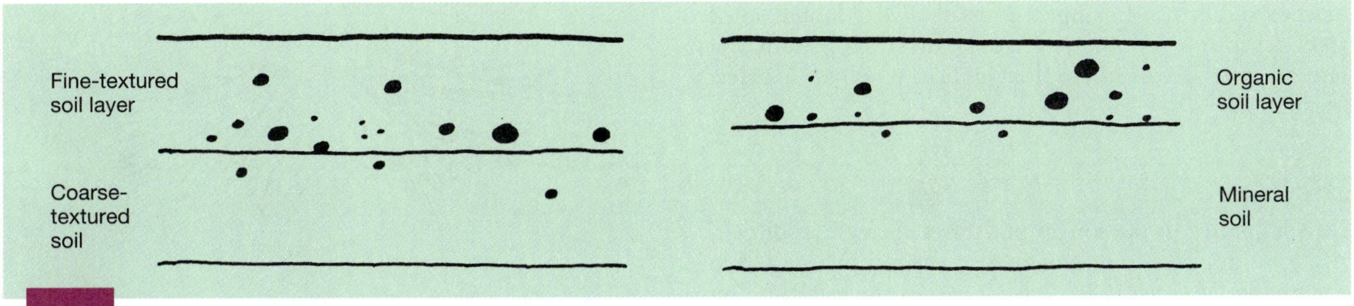

7-21 Roots are often concentrated just above the boundary between two soil types. Roots occasionally penetrate the boundary and grow in the lower soil layer.

Many urban sites have loose top soil with a compacted soil layer underneath. Root growth in this situation is fairly predictable. The root system will be predominantly located in the layer of loose soil on top of the compacted layer (Figure 7-4).

Effects of soil layering

When a fine-textured soil layer is placed over a coarse-textured soil, water movement from the fine soil into the coarse soil below is slow (Harris 1992). This can have a dramatic impact on water and root penetration into the soil. Roots often concentrate just above the boundary between the two soil types (Figure 7-21).

Species differences

Natural rooting depth varies among tree species, but the influences of soil, cultural, and environmental conditions previously mentioned often override the tree's natural rooting ability (Figure 7-22a–c). Seeded-in-place Monterey pine and other trees growing in a well-drained soil develop descending roots as well as shallow root systems (Somerville 1979). Red maple (and other trees) produce mostly shallow roots with few deep roots, even in well-drained, sandy soil. However, if pines and maples were grown in the same high-water-table or compacted soil, the root systems would be very similar. Descending roots would be aborted, twisted, or deflected by the water table or low soil-oxygen content,

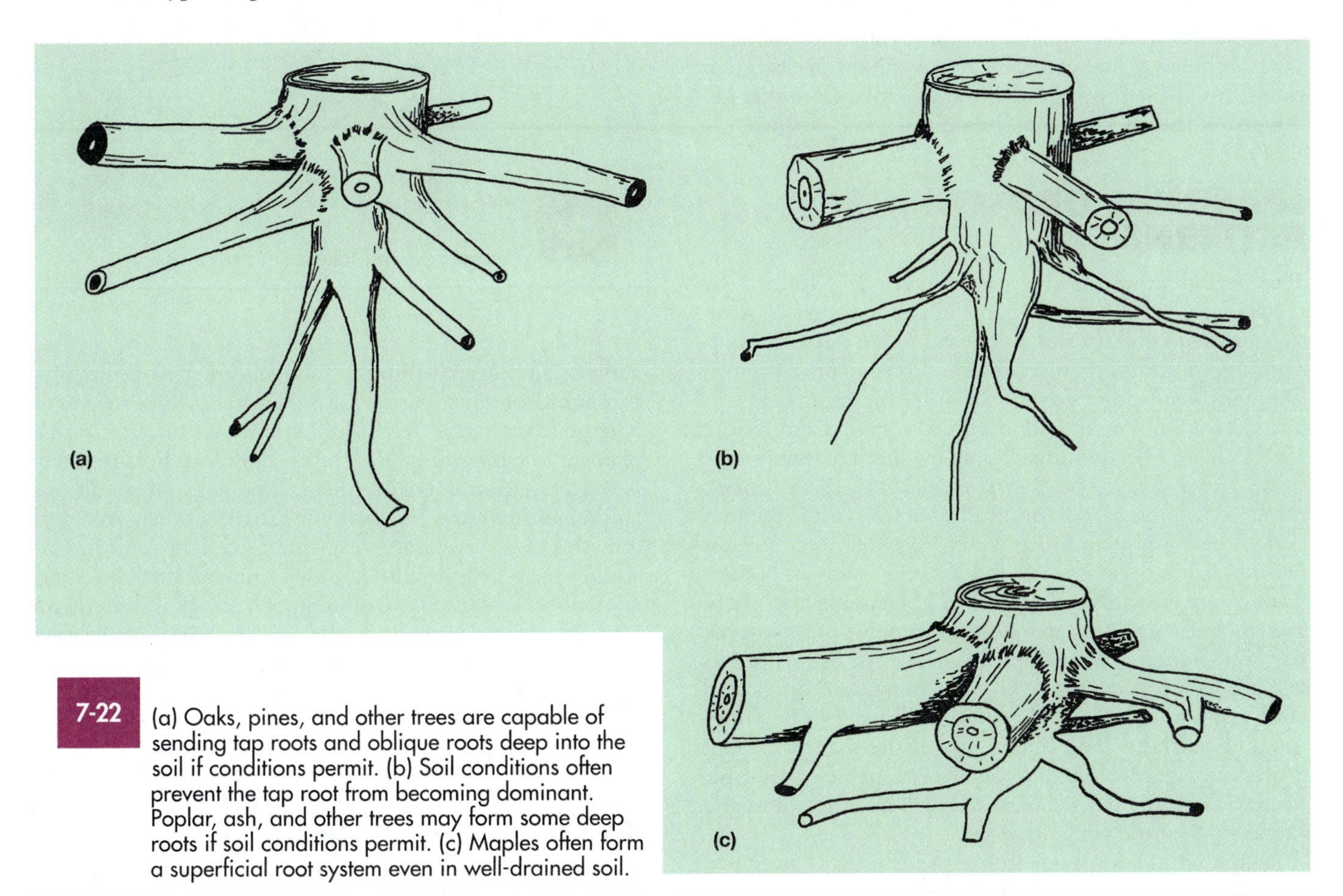

7-22 (a) Oaks, pines, and other trees are capable of sending tap roots and oblique roots deep into the soil if conditions permit. (b) Soil conditions often prevent the tap root from becoming dominant. Poplar, ash, and other trees may form some deep roots if soil conditions permit. (c) Maples often form a superficial root system even in well-drained soil.

or they might not be apparent at all. Even though some trees can produce some deep roots, most of the root system on these trees is still located in the top two or three feet of soil.

Effects of mulch

Mulch applied to the soil around a tree increases root density by stimulating root growth near the soil surface and in the mulch (Watson 1988). This increases the total amount of roots on the tree and may enhance drought resistance. Root growth deeper in the soil appears to be unaffected by a layer of mulch.

ROOT FORM VARIATION WITH SPECIES

Palm trees, young crape myrtle, and probably other trees have many small-diameter roots originating from the base of the trunk and no tap root (Figure 7-23). Oaks and pines have a coarser root structure, with several larger-diameter roots dominating the permanent root system. Young maples have many small-diameter roots, making them easy to transplant. Many roots on screw pine emerge from above the ground, eventually reaching the soil surface and rooting into the ground. Several publications describe root structure, depth, and spread on many tree species (Burns and Honkala 1990a and 1990b), but remember that most of these studies were conducted on undisturbed forest land. We now know (as the preceding discussion made clear) that root form in the forest bears little resemblance to that developed in the urban or suburban landscape.

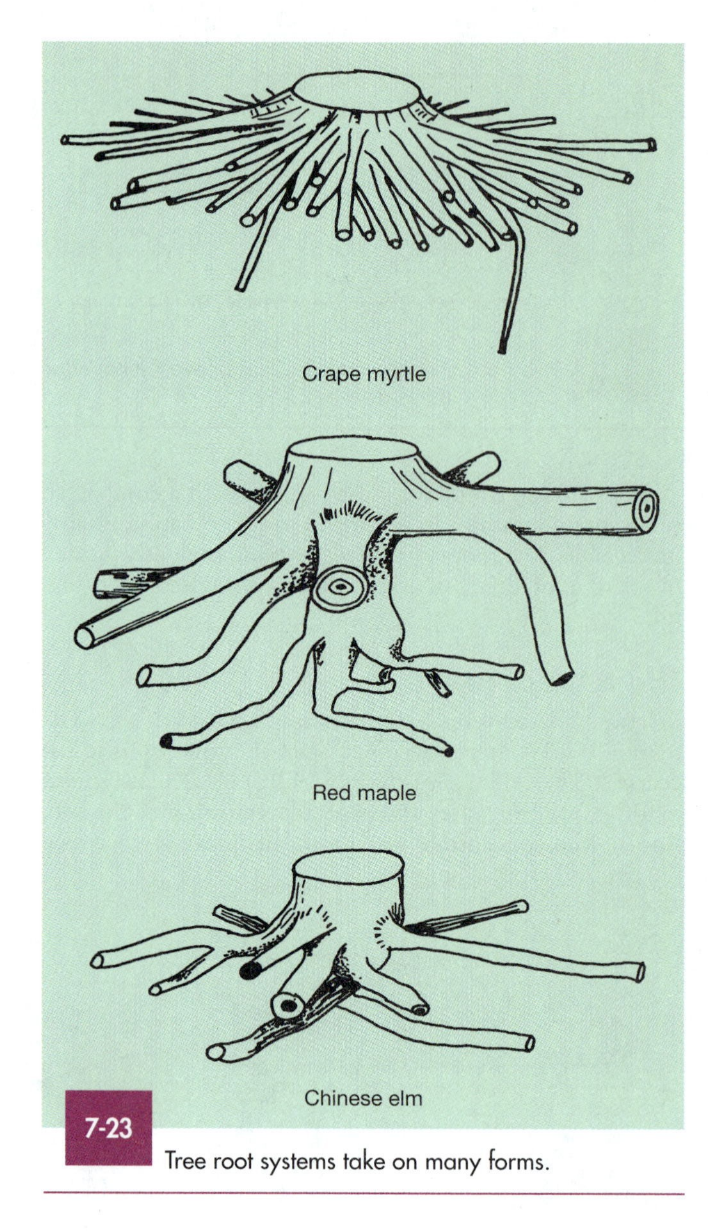

7-23 Tree root systems take on many forms.

SEASONAL GROWTH PATTERNS

The period of maximum root growth is governed by species, stage of shoot development, soil moisture content, soil nutrient content, soil temperature, time of year, root-pruning, top pruning, tree age, and probably other factors. On many plants, root growth begins in the spring before or just after shoot growth (Atkinson 1980; Harris and Bassuk 1994). In temperate climates, many species, such as the stone fruits, apple, blueberry (*Vaccinium corymbosum*), raspberry (*Rubus idaeus*), and some oaks, have a maximum period of root growth in April and May, followed by a rest period during shoot growth and leaf expansion and then a second root growth period in the fall, beginning after shoot growth ends. Some research does not support this two-peak pattern, though. For example, root growth begins just as leaf expansion is complete on pear, plum, and birch, and ends just before leaf-fall (Lyr and Hoffman 1967; Ovington and Murray 1968). Shoot growth alternates with root growth in vigorous sapling-sized trees capable of multiple flushes. On some species, however, such as citrus, roots and shoots grow at the same time.

Mugho pine roots grew from April through November, with peak activity in the summer. Scots pine had highest activity from March through August, earlier than the July-through-September period for Sitka spruce. Roots began to emerge from fringe tree in July, seven months after transplanting the preceding November. However, because there is wide variation in environmental and cultural conditions among research sites, some of this variation in reported root growth activity may not reflect true species differences. Few studies show a peak in root activity during the winter, attributable largely to low soil temperature. Peak activity on most species was usually found in the spring or summer, though occasionally in fall. Root growth seems to be uniformly retarded during active shoot growth, although one study on spruce found no reduction in root activity during the period of shoot elongation (Ford and Deans 1977).

Of all environmental conditions, soil moisture and soil temperature have the largest influence on root growth. Some estimate that these two factors account for more than three-quarters of the variability in root growth activity (Singh and Srivastava 1985). When soil temperature is not limiting (in summer), dry soil readily reduces the rate of root growth, regardless of the stage of shoot activity. However, roots are capable of quickly resuming growth several days after dry soil is rewet.

The most intense root growth occurs when soil temperatures are between 68 to 84°F, depending on species (Bevington and Castle 1985; Richardson 1956). Intense root growth during the summer is probably a result of adequate soil moisture and suitable soil temperatures. Soil temperatures in warm climates are within this optimum range longer than in cooler regions. As a result, root and shoot growth appear to be much faster in the warmest climates, because roots grow over a longer period.

In short, the period of maximum root growth varies. There appears to be too much variability among species to make general statements about the maximum root growth period, although it rarely occurs in winter or during shoot elongation.

SUMMARY

One of the most revealing studies of root systems in recent history resulted from a storm that struck Europe. Thousands of trees were uprooted from clayey and gravelly soil and hundreds were studied (Cutler 1990). Fifty-three percent of the root systems were described as having lateral roots with sinkers (Figures 7-1, 7-13); 27 percent had only lateral roots (Figure 7-17b); 12 percent had mostly oblique roots (Figure 7-8a); 2.4 percent had lateral roots and a tap root (Figure 7-22d); 2 percent had only a tap root, and 2 percent had a double-layered root system (Figure 7-4).

Chapter 8

Fertilizing Trees

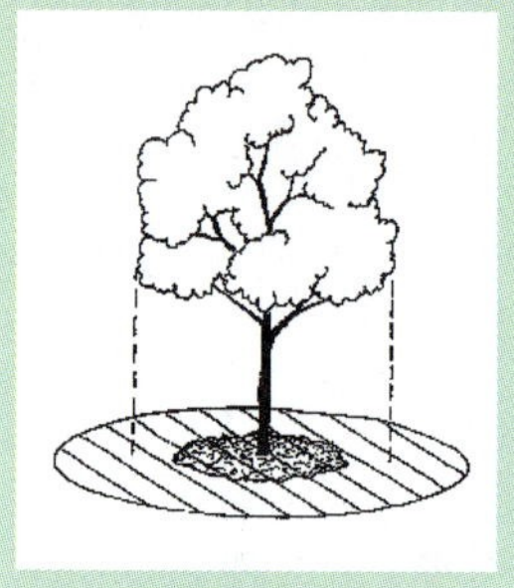

INTRODUCTION

Additional fertilizer is usually not necessary for established trees growing in or near lawns, ground covers, and shrub beds that receive regular fertilizer applications. Their root systems extend throughout the lawn and landscape and receive essential elements when these areas are fertilized. In contrast, trees growing in confined soil spaces such as parking lot islands can benefit from a regular fertilization program. Fertilizing some urban trees in confined spaces can be difficult if there is little open soil on which to apply fertilizer, so this is best done by an arborist or urban forester.

Occasionally, the need may arise, for some species, to apply specific formulations in addition to what the lawn and landscape are receiving. For example, in south Florida, supplemental applications are needed beneath the canopy for some trees, especially palms, or nutrient deficiencies can develop. The same is true in cooler climates for trees like pin oak, maples, and some other species. These trees often show symptoms typical of manganese or iron deficiency when improperly planted in alkaline soil. Supplemental nitrogen fertilizer may also be applied to young trees to encourage faster growth. Recently transplanted trees usually do not need to be fertilized in order to survive and grow.

WHEN TO APPLY FERTILIZER

According to scientific studies conducted during the last several decades on a variety of soil types, there is no best time of year to fertilize (Neely, Himelick, and Crowley 1970). Unfortunately, you might read differently in popular literature. If one season gives a better tree response than another, the difference is usually negligible. Fertilizing in the fall does not help "winterize" a tree, as advertized on some fertilizer tags.

For applications made in late summer to early fall, use an insoluble, slow-release fertilizer. A soluble fertilizer applied at this time, combined with unusually abundant rainfall and warm weather, could stimulate a late-season flush of growth or delay acclimation on some species. Such trees could be damaged by an early freeze.

HOW AND WHERE TO APPLY FERTILIZER

Fertilizer can be broadcast over the surface of the landscape—for lawn, trees, and shrubs alike. It can be applied directly on top of the mulch; there is no need to place it below the mulch because most leaches through quickly (Gilman, Yeager, and Weigle 1990). For many people, it is easier to apply fertilizer with a rotary spreader than with a drop spreader. Apply it evenly out to about 1½ times the canopy diameter (Figure 8-1).

Because most smaller-diameter roots (commonly referred to as feeder roots) on trees are shallow (within the top 12 inches of soil), there is no need to inject or place fertilizer deep into soil. Studies consistently show that trees respond about the same to injected fertilizer treatments as they do to surface applications (Davey 1930; Gilman 1989a; van de Werken 1981). However, shallow soil injections (four to six inches deep) on mounds, berms, and

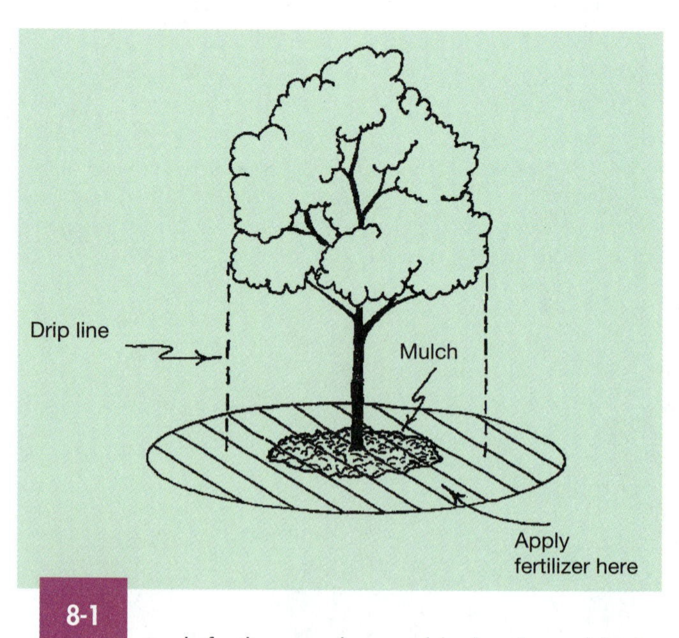

8-1 Apply fertilizer evenly on mulched and unmulched surfaces out to about 1.5 times the canopy diameter.

slopes, in compacted soil, and in other areas where runoff is likely, could reduce the amount of fertilizer washing off the landscape. To minimize the amount of fertilizer leaching below the root zone, use a water-*insoluble* (slow-release) fertilizer when injecting into the soil.

Trunk injection of micronutrients is another method of tree fertilization best used by a professional arborist certified by the International Society of Arboriculture (Savoy, IL), and then only as a last resort when conventional fertilization is not possible or is ineffective. One such situation would be an unhealthy tree showing nutrient deficiency symptoms in a space where most of the root system is covered with asphalt. Trunk injection with macronutrients (nitrogen, potassium, and phosphorus) is rarely warranted. I cannot think of a situation in which young, establishing trees need to be injected with fertilizer. Trees are permanently damaged by trunk injections and the potential benefits must outweigh this damage. Consult an arborist or urban forester for more information.

THE BEST TYPE OF FERTILIZER

Unromantic as it may seem, plants cannot differentiate among lawn, shrub, and tree fertilizer. They utilize it all the same. It does not matter to most trees what type of fertilizer is applied, as long as it contains nitrogen. In most cases, *nitrogen is the only element that consistently causes a shoot or root growth response in trees* (Davey 1930; Neely, Himelick, and Crowley 1970; Watson 1994). Despite claims made in various popular literature, *phosphorus and potassium do not stimulate root or shoot growth* unless there is a serious deficiency in the soil. This is not common. There are cases in south Florida and some other regions where potassium appears to be deficient in the soil. Here, regular potassium applications can prevent and correct deficiency symptoms. Applying potassium in slow-release formulations should help minimize leaching, which occurs in sandy soils. Chlorotic maples responded to manganese applications by increasing shoot growth in one study in hardiness zone 5 (Messenger and Hurby 1990). If soil samples include a low level of phosphorus or potassium, apply it according to recommendations from the lab.

Fertilizers listed as organic, inorganic, synthetic, natural, composted sludge, coated, slow-release, controlled-release, granular, or liquid all give similar responses because they all contain nitrogen. Studies confirming this have been conducted in sand, loam, and clay in different parts of the country (Conover and Joiner 1974; Corley, Goodroad, and Robacker 1988; Gilman 1987; Gilman and Yeager 1990; Smith 1965; van de Werken 1981).

Selection of fertilizer to minimize leaching may have some merit, although there is little research to help select which one. Look for terms like "slow-release," "controlled-release," "sulfur coated urea," "resin coated," "plastic coated," "IBDU," "water insoluble," and "ureaformaldahyde." All these forms of fertilizer are released slowly. This is thought to be beneficial because nitrogen is available to plants over a long period of time and less nitrogen is probably leached (Yeager et al. 1993). You will pay more for these types of fertilizers and the benefits they provide compared to water-soluble materials.

Fertilizer containing water-soluble nitrogen (either granular or liquid) is less expensive, but it may leach quickly through the soil. In sandy soils, most of it may leach past the root system after only several inches of rainfall or irrigation. In more fertile marl, clay, loam, or muck soils, leaching will be slower but runoff may be greater. Apply soluble fertilizers more frequently than slow-release forms, but use less of it each time you fertilize. (For example, make two applications of one pound of soluble nitrogen per thousand square feet during the year instead of one two-pound application of water-insoluble fertilizer.)

Fertilizers containing a ratio of 3:1:3:1 nitrogen, phosphorus, potassium, magnesium provide a good program for palms. In alkaline soils, it is also a good idea to include manganese and iron at about 0.5. Manganese sulfate is often applied to the soil to prevent or correct severe iron or manganese deficiencies in soils with a pH less than 8. In more alkaline soils, it can also be applied directly to the bud at the top of the palm for a quick response.

HOW MUCH FERTILIZER TO USE

Trees generally do not respond to applications in excess of about four pounds of actual nitrogen per thousand square feet per year. However, most trees, especially those that are mature, appear to do fine with much less. If you fertilize the lawn, shrubs, and ground covers near a tree once or twice each year, the tree is probably receiving enough nutrients to remain healthy. It probably does not need additional fertilizer. Too much nitrogen invites pest problems and can disrupt the shoot to root ratio. However, many palms perform best with three or four applications each year during their life.

Each time you fertilize, apply a maximum of one pound of "actual" nitrogen per thousand square feet of ground area. Up to two pounds can be applied if using slow-release material. This sounds complicated, but it is easy to calculate from the information given on every fertilizer bag.

Example: You have purchased a 10-5-5 (N-P-K) fertilizer. Divide the nitrogen (N) content (10) into 100.

$$\frac{100}{10} = 10 \text{ pounds}$$

To apply the correct amount, spread 10 pounds of 10-5-5 per 1,000 square feet of ground area. Apply it to the area indicated in Figure 8-1.

SECTION 2

Tree Use and Management

This section is meant as a reference guide for trees used in urban and suburban landscapes. Pages contain color photographs, data, and text. This should provide all the information you need to select and care for the correct tree for any landscape situation.

The following glossary defines or explains terms used in this section's listings.

GLOSSARY OF TERMS

HARDINESS ZONES

Trees are adapted to regions of the country represented by their hardiness zones. The hardiness zone map (see page 110) was developed to specify the lowest winter temperature a tree will tolerate. It is a fairly reliable guide in this respect. This section includes a range of hardiness numbers for each tree. The warmest zone, represented by the largest number in this range, is less reliable. In other words, some trees will be able to tolerate the climate in an higher zone number than indicated in the range.

A number of trees will not grow in hardiness zones 9 and 10 in Florida and other subtropical areas in the Caribbean region, but do nicely in these same zones in California. This may be due to cooler summer temperatures in California coastal areas and lower humidity. The warmest zone indicated for these trees is usually 8 or 9, indicating that they are not suited for planting in parts of Texas and Florida. A note in parentheses is included for these trees if they grow well in southern California. (e.g. see *Cedrus deodara*).

HEIGHT

The height indicated for each tree is the size expected in most urban or suburban landscapes and is to be used as a guide only. The site and climate have a large impact on the ultimate size of a tree. Trees in the forest will typically grow taller than indicated in this book.

WIDTH

The width indicated for each tree is the canopy diameter expected in most urban or suburban landscapes and is to be used as a guide only. The site and climate have a large impact on the ultimate size of a tree. Trees in the forest will typically be narrower than indicated in this book due to competition for light and space from adjacent trees.

FRUIT

Causes litter—The fruits (e.g., Persimmon, Cornelian Cherry) create a mess when they fall.

No significant litter—Fruit may develop but will not usually cause enough of a mess to be a safety hazard.

Hard—Hard enough for a person to roll on if stepped on.

Fleshy—Soft and mushy when stepped on or as the fruit rots.

GROWTH RATE

A slow rate of growth is usually less than 12 inches per year. Trees growing 12 to 24 inches each year are considered to have a moderate growth rate. Trees rated as rapid-growing usually grow more than 24 inches each year.

Life span—A tree with a short life span will usually live no longer than 25 to 30 years in an urban or suburban landscape. One with a moderate life span usually lives no longer than about 50 years. A long-lived tree will usually live for longer than 50 years in urban or suburban landscapes.

HABIT

Canopy Density

Open—A branch canopy that is open and allows filtered light to penetrate to the ground below (e.g., pine).

Moderate density—This branch canopy has a medium density compared to other trees (e.g., many maples).

Dense—A branch canopy that allows little light to penetrate and reach the ground below (e.g., Callery Pear).

Canopy Symmetry

Symmetrical—A branch canopy with a regular or smooth outline; individuals of the species or cultivar have more or less identical and predictable forms. Many cultivars, such as Bradford Callery Pear, have symmetrical canopies.

Irregular—A branch canopy with an irregular or uneven outline. Individuals of the species or cultivar do not have identical shapes.

EXPOSURE/CULTURE

Light Requirement

Full sun—Tree requires full sun to develop to its greatest potential.

Part shade—Tree grows well with 2 to 5 hours of sun, or in filtered sun.

Shade—Tree grows well with less than 2 hours of sun.

Soil Tolerances

Acidic—Tree grows in soil with a pH of less than 7.0.

Slightly alkaline—Tree grows in soil with a pH between 7.0 and 7.5.

Alkaline—Tree grows in soil with a pH greater than 7.5. However, some of these trees become chlorotic or grow poorly at soil pH above about 8.2.

Occasionally wet—Tree grows in soil that might become saturated for several days during the growing season.

Wet—Tree grows in soil that is saturated more than several days during the growing season.

Drought tolerant—Tree needs no irrigation once it is well established.

Moderately drought tolerant—Tree could benefit from or require occasional irrigation during extended dry periods, especially if this occurs in the warm season.

Aerosol Salt Tolerance

Trees are rated as to the level of salt exposure they can survive. Although trees may survive and grow with a certain level of salt exposure, their canopies may become deformed.

High salt tolerance—Tolerant of direct exposure to salt spray; typical of trees along the first dune of a beach or next to a heavily salted roadway.

Moderate salt tolerance—Tolerant of moderate exposure to aerosol salt; typical of the second or third dune back from the beach.

Low salt tolerance—Tolerant to some aerosol salt exposure.

Salt-sensitive—Not tolerant of salt exposure.

PEST PROBLEMS

The word pest is used generically for problems caused by diseases, insects, and mites.

Pest-free—Tree is generally free of pests.

Resistant—Tree may get pests, but none are usually serious enough to warrant control measures.

Sensitive—Tree is sensitive to at least one serious pest or disease that can affect tree health or aesthetics. This pest or disease could be restricted to one region of the country, or it could be spread over a large geographical area.

BARK/STEMS

Drooping branches—Branches droop as the tree grows and will require pruning for vehicular or pedestrian clearance beneath. The branches on Live Oak, Wax Myrtle, Weeping Willow, Tree Lilac, most maples, and many other trees droop as the branches increase in size and weight. Upright trees like American Elm and columnar-shaped plants have few low, drooping branches.

Multiple-trunk or multitrunked—A tree routinely grown or trainable to grow, with more than one trunk.

Showy trunk—The trunk has prominent multiple colors, is unusually smooth or textured, or is otherwise outstanding and ornamental.

Thin bark—Trees (such as Red Maple) with thin bark are easily damaged by mechanical injury. The green cambium can often be exposed easily with a fingernail.

Thorns—Thorns are present on the trunk or branches.

Trainable to a single trunk—A tree that typically grows with multiple trunks but can be trained to a single trunk in the nursery (e.g., Crape Myrtle, Amur Maple).

Upright branch habit—Branches grow mostly upright and do not droop, so they will not require regular pruning for pedestrian or vehicular clearance beneath.

Pruning Requirement

Little pruning required—A tree that needs little pruning to develop a strong trunk and branch structure (e.g., Sycamore and many conifers).

Requires pruning for strong structure—Pruning is usually needed, especially when the tree is less than 15 years old, to develop strong, durable structure.

Limb Breakage

Resists breakage—The tree has strong (e.g., many oaks) or flexible (e.g., Gumbo Limbo) wood.

Susceptible to breakage—Branches are susceptible to breakage either at the crotch, due to formation of included bark, or the wood itself is brittle and tends to break easily.

Current Year Twig

Twig color—This usually indicates the color of the twig at the beginning or middle of the growing season. Color may change as the growing season progresses.

Roots

Invasive roots—Roots that are very aggressive. Large-diameter surface roots often develop which can lift walks, curbs, and driveways. They also can interfere with mowing the lawn.

LANDSCAPE NOTES

Origin

This indicates where the tree comes from, that is, where it is native.

Usage

Above-ground planter—Usually a small tree. Trees in planters and containers need regular irrigation and fertilization and require frequent replacement.

Bonsai—Trees that can be trained into a bonsai.

Espalier—Trees (such as Southern Magnolia or Apple) that can be trained to grow flat against a wall or trellis.

Fruit tree—A tree (such as Loquat or Apple) that may be planted for its edible fruit.

Hedge—These trees (such as American Hornbeam, Stoppers, and Orange Jessamine) can be clipped regularly and make respectable hedges if well maintained.

Highway—A tree suited for planting along an interstate highway, interchange, or other low-maintenance area that receives little care after planting. Trees establish quickest if planted when small.

Median or buffer strip—A tree suited for a median strip in a highway or buffer strip around the edge of a parking lot. Warning: The Department of Transportation or local government may regulate planting in these areas. Check with appropriate authorities before planting.

Not recommended for planting—Due to weak branches, invasive habit, or other recognized problems, the tree is not a good choice for planting.

Patio—These trees are useful near a deck or patio and generally do not have messy fruit. They are usually small or medium-sized trees with ornamental interest without thorns on the trunk or branches.

Parking lot island—These are usually drought-tolerant trees without messy fruit that grow or can be pruned to a more-or-less upright position.

Reclamation—Typically a native (such as Black Locust, Wax Myrtle, or Pinyon Pine) that grows well on poor sites. Some non-native trees are also suited for reclamation sites as long as they are not invasive.

Shade tree—A tree (such as an oak) with a rounded, oval, or spreading habit that usually casts significant shade.

Sidewalk cutout—These trees have been successfully planted or could be planted in sidewalk cutouts. This is a very stressful environment that tests the tolerances of trees to city conditions. These trees will not grow in all sidewalk cutouts as the soil environment beneath the sidewalk varies considerably from one site to the next. Be sure to test the soil compaction (aeration) and drainage at the planting site and choose a tree that tolerates these conditions. Consult Chapter 1 for advice on selecting the best tree for your site.

Specimen—A tree (such as Crape-myrtle, Magnolia, or a birch) that is worth planting by itself and is featured in the landscape due to one or several attractive traits, such as glossy or unusual foliage, striking habit, nice bark, interesting multiple-trunked arrangement, attractive, long-lasting fruit or flower display, or showy fall color.

Standard—A tree trained with one, straight trunk and a head of foliage growing from branches that usually originate at the top of the trunk. A tree trained in this manner should be maintained in this form. If a tree purchased as a standard is allowed to grow to a larger size, branches often break off from the tree at the top of the trunk.

Street tree—Recommended for planting along a street. Trees maturing at more than 50-feet-tall should be planted in a soil space at least 7 feet wide. If there is no sidewalk, plant at least 6 feet from the curb or sidewalk, preferably farther if space is available. Some of these trees may not be suited for planting close to the street where pedestrians walk or vehicles park, because of messy fruit or leaves, or some other characteristic.

Urban tolerant—A tree with a proven track record in many urban sites. These trees are usually tolerant of air pollution, poor drainage, compacted soil, and/or drought. They might not be tolerant of all these adversities, so check the data for the tree carefully.

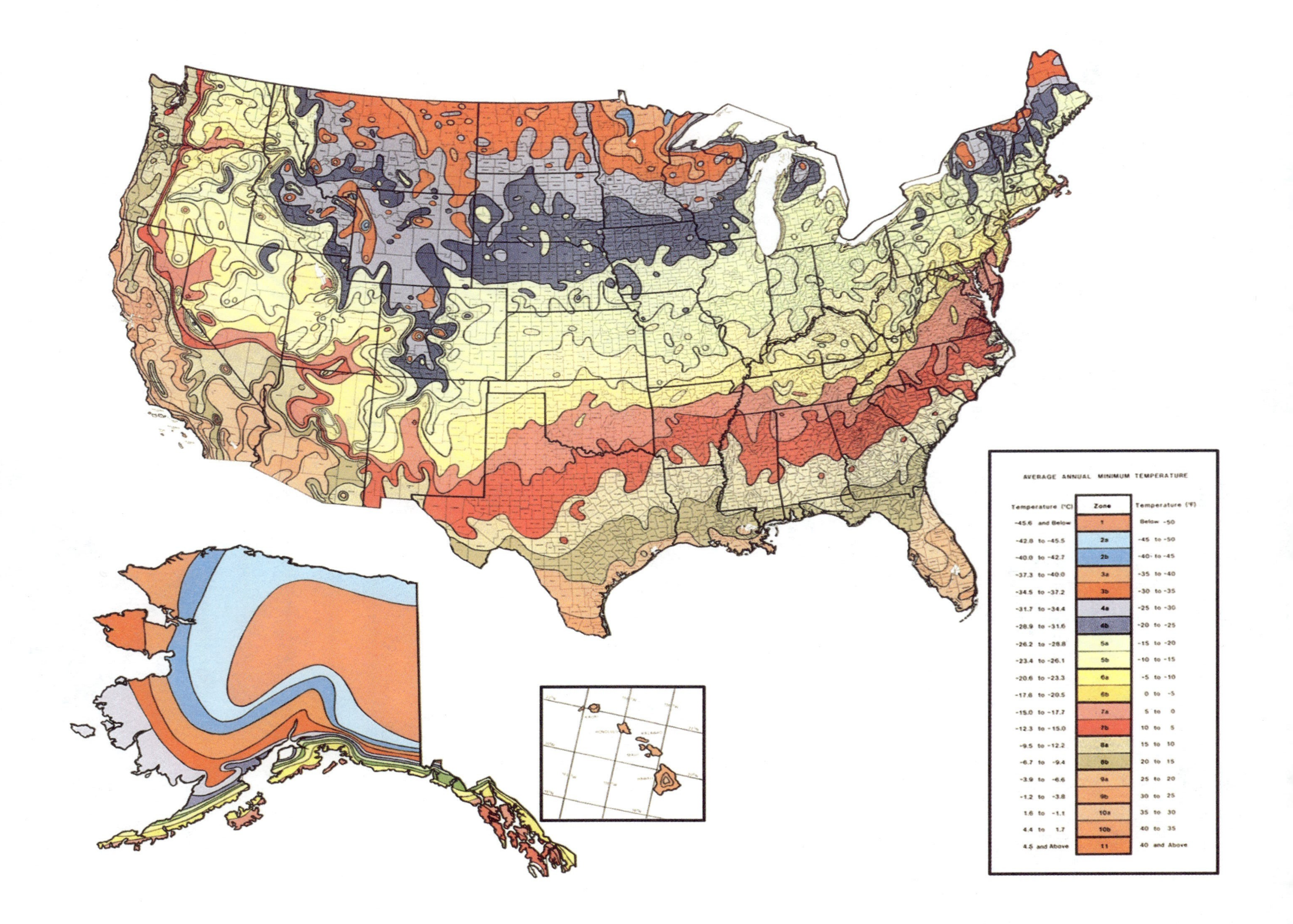
AVERAGE ANNUAL MINIMUM TEMPERATURE
Temperature (°C) Zone Temperature (°F)
-45.6 and Below 1 Below -50
-42.8 to -45.5 2a -45 to -50
-40.0 to -42.7 2b -40 to -45
-37.3 to -40.0 3a -35 to -40
-34.5 to -37.2 3b -30 to -35
-31.7 to -34.4 4a -25 to -30
-28.9 to -31.6 4b -20 to -25
-26.2 to -28.8 5a -15 to -20
-23.4 to -26.1 5b -10 to -15
-20.6 to -23.3 6a -5 to -10
-17.8 to -20.5 6b 0 to -5
-15.0 to -17.7 7a 5 to 0
-12.3 to -15.0 7b 10 to 5
-9.5 to -12.2 8a 15 to 10
-6.7 to -9.4 8b 20 to 15
-3.9 to -6.6 9a 25 to 20
-1.2 to -3.8 9b 30 to 25
1.6 to -1.1 10a 35 to 30
4.4 to 1.7 10b 40 to 35
4.5 and Above 11 40 and Above

Abies cilicica Cilician Fir

Pronunciation: AY-beez sill-LISS-ih-cah

ZONES: 6A, 6B, 7A

HEIGHT: 60 to 80 feet

WIDTH: 15 to 20 feet

FRUIT: elongated cone; 6 to 8"; dry; brown; no significant litter

GROWTH RATE: slow-growing; long-lived

HABIT: narrow pyramidal to columnar; dense; symmetrical; fine texture

FAMILY: *Pinaceae*

LEAF COLOR: green

LEAF DESCRIPTION: 1- to 2-inch-long needles

FLOWER DESCRIPTION: not showy

EXPOSURE/CULTURE:

Light Requirement: part shade to full sun

Soil Tolerances: all textures; acidic to slightly alkaline; well drained; drought

PEST PROBLEMS: resistant

BARK/STEMS:

Branches droop as tree grows and will require pruning for pedestrian clearance beneath canopy; normally grows with a single leader; no thorns

Pruning Requirement: needs little, if any, pruning

Limb Breakage: resistant

CULTIVARS AND SPECIES:

A. fraseri and *A. balsamea* are popular as Christmas trees and can be planted as landscape plants in cooler, mountainous areas. They probably look best planted in a row to form a screen or windbreak. *A. balsamea* var. *phanerolipsis* looks like Fraser Fir and grows in wetter soils than Fraser or Balsam Fir. *A. procera,* Noble Fir, a native of western Washington state and Oregon, is popular as a Christmas tree and is suited for planting in landscapes with plenty of soil space for root expansion. It is a beautiful fir well deserving of specimen use.

LANDSCAPE NOTES:

Origin: Asia Minor, Cilician Taurus Mountains

Usage: buffer strip; specimen; possible Christmas tree; screen

One of the nicest firs for the east, Cilician Fir reaches a mature height of about 100 feet in its native range, but is usually smaller in the landscape. It has horizontal branching with the lower branches drooping toward the ground. The tree should be grown in an open area so lower branches can touch the ground. This will display the soft, beautiful form of Cilician Fir. Cilician Fir may grow 1.5 feet per year with good growing conditions. This tree can take exposure and will withstand some heat and drought better than most firs.

A moist, well-drained loam provides the best growing conditions. The root system can adapt to shallow or rocky soil conditions by growing close to the surface of the soil. It will not tolerate constantly wet soil, but grows nicely in alkaline soil. It could be tried as a Christmas tree. Cilician Fir might be an excellent landscape substitute for the disease-sensitive Colorado Blue Spruce, although the foliage color is different from Blue Spruce.

Generally no insects or diseases limit growth, but a few cause some damage, including the twig aphid, bagworm, scales, and spider mites. Look for spider mites in hot weather. Diseases include needle and twig blight, pine twig blight, and cankers, but these are usually not severe. *Phytophthora* can infect and kill roots.

Abies concolor White Fir

Pronunciation: *AY-beez CON-cull-er*

ZONES: 3A, 3B, 4A, 4B, 5A, 5B, 6A, 6B, 7A, 7B

HEIGHT: 70 to 100 feet

WIDTH: 15 to 25 feet

FRUIT: elongated; upright; 3 to 6"; dry or hard; brown; not showy; no significant litter

GROWTH RATE: slow-growing; long-lived

HABIT: pyramidal to columnar; dense; symmetrical; fine texture

FAMILY: *Pinaceae*

LEAF COLOR: ranges from green to blue-gray and silver-gray

LEAF DESCRIPTION: 2-inch-long blunt needles

FLOWER DESCRIPTION: red; not showy

EXPOSURE/CULTURE:

Light Requirement: part shade to full sun

Soil Tolerances: all textures; acidic to slightly alkaline; drought; some salt

PEST PROBLEMS: resistant

BARK/STEMS:

Branches droop as tree grows and will require pruning for pedestrian clearance beneath canopy; should be grown with a single leader; no thorns

Pruning Requirement: needs little pruning except for removing double leaders

Limb Breakage: resistant

CULTIVARS AND SPECIES:

White Fir is not commonly known. Cultivars include 'Compacta' and 'Conica', which are more dwarf and conical-shaped; and 'Argentea' and 'Violacea', which have beautiful silver-gray foliage. *A. firma* is probably more heat-tolerant, has green glossy foliage, and grows even further south into zone 8. *A. amabilis,* Pacific Silver Fir, is a tall northwest American native tree suited for a large area, and grows best with irrigation. *A. lasiocarpa,* Subalpine Fir, has a narrow canopy and grows at high altitudes in the Rocky Mountains.

LANDSCAPE NOTES:

Origin: throughout the mountains of California and other western states

Usage: buffer strip; specimen; Christmas tree; bonsai

One of the best firs for the east, White Fir reaches a mature height of 100 to 180 feet in its native range, but is often much smaller in the landscape. It has horizontal branching with the lower branches drooping toward the ground. The tree should be grown in an open area so the lower branches can touch the ground. White Fir may grow 1.5 feet per year with good growing conditions. This tree can take exposure and will withstand some heat and drought better than most firs.

White Fir transplants well balled and burlapped, if properly root-pruned. The tree prefers a moist, well-drained loam. The root system can adapt to shallow or rocky soil conditions by growing close to the surface of the soil. It will not tolerate wet soil and often declines as a result of it in the spring unless located in a well-drained site. As a Christmas tree, White Fir remains fresh and retains its needles indoors for two weeks or more if provided with water. White Fir is an excellent landscape substitute for the disease-sensitive Colorado Blue Spruce because it is less prone to diseases.

Generally no insects or diseases limit growth, but a few cause some damage, including the balsam twig aphid, bagworm, scales, and spider mites. Look for spider mites in hot weather. Diseases include needle and twig blight, pine twig blight, and cankers. *Phytophthora* can infect and kill the roots. Roots decline in wet soil.

Abies firma Japanese Fir

Pronunciation: AY-beez FUR-mah

ZONES: 5B, 6A, 6B, 7A, 7B, 8A, 8B

HEIGHT: 20 to 30 feet

WIDTH: 10 to 15 feet

FRUIT: elongated; 3 to 6"; dry; brown; no significant litter; persistent

GROWTH RATE: slow-growing; long-lived

HABIT: pyramidal; dense; symmetrical; fine texture

FAMILY: *Pinaceae*

LEAF COLOR: light- to medium-green

LEAF DESCRIPTION: 1- to 2-inch-long blunt needles

FLOWER DESCRIPTION: not showy; spring

EXPOSURE/CULTURE:

Light Requirement: full sun

Soil Tolerances: all textures; acidic; drought

PEST PROBLEMS: pest-free

BARK/STEMS:

Branches droop as tree grows and will require pruning for pedestrian clearance beneath canopy; should be grown with a single leader; no thorns

Pruning Requirement: needs little pruning

Limb Breakage: resistant

LANDSCAPE NOTES:

Origin: Asia

Usage: bonsai; buffer strip; specimen

Japanese Fir has a form similar to White Fir, but has stiffer needles. The trunk grows straight up the center of the tree and the crown maintains a soft, tight, pyramidal shape without pruning. Branches are held upright on young trees but give way to a more horizontal form as the tree grows older. The tree looks best with lower branches left on the tree so they sweep the ground. Growth is very slow in the seedling stage and after transplanting, but once established the tree will grow about 12 inches each year. It grows about 20 feet in 30 years.

It is used as a Christmas tree in the western United States, and should be tried in the east as a landscape plant and Christmas tree. Although rare in the nursery trade, Japanese Fir is a beautiful plant that should be planted more often. It can be seen in a number of arboreta in the south and central part of the country. Use it to create a slow-growing screen planted on 10-foot centers, or as a specimen.

Grown best in acid soil in full sun, Japanese Fir is surprisingly tolerant of heat and drought, even in clay soil, but allow for good drainage. It has not grown well in alkaline soil. It should be a low-maintenance tree requiring little or no fertilizer or irrigation once established. It is probably one of the best (if not the best) firs to grow in the southeast United States where summers are hot and long.

There are no reports of serious pest problems, although the tree has not been grown much or extensively tested. *Phytophthora* can rot roots on other fir trees.

Abies nordmanniana Nordmann Fir

Pronunciation: AY-beez nord-man-ee-AY-nah

ZONES: 5A, 5B, 6A, 6B, (7A)

HEIGHT: 60 to 80 feet

WIDTH: 15 to 20 feet

FRUIT: cone 5 to 6" long, held upright on upper branches; dry; reddish brown; showy; no significant litter

GROWTH RATE: slow-growing; long-lived

HABIT: narrow pyramidal to columnar; dense; symmetrical; fine texture

FAMILY: *Pinaceae*

LEAF COLOR: dark green

LEAF DESCRIPTION: 1-inch-long needles

FLOWER DESCRIPTION: not showy

EXPOSURE/CULTURE:

Light Requirement: part shade to full sun

Soil Tolerances: all textures; acidic to slightly alkaline; drought

PEST PROBLEMS: resistant

BARK/STEMS:

Branches droop as tree grows and will require pruning for pedestrian clearance beneath canopy; normally grows with a single leader; no thorns

Pruning Requirement: needs little, if any pruning

Limb Breakage: resistant

CULTIVARS AND SPECIES:

A. procera, Noble Fir, a native of western Washington state and Oregon, is popular as a Christmas tree and is suited for planting in landscapes with plenty of soil space for root expansion. It is a beautiful fir well deserving of specimen use.

LANDSCAPE NOTES:

Origin: Asia Minor, Caucasus

Usage: buffer strip; specimen; Christmas tree; screen

One of the best-looking firs anywhere, Nordmann Fir can reach a mature height of about 80 feet in a landscape. It has horizontal branching with lower branches drooping toward the ground. The tree should be grown in an open area so lower branches can touch the ground. Nordmann Fir may grow 1.5 feet per year with good growing conditions. This tree can take exposure and will withstand some heat and drought.

Plant Nordmann Fir as a screen in an area protected from direct winter wind and it will reward you with rich color year-round. Its beautiful needles and soft texture make it nicely suited for planting as a specimen in any landscape. The cultivar 'Tortifolia' is better suited for the smaller, residential landscape.

The tree prefers a moist, well-drained loam. The root system can adapt to shallow or rocky soil conditions by growing close to the surface of the soil. Roots branch infrequently, forming a spreading root system. It will not tolerate constantly wet soil, but grows nicely in alkaline soil. It could be tried as a Christmas tree. Nordmann Fir might be an excellent landscape substitute for the disease-sensitive Colorado Blue Spruce, although the foliage color is different.

Generally no insects or diseases limit growth, but a few cause some damage, including woolly adelgids, twig aphids, bagworm, scales, and spider mites. Look for spider mites in hot weather. Diseases include needle and twig blight, pine twig blight, and cankers. *Phytophthora* can rot roots.

Acacia auriculiformis Earleaf Acacia

Pronunciation: ah-CAY-shaw ah-rick-you-lih-FOR-miss

ZONES: 10A, 10B, 11

HEIGHT: 35 to 40 feet

WIDTH: 25 to 35 feet

FRUIT: irregular; 1 to 2" long; fleshy; green turning brown; fruit and twigs cause significant litter; persistent on tree; showy

GROWTH RATE: rapid growth; short life

HABIT: round; pyramidal; moderate density; irregular silhouette; medium texture

FAMILY: *Leguminosae*

LEAF COLOR: evergreen; summer–medium green

LEAF DESCRIPTION: alternate; simple; entire; 4 to 6" long; not a true leaf

FLOWER DESCRIPTION: yellow; showy; spring flowering

EXPOSURE/CULTURE:

Light Requirement: full sun

Soil Tolerances: all textures; alkaline or acidic; wet soil; drought

Aerosol Salt Tolerance: moderate

PEST PROBLEMS: resistant; occasionally anthracnose infects leaves

BARK/STEMS:

Branches droop as tree grows and will require pruning for vehicular or pedestrian clearance beneath canopy; should be grown with a single leader but seldom is; no thorns

Pruning Requirement: requires regular pruning to prevent back inclusions

Limb Breakage: susceptible to breakage at crotch due to poor collar formation; wood itself is weak and tends to break

Current Year Twig: green; slender

LANDSCAPE NOTES:

Origin: Australia

Usage: not recommended for extensive planting

Quickly reaching mature height and width, and growing 6 to 8 feet per year, Earleaf Acacia becomes a loose, rounded, evergreen, open shade tree. It is often planted for its abundance of small, beautiful, bright yellow flowers. The flattened, curved branchlets, which look like leaves, are joined by twisted, brown, ear-shaped seed pods. It has brittle wood and weak branch crotches, and the tree can be badly damaged during thunderstorms.

The tree has been used extensively for planting in parking lot islands and in other urban areas, but it breaks apart too easily to recommend it for planting. Seeds germinate in the landscape and it has escaped cultivation in south Florida, where it is becoming a mildly invasive weed in some areas. For these reasons, many people consider this an undesirable tree. Some communities prohibit the planting of this tree.

Prune branches so there is a wide angle of attachment, to help keep them from splitting from the tree. Also be sure to keep the major branches thinned so they stay less than half the diameter of the trunk and to prevent bark inclusions. These techniques might increase the longevity of existing trees.

Earleaf Acacia grows in full sun on almost any soil, including alkaline, and is moderately salt-tolerant. It will withstand periods of water inundation but is also very tolerant of drought.

Acacia farnesiana Sweet Acacia, Huisache

Pronunciation: ah-CAY-shaw far-nee-she-AY-nah

ZONES: 9A, 9B, 10A, 10B, 11

HEIGHT: 20 feet

WIDTH: 25 feet

FRUIT: elongated; pod; 3 to 6"; dry; brown; attracts wildlife; no significant litter; persistent; showy

GROWTH RATE: slow-growing; short-lived

HABIT: spreading vase; open; symmetrical on older trees; fine texture

FAMILY: *Leguminosae*

LEAF COLOR: soft, medium-green; no fall color change

LEAF DESCRIPTION: alternate; bipinnately compound; entire margins; semi-evergreen; leaflets less than 2 inches long

FLOWER DESCRIPTION: yellow; slight fragrance; showy; year-round

EXPOSURE/CULTURE:

Light Requirement: full sun

Soil Tolerances: all textures; alkaline to acidic; occasionally wet soil; drought

Aerosol Salt Tolerance: moderate

PEST PROBLEMS: resistant; occasionally anthracnose can infect leaves

BARK/STEMS:

Low branching; branches droop as tree grows and will require pruning for pedestrian clearance beneath canopy; routinely grown with multiple trunks; thorns are present on trunk or branches

Pruning Requirement: requires regular pruning to develop a short, single trunk; main trunks on older trees need cabling to hold them together, as they may split apart at ground level.

Limb Breakage: trunks often split apart at crotches

Current Year Twig: brown; slender

LANDSCAPE NOTES:

Origin: extreme southern United States south into northern South America

Usage: bonsai; container; buffer strip; reclamation; specimen; highway

This tall, semi-evergreen, native shrub or small tree has feathery, finely divided leaflets. The slightly rough stems are a rich chocolate brown or grey, possessing long, sharp, multiple thorns. The small, yellow, puff-like flowers are very fragrant and appear in clusters in late winter and spring, then sporadically after each new flush of growth, providing nearly year-round bloom.

It can be trained in the nursery into a tree for use in highway median strips, or in other urban areas where there is no need for tall-vehicle clearance beneath the crown. The small stature and low, spreading branching habit make pruning for pedestrian clearance difficult unless the tree is properly trained from an early age. Head back aggressive branches on the lower trunk to force growth in the upper canopy, and space branches along a short, single trunk. The hours required for early training may be offset by the high drought, pest, and insect resistance of this tree. Do not locate the tree too close to where people can be injured by the sharp thorns on the branches.

Although easy to grow, the leaves will drop if the soil is allowed to dry out. This drought-avoidance mechanism allows the plant to grow well with no irrigation, once established. Growing best in full sun, this thorny, well-branched shrub makes an excellent barrier planting or nesting cover for wildlife. When trained as a small tree and used as a freestanding specimen, it is likely to provoke many positive comments. Sweet Acacia has its place in any sunny shrub border or as an accent plant in any garden, but locate it away from areas where children frequent, because the thorns can inflict severe pain. It is well suited for dry climates with little rainfall.

Acacia wrightii Wright Acacia, Wright Catclaw

Pronunciation: ah-CAY-shaw RIGHT-ee-eye

ZONES: 7A, 7B, 8A, 8B, 9A, 9B, 10A, 10B

HEIGHT: 25 to 30 feet

WIDTH: 20 to 30 feet

FRUIT: pod; 2 to 4" long; dry; turns brown; no significant litter; showy

GROWTH RATE: slow-growing; moderate life span

HABIT: round to upright, vase; moderate density; somewhat irregular; fine texture

FAMILY: *Leguminosae*

Photo Courtesy Benny Simpson

LEAF COLOR: bright green; no fall color change

LEAF DESCRIPTION: tiny; alternate; even pinnately compound; entire margin; oblong to obovate; semi-evergreen

FLOWER DESCRIPTION: white; pleasant fragrance; showy; spring

EXPOSURE/CULTURE:

Light Requirement: full sun

Soil Tolerances: all textures; alkaline to acidic; drought

PEST PROBLEMS: resistant

BARK/STEMS:

Branches droop as tree grows and will require pruning for vehicular or pedestrian clearance beneath canopy; routinely grown with multiple trunks; small thorns

Pruning Requirement: requires pruning to develop strong structure and uniform canopy

Limb Breakage: resistant

Current Year Twig: brown or gray; slender

CULTIVARS AND SPECIES:

Other species include *A. abyssinica* (zone 9A), which is deciduous; *A. saligna* (zone 9A) is evergreen; both are without thorns and suited for arid climates. *A. baileyana* (zone 9A) is commonly planted in southern California and has beautiful yellow flowers.

LANDSCAPE NOTES:

Origin: southwestern United States and into Mexico

Usage: container or above-ground planter; parking lot island; buffer strip; near a deck or patio; reclamation; small shade tree; specimen

Wright Acacia forms a rounded, open canopy composed of tiny leaflets, and remains semi-evergreen. The showy springtime displays of 1.5-inch-long spikes of yellow to white blossoms are abundantly produced over the slightly drooping branches. The blooms are followed by 2- to 4-inch-long, brown, compressed pods that contain small seeds. Multiple trunks arise from the ground, growing into the rounded canopy which provides moderately dense shade.

The tree is well suited for planting near a patio or deck, or it can make a nice street tree or parking lot tree for hot, dry sites. The spreading crown can cover an area rather quickly, producing ample shade in a small landscape. It is well suited for planting in small landscapes where space is at a premium. Plant trees on 20-foot centers to form an arcade of fine-textured small trees. Little irrigation is required after the tree is well established in the landscape.

Wright Acacia should be grown in full sun on well-drained soil. Acacias often develop a thin canopy if grown in partial sun. The somewhat-drooping branches may require regular pruning to allow passage of vehicles and pedestrians if planted close to sidewalks or streets. Due to the open habit, young trees benefit from training to guide the multiple trunks to grow upright. Without this training, plants become more of a shrub than a tree.

Propagation is by seed, which germinates easily, or by collection of small plants.

Acer barbatum Florida Maple, Southern Sugar Maple

Pronunciation: AY-sir bar-BAY-tum

ZONES: 6B, 7A, 7B, 8A, 8B, 9A

HEIGHT: 50 to 60 feet

WIDTH: 25 to 40 feet

FRUIT: samara; 1 to 3" long; dry; green turning brown; no significant litter

GROWTH RATE: moderate growth rate; long-lived on suitable site

HABIT: oval to round; moderate density; symmetrical; medium texture

FAMILY: *Aceraceae*

LEAF COLOR: summer–medium green; showy orange to yellow fall color

LEAF DESCRIPTION: opposite; simple; lobed; entire margin; deciduous; 2 to 4" long

FLOWER DESCRIPTION: red; not showy; spring flowering

EXPOSURE/CULTURE:

Light Requirement: part shade to full sun

Soil Tolerances: all textures; slightly alkaline to acidic; moderate drought

Aerosol Salt Tolerances: none

PEST PROBLEMS: resistant

BARK/STEMS:

Branches grow mostly upright and will not droop; showy trunk; should be grown with a single leader; no thorns

Pruning Requirement: needs minimum pruning to develop a strong structure; eliminate double leaders

Limb Breakage: resistant

Current Year Twig: brown; medium thickness

CULTIVARS AND SPECIES:

The Northern Sugar Maple is similar to Florida Maple. The cultivar 'Caddo' grows in alkaline soil at a moderate to slow rate—about 1 foot each year in most soils—but is fairly tolerant of drought. It is most popular in the central and southern states, but has grown poorly in zone 9. Leaf scorch will be less than on other maples where the root system is restricted to a small soil area, such as a street tree planting. It is most drought-tolerant in open areas where the roots can proliferate into a large soil space.

LANDSCAPE NOTES:

Origin: southeastern North America

Usage: bonsai; wide tree lawns; buffer strips; deck or patio; shade tree; street tree

The Florida Maple is still considered to be *Acer saccharum* var. *floridanum* to some. Others call it *A. floridanum*. Florida Maple is smaller than Northern Sugar Maple. The edges of the leaves turn under slightly, giving them a distinct appearance. The trunk on older specimens resembles that on the Northern Sugar Maple, which is an attractive gray with longitudinal ribs.

Growing in full sun or partial shade, Florida Maple will tolerate a wide variety of soil types. Established trees look better when given some irrigation during dry weather, particularly in the south. Although leaves will eventually fall, in the south many remain in the central portion of the canopy for much of the winter, giving the tree a somewhat unkempt appearance. Sensitivity to compaction, heat, drought, and road salt limit usage of Sugar Maple as an urban street tree. It is still recommended for parks and other areas away from roads where soil is loose and well-drained. *Acer nigrum* (zone 5), a similar species, is more tolerant of heat and drought.

The limbs of Florida Maple are strong and not susceptible to wind damage. Roots are often shallow and reach the surface at an early age, even in sandy soil. The trees appear to suffer when the roots are confined to a small area. Allow roots to expand by planting the tree in an open soil area.

See *Acer saccharum*, Sugar Maple, for more discussion. Some potential problems include cottony maple scale, borers, aphids, gall mites, and a vascular wilt disease.

Acer buergeranum Trident Maple

Pronunciation: AY-sir burr-jurr-AY-num

ZONES: 4B, 5A, 5B, 6A, 6B, 7A, 7B, 8A, 8B, 9A, 9B

HEIGHT: 30 to 40 feet

WIDTH: 25 feet

FRUIT: samara; 1"; dry; red; no significant litter

GROWTH RATE: medium growth rate and life span

HABIT: oval; round; upright; moderate density; irregular silhouette; medium texture

FAMILY: *Aceraceae*

LEAF COLOR: summer–medium green; orange, red, or yellow fall color

LEAF DESCRIPTION: opposite; simple; serrate margin; deciduous; 2 to 3" long

FLOWER DESCRIPTION: yellow; showy; spring flowering

EXPOSURE/CULTURE:

Light Requirement: part shade to full sun

Soil Tolerances: all textures; slightly alkaline to acidic; occasionally wet; moderate drought and salinity

Aerosol Salt Tolerance: moderate

PEST PROBLEMS: resistant

BARK/STEMS:

Trainable to be grown with single or multiple trunks; branches droop; showy trunk with flaking, orange-brown bark; no thorns

Pruning Requirement: requires pruning to develop single trunk structure

Limb Breakage: resistant

Current Year Twig: brown or gray; medium thickness

CULTIVARS AND SPECIES:

Several cultivars are known but rare, with trees having dwarf growth, corky bark, variegated leaves, and a variety of leaf shapes. Cultivars include: 'Akebono', 'Goshiki Kaede', 'Maruba', and 'Mino Yatsubusa'. 'Redman' has nice fall color and a fairly straight trunk.

LANDSCAPE NOTES:

Origin: China

Usage: bonsai; container; parking lot island; highway; buffer strip; deck or patio; shade tree; specimen; street tree; urban tolerant

Trident Maple naturally exhibits low spreading growth and multiple stems, but can be trained to a single trunk and pruned to make it branch higher, allowing passage below its broad, oval-to-rounded canopy. With its moderate growth rate, attractive bark, and easy maintenance, Trident Maple is popular as a patio or street tree and is also highly valued as a bonsai subject. Crown form is often variable and selection of a uniformly shaped, vigorous cultivar is needed.

Trees in partial shade can grow much taller (up to 50 feet tall), especially when the crown is touching adjacent trees that prevent branches from spreading. The tree is reported to be weak-wooded in North Carolina, but some of this may be due to poor structure, not weak wood. This can be at least partially prevented by pruning major lateral branches so they grow no larger than half the diameter of the main trunk. Be sure that there are no weak crotches with included bark, nor double or multiple leaders that could cause the tree to split apart. Specify single-leadered trees when planting along streets, in parking lots, or in other commercial landscapes.

Trident Maple has not been extensively used as a street tree, probably due to its unavailability, but the cultural requirements, size, and form make it a great candidate. It should also be planted more around residences and commercial landscapes due to its pleasing form and small size.

Plant in full sun or partial shade on any well-drained, acid soil. Trident Maple is quite tolerant of salt, air pollution, wind, and drought. Like other maples, some chlorosis can develop in soils with pH over 7, but it is moderately tolerant of soil salt. It performs well in urban areas where soils are often poor and compacted. Trees are easily transplanted due to their shallow root system. They are fairly clean trees, as they do not drop messy leaves, fruit, or flowers.

Acer campestre Hedge Maple

Pronunciation: AY-sir cam-PES-tree

ZONES: 5A, 5B, 6A, 6B, 7A, 7B, 8A, (9A)

HEIGHT: 30 to 35 feet

WIDTH: 30 to 35 feet

FRUIT: samara; 1 to 3"; dry; green turning brown; not showy; no significant litter

GROWTH RATE: slow-growing; moderate longevity

HABIT: oval to round; dense; symmetrical; fine texture

FAMILY: *Aceraceae*

LEAF COLOR: summer–dark green; yellow, showy fall color

LEAF DESCRIPTION: opposite; simple; lobed; entire margin; deciduous; 2 to 4" long

FLOWER DESCRIPTION: green; spring

EXPOSURE/CULTURE:

Light Requirement: part shade to full sun

Soil Tolerances: all textures; slightly alkaline to acidic; moderate salt; drought

Aerosol Salt Tolerance: moderate to high

PEST PROBLEMS: resistant

BARK/STEMS:

Bark is thin and easily damaged by mechanical impact; branches droop as tree grows and will require pruning for vehicular or pedestrian clearance beneath canopy; usually has multiple trunks, but can be trained to a single leader; no thorns

Pruning Requirement: requires pruning to develop a single leader

Limb Breakage: resistant

Current Year Twig: brown; medium to thin

CULTIVARS AND SPECIES:

'Evelyn' may be more vigorous, and has an upright branching habit, but is cold-tolerant only to zone 6. 'Compactum' is dwarf; 'Postelense' has golden leaves that gradually change to green during the summer; 'Queen Elizabeth' is more upright-formed than the species and makes a good street tree.

LANDSCAPE NOTES:

Origin: Europe, Africa

Usage: above-ground planter; bonsai; tall hedge; parking lot island; buffer strip; highway; deck or patio; standard; shade tree; specimen; sidewalk cutout; street tree; urban tolerant

Hedge Maple is usually low-branched with a rounded form, but there is considerable variability among trees. The branches are slender and branch profusely, lending a fine texture to the landscape, particularly during winter. Lower branches can be removed to create clearance beneath the crown for vehicles and pedestrians. The small stature and vigorous growth make this an excellent street tree for residential areas or downtown urban sites. It is also suitable as a patio or yard shade tree because it stays moderately small and creates dense shade.

The tree excels in its ability to tolerate dry, alkaline soil, but some protection from open winds is helpful, and it is not for highly compacted soil. It is well-suited for and looks great during drought in a partially shaded location or on the north side of a building. The main ornamental feature is the bright yellow fall color. The common name alludes to the plant's tolerance of severe pruning, and it will make a dense, tall screen whether pruned or not. Branches are arranged closely on the trunk and some pruning is usually desirable to create a well-formed tree. Prune early in the life of the tree to develop several major branches well-spaced along a central trunk. This will improve the durability of the tree compared to trees with many upright and spreading branches originating from one point on the trunk.

Pests are usually not serious, but some potential problems include leaf stalk borer, aphids, scales, and other borers. It is susceptible to verticillium wilt, which can kill the tree. Tar spot and a variety of leaf spots cause some concern among homeowners but are rarely serious enough for control. Seeds germinate in the landscape and can become weedy.

Acer cissifolium Ivy-Leaf Maple

Pronunciation: AY-sir siss-uh-FOAL-ee-um

ZONES: 5B, 6A, 6B, 7A, 7B, (8A)

HEIGHT: 20 to 30 feet

WIDTH: 20 to 30 feet

FRUIT: samara; 1"; dry; green to brown; no significant litter

GROWTH RATE: slow-growing; moderate life span

HABIT: round; upright; dense; symmetrical; fine texture

FAMILY: *Aceraceae*

LEAF COLOR: summer–light green; red to yellow showy fall color

LEAF DESCRIPTION: opposite; trifoliate; serrate margin; obovate to ovate; deciduous; 2" long

FLOWER DESCRIPTION: yellow; not showy; spring

EXPOSURE/CULTURE:

Light Requirement: part shade to full sun

Soil Tolerances: all textures; slightly alkaline to acidic; moderate drought

PEST PROBLEMS: resistant

BARK/STEMS:

Routinely grown with, or trainable to be grown with, multiple trunks; branches grow mostly upright; showy trunk; no thorns

Pruning Requirement: requires pruning to develop strong structure

Limb Breakage: may be susceptible on trees with included bark

Current Year Twig: reddish; slender

LANDSCAPE NOTES:

Origin: Japan

Usage: bonsai; container or above-ground planter; hedge; near a deck or patio; specimen; sidewalk cutout; street tree

Ivy-Leaf Maple reaches 10 to 20 feet in height (occasionally taller) and is equally wide, with a broad, rounded, very dense, fine-textured canopy. The branches have a twisted and contorted growth habit which, along with the smooth, gray bark, create an attractive winter silhouette. Upper branches are upright, middle ones horizontal, and lower branches slightly pendulous. The crown is made up of a large number of finely-divided, small-diameter branches, with none really dominant. The divided leaflets turn muted shades of red and yellow before dropping in fall.

Although it is probably best used as a patio tree or specimen, it could be tried as a street tree, particularly in areas where overhead space is restricted. It would make a lovely tree for planting on 20-foot centers along an entry driveway to a commercial landscape, or along a suburban street. Set them back from the street 8 to 12 feet if large trucks use the street regularly, as the tree loses its attractiveness when lower branches are removed for vehicle clearance. They can be planted closer if the street is residential and predominantly travelled by automobiles.

Ivy-Leaf Maple prefers a partially shaded location (particularly in the southern part of its range), but will grow well in full sun when grown on well-drained, moist, acid soil. Tolerating drought without leaf scorch except in the driest, sandy soils, Ivy-Leaf Maple appears to be a tough maple deserving greater usage in urban and suburban landscapes, though it is not really grown enough to determine.

Acer ginnala Amur Maple

Pronunciation: AY-sir gin-NAY-lah

ZONES: 3A, 3B, 4A, 4B, 5A, 5B, 6A, 6B, 7A, 7B, 8A, 8B

HEIGHT: 15 to 20 feet

WIDTH: 15 to 25 feet

FRUIT: samara; oval; 1"; dry; pinkish to bright red; abundant; causes some litter; very showy

GROWTH RATE: moderate growth rate; moderate life span

HABIT: round and spreading; dense; symmetrical; fine texture

FAMILY: *Aceraceae*

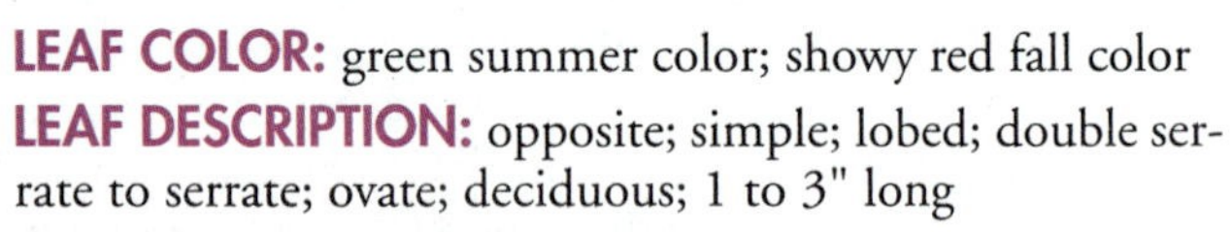

LEAF COLOR: green summer color; showy red fall color

LEAF DESCRIPTION: opposite; simple; lobed; double serrate to serrate; ovate; deciduous; 1 to 3" long

FLOWER DESCRIPTION: white to yellow; pleasant fragrance; showy; spring flowering for 2 weeks

EXPOSURE/CULTURE:

Light Requirement: part shade to full sun

Soil Tolerances: all textures; alkaline to acidic; well drained; moderate salt; drought

Aerosol Salt Tolerance: moderate

PEST PROBLEMS: resistant

BARK/STEMS:

Bark is thin and easily damaged by mechanical impact; branches droop as tree grows and will require pruning for vehicular or pedestrian clearance beneath canopy; routinely grown with multiple trunks but can be trained to grow with a single trunk; no thorns

Pruning Requirement: requires regular pruning to develop and maintain one trunk and strong structure

Limb Breakage: resistant

Current Year Twig: brown or gray; medium to thin

CULTIVARS AND SPECIES:

Several cultivars, including 'Compactum', which is smaller; 'Embers', 'Red Fruit', and 'Durland Dwarf'. 'Bailey Compact' requires little, if any, pruning to develop a neat, compact, dark green, round crown and is pest-free. This group of cultivars is grown for their bright red fruit, which is very showy in summer. 'Summer Splendor' may produce less fruit.

LANDSCAPE NOTES:

Origin: Japan and China

Usage: bonsai; container or above-ground planter; hedge; near a deck or patio; specimen; street tree

Amur Maple is an excellent, low-growing tree for small yards, patios, and other small-scale landscapes. It can be grown as a multi-stemmed clump or trained into a small tree with a single trunk up to 4 to 6 feet tall. The tree eventually grows about 15 to 20 feet tall and has spreading, rounded, finely branched growth habit on several thick main branches, which creates dense shade under the crown. Trunks droop and spread as they grow older. Amur Maple can grow rapidly when it is young if it receives water and fertilizer, but it is well suited for planting close to power lines because it slows down and remains small at maturity.

The main ornamental value of Amur Maple is its brilliant red fall foliage color and fruit that sports bright pink or red wings. It is a durable tree, tolerating poor soil, but will grow less vigorously in the southern end of its range. It will leaf scorch in dry summers in full sun, but is very drought-tolerant and will not die back. It is most drought-tolerant in partial shade. The plant is sometimes used in hedges or screens, and can be used for planting along streets beneath power lines if it is set back from the road. Without early thinning and training, the multiple trunks droop considerably. Main branches and trunks can be headed back and/or drop-crotched several times when the tree is young to create a more upright form. The tree makes a nice specimen or patio tree.

Amur Maple is usually pest-free. The drooping, spreading form should be taken into account when locating the tree close to walks. Typical maple diseases such as verticillium wilt and bacterial blight can be a problem. Seeds germinate in the landscape and can become a pesty weed. Chlorosis occurs on highly alkaline soil.

Acer grandidentatum Bigtooth Maple

Pronunciation: AY-sir gran-dih-den-TAY-tum

ZONES: 5A, 5B, 6A, 6B, 7A, 7B, 8A, 8B

HEIGHT: 40 to 50 feet

WIDTH: 25 to 35 feet

FRUIT: samara; 1"; dry; green to brown; attracts birds; no significant litter

GROWTH RATE: slow to moderate growth rate; long-lived

HABIT: oval to round; moderate density; irregular; medium texture

FAMILY: *Aceraceae*

LEAF COLOR: dark green; orange, red, or yellow fall color

LEAF DESCRIPTION: opposite; simple; 3 to 5 lobes; star-shaped; deciduous; 2 to 5" long

FLOWER DESCRIPTION: yellow; spring

EXPOSURE/CULTURE:

Light Requirement: part shade to full sun

Soil Tolerances: loam; sand; alkaline to acidic; salt-sensitive; moderate drought

PEST PROBLEMS: resistant

BARK/STEMS:

Branches droop as tree grows and will require pruning for vehicular or pedestrian clearance beneath canopy; should be grown with a single leader; no thorns

Pruning Requirement: requires some pruning to develop strong structure

Limb Breakage: resistant

Current Year Twig: reddish-brown; slender

CULTIVARS AND SPECIES:

'Schmidt' (Rocky Mountain Glow™) grows quickly in the nursery; *A. glabrum,* Rocky Mountain Maple, is native to the northwestern United States and is a smaller, multitrunked tree without the brilliant fall color. *A. macrophyllum,* Bigleaf Maple, is a large shade tree used for large-scale landscapes in the western United States.

LANDSCAPE NOTES:

Origin: western North America, Rocky Mountains

Usage: buffer strip; reclamation; shade tree; specimen; street tree

This Rocky Mountain North American native slowly reaches 50 feet in height with a broad, spreading canopy and grayish-brown bark that may be either smooth or scaly. The 2- to 5-inch-diameter, lustrous, lobed leaves, which have a pale underside, are noted for their striking brilliance in fall, when they change into beautiful shades of red, orange, and yellow before dropping. This is one of the best maples to plant in the drier parts of the west to create a suburban fall color show. Color resembles the Sugar Maple (it is considered by some to be a western variety of Sugar Maple). The canopy is fairly dense with billowing foliage displaying a finer texture than most other maples.

Plants in the wild often begin with a multitrunked habit. Some training is needed to develop a dominant central leader. Once selected, this leader should more or less stay dominant. This makes Bigtooth Maple a good candidate for planting along streets in an area with plenty of soil space for root expansion. Roots will raise sidewalks if planted too close. It is not adapted to the harsh conditions along city streets or parking lots where soil space is often at a premium. Its best use is for residential shade tree planting. Plant it in masses in parks for a canopy of shade.

Bigtooth Maple will grow in full sun or partial shade and is found most often in its natural habitat in moist, well-drained soils along stream banks and in canyons. Some refer to it as Canyon Maple. It tolerates limestone soils well, which is uncharacteristic of most maples. Plants in the wild grown in open areas have withstood long periods of drought.

Acer griseum Paperbark Maple

Pronunciation: AY-sir GRIZZ-ee-um

ZONES: 5A, 5B, 6A, 6B, 7A, 7B

HEIGHT: 20 to 30 feet

WIDTH: 15 to 25 feet

FRUIT: samara; oval; 1 to 3"; dry; brown; no significant litter; showy

GROWTH RATE: very slow-growing; moderate life span

HABIT: oval to vase; moderate density; slightly irregular silhouette; fine texture

FAMILY: *Aceraceae*

LEAF COLOR: dark green summer color; showy yellow, red, or orange fall color

LEAF DESCRIPTION: opposite; trifoliate; serrate margin; elliptic to ovate; hairy; deciduous; 3 to 5" long

FLOWER DESCRIPTION: green; not showy; spring flowering

EXPOSURE/CULTURE:

Light Requirement: part shade to full sun

Soil Tolerances: all textures; slightly alkaline to acidic; some drought; moderate salt

PEST PROBLEMS: resistant; mostly pest-free

BARK/STEMS:

Outstanding color and texture, one of the best; trainable to be grown with multiple trunks or with a single, short trunk; branches grow mostly upright and will not droop; no thorns

Pruning Requirement: needs little pruning to develop a strong trunk and branch structure

Limb Breakage: resistant

Current Year Twig: brown; slender

CULTIVARS AND SPECIES:

Cultivars are being developed. Only 'Gingerbread' is currently listed.

LANDSCAPE NOTES:

Origin: China

Usage: bonsai; container or above-ground planter; near a deck or patio; specimen; possible street tree

Perhaps the most beautiful maple, Paperbark Maple has trifoliate leaves and wonderful orange to bronze, peeling, papery bark that provides year-round interest. The bark begins peeling on the sculptured trunk and on 2- or 3-year-old branches. It may be cinnamon brown or orange but is usually a dark reddish-brown that looks particularly striking in the snow. Even small branches display exfoliating bark making this a true specimen tree even at a young age. The bark on the lower trunk of old trees becomes darkened and platy. Most specimens have multiple trunks that branch close to the ground, but proper training when young can create a single, short trunk. Without irrigation, 40-year-old specimens may only be 20 feet tall.

The multistemmed habit, unusual leaves, and wonderful bark makes this a prime candidate for specimen planting in any commercial, institutional, or residential landscape. Plant it by a patio or other prime location and light it from below for nighttime enjoyment. Transplant in spring for best establishment.

The tree is hardy, grows very slowly to 25 or 30 feet tall, but, unfortunately, is difficult and expensive to propagate. It does not tolerate extended drought or other environmental stresses, especially in poor soil (perhaps moderate drought tolerance in sandy loam), but will grow in sun or shade. Leaves will scorch during dry summers unless provided with some irrigation. Not for the drier parts of the country, probably due to high temperature stress. The beauty of this tree makes up for the extra effort required to find it.

Acer japonicum Fullmoon Maple

Pronunciation: AY-sir juh-PON-eh-kum

ZONES: 5A, 5B, 6A, 6B, 7A, 7B

HEIGHT: 10 to 15 feet

WIDTH: 6 to 10 feet

FRUIT: samara; oval; 0.5 to 1"; dry; green; not showy; no significant litter

GROWTH RATE: slow-growing; short to moderate life span

HABIT: oval; round; vase; dense; symmetrical; medium to coarse texture

FAMILY: *Aceraceae*

LEAF COLOR: soft green in summer; good red fall color

LEAF DESCRIPTION: opposite; simple; lobed; incised margin; star-shaped; deciduous; 2 to 4" long

FLOWER DESCRIPTION: red; spring; surprisingly abundant and showy, but small in size

EXPOSURE/CULTURE:

Light Requirement: part shade to shade for best growth

Soil Tolerances: all textures; slightly alkaline to acidic; sensitive to drought

Aerosol Salt Tolerance: none

PEST PROBLEMS: resistant; may be susceptible to verticillium wilt

BARK/STEMS:

Branches droop as tree grows; routinely grown with multiple trunks; no thorns

Pruning Requirement: needs little pruning to develop a strong trunk and branch structure; space close branches along trunk

Limb Breakage: resistant

Current Year Twig: green; slender

LANDSCAPE NOTES:

Origin: Japan

Usage: bonsai; near a deck or patio; standard; specimen

Full-Moon (or Fernleaf) Maple is a small, deciduous tree that creates a smooth, rounded canopy. It fits well into the oriental garden due to its exotic silhouette. The cultivar 'Acontifolium' is exceptionally cold-hardy, having survived temperatures of -25°F. The deeply divided leaves have 9 to 11 lobes and are delicately displayed on thin, drooping branches. The cultivar 'Vitifolium' leaves are less divided, providing a coarse texture in the landscape. Leaves take on a beautiful yellow to red coloration in the fall before dropping, making this small, dense plant really stand out in the landscape. The hanging clusters of showy, purple/red flowers appear in late spring and are followed by the production of winged seeds. The flowers stand out among the maples.

This maple would be at home in the residential landscape, planted near the patio or deck, as well as the commercial setting. It is probably best used as a specimen, planted to attract attention to an area. It should live for at least 20 years. Nice specimens can be viewed at arboreta, but few nurseries currently offer these cultivars for sale.

Full-Moon Maple can be grown in sun to almost full shade. Nice specimens can be seen growing in the filtered shade of tall, overstory trees, or with 2 to 6 hours of direct sun. Where the sunlight is intense, the tree will benefit from having its roots shaded to help keep the soil cool. A generous helping of mulch out to the edge of the canopy is beneficial.

Acer macrophyllum Bigleaf or Oregon Maple

Pronunciation: AY-sir mack-row-FIE-lum

ZONES: 7A, 7B, 8A, 8B, 9A, 9B, 10A, 10B

HEIGHT: 50 to 75 feet

WIDTH: 40 to 60 feet

FRUIT: elongated; 1.5 to 2"; dry; green turning brown; attracts wildlife; some litter; showy

GROWTH RATE: rapid growth; moderate life span

HABIT: round; very thick and dense; nearly symmetrical; coarse-textured

FAMILY: *Aceraceae*

LEAF COLOR: dark green in summer; bright yellow to orange fall color

LEAF DESCRIPTION: opposite; simple; 5 lobes; star-shaped; deciduous; 6 to 12" long

FLOWER DESCRIPTION: greenish-yellow; somewhat showy and fragrant; spring flowering with emerging leaves

EXPOSURE/CULTURE:

Light Requirement: part shade to full sun

Soil Tolerances: all textures; alkaline to acidic; occasionally wet soil; moderate drought

PEST PROBLEMS: resistant

BARK/STEMS:

Buttressed trunk; thin bark is easily damaged; branches droop as tree grows and will require pruning for vehicular or pedestrian clearance beneath canopy; should be grown with a single leader; no thorns

Pruning Requirement: requires early pruning to develop strong structure and to direct or slow growth on aggressive branches; prune in late summer to avoid sap flow from cuts

Limb Breakage: susceptible

Current Year Twig: brown; thick

CULTIVARS AND SPECIES:

'Seattle Sentinel' has a more upright form than the species.

LANDSCAPE NOTES:

Origin: Oregon to southern California

Usage: shade tree; street tree at least 8 feet from pavement

Bigleaf Maple in cultivation has a height of 50 feet but can grow taller when shaded on the sides on a good site. The dense shade it creates and its shallow root system compete with lawn grasses, and the shallow roots can make mowing under the tree difficult. Surface roots often form in urban soils. The shallow roots can heave sidewalks, so be certain to locate the tree 6 to 10 feet away. The tree is easily transplanted, grows quickly, and is adapted to a wide variety of soils (including alkaline).

Bigleaf Maple is a large tree suited for planting in a large landscape with plenty of soil space for root expansion. The lower trunk swells as it meets the root system, forming a wide, buttressed trunk. This limits its usefulness in most tree lawns and other small, narrow soil spaces. When leaves drop, they decompose slowly and blow around in the landscape. This makes a messy tree during the late summer and fall.

Bigleaf Maple is native to a variety of habitats along the west coast of the United States from stream bank to sloping hillside, and thus is adapted to a number of soil types. Although it tolerates poor drainage for a short period, prolonged root exposure to standing water will cause poor growth and decline. Storms frequently break branches, which shortens the life of the tree.

Dense surface rooting in compacted soils. Roots rot if soil is kept too wet. Aphid infestations produce copious honeydew, which drips on sidewalks and cars making them sticky. Seeds germinate readily in the landscape. Trees often break apart in a major storm.

Acer miyabei Miyabe Maple

Pronunciation: AY-sir me-YAH-bee-eye

ZONES: 4A, 4B, 5A, 5B, 6A, 6B

HEIGHT: 30 to 60 feet

WIDTH: 35 to 50 feet

FRUIT: samara; 1 to 2"; dry; green turning brown; no significant litter

GROWTH RATE: slow to moderate growth rate, depending on soil moisture; long-lived

HABIT: young trees distinctly upright or pyramidal, older ones rounded; dense when young, more open later; symmetrical; medium texture

FAMILY: *Aceraceae*

LEAF COLOR: medium green summer color; striking orange, red, or yellow fall color

LEAF DESCRIPTION: opposite; simple; 5 lobes; star-shaped; deciduous; 2 to 4" long

FLOWER DESCRIPTION: green to yellow; spring; not showy

EXPOSURE/CULTURE:

Light Requirement: partial shade to full sun

Soil Tolerances: all textures; alkaline to acidic; some salt and drought

Aerosol Salt Tolerance: moderate

PEST PROBLEMS: resistant

BARK/STEMS:

Branches on young trees are upright and do not droop; showy, orange-brown bark; should be grown with a single leader for street tree use, or can be trained with several trunks for specimen use to show the attractive bark; no thorns

Pruning Requirement: thin or remove aggressive lower limbs for street tree planting

Limb Breakage: resistant

Current Year Twig: brown; medium to slender

CULTIVARS AND SPECIES:

'State Street' is upright and coming into production at a number of nurseries.

LANDSCAPE NOTES:

Origin: Japan

Usage: bonsai; wide tree lawns; shade tree; street tree; specimen

Miyabe Maple is not common, except at selected arboreta in the northern United States. It was brought to this country about 100 years ago, but never made it out of botanical gardens and arboreta. The tree will be 30 to 60 feet tall at maturity, with a low-branching, spreading canopy unless pruned to remove lower branches. With lower limbs removed, an upright habit is maintained, which is ideal for some street tree planting sites. It grows less than 1 foot each year unless near water, but is sensitive to reflected heat and to drought; leaves turn brown (scorch) along their edges in dry weather in urban sites. Leaf scorch from dry soil will probably be evident in areas where the root system is restricted to a small soil area. Miyabe Maple is quite drought-tolerant in open areas where the roots can expand into a large soil space.

The thick, branched canopy creates dense shade and will prevent good lawn growth unless lower branches are removed. Branches are usually small in diameter and well attached to trunks. One of the main ornamental features of the tree is the brilliant yellow fall color. The bark is also strikingly showy and adds to the appeal of this underused tree.

The limbs of Miyabe Maple are usually strong and not susceptible to wind damage. The tree tends to develop several trunks early in its life with bark included between them. Remove all but one to develop a strong tree that will be well suited for planting in urban landscapes. A wider-spreading tree develops when low branches are left intact. They eventually droop and sweep the ground, creating a nice place under the tree to sit and enjoy the shade. Plant in an area where grass below it need not be mowed, so the shallow roots will not be damaged by mowing.

It is supposedly difficult to propagate. Grafting has met with some success.

Acer negundo Boxelder

Pronunciation: AY-sir nay-GHUN-doe

ZONES: 3A, 3B, 4A, 4B, 5A, 5B, 6A, 6B, 7A, 7B, 8A, 8B, 9A

HEIGHT: 40 to 50 feet

WIDTH: 35 to 40 feet

FRUIT: samara; produced in copious amounts; 0.5 to 1"; dry; green turning brown; attracts wildlife; causes significant litter; persistent; showy

GROWTH RATE: rapid growth; moderate life span

HABIT: oval; round; dense; symmetrical; medium texture

FAMILY: *Aceraceae*

LEAF COLOR: very light green in summer; showy orange to yellow fall color in the north

LEAF DESCRIPTION: opposite; odd pinnately compound; lobed; serrate margin; lanceolate to ovate; deciduous; 2 to 4" long leaflets

FLOWER DESCRIPTION: yellowish; not showy; spring flowering

EXPOSURE/CULTURE:

Light Requirement: part shade to full sun

Soil Tolerances: all textures; alkaline to acidic; wet soil; moderate salt; drought

Aerosol Salt Tolerance: none

PEST PROBLEMS: sensitive

BARK/STEMS:

Branches droop as tree grows and will require pruning for vehicular or pedestrian clearance beneath canopy; should be grown with a single leader; no thorns

Pruning Requirement: requires pruning to develop good structure

Limb Breakage: susceptible to breakage either at crotch, due to poor collar formation, or wood itself is weak and tends to break

Current Year Twig: brown to green; medium thickness

CULTIVARS AND SPECIES:

The cultivars of Boxelder are more ornamental but still share the tree's undesirable characteristics: 'Aureo-variegatum' has leaves bordered in gold; 'Flamingo' has variegated leaves with pink margins and is somewhat available; 'Variegatum' has leaves bordered in white; 'Auratum' has gold leaves; 'Elegantissima' has yellow leaf variegations. 'Sensation' has spectacular red-orange fall color and is seedless.

LANDSCAPE NOTES:

Origin: native throughout North America

Usage: bonsai; reclamation plant

Sometimes listed in catalogs as Ash-Leaved Maple. The Boxelder is an undesirable tree for many urban situations; in some cities, planting it may be illegal. The undesirable characteristics are brittle, weak wood and susceptibility to boxelder bug and trunk decay. The best thing about the tree is that it will grow on adverse sites where more desirable trees may not. Therefore, it may have uses due to its adaptability. Boxelder is native along streambanks over a wide area of North America, grows along flood plains, and naturalizes quickly on disturbed sites. It may be best to restrict planting to these areas, where it would make a good soil stabilizer. It is tolerant of drought and was planted as a shelter-belt tree.

The numerous, attractive seeds are very popular with squirrels and birds. A nice plant for naturalized areas, particularly if the soil is wet or the pH is alkaline. It is probably best used in these areas to help stabilize stream beds and colonize reclaimed land. Do not plant it as a street tree. If you use the tree, plant it for its quick growth, but interplant with more desirable trees to provide a lasting tree canopy. There are so many other trees to plant that there should be no need to use this tree for anything other than a reclamation site or for stream bank stabilization.

Boxelder harbor a variety of insects and diseases. It also is susceptible to breakage in storms and produces a copious amount of fruit.

Acer palmatum Japanese Maple

Pronunciation: AY-sir paul-MAY-tum

'Atropurpureum'　　'Dissectum'

ZONES: 5B, 6A, 6B, 7A, 7B, 8A, 8B, (10A)

HEIGHT: 15 to 25 feet

WIDTH: 15 to 25 feet

FRUIT: samara; 0.5 to 1"; dry; red; no significant litter

GROWTH RATE: slow-growing; moderate life span

HABIT: rounded vase; moderate density; symmetrical; medium to fine texture

FAMILY: *Aceraceae*

LEAF COLOR: green or red, depending on cultivar; showy copper, orange, red, or yellow fall color

LEAF DESCRIPTION: opposite; simple; lobed; serrate margin; star-shaped; deciduous; 2 to 4" long

FLOWER DESCRIPTION: red; not showy; spring

EXPOSURE/CULTURE:

Light Requirement: part shade to shade; protect from afternoon sun; full sun if irrigated

Soil Tolerances: sand to loam, clay if well-drained; neutral to acidic; well-drained soil is essential; some salt and drought

Aerosol Salt Tolerance: none

PEST PROBLEMS: resistant; verticillium wilt can kill plants; may have aphids, scales, and borers

BARK/STEMS:

Bark is thin and easily damaged by mechanical impact; branches droop as tree grows and will require pruning for vehicular or pedestrian clearance beneath canopy; routinely grown with multiple trunks; showy trunk and branch structure; no thorns

Pruning Requirement: requires pruning to develop strong structure; choose branches with wide angles of attachment for main limbs

Limb Breakage: resistant, except for those with included bark

Current Year Twig: green or red; slender

CULTIVARS AND SPECIES:

Variegated types are a bit more difficult to grow and are subject to leaf scorch. Many cultivars of Japanese maple exist, with a wide variety of leaf shapes and colors, growth habits, and sizes. 'Atropurpureum' can enhance any landscape with its delightful spring flush of red foliage, but it will fade to bronze-green in the beginning of the summer (better red color in the sun). 'Bloodgood's' spring foliage is bright red, changing to dark green, and holds leaves well

Continued

into fall. 'Burgundy Lace' is grown for its purple-red colored leaves, interesting growth habit, and fine leaf texture; leaves are dissected almost to the petiole; red leaf color is best as new leaves emerge in spring and fall; leaves turn almost green during summer heat; growth habit is more like a large shrub with branches to the ground. 'Dissectum' has finely dissected leaves in green or red and grows 10 to 12 feet tall. 'Elegans' has leaves with rose-colored margins when they first unfold. 'Ornatum' foliage is cut and reddish.

LANDSCAPE NOTES:

Origin: China, Korea, Japan

Usage: bonsai; container or above-ground planter; near a deck or patio; trainable as a standard; specimen

Japanese Maple has a height and spread of about 20 feet, but much smaller selections are available. The multiple trunks are muscular-looking, picturesque, gray, and show nicely when lit up at night. Grown for its colored leaves, interesting growth habit, and fine leaf texture. Fall color is often striking, even on trees grown in total shade. Growth habit varies widely depending on cultivar, from globose, branching to the ground, to upright, vase-shaped. Globose selections look best when they are allowed to branch to the ground.

This large shrub or small tree tends to leaf out early, so it may be injured by spring frosts. Protect them from drying winds and direct sun by providing partial or filtered shade and well-drained, acid soil with plenty of organic matter, particularly in the southern part of its range. Leaves often scorch in hot summer weather unless they are in some shade or irrigated during dry weather. Trees with diseased or inadequate root systems will also show scorching. More direct sun can be tolerated in the northern part of the range. Be sure drainage is maintained and never allow water to stand around the roots. Grows fine on well-drained clay soils; often planted on raised beds or on high ground in clay soil. Responds well to several inches of mulch placed beneath the canopy. Be sure to clear all turf away from beneath the branches of low-growing types so lawn mowers will not damage the tree.

The more upright selections, such as 'Atropurpureum', make nice patio or small shade trees for residential lots, and, with pruning to remove drooping branches, provide adequate clearance for pedestrian traffic to pass close to the tree. More compact cultivars make wonderful accents for any landscape.

Train trunks and branches so they will not touch each other. Eliminate branches with included bark, or those which are likely to develop it, as soon as possible to reduce the likelihood of one splitting from the tree later. Remove small twigs to enhance the showy trunk and bark structure. Locate the tree properly, taking into account the ultimate size; the tree looks best if it is not pruned to control size.

Acer platanoides Norway Maple

Pronunciation: AY-sir plah-tah-NOY-dees

'Erectum'

'Crimson King'

'Globosum'

ZONES: 4A, 4B, 5A, 5B, 6A, 6B, 7A

HEIGHT: 40 to 50 feet

WIDTH: 35 to 40 feet

FRUIT: elongated; 1 to 3"; dry; green turning brown; attracts birds; some litter; showy

GROWTH RATE: moderate to rapid growth; long-lived

HABIT: round; dense; nearly symmetrical; coarse-textured

FAMILY: *Aceraceae*

LEAF COLOR: dark green in summer; striking yellow fall color

LEAF DESCRIPTION: opposite; simple; 5 lobes; star-shaped; deciduous; bleeds milky sap; 4 to 5" long

FLOWER DESCRIPTION: greenish-yellow; somewhat showy; spring flowering with emerging leaves

EXPOSURE/CULTURE:

Light Requirement: part shade to full sun

Soil Tolerances: all textures; alkaline to acidic; occasionally wet soil; salt; moderate drought

Aerosol Salt Tolerance: moderate

PEST PROBLEMS: sensitive

BARK/STEMS:

Branches droop as tree grows and will require pruning for vehicular or pedestrian clearance beneath canopy; should be grown with a single leader; no thorns

Pruning Requirement: requires early pruning to develop strong structure

Limb Breakage: resistant

Current Year Twig: brown; thick

CULTIVARS AND SPECIES:

A large number of cultivars are available and are better suited for urban planting than the species. Those having colored summer foliage are: 'Crimson King'—oval, 45

Continued

feet tall, good compartmentalizer, foliage purple during the entire summer, well-adapted to street tree use but (due to its bold leaf color) may produce an overpowering, negative effect if planted along an entire street; 'Drummondii'—leaves edged in white; 'Schwedleri'—hardy to zone 3, oval, 50 feet tall and wide, foliage reddish in spring, then becoming greenish-bronze. Other cultivars include: 'Almira'—round-headed, mature height about 20 feet; 'Cleveland'—50 feet tall, branches on young and medium-aged trees are more upright, reducing the pruning requirement when planted close to buildings or other structures compared to the species; 'Columnare'—columnar or upright growth habit, 35 feet tall, narrow crown well suited to street tree plantings where vertical space is limited; 'Deborah'—new leaves appear as a deep red; 'Emerald Queen'—crown oval, growth rate faster, 60 feet tall; 'Erectum'—upright growth habit; 'Globosum'—compact rounded head, 20 feet tall, slow-growing, meatball-on-a-stick tree that requires no pruning; 'Greenlace'—cutleaf cultivar with rapid growth rate; 'Jade Glen' and 'Parkway'—supposedly resistant to verticillium wilt; 'Olmstead'—upright growth habit, 45 feet tall; 'Summershade'—oval crown with a central leader, growth rate faster than species, compartmentalizes decay well; 'Superform'—round, 45 feet tall, may show more resistance to frost cracks.

LANDSCAPE NOTES:

Origin: Europe

Usage: bonsai; parking lot island; buffer strip; shade tree; street tree; urban tolerant

Norway Maple in cultivation has a height of 40 to 50 feet but can grow much taller when shaded on the sides on a good site. The rounded crown fills with greenish-yellow flowers in the spring. Norway Maple's dense shade and shallow root system competes with lawn grasses, and the shallow roots can make mowing under the tree difficult. The shallow roots can heave sidewalks, so be certain to locate the tree 4 to 6 feet away. The tree is easily transplanted, grows quickly, is adapted to a wide variety of soils (including alkaline), and has brilliant fall color. It can also tolerate coastal conditions. Well adapted to street tree plantings, it is often overused.

Trunks can crack on the southern side during the winter, initiating some trunk decay, but the tree usually remains intact. Branches sometimes fail from too much weight toward the end of the branch. Although this is rare, it can be prevented by thinning the canopy to distribute the weight evenly along the main branches. A variety of birds are known to use seeds as a food source. Seeds germinate readily in the landscape and adjacent woodlands, and could become a nuisance.

Dense surface rooting in compacted soils. Verticillium wilt can kill trees. Aphid infestations produce copious honeydew, which drips on sidewalks and cars, making them sticky. Japanese beetles attack purple-leaf cultivars. Girdling roots are often found around the trunk near the soil surface. These can originate from the roots cut when the tree was dug from the nursery. Thin bark on young trees makes them very suscptible to sunscald.

Acer pseudoplatanus Sycamore Maple

Pronunciation: AY-sir sue-doe-PLAH-tah-nuss

ZONES: 5A, 5B, 6A, 6B, 7A, 7B

HEIGHT: 60 to 70 feet

WIDTH: 40 to 60 feet

FRUIT: samara; 1 to 3"; dry; green turning red; showy; no significant litter

GROWTH RATE: moderately fast-growing; long-lived

HABIT: round; moderate to dense; symmetrical; coarse texture

FAMILY: *Aceraceae*

LEAF COLOR: dark green in summer; muted yellow fall color

LEAF DESCRIPTION: opposite; simple; 3 to 5 lobes; serrate margin; ovate; deciduous; 3 to 7" long and wide

FLOWER DESCRIPTION: green; somewhat showy, pendulous; spring

EXPOSURE/CULTURE:

Light Requirement: part shade to full sun

Soil Tolerances: all textures; alkaline to acidic; occasionally wet soil; salt-tolerant by some accounts, sensitive by others; drought

Aerosol Salt Tolerance: high

PEST PROBLEMS: resistant

BARK/STEMS:

Branches grow mostly upright and spreading; showy trunk; should be grown with a single leader; no thorns

Pruning Requirement: needs little pruning to develop a strong structure

Limb Breakage: very resistant

Current Year Twig: green-brown to gray; stout

CULTIVARS AND SPECIES:

'Atropurpureum' ('Spaethii') has dark purple lower leaf surfaces; other cultivars have different leaf and fruit colors.

LANDSCAPE NOTES:

Origin: Europe

Usage: shade tree; street tree; parking lot buffer strips

Occasionally called Planetree Maple, this large deciduous tree is capable of reaching over 100 feet in height. The spreading branches form an oval or rounded canopy. The lobed leaves do not ordinarily become showy in the fall, but this will vary. The gray to reddish-brown bark flakes off in small scales to reveal the showy, orange inner bark. Flowers appear in 3- to 6-inch-long hanging panicles among the leaves in late spring and are followed by 1- to 2-inch-long, winged seeds.

This large tree requires space to spread. Not for the small landscape, its large, falling leaves and early defoliation in the fall can create a maintenance challenge. This tree may be best saved for the park or other large open-space planting site, as its coarse texture blends poorly in the small residential lots. It is used for a street tree throughout Europe.

Sycamore Maple grows in full sun or partial shade on almost any well-drained soil, acid or alkaline. Sycamore Maple is quite adaptable to various soils and is also highly salt-tolerant. Little pruning is needed to develop a good trunk and branch structure. Develop one leader early in the life of the tree and it will serve you well for many years.

Trunk and branch cankers. Aphids drop honeydew on anything beneath the canopy. Tar-spot fungi cause homeowners much concern, but they are of no consequence.

Acer rubrum Red Maple

Pronunciation: AY-sir ROO-brum

ZONES: 4A, 4B, 5A, 5B, 6A, 6B, 7A, 7B, 8A, 8B, 9A, 9B, (10A)

HEIGHT: 60 to 75 feet

WIDTH: 25 to 35 feet

FRUIT: samara; 1 to 3"; dry; red; attracts wildlife; no significant litter; showy

GROWTH RATE: rapid-growing; moderate life span in urban landscapes, longer-lived on wet sites

HABIT: oval; round; upright; moderate density; irregular silhouette; medium texture

FAMILY: *Aceraceae*

Acer x freemanii

LEAF COLOR: medium green summer color; very showy orange, red, or yellow fall color

LEAF DESCRIPTION: opposite; simple; 3 to 5 lobes; serrate margins; ovate; deciduous; 2 to 4" long

FLOWER DESCRIPTION: red; showy; winter to spring

EXPOSURE/CULTURE:

Light Requirement: part shade to full sun

Soil Tolerances: all textures; acidic; wet soil; salt-sensitive; moderate drought

Aerosol Salt Tolerance: none

PEST PROBLEMS: resistant

BARK/STEMS:

Bark is thin and easily damaged by mechanical impact; branches droop as tree grows and will require early pruning for vehicular or pedestrian clearance beneath canopy; should be grown with a single leader; no thorns

Pruning Requirement: requires pruning to develop strong structure

Limb Breakage: some trees are susceptible to breakage at crotch due to poor collar formation

Current Year Twig: reddish turning gray; medium thickness

CULTIVARS AND SPECIES:

Due to graft incompatibility problems, which cause trees to break apart, preference should be given to cultivars produced on their own roots. In the northern and southern end of the range, choose cultivars with regional adaptation. The cultivars are: 'Armstrong'—upright growth habit, almost columnar, somewhat prone to splitting branches due to tight crotches, 50 feet tall, weak compartmentalizer of decay; 'Autumn Flame'—45 feet tall, dense, uniform, round crown, above-average fall color, good decay compartmentalizer, slower growing than other cultivars; 'Autumn Spire'—broad columnar from Minnesota, great red fall color; 'Bowhall' and 'Karpick'—upright, broadly columnar growth habit, branches form included bark, graft incompatibility, weak compartmentalizer of decay; 'Gerling'—slow-growing, densely branched, broadly pyramidal, about 35 feet tall when mature; 'Northwood'—from northern Minnosota, well adapted for Rocky Mountains and northern part of hardiness range; 'October Glory'—above-average, dependable red to orange fall color, retains leaves late, 60 feet tall, good decay compartmentalizer, best in zones 4 through 7 in the east; 'Red Sunset'—above-average, dependable orange to red fall color, does well in the south into zone 8A, good central leader dominance with small-diameter lateral branches, oval, 50 feet tall, good compartmentalizer; 'Scanlon'—upright growth habit, weak decay compartmentalizer; 'Schlesinger'—good fall color lasting several weeks, rapid growth rate, good heat tolerance, good decay compartmentalizer; 'Silhouette' (zone 5)—brilliant red fall color; 'Tilford'—upright oval-shaped crown. Variety *drummondii* is well suited for the deep south and tolerates saline irrigation and soil.

There is a hybrid cross between Red and Silver Maple called Freeman Maple (*Acer x freemanii*) (zone 4A). Cultivars of this hybrid include 'Armstrong', with a narrow columnar crown to 60 feet tall; 'Autumn Blaze', with an oval crown to 50 feet tall; 'Celebration', with a narrow upright crown and a strong central leader to 50 or 60 feet tall; 'Celzam', with a narrow oval crown to 50 feet tall; 'Firedance', with outstanding red fall color; 'Jeffersred', with good heat tolerance; and 'Scarlet Sentenial', with great fall color and oval crown to 40 feet tall. The culture of these trees is similar to Red Maple.

LANDSCAPE NOTES:

Origin: native to eastern North America

Usage: buffer strip; highway swales; near a deck or patio; reclamation; shade tree; street tree

Continued

Acer rubrum *Continued*

The outstanding ornamental characteristic of Red Maple is red, orange, or yellow fall color (sometimes on the same tree) lasting several weeks. Red Maple is often one of the first trees to color up in autumn, and puts on one of the most brilliant displays of any tree, but trees vary greatly in fall color and intensity. Cultivars are more consistently colored.

Grows best in wet places; has no other particular soil preference, except chlorosis may develop on alkaline soil. Unless irrigated or on a wet site, Red Maple is best used north of zone 9. Well-suited as a street tree in northern and mid-south climates in residential and other suburban areas, but bark is thin and easily damaged by mowers. Susceptible to trunk decay. Irrigation is often needed to support street tree plantings in well-drained soil in the south. Roots can raise sidewalks, as with Silver Maples, but Red Maples have a less aggressive root system and so make a good street tree. Surface roots beneath the canopy can make mowing difficult.

Red Maple is easily transplanted and usually develops surface roots in soil ranging from well-drained sand to clay. Roots may regenerate slowly on fall-transplanted trees. It is not especially drought-tolerant, particularly in the southern part of the range, but selected individual trees can be found growing on dry sites. Branches often grow upright through the crown, forming poor attachments to the trunk. These should be removed as soon as they are seen in the nursery or after planting in the landscape to help prevent branch failure in older trees during storms. Select branches with a wide angle to the trunk and prevent branches from growing larger than half the diameter of the trunk. Some cultivars have better structure.

Roots occasionally girdle the trunk. Flathead borer attacks young, stressed trees. Leafhoppers crinkle leaves and can become serious. Twig borers cause die-back on small twigs. Nectria cankers may cause deformities in the trunk and branches. Nutrient deficiency symptoms appear in soil with a pH above 7. Bacterial leaf scorch can cause serious damage. Drought, pollution, and salt injures Red Maples in the central United States.

Acer saccharinum Silver Maple

Pronunciation: AY-sir sack-ah-RYE-num

ZONES: 3A, 3B, 4A, 4B, 5A, 5B, 6A, 6B, 7A, 7B, 8A, 8B, 9A, 9B

HEIGHT: 60 to 80 feet

WIDTH: 40 to 50 feet

FRUIT: copious; elongated; 1 to 2"; dry; green turning brown; fruit, twigs, and foliage cause significant litter; showy

GROWTH RATE: rapid-growing; long-lived, especially on moist sites

HABIT: upright; vase shape; moderate density; irregular silhouette; medium texture

FAMILY: *Aceraceae*

LEAF COLOR: green summer color; showy yellow fall color

LEAF DESCRIPTION: opposite; simple; star-shaped; deciduous; 3 to 7" long; some leaves drop in drought

FLOWER DESCRIPTION: yellow to red; showy; spring

EXPOSURE/CULTURE:

Light Requirement: part shade to full sun

Soil Tolerances: all textures; slightly alkaline to acidic (can tolerate very low pH); wet soil; moderate salt; drought

Aerosol Salt Tolerance: moderate

PEST PROBLEMS: sensitive

BARK/STEMS:

Branches on young trees droop as tree grows and require pruning for vehicular or pedestrian clearance beneath canopy; bark not particularly showy; should be grown with a single leader; no thorns

Pruning Requirement: requires pruning to develop good structure; thin the most aggressive, upright branches to prevent formation of included bark; prune in dormant season to minimize bleeding

Limb Breakage: susceptible to breakage, especially in ice storms

Current Year Twig: brown to reddish; medium to slender

CULTIVARS AND SPECIES:

'Pyramidale'—narrow crown; 'Silver Queen'—lower leaf surfaces silvery; 'Skinneri'—somewhat weeping, dissected leaves with a central leader and a better branching habit (may be the best Silver Maple); 'Weiri'—cutleaved form with pendulous branches. See *Acer rubrum,* Red Maple, for information on a hybrid cross between Red and Silver Maple.

LANDSCAPE NOTES:

Origin: native to eastern North America, except the coastal plain

Usage: reclamation and soil stabilization; shade tree; urban tolerant

Silver Maple grows best along streambanks and flood plains and is easy to transplant. Fast growth and large size makes this tree popular in the high plains region. Save for planting in areas where nothing else will thrive. Roots often grow on the surface of the soil, making mowing grass difficult under the canopy. They also are aggressive and invasive, growing into septic tank drain fields and into broken water and sewer pipes. It is hard to plant shrubs and other plants beneath the branches due the dense root system. Branches tend to droop, almost weep, forming a graceful canopy outline as the tree grows older. The attraction of bright yellow fall color may be offset by the abundant number of leaves to rake.

Silver Maple will grow in areas that have standing water for several weeks at a time. It grows best on acid soil that remains moist, but adapts to very dry, alkaline soil. Leaves may scorch in areas with restricted soil space during dry spells in the summer, but this will seldom harm the tree. There are too many other superior trees to warrant wide use of this species, but it has a place in tough sites away from buildings, pavement, and people. Trees suffer in hot climates.

Weak structure, messy habit, and aggressive roots. Chlorosis appears on trees in alkaline soil. Seedlings sprout everywhere. Trees are susceptible to many pest problems, including verticillium wilt, some of which may cause severe problems. Some communities prohibit planting Silver Maple.

Acer saccharum Sugar Maple

Pronunciation: AY-sir SACK-ah-rum

ZONES: 3A, 3B, 4A, 4B, 5A, 5B, 6A, 6B, 7A, 7B, 8A

HEIGHT: 50 to 80 feet

WIDTH: 35 to 50 feet

FRUIT: samara; 1 to 3"; dry; green turning brown; no significant litter

GROWTH RATE: moderate growth rate, long-lived

HABIT: oval; round; dense; symmetrical; medium to coarse texture

FAMILY: *Aceraceae*

LEAF COLOR: medium green summer color; striking orange, red, or yellow fall color

LEAF DESCRIPTION: opposite; simple; 5 lobes; star-shaped; deciduous; 4 to 6"

FLOWER DESCRIPTION: green to yellow; spring

EXPOSURE/CULTURE:

Light Requirement: shade to full sun

Soil Tolerances: all textures; slightly alkaline to acidic; salt-sensitive; moderate drought

Aerosol Salt Tolerance: none

PEST PROBLEMS: verticillium wilt, aphids, bacterial leaf scorch, leaf hoppers (although more resistant than Red Maples)

BARK/STEMS:

Branches droop slightly as tree grows and will require early pruning for vehicular or pedestrian clearance beneath canopy; showy, gray trunk; should be grown with a single leader; no thorns

Pruning Requirement: needs little pruning to develop a strong trunk and branch structure

Limb Breakage: resistant

Current Year Twig: brown; medium to slender

CULTIVARS AND SPECIES:

Nurseries may offer one or several cultivars of Sugar Maple: 'Bonfire'—brilliant orange-red fall color, resists leaf hoppers; 'Caddo'—superior drought tolerance; 'Endowment Columnar'—50 feet tall, columnar form, red and yellow fall color, reportedly scorches less than the species and is more drought-tolerant; 'Globosum'—slow grower with dense round crown and mature height of about 10 feet; 'Goldspire'—dense, compact, pyramidal form, 10 to 12 feet wide, gold fall color; 'Green Mountain' (a specific cross)—upright oval crown and consistently good scarlet fall color, may be more resistant to leaf scorch, compartmentalizes wounds well; 'Majesty'—ovate form, resistant to frost crack and sunscald; 'Newton Sentry'—an old selection, upright growth habit, 15 feet wide; 'Sweet Shadow'—cut leaves with an open habit when young, filling in later; 'Temple's Upright'—upright, pole-like growth habit, 4 to 10 feet wide. There are other cultivars, including 'Commemoration' and 'Legacy', which are reportedly resistant to leaf tatter. *A. leucoderme* is a small, drought-tolerant tree native to the southern United States. 'Legacy' and 'Green Mountain' are popular in the Rocky Mountains, and are very hardy.

LANDSCAPE NOTES:

Origin: north central to northeastern North America

Usage: bonsai; wide tree lawns; shade tree; street tree

Sugar Maple is one of the most common maples in the east and is a hard-wooded tree with a moderate to slow growth rate. Sugar Maple grows about 1 foot each year in most soils but is sensitive to reflected heat and drought; leaves often scorch along their edges. Leaf scorch from dry soil is often evident in areas where the root system is restricted to a small soil area, such as a street tree planting. It is more drought-tolerant in open areas, so plant where the roots can expand into a large soil space.

The thick, branched canopy creates dense shade and will prevent good lawn growth unless lower branches are removed. Branches are usually well attached to trunks, resulting in a low branch failure rate. The main ornamental feature of the tree is the brilliant fall color. The tree transplants fairly easily but may develop girdling roots, which can reduce growth or even kill the tree.

The limbs of Sugar Maple are usually strong and not susceptible to wind damage. The bark forms attractive bright gray plates that stand out, especially during the winter. Roots are often shallow and reach the surface at an early age, even in sandy soil. Plant where grass below need not be mowed, so the roots will not be damaged by the mower. A variety of birds use the tree for food, nesting, and cover, and the fruits are especially popular with squirrels. Thirty-five to forty gallons of sap are required to produce a gallon of maple syrup.

Acer tataricum Tatarian Maple

Pronunciation: AY-sir ta-TAR-ih-kum

ZONES: 3A, 3B, 4A, 4B, 5A, 5B, 6A, 6B, 7A, 7B

HEIGHT: 20 to 25 feet

WIDTH: 25 to 35 feet

FRUIT: samara; showy, oval; 1"; dry; green turning red or pink; no significant litter; very showy

GROWTH RATE: moderate growth rate; long-lived

HABIT: round and spreading; dense; symmetrical; fine texture

FAMILY: *Aceraceae*

LEAF COLOR: medium green summer color; showy, yellow to red fall color

LEAF DESCRIPTION: opposite; simple; lobed; double serrate to serrate; ovate; deciduous; 1 to 3" long

FLOWER DESCRIPTION: greenish to yellow; pleasant fragrance; showy; spring flowering for 2 weeks

EXPOSURE/CULTURE:

Light Requirement: part shade to full sun

Soil Tolerances: all textures; alkaline to acidic; occasionally wet soil; some salt and drought

PEST PROBLEMS: nearly pest-free; resists verticillium wilt

BARK/STEMS:

Bark is thin and easily damaged by mechanical impact; branches droop as tree grows and will require pruning for vehicular or pedestrian clearance beneath canopy; routinely grown with multiple trunks; can be trained to grow with a single trunk; no thorns

Pruning Requirement: requires pruning to develop one trunk and strong structure

Limb Breakage: resistant

Current Year Twig: brown or gray; medium to slender

CULTIVARS AND SPECIES:

A. ginnala is very similar to Tatarian Maple and can be used in much the same manner.

LANDSCAPE NOTES:

Origin: Europe, Asia Minor, southern Russia

Usage: bonsai; container or above-ground planter; hedge; near a deck or patio; specimen; street tree; urban tolerant

Tatarian Maple is an excellent, low-growing tree for small yards, patios, and other small-scale landscapes. It can be grown as a multi-stemmed clump or can be trained into a small tree with a single trunk up to 6 to 10 feet tall. The tree has an upright, rounded, finely branched, growth habit on several thick main branches, which creates dense shade under the crown. Trunks droop and spread as they grow older, forming a canopy that is wider than tall; consider this when planting close to walks and buildings. Tatarian Maple can grow rapidly when young if it receives water and fertilizer, but it is well-suited for planting close to power lines, as it slows down and remains small at maturity.

The main ornamental value is the brilliant red fall foliage color and fruit that sports bright pink wings. It is a durable tree, tolerating poor soil, but will grow less vigorously in the southern end of its range. It will leaf-scorch in dry summers in full sun, but is very drought-tolerant and will not die back. Most drought-tolerant in partial shade.

The plant is sometimes used in hedges or screens, and can be used for planting along streets beneath power lines, provided it is set back from the road. Without early thinning and training, the multiple trunks droop considerably. Main branches and trunks can be headed back several times when the tree is young to create a more upright form. Remove lower branches early to force growth in the upper canopy for planting along streets and for other areas where clearance is needed beneath the canopy. The tree makes a nice specimen or patio tree.

Acer triflorum Three-Flowered Maple

Pronunciation: AY-sir try-FLOOR-um

ZONES: 3B, 4A, 4B, 5A, 5B, 6A, 6B, 7A, 7B

HEIGHT: 15 to 20 feet

WIDTH: 15 to 25 feet

FRUIT: samara; 1 to 1.5"; dry; no significant litter; showy

GROWTH RATE: slow-growing; long-lived

HABIT: round; dense; symmetrical; fine texture

FAMILY: *Aceraceae*

LEAF COLOR: light green summer color; exceptionally showy orange or red fall color

LEAF DESCRIPTION: opposite; trifoliate; entire margin; deciduous; 2 to 3.5" long

FLOWER DESCRIPTION: green; not showy; spring

EXPOSURE/CULTURE:

Light Requirement: part to full shade

Soil Tolerances: loam; sand; slightly alkaline to acidic; moderate salt; drought-sensitive

Aerosol Salt Tolerance: moderate

PEST PROBLEMS: resistant

BARK/STEMS:

Drooping low branches require pruning to create clearance beneath canopy; very showy trunk form and papery bark; can be grown with single or multiple trunk; no thorns

Pruning Requirement: requires little pruning to develop strong structure

Limb Breakage: resistant

Current Year Twig: brown; medium to slender

LANDSCAPE NOTES:

Origin: Manchuria, Korea

Usage: bonsai, container or above-ground planter; near a deck or patio; standard; specimen; street tree

Three-Flowered Maple slowly grows to mature height with an equal width. The dense, rounded canopy casts dense shade below. The 2- to 3.5-inch-long, compound leaves are slightly hairy and turn attractive shades in the fall before dropping. True to its name, Three-Flowered Maple produces three, greenish-yellow flowers clustered together in springtime; these are followed by 1- to 1.5-inch-long winged seeds that persist on the tree. The attractive red/brown to tan bark, which peels off in long, thin strips, is one of the most striking attributes of this tree. Although hardy in zone 3B, some seed sources may only be hardy to zone 4.

This handsome tree is well suited for use as a specimen planted in the lawn or in a low ground cover bed. Its shade tolerance gives it a special place in an established landscape. The showy bark and fine texture create a striking display during the growing season. It provides interest in the winter when branches are bare, showing the darkened bark, which contrasts nicely against snow. Once the main structure of the tree has been developed, lower branches can be thinned to allow the bark to show its true beauty. Thin some small branches on the outside of the canopy so you can see the inside of the canopy.

Three-Flowered Maple grows best in partial shade on well-drained soil. It does best when protected from sun in the afternoon. For this reason, it is usually not well suited for planting in new subdivisions without an overstory of existing trees. Due to its shallow root system, Three-Flowered Maple will not tolerate soil compaction or drought. It does not tolerate soil that remains wet for long periods of time. It is hard to come by in nurseries.

Acer truncatum Purpleblow or Shantung Maple

Pronunciation: AY-sir trun-KAY-tum

ZONES: (3A), 3B, 4A, 4B, 5A, 5B, 6A, 6B, 7A, 7B, 8A

HEIGHT: 20 to 35 feet

WIDTH: 20 to 35 feet

FRUIT: samara; oval; 1"; dry; green; no significant litter

GROWTH RATE: moderate growth rate; moderate life span

HABIT: round and spreading; dense; symmetrical; fine texture

FAMILY: *Aceraceae*

LEAF COLOR: green summer color; showy red, orange, or yellow fall color

LEAF DESCRIPTION: opposite; simple; 5 lobes; entire margin; star-shaped and glossy; deciduous; 3 to 5" long

FLOWER DESCRIPTION: yellow; showy; spring flowering for 2 weeks

EXPOSURE/CULTURE:

Light Requirement: part shade to full sun

Soil Tolerances: all textures; slightly alkaline to acidic; moderate salt; drought

PEST PROBLEMS: nearly pest-free

BARK/STEMS:

Bark is thin and easily damaged by mechanical impact; branches droop as tree grows and will require pruning for vehicular or pedestrian clearance beneath canopy; routinely grown with multiple trunks; can be trained to grow with a single trunk; no thorns

Pruning Requirement: requires pruning to develop one trunk and strong structure

Limb Breakage: resistant

Current Year Twig: brown or gray; medium to thin

CULTIVARS AND SPECIES:

A. ginnala and *A. tataricum* are very similar to Purpleblow Maple and can be used in much the same manner. *A. platanoides* is related to and hybridizes with *A. truncatum*. Some hybrids are marketed.

LANDSCAPE NOTES:

Origin: China

Usage: container or above-ground planter; hedge; near a deck or patio; large parking lot island; buffer strip; specimen; street tree

Purpleblow or Shantung Maple is an excellent, low-growing tree for small yards, patios, and other small-scale landscapes. It can be grown as a multi-stemmed clump or can be trained into a small tree with a single trunk up to about 6 feet tall. The tree has an upright, rounded, finely branched growth habit on several thick main branches, which creates dense shade under the crown. Trunks droop and spread as they grow older, forming a canopy that is wider than tall. The drooping, spreading form should be considered when locating the tree close to walks. Purpleblow Maple can grow rapidly when it is young if it receives water and fertilizer, but it is well suited for planting close to power lines because it slows down and remains small at maturity.

The main ornamental value of Purpleblow Maple is the brilliant fall foliage color. It is a durable tree, tolerating poor soil, but will grow less vigorously in the southern end of its range. It will leaf-scorch very little in dry summers in full sun and is very drought-tolerant for a maple. It appears to tolerate the reflected light typical of urban areas very well.

The plant is sometimes used in hedges or screens, and can be used for planting along streets beneath power lines, provided it is set back from the road. Without early thinning and training, the multiple trunks droop considerably. Main low branches and trunks can be headed back several times when the tree is young to create a more upright form. Remove lower branches early to force growth in the upper canopy for planting along streets and for other areas where clearance is needed beneath the canopy. Makes a nice specimen or patio tree.

Acoelorrhaphe wrightii Paurotis Palm

Pronunciation: ah-see-leh-RAY-fee RYE-tee-eye

ZONES: 9B, 10A, 10B, 11

HEIGHT: 15 to 25 feet

WIDTH: 10 to 15 feet

FRUIT: round; less than ½"; fleshy; black; no significant litter; showy

GROWTH RATE: slow-growing; long-lived

HABIT: palm; upright; open; irregular silhouette; fine texture

FAMILY: *Arecaceae*

LEAF COLOR: blue-green with silver backs

LEAF DESCRIPTION: alternate; star-shaped; 24 to 36" long blade; sharp spines on petiole

FLOWER DESCRIPTION: creamy yellow; showy, 4' long inflorescence; spring and summer

EXPOSURE/CULTURE:

Light Requirement: shade to full sun

Soil Tolerances: all textures; slightly alkaline to acidic; wet soil; moderate drought

Aerosol Salt Tolerance: moderate

PEST PROBLEMS: resistant

BARK/STEMS:

Routinely grown with multiple, slender trunks 6–8" in diameter; trunks grow mostly upright and will not droop; showy trunk; thorns on petioles

Pruning Requirement: needs little pruning to develop a strong trunk

Limb Breakage: resistant

LANDSCAPE NOTES:

Origin: native to south Florida

Usage: container or above-ground planter; buffer strip; near a deck or patio; reclamation; specimen; wet sites

This striking fan palm has several to many showy, upright, slender trunks that form attractive tight clumps with multiple suckers clustered at the base. This ensures that there are trunks of different heights on the palm at all times if suckers are not pruned off. Remove the suckers to prevent formation of additional trunks. Removing suckers also keeps the spines on the palm out of reach by preventing new fronds from developing on the lower trunks. Fruits are borne on bright orange stalks that can be quite showy in the fall. Eventually reaching mature height with a variable spread, the Paurotis Palm is highly desirable in the landscape. It is exceptionally attractive with nighttime lighting from below.

A native of the Florida Everglades, Paurotis Palm is hardy to about 25 to 28°F and prefers rich, moist locations. In many years, it can be grown as far north as zone 9B. Growing in full sun or partial shade, growth is considerably slower in drier soils; this tree could decline and die without irrigation on a dry, well-drained site. Paurotis Palm is moderately tolerant of salt spray, so it can be located in a protected spot along the coast. This beautiful palm is popular as an accent or specimen for large residential, commercial, or municipal landscapes where it can be seen in an open setting. Set it in a bed of Plumbago or other sprawling shrub or groundcover to create a striking accent. Unfortunately, it grows slowly, and it is hard to come by in quantity in the trade.

Paurotis Palm is susceptible to chlorosis from micronutrient deficiencies (especially manganese) when grown in soil with a high pH. It may also show deficiencies of potassium in sandy soils unless appropriate fertilizers are applied regularly. The spines on the petioles can inflict pain, so locate the trees where children will not contact them, or remove the lower fronds and developing trunks.

Aesculus x carnea Red Horsechestnut

Pronunciation: ESS-kew-luss x CAR-nee-ah

ZONES: 5A, 5B, 6A, 6B, 7A, 7B

HEIGHT: 30 to 45 feet

WIDTH: 30 to 45 feet

FRUIT: 1.5"; causes significant litter

GROWTH RATE: slow-growing, long-lived

HABIT: pyramidal, then rounded; dense; symmetrical; coarse texture

FAMILY: *Hippocastanaceae*

LEAF COLOR: dark green in summer, no significant fall color change

LEAF DESCRIPTION: opposite; palmate; serrate margin; oblanceolate; deciduous; 4 to 8" long

FLOWER DESCRIPTION: pink or red; very showy and delightful large clusters; spring flowering with foliage

EXPOSURE/CULTURE:

Light Requirement: full sun for best growth and flowering

Soil Tolerances: all textures; slightly alkaline to acidic; occasionally wet soil; moderate salt and drought

Aerosol Salt Tolerance: moderate

PEST PROBLEMS: resistant

BARK/STEMS:

Branches droop somewhat as tree grows and will require some pruning for vehicular or pedestrian clearance beneath canopy; tree grows with several trunks but should be trained to grow with a single trunk; no thorns

Pruning Requirement: needs some pruning to develop a strong central trunk and good branch structure

Limb Breakage: some

Current Year Twig: brown; stout

CULTIVARS AND SPECIES:

'Briotii' has bright red flowers, is nearly fruitless, and is usually planted in place of the species; 'O'Neil' has red flowers; 'Rosea' has pink flowers.

LANDSCAPE NOTES:

Origin: hybrid

Usage: large container; parking lot island; buffer strip; highway; shade tree; specimen; sidewalk cutout; street tree; urban tolerant

This old hybrid of *Aesculus hippocastanum* and *Aesculus pavia* has leaves composed of 5 to 7 leaflets, and is popular for the showy red flowers and interesting leaf texture. Specimens exist in the United States up to 50 feet tall. The multitude of pink to bright scarlet blooms appear on erect, 8-inch-long panicles at each branch tip and are quite attractive to bees and hummingbirds. The tree is a wonderful sight in full bloom.

Red Horsechestnut will grow in full sun or light shade; it prefers moist, well-drained, acid soils, but also grows in slightly alkaline soil. Plants are moderately tolerant to drought, wind, and salt and resist the heat of the south very well. Holds up well in urban areas, even in restricted and compacted soil spaces. Trunk bark may crack when exposed to direct sun, so keep it shaded as much as possible by leaving lower branches on small trees, and do not overprune the tree, exposing the trunk suddenly to direct sun. Leaf litter in the fall may be objectionable, as the leaves are large and decompose slowly.

Trees in open soil look very nice throughout the summer. Makes a great street tree but will leaf scorch and drop leaves in extended drought. Much less susceptible (not immune) to leaf scorch and leaf blotch than *Aesculus hippocastanum*, and should be planted in its place, although leaves drop in prolonged drought nearly every year. Susceptible to rust, mildew, and Japanese beetle. Continuously dropping something, whether twigs, foliage, fruit, and/or flowers. Recovers slowly from transplanting.

Aesculus flava (octandra) Yellow Buckeye

Pronunciation: ESS-kew-luss FLAY-vah (okt-AN-druh)

ZONES: 4A, 4B, 5A, 5B, 6A, 6B, 7A, 7B, 8A

HEIGHT: 60 to 75 feet

WIDTH: 25 to 35 feet

FRUIT: oval; 2 to 3"; dry and hard; brown, smooth, shiny; attracts mammals; significant litter

GROWTH RATE: moderate; long-lived

HABIT: oval; dense; symmetrical; coarse texture

FAMILY: *Hippocastanaceae*

LEAF COLOR: dark green in summer; brilliant pumpkin yellow fall color

LEAF DESCRIPTION: opposite; palmate; serrate margins; oval; deciduous; 4 to 7" long; showy

FLOWER DESCRIPTION: yellow/green; very showy; spring

EXPOSURE/CULTURE:

Light Requirement: grows best in full sun

Soil Tolerances: all textures; slightly alkaline to acidic; occasionally wet soil; moderate drought

Aerosol Salt Tolerance: moderate

PEST PROBLEMS: resistant

BARK/STEMS:

Branches droop as tree grows and will require pruning for vehicular or pedestrian clearance beneath canopy; grows with several trunks but can be trained to grow with a single trunk; no thorns

Pruning Requirement: requires pruning to develop strong structure

Limb Breakage: some minor branch breakage can be expected

Current Year Twig: brownish-green; very stout

LANDSCAPE NOTES:

Origin: Appalachian mountains of North America

Usage: buffer strip; highway; shade tree; specimen

Yellow Buckeye has an oval to slightly spreading canopy. The leaves, composed of five leaflets, cast dense shade below. The thick canopy makes this well suited for a tall screen or shade tree. Small yellow/green flowers appear in dense, upward-pointing, 6- to 7-inch-long terminal panicles in early spring and are followed by 2.5-inch-long, smooth, pear-shaped capsules containing bitter, poisonous seeds. A tree in full flower is wonderful to behold.

The tree has a striking, coarse texture and leaves that are as dark as any other tree, attracting attention as a specimen plant. Save it for large areas so the wonderful form of this medium-large tree can be appreciated. Not suited for small residences, due to its overpowering size and texture. Leaf and flower litter in summer and fall may be objectionable to some. The nuts make good food for wildlife but are not desirable scattered along city streets.

A North American native, Yellow Buckeye grows best along stream beds in full sun or partial shade and should be planted in moist, well-drained soil rich in organic matter.

This and the Red Horsechestnut are less sensitive to diseases and scorch than other buckeyes. Mildew and lacebugs are usually minor problems compared to other buckeyes. Leaves often drop in the summer. Fruit, twigs, and foliage cause significant litter.

Aesculus glabra Ohio Buckeye

Pronunciation: ESS-kew-luss GLAY-brah

ZONES: 4A, 4B, 5A, 5B, 6A, 6B, 7A

HEIGHT: 50 to 70 feet

WIDTH: 40 to 50 feet

FRUIT: oval to round; 1 to 2"; dry and hard; brown; attracts mammals; significant litter; showy

GROWTH RATE: moderate; long-lived

HABIT: oval; round; dense; symmetrical; coarse texture

FAMILY: *Hippocastanaceae*

LEAF COLOR: bright green; good orange to yellow fall color if the summer was not very dry

LEAF DESCRIPTION: opposite; palmate; serrate margin; oval to ovate; deciduous; 3 to 8" long

FLOWER DESCRIPTION: yellow; showy; spring

EXPOSURE/CULTURE:

Light Requirement: part shade/part sun; grows best in full sun

Soil Tolerances: all textures; slightly alkaline to acidic; wet soil; moderate drought

Aerosol Salt Tolerance: none

PEST PROBLEMS: nonlethal foliage pests are troublesome; otherwise none

BARK/STEMS:

Branches droop as tree grows and will require pruning for vehicular or pedestrian clearance beneath canopy; should be grown with a single leader; no thorns

Pruning Requirement: requires early pruning to develop strong structure

Limb Breakage: mostly resistant

Current Year Twig: reddish-brown; stout

CULTIVARS AND SPECIES:

'Autumn Splendor' resists leaf scorch and has clean, dark green foliage with splendid purple fall color; 'Homestead' has wonderfully bright red fall color.

This is a variable species with a number of named, naturally occurring varieties.

LANDSCAPE NOTES:

Origin: western Pennsylvania to Texas

Usage: shade tree

Ohio Buckeye forms a short trunk with a low branching structure, creating an oval to rounded canopy. The coarse, palmately compound leaves have an unpleasant odor when crushed, as do the twigs, giving this tree its other common name of Fetid Buckeye. The fragrance should be a warning sign for this plant, as all parts of it are poisonous, a factor that should be considered when placing it in the landscape. One of the first trees to shed its leaves in autumn, Ohio Buckeye puts on a vivid display of bright orange and yellow fall foliage. The spring flowers are also quite showy, with erect, 6-inch-high panicles decorating the branches. The fruits ripen in late summer and may become a litter problem because they are very prickly. These fruits are quite popular with squirrels and other wildlife.

Ohio Buckeye is naturally found on moist stream banks, occasionally on drier land where it becomes a shrub. It will grow in full sun or partial shade on well-drained, fertile soil but should not be exposed to extended drought or excessive heat, two factors that make this tree unsuitable for use as a street tree. Falling fruit, twigs, and foliage would also be a problem along streets. It is suited for naturalized plantings or for establishing a native grove in a park.

Subject to leaf blotch, powdery mildew, and leaf scorch. Leaves dry up, turn brown, and fall from the tree in summer drought. This usually does not hurt the tree, but creates a mess in the landscape. It almost seems as though one is always picking something up from under this tree.

Aesculus hippocastanum Horsechestnut

Pronunciation: ESS-kew-luss hip-poe-CAS-tah-num

ZONES: 4A, 4B, 5A, 5B, 6A, 6B, 7A, 7B

HEIGHT: 50 to 80 feet

WIDTH: 40 to 50 feet

FRUIT: oval; round; 1 to 2"; dry and hard; brown; attracts mammals; significant litter; showy

GROWTH RATE: moderate growth rate; long-lived

HABIT: oval; dense; symmetrical; coarse texture

FAMILY: *Hippocastanaceae*

LEAF COLOR: dark green in summer; yellow fall color

LEAF DESCRIPTION: opposite; palmate; double serrate margin; obovate; deciduous; 6 to 10" long

FLOWER DESCRIPTION: white; showy; spring

EXPOSURE/CULTURE:

Light Requirement: full sun

Soil Tolerances: loam; sand; slightly alkaline to acidic; occasionally wet soil; salt; moderate drought

Aerosol Salt Tolerance: moderate

PEST PROBLEMS: sensitive to nonlethal pests including leaf blotch disease which causes significant defoliation

BARK/STEMS:

Exfoliating bark on a thick trunk; branches droop as tree grows and will require pruning for vehicular or pedestrian clearance beneath canopy; should be grown with a single leader; no thorns

Pruning Requirement: requires early pruning to develop strong structure; thin branches with included bark

Limb Breakage: somewhat sensitive

Current Year Twig: brown; stout

CULTIVARS AND SPECIES:

Several cultivars can be found, but also have leaf scorch problems. 'Baumannii' grows 70 feet tall with an oval canopy, has double flowers, grows slower, and does not produce any nuts. This cultivar is recommended over the species.

LANDSCAPE NOTES:

Origin: Albania, northern Greece and Bulgaria

Usage: shade tree; specimen; street tree; urban tolerant

Horsechestnut can grow 90 or 100 feet tall, but is often 50 to 80 feet in the landscape. The prominent white flowers, occurring in panicles at the branch tips, are the main ornamental feature of Horsechestnut. The large brown nuts covered with spiny husks fall and can dent cars and create a hazard on hard surfaces if people roll on the golfball-sized fruit.

The growth rate is rapid while the tree is young but slows down with age. Horsechestnut is weak-wooded and some branches break from the trunk under ice or snow loads. Japanese beetles eat the leaves down to the veins. The tree usually develops leaf scorch in dry soil or summer drought, and powdery mildew during the summer, causing leaves to drop (though this will not kill the tree). Leaf blotch disease is also common, and turns leaves brown in summer. This limits its wide use in any one area, but any large landscape would benefit from a Horsechestnut or two. It is also very tolerant of city conditions, does well in small, restricted root zone areas, such as along streets, and is well suited to pollarding, a practice used to keep trees small.

Horsechestnut prefers a sunny exposure sheltered from wind, and casts dense shade with its coarse-textured leaves. The tree is easily transplanted and grows in almost any urban soil, including alkaline. The large leaves, nuts, and dropping twigs create litter that is considered objectionable by many. It is probably best located away from hard surfaces, where people can enjoy it from a distance.

Aesculus indica Indian Horsechestnut

Pronunciation: ESS-kew-luss IN-duh-cah

ZONES: 7A, 7B, 8A, 8B, 9A, 9B, 10A

HEIGHT: 40 to 60 feet

WIDTH: 35 to 50 feet

FRUIT: round; 1 to 2"; dry and hard; attracts mammals; significant litter; showy

GROWTH RATE: slow-growing; long-lived

HABIT: oval; round; dense; symmetrical; coarse texture

FAMILY: *Hippocastanaceae*

LEAF COLOR: dark green in summer; orange fall color

LEAF DESCRIPTION: opposite; palmate; serrated, undulating margin; lanceolate to obovate; deciduous; 6 to 8" long

FLOWER DESCRIPTION: white; showy; summer

EXPOSURE/CULTURE:

Light Requirement: full sun

Soil Tolerances: all textures; alkaline to acidic; moderate drought

Aerosol Salt Tolerance: none

PEST PROBLEMS: resistant

BARK/STEMS:

Branches droop as tree grows and will require pruning for vehicular or pedestrian clearance beneath canopy; often grows with many trunks, but should be grown with a single leader; no thorns

Pruning Requirement: needs some pruning to develop a strong trunk and branch structure

Limb Breakage: mostly resistant

Current Year Twig: brown to gray; stout

CULTIVARS AND SPECIES:

The cultivar 'Sydney Pearce' has richer green leaves and flower spikes 12 inches high. *A. californica* is a California native with showy spring flowers that drops leaves in summer to cope with the dry season.

LANDSCAPE NOTES:

Origin: Himalayas

Usage: buffer strip; highway; shade tree; specimen

Indian Horsechestnut is a large, rounded tree, reaching up to 100 feet in height, with smooth, grayish-red bark. Trees grown in an open landscape setting probably reach about 40 to 60 feet tall. In June and July, the tree is decorated with upright panicles of white blooms; the flowers stalks most often seen are 4 to 6 inches high, but they may be much larger. These blooms are followed by the production of a spiny, green fruit that holds several brown seeds.

The tree has not been made extensively available, but could make a nice park or landscape tree for a large commercial landscape or an estate. It has been successfully grown in California and in the northwest part of the United States, and could be tried in the east. The cold-hardiness of the plant is uncertain. The coarse texture, low branching habit, and uniformly round canopy make it stand out among other trees. Children would enjoy climbing this well-branched tree. The beauty of this tree calls for some trials in the eastern part of the country.

The pests and diseases affecting this tree are not well understood, due to the limited experience with this tree.

Aesculus pavia Red Buckeye

Pronunciation: ESS-kew-luss PAY-vee-ah

ZONES: 6A, 6B, 7A, 7B, 8A, 8B, 9A

HEIGHT: 15 to 20 feet

WIDTH: 15 to 25 feet

FRUIT: round; 1 to 2"; dry and hard; brown; attracts mammals; significant litter; showy

GROWTH RATE: slow to moderate; moderate life span

HABIT: oval; round; moderate density; irregular when young, becoming symmetrical; coarse texture

FAMILY: *Hippocastanaceae*

LEAF COLOR: dark green in summer; no fall color change

LEAF DESCRIPTION: opposite; palmate; double serrate margin; oval to obovate; deciduous; 4 to 8" long

FLOWER DESCRIPTION: red; showy; spring; pollinated by hummingbirds

EXPOSURE/CULTURE:

Light Requirement: shade to full sun; growth is shrubby and open in shade

Soil Tolerances: all textures; slightly alkaline to acidic; wet soil; moderate salt, moderate drought

Aerosol Salt Tolerance: moderate

PEST PROBLEMS: nearly pest-free

BARK/STEMS:

Branches on a young tree droop as tree grows, and will require pruning for vehicular or pedestrian clearance beneath canopy; once tree is trained, trunks and branches remain upright; routinely grown with, or trainable to be grown with, multiple trunks; can be trained to grow with a single trunk; no thorns

Pruning Requirement: requires pruning to develop a single trunk, then very little is needed

Limb Breakage: resistant

Current Year Twig: brown; stout

CULTIVARS AND SPECIES:

'Atrosanguinea' has deeper red flowers; var. *splendens* has bright red flowers; yellow-flowering forms exist in Texas.

LANDSCAPE NOTES:

Origin: southeastern United States

Usage: container; buffer strip; highway; near a deck or patio; reclamation; small shade tree; specimen; street tree

Red Buckeye is capable of reaching 25 to 30 feet tall in the wild, but most often grows 15 to 20 feet high in cultivation. Red Buckeye is most popular for its springtime display of 3- to 6-inch-long, upright, terminal panicles composed of 1.5-inch-wide, red flowers, which are quite attractive to hummingbirds. Trees flower at 3 years old. These blooms are followed by flat, round capsules that contain bitter and poisonous seeds. The large palmate leaves usually offer no great color change in fall and often drop as early as late September.

The coarse, open structure and the light brown, flaky bark are quite attractive and offer great winter landscape interest. Several trunks typically arise from the ground, forming a shrub unless pruned into a tree, but this effort is rewarded with rich red flowers in the spring along with the new growth. Main branches begin forming low on the trunk and remain there when grown in full sun.

Best used as a novelty patio tree or as part of a shrubbery border to add bright red color for several weeks in the spring. Plant it in a medium- to large-sized residential landscape as a very coarse accent. Lower branches can be removed to allow clearance beneath the crown for use as a street tree. When planted in the open, allow branches to fully develop to the ground. Red Buckeye will flower well in rather dense shade, but does best in full sun. It is native along moist stream banks, but is moderately drought-tolerant.

The seeds are poisonous. Fruit is abundant in some years and can litter the ground, as can twigs and foliage. Unfortunately, asexual propagation is often challenging.

Ailanthus altissima Tree-of-Heaven

Pronunciation: ale-AN-thuss al-TISS-sih-mah

ZONES: 5A, 5B, 6A, 6B, 7A, 7B, 8A, 8B, 9A, 9B, 10A

HEIGHT: 60 to 75 feet

WIDTH: 35 to 50 feet

FRUIT: elongated; 1 to 2"; dry; orange or yellow; causes significant litter; showy

GROWTH RATE: rapid-growing; moderate life span

HABIT: irregular; round; vase; open; coarse texture

FAMILY: *Simaroubaceae*

LEAF COLOR: dark green in summer; no fall color change

LEAF DESCRIPTION: alternate; even pinnately compound; ciliate margin; ovate; deciduous; 2 to 6" long leaflets

FLOWER DESCRIPTION: green; showy; spring; offensive odor

EXPOSURE/CULTURE:

Light Requirement: part shade to full sun

Soil Tolerances: tolerant of any soil

Aerosol Salt Tolerance: high

PEST PROBLEMS: pest-resistant

BARK/STEMS:

Bark is thin and easily damaged by mechanical impact; branches grow mostly upright once trained; should be grown with a single leader; no thorns

Pruning Requirement: head back laterals to develop sturdy branches without included bark

Limb Breakage: susceptible to breakage

Current Year Twig: reddish-brown; stout

LANDSCAPE NOTES:

Origin: northern China

Usage: reclamation; urban tolerant

This nonnative deciduous tree will rapidly grow in height and produces an open canopy of stout branches. Leaves can be 1 to 3-foot-long; leaflets 2 to 6-inches-long. Broken stems smell of rancid peanut butter, and males reportedly smell worse than female trees. The leaves turn only slightly yellow in fall before dropping. The small, green, male and female flowers are produced on separate trees and appear in dense terminal clusters.

The 1.5-inch-long, yellow to red/brown, winged fruits that follow the blossoms will persist on the tree in dense clusters throughout the fall and into the winter months, and are quite showy. They can create a crunchy mess when they fall to the ground. Seeds sprout easily and seedlings usually invade surrounding land. Suckering on the trunk can be a maintenance nuisance.

Tree-of-Heaven performs best in full sun on well-drained, moist soil, but will survive almost anywhere, under any cultural conditions—smoke, dust, hot, cold, wet, dry, or salt. Growth may be poor in wet soil. It has been known to appear in cracks of pavement or even trash piles, and it will survive where no other trees will grow. Spreading rapidly by seed and suckers, Tree-of-Heaven is viewed by many as a pest- and weed-tree. If well cared for, this tree can persist for a long time. Large specimens may grow trunks up to 5 feet in diameter.

Weedy, sprouting, thicket-forming, open habit is objectionable to some. Weak branch crotches can cause limb breakage. It may be hard to believe, but the tree is occasionally used as a unique, almost tropical specimen. Tree-of-Heaven can actually be quite attractive used in this manner. Extracts from root bark and leaflets are being tested for herbicidal activity.

It is reported to be difficult to propagate but this could be overcome with research.

Albizia julibrissin Mimosa

Pronunciation: al-BIZZ-ee-ah jew-lee-BRISS-in

ZONES: 6B, 7A, 7B, 8A, 8B, 9A, 9B

HEIGHT: 15 to 25 feet

WIDTH: 25 to 35 feet

FRUIT: elongated pod; 3 to 6"; dry and hard; brown; causes significant litter; persistent; showy

GROWTH RATE: rapid growth; short-lived

HABIT: upright or spreading vase with a flat top; open; irregular silhouette; fine texture

FAMILY: *Leguminosae*

LEAF COLOR: medium green in summer; no fall color change

LEAF DESCRIPTION: alternate; bipinnately compound; ciliate margin; lanceolate to oblong; deciduous; less than 2" long leaflets

FLOWER DESCRIPTION: pink; pleasant fragrance; showy; spring or summer

EXPOSURE/CULTURE:

Light Requirement: full sun

Soil Tolerances: all textures; slightly alkaline to acidic; wet soil; salt; drought

Aerosol Salt Tolerance: moderate

PEST PROBLEMS: sensitive

BARK/STEMS:

Bark is thin and easily damaged by mechanical impact; branches droop as tree grows and will require pruning for vehicular or pedestrian clearance beneath canopy; routinely grown with multiple trunks; included bark routinely forms in crotches; no thorns

Pruning Requirement: requires pruning to develop good structure and upright habit

Limb Breakage: susceptible to breakage at crotch due to poor collar formation

Current Year Twig: gray; stout

CULTIVARS AND SPECIES:

Purchase wilt-resistant cultivars: 'Charlotte', 'Tyron', and 'Union' are reportedly wilt-resistant, may be coming into production in selected nurseries, and test results of supposed resistance have not been well publicized.

LANDSCAPE NOTES:

Origin: Iran to central China

Usage: near a deck or patio; reclamation; specimen

This fast-growing, deciduous tree has a low-branching, open, spreading habit and delicate, lacy, almost fern-like foliage. Fragrant, silky, pink puffy pompom blooms, 2 inches in diameter, appear in abundance from late April to early July, creating a spectacular sight. But the tree produces numerous seed pods and harbors insect (webworm) and disease (vascular wilt) problems. Although rather short-lived (10 to 20 years), Mimosa is popular for use as a terrace or patio tree for its light, dappled shade and tropical effect.

Mimosa tolerates drought conditions well but has a deeper green color and more lush appearance when given adequate moisture. The litter problem of the blooms, leaves, and especially the long seed pods requires consideration when planting this tree. Also, the wood is brittle and has a tendency to break during storms, though usually the wood is not heavy enough to cause damage. Typically, most of the root system grows from only 2 or 3 large-diameter roots originating at the base of the trunk. Surface roots can raise walks and patios as they grow in diameter and make for poor transplanting success.

Unfortunately, fusarium (vascular) wilt is becoming a very widespread problem in many areas of the country and has killed many roadside trees. Despite its picturesque growth habit and its beauty when in bloom, some cities have passed ordinances outlawing further planting of this species due to its weed potential and wilt disease problem. Other problems include cottony cushion scale, mites, and mimosa webworm. Chlorosis occurs in alkaline soil.

Alnus glutinosa Black Alder

Pronunciation: ALL-nuss glue-teh-NO-sah

ZONES: 3A, 3B, 4A, 4B, 5A, 5B, 6A, 6B, 7A, 7B

HEIGHT: 50 to 80 feet

WIDTH: 20 to 40 feet

FRUIT: elongated oval; less than ½"; dry and hard; brown; no significant litter; persistent and showy

GROWTH RATE: rapid growth; moderate life span

HABIT: oval; pyramidal; dense; nearly symmetrical; coarse texture

FAMILY: *Betulaceae*

LEAF COLOR: dark green in summer; no fall color change

LEAF DESCRIPTION: alternate; simple; double serrate; round; deciduous; 2 to 4" long

FLOWER DESCRIPTION: purple; red; spring; not showy

EXPOSURE/CULTURE:

Light Requirement: part shade to full sun

Soil Tolerances: all textures; alkaline to acidic; wet soil; moderate salt; drought-sensitive on wet sites, tolerant on dry sites where roots can penetrate well into soil

Aerosol Salt Tolerance: moderate

PEST PROBLEMS: resistant

BARK/STEMS:

Trainable to be grown with multiple trunks, but often grows with a single trunk; trunks and branches grow mostly upright or horizontal and will not droop; showy trunk; no thorns

Pruning Requirement: requires little pruning to develop one trunk

Limb Breakage: resistant

Current Year Twig: brown; gray; medium to slender

CULTIVARS AND SPECIES:

'Aurea', golden yellow leaves; 'Fastigiata', very narrow, upright form to 6 feet wide; 'Laciniata', leaves deeply lobed, vigorous growth; 'Pyramidalis', upright or columnar form to 50 feet tall.

LANDSCAPE NOTES:

Origin: Europe and Asia, naturalized in eastern North America

Usage: reclamation; specimen

A popular tree for moist to wet soils, Common, European, or Black Alder is a moderate- to fast-growing (2 feet per year) deciduous tree that usually grows to 40 to 50 feet in height with a 20- to 40-foot spread and a 12- to 18-inch trunk, but is capable of reaching 80 feet in height. It is not native but has escaped from cultivation and will form pure stands or thickets on disturbed wet sites. Pyramidal when young, Alder often has multiple stems, making it ideal for use as a screen or specimen; trees eventually become more rounded or oval as they mature. The 2- to 4-inch-wide, roundish leaves with toothed edges and pale undersides are joined in spring by rather insignificant male and female flowers. Foliage remains green well into the fall. The fruits are more interesting—small, nutlike "cones" that persist throughout the fall and winter, long after the darkening leaves have fallen. The fruits are food for a variety of wildlife.

A good plant for establishing along stream banks to stabilize soil and add interest, Alder can also be used as a specimen in a more formal landscape where wet soil challenges most other plants. It can be pruned to one central leader or used as a multi-stemmed specimen. Branches on central-leadered trees form attractive horizontal layers, unlike most other trees. It is similar to the dogwoods in this respect.

Common Alder will grow easily in full sun or partial shade in almost any landscape setting because the trees are able to fix nitrogen (take it out of the soil atmosphere), enabling them to grow in the poorest soils where other trees might fail. Common Alder transplants easily and will seed itself into an area, creating a thicket, if it is planted and left alone in an unmaintained area. Trees on dry sites appear to be drought-tolerant if roots can expand into a large soil volume. Alders adapt to a variety of sites and they could be planted more often.

Subject to leaf miners, tent caterpillars, and powdery mildew.

Alnus rhombifolia White Alder

Pronunciation: ALL-nuss rom-beh-FOE-lee-ah

ZONES: 8A, 8B, 9A, 9B, 10A, 10B, 11

HEIGHT: 50 to 75 feet

WIDTH: 30 to 40 feet

FRUIT: elongated; ½ to 1"; dry; brown; no significant litter

GROWTH RATE: rapid growth; moderate life span

HABIT: oval; pyramidal; open; symmetrical; medium texture

FAMILY: *Betulaceae*

LEAF COLOR: dark green above, lighter below in summer; no fall color change

LEAF DESCRIPTION: alternate; simple; finely serrate to entire margin; oval to ovate; deciduous; 2 to 4" long

FLOWER DESCRIPTION: yellow; not showy; spring

EXPOSURE/CULTURE:

Light Requirement: part shade to full sun

Soil Tolerances: all textures; alkaline to acidic; wet soil; drought-sensitive

PEST PROBLEMS: sensitive

BARK/STEMS:

Thin bark; branches grow mostly upright or horizontal and will not droop; showy light-colored trunk; should be grown with a single leader; no thorns

Pruning Requirement: needs little pruning to develop a strong structure

Limb Breakage: resistant

Current Year Twig: brown; slender

LANDSCAPE NOTES:

Origin: California north along the coast to British Columbia

Usage: parking lot islands; street tree; reclamation, shade

A North American native commonly found along streams, White Alder is a quick-growing tree (to 30 inches per year) reaching sometimes up to 100 feet. The glossy leaves appear just after the springtime display of the 6-inch-long, greenish-yellow catkins. These flowers are followed by the appearance of brown, cone-like fruits that persist on the tree throughout most of the year and are quite popular for use in dried flower arrangements.

White Alder creates an interesting specimen during the winter with a straight trunk, persistent fruits, attractive silver/grey bark, and slender, horizontal branches with pendulous tips. Bark becomes smooth, gray to green and mottled, especially along the lower half of the tree. Once chosen, the central leader dominates the tree, forming a pyramidal crown shape, but becomes more oval with age.

White Alder will grow easily in full sun or partial shade in any moist soil and will tolerate the wettest conditions. For this reason it grows well in irrigated landscapes. It is native to stream banks. It is usually not grown or offered by nurseries in the eastern United States, but is common in some areas of California. It has been used in parking lots as a shade tree and along streets. Leave plenty of soil space for root development. Keep the tree away from places where it could suffer mechanical damage to the trunk; the thin bark is easily damaged.

White Alder is attacked and killed by the flatheaded borer in California. This borer is closely related to the bronze birch borer, which devastates white-barked birches in the eastern United States. Prune infested branches fall through March when the infestation is visible and easy to detect and the beetle does not fly. Avoid pruning in the spring. Tent caterpillars occasionally eat foliage, but cause no lasting harm.

Amelanchier canadensis Shadblow, or Downy Serviceberry

Pronunciation: am-eh-LANG-key-er can-ah-DEN-siss

ZONES: 4A, 4B, 5A, 5B, 6A, 6B, 7A, 7B

HEIGHT: 20 to 25 feet

WIDTH: 15 to 20 feet

FRUIT: round; less than ½"; fleshy; purple to red; attracts wildlife; edible; no significant litter; showy

GROWTH RATE: moderate; moderate life span

HABIT: rounded, upright vase; open; nearly symmetrical; fine texture

FAMILY: *Rosaceae*

LEAF COLOR: dark green in summer; very showy orange, red, or yellow fall color

LEAF DESCRIPTION: alternate; simple; serrate margin; oval to oblong; deciduous; 2" long

FLOWER DESCRIPTION: white; very showy; spring

EXPOSURE/CULTURE:

Light Requirement: part shade to full sun

Soil Tolerances: loam; sand; acidic; occasionally wet; moderate salt and drought

PEST PROBLEMS: resistant

BARK/STEMS:

Bark is thin and easily damaged by mechanical impact; branches droop as tree grows, and will require pruning for vehicular or pedestrian clearance beneath canopy; routinely grown with multiple trunks; can be trained to grow with a single trunk; no thorns

Pruning Requirement: requires little pruning to develop strong structure; prune lightly to minimize suckering

Limb Breakage: brittle branches

Current Year Twig: brown; slender

LANDSCAPE NOTES:

Origin: eastern North America

Usage: container; near a deck or patio; specimen; street tree

Downy Serviceberry is an upright, multi-stemmed large shrub. This North American native is usually the first to be noticed in the forest or garden at springtime, as its pure white, glistening flowers are some of the earliest to appear. The small white flowers are produced in dense, erect, 2- to 3-inch-long racemes, opening up to a delicate display before the attractive reddish-purple buds unfold into small, rounded leaves. Leaves are covered with a fine, soft gray fuzz when young, giving the plant its common name. Following the blooms are many small, dark red/purple, luscious, sweet and juicy, apple-shaped fruits, often well-hidden by the leaves, which are quickly consumed by birds and other wildlife that seem to find their flavor irresistible.

When the shortened days of autumn arrive, Downy Serviceberry is alive with a variety of colorful hues, from yellow and gold to orange and deep red. Downy Serviceberry is ideal for planting in the naturalized garden, where it can be allowed to spread by its natural suckering habit, or is striking when placed in a mixed shrubbery border, where its brilliant white blooms and fall color stand out nicely against a background of evergreen shrubs. The light shade cast by the open crown makes the tree well-suited for planting as a specimen near the deck or patio.

With a native habitat of wet bogs and swamps, Downy Serviceberry should be grown in full sun or light shade on moist, well-drained, acid soil. Plants will rarely require any pruning or fertilizing, except if thinning of the multiple stems is desired to clean up or display the nice trunk form.

Trees often sucker from the base of the trunk. Recovers slowly from field transplanting. Fire blight occurs with too much soil nitrogen.

Amelanchier x grandiflora Apple Serviceberry

Pronunciation: am-eh-LANG-key-er x gran-deh-FLOOR-ah

ZONES: 4A, 4B, 5A, 5B, 6A, 6B, 7A, 7B

HEIGHT: 20 to 35 feet

WIDTH: 15 to 25 feet

FRUIT: round; less than ½"; fleshy; red; edible; no significant litter; showy; quickly eaten by birds

GROWTH RATE: slow-growing; moderate life span

HABIT: upright vase; moderate density; slightly irregular silhouette; fine texture

FAMILY: *Rosaceae*

LEAF COLOR: purplish, turning to flat green in summer; stunning orange, red, or yellow fall color

LEAF DESCRIPTION: alternate; simple; serrate margins; oval to oblong; deciduous; 1 to 3" long

FLOWER DESCRIPTION: white; very showy; spring

EXPOSURE/CULTURE:

Light Requirement: shade to full sun

Soil Tolerances: loam; sand; slightly alkaline to acidic; some salt; moderate drought; good drainage is essential

Aerosol Salt Tolerance: little

PEST PROBLEMS: resists most lethal pests

BARK/STEMS:

Bark is thin and easily damaged by mechanical impact; routinely grown with multiple trunks; branches grow mostly upright and will not droop; showy trunk; can be trained to grow with a single trunk; no thorns

Pruning Requirement: needs little pruning to develop a strong trunk and branch structure

Limb Breakage: brittle branches

Current Year Twig: brown; slender

CULTIVARS AND SPECIES:

'Autumn Brilliance' is supposedly resistant to leaf spot; 'Cumulus' is moderately columnar, may sucker at the root collar; 'Robin Hill' has an upright habit, 20 to 25 feet tall, sensitive to drought; 'Strata' is 35 feet tall and wide with beautiful orange-red fall color.

LANDSCAPE NOTES:

Origin: hybrid

Usage: container or above-ground planter; buffer strip; highway; near a deck or patio; specimen; street tree

Apple Serviceberry is a hybrid between *A. canadensis* and *A. laevis,* which grow 15 to 25 feet tall. Multiple stems are upright and highly branched, forming a dense shrub, or, if properly pruned in the nursery, a small tree. It is superior to either species in that it suckers less, is adapted to a wide range of soils, and tolerates some drought. The main ornamental feature is the white flowers, which are larger than those of other Amelanchiers. The flowers are borne in early spring and are at first tinged with pink but later turn to bright white. Well-adapted for planting beneath power lines due to its small size.

Some early training is needed to create an upright tree. The tree naturally forms many trunks originating from the ground level, but with staking and pruning can develop one trunk. Branching can begin at 3 to 6 feet from the ground to form a uniform canopy of feathery foliage. Little pruning is needed beyond this initial training to maintain a tree form. One of the nicest uses for Serviceberry is in a container on the patio or deck. It can be moved closer when it is in flower and fall color to better enjoy this special tree.

Nonlethal leaf diseases and insects can nearly defoliate trees. Borers can infest drought-stressed trees. Spider mites and fireblight can be troublesome. Recovers slowly from transplanting. Trees sucker less than other Amelanchiers. Fertilize and prune lightly to minimize fireblight disease.

Amelanchier laevis Allegheny Serviceberry

Pronunciation: am-eh-LANG-key-er LAY-viss

ZONES: 5A, 5B, 6A, 6B, 7A, 7B, 8A, 8B

HEIGHT: 30 to 40 feet

WIDTH: 15 to 20 feet

FRUIT: round; less than ½"; fleshy; black to purple; attracts birds and mammals; edible; no significant litter; showy

GROWTH RATE: moderate; moderate life span

HABIT: oval; upright vase; moderate density; slightly irregular silhouette; fine texture

FAMILY: *Rosaceae*

LEAF COLOR: blue-green in summer; showy red, orange, or yellow fall color

LEAF DESCRIPTION: alternate; simple; serrate margin; oval to oblong or ovate; deciduous; 1 to 3" long

FLOWER DESCRIPTION: white; showy; spring

EXPOSURE/CULTURE:

Light Requirement: partial shade to partial sun; becomes stressed in full sun in dry soil

Soil Tolerances: loam; sand; slightly alkaline to acidic; some salt; moderate drought; good drainage is essential

PEST PROBLEMS: resistant to most lethal pests

BARK/STEMS:

Bark is thin and easily damaged by mechanical impact; routinely grown with multiple trunks; branches grow mostly upright and will not droop; can be trained to grow with a single trunk; no thorns

Pruning Requirement: needs little pruning to develop a strong trunk and branch structure

Limb Breakage: resistant

Current Year Twig: brown; reddish; slender

CULTIVARS AND SPECIES:

A. alnifolia is common in the western United States at elevations from 4,500 to 9,000 feet and in the Rocky Mountains where annual rainfall is as low as 16 inches. They show fabulous yellow, orange, and red fall color and grow 15 feet tall.

LANDSCAPE NOTES:

Origin: native to eastern North America

Usage: container or above-ground planter; near a deck or patio; specimen; street tree

The Allegheny Serviceberry is similar in most ways to other Serviceberries. It grows naturally in shade or partial shade as an understory tree. Multiple stems are upright and highly branched, forming a dense shrub, or, if properly pruned, a small tree. The tree is fairly short-lived, but can be used as a filler plant or to attract birds. The main ornamental feature is the white flowers borne in drooping clusters in mid-spring. They brighten up any shady area of the landscape and signal that spring has arrived. The purplish-black berries are sweet and juicy but are soon eaten by birds. The fall color is a striking yellow to red that is rivalled by few other trees.

Amelanchier is well-adapted for planting beneath power lines, because of its small size. It can be placed in a container for use on the patio or deck, or planted in a narrow space. Plant a low, uniform ground cover beneath the canopy to display the nice trunk form. The canopy looks wonderful in the fall with a background of green foliage, so try to locate it with evergreens behind.

May suffer nonlethal leaf diseases and insects. Borers can infest drought-stressed trees. Spider mites and fireblight can be troublesome. Minimize fireblight disease by applying nitrogen in moderation.

Aralia spinosa Devil's-Walkingstick

Pronunciation: ah-RAY-lee-ah spy-NO-sah

ZONES: 5A, 5B, 6A, 6B, 7A, 7B, 8A, 8B, 9A

HEIGHT: 10 to 15 feet

WIDTH: 6 to 10 feet

FRUIT: round; less than ½"; fleshy; black to purple; attracts birds; creates some litter

GROWTH RATE: moderate; short life span

HABIT: irregular; upright; open; fine texture

FAMILY: *Araliaceae*

LEAF COLOR: young leaves bronze, turn to green in summer; showy copper to red fall color

LEAF DESCRIPTION: alternate; bipinnately compound; serrate margins; ovate; deciduous

FLOWER DESCRIPTION: white; very showy; summer

EXPOSURE/CULTURE:

Light Requirement: part shade to full sun

Soil Tolerances: all textures; alkaline to acidic; wet soil; moderate drought

PEST PROBLEMS: pest-free

BARK/STEMS:

Branches grow mostly upright; can be trained to grow with a single, short trunk or with multiple trunks; thorns are present on trunk, branches, or leaves

Pruning Requirement: requires pruning to develop strong structure

Limb Breakage: susceptible

Current Year Twig: brown or gray; stout

LANDSCAPE NOTES:

Origin: native to eastern North America

Usage: reclamation; specimen; ditch banks; living fence

Anyone who has accidentally brushed against a Devil's-Walkingstick does not soon forget the experience, for this tall, spindly native shrub or small tree is armed up and down its thin trunk with extremely sharp, treacherous spines. Even the huge, much-divided leaves, which can reach 4 feet long and 3 feet wide, are armed with pointed prickles, ready to scratch anyone who comes within range. But when placed in an area where they can do no harm, Devil's-Walkingsticks add a tropical effect to a mixed shrubbery border or other naturalized setting, where the large leaves can easily spread out to their full length atop the slender, 10-foot-tall trunks. Under ideal conditions, these small trees can even reach 20 feet tall.

As striking as the thorns are the large, summertime panicles of bloom, 12 to 18 inches in diameter and up to 3 feet long, are held above the crown of leaves for 5 to 10 days and gently drape outward under the weight of the flower head. Following these blooms comes the production of a great quantity of dark, purple-black, juicy berries that are exceptionally popular with birds.

Devil's-Walkingstick is quite easy to grow and literally thrives on neglect. Invasive habit; it usually spreads by sprouts produced from its base, eventually creating an impenetrable thicket if left to its own devices. Although the temptation is to use this plant as a barrier planting, the effect of the thorned, naked trunks during wintertime is quite harsh and perhaps unattractive to some. The plant can be trained into a small, single or multi-stemmed tree and used in a shrubbery border to add height and interest. It can be transplanted during winter and used as a specimen in an out-of-the-way place where it is sure to capture the curiosity of many visitors, particularly when it is in flower. *A. elata* is more commonly cultivated, especially the strikingly beautiful cultivar 'variegata'.

Araucaria bidwillii False Monkey-Puzzletree

Pronunciation: ah-row-CARE-ee-yah bid-WILL-ee-eye

ZONES: 9A, 9B, 10A, 10B, 11

HEIGHT: 60 to 80 feet

WIDTH: 15 to 20 feet

FRUIT: rare; oval; 6 to 12"; dry and hard; brown; causes significant litter but fruits are formed infrequently in humid climates such as Florida; showy

GROWTH RATE: moderate; long-lived

HABIT: columnar; oval; pyramidal; moderate density; symmetrical; fine texture

FAMILY: *Araucariaceae*

LEAF COLOR: medium to dark green; evergreen

LEAF DESCRIPTION: spiral arrangement; simple; entire margin; lanceolate to oval 1.5 to 2" long; sharp tip

FLOWER DESCRIPTION: not showy

EXPOSURE/CULTURE:

Light Requirement: grows best in full sun

Soil Tolerances: all textures; alkaline to acidic; moderate salt and drought

Aerosol Salt Tolerance: moderate

PEST PROBLEMS: resistant

BARK/STEMS:

Branches droop; should be grown with a single leader; persistent leaves are spiny

Pruning Requirement: needs little pruning to develop a strong structure

Limb Breakage: resistant

Current Year Twig: brown; medium thickness

LANDSCAPE NOTES:

Origin: Australia

Usage: specimen for large landscapes; windbreak

This large evergreen has a single upright trunk, tiered branching habit, and a pyramidal or columnar canopy with a rounded top. The tree could grow taller than 80 feet, but lightning frequently limits height growth in humid climates. The individual leaves are lanceolate when young becoming ovals at maturity. Both leaf types appear on the tree at the same time. The large, spiny, 10- to 15-pound cones are rare in cultivation.

Although they provide some shade, they are not suitable for patios or terraces because they are too large and large surface roots are common. In addition, columnar-formed trees generally cast limited shade due to the narrow crown. Many people forget how tall these trees grow. They often have an attractive pyramidal form (like a conifer) when they are small, but they quickly grow too tall for most residential sites. They look entirely out of place in most residential landscapes. This tree is best saved for large-scale landscapes like parks and municipal buildings. This tree is probably better suited than Norfolk Island Pine for planting in the residential landscape. Its canopy is wider and it casts more shade.

Growing best in full sun locations, this tree thrives on a variety of soils and is suited for coastal planting sites. Young trees should be watered well, especially during periods of drought. Be sure to prune out multiple trunks or leaders, as the tree is strongest when grown with one central leader. Small branches periodically litter the ground beneath the canopy, and they decompose slowly.

Scale, leaf spots, and sooty mold are minor problems. Leaves are spiny and they inflict pain when they meet flesh.

Araucaria heterophylla Norfolk Island Pine

Pronunciation: ah-row-CARE-ee-yah het-er-o-FYE-lah

ZONES: 10A, 10B, 11

HEIGHT: 60 to 80 feet

WIDTH: 10 to 12 feet

FRUIT: rare; oval; 6 to 12"; dry and hard; green

GROWTH RATE: rapid growth; long-lived

HABIT: columnar; pyramidal; open; symmetrical; fine texture

FAMILY: *Araucariaceae*

LEAF COLOR: evergreen; dark green

LEAF DESCRIPTION: spiral; simple; entire margin; needle-like

FLOWER DESCRIPTION: not showy

EXPOSURE/CULTURE:

Light Requirement: grows best in full sun

Soil Tolerances: all textures; alkaline to acidic; moderate salt; drought

Aerosol Salt Tolerance: moderate

PEST PROBLEMS: resistant

BARK/STEMS:

Branches grow mostly horizontal and will not droop; should be grown with a single leader; no thorns

Pruning Requirement: needs little pruning to develop a strong structure; remove upright trunks that compete with leader as soon as they form

Limb Breakage: resistant

Current Year Twig: brown or green; medium thickness

LANDSCAPE NOTES:

Origin: Australia (Norfolk Island)

Usage: specimen for large landscapes

This large evergreen has a single upright trunk, tiered branching habit, and a narrow pyramidal or columnar shape. The tree could grow taller than 80 feet, but lightning frequently limits height growth in humid climates. The ½-inch-long, individual leaves on young trees are lanceolate and look somewhat like conifer needles at first glance. Mature leaves are somewhat contorted on twisted branches. Both leaf types can appear on the tree at the same time. The trunk is often black, curved, and swollen at the base. It contrasts well with everything else in the landscape. The large, spiny, 10- to 15-pound cones are rare in cultivation.

They provide very little shade and are not suitable for patios or terraces because they are too large. Large surface roots are common. Many people forget how tall these trees grow. They often display an attractive pyramidal form (like a conifer) when they are small, but they quickly grow too tall for most residential sites. There are many other trees that can provide more benefits to the landscape. They are well suited for planting in parks and other large, open spaces. Planting them in groups of 4 or 5 would help protect them from wind damage. They can live as a house plant for a long time if not overwatered.

Growing best in full sun locations, this tree thrives on a variety of soils and is suited for coastal planting sites. Young plants should be watered well, especially during periods of drought. Be sure to prune out multiple trunks or leaders, as these trees should be grown with one central leader.

Scale, sooty mold, and leaf spot are minor problems.

Arbutus menziesii Pacific Madrone

Pronunciation: are-BEW-tuss men-ZEE-see-eye

ZONES: (7A), (7B), 8A, 8B, 9A, 9B, 10A, 10B

HEIGHT: 40 to 80 feet

WIDTH: 30 to 40 feet

FRUIT: round; ½ to 1"; fleshy; red or orange; attracts birds; can be a litter problem; persistent

GROWTH RATE: slow-growing; moderate to long life span

HABIT: upright to rounded vase; dense; symmetrical; medium texture

FAMILY: *Ericaceae*

LEAF COLOR: glossy; evergreen; dark green; some yellow fall color due to shedding of last year's older leaves

LEAF DESCRIPTION: alternate; simple; serrate margin; oblanceolate to obovate; 3 to 4" long;

FLOWER DESCRIPTION: showy; white; fragrant; spring

EXPOSURE/CULTURE:

Light Requirement: part shade to full sun

Soil Tolerances: all textures; acidic; drought

PEST PROBLEMS: resistant

BARK/STEMS:

Outstanding exfoliating bark, one of the showiest; branches droop somewhat as tree grows, and will require pruning for vehicular or pedestrian clearance beneath canopy; can be trained to grow with a single trunk or with multiple trunks; no thorns

Pruning Requirement: requires pruning to space main branches along the trunk to develop strong structure

Limb Breakage: resistant

Current Year Twig: red and green; medium thickness

LANDSCAPE NOTES:

Origin: Vancouver to southern California

Usage: parking lot island; buffer strip; highway; near a deck or patio; specimen

Pacific Madrone is most often seen as a multi-stemmed, rounded, evergreen shrub or small tree reaching 20 feet in height on a dry site, but is capable of reaching more than 100 feet in height on a good site in the forest. The trees take on a picturesque, somewhat twisted appearance over time, and exhibit red-brown, flaking and shreddy bark that reveals a smooth green, yellow, or reddish inner bark. The tree could be grown for the bark characteristic alone.

The small, white blooms appear in late spring to cover the tree in white. The red to orange fruits are especially attractive to wildlife and add to the tree's overall attractiveness.

Pacific Madrone makes an ideal shade tree, screen, hedge, or patio tree for dry sites. The showiness of both fruits and flowers also makes this plant very popular for accent plantings. It is one of the most attractive trees available for residential use in the western United States. It could be grown and planted more in dry climates. Expect trees to grow taller than 50 feet on an irrigated landscape site with plenty of soil space for root expansion. These trees should be trained to grow with branches well spaced along a central trunk.

Pacific Madrone will grow slowly when planted in either full sun or partial shade on well-drained, acid soil. Plants are tolerant of wind and moderate drought once established and can grow in well-drained clay. The wood is valuable for veneers and for flooring, paneling, interior trim, charcoal, and many other uses.

Madrone canker is the biggest problem and has killed many trees in parts of the natural range. It appears to prefer weakened trees. Keep trees out of irrigated turf areas to minimize root rot. A leaf spot can defoliate the tree. A root rot has also killed trees and could be potentially troublesome. Borers, beetles, and foliage insects usually cause only minor damage. Trees are notoriously hard to transplant.

Arbutus texana Texas Madrone

Pronunciation: are-BEW-tuss tex-A-nah

ZONES: 7A, 7B, 8A, 8B

HEIGHT: 25 to 40 feet

WIDTH: 15 to 25 feet

FRUIT: oval; round; ½ to 1"; fleshy; orange to red; very showy

GROWTH RATE: moderate; long-lived

HABIT: pyramidal when young; upright vase eventually; moderate density; irregular silhouette; medium texture

FAMILY: *Ericaceae*

Photo Courtesy of Benny Simpson

LEAF COLOR: dark green; evergreen

LEAF DESCRIPTION: alternate; simple; crenate to entire margin; oblong to ovate; 2 to 4" long

FLOWER DESCRIPTION: white; showy; spring

EXPOSURE/CULTURE:

Light Requirement: grows best in full sun

Soil Tolerances: all textures; alkaline to acidic; drought

PEST PROBLEMS: resistant

BARK/STEMS:

Bark is thin and easily damaged by mechanical impact; routinely grown with multiple trunks; branches grow mostly upright and will not droop; uniquely showy bark; no thorns

Pruning Requirement: needs little pruning to develop a strong trunk and branch structure

Limb Breakage: resistant

Current Year Twig: green to gray; medium thickness

LANDSCAPE NOTES:

Origin: southwestern North America

Usage: container or above-ground planter; parking lot island; buffer strip; highway; near a deck or patio; small shade tree; specimen; street tree

This native North American evergreen tree has beautiful peeling bark. Although it can reach a height of 40 feet, many trees are much smaller. As the tree ages, the outer bark drops off to reveal smooth, new bark, which can range in color from white to apricot, tan, or dark red. The mixture of colors between old and new bark is quite striking. The dark green leaves, with paler undersides, are joined in springtime with small, white flowers that have an interesting shape, almost like small lanterns. These blooms are followed by the production of orange or red berries that ripen in fall. Berries are very brightly colored and will attract attention in any landscape.

Multiple trunks arise from the ground, much like those of Crapemyrtle or Tree Lilacs. Lower foliage and branches are often removed to show off this feature. Plant this tree to display the trunks near a patio, deck, walk, or other area where people come close to the tree. A row of Texas Madrone planted on 15- to 20-foot centers can enliven the entryway to an office park or condominium complex, or add character to a residential street.

Texas Madrone grows in full sun on any well-drained soil, acid or alkaline. It appears to be a very adaptable tree and should do well in a variety of landscape sites in regions receiving less than 30 inches annual rainfall. Propagation is by seed or micropropagation. Improved selections can probably be developed through micropropagation techniques. Some may be available soon.

Arbutus unedo Strawberry-Tree

Pronunciation: are-BEW-tuss YOU-nee-doe

ZONES: 8B, 9A, 9B, 10A, 10B, (11)

HEIGHT: 8 to 20 feet

WIDTH: 8 to 15 feet

FRUIT: round; ½ to 1"; fleshy; yellow turning red; attracts birds; edible; can be a litter problem; persistent through the winter

GROWTH RATE: slow-growing; moderate life span

HABIT: rounded vase; dense; somewhat symmetrical; fine texture

FAMILY: *Ericaceae*

LEAF COLOR: dark green; evergreen

LEAF DESCRIPTION: alternate; simple; serrate; oblanceolate to obovate; 2 to 3" long

FLOWER DESCRIPTION: showy pink or white; fall to winter

EXPOSURE/CULTURE:

Light Requirement: part shade to full sun

Soil Tolerances: all textures; alkaline to acidic; salt-sensitive; moderate drought

PEST PROBLEMS: pest-free

BARK/STEMS:

Branches droop as tree grows and will require pruning for vehicular or pedestrian clearance beneath canopy; routinely grown with multiple trunks; showy trunk; can be trained to grow with a short, single trunk; no thorns

Pruning Requirement: requires pruning to space main branches along trunk to develop strong structure and single trunk

Limb Breakage: resistant

Current Year Twig: red and green; medium thickness

CULTIVARS AND SPECIES:

The cultivar 'Compacta' is a smaller shrub, 6 to 10 feet tall and wide; 'Elfin King' has a contorted, dwarf form, flowers and fruits year-round; 'Rubra' has deep pink flowers.

LANDSCAPE NOTES:

Origin: Mediterranean and southern Europe

Usage: container or above-ground planter; hedge; parking lot island; buffer strip; highway; near a deck or patio; specimen

Most often seen as a multi-stemmed, rounded, evergreen shrub or small tree, Strawberry-Tree is capable of reaching 20 to 25 feet in height. The trees take on a picturesque, somewhat twisted appearance over time, and exhibit dark, red/brown, flaking, and shreddy bark accompanied by the lush, dark green, leathery, red-stemmed leaves.

In fall and winter, small white or pink blooms in 2-inch-long panicles appear at the same time the previous year's fruits are ripening. These unusual fruits have a rough, pebbled outer surface that add much to the tree's overall attractiveness.

Strawberry-Tree spreads quite wide as it grows and produces dense shade, making it ideal for use as a small, screen, hedge, or patio tree. The appearance of both fruits and flowers during the winter months also makes this plant very popular for accent plantings. It is one of the most attractive small trees available for residential use in the western United States. It could be grown and planted more in dry climates.

Strawberry-Tree will grow at a slow pace when planted in either full sun or partial shade on well-drained, acid soil. Plants are tolerant of wind and moderate drought once established and grow in well-drained clay.

Asimina triloba Pawpaw

Pronunciation: ah-SIM-in-ah try-LOW-bah

ZONES: 5A, 5B, 6A, 6B, 7A, 7B, 8A, 8B

HEIGHT: 15 to 20 feet

WIDTH: 15 to 20 feet

FRUIT: elongated; 3 to 5"; fleshy; black or brown; attracts wildlife; edible and tasty; fruit and foliage cause significant litter; showy

GROWTH RATE: moderate; moderate life span

HABIT: round; upright pyramid; moderate density; irregular or symmetrical silhouette; coarse texture

FAMILY: *Annonaceae*

LEAF COLOR: dark green in summer; very showy yellow fall color

LEAF DESCRIPTION: alternate; simple; entire; oblong to obovate; deciduous; 4 to 10" long

FLOWER DESCRIPTION: unusually dark purple; somewhat showy; spring flowering

EXPOSURE/CULTURE:

Light Requirement: shade to mostly sunny

Soil Tolerances: all textures; slightly alkaline to acidic; occasionally wet soil; some salt; drought-sensitive in sunny location

Aerosol Salt Tolerance: moderate

PEST PROBLEMS: pest-free

BARK/STEMS:

Branches droop as tree grows, and will require some pruning for vehicular or pedestrian clearance beneath canopy; can be trained to grow with multiple trunks; frequently grows with a central leader; no thorns

Pruning Requirement: requires little pruning to develop strong structure

Limb Breakage: resistant

Current Year Twig: brown; medium thickness

LANDSCAPE NOTES:

Origin: eastern North America

Usage: reclamation; specimen; fruit; medicinal

A native deciduous tree, the coarse-textured Pawpaw may ultimately reach 30 feet in height with an equal spread, and creates an upright, wide pyramidal silhouette. The large leaves, 4 to 10 inches in length and 3 to 5 inches wide, seem to droop from their weight at branch tips, giving the plant a distinctive, almost wilted appearance. The 2-inch-wide purple flowers with their less-than-pleasant fragrance appear before the leaves unfurl in springtime, and are followed by the production of unusual, fleshy, 3- to 5-inch-long, oval to irregular-shaped fruits, green when young but ripening to a brown/black, wrinkled texture. When fully ripe, the edible flesh becomes soft (almost custard-like), has a sweet, rich taste similar to bananas, and is very nutritious. The fruits are popular with humans and wildlife, especially raccoons and birds.

The Pawpaw tree will grow in full sun or dense shade but will have denser growth in the sun. Branches arch and reach for the sun in shaded sites, often creating an open, irregularly shaped canopy. The soil should be rich, moist, and slightly acid; these trees will even tolerate wet, soggy soils for a period. It can be found in multi-stemmed thickets along stream banks and on flood plains in the wild. The tree is probably best used in a natural area for stabilizing stream banks and to add yellow fall color to a shaded landscape. Ground-up twigs and bark contain a potentially useful insecticide.

This very nice tree could be used more often. Nurseries have not been able to keep up with recent demand for it.

Bauhinia spp. Orchid-Tree

Pronunciation: baw-HINN-ee-ah species

ZONES: 9B, 10A, 10B, 11

HEIGHT: 25 to 40 feet

WIDTH: 25 to 35 feet

FRUIT: elongated pod; 6 to 12"; dry; brown; significant litter; persistent; showy

GROWTH RATE: rapid-growing; short-lived

HABIT: rounded vase; moderate density; irregular silhouette when young, more symmetrical with age; coarse texture

FAMILY: *Leguminosae*

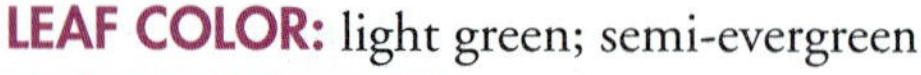

LEAF COLOR: light green; semi-evergreen

LEAF DESCRIPTION: alternate; simple; bi-lobed; orbiculate; 2 to 4" long

FLOWER DESCRIPTION: pink, purple, red, white, yellow; showy; year-round

EXPOSURE/CULTURE:

Light Requirement: part shade to full sun; best flowering and canopy form in full sun

Soil Tolerances: all textures; slightly alkaline to acidic; moderate salt; drought

Aerosol Salt Tolerance: moderate

PEST PROBLEMS: resistant

BARK/STEMS:

Branches droop as tree grows, and will require pruning for vehicular or pedestrian clearance beneath canopy; routinely grown with multiple trunks; can be trained to grow with a short, single trunk; no thorns

Pruning Requirement: requires pruning to develop strong structure or a central leader

Limb Breakage: susceptible to breakage at crotch, due to poor collar formation; wood itself is weak and tends to break

Current Year Twig: brown; medium to slender

CULTIVARS AND SPECIES:

B. blakeana, the Hong Kong Orchid-Tree, is seedless and would not present such a litter problem, bearing 6-inch, orchid-like flowers of rich reddish or rose-purple during the winter, but it is very tender in freezing temperatures. *B. variegata*, produces in winter and spring orchid-like blossoms of purplish casts or pure white in cultivar 'Candida'. *B. purpurea*, produces narrow-petaled, red-purple to blue-purple flowers in late fall and early winter while leaves are on the trees. *B. monandra* produces pink, single-stamened flowers all summer. *B. acuminata* also blooms all summer, but with white flowers. *B. aculeata*, with white flowers, is hardy as far north as the southern part of zone 8B, but tends to produce many root suckers and thorny branches. *B. forficata* has white flowers and no thorns.

LANDSCAPE NOTES:

Origin: Asia and South America

Usage: parking lot island; buffer strip; highway; near a deck or patio; reclamation; shade tree; specimen; street tree

This deciduous to semi-evergreen tree has a vase-shaped canopy (if lower branches are removed) made up of large, papery leaves. The orchid-like blooms are 3 to 4 inches across and are produced in abundance at various times of the year, depending upon species and cultivar. Orchid-Tree makes a spectacular specimen or shade tree or fits well into mixed shrubbery borders.

Growing best in full sun or high, shifting pine shade, Orchid-Tree thrives in any well-drained soil but in alkaline soils will show interveinal chlorosis (yellowing) on the leaves. The flowers are followed by many brown, woody, 12-inch-long seed pods, which are unattractive on the tree and a nuisance when they drop. The wood tends to be weak. Some consider the fallen leaves messy because they are large and decompose slowly.

In flower, Orchid-Tree makes a beautiful street tree, with foliage and flowers arching over the road. However, drooping branches must be removed as they develop to allow for vehicle clearance. Because sprouts will have to be removed regularly and the tree is bare for a month or two in winter, this is considered by many to be a high-maintenance tree not suited for large-scale street-tree planting. But it is a tough tree growing in most soils with pH below 7.5.

Chewing insects and borers may be a problem. Interveinal chlorosis occurs on high pH soil from micronutrient deficiency. Regular fertilization helps prevent potassium deficiency and maintains good foliage color. Flowers, fruit, twigs, and foliage regularly litter the ground. Be careful pruning because bark rips and is injured easily.

Beaucarnea recurvata Ponytail

Pronunciation: bo-CAR-nee-ah re-kurr-VAY-tah

ZONES: 10A, 10B, 11

HEIGHT: 12 to 18 feet

WIDTH: 10 to 15 feet

FRUIT: elongated; less than ½"; dry; not showy; no significant litter

GROWTH RATE: slow-growing; long-lived

HABIT: palm shape; upright; open; irregular silhouette; fine texture

FAMILY: *Agavacea*

LEAF COLOR: light green; evergreen

LEAF DESCRIPTION: spiral; simple; serrate margin; 18 to 36" long

FLOWER DESCRIPTION: whitish-yellow; very showy and dense; spring and summer

EXPOSURE/CULTURE:

Light Requirement: shade to full sun; best growth is in sun

Soil Tolerances: all textures; alkaline to acidic; drought

Aerosol Salt Tolerance: moderate

PEST PROBLEMS: resistant

BARK/STEMS:

Older plants eventually branch; branches grow mostly upright and will not droop; showy, swollen trunk base supposedly functions in water storage; no thorns

Pruning Requirement: needs little pruning to develop a strong structure

Limb Breakage: resistant

Current Year Twig: gray; stout

LANDSCAPE NOTES:

Origin: southern Mexico

Usage: container or above-ground planter; near a deck or patio; specimen; house plant

This upright evergreen tree grows very slowly up to 30 feet in height, but rarely exceeds 10 feet in most landscapes. Despite its name, it is not a palm. A distinctive plant, Ponytail Palm has a greatly swollen trunk base (sometimes to 7 feet across) that quickly tapers and eventually branches in older specimens. The leaves, up to 5 feet long and ¾ inch wide, are produced in tufts clustered at the tips of branches. The cascading nature of the leaves gives the appearance of a pony's tail. Creamy yellow flowers are quite showy, as they are held above the foliage in spring or summer for several weeks. The tree will occasionally flower two or even three times a year. This plant makes a great conversation piece, whether grown as a specimen, a container plant, near patios, or placed in rock gardens. It can also be used as a houseplant in a sunny location.

Ponytail Palm is nicely suited for maintaining in a container. Regular watering and occasional fertilizer will keep it going for years with few problems. It grows slowly but the trunk expands quickly. Transplant to a bigger pot before it becomes pot-bound to prevent the pot from cracking. In subtropical areas, it makes a striking accent plant near a deck or patio, or adds texture to a shrub border.

Ponytail Palm grows in full sun or partial shade on a wide range of soils. Soil must have good drainage, as plants have a tendency to develop root rot on poorly drained soils. Plants moved from indoors to permanent outside locations should be gradually exposed to the increase in light and temperature change.

Chewing insects may disfigure the leaves.

Betula nigra River Birch

Pronunciation: BET-chew-lah NYE-grah

ZONES: 4A, 4B, 5A, 5B, 6A, 6B, 7A, 7B, 8A, 8B, 9A

HEIGHT: 40 to 50 feet

WIDTH: 40 to 50 feet

FRUIT: elongated; 1 to 3"; dry; brown; no significant litter

GROWTH RATE: rapid-growing; moderate longevity on urban sites, long-lived on moist sites

HABIT: oval; pyramidal; dense; somewhat symmetrical when young; medium texture

FAMILY: *Betulaceae*

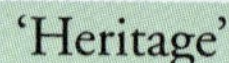

'Heritage'

LEAF COLOR: light green in summer; showy yellow fall color

LEAF DESCRIPTION: alternate; simple; double serrate margins; ovate to rhomboid; deciduous; 2 to 4" long

FLOWER DESCRIPTION: brown; not showy; spring and winter

EXPOSURE/CULTURE:

Light Requirement: part shade to full sun

Soil Tolerances: all textures; acidic; wet soil; some salt; drought-sensitive in confined soil spaces

Aerosol Salt Tolerance: low

PEST PROBLEMS: resistant

BARK/STEMS:

Very showy trunk; branches droop as tree grows, and will require pruning for vehicular or pedestrian clearance beneath canopy; trainable to be grown with a single or multiple trunks; no thorns

Pruning Requirement: needs little pruning to develop a strong trunk and branch structure

Limb Breakage: susceptible to ice storms; otherwise resistant

Current Year Twig: reddish-brown; slender

CULTIVARS AND SPECIES:

'Heritage', a patented cultivar, has scaly, light beige bark and is the closest to a paper-white birch that will survive in hot areas. It is reportedly more vigorous than the species.

LANDSCAPE NOTES:

Origin: eastern North America

Usage: screen; near a deck or patio; shade tree; specimen; possible street tree

River Birch can grow 50 to 90 feet tall. In its native habitat, it has a central leader and small-diameter, dark-colored lateral branches. It has a narrow, oval to pyramidal crown when young, but spreads wider with age as several branches become dominant. The trunk is distinguished by reddish-brown bark that peels off in film-like papery curls, providing interest all year. The bark darkens and becomes less showy as the tree matures.

Requires acid soil; otherwise leaves become chlorotic. Very well-suited for planting along stream banks (where it is native) and in other wet areas. River Birch tends to be short-lived (30 to 40 years) in many urban settings, possibly due to inadequate water supply in restricted soil spaces. Situate the tree to allow roots to spread for best growth. Though not recommended as a street tree (unless irrigated), it makes a nice specimen.

Tolerates low soil oxygen, flooding, and clay soil and needs moist conditions. Surface roots develop in this environment. Can easily be trained with one central leader or as a multi-stemmed clump. Some nurseries plant two or three trees together to form a clump, but one often outgrows the others, and these trunks will not fuse into one strong trunk. It should be grown more as a single-trunked specimen.

Not susceptible to bronze birch borer, but gets leaf miner, as do other birches. It is not especially drought-tolerant and should be irrigated if used in confined soil spaces. It is not suited for parking lots or other hot, exposed urban sites unless carefully managed. Aphids can infest the foliage, forming sticky honeydew on surfaces below the canopy. Chlorosis develops on high pH soils. Small leaves sometimes develop from an unknown cause on trees in containers.

Betula papyrifera Paper Birch

Pronunciation: BET-chew-lah pap-eh-RIFF-er-ah

ZONES: 3A, 3B, 4A, 4B, 5A, 5B, 6A

HEIGHT: 45 to 50 feet

WIDTH: 25 to 30 feet

FRUIT: elongated; 1 to 3"; dry; brown; relished by wildlife; no significant litter

GROWTH RATE: moderate; short-lived in urban areas

HABIT: oval; pyramidal; open; irregular silhouette; fine texture

FAMILY: *Betulaceae*

LEAF COLOR: medium green in summer; bright yellow fall color

LEAF DESCRIPTION: alternate; simple; double serrate margin; ovate; deciduous; 2 to 4" long

FLOWER DESCRIPTION: brown or green; showy; spring

EXPOSURE/CULTURE:

Light Requirement: part shade to full sun

Soil Tolerances: all textures; slightly alkaline to acidic; salt; drought sensitive in urban sites

Aerosol Salt Tolerance: moderate to high

PEST PROBLEMS: sensitive

BARK/STEMS:

Can be trained with a single or multiple trunks; major branches grow mostly upright, lower branches droop; showy trunk; no thorns

Pruning Requirement: do not prune from February through July

Limb Breakage: susceptible to ice storms

Current Year Twig: reddish-brown; slender

LANDSCAPE NOTES:

Origin: native to Canada and northern United States

Usage: specimen on a site with plenty of soil

This birch, a native to northern areas, is grown for its beautiful white bark and airy habit. When all cultural requirements are met, the tree will reach 50 feet or more and spread about half that amount in landscapes with plenty of soil space. Trees planted in urban areas rarely live long enough to reach this size. Paper Birch has excellent cold tolerance and will grow in zone 3. It is less adapted to zones 5B and 6, but is planted there frequently because of its beautiful bark. It is hard to resist the temptation of planting large numbers of this tree. It may be grown as a single-stemmed tree or in a multi-stemmed clump to show off the bark.

The striking yellow fall leaf color and white bark look best when displayed against a dark background of evergreen trees and shrubs. Often planted in clumps or groves on commercial landscapes to mass the white bark effect. The trunks contrast nicely with dark mulch placed beneath the canopy. Expand the mulched area each year to eliminate grass and other plants from beneath the canopy.

The tree is best adapted to the coolest parts of the country on moist sites with deep, loose soil with irrigation provided in times of drought. Few urban sites fit these requirements. Watch for pests if grown on other sites. Quick action is essential if pests are noticed. Recovers best from spring transplanting. Pruning is best performed in the dormant season or late summer, not in the spring or early summer.

Short life on many sites due to the bronze birch borer. Control leaf miners to help reduce borer infestation, which is lethal. Cankers can cause problems in the trunk and main branches. Walnut scale can be a big problem in California. This is a high-maintenance tree, but you may get lucky and enjoy many years with one. In the southern part of its range, protect from all-day sun by locating the tree so it is shaded for part of the afternoon in the summer. Be sure to pick local seed sources for planting in your area.

Betula pendula European Birch

Pronunciation: BET-chew-lah PEN-jew-lah

ZONES: 3A, 3B, 4A, 4B, 5A, 5B, 6A, 6B, (8)

HEIGHT: 30 to 45 feet

WIDTH: 20 to 35 feet

FRUIT: elongated; 1 to 3"; dry; brown; attracts wildlife; no significant litter

GROWTH RATE: rapid-growing, short life

HABIT: weeping oval; open; irregular silhouette; fine texture

FAMILY: *Betulaceae*

LEAF COLOR: dark green in summer; showy yellow fall color

LEAF DESCRIPTION: alternate; simple; double serrate margin; ovate to rhomboid; deciduous; 1 to 2" long

FLOWER DESCRIPTION: brown; showy; spring

EXPOSURE/CULTURE:

Light Requirement: full sun

Soil Tolerances: all textures; acidic; occasionally wet soil; moderate salt; drought-sensitive

Aerosol Salt Tolerance: moderate

PEST PROBLEMS: sensitive

BARK/STEMS:

Branches droop as tree grows, and will require pruning for vehicular or pedestrian clearance beneath canopy; trainable to be grown with a single or multiple trunks; no thorns

Pruning Requirement: requires some pruning to develop strong structure; do not prune February through July

Limb Breakage: susceptible to ice storms

Current Year Twig: brown; slender

CULTIVARS AND SPECIES:

Several cultivars are available, but these too will have pest problems: 'Dalecarlica' has deeply lobed, lacy leaves on pendulous branches; 'Laciniata' has cut leaves, very delicate; 'Fastigiata' has upright growth habit; 'Purple Splendor' (*purpurea*) and 'Scarlet Glory' have purple leaves; 'Tristis' and 'Youngii' have weeping habit. The beauty of these trees make them worthy of planting, even if they have a short life.

LANDSCAPE NOTES:

Origin: Europe and Asia Minor

Usage: near a deck or patio; specimen on a site with plenty of soil for root growth; street tree in cool climates

European Birch is graceful and ornamental, but is susceptible to fatal attacks of bronze birch borer. The delicate leaves are often fried by birch leaf miner. Plan to provide the necessary insect control and cultural conditions for best growth. Lawns grow well in its light shade, but should be eliminated with a layer of mulch to the edge of the canopy.

The striking fall leaf color and white bark look best when displayed against a dark background of evergreen trees and shrubs. Often planted in clumps or groves on commercial landscapes to mass the white bark effect. The trunks contrast nicely with dark mulch placed beneath the canopy. This also helps provide needed moisture to the roots and reduces competition from other plants. Expand the mulched area each year to eliminate grass and other plants from beneath the canopy.

Best adapted to moist sites with deep, loose soil with irrigation provided in times of drought. Few urban sites fit these requirements. Watch for pests if grown on other sites. Quick action is essential if pests are noticed. Recovers best from spring transplanting. Short life on most sites due to the bronze birch borer, except for the northwest part of the country. See *B. papyrifera* for discussion.

Betula utilis var. jacquemontii
Jacquemontii Birch, Whitebarked Himalayan Birch

ZONES: 5A, 5B, 6A, 6B, 7A, 7B, (8)

May have some resistance to borers and miners. Japanese beetles have been a big problem. Tolerates slightly alkaline soil conditions; grows rapidly but is short-lived.

Betula ermanii Erman Birch

ZONES: 5A, 5B, 6A, 6B, 7A, 7B, (8)

Susceptible to leaf miner and borers; short-lived in urban landscapes; tolerates slightly alkaline soil conditions.

Betula platyphylla 'Whitespire' Whitespire Birch

ZONES: 4A, 4B, 5A, 5B, 6A, 6B, 7A, 7B, (8)

One of the most pest resistant white-barked birches. It must be grown from cuttings or from tissue culture of the original tree in Madison, Wisconsin, to ensure borer resistance. The cultivar 'Whitespire Senior' is micropropagated from the original tree. Tolerates slightly alkaline soil conditions. 'Whitespire' propagated on its own roots grows poorly in clay, but grows vigorously when grafted onto *B. pendula* or *B. nigra*.

Betula populifolia **Gray Birch**

Pronunciation: BET-chew-lah pop-you-leh-FOE-lee-ah

ZONES: 4A, 4B, 5A, 5B, 6A, 6B

HEIGHT: 20 to 30 feet

WIDTH: 10 to 20 feet

FRUIT: elongated; 1 to 3"; dry; brown; attracts wildlife; no significant litter; persistent

GROWTH RATE: moderate; short-lived

HABIT: pyramidal; upright; weeping; open; symmetrical; fine texture

FAMILY: *Betulaceae*

LEAF COLOR: glossy dark green in summer; showy yellow fall color

LEAF DESCRIPTION: alternate; simple; double serrate margin; ovate; deciduous; 2 to 3" long

FLOWER DESCRIPTION: brown; showy; spring

EXPOSURE/CULTURE:

Light Requirement: part shade to full sun; afternoon shade beneficial in confined soil spaces

Soil Tolerances: all textures; slightly alkaline to acidic; wet soil; salt; drought-tolerant if planted where roots can expand unimpeded by urban structures

Aerosol Salt Tolerance: high

PEST PROBLEMS: sensitive

BARK/STEMS:

Trainable to be grown with a single or multiple trunks; showy trunk; no thorns

Pruning Requirement: needs little pruning to develop a strong structure; do not prune February through July

Limb Breakage: resistant

Current Year Twig: brown to gray; slender

CULTIVARS AND SPECIES:

'Laciniata', with pinnately-lobed leaves; 'Pendula', with drooping branches; and 'Purpurea', with purple young leaves.

LANDSCAPE NOTES:

Origin: native to northern North America

Usage: container or above-ground planter; reclamation; highway; near a patio or deck; specimen on a site with plenty of soil

Gray Birch forms loose, open thickets in the wild but is easily trained to a single, slender trunk with an irregular, upright, pyramidal silhouette. The glossy leaves on reddish-brown twigs are triangular-shaped and turn a lovely yellow in autumn before dropping. Both male and female catkins, or blooms, appear on the same tree, eventually producing a small cylindrical cone, with the male catkins persisting on the trees well into the winter. For the first four or five years, the bark of Gray Birch is dark brown, but later it takes on a smooth, chalky-white appearance, though it does not peel as readily as the bark of European and Paper Birch.

With the attractive bark, persistent catkins, and interesting, finely branched silhouette, Gray Birch makes a striking winter landscape plant, especially against a backdrop of dark green evergreens. It could be used as a specimen tree where a small- to medium-sized, fine-textured, upright plant is needed for a short time.

Gray Birch will grow easily in full sun or partial shade on almost any soil and location. It is an early colonizer of recently disturbed sites and suckers from the roots. It may suffer if interplanted with other competing shrubs and ground covers, so mulch generously beneath the canopy. Any necessary pruning should be done in summer or fall, as trees pruned in late winter or early spring will bleed excessively.

Leaf miners turn trees brown in a hurry. Gray birch may be more resistant to bronze birch borer than other white-barked birches.

Bischofia javanica Toog Tree

Pronunciation: bih-SHOW-fee-ah jah-VAY-nee-kah

ZONES: 10A, 10B, 11

HEIGHT: 30 to 50 feet

WIDTH: 25 to 35 feet

FRUIT: round; less than .5"; fleshy; blackish-red; causes messy litter on pavement; showy

GROWTH RATE: rapid-growing; moderate life span

HABIT: pyramidal when young, then round; dense; symmetrical; medium texture

FAMILY: *Euphorbiaceae*

Photo Courtesy of G. Joyner

LEAF COLOR: bronze-toned green; evergreen

LEAF DESCRIPTION: alternate; trifoliate; serrulate margin; oval to ovate; 2 to 5" long leaflets

FLOWER DESCRIPTION: green; not showy; spring

EXPOSURE/CULTURE:

Light Requirement: part shade to full sun

Soil Tolerances: all textures; alkaline to acidic; wet soil; moderate drought

Aerosol Salt Tolerance: moderate

PEST PROBLEMS: resists most lethal pests

BARK/STEMS:

Bark is thin and easily damaged by mechanical impact; branches droop as tree grows, and will require pruning for vehicular or pedestrian clearance beneath canopy; should be grown with a single leader; no thorns

Pruning Requirement: needed regular thinning to open up the canopy

Limb Breakage: somewhat susceptible

Current Year Twig: green; medium thickness

LANDSCAPE NOTES:

Origin: tropical Asia and Pacific islands

Usage: some communities have outlawed this tree because it is invasive; many homeowners like it as a shade tree

This rapidly growing tree can reach a height of 65 feet in California, but usually is seen 30 to 50 feet tall in Florida, Puerto Rico, and the Virgin Islands. The dense rounded crown makes Toog Tree a popular shade tree. Not enough light will penetrate for a lawn to grow underneath the canopy, but a groundcover will help to cover the exposed tree roots that commonly develop.

Branching is typically coarse, with several large-diameter laterals originating fairly close to the ground. The shiny trifoliate leaves are especially attractive when young. The stem will exude a milky sap when wounded. Small blue-black or reddish berries are produced in copious drooping clusters and drop to the ground, creating a mess on female trees. Unfortunately, the sex of the tree cannot be determined on young plants.

Toog Tree grows in full sun on various soil types. It appears to grow well in confined urban soil spaces; however, the fruit is considered messy and stains walks when it drops to the ground. Seeds often germinate in the landscape and could become a nuisance. It has escaped cultivation in parts of southern Florida. Some communities prohibit or discourage planting this tree due to its invasive habit. Severe scale infestations, especially false Oleander scale, are followed by sooty mold.

Aggressive roots can lift sidewalks if planted within 5 or 6 feet of the walk. Locate this tree in a lawn area where regular mowing will kill the sprouting seedlings, not in a landscape bed. The tree is not generally recommended for street tree planting and can be a nuisance in lawns, as surface roots make mowing difficult close to the trunk. Branches reportedly break from the tree on occasion. There are too many other high-quality trees available in zones 10 and 11 to encourage planting this tree.

Bismarckia nobilis Bismarck Palm

Pronunciation: biz-MAR-kee-ah NO-bill-iss

ZONES: 10A, 10B, 11

HEIGHT: 60 to 100 feet (200 feet in the wild)

WIDTH: 10 feet

FRUIT: round; 1.5"; no significant litter

GROWTH RATE: slow-growing; long-lived

HABIT: palm; round; open; symmetrical; coarse texture

FAMILY: *Arecaceae*

LEAF COLOR: blue, blue-green, blue-gray

LEAF DESCRIPTION: simple; lobed; star-shaped; 4 to 10' long fronds

FLOWER DESCRIPTION: white; not showy; spring

EXPOSURE/CULTURE:

Light Requirement: part shade to full sun

Soil Tolerances: all textures; alkaline to acidic; occasionally wet soil; salt; drought

Aerosol Salt Tolerance: moderate

PEST PROBLEMS: resistant

BARK/STEMS:

Showy; produces one trunk up to 2' thick; drops older fronds; no crown shaft; no thorns

Pruning Requirement: remove dead or dying fronds only

Limb Breakage: resistant

LANDSCAPE NOTES:

Origin: Madagascar

Usage: buffer strips; specimen

Lending a tropical flair to the landscape, Bismarck Palm slowly reaches 35 to 50 feet or more in height and is topped with gorgeous stiff, waxy leaves. These palms cannot be missed in the landscape, because of the very striking frond color and texture. The flower stalks are 4 feet long and produce many fruits.

Several of these palms placed together in a commercial or large residential setting contrast dramatically with existing vegetation, providing wonderful relief from the greens so common in most landscapes. Single specimens are also attractive and well suited for most residential-sized landscapes. A row of Bismarck Palms spaced 10 to 15 feet apart along each side of an entry road or wide walkway can create a dramatic impact. Even if you do not collect palms, this one should be considered when planning a landscape in zones 10 and 11.

Bismarck Palm should be grown in full sun or partial shade on well-drained soil. This palm is highly drought- and salt-tolerant and is becoming more popular. As with a number of palms, it cannot be transplanted until a trunk develops and is visible at the base of the plant. Container-grown Bismarck Palms can be planted any time of year. Most nurseries routinely root-prune this palm to help ensure transplantability.

There are usually no major pest problems on this palm, but watch for scale infestations.

Brahea armata Mexican Blue Palm

Pronunciation: BRAY-ya are-MAH-tah

ZONES: 9A, 9B, 10A, 10B, 11

HEIGHT: 30 to 40 feet

WIDTH: 6 to 10 feet

FRUIT: round; ⅓"; dry; black; attracts wildlife; causes some litter

GROWTH RATE: slow-growing; long-lived

HABIT: palm; open; symmetrical; coarse

FAMILY: *Arecaceae*

LEAF COLOR: silvery blue

LEAF DESCRIPTION: costapalmate; 5 to 6' leaves; some persist, others drop from trunk; thorns on the petiole

FLOWER DESCRIPTION: white; showy; summer; borne on long, hanging clusters up to 10' long

EXPOSURE/CULTURE:

Light Requirement: full sun

Soil Tolerances: all textures; alkaline to acidic; extended drought

PEST PROBLEMS: resistant

BARK/STEMS:

Single trunk about 24" thick; some have persistent fronds, others are smooth with old leaf scars; no crown shaft

Pruning Requirement: prune only drooping lower fronds, not horizontal or upright ones

Limb Breakage: very resistant

CULTIVARS AND SPECIES:

Other Mexican natives include *B. brandegeei,* a tall palm with a slender trunk; *B. edulis,* with green fronds instead of blue; and *B. elegans*, a slow-growing short palm suited for residential and commercial landscapes.

LANDSCAPE NOTES:

Origin: Mexico

Usage: parking lot island; buffer strip; highway; reclamation; specimen; sidewalk cutout; street tree; urban tolerant

Mexican Blue Palm is capable of reaching 40 feet in height, but takes 50 or more years to do so. It is topped with a dense, 8- to 10-foot-diameter, round crown comprised of deeply cut, nearly flat, costapalmate leaves. The bright fronds complement dry desert areas, yet fit extremely well into the greens so common in irrigated landscapes. This is one of Mexico's desert palms, and unlike the Washington Palms, it is tolerant of extended drought. The 6- to 10-foot-long, creamy white flower stalks in summer are followed by small fruits that could create a short-term mess on a nearby sidewalk. Although one of the hardier palms, 18°F will kill the plant.

Mexican Blue Palm is about as wind-proof as a tree can be, still standing after other trees have blown over or snapped in two. The leaf bases on some palms are persistent for several years, but they can be cleaned off to make the palm more appealing in the home or commercial landscape. Blue Palm is easy to transplant and will thrive in full sun or partial shade. It will adapt to dry or sandy locations and requires no special care once established. Do not overirrigate.

It is well suited to use as a street planting or clustered in informal groupings of varying size. It adapts well to small cutouts in the sidewalk, and can even create shade if planted on 6- to 10-foot centers. It has been popular in parts of the southwestern United States for a number of years, where it is used as a street tree. It is better adapted for this use than Washingtonia Palms because it is more drought-tolerant, grows slowly, and does not form the tall "telephone pole with a green top" effect along the street. It needs to be watered regularly until well established.

Bucida buceras Black Olive

Pronunciation: bew-SEE-dah bew-SAIR-us

ZONES: 10B, 11

HEIGHT: 40 to 50 feet

WIDTH: 35 to 50 feet

FRUIT: oval; less than .5"; fleshy; black; somewhat showy; occasionally distorted

GROWTH RATE: moderate; moderately long-lived

HABIT: oval; round; dense; irregular silhouette; fine texture

FAMILY: *Combretaceae*

LEAF COLOR: dark bluish-green to yellowish-green; evergreen

LEAF DESCRIPTION: alternate; whorled; simple; entire margin; oblanceolate to obovate; 2 to 4" long

FLOWER DESCRIPTION: greenish-yellow; not showy; spring and summer

EXPOSURE/CULTURE:

Light Requirement: part shade to full sun

Soil Tolerances: all textures; slightly alkaline to acidic; moderate salt; drought

Aerosol Salt Tolerance: high

PEST PROBLEMS: resistant

BARK/STEMS:

Branches droop as tree grows, and will require pruning for vehicular or pedestrian clearance beneath canopy; should be grown with a single leader; some thorns on branches

Pruning Requirement: requires early pruning to develop strong structure; unfortunately, many trees are topped in the nursery, spoiling the central leader that often develops naturally

Limb Breakage: resistant if pruned to one central leader, susceptible otherwise

Current Year Twig: gray; medium to slender

CULTIVARS AND SPECIES:

'Shady Lady' has a compact canopy. *Bucida spinosa* is small, slow-growing with tiny leaves, and has a fine-textured branching habit. It is a superior tree in most ways.

LANDSCAPE NOTES:

Origin: Puerto Rico, Caribbean

Usage: hedge; buffer strip; highway; reclamation; shade tree; specimen; street tree; urban tolerant

Black Olive is a nice evergreen tree, but is overused. It grows with a smooth trunk holding up strong, moderately wind-resistant branches, forming a pyramidal shape when young but developing a very dense, full, oval to rounded crown with age. Sometimes the top of the crown will flatten on an older tree so that the tree grows horizontally. The lush, leathery leaves are clustered at branch tips, sometimes mixed with the 0.5- to 1.5-inch-long spines found along the branches. The trunk can grow to at least 3 feet in diameter.

The inconspicuous flowers are produced in 4-inch-long spikes during spring and summer and eventually form the black fruits, which, unfortunately, exude a staining tannic acid material that stains patios, sidewalks, or vehicles parked below. For this reason it is probably best to locate the tree away from cars and sidewalks. Some communities discourage planting this tree.

A straight central trunk often develops when the tree is not topped. This structure should be maintained if trees will be allowed to grow to a large size. Multiple-trunked trees are susceptible to splitting apart.

Unfortunately, Black Olive is often planted in parking lot islands. When the tree begins to produce the staining fruit, it is usually topped or "hat-racked" (improper pruning) to prevent the branches from growing over the cars. Many native trees could be used in its place, including Satin Leaf, Gumbo-Limbo, and Wild Tamarind.

Occasionally bothered by sooty mold and bark borer. Eryphide mites cause galls, but no control is needed.

Bulnesia arborea Bulnesia

Pronunciation: bull-NEES-ee-ah are-BORE-ee-ah

ZONES: 10A, 10B, 11

HEIGHT: 30 to 40 feet

WIDTH: 25 to 35 feet

FRUIT: little seed set; no significant litter

GROWTH RATE: moderate; moderate life span

HABIT: round and spreading; dense; symmetrical; fine texture

FAMILY: *Zygophyllaceae*

Photo Courtesy G. Joyner

LEAF COLOR: medium green

LEAF DESCRIPTION: pinnately compound; ovate, 1" long leaflets; deciduous

FLOWER DESCRIPTION: yellow; very showy; spring and summer

EXPOSURE/CULTURE:

Light Requirement: full sun for best flowering

Soil Tolerances: all textures; alkaline to acidic; salt-sensitive; drought

Aerosol Salt Tolerance: high

PEST PROBLEMS: nearly pest-free

BARK/STEMS:

Bark is thin and easily damaged by mechanical impact; branches droop as tree grows, and will require pruning for vehicular or pedestrian clearance beneath canopy; routinely grown with multiple trunks; can be trained to grow with a short, single trunk; no thorns

Pruning Requirement: requires pruning to develop one trunk and strong structure; space branches along trunk and head back or thin aggressive lower branches to prevent formation of included bark

Limb Breakage: resistant, once good structure has been established

Current Year Twig: brown or gray; medium to slender

LANDSCAPE NOTES:

Origin: Colombia and Venezuela

Usage: container or above-ground planter; near a deck or patio; specimen; buffer strip; parking lot island

Bulnesia is an excellent, low-growing tree for small yards, patios, and other small-scale landscapes. It can be grown as a multi-stemmed clump or can be trained into a small to medium-sized tree with a single trunk up to 15 to 20 feet tall. The tree eventually has a spreading, rounded, finely branched growth habit on several thick main branches, which create dense shade under the crown. Trunks droop and spread as they grow older, forming a canopy that is wider than tall. Bulnesia can grow rapidly when it is young if it receives water and fertilizer.

The main ornamental value of Bulnesia is the brilliant yellow flowers displayed for 2 or 3 months during the warm season. It is a durable tree, tolerating poor soil. Like any other tree, it needs irrigation during the establishment period, but then survives on rainfall alone in humid climates.

Bulnesia can be planted along streets beneath tall power lines if it is set back from the road to allow for the spreading canopy. Without early thinning and training, the multiple trunks droop considerably. Main branches and trunks can be pruned back several times when the tree is young to create a more upright form. Remove lower branches early to force growth in the upper canopy for planting along streets and other areas where clearance is needed beneath the canopy. The tree makes a nice specimen or patio tree.

Bulnesia is usually pest-free. The tree has been difficult to propagate and so is not readily available in nurseries.

Bumelia lanuginosa Chittamwood

Pronunciation: boo-MEAL-ee-ah lan-you-gin-OH-sah

ZONES: 5A, 5B, 6A, 6B, 7A, 7B, 8A, 8B, 9A, 9B

HEIGHT: 30 to 40 feet

WIDTH: 25 to 35 feet

FRUIT: round; less than ½"; fleshy; black or blue; attracts wildlife; edible; no significant litter; persistent

GROWTH RATE: moderate; long-lived

HABIT: oval; round; open; irregular silhouette; symmetrical; medium texture

FAMILY: *Sapotaceae*

LEAF COLOR: green upper; red-brown to gray under; no fall color change

LEAF DESCRIPTION: whorled or alternate; simple; entire margin; oblanceolate to obovate; deciduous; 1 to 3" long; pubescent

FLOWER DESCRIPTION: white; pleasant fragrance; not showy; spring and summer

EXPOSURE/CULTURE:

Light Requirement: grows best in full sun

Soil Tolerances: clay; loam; sand; acidic; alkaline; drought

PEST PROBLEMS: pest-free

BARK/STEMS:

Routinely grown with multiple trunks; branches droop on older trees; small thorns present on trunk or branches

Pruning Requirement: requires pruning to develop good structure and an upright, tree-form habit

Limb Breakage: resistant

Current Year Twig: brown or gray; slender

LANDSCAPE NOTES:

Origin: southeastern North America

Usage: buffer strip; highway; reclamation; shade; specimen; street tree

This native North American deciduous tree also goes by the names of Gum Bumelia, Gum-Elastic Buckthorn, and Wollybucket Bumelia. Because there appear to be many forms of the plant in nature, from shrubby to tree form, nursery operators could make superior selections.

The bark varies considerably from tree to tree, making this a potential selection criterion for cultivar development. The leathery leaves are smooth on the upper side and fuzzy beneath. They drop in late fall without a show. Small, fragrant white flowers appear from June to July and are followed in fall by large, shiny, blue/black, fleshy fruits that are extremely popular with birds and other wildlife. Although the fruits are edible by humans, they have been known to produce unpleasant side effects if eaten in quantity.

Young trees require training to display a tree-like form. A shrubby, spreading, rounded ball of foliage often develops without pruning. Aggressive lateral branches must be reduced in length on the lower part of the trunk to help develop a tree form. It could be planted in urban and suburban landscapes, especially in areas that receive minimum maintenance. It is used quite extensively in Texas landscapes.

The common names of Gum Bumelia and Gum-Elastic are derived from the sap that quickly oozes from cuts or cracks to the bark. Youngsters in pioneer days were known to chew this sap as a gum.

Chittamwood should be grown in full sun or partial shade on well-drained soils. Trees found on poor soils in the wild grow slowly and are stunted, but with normal care they will grow well in a variety of landscapes. It is well suited for a reclamation site because of its adaptability to a wide range of soil types.

Bursera simaruba Gumbo-Limbo

Pronunciation: BURR-sir-ah sim-ah-ROO-bah

ZONES: 10B, 11

HEIGHT: 25 to 40 feet

WIDTH: 35 to 50 feet

FRUIT: oval; ½ to 1"; fleshy; red; no significant litter

GROWTH RATE: moderate; long-lived

HABIT: round; spreading; variable; open; irregular silhouette; medium texture

FAMILY: *Burseraceae*

LEAF COLOR: dark green; semievergreen

LEAF DESCRIPTION: alternate; odd pinnately compound; entire margin; oval to ovate; 2 to 4" long

FLOWER DESCRIPTION: green; not showy; spring

EXPOSURE/CULTURE:

Light Requirement: part shade to full sun

Soil Tolerances: all textures; alkaline to acidic; drought

Aerosol Salt Tolerance: high

PEST PROBLEMS: pest-free

BARK/STEMS:

Showy bronze, exfoliating; bark on young trees is thin and easily damaged by mechanical impact; branches droop as tree grows, and will require early training and pruning for vehicular or pedestrian clearance beneath canopy; routinely grown with multiple trunks; should be trained to grow with a single trunk; no thorns

Pruning Requirement: needs pruning to develop a strong trunk and branch structure

Limb Breakage: resistant; tree defoliates in hurricanes, protecting it from severe damage

Current Year Twig: brownish-green to reddish; stout

LANDSCAPE NOTES:

Origin: South Florida to Puerto Rico and Virgin Islands

Usage: parking lot island; buffer strip; highway; deck or patio; shade; specimen; street tree; urban tolerant

This large, native tree, with an open, irregular to rounded crown, may reach 60 feet in height, with an equal or wider spread, but is usually seen smaller in landscape plantings. The trunk and branches are thick and covered with resinous, smooth, peeling coppery bark with an attractive, shiny, freshly varnished appearance. Typically develops from 2 to 4 large-diameter limbs originating close to the ground. A salve is made from the bark to treat dermatitis.

Although growth rate is moderate to rapid and wood is soft, Gumbo-Limbo trees have good resistance to strong winds, drought, and neglect. Drought avoidance is accomplished by leaf drop and growth is often best in drier locations not receiving irrigation.

Gumbo-Limbo grows in full sun or partial shade on a wide range of well-drained soils. Gumbo-Limbo adapts to alkaline or poor, deep white sands but will also grow quickly on more fertile soil. Once established with good branch structure, Gumbo-Limbo requires little attention other than occasional pruning to remove lower branches that may droop close to the ground.

Gumbo-Limbo is ideal as a freestanding specimen on a large property or as a street tree, but does need room to grow. Lower branches will naturally grow close to the ground, so street trees will have to be trained early to develop a clear trunk with branches high enough for vehicle clearance. Locate the lowest permanent branch about 13 feet off the ground to provide enough clearance for a street tree planting. Specimen trees are often grown with branches beginning much closer to the ground, providing a beautiful plant with wonderful bark. Prune young trees to prevent formation of included bark.

Occasionally caterpillars will chew holes in the leaves, but control is usually not neccesary. Ganoderma fungus can infect the trunk and kill the tree.

Butia capitata Pindo Palm

Pronunciation: BEW-tee-ah cap-eh-TAY-tah

ZONES: 8B, 9A, 9B, 10A, 10B, 11

HEIGHT: 15 to 25 feet

WIDTH: 10 to 15 feet

FRUIT: round; ¾ to 1"; fleshy; orange or yellow; attracts squirrels and other mammals; suited for human consumption; fruit causes significant litter; showy

GROWTH RATE: slow-growing; long-lived

HABIT: palm; open; symmetrical; coarse texture

FAMILY: *Arecaceae*

LEAF COLOR: blue or blue-green to silver, blue-gray

LEAF DESCRIPTION: odd pinnately compound; entire margins; 18 to 36" long leaflets; 8 to 12' long, recurved fronds

FLOWER DESCRIPTION: white; showy; spring flowering and sporadically throughout the year

EXPOSURE/CULTURE:

Light Requirement: part shade to full sun

Soil Tolerances: all textures; slightly alkaline to acidic; drought

Aerosol Salt Tolerance: moderate

PEST PROBLEMS: resistant

BARK/STEMS:

Single trunk 2' thick with persistent leaf bases; no crown shaft; no thorns

Pruning Requirement: remove dead and dying fronds

Limb Breakage: resistant

LANDSCAPE NOTES:

Origin: Brazil

Usage: parking lot island; buffer strip; highway; near a deck or patio; specimen; street tree; urban tolerant

This cold-hardy, single-trunked palm is easily recognized by its rounded canopy of graceful fronds that curve in toward the trunk. Frond color varies from green through blue and silver. The heavy, stocky trunks are covered with persistent leaf bases. Large, showy clusters of orange-yellow, juicy, edible fruits, the size of large dates, are produced and often are used to make jams or jellies. The fruit, ripening in summer, can be messy, so plant 5 feet away from walks or patios. This slow-growing palm is attractive as a freestanding specimen or grouped with other palms. Most are seen small, as growth rate is very slow.

Growing in full sun or part shade on a wide variety of soils, including alkaline, Pindo Palm is moderately salt-tolerant. Pindo Palm can survive hot, windy conditions, asphalt, and concrete areas, but looks better in good soil with adequate moisture. Some people do not consider this a pretty palm, but it certainly will grow just about anywhere.

Plant 10 feet apart as a street tree to create a wall of blue foliage. They can be planted beneath power lines due to slow growth and small size. Locate one or a group in a low-growing ground cover to set them off from the rest of the landscape and to protect the trunks from damage. The blue foliage stands out in a shaded garden, adding a certain warmth.

Few nurseries have large numbers available. Palm leaf skeletonizer, scale, and micronutrient deficiencies (especially manganese and iron) are occasional problems for Pindo Palm. Micronutrient deficiencies only show up in soil with a high pH. Roots and lower trunk can rot if soil is kept too moist.

Caesalpinia granadillo Bridalveil Tree

Pronunciation: see-sal-PIH-nee-ah gran-ah-DILL-oh

ZONES: 10B, 11

HEIGHT: 30 to 35 feet

WIDTH: 25 to 35 feet

FRUIT: pod; dry; no significant litter

GROWTH RATE: moderate; long-lived

HABIT: vase; moderate density; nearly symmetrical; fine texture

FAMILY: *Leguminosae*

LEAF COLOR: dark green; evergreen

LEAF DESCRIPTION: alternate; bipinnately compound; entire margin; obovate; leaflets less than 2" long

FLOWER DESCRIPTION: yellow; summer and fall flowering; showy

EXPOSURE/CULTURE:

Light Requirement: grows best in full sun

Soil Tolerances: loam; sand; alkaline to acidic; moderate drought

Aerosol Salt Tolerance: low

PEST PROBLEMS: resistant

BARK/STEMS:

Very showy; bark is thin and easily damaged by mechanical impact; routinely grown with multiple trunks; branches grow mostly upright and will not droop until tree is older; no thorns

Pruning Requirement: needs some early pruning to develop strong trunk and branch structure to prevent formation of included bark

Limb Breakage: resistant if there is no included bark

Current Year Twig: green; slender

LANDSCAPE NOTES:

Origin: northern South America

Usage: container or above-ground planter; parking lot island; buffer strip; highway; shade; specimen; sidewalk cutout; street tree

Bridalveil Tree makes a wonderful shade tree, with its high canopy clothed with finely textured leaves. In summer and fall, Bridalveil Tree is decorated with showy yellow blossoms. The bark is also quite striking, peeling off in thin strips showing an unusual yellow, green, and gray mottling. The tree is usually found growing with several trunks originating from the lower four feet of the tree. This feature, along with the unusual bark traits, make this a highly desirable tree for planting in almost any landscape.

Bridalveil Tree will increase in popularity once people discover its outstanding characteristics. The foliage combines with an upright-vase shape to form a canopy tree with few equals. It is well suited for a residence, staying small enough to keep it from overtaking a property. It can be planted on 25-foot centers along a road or placed in a parking lot buffer strip to create a nice canopy of soft foliage. Its only limitation is lack of availability.

Bridalveil Tree should be grown in full sun on well-drained soil. Early pruning is needed to prevent bark from pinching or becoming included in the crotches. Branches with included bark can split from the tree. Aggressive, low branches can be trimmed back when the tree is young to help force growth in the upper part of the tree. Choose branches that are spaced 12 inches or more along the trunk as the main scaffold branches. Those forming a narrow crotch should be removed.

Caesalpinia pulcherrima Dwarf Poinciana

Pronunciation: see-sal-PIH-nee-ah pull-KEH-reh-mah

ZONES: 9B, 10A, 10B, 11

HEIGHT: 8 to 10 feet

WIDTH: 10 to 12 feet

FRUIT: pod; 3 to 6"; dry; brown; no significant litter

GROWTH RATE: moderate; short-lived

HABIT: round to vase; moderate density; symmetrical; fine texture

FAMILY: *Leguminosae*

LEAF COLOR: dark green; evergreen

LEAF DESCRIPTION: opposite/subopposite; bipinnately compound; entire margin; oval to oblong; leaflets less than 2" long

FLOWER DESCRIPTION: orange to red or yellow; showy; year-round

EXPOSURE/CULTURE:

Light Requirement: part shade to full sun; best flowering in full sun

Soil Tolerances: all textures; alkaline to acidic; drought

PEST PROBLEMS: resistant

BARK/STEMS:

Branches droop as tree grows; routinely grown with multiple trunks; can be trained to grow with short, single trunk; thorns are present on trunk or branches

Pruning Requirement: requires pruning to develop strong structure or single trunk

Limb Breakage: susceptible to breakage at crotch due to poor collar formation; wood itself is weak and tends to break

Current Year Twig: brown to green; stout

LANDSCAPE NOTES:

Origin: South America

Usage: reclamation; specimen; highway; standard

Brilliant flowers, feathery foliage, and quick growth make Dwarf Poinciana a popular shrub or small tree. It is hard to find a more attractive flower. Also known as Barbados Flower-Fence, this open-branched shrub will tolerate hot, dry areas and forms an effective thorny barrier. It flowers year-round with peak displays in spring and fall. Numerous trunks ascend from the ground, forming a mound of flowers and foliage.

Full sun is preferred for best flowering, but some shade is tolerated. Any well-drained soil is suitable. Dwarf Poinciana is perfectly suited to informal plantings. This is a beautiful, refreshing addition to any garden or yard as a specimen or as an accent toward the middle or back of a shrub border. Tipping the branches during the growing season creates a fuller shrub and more flowers. With some training and pruning, you can create a small, 12- to 15-foot-tall multi-stemmed tree, but the natural form is a low-branched, full, wide-spreading shrub about 10 feet tall and wide. The trunk must be staked to develop a small tree, but the wonderful flower display makes it worth extra effort.

Allow plenty of room for this plant to develop as a shrub. Set it off by itself in the lawn, or plant it as a tall component of a shrub border. It is sure to attract attention wherever it is planted. Regularly thin older stems to help keep it looking neat.

Scale presents an occasional problem. Dwarf Poinciana is susceptible to mushroom root rot, especially in poorly drained soil. This is not a problem on most soils.

Calliandra haematocephala Powderpuff

Pronunciation: cal-lee-AN-drah hee-mat-o-SEFF-ah-lah

ZONES: 9B, 10A, 10B, 11

HEIGHT: 12 to 15 feet

WIDTH: 10 to 15 feet

FRUIT: pod; 3 to 6"; dry; brown; no significant litter

GROWTH RATE: rapid-growing; short-lived

HABIT: rounded vase; open when young, filling in later; fine texture

FAMILY: *Leguminosae*

LEAF COLOR: evergreen; glossy copper when new, dark metallic green when older

LEAF DESCRIPTION: alternate; bipinnately compound; entire margin; oblong; bowed venation; 1 to 2" long leaflets

FLOWER DESCRIPTION: pink to red; pleasant fragrance; showy; year-round

EXPOSURE/CULTURE:

Light Requirement: flowers best in full sun

Soil Tolerances: all textures; slightly alkaline to acidic; drought

PEST PROBLEMS: resistant

BARK/STEMS:

Bark is thin and easily damaged by mechanical impact; branches droop as tree grows; routinely grown with multiple trunks; can be trained to a single trunk; no thorns

Pruning Requirement: requires pruning to develop strong structure and produce upright growth

Limb Breakage: resistant

Current Year Twig: brown to green or gray; medium to slender

CULTIVARS AND SPECIES:

'White' has snow-white flowers.

LANDSCAPE NOTES:

Origin: Boliva

Usage: container or above-ground planter; espalier; hedge; buffer strip; highway; near a deck or patio; standard; specimen; street tree

This large, multi-trunked, low-branching shrub has silky leaflets. The profuse, fragrant bloom is the main reason for its popularity, with big puffs of watermelon pink, deep red, or white silky stamens, produced during warm months. Many stems ascend from the ground on older plants, giving rise to a gorgeous, round canopy of pink flowers. Powderpuff can become quite bushy and overpowering in a small landscape unless it is trained to a small tree or thinned regularly.

With rapid growth in sandy soils and full sun, Powderpuff Bush will respond favorably to regular watering while young, but should require no special care once established except an occasional pruning to keep it within bounds. Powderpuff may be maintained as a tall (5- to 6-foot) flowering, clipped hedge. It is often seen as a small, flowering specimen tree with the lower branches pruned off. It makes an effective, colorful windbreak as it grows denser.

Powderpuff can grow to about 15 feet tall when trained and pruned into a small tree. Develop a short trunk by staking or heading back lateral branches until a sturdy trunk can support the developing canopy. The long, arching branches form an attractive canopy suitable for patio or container plantings. Pinching the new growth increases branch number and produces more flowers on a more compact plant. Although plants are damaged by freezing temperatures, they grow back from the base in the spring throughout zone 9.

Though usually pest-free, Powderpuff foliage can occasionally be infested by mites and eaten by caterpillars or other chewing insects.

Callistemon citrinus (lanceolatus)

Red Bottlebrush, Lemon Bottlebrush

Pronunciation: cal-eh-STEE-mon sih-TRINE-us

ZONES: 9B, 10A, 10B, 11

HEIGHT: 10 to 15 feet

WIDTH: 10 to 15 feet

FRUIT: round; less than 0.5"; dry or hard in clusters; brown; no significant litter; persistent on tree for several years; showy

GROWTH RATE: slow-growing; short-lived

HABIT: oval; round; moderate density; symmetrical; fine texture

FAMILY: *Myrtaceae*

LEAF COLOR: evergreen; medium green

LEAF DESCRIPTION: alternate; simple; entire margin; hairy below; lanceolate; 1 to 2" long

FLOWER DESCRIPTION: bright red; very showy; spring and summer; attracts hummingbirds

EXPOSURE/CULTURE:

Light Requirement: flowers best in full sun

Soil Tolerances: all textures; slightly alkaline to acidic; salt; drought

Aerosol Salt Tolerance: moderate

PEST PROBLEMS: resistant

BARK/STEMS:

Branches droop as tree grows; routinely grown with multiple trunks; no thorns

Pruning Requirement: needs little pruning to develop strong trunk and branch structure

Limb Breakage: resistant

Current Year Twig: gray; medium thickness

CULTIVARS AND SPECIES:

There are a variety of other species including *viminalis,* which grows larger, and *acuminatus,* which has deep red flowers. 'Red Cascade' makes a nice pinkish-red accent plant.

LANDSCAPE NOTES:

Origin: western Australia

Usage: container or above-ground planter; espalier; hedge; buffer strip; highway; near a deck or patio; parking lot island; specimen; standard; street tree

The common name, Bottlebrush, perfectly describes this plant's flower spikes, which are borne several inches back from the branch tips. Hummingbirds love the flowers, which are followed by small, woody capsules that look like bead bracelets on the bark.

Usually offered as a shrub, Bottlebrush can be trained as a tree to 15 feet or espaliered as a quick wall cover. It makes a nice screen or tall unclipped hedge. Pruning to develop several trunks and removing some lower branches can create a fine small specimen tree.

A good choice for a sunny spot, it will adapt to a variety of soils. Very drought-tolerant once established, Bottlebrush tolerates any soil except very poor, alkaline, or poorly drained. Fertilize regularly to maintain good flower color and dark green foliage. It makes a great small tree pruned to a short trunk in a container. Keep it in a prominent spot in the sun and enjoy the long flower display. It can be moved indoors for a short period in full flower.

No particular insect pests are listed for Callistemon. If the soil is too moist, root and trunk diseases can be a problem. Prevention is the best defense. Keep the plant on the dry side with good air circulation. A fungal-induced twig gall can disfigure the tree. Chlorosis, a systemic condition in alkaline soil that causes new leaves to turn yellow, can be temporarily corrected with treatment of the soil using iron sulfate or iron chelate. Suckers from the trunk should be removed periodically to maintain tree form.

Callistemon viminalis Weeping Bottlebrush

Pronunciation: cal-eh-STEE-mon vih-min-AL-iss

ZONES: 9B, 10A, 10B, 11

HEIGHT: 15 to 20 feet

WIDTH: 15 to 25 feet

FRUIT: round; less than 0.5"; dry; brown clusters; somewhat showy; no significant litter; persistent

GROWTH RATE: moderate; short to moderate life

HABIT: rounded, weeping; open; irregular silhouette; fine texture

FAMILY: *Myrtaceae*

LEAF COLOR: evergreen; light green

LEAF DESCRIPTION: alternate; simple; entire, hairy margins; narrow; lanceolate to linear; 2 to 3" long

FLOWER DESCRIPTION: bright scarlet; very showy; spring and summer

EXPOSURE/CULTURE:

Light Requirement: flowers best in full sun

Soil Tolerances: all textures; slightly alkaline to acidic; moderate salt; moderate drought

Aerosol Salt Tolerance: moderate

PEST PROBLEMS: sensitive in Florida

BARK/STEMS:

Branches droop as tree grows and will require pruning for vehicular or pedestrian clearance beneath canopy; routinely grown with, or trainable to be grown with, multiple trunks; can be trained to grow with short, single trunk; no thorns

Pruning Requirement: requires pruning to develop upright structure and single trunk

Limb Breakage: moderately sensitive

Current Year Twig: gray; medium thickness

CULTIVARS AND SPECIES:

The cultivar 'Red Cascade' has large red flowers in spring and fall. 'Texas' leaves have bright silver backs. *C. citrinis* and *C. rigidus* are hardy, growing well into zone 9.

LANDSCAPE NOTES:

Origin: Australia

Usage: container or above-ground planter; screen; parking lot island; buffer strip; highway median; near a deck or patio; specimen

This popular tree has a dense, multi-trunked, low-branching, pendulous growth habit. Mature specimens can reach 25 to 30 feet tall in 30 years, but most trees are seen 15 to 20 feet high and wide. The leaves tend to grow only at the ends of the long, hanging branches, creating a weeping effect. The cylindrical bloom clusters, 3 to 5 inches long and 1 inch wide, are composed of multiple, long, bristlelike stamens. These clusters appear in great abundance during March through July, less so throughout the year. The flowers are followed by persistent woody capsules that are not noticeable unless you are close to the tree.

Weeping Bottlebrushes should be grown in full sun, preferably on moist, well-drained soil. Although they can tolerate some drought, best flowering and growth are obtained with ample moisture and regular fertilization. Be sure the soil drains well, as roots often rot in wet soil. The brittle wood of Weeping Bottlebrush may make it unsuitable for windy areas. They are not suitable for street tree planting, because of the weeping growth habit, but make nice plantings in wide medians. Lower branches can be removed so cars can fit beneath in parking lots. Bottlebrushes grow well in restricted soil space. Occasional pruning of pendulous branches will be required for vehicle clearance. One of the best uses is for lawn specimens, or screens on large properties, with a regular maintenance program.

Mites and witches broom can be troublesome, as can root rot in wet soil and canker. A fungal-induced twig gall can severely disfigure the tree. The tree is often short-lived in Florida due to disease.

Calocedrus decurrens, California Incense-Cedar

Pronunciation: cal-o-SEED-rus dee-CURR-enz

ZONES: 5A, 5B, 6A, 6B, 7A, 7B, 8A, 8B, 9A, (9B), (10A)

HEIGHT: 40 to 60 feet

WIDTH: 8 to 12 feet

FRUIT: elongated; oval; 0.5 to 1"; dry; brown; attracts wildlife; no significant litter; persistent

GROWTH RATE: slow-growing; long-lived

HABIT: columnar; pyramidal; moderate density; symmetrical; fine texture

FAMILY: *Cupressaceae*

LEAF COLOR: evergreen; light green

LEAF DESCRIPTION: in fours; simple; scale-like

FLOWER DESCRIPTION: not showy

EXPOSURE/CULTURE:

Light Requirement: part shade to full sun; full sun for best growth

Soil Tolerances: all textures; slightly alkaline to acidic; drought

Aerosol Salt Tolerance: high

PEST PROBLEMS: resistant

BARK/STEMS:

Gorgeous, thick, orange-brown trunk; branches grow mostly upright and will not droop; usually grown with a single leader; no thorns

Pruning Requirement: needs little or no pruning to develop strong structure

Limb Breakage: resistant

Current Year Twig: brown; green; slender

CULTIVARS AND SPECIES:

Nurseries frequently offer the cultivar 'Columnaris', which has a formal landscape effect. The cultivar 'Compacta' has a dwarf, compact growth habit; 'Aureovariegata' has interspersed sprays of bright yellow foliage.

LANDSCAPE NOTES:

Origin: California and Oregon

Usage: hedge; specimen; screen; possible street tree

This stately native evergreen tree slowly grows straight up in a very narrow cone or columnar shape to a height of 70 to 150 feet or more in the wild, 50 to 60 feet in landscapes, yet is only 8 to 15 feet wide at maturity. The branches, which are densely clothed with flat, shiny, aromatic needles, extend to the ground and remain on the tree unless shaded out by other trees, making California Incense-Cedar ideal for use as a screen, hedge, or windbreak. The small cones produced at branch tips are quite persistent. The seeds provide a welcome treat for many varieties of birds and wildlife.

Although growth is slow, California Incense-Cedar trees are extremely long-lived, surviving 500 to 1,000 years in the wild. Mature specimens have outstanding, brick red, flaky, furrowed bark. Bark is more than six inches thick on mature trees. You cannot miss this tree in the woods.

California Incense-Cedar is a bit particular as to its growing requirements, being very sensitive to wet conditions. It can be found growing in canyons and on steep, well-drained slopes down to the edge of a stream, but not on the flood plain where water would stand for a period of time. Some forms are tolerant of heat and drought; it grows in areas receiving only 20 inches of rainfall each year. Pruning is rarely necessary, but trees may be sheared, if desired, to maintain a shorter, denser screen. Beautiful specimens of this tree can be seen growing in zones 6, 7, 8, and 9. Plants withstand ice and snow loads well due to wide branch angles and small-diameter branches.

Heart rot is a serious problem on mature trees, but this is not cause to avoid planting this beautiful tree. Armillaria and other root diseases kill more trees than any other agent. Rust and leafy mistletoe are less serious.

Calodendrum capense Cape Chestnut

Pronunciation: kal-oh-DEN-drum ka-PEN-sa

ZONES: 9B, 10A, 10B, 11

HEIGHT: 35 to 50 feet

WIDTH: 35 to 50 feet; can grow larger

FRUIT: oval pod; 1.5"; dry; gray-brown; no significant litter; persistent

GROWTH RATE: moderate; long-lived

HABIT: pyramidal when young, spreading and rounded later; dense; somewhat irregular, billowing silhouette; medium texture

FAMILY: *Rutaceae*

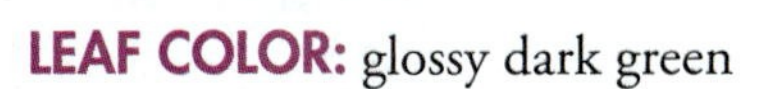

LEAF COLOR: glossy dark green

LEAF DESCRIPTION: simple; ovate; briefly deciduous; 3 to 6" long

FLOWER DESCRIPTION: white to pink, variable; very showy and long-lasting; spring through early summer

EXPOSURE/CULTURE:

Light Requirement: flowers best in full sun

Soil Tolerances: clay or loam is best; slightly alkaline to acidic; moderate drought

PEST PROBLEMS: resistant

BARK/STEMS:

Branches droop and will need removal or pruning for vehicle and pedestrian clearance beneath canopy; can be grown with a short, single leader or with several trunks; no thorns

Pruning Requirement: requires some early pruning to develop one leader; thinning canopy enhances wind firmness

Limb Breakage: resistant; old trees with multiple trunks could split apart

Current Year Twig: thick

LANDSCAPE NOTES:

Origin: South Africa to Kenya

Usage: buffer strip; parking lot islands; street tree; highway; specimen; shade

A native of South Africa, Cape Chestnut grows best on the western U.S. coast and is seldom successful in the interior, primarily due to cold sensitivity. Its flowers are spectacular, completely covering the canopy for two months. The leaves are almost completely hidden by the dense flower display. The largest trees in California have a canopy spread of about 80 feet.

This is an especially beautiful tree used along a street, as a large specimen, or as a shade tree in any number of landscape settings. Place it in the front lawn area or in a landscape bed along with low-growing ground covers or shrubs. The flower display shows off nicely along highways and at interchanges. It adapts well to the dry heat of parking lot buffer strips if it is irrigated regularly. It is well suited for planting near curbs and sidewalks, as roots are less aggressive than those on larger-maturing trees.

Trees need to be regularly irrigated in the western U.S. to maintain the full canopy and moderate growth rate. Fertilizer applications each year or two can also help to maintain vigor.

Calophyllum brasiliense (calaba) Santa Maria

Pronunciation: cal-o-FILL-um brah-sih-lee-EN-see

ZONES: 10B, 11

HEIGHT: 30 to 45 feet

WIDTH: 30 to 50 feet

FRUIT: round; 1 to 2"; dry and hard, then fleshy; green; fruit causes significant litter; showy

GROWTH RATE: slow to moderate; long-lived

HABIT: round and spreading; dense; somewhat irregular silhouette; coarse texture

FAMILY: *Clusiaceae*

LEAF COLOR: evergreen; medium green

LEAF DESCRIPTION: opposite; simple; entire margin; oval to oblong; 2 to 4" long

FLOWER DESCRIPTION: white; pleasant fragrance; briefly showy; summer

EXPOSURE/CULTURE:

Light Requirement: part shade to full sun

Soil Tolerances: adaptable; all textures; alkaline to acidic; salt; drought

Aerosol Salt Tolerance: high

PEST PROBLEMS: pest-free

BARK/STEMS:

Branches droop as tree grows and will require pruning for vehicular or pedestrian clearance beneath canopy; showy, black trunk; often grows with several trunks but can be trained with a short, single leader; no thorns

Pruning Requirement: needs little pruning to develop strong structure

Limb Breakage: resistant

Current Year Twig: green; moderate to stout

CULTIVARS AND SPECIES:

C. inophyllum has larger leaves and flower racemes and may grow larger. The plant can invade native habitat in Florida.

LANDSCAPE NOTES:

Origin: Puerto Rico to Mexico south to Brazil

Usage: container; espalier; hedge; parking lot island; buffer strip; highway; deck or patio; reclamation; shade; specimen; sidewalk cutout; street tree; urban tolerant

This rounded tree is densely foliated with four-inch-long, leathery leaves. Although able to reach 50 feet in height in the forest, Santa Maria tends to be a slow-growing, moderately sized tree. Small, white, fragrant flowers appear on 1- to 2-inch-long racemes among the 4- to 6-inch-long, glossy leaves. The bark on older trees is almost black and longitudinally furrowed and is quite attractive.

Santa Maria has good salt tolerance and is often seen along the beach. Grows well in confined soil spaces, such as along a street. The tree should be propagated, sold, and planted in urban areas much more often. The golfball-sized fruit is poisonous and hard; this could be undesirable or a nuisance if the tree is planted near sidewalks and pavements. The trunk grows to about three feet in diameter.

Well-suited as a street, parking lot, patio, or small shade tree, especially for coastal areas, Santa Maria can also be used as a screen and can be maintained as a shrub. The tree can be trained to a single trunk by reducing the length of aggressive lateral branches when the tree is young. Choose single-trunked trees for planting along streets and in parking lots. The stems bleed a yellow latex when injured or pruned. The wood of the Santa Maria tree is valuable for ship building and cabinet work.

Camellia oleifera Tea-Oil Camellia

Pronunciation: cah-MEAL-lee-ah oh-lee-IF-er-ah

ZONES: 5B, 6A, 6B, 7A, 7B, 8A, 8B, 9A, 9B, (10)

HEIGHT: 20 feet

WIDTH: 10 to 15 feet

FRUIT: round; 0.5 to 1"; dry; no significant litter

GROWTH RATE: slow-growing; moderate life span

HABIT: round; pyramidal to vase; dense; symmetrical; medium texture

FAMILY: *Theaceae*

LEAF COLOR: dark green; evergreen

LEAF DESCRIPTION: alternate; simple; serrate margins; oblanceolate to obovate; broadleaf evergreen; 3 to 5" long

FLOWER DESCRIPTION: white; showy; winter

EXPOSURE/CULTURE:

Light Requirement: part shade to full sun

Soil Tolerances: all textures; slightly alkaline to acidic; moderate drought

PEST PROBLEMS: resistant

BARK/STEMS:

Smooth and showy branches grow mostly upright; routinely grown with, or trainable to be grown with, multiple trunks; can be trained to grow with a short, single trunk; no thorns

Pruning Requirement: requires pruning to develop single trunk; grows nicely with multiple trunks

Limb Breakage: resistant

Current Year Twig: brown; medium thickness

LANDSCAPE NOTES:

Origin: China

Usage: bonsai; container or above-ground planter; hedge or screen; parking lot islands; buffer strip; near a deck or patio; trainable as a standard; specimen; sidewalk cutout; street tree

This large shrub or small tree will reach a height of 20 feet with thin, upright, multiple trunks and branches. The crown forms a rounded or oval vase with lower branches removed. Tea-Oil Camellia is so named because the trees are cultivated in their native China specifically for the seeds, from which is extracted commercial tea oil. This particular camellia species looks much like *Camellia sasanqua*, the more familiar Camellia, except the leaves are a bit larger, 3 to 5 inches long and 2 to 3 inches wide. Single, white, fragrant flowers are produced in late winter.

Plants should require little pruning. Their dense, compact crown makes them ideally suited for borders, specimens, accents, and sheared or natural hedges or screens. Large specimens may be trimmed to multi- or single-trunked small trees. This is a great tree for planting along a street beneath power lines, as it will not require pruning by the utility company, but unfortunately it is not yet readily available. Nurseries could grow and market this plant in multi-stemmed or single-trunked forms for this purpose.

Smooth exfoliating bark combines with a showy trunk form to make this an ideal small tree for accenting in a garden. Light the canopy from below to feature it in a nighttime landscape.

Tea-Oil Camellia should be grown in full sun or partial shade on rich, moist, acid soils. It will tolerate drought once established and grows well in clay soil. Watch for scale infestation.

Carpentaria acuminata Carpentaria Palm

Pronunciation: car-pen-TEAR-ee-ah ah-coo-mih-NAY-tah

ZONES: 10B, 11

HEIGHT: 35 to 40 feet

WIDTH: 8 to 15 feet

FRUIT: very showy; red; round; 0.5 to 1"; no significant litter

GROWTH RATE: rapid; moderate life span

HABIT: open; symmetrical; medium texture

FAMILY: *Arecaceae*

LEAF COLOR: green

LEAF DESCRIPTION: odd pinnately compound; entire margins; fronds are 6 to 10' long

FLOWER DESCRIPTION: white; somewhat showy; spring and summer

EXPOSURE/CULTURE:

Light Requirement: best in full sun

Soil Tolerances: loam; slightly alkaline to acidic; occasionally wet soil; drought sensitive

Aerosol Salt Tolerance: low

PEST PROBLEMS: resistant

BARK/STEMS:

Single, slender trunk; no thorns

Pruning Requirement: remove dead fronds

Limb Breakage: resistant

LANDSCAPE NOTES:

Origin: Australia

Usage: container or above-ground planter; near a deck or patio; specimen

This palm tree quickly grows to a height of 40 feet, the smooth gray trunk topped with 10-foot-long, spreading fronds. The inconspicuous white flowers appear from spring through fall and are followed by the production of bright red fruits that are less than one inch long. The juice from these fruits can cause skin irritation. Once highly recommended as a replacement for the Christmas Palm, which is very susceptible to lethal yellowing disease, Carpentaria Palm apparently requires a richer soil than many landscapes can provide. It is also susceptible to thrips and trunk cracks in freezing weather, a condition that opens the trunk to decay organisms.

Carpentaria Palm is probably best suited for an occasional accent or specimen planting where temperatures stay warm in the winter. A number of them grouped together can be attractive. Choose the best soil on your site for planting this palm. Carpentaria Palm should be grown in full sun on rich, moist but well-drained, fertile soil; it has a low tolerance for salt and drought. It tolerates a week or more of flooding or wet soil on a regular basis.

Carpinus betulus 'Fastigiata' 'Fastigiata' European Hornbeam

Pronunciation: car-PINE-us BET-u-luss

ZONES: 5A, 5B, 6A, 6B, 7A, 7B, 8A, 8B

HEIGHT: 30 to 40 feet

WIDTH: 15 to 20 feet

FRUIT: oval; less than 0.5"; dry; brown; attracts birds; no significant litter; persistent

GROWTH RATE: slow-growing; moderate life span

HABIT: columnar; oval; pyramidal; upright; dense; perfectly symmetrical; unusually fine texture

FAMILY: *Betulaceae*

LEAF COLOR: green in summer; somewhat showy yellow fall color

LEAF DESCRIPTION: alternate; simple; double serrate margin; oblong to ovate; deciduous; 2 to 3" long

FLOWER DESCRIPTION: white; not showy; spring

EXPOSURE/CULTURE:

Light Requirement: part shade to full sun; stressed in full sun in confined soil spaces

Soil Tolerances: all textures; alkaline to acidic; occasionally wet soil; salt-sensitive; moderate drought

Aerosol Salt Tolerance: none

PEST PROBLEMS: mostly resistant

BARK/STEMS:

Branches grow mostly upright and will not droop; should be grown with a single leader as high into canopy as possible; no thorns

Pruning Requirement: needs a little pruning when young to prevent formation of low, double leaders

Limb Breakage: mostly resistant

Current Year Twig: brown; slender

LANDSCAPE NOTES:

Origin: Europe to Iran

Usage: a favorite for bonsai; container or above-ground planter; espalier; hedge; parking lot island; buffer strip; specimen; sidewalk cutout; street tree; urban tolerant

'Fastigiata' European Hornbeam, the most common Hornbeam cultivar sold, grows into a very densely foliated, columnar or oval-shaped tree, making it ideal for use as a hedge, screen, or windbreak. The smooth, gray, rippling bark shields the extremely hard, strong wood.

The tree is sold as tree-form with a short trunk for street tree use, or low-branched for use as specimens and screens. Young trees will be quite narrow, but they will broaden some with age, making them well-adapted for planting in areas with limited horizontal space for crown development. It is an excellent tree well adapted to urban life. European Hornbeam tolerates clipping extremely well and has been used as a screening plant for centuries. The species is rare in the trade, but it forms a beautiful vase-shaped canopy with buttressed roots.

'Fastigiata' European Hornbeam will grow in full sun or light shade on almost any well-drained soil, from dry and rocky to wet. Because of the dense arrangement of leaves on the fine-textured, upright stems, pruning is seldom required once the proper branch-height clearance is established early in the life of the tree. The upright branching habit will not form a canopy over a street as do broad-spreading trees. Usually pest-free, it is considered a low-maintenance tree.

Occasionally bothered by two-lined chestnut borer or trunk canker. Japanese beetle can eat significant quantities of foliage. It may die back in severe drought in areas with limited soil space, such as narrow tree lawns or small parking lot islands, but it performs well without irrigation in areas with open soil where roots can grow unrestricted. Trees recover quickest with spring transplanting.

Carpinus caroliniana American Hornbeam

Pronunciation: car-PINE-us care-o-lin-ee-A-nah

ZONES: 3A, 3B, 4A, 4B, 5A, 5B, 6A, 6B, 7A, 7B, 8A, 8B, 9A

HEIGHT: 20 to 40 feet

WIDTH: 20 to 30 feet

FRUIT: elongated; oval; 0.5 to 1"; dry; brown; attracts wildlife; no significant litter

GROWTH RATE: slow-growing; moderate life span

HABIT: columnar to oval when young, pyramidal to vase later; dense; symmetrical; medium texture

FAMILY: *Betulaceae*

LEAF COLOR: medium green in summer; exceptionally showy burnt-orange or red to yellow fall color

LEAF DESCRIPTION: alternate; simple; double serrate margin; oblong to ovate; deciduous; 2 to 4" long

FLOWER DESCRIPTION: orange; yellow; not showy; spring

EXPOSURE/CULTURE:

Light Requirement: part shade to shade for best growth; full sun with irrigation or in open soil

Soil Tolerances: all textures; slightly alkaline to acidic; occasionally wet soil; salt-sensitive; moderate drought

Aerosol Salt Tolerance: none

PEST PROBLEMS: nearly pest-free

BARK/STEMS:

Showy, gray trunk; bark is thin and easily damaged by mechanical impact; branches droop slightly and will require some pruning for vehicular or pedestrian clearance beneath canopy; routinely grown with multiple trunks, but can be trained to grow with a single trunk; no thorns

Pruning Requirement: needs little or no pruning to develop strong structure; prune to one trunk for street tree use

Limb Breakage: very resistant

Current Year Twig: brown; reddish; slender

LANDSCAPE NOTES:

Origin: eastern North America along streams and moist soils

Usage: bonsai; hedge or screen; near a deck or patio; small shade tree; specimen; street tree

A handsome tree in many locations, the tree slowly reaches mature height and spread. Narrow-crowned cultivars may be introduced soon. The variety *virginiana* maintains a neat, compact size. It will grow with an attractive open habit in total shade, but is dense in full sun. The muscle-like bark is smooth, gray, and fluted. American Hornbeam, also called Blue-Beech and Ironwood, has a slow growth rate and is reportedly difficult to transplant from a field nursery, but is easy from containers. Brown leaves occasionally hang on the tree into the winter.

With age, a multiple-trunked, low-branching specimen can be very attractive, showing off the bark and trunk form particularly well when lit at night. Tolerant of pruning, the tree can be used as a hedge plant or as a screen due to the densely foliated crown. It can also be trained for street tree use by pruning to one central leader with small-diameter horizontal branches forming layers of foliage in the crown. Some nurseries offer single-stemmed specimens. Well-suited for small overhead spaces in the shade or sun, Hornbeam is tolerant of occasional flooding. The wood is very hard and strong and makes a great tree for climbing if allowed to grow with low branches. If transplanting from the field, do so in spring.

This tree performs well even in areas inundated with water for several days to a week or two once it is established. It is probably best to provide even established trees with some irrigation during dry spells in the south. Ironwood grows in sun or shade (as an understory tree in the woods) and tolerates most soils, except alkaline.

Relatively few insects attack Hornbeam, though borers can shorten their life. Several fungi cause leaf spots. Leaf spots are not serious, so control measures are usually not needed. Cankers infect stressed trees. Trees recover slowly from transplanting probably due to their natural slow-growing habit.

Carya glabra Pignut Hickory

Pronunciation: CARE-yah GLAY-bra

ZONES: 5A, 5B, 6A, 6B, 7A, 7B, 8A, 8B, 9A, 9B

HEIGHT: 50 to 65 feet

WIDTH: 30 to 40 feet

FRUIT: oval; round; 1 to 3"; dry and hard; green turning brown; attracts squirrels and other mammals; fruit and foliage cause significant litter

GROWTH RATE: rapid-growing; long-lived

HABIT: oval; dense; nearly symmetrical; medium texture

FAMILY: *Juglandaceae*

LEAF COLOR: medium green in summer; showy yellow to copper fall color for several days

LEAF DESCRIPTION: alternate; odd pinnately compound; serrate margin; lanceolate to obovate leaflets; deciduous; 6 to 12" long

FLOWER DESCRIPTION: yellow; not showy; spring

EXPOSURE/CULTURE:

Light Requirement: part shade to full sun

Soil Tolerances: all textures; slightly alkaline to acidic; occasionally wet soil; salt-sensitive; drought

Aerosol Salt Tolerance: moderate

PEST PROBLEMS: resistant

BARK/STEMS:

Branches grow mostly upright and will not droop; should be grown with a single leader; no thorns

Pruning Requirement: needs little pruning to develop a strong structure

Limb Breakage: very resistant

Current Year Twig: brown; stout

CULTIVARS AND SPECIES:

Variety *megacarpa* has larger fruit, is salt tolerant, and grows along the coastal plain from Virginia to Louisiana. *C. tomentosa* grows 40 to 50 feet tall and is found on dry, sandy soils in pine forests. It also tolerates periodic flooding. *C. aquatica* is found on poor, bottomland soils and tolerates flooding.

LANDSCAPE NOTES:

Origin: eastern North America

Usage: shade; specimen

Pignut Hickory is capable of reaching 100 feet in the forest, but is usually much smaller in the landscape. The leaves create a coarse, oval canopy and the strong but irregularly spaced branches resist breakage in storms, making it useful as a shade tree. The green fruits are quite bitter and are popular with various forms of wildlife, but not humans. Hickory usually grows with branches well spaced along one, straight central leader, with little or no pruning required.

Fruits may damage cars as they fall and people could slip on fallen fruit, so it may be best to locate the tree away from streets, parking lots, sidewalks, and other areas where cars regularly park. It makes a nice shade tree or median strip tree planted on 25- to 30-foot-centers and turns a striking color in the fall. This is an underutilized native tree with potential for much wider use.

Pignut Hickory grows best in sun or partial shade on well-drained, acid soils and is very drought-tolerant. It grows well in sand or clay, sending deep roots down below the trunk in well-drained soil. Tap roots are produced on trees grown in the forest on deeper soils. Natural habitat ranges from stream banks to dry hillsides.

Hickory bark beetle is a problem, particularly during droughts. Trees show minor element deficiencies on alkaline soils. Borers, bagworms, and fall webworms can be troublesome, but none are normally serious. Galls are common on the leaves but cause no real damage.

Carya illinoensis Pecan

Pronunciation: CARE-yah ill-eh-noy-EN-siss

ZONES: 6A, 6B, 7A, 7B, 8A, 8B, 9A, 9B, (10)

HEIGHT: 70 to 100 feet

WIDTH: 50 to 70 feet

FRUIT: oval; round; 1 to 2"; dry and hard; brown; attracts squirrels and other mammals; edible; fruit, twigs, and foliage cause significant litter; persistent

GROWTH RATE: moderate; long-lived

HABIT: oval; round; spreading; open; nearly symmetrical; coarse texture

FAMILY: *Juglandaceae*

LEAF COLOR: medium green in summer; somewhat yellow fall color

LEAF DESCRIPTION: alternate; odd pinnately compound; serrate margin; lanceolate; deciduous; 4 to 6" long leaflets

FLOWER DESCRIPTION: yellow; not showy; spring

EXPOSURE/CULTURE:

Light Requirement: grows best in full sun

Soil Tolerances: all textures; alkaline to acidic; wet soil; salt-sensitive; drought

Aerosol Salt Tolerance: low to moderate

PEST PROBLEMS: pest-sensitive

BARK/STEMS:

Branches grow mostly upright and will not droop; showy trunk; should be grown with a single leader; single trunk with well-spaced branches forms if tree is not topped; no thorns

Pruning Requirement: needs some early pruning to develop strong structure

Limb Breakage: susceptible to occasional breakage either at crotch or wood itself tends to break

Current Year Twig: brown to gray; thick

LANDSCAPE NOTES:

Origin: Illinois to Texas and into northern Mexico

Usage: shade; nuts

The state tree of Texas, this tree needs plenty of room, above and below ground, to grow. The largest of the hickories, this deciduous American native tree has a uniform, symmetrical, broadly oval crown and is massively branched especially if the main trunk is headed. Large major limbs on some trees grow up and out from the trunk in a distinctive upright, spreading fashion. On older trees, lower branches become wide-sweeping, with their tips almost touching the ground. Trunks can grow to 6 feet in diameter.

Pecan trees are well suited to large landscapes and natural settings, where wildlife is greatly attracted by the fruit. Branches on the tree are sometimes rather brittle, breaking toward the base. Be sure to space major limbs along the trunk so they do not all originate from the same point on the trunk. The central leader can usually be selected easily if the tree has not been topped, because Pecan naturally grows with one central trunk. The weaker, spreading habit familiar to Pecan orchards results from early topping. This is not recommended for shade tree plantings. Leaves sometimes drop during the summer due to insects or disease, stimulating the tree to initiate a new set of leaves in late summer.

Care must be taken in locating this tree, not only due to its size and fruit drop but also because the leaves and husks release a substance that can stain clothes, pavements, and cars during rains. The falling nuts can also damage vehicles. Pecan is subject to sudden summer limb drop, a phenomenon characterized by failure in the middle of a large-diameter branch. This typically occurs on a very hot summer day with no wind. The best treatment is preventive thinning to eliminate some of the weight along major horizontal limbs.

There are many pests of Pecan, making it a high-maintenance tree. Zinc deficiencies often occur on alkaline soils. Be sure to select a cultivar suitable for your particular area of the country.

Carya ovata Shagbark Hickory

Pronunciation: CARE-yah oh-VAY-tah

ZONES: 5A, 5B, 6A, 6B, 7A, 7B, 8A

HEIGHT: 60 to 80 feet

WIDTH: 25 to 35 feet

FRUIT: oval; round; 1 to 2"; dry and hard; green turning brown; attracts squirrels and other mammals; causes significant litter; showy

GROWTH RATE: slow-growing; long-lived

HABIT: oval; moderate density; symmetrical when young; coarse texture

FAMILY: *Juglandaceae*

LEAF COLOR: green in summer; showy, burnt-yellow fall color

LEAF DESCRIPTION: alternate; odd pinnately compound; serrate margin; lanceolate to oblong or oblanceolate; deciduous; 4 to 7" long leaflets

FLOWER DESCRIPTION: calkins to 6" long; showy; spring

EXPOSURE/CULTURE:

Light Requirement: part shade to full sun

Soil Tolerances: all textures; slightly alkaline to acidic; occasionally wet soil; some salt; drought

Aerosol Salt Tolerance: low

PEST PROBLEMS: resistant

BARK/STEMS:

Showy, shaggy bark; branches mostly upright on larger trees; should be grown with a single leader; no thorns

Pruning Requirement: needs little pruning to develop strong structure

Limb Breakage: resistant

Current Year Twig: brown to gray; thick

LANDSCAPE NOTES:

Origin: Eastern United States

Usage: shade tree; specimen; highway

Shagbark Hickory often towers over the forest from a straight trunk with branches high in the canopy. The young tree has a picturesque, oval outline with the lower branches somewhat drooping, the upper branches upright, and the middle branches just about horizontal. It may be the best ornamental hickory due to the open branching habit and shaggy, gray bark that stands out in any landscape. It grows naturally in moist river bottoms and slopes, though not inundated flood plains.

Growth rate is moderately slow and the tree is somewhat difficult to transplant due to a coarse root system, but is adaptable to many different soils. Transplant in the spring for best establishment. There is usually a tap root on trees grown in well-drained soil. Shagbark Hickory grows best in a sunny location in light, well-drained soil.

The tree has two ornamental characteristics: shaggy bark, which peels off in long strips, and striking fall color. Although not a tree for general use, Shagbark Hickory is nice as an occasional specimen in parks and in other large, open areas. It may be best planted in a landscape bed or in an unmowed area where the nuts can drop unnoticed. It may not be suited for the lawn because a lawn mower can shoot the nuts into people and cars. Nuts can damage cars when they fall from the tree, so do not locate this tree near streets or parking areas. The nuts, which are produced in abundance, are edible, eaten by birds and mammals, but must also be cleaned up from the ground in a maintained lawn.

The Hickory bark beetle is probably the most serious pest, killing trees in drought. The twig girdler larva girdles twigs, causing weakened twigs to break off and drop. A variety of other insects can infest the tree, but they rarely cause serious damage. Diseases can cause some minor damage.

Caryota mitis Fishtail Palm

Pronunciation: care-ee-O-tah MY-tiss

ZONES: 10B, 11

HEIGHT: 15 to 25 feet

WIDTH: 10 to 15 feet

FRUIT: brown; not showy; no significant litter

GROWTH RATE: moderate; short-lived

HABIT: moderate density; irregular silhouette; coarse texture

FAMILY: *Arecaceae*

LEAF COLOR: evergreen; medium-green

LEAF DESCRIPTION: odd pinnately compound; incised; 3 to 6' long fronds

FLOWER DESCRIPTION: white; spring and summer

EXPOSURE/CULTURE:

Light Requirement: shade to full sun

Soil Tolerances: all textures; alkaline to acidic; drought

Aerosol Salt Tolerance: low

PEST PROBLEMS: sensitive

BARK/STEMS:

Grows with multiple, upright trunks; no thorns

Pruning Requirement: needs pruning to remove dead trunks

Limb Breakage: resistant

LANDSCAPE NOTES:

Origin: southeast Asia

Usage: container or above-ground planter; near a deck or patio; specimen; interiorscapes; house plant

This clump-growing palm's leaf blades are divided into many segments, each of which resembles the tail of a fancy goldfish. Rarely exceeding 25 feet in height, Fishtail Palm produces many suckers from its base, creating a very attractive, dense specimen palm. The foliage is displayed in layers because new foliage is continually arising from young trunks. The multi-divided fronds rustle in a slight breeze, adding refreshing sound to the landscape.

Its exotic habit makes it ideal for use at poolside and in urns or other containers. It is often seen in well-lit interiorscapes where its distinct form lends a tropical effect. It can be used as a house plant in large homes with plenty of light. Set it off by planting in a low ground cover to display the nice trunk habit and beautiful foliage. It is well suited for a garden that is shaded by overstory trees.

Fishtail Palms can thrive in light conditions from full sun to deep shade, requiring only good drainage and reasonably fertility. It has a moderate to rapid growth rate and should be located outdoors in a sheltered location protected from cold. However, it is monocarpic, so a trunk, with its fronds, completely dies to the ground after it flowers. This stem must be removed to maintain a neat appearance. Allow emerging trunks to continue to grow from the base of the palm to take the place of the dying ones. The removed, dead trunk usually does not leave a large gap in the canopy because there are many other trunks clumped together. The small gap fills in quickly. New trunks can be pinched off the palm as they emerge, creating a neater lower trunk.

Unfortunately, the palm is susceptible to lethal yellowing disease, which is terminal. Other *Caryota* species such as *C. urens* make nice specimens. Red spider mites and scales are serious problems, especially when Fishtail Palms are grown indoors.

Cassia fistula Golden-Shower

Pronunciation: CASS-ee-ah FISS-tew-lah

ZONES: 10A, 10B, 11

HEIGHT: 30 to 40 feet

WIDTH: 30 to 40 feet

FRUIT: pod; 12 to 24"; dry; purple; causes some litter; showy; seeds poisonous

GROWTH RATE: rapid; moderate life span

HABIT: oval vase; moderate density; irregular silhouette; coarse texture

FAMILY: *Leguminosae*

LEAF COLOR: medium green; no fall color change

LEAF DESCRIPTION: alternate; even pinnately compound; entire to undulate margin; oval; briefly deciduous; 4 to 8" long leaflets

FLOWER DESCRIPTION: yellow; strikingly showy; summer

EXPOSURE/CULTURE:

Light Requirement: flowers best in full sun

Soil Tolerances: all textures; alkaline to acidic; moderate salt; drought

Aerosol Salt Tolerance: moderate

PEST PROBLEMS: resistant

BARK/STEMS:

Branches droop as tree grows and will require some pruning when tree is young for vehicular or pedestrian clearance beneath canopy; should be grown with a single leader; no thorns

Pruning Requirement: requires pruning to develop strong structure; be sure major limbs are spaced along trunk, not clustered together; thin aggressive limbs to prevent formation of included bark

Limb Breakage: resistant, if properly pruned

Current Year Twig: brown; stout

LANDSCAPE NOTES:

Origin: India

Usage: shade tree; specimen; street tree; large parking lot island; buffer strip

Golden-Shower is a fast-growing tree. Older specimens are capable of growing more than 40 feet high. The well-spaced, large-diameter branches are clothed with enormous leaves, with leaflets up to 8 inches long and 2.5 inches wide. These leaves drop from the tree for a short period of time, but are quickly replaced by new ones. In summer, Golden-Shower is decorated with thick clusters of showy yellow blooms that cover the slightly drooping branches. The blooms are followed by the production of two-foot-long, dark brown, cylindrical seedpods which persist on the tree throughout the winter before falling to litter the ground. The seeds within the pods are poisonous.

Golden-Shower is ideal for use as a specimen planting. It can look a bit coarse and unkempt for short periods when the leaves drop, but the vibrant flower display more than makes up for this. Some communities have planted this as a street tree, where it has held up quite well.

Golden-Shower should be grown in full sun on well-drained soil. Although Golden-Shower is damaged by temperatures falling slightly below freezing, it will come back with warmer weather. Trees will need occasional pruning when they are young to control shape and develop a uniform crown. A central leader should be encouraged to develop on young trees by reducing the length of competing branches. Remove or thin branches with bark included in the crotch. Young trees grow asymmetrically, with branches often drooping toward the ground. Staking and proper pruning will help develop a well-shaped and well-structured crown.

No limitations of major concern, but occasionally Golden-Shower is bothered by caterpillars, mildew, leaf spot, and root rot diseases.

Cassia javanica Pink and White Shower

Pronunciation: CASS-ee-ah jah-VAN-ih-kah

ZONES: 10B, 11

HEIGHT: 40 to 45 feet

WIDTH: 40 to 45 feet

FRUIT: pod; 12" long; dry; causes some litter; showy

GROWTH RATE: rapid; moderate life span

HABIT: spreading, vase; open; irregular; medium to fine texture

FAMILY: *Leguminosae*

LEAF COLOR: dark green in summer; no fall color change

LEAF DESCRIPTION: alternate; even pinnately compound; entire margin; oval; deciduous; 4 to 5" long leaflets

FLOWER DESCRIPTION: pink; strikingly showy; summer

EXPOSURE/CULTURE:

Light Requirement: flowers best in full sun

Soil Tolerances: all textures; alkaline to acidic; moderate salt and drought

Aerosol Salt Tolerance: moderate

PEST PROBLEMS: resistant

BARK/STEMS:

Branches droop somewhat as tree grows and will require some pruning and training when tree is young for vehicular or pedestrian clearance beneath canopy; should be grown with at least a short, single leader; no thorns

Pruning Requirement: requires pruning to develop strong structure; long, young, unbranched limbs can be headed to encourage branching and increase strength

Limb Breakage: somewhat susceptible

Current Year Twig: brown; stout

LANDSCAPE NOTES:

Origin: Indonesia

Usage: shade tree; specimen; street tree; large parking lot island; buffer strip

This cassia is a very vigorous tree that grows equally tall and wide. The large-diameter branches are clothed with large leaves, with leaflets up to four inches long and one inch wide. These leaves drop from the tree for a short period of time, but are quickly replaced by new ones. In summer, Pink and White Shower is decorated with thick clusters of showy pinkish-red blooms that cover the branches. Flowers are borne in an open canopy that provides no green background as on *C. leptophylla* and some other *Cassia*. Nonetheless, the tree is beautiful in flower and is popular in Hawaii and other tropical areas.

This tree is ideal for use as a specimen planting or medium-sized shade tree. It looks a bit coarse and unkempt for short periods when the leaves drop, but the vibrant flower display more than makes up for this. Some communities have planted this as a street tree, where it has held up quite well because it maintains a more upright canopy than many other cassias. It is similar to *C. fistula* and is well suited for planting in a wide highway median, or along the edge of a highway or turnpike where there is plenty of space for root development.

Pink and White Shower should be grown in full sun on well-drained soil. Although it is damaged by temperatures falling slightly below freezing, it will come back with warmer weather. Trees will need pruning when they are young to control shape and develop a uniform crown. A central leader should be encouraged to develop on young trees by reducing the length of competing branches. Increase the strength of young, aggressive, main limbs by heading them back. This slows their growth, increases branching along the limb, and allows the branch collar to develop properly. Remove or shorten branches with bark included in the crotch. Staking and regular pruning will help develop a well-shaped and well-structured crown.

No limitations of major concern, but occasionally bothered by caterpillars, mildew, leaf spot, and root rot diseases.

Cassia leptophylla Gold Medallion Tree

Pronunciation: CASS-ee-ah lep-toe-FIE-la

ZONES: 10A, 10B, 11

HEIGHT: 30 to 40 feet

WIDTH: 30 to 40 feet

FRUIT: pod; 12" long; dry; causes some litter; showy

GROWTH RATE: rapid; moderate life span

HABIT: rounded; moderate density; nearly symmetrical; medium to fine texture

FAMILY: *Leguminosae*

LEAF COLOR: dark green; no fall color change

LEAF DESCRIPTION: alternate; even pinnately compound; entire margin; oval; briefly deciduous; 2 to 3" long leaflets

FLOWER DESCRIPTION: yellow; strikingly showy; summer

EXPOSURE/CULTURE:

Light Requirement: flowers best in full sun

Soil Tolerances: all textures; alkaline to acidic; moderate drought

PEST PROBLEMS: resistant

BARK/STEMS:

Branches droop as tree grows, and will require some pruning when tree is young for vehicular or pedestrian clearance beneath canopy; should be grown with a single leader; no thorns

Pruning Requirement: requires pruning to develop strong structure; space major branches along a central trunk; do not cluster them together

Limb Breakage: mostly resistant

Current Year Twig: brown; stout

LANDSCAPE NOTES:

Origin: Brazil

Usage: above-ground planter; shade tree; specimen; street tree; large parking lot island; buffer strip

Gold Medallion Tree is a fast grower. The large-diameter branches are clothed with large, pendulous leaves, with leaflets up to 4 inches long and ½ inch wide. These leaves drop from the tree for a short period of time, but are quickly replaced by new ones. In summer, Gold Medallion Tree is decorated with thick clusters of showy yellow blooms that cover the drooping branches. The thick canopy makes an especially nice background for the bright yellow flowers, which are displayed on the outer canopy as on no other yellow flowering tree.

Gold Medallion is ideal for use as a specimen planting, or small to medium shade tree. It can look a bit coarse and unkempt for short periods when the leaves drop, but the vibrant flower display more than makes up for this. Some communities have planted this as a street tree, where it has held up quite well. It is well suited for planting in a wide highway median, or along the edge of a highway or turnpike where there is plenty of space for the low, spreading canopy to develop.

Gold Medallion Tree should be grown in full sun on well-drained soil. Although it is damaged by temperatures falling slightly below freezing, it will come back with warmer weather. Trees will need pruning when they are young to control shape and develop a uniform crown. A central leader should be encouraged to develop on young trees by reducing the length of competing branches. Increase the strength of young branches by heading them back or thinning them. This slows their growth and allows the branch collar to develop properly. Remove branches with bark included in the crotch. Young trees grow asymmetrically, with branches often drooping. Staking and proper pruning will help develop a well-shaped and well-structured crown.

No major concerns, but occasionally bothered by caterpillars, mildew, leaf spot, and root rot diseases.

Cassia surattensis Glaucous Cassia

Pronunciation: CASS-ee-ah sir-ah-TEN-siss

ZONES: 10B, 11

HEIGHT: 15 to 20 feet

WIDTH: 15 to 20 feet

FRUIT: pod; 4 to 6" long; dry; brown; no significant litter

GROWTH RATE: moderate; short-lived

HABIT: upright, vase; open; irregular silhouette; nearly symmetrical; fine texture

FAMILY: *Leguminosae*

LEAF COLOR: green; evergreen

LEAF DESCRIPTION: alternate; even pinnately compound; entire margin; oval; 1 to 2" leaflets

FLOWER DESCRIPTION: very striking yellow; spring and fall flowering

EXPOSURE/CULTURE:

Light Requirement: flowers best in full sun

Soil Tolerances: all textures; alkaline to acidic; moderate drought

PEST PROBLEMS: resistant

BARK/STEMS:

Bark is thin and easily damaged by mechanical impact; branches droop as tree grows and will require pruning for vehicular or pedestrian clearance beneath canopy; routinely grown with multiple trunks; can be trained to a single trunk; no thorns

Pruning Requirement: requires pruning and staking to develop strong, upright structure

Limb Breakage: susceptible to breakage either at crotch, due to poor collar formation, or wood itself is weak and tends to break

Current Year Twig: green; moderate thickness

CULTIVARS AND SPECIES:

There are many cassias suited for tropical areas, including *C. alata* (10 to 15 feet tall) with large yellow flowers and *C. marginata* (10 to 15 feet tall) with reddish flowers. *C. splendida* grows 15 feet tall and wide and flowers in late summer.

LANDSCAPE NOTES:

Origin: Tropical Asia, Australia

Usage: container or above-ground planter; buffer strip; highway; specimen

This increasingly popular, sprawling evergreen shrub or small tree is composed of many delicate leaflets. Trees grow very quickly into an upright, spreading form, and are decorated twice a year, in spring and fall, with short clusters of deep yellow blooms, appearing at each branch tip. A tree in full bloom is a site for sore eyes. The thick, flat seedpod, 4 to 6 inches long and 2.5 inches wide, explodes when ripe, dispersing many seeds that readily germinate in the landscape.

Trees must to be trained and pruned to develop into a small tree; otherwise a rounded, sprawling shrub develops. The trunk should be staked and lateral branches kept small by pruning back several times during the year to develop adequate trunk diameter to hold the crown erect. This will also help to force growth in the upper branches, which should be developed to create a tree form. Branches naturally droop as the weight of flowers and foliage at the ends of branches increases. But pruning and training is well worth the effort, as the bright yellow flowers are spectacular.

Glaucous Cassia becomes an outstanding small tree in full-sun, frost-free locations and will tolerate almost any soil with average moisture. It is very appropriate for planting in commercial and residential landscapes where the roots can spread out without competition from sidewalks and other urban structures. The fallen flowers blanket the ground, forming a solid yellow mass beneath the canopy, and look especially nice on turf.

Reportedly has a weak root system and blows over on exposed sites. Occasional caterpillars. Seeds germinate readily in the landscape.

Castanea mollissima Chinese Chestnut

Pronunciation: cass-TAIN-ee-ah mol-LISS-eh-mah

ZONES: 5A, 5B, 6A, 6B, 7A, 7B, 8A, 8B

HEIGHT: 35 to 40 feet

WIDTH: 40 to 50 feet

FRUIT: oval; round; 1 to 3"; dry; green; attracts squirrels and other mammals; edible; causes significant litter; showy

GROWTH RATE: moderate; moderate life span

HABIT: round, spreading vase; dense; nearly symmetrical; coarse texture

FAMILY: *Fagaceae*

LEAF COLOR: green in summer; copper to yellow fall color

LEAF DESCRIPTION: alternate; simple; serrate margin; lanceolate to oblong; deciduous; 4 to 8" long

FLOWER DESCRIPTION: yellowish-white; pleasant fragrance; showy; summer

EXPOSURE/CULTURE:

Light Requirement: grows best in full sun

Soil Tolerances: all textures; slightly alkaline to acidic; moderate drought

Aerosol Salt Tolerance: moderate

PEST PROBLEMS: resistant

BARK/STEMS:

Branches droop as tree grows and will require pruning for vehicular or pedestrian clearance beneath canopy; routinely grown with multiple trunks; can be trained to grow with a short, single trunk; no thorns

Pruning Requirement: little is needed except if a central leader is desired

Limb Breakage: resistant

Current Year Twig: brown; moderate thickness

CULTIVARS AND SPECIES:

The following four cultivars have been selected for their nut production: 'Abundance', 'Meiling', 'Nanking', and 'Kuling'. Others include: 'Estate-jap'—highly resistant to chestnut blight; 'Sleeping Giant'—grows larger than species; 'Kelsey'—smaller tree with good nut quality.

LANDSCAPE NOTES:

Origin: China, Korea

Usage: nuts; shade tree; specimen; urban tolerant

Chinese Chestnut usually branches close to the ground, making it a good candidate for a specimen or a tree to climb. The chief ornamental features are yellowish white-catkins, present in early summer, and coarse foliage. The flower odor may be offensive to some people. The nuts are edible but not as sweet as the American Chestnut. The soft, spiny fruit could become a nuisance on sidewalks for a short period, so locate these trees accordingly. In cold climates, the growing season may not be long enough for nuts to mature.

This urban-tough tree may not be suitable for street or parking lot locations, but it can make a nice shade tree. Any advantages of this tree may be overshadowed by potential disease problems (although it is moderately resistant to chestnut blight), so plant it in limited numbers. Makes a nice tree to line entry roads or along walks to create a low canopy. This novelty tree should be planted occasionally as a specimen. The tree is hard to transplant, perhaps due to a coarse root system. Spring transplanting appears to be the most successful.

Blight has virtually eliminated the American Chestnut from the landscape. The disease causes cankers on the branches, then moves into the trunk, killing the tree. There is no chemical control for the disease. Most chestnuts now grown are asiatic types and, like Chinese Chestnut, are moderately resistant (but not immune) to this disease, caused by the chestnut blight fungus.

Catalpa spp. Catalpa

Pronunciation: cah-TOWEL-pah species

ZONES: 5A, 5B, 6A, 6B, 7A, 7B, 8A, 8B, 9A

HEIGHT: 50 to 60 feet

WIDTH: 40 to 50 feet

FRUIT: elongated string-bean-like; 12" or more; dry; brown; fruit, twigs, and foliage cause significant litter; showy; persistent

GROWTH RATE: rapid; moderate life span

HABIT: oval; upright; moderately dense; irregular silhouette; coarse texture

FAMILY: *Bignoniaceae*

LEAF COLOR: light green in summer; showy, yellow fall color

LEAF DESCRIPTION: opposite/subopposite; whorled; simple; entire margin; ovate; deciduous; 8 to 12" long

FLOWER DESCRIPTION: white with purple markings; very showy; spring and early summer

EXPOSURE/CULTURE:

Light Requirement: part shade to full sun; best in full sun

Soil Tolerances: all textures; alkaline to acidic; occasionally wet soil; moderate salt; drought

Aerosol Salt Tolerance: moderate

PEST PROBLEMS: resistant

BARK/STEMS:

Bark is thin and easily damaged by mechanical impact; branches droop as tree grows and will require some pruning when tree is young for vehicular or pedestrian clearance beneath canopy; should be grown with a single leader; no thorns

Pruning Requirement: thin main lateral branches to slow their growth rate and encourage the main trunk to dominate

Limb Breakage: resistant

Current Year Twig: brown; green; very stout

LANDSCAPE NOTES:

Origin: native to North America

Usage: reclamation; shade tree; urban tolerant

Catalpa speciosa (Northern or Western Catalpa) grows in a loose oval, occasionally to 90 feet. This coarse, large-leaved tree spreads wide and tolerates hot, dry weather, but leaves may scorch and some drop from the tree in very dry summers. *Catalpa bignonioides* (Southern Catalpa) is somewhat smaller, reaching about 30 to 40 feet tall. Leaves are arranged opposite or in whorls (*speciosa* leaves are opposite). Both trees have a coarse growth habit forming an irregularly shaped crown. *C. ovata* is more durable, nearly pest-free, and has a more delicate habit than the other, more popular catalpas. *C. bignoniodes* 'Bungei' bears no fruit and grows to 15 feet tall; 'Nana' is umbrella shaped and grows to about 15 feet tall.

Leaves fall throughout the summer, creating a mess and ragged-appearing tree. Flowers make somewhat of a slimy mess for a short period when they drop on a sidewalk, but are no problem falling into shrubs, groundcovers, or turf. The fruit is a long pod (up to two feet long) resembling a string bean that can be a slight litter problem to some, but it is quite interesting.

The tree is useful in areas where quick growth is desired, but better, more durable trees are available for street and parking lot plantings. Sixty-year-old trees in Williamsburg, Virginia, have 3- to 4-foot-diameter trunks and are 40 feet tall. A sunny exposure and well-drained, moist, rich soil are preferred for best growth of Catalpa, but it will tolerate a range of soils from acid to calcareous. It is a tough tree suited for planting in large-scale landscapes. Not for downtown areas. Catalpas are planted to attract catalpa worms, a large caterpillar prized for fish bait because the skin is very tough and the caterpillar is juicy.

The larva of the Catalpa sphinx moth can eat large quantities of leaves. The caterpillar is yellow with black markings. The tree is regularly defoliated and often looks terrible by the end of the summer. Verticillium wilt can kill the tree. Leaves scorch in hot, dry weather.

Cedrus atlantica 'Glauca' Blue Atlas Cedar

Pronunciation: SEE-druss at-LAN-teh-cah

ZONES: 6A, 6B, 7A, 7B, 8A, 8B, (9), (10A)

HEIGHT: 40 to 60 feet

WIDTH: 30 to 40 feet

FRUIT: oval; upright; 3 to 5"; dry; brown; no significant litter

GROWTH RATE: moderate; long-lived

HABIT: pyramidal; weeping; open; nearly symmetrical when young, irregular later; fine texture

FAMILY: *Pinaceae*

LEAF COLOR: blue or blue-green to silver; needle-leaf evergreen

LEAF DESCRIPTION: spiral; simple; needle-like; less than 2" long needles

FLOWER DESCRIPTION: not showy; spring

EXPOSURE/CULTURE:

Light Requirement: part shade to full sun

Soil Tolerances: all textures; slightly alkaline to acidic; drought

Aerosol Salt Tolerance: moderate

PEST PROBLEMS: resistant

BARK/STEMS:

Branches droop as tree grows and will require pruning for vehicular or pedestrian clearance beneath canopy; should be grown with a single leader; a lower branch occasionally outgrows the leader; no thorns

Pruning Requirement: needs little or no pruning to develop strong trunk and branch structure

Limb Breakage: somewhat susceptible

Current Year Twig: brown; green; medium thickness

CULTIVARS AND SPECIES:

'Pendula' has a very strong weeping habit but must be staked and trained to make a tree, 15 feet tall, depending on the staking; grafted onto *C. deodara* for best root system; 'Argentea' has silver, almost white foliage.

LANDSCAPE NOTES:

Origin: North Africa

Usage: bonsai; specimen

C. atlantica is now technically a subspecies of *C. libani* but is still commonly referred to as *C. atlantica.* A handsome evergreen, Blue Atlas Cedar is perfect for specimen planting where it can grow without being crowded. The tree looks its best when branches are left on the tree to the ground; this shows off the wonderful irregular, open pyramidal form with lower branches spreading about half the height. It grows rapidly when young, then more slowly. The trunk stays fairly straight with lower, lateral branches nearly horizontal. Allow plenty of room for these trees to spread. They are best located as a lawn specimen away from walks, streets, and sidewalks so branches will not have to be pruned. It looks odd if lower branches are removed. Older trees become flat-topped and are a beautiful sight to behold.

Difficult to transplant, Blue Atlas Cedar is often planted from a container. Soil preference is for well drained deep loam, on the acid side, but it can tolerate sandy or clay soils and alkaline soil if they are very well drained. The tree grows best when sheltered from strong winds, but tolerates open conditions, and will grow in full sun or partial shade. Growth will be poor in restricted soil space. Tolerates extensive drought only when grown in an area where roots can explore a large soil area. Performs well in all areas within its hardiness range.

Generally free of insect pests and resistant to diseases, they may occasionally fall prey to tip blight, root rots (in poorly drained soil), or black scale and the deodar weevil. Usually no pest protection or control is necessary. Sapsuckers are attracted to the trunk and often riddle it with small holes. This usually does little lasting harm to the tree.

Cedrus deodara Deodar Cedar

Pronunciation: SEE-druss dee-o-DARE-ah

ZONES: 7A, 7B, 8A, 8B, (9)

HEIGHT: 40 to 60 feet

WIDTH: 25 to 30 feet

FRUIT: oval; 3 to 6"; dry and hard; upright; green turning brown; no significant litter; persistent; showy

GROWTH RATE: moderately fast; long lived

HABIT: pyramidal; open; nearly symmetrical; very fine texture

FAMILY: *Pinaceae*

LEAF COLOR: needle-leaf evergreen; blue-green to silver

LEAF DESCRIPTION: spiral; simple; needle-like; less than 2"

FLOWER DESCRIPTION: not showy; spring

EXPOSURE/CULTURE:

Light Requirement: grows best in full sun

Soil Tolerances: all textures; slightly alkaline to acidic; drought

PEST PROBLEMS: resistant

BARK/STEMS:

Branches droop as tree grows and will require pruning for vehicular or pedestrian clearance beneath canopy; should be grown with a single leader; no thorns

Pruning Requirement: needs little or no pruning to develop strong trunk and branch structure

Limb Breakage: branches break in ice storms

Current Year Twig: brown; green; medium thickness

CULTIVARS AND SPECIES:

'Kashmir' has silvery foliage and is hardy in zone 6; 'Aurea' has yellow leaves (looks ill); 'Pendula' has long, drooping leaves; 'Robusta' has stiffer twigs.

LANDSCAPE NOTES:

Origin: Himalayas in India

Usage: buffer strip; highway; specimen; street tree

With its pyramidal shape and drooping branches, this cedar makes a graceful specimen or accent tree. Growing rapidly, it also works well as a soft screen. The trunk stays fairly straight with lateral branches nearly horizontal and drooping. The top of the central leader droops characteristically on every tree. Lower branches should be left on the tree to display its real beauty. Allow plenty of room for this tree to spread. It is best located as a lawn specimen away from walks, streets, and sidewalks so branches will not have to be pruned. Large specimens have trunks almost three feet in diameter. Lower branches can grow to 25 feet long as they sweep the ground.

The tree has been successfully used as a street or median planting with lower branches removed. It appears to tolerate compacted, poor soil but declines in areas where smog is a problem. Plant on 20-foot-centers to create a canopy of blue foliage over a small residential street or entrance to a park. This is probably the best true cedar for the South, but it adapts well to all areas within its hardiness zone. It is widely planted in California, the Northwest, and the southeastern United States.

Transplants easily if root-pruned or from a container and protected from sweeping winds. It does well in dry, sunny spots and will tolerate slightly alkaline pH and clay soil. Keep turf from the root system by mulching beneath the canopy; this cedar has shallow surface roots.

May have scales, borers, deodar weevils, and bagworms. Following a cold winter, tops often decline and die back in zone 7. Secondary fungi can sometimes be associated with this decline. Chlorosis develops in alkaline soil.

Cedrus libani Lebanon Cedar

Pronunciation: SEE-druss LIB-ann-eye

ZONES: 6A, 6B, 7A, 7B, 8A, 8B, 9A, 9B, (10A)

HEIGHT: 40 to 50 feet

WIDTH: 20 to 30 feet

FRUIT: oval; 3 to 6"; dry and hard; upright; blue to green; no significant litter; showy; persistent

GROWTH RATE: slow-growing; long-lived

HABIT: pyramidal; open; nearly symmetrical when young, irregular later; medium texture

FAMILY: *Pinaceae*

LEAF COLOR: needle-leaf evergreen; dark green

LEAF DESCRIPTION: spiral; simple; about 1" long

FLOWER DESCRIPTION: brown; purple; not showy

EXPOSURE/CULTURE:

Light Requirement: grows best in full sun

Soil Tolerances: all textures; alkaline to acidic; drought

PEST PROBLEMS: resistant

BARK/STEMS:

Branches droop as tree grows and will require pruning for vehicular or pedestrian clearance beneath canopy; should be grown with a single leader; no thorns; large lower branches often outgrow the leader, forming an interesting silhouette

Pruning Requirement: needs little or no pruning to develop strong structure

Limb Breakage: resistant

Current Year Twig: brown; green; medium thickness

CULTIVARS AND SPECIES:

The subspecies *stenocoma* is hardy into zone 5B.

LANDSCAPE NOTES:

Origin: Lebanon and Turkey

Usage: buffer strip; highway; specimen; street tree

This is a large stately evergreen, with a massive trunk when mature, and wide-sweeping, sometimes upright branches (more often horizontal) that originate on the lower trunk. Allow plenty of space for proper development. Dark green needles and cones, which are held upright above the foliage, add to the impressive appearance. Young specimens retain a pyramidal shape, but the tree takes on a more open form with age. Like most true cedars, it does not like to be transplanted, and prefers a pollution-free, sunny environment.

Cedars are not well-suited for street tree planting in downtown situations, but are unrivaled as specimens, even for hot, dry locations. There are examples of residential street tree plantings on 20-foot-centers which look rather striking. They have not been tried extensively for street-side planting, but should be adaptable if cultural conditions are met. Provide as much open soil space as possible to allow the root system to expand unimpeded by urban structures. Most specimens can be seen growing in parks and other open areas where there is plenty of soil space.

The grand stature of the cedars is not rivalled by any other tree. Although they require space to grow properly, you can enjoy the tree as it develops on a small residential lot. When the lower branches get in the way, cut them back to the trunk. Encourage lower branches to grow in a sweeping, upright fashion for a truly picturesque specimen. Old trees often develop low branches nearly as large as the main trunk, and these branches appear if as they have developed into a separate tree.

The tree is well adapted to high pH and dry soil once it is established. Generally free of insect pests and resistant to diseases, they may occasionally fall prey to tip blight, root rots (in poorly drained soil), or black scale and the deodar weevil. Usually no pest protection or control is necessary. Sap-suckers are attracted to the trunk and often riddle it with small holes. This usually does little lasting harm to the tree.

Celtis australis Mediterranean Hackberry

Pronunciation: SELL-tiss auss-TRAL-iss

ZONES: 7B, 8A, 8B, 9A, 9B, (10A)

HEIGHT: 40 to 70 feet

WIDTH: 40 to 50 feet

FRUIT: round; less than 0.5"; fleshy; purple; attracts wildlife; causes messy litter

GROWTH RATE: rapid-growing; long-lived

HABIT: oval; round; pyramidal when young; moderate density; somewhat symmetrical; medium texture

FAMILY: *Ulmaceae*

LEAF COLOR: dark green in summer; yellow fall color

LEAF DESCRIPTION: alternate; simple; serrate margin; ovate; deciduous; 4 to 6" long

FLOWER DESCRIPTION: not showy; spring

EXPOSURE/CULTURE:

Light Requirement: full sun to partial shade

Soil Tolerances: all textures; slightly alkaline to acidic; moderate salt; drought

PEST PROBLEMS: resistant

BARK/STEMS:

Bark is thin and easily damaged by mechanical impact; branches droop as tree grows and will require pruning for vehicular or pedestrian clearance beneath canopy; showy trunk; should be grown with a single leader; no thorns

Pruning Requirement: select major branches with a wide crotch angle; thin major limbs so they remain less than half the trunk diameter

Limb Breakage: susceptible to breakage either at crotch, due to poor collar formation, or wood itself occasionally breaks

Current Year Twig: brown; green; slender

CULTIVARS AND SPECIES:

C. reticulata (zone 5) is a smaller tree suited for dry climates.

LANDSCAPE NOTES:

Origin: Europe

Usage: bonsai; buffer strip; highway; reclamation; shade tree; street tree; specimen

European or Mediterranean Hackberry is a deciduous tree with smooth, light gray, somewhat warty bark and a wide, broad, rounded canopy, making it a good potential shade tree. Once trained to a central leader, habit is more upright than other hackberries, making it a good candidate for street tree usage.

Trees benefit from some pruning in the nursery and landscape. The canopy grows from only a few, large-diameter branches, and these can develop included bark in the crotch. This is common if they are allowed to grow upright and become larger than about one-half the trunk size. This problem can be avoided by pruning back branches to slow their growth so more, smaller-diameter dominant branches develop in the crown. Be sure that branches arise from the trunk at a wide angle, and slow the growth of these branches by pruning. This will help the tree develop a strong branch structure. Large-diameter surface roots can form (particularly in poorly drained soil), raising sidewalks and making mowing grass difficult. Locate the tree eight feet or more from a sidewalk or street to minimize damage from roots.

Hackberry has a reputation for internal trunk rot, particularly following mechanical injury to the trunk. Locate the tree so it will not be injured by mowing equipment, string trimmers, or other vehicles, and keep grass away from the base of the trunk. European Hackberry will display quickest growth in full sun on moist soil, but will tolerate poorer soil conditions very well with slower growth.

Twigs may occasionally die from the parasitic witches' broom fungus. Affected wood should simply be removed when noticed. Not often seen with the leaf gall that is so common on *C. occidentalis.*

Celtis laevigata Sugarberry, Hackberry

Pronunciation: SELL-tiss lev-eh-GAY-tah

ZONES: 5B, 6A, 6B, 7A, 7B, 8A, 8B, 9A, 9B, 10A, 10B

HEIGHT: 50 to 70 feet

WIDTH: 50 to 70 feet

FRUIT: round; less than 0.5"; fleshy; red to black; attracts wildlife; causes a mushy mess for a short time

GROWTH RATE: rapid-growing; moderate life span

HABIT: rounded vase; moderate density; irregular silhouette; medium texture

FAMILY: *Ulmaceae*

LEAF COLOR: medium green in summer; showy yellow fall color

LEAF DESCRIPTION: alternate; simple; serrate margin; lanceolate to ovate; deciduous; 2 to 4" long

FLOWER DESCRIPTION: green; not showy; spring

EXPOSURE/CULTURE:

Light Requirement: part shade to full sun

Soil Tolerances: all textures; slightly alkaline to acidic; wet soil; salt and drought

Aerosol Salt Tolerance: moderate

PEST PROBLEMS: resistant

BARK/STEMS:

Bark is thin and easily damaged by mechanical impact; branches droop as tree grows and will require early pruning for vehicular or pedestrian clearance beneath canopy; showy trunk; should be grown with a single leader; no thorns; susceptible to decay

Pruning Requirement: thin major limbs to keep them less than half the trunk diameter

Limb Breakage: susceptible to breakage at crotch due to poor collar formation; also breaks in ice storms

Current Year Twig: brown; green; slender

CULTIVARS AND SPECIES:

'All Seasons' has a rounded crown, does not shed twigs as much as the species, and is very hardy (zone 5). The canopy is supposed to be tighter and better formed than the species, but apparently this is not always the case.

LANDSCAPE NOTES:

Origin: Missouri south to Texas and Florida

Usage: large parking lot island; buffer strip; highway; reclamation; shade tree; street tree; urban tolerant

This very large, broad, fast-growing deciduous North American native tree has a rounded vase crown with spreading, pendulous branches. The crown is dominated by several large-diameter branches that divide many times before ending in fine-textured, thin twigs. Tiny, berry-like, sweet fruits attract many birds, but create a mess on sidewalks for several weeks.

Skilled pruning is required several times during the first 15 years of life to shape the crown and prevent formation of weak branch crotches and multiple trunks. Remove, shorten, or thin branches if they develop included bark in the branch crotches, and keep major branches less than one-half the diameter of the trunk. Avoid pruning large-diameter branches from the trunk, as the tree compartmentalizes decay poorly and pruning often leads to decay in the trunk and branches. Branches damaged by ice also begin to decay.

Sugarberry transplants best in the spring. A number of southern cities used Sugarberry as a street tree, but others ban it because of the potential for trunk rot. Lack of popularity may be due to the open, awkward appearance of young trees; however, trees fill in and become more uniformly shaped as they grow older. They make a beautiful street tree planted on 30- to 40-foot centers.

Chlorosis develops on highly alkaline soil, but witches' broom and nipple gall are not a problem as they are on *C. occidentalis*. Older trees often develop trunk decay. Mistletoe infects the branches, occasionally causing tip die-back. Large surface roots form on poorly drained soil. Seedlings germinate readily creating a potential weed problem. Root rot can shorten the life of the tree.

Celtis occidentalis Common Hackberry

Pronunciation: SELL-tiss ox-see-den-TAY-liss

ZONES: 3A, 3B, 4A, 4B, 5A, 5B, 6A, 6B, 7A, 7B, 8A, 8B, 9A, 9B

HEIGHT: 55 to 80 feet

WIDTH: 40 to 60 feet

FRUIT: round; less than 0.5"; fleshy; purple; red; attracts wildlife; causes messy litter for several weeks

GROWTH RATE: rapid-growing; long-lived

HABIT: rounded vase; moderate density; irregular silhouette; medium texture

FAMILY: *Ulmaceae*

LEAF COLOR: medium green in summer; showy yellow fall color

LEAF DESCRIPTION: alternate; simple; serrate margin; oval to ovate; deciduous; 2 to 4" long

FLOWER DESCRIPTION: green; not showy; spring

EXPOSURE/CULTURE:

Light Requirement: part shade to full sun

Soil Tolerances: all textures; alkaline to acidic; wet soil; salt; drought

Aerosol Salt Tolerance: moderate to low

PEST PROBLEMS: sensitive to non-lethal pests

BARK/STEMS:

Showy trunk; bark is thin and easily damaged by mechanical impact; branches grow mostly upright following proper early training; should be grown with a single leader; no thorns

Pruning Requirement: requires pruning to develop strong structure, uniform canopy, and a central trunk

Limb Breakage: susceptible to breakage at crotch due to formation of included bark; ice breaks branches

Current Year Twig: brownish-green; slender

CULTIVARS AND SPECIES:

'Prairie Pride' has a more uniform, upright, compact crown that is easier to prune into a strong, well-formed tree. Thin the major limbs to prevent formation of included bark.

LANDSCAPE NOTES:

Origin: north central to northeastern United States

Usage: bonsai; buffer strip; highway; reclamation; shade tree; street tree; urban tolerant

Trees in the woods often grow with one trunk, but open-grown trees often develop from several large-diameter branches, forming a tall, wide canopy.

Hackberry grows naturally in moist bottomland soil and mesic sites, but will grow rapidly in a variety of soil types, from moist, fertile soils to hot, dry, rocky locations in the full sun. Extensive surface roots can develop in poorly drained soil. Hackberry is tolerant of alkaline soil; Sugarberry is less tolerant. Skilled pruning is required several times during the first 15 years of life to prevent formation of weak branch crotches and weak multiple trunks. Space major branches along the trunk, and keep them less than one-half the diameter of the trunk to ensure strong attachments.

It was extensively used in street plantings in parts of Texas and other cities, as it tolerates most soils except extremely alkaline (pH > 8), but branches may break out from the trunk if proper pruning and training are not conducted early in the life of the tree. If you use this tree, locate it where it will be protected from mechanical injury. Use cautiously along streets in the ice belt due to suceptibility to ice damage.

The fruit temporarily stains walks and ice can cause extensive damage. Hackberry nipple gall is a bit unsightly, but causes no harm to the tree. Witches' broom can disfigure portions of the canopy. Other foliage insects or diseases eat or disfigure the leaves, but none are lethal. Hackberry may recover slowly from transplanting from a field nursery due to its extensive, coarsely branched root system, but this can be overcome by planting from containers. Transplant B&B material in the spring.

Celtis sinensis Japanese Hackberry

Pronunciation: SELL-tiss sye-NEN-siss

ZONES: 7B, 8A, 8B, 9A, (10)

HEIGHT: 40 to 60 feet

WIDTH: 35 to 50 feet

FRUIT: round; less than 0.5"; fleshy; orange; attracts wildlife; causes a messy litter

GROWTH RATE: rapid-growing; long-lived

HABIT: vase; moderate density; irregular when young, becoming symmetrical; medium texture

FAMILY: *Ulmaceae*

LEAF COLOR: medium green in summer; yellow fall color

LEAF DESCRIPTION: alternate; simple; serrate margin; oval; deciduous; 2 to 3" long

FLOWER DESCRIPTION: not showy; spring

EXPOSURE/CULTURE:

Light Requirement: part shade to full sun

Soil Tolerances: all textures; slightly alkaline to acidic; wet soil; drought

PEST PROBLEMS: resistant

BARK/STEMS:

Bark is thin and easily damaged by mechanical impact; branches droop when tree is young and will require pruning for vehicular or pedestrian clearance beneath canopy; should be grown with a single leader; no thorns

Pruning Requirement: thin major limbs to keep them less than half the trunk diameter

Limb Breakage: susceptible to breakage at crotch due to poor collar formation

Current Year Twig: brownish-green; slender

LANDSCAPE NOTES:

Origin: Asia

Usage: buffer strip; highway; shade tree; street tree

Japanese Hackberry transplants easily in spring. It can be 20 to 25 feet tall and wide 10 years after planting. Trees are very similar to the Common Hackberry, *C. occidentalis* with four-inch-long leaves have wavy, toothed margins. The mature bark is light gray, rough, and corky, and the small, dark-orange-colored fruits are relished by birds. The seeds are hard and people could slip and fall on them for a short period if fruits drop on a hard surface.

Skilled pruning is required several times during the first 15 years of life to prevent formation of weak branch crotches, multiple trunks, and drooping branches (see *C. occidentalis*). Japanese Hackberry is used in street plantings in California where there is plenty of soil space, as it tolerates most soils and grows in sun or partial shade, but branches may break out from the trunk if proper pruning and training are not conducted early in the life of the tree.

The trunk is often injured if the tree is planted too close to the street in a restricted soil space. Some Hackberries have a reputation for internal trunk rot, particularly following mechanical injury to the trunk. Locate the tree so it will not be injured by mowing equipment or other vehicles, and keep grass away from the base of the trunk so string trimmers will not cause injury.

Japanese Hackberry will grow rapidly in a variety of soil types from moist, fertile soils to hot, dry locations in the full sun. It is wind- and drought-tolerant once established. Surface roots can form in compacted soils.

If this tree responds similar to Common Hackberry, the trunk will rot following mechanical injury. Leaf gall does not seem to be a problem, as is so common on *C. occidentalis*.

Cercidiphyllum japonicum Katsuratree

Pronunciation: sir-sid-eh-FILE-lum jah-PON-eh-cum

ZONES: 4B, 5A, 5B, 6A, 6B, 7A, 7B, 8A, 8B, (9)

HEIGHT: 40 to 60 feet

WIDTH: 35 to 60 feet

FRUIT: elongated; pod; 0.5 to 1"; dry; no significant litter

GROWTH RATE: rapid-growing; long-lived

HABIT: oval; pyramidal; moderate to dense; symmetrical; medium texture

FAMILY: *Cercidiphyllaceae*

LEAF COLOR: dark blue-green in summer; stunning yellow fall color

LEAF DESCRIPTION: opposite to subopposite; simple; crenate margin; orbiculate to ovate; deciduous; 2 to 4" long

FLOWER DESCRIPTION: green; not showy; spring

EXPOSURE/CULTURE:

Light Requirement: part shade to full sun; afternoon shade or regular irrigation beneficial during establishment

Soil Tolerances: all textures; slightly alkaline to acidic; low to moderate salt; moderate drought

Aerosol Salt Tolerance: low to moderate

PEST PROBLEMS: mostly pest-free

BARK/STEMS:

Showy trunk; branches droop as tree grows; often grown with multiple trunks; easily trained to grow with a single trunk for streets; no thorns; leave shortened lower branches on trunk of young trees to prevent sunburn

Pruning Requirement: requires some early pruning to develop strong structure and single trunk; then little pruning is required

Limb Breakage: susceptible to breakage in ice storms, especially trees with included bark

Current Year Twig: brown; slender

CULTIVARS AND SPECIES:

'Pendula' grows 15 to 25 feet tall and displays a beautifully graceful, weeping form, but it is hard to find.

LANDSCAPE NOTES:

Origin: China and Japan

Usage: large parking lot island; buffer strip; shade tree; specimen; street tree

Katsuratree has an oval to pyramidal form in youth, becoming more upright and spreading with age; it makes a good shade tree for residential property due to its medium stature. Fall color is a spectacular yellow, with some red. The trunk normally flares out at the base, gracefully dividing into the numerous shallow roots often prominent at the soil surface. Large surface roots can form on older trees. Without pruning, most trees develop multiple trunks originating at the soil line. These can split apart at the soil line.

Katsuratree grows best in a sunny exposure and moist soil, but is considered drought-tolerant once established. It may also have uses as a street tree where there is adequate soil space to prevent surface roots from raising walks and curbing. Select single-stemmed specimens for street tree use. Lateral branches on single-trunked trees are small in diameter and well secured to the trunk. Remove upright-growing branches, because these often develop weak branch attachments. Multi-stemmed trees are also sold and they make nice specimens for lawn and park areas, though not street trees.

Leaves often drop in middle to late summer in response to dry weather. Not suited for compacted soil, as the shallow roots will be a nuisance for lawn and sidewalk maintenance. Provide irrigation and keep the soil beneath the canopy mulched. The coarse root system calls for production in fabric, copper treated, or air-root-pruned containers, or frequent root-pruning of field-grown stock.

Occasionally chewing insects eat foliage. Transplant in spring for best recovery.

Cercidium floridum Blue Palo Verde

Pronunciation: cer-SID-e-um FLOOR-ah-dum

ZONES: 8B, 9A, 9B, 10A, 10B, 11

HEIGHT: 15 to 20 feet

WIDTH: 20 to 25 feet

FRUIT: pod; 3 to 6"; dry brown; some litter; persistent; showy

GROWTH RATE: rapid growth; moderate life span

HABIT: spreading vase; open; irregular; very fine texture

FAMILY: *Leguminosae*

LEAF COLOR: light green; no fall color change

LEAF DESCRIPTION: alternate; pinnately compound; entire margins; lanceolate; deciduous; leaflets less than 2" long

FLOWER DESCRIPTION: yellow; slight pleasant fragrance; very showy; spring and summer

EXPOSURE/CULTURE:

Light Requirement: grows best in full sun

Soil Tolerances: all textures; alkaline to acidic; salt and drought

PEST PROBLEMS: resistant

BARK/STEMS:

Bark is thin and easily damaged by mechanical impact; branches droop as tree grows, and will require pruning for vehicular or pedestrian clearance beneath canopy; routinely grown with, or trainable to be grown with, multiple trunks; should be trained to a short central leader for planting close to walks and pavement; showy trunk; thorns are present on the trunk or branches

Pruning Requirement: requires pruning to develop strong structure, to direct growth on low branches, and to remove dead wood

Limb Breakage: resistant

Current Year Twig: green; medium

LANDSCAPE NOTES:

Origin: Mexico, California, and Arizona

Usage: parking lot island; buffer strip; highway; specimen; street; urban tolerant

Pendulous, delicate leaflets, low-branching growth habit, and a profusion of small, but very striking, slightly fragrant yellow blooms combine to create this popular, small landscape tree. Open-grown trees are beautiful if left unpruned forming a fountain of fine texture. Stems are armed with short, sharp spines, and the trees should be located where they will not injure passersby.

Trees have a short-life span in many urban sites. Poor drainage may account for this on many sites, so locate the tree accordingly in areas where soil is never saturated for more than an hour or two. Locate the tree properly, and design and maintain the site to minimize trunk injury. It is easily injured from mowers, string trimmers, and bumps from other equipment.

Adapted to arid regions, Blue Palo Verde is one of the best choices for hot, dry locations. The light shade afforded by the fine-textured foliage allows lawns to thrive beneath this tree. Be sure to purchase trees with a 6- to 8-foot clear trunk and upright branches for street and parking lot plantings to allow for passage of vehicles. This can be hard to find because branches weep toward the ground as they spread from the tree.

Scales and thorn bugs can cause problems but not serious ones. Root rot will occur if planted in wet soils. Witches' broom occasionally cause a proliferation of branches forming tight heads of foliage. Sharp spines may limit planting close to streets.

Cercis canadensis Eastern Redbud

Pronunciation: SIR-siss can-a-DEN-siss

Cercis canadensis forma alba

Cercis reniformis 'Oklahoma'

Cercis canadensis var. *texensis*

ZONES: 4B, 5A, 5B, 6A, 6B, 7A, 7B, 8A, 8B, 9A

HEIGHT: 20 to 30 feet

WIDTH: 15 to 30 feet

FRUIT: pod; 2 to 3"; dry; brown; attracts birds; no significant litter; persistent; showy

GROWTH RATE: rapid when young; short-lived

HABIT: rounded vase; moderate density; irregular silhouette; moderate to coarse texture

FAMILY: *Leguminosae*

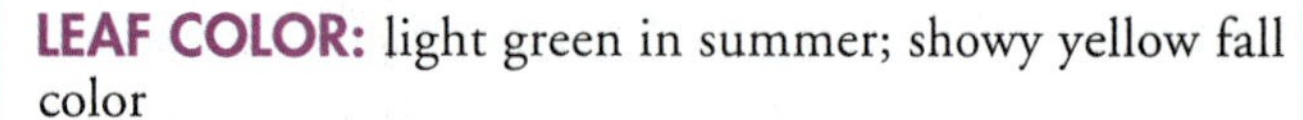

LEAF COLOR: light green in summer; showy yellow fall color

LEAF DESCRIPTION: alternate; simple; orbiculate to ovate; deciduous; 2 to 5" long

FLOWER DESCRIPTION: lavender; pink; purple; showy; spring

EXPOSURE/CULTURE:

Light Requirement: part shade to full sun

Soil Tolerances: all textures; alkaline to acidic; salt-sensitive; moderate to high drought

Aerosol Salt Tolerance: none

PEST PROBLEMS: resistant when young, sensitive later

BARK/STEMS:

Bark is thin and easily damaged by mechanical impact; branches droop as tree grows and will require pruning for vehicular or pedestrian clearance beneath canopy; trainable to be grown with a single or multiple trunks; no thorns

Pruning Requirement: requires pruning to develop durable structure and remove drooping branches; remove or shorten young branches with included bark

Limb Breakage: old trees are susceptible to breakage at crotch due to poor collar formation

Current Year Twig: brown; medium thickness

Continued

CULTIVARS AND SPECIES:

'Forest Pansy' is a particularly attractive cultivar with purple-red leaves in the spring fading to green in late summer. The dark maroon leaf veins are especially prominent on the underside of the leaf. *C. mexicana* and *C. reniformis* 'Oklahoma' (deep purple flowers) and 'Texas White' have superior, glossy foliage and make wonderful substitutes for Eastern Redbud, particularly in dry climates. The 'Oklahoma' Rebud has incredibly shiny, thick, leathery, dark green leaves with rounded or notched tips, 2 to 3 inches wide. The deep pink to lavender flowers appear in profusion up and down the tree limbs in springtime, well before the leaves begin to emerge, creating unquestionably the best Redbud display. The Redbud *forma alba* has bright white flowers and is exquisite in full bloom. 'Appalachian Red' has reddish foliage.

LANDSCAPE NOTES:

Origin: Missouri to New Jersey, south to Mexico and Florida

Usage: container; parking lot island; buffer strip; highway; near a deck or patio; reclamation; small shade tree; specimen; sidewalk cutout; street tree

The state tree of Oklahoma, Eastern Redbud has an irregular growth habit when young but forms a graceful flat-topped vase-shape as it gets older. The splendid purple-pink flowers appear all over the tree in spring, just before the leaves emerge. Thirty-year-old specimens are rare but they can reach 35 feet in height; trees of this size are often found on moist sites. The tree usually branches low on the trunk, and if left intact forms a graceful multi-trunked habit.

To increase longevity, thin or head aggressive young branches that have V-shaped crotches with included bark, and space branches along the trunk. Save the branches that have a U-shaped crotch and thin them so they remain less than two-thirds the diameter of the trunk to prevent branch splitting.

Eastern Redbud grows well in full sun in the northern part of its range but will benefit from some shade in the southern zones, particularly where summers are hot. Best growth occurs in a light, rich, moist soil, but this tree adapts well to a variety of soils, including sandy or alkaline. Trees from west Texas adapt best to alkaline soil. Although native habitat ranges from streambank to dry ridge, demonstrating genetic diversity and adaptability, trees look better when they receive some irrigation in summer dry spells. Trees are short-lived but provide a wonderful show in the spring and fall.

Best not used extensively as a street tree, due to low disease resistance (*Botryosphaeria canker*) and short life, but they are nice in commercial and residential landscapes where trunks can be protected from mechanical injury. Regular pruning in the first several years improves form.

Cercis mexicana Mexican Redbud

Pronunciation: SIR-siss mex-eh-CAY-nah

ZONES: 7A, 7B, 8A, 8B, 9A, 9B, 10A, 10B

HEIGHT: 18 to 25 feet

WIDTH: 18 to 25 feet

FRUIT: pod; 2 to 3"; dry; brown; attracts wildlife; no significant litter; persistent; showy

GROWTH RATE: slow; short to moderate life span

HABIT: rounded vase; slow to moderate density; irregular silhouette; coarse texture

FAMILY: *Leguminosae*

LEAF COLOR: green in summer; showy yellow fall color

LEAF DESCRIPTION: alternate; simple; entire, undulating margin; cordate to orbiculate; deciduous; 1 to 3" long

FLOWER DESCRIPTION: pink; very showy; spring

EXPOSURE/CULTURE:

Light Requirement: part shade to full sun

Soil Tolerances: all textures; alkaline to acidic; salt-sensitive; drought; good drainage is essential

Aerosol Salt Tolerance: none

PEST PROBLEMS: resistant when young

BARK/STEMS:

Bark is thin and easily damaged by mechanical impact; branches droop as tree grows and will require pruning for vehicular or pedestrian clearance beneath canopy; routinely grown with multiple trunks; can be trained to grow with a short, single trunk; no thorns

Pruning Requirement: requires pruning to develop strong structure and single leader

Limb Breakage: susceptible to breakage at crotch due to poor collar formation

Current Year Twig: brown; medium thickness

LANDSCAPE NOTES:

Origin: western Texas into Mexico

Usage: container or above-ground planter; parking lot island; buffer strip; highway; near a deck or patio; reclamation; specimen; sidewalk cutout; street tree

The splendid purple-pink flowers appear all over the tree in spring, just before the leaves emerge. Leaves are considerably smaller than those on Texas Redbud and Eastern Redbud, they are shiney, and the edges are distinctly undulating. Mexican Redbud has an irregular, upright growth habit when young but forms a graceful flat-topped vase shape as it gets older. The tree usually branches low on the trunk, and if left intact forms a graceful multi-trunked habit. Mexican and Texas Redbud are best suited for drier climates with less than 30 inches of rainfall.

Avoid weak branch crotches by pruning to keep main lateral branches from growing larger than two-thirds the diameter of the trunk. Save those that form a U-shaped crotch; remove those with a V shape. Do not allow multiple trunks to grow with tight crotches; instead, space branches about 6 to 10 inches apart along a main trunk.

Yellow fall color and tolerance to partial shade make this a suitable, attractive tree for understory or specimen planting in the shaded garden. Select single-trunked specimens for planting along streets. Plant the multi-trunked version in a shrub border for a spring and fall color display, or in commercial and residential landscapes.

Redbud grows well in its native habitat on limestone soils. Trees might look better when they receive some irrigation in summer dry spells, especially if they are in a restricted soil space. Young trees are easiest to transplant and survive best when planted in spring or fall. Containerized trees can be planted anytime. Trees are often short-lived but provide a wonderful show in the spring and fall, and they should be planted.

Borers and cankers attack the trunk of older and stressed Eastern Redbuds, and may do so on Mexican Redbud as well.

Chamaecyparis lawsoniana
Lawson Falsecypress, Port-Orford-Cedar
Pronunciation: cam-eh-SIP-ah-riss law-so-nee-A-nah

ZONES: 5B, 6A, 6B, 7A, 7B

HEIGHT: 40 to 60 feet

WIDTH: 10 to 15 feet

FRUIT: round; less than 0.5"; dry; blue turning brown; attracts wildlife; no significant litter

GROWTH RATE: moderate; long-lived on well-drained sites

HABIT: columnar; pyramidal; upright; dense; symmetrical; fine texture

FAMILY: *Cupressaceae*

LEAF COLOR: evergreen; blue or blue-green to green

LEAF DESCRIPTION: opposite; scale-like

FLOWER DESCRIPTION: blueish; not showy

EXPOSURE/CULTURE:

Light Requirement: grows best in full sun; will tolerate partial sun

Soil Tolerances: all textures; slightly alkaline to acidic; some salt; moderate drought

Aerosol Salt Tolerance: may have some tolerance

PEST PROBLEMS: resistant

BARK/STEMS:

Reddish-brown, deeply furrowed bark; lower branches on older trees droop; showy trunk; occasionally seen with multiple trunks, but should be grown with a single leader; no thorns

Pruning Requirement: needs little pruning to develop strong trunk and branch structure

Limb Breakage: resistant

Current Year Twig: brownish-green; slender

CULTIVARS AND SPECIES:

Various cultivars are available with different foliage colors (golden, blue, or silver-variegated) and different growth forms (dwarf, columnar, and low-spreading).

LANDSCAPE NOTES:

Origin: Southwest Oregon to northwest California

Usage: bonsai; specimen; screen

Often seen shorter in its cultivated form, this North American native can soar to heights of 100 to 150 feet in the wild. The massive, thick trunk and formal, upright, conical silhouette is softened by the gently weeping tips of the short, upright branches. Available in a wide variety of forms and bluish foliage colors, Lawson Falsecypress remains an important timber tree in the Pacific Northwest, but it is rare in the nursery trade and probably not well adapted to most landscapes. It grows nicely where its cultural requirements are met.

Lawson Falsecypress is only suited for the largest landscapes, such as in parks, golf courses, large industrial or commercial landscapes or estates. The unusually blue foliage and dense, symmetrical growth habit make it ideally suited as a screen in a sunny spot protected from constant wind. It has somewhat picky cultural requirements and should be grown in full sun in moist, well-drained soil (not clay), in areas of moderate to high humidity, preferably where it can be protected from harsh winds. These sites may be hard to find except in the Pacific northwestern United States.

A fungus (*Phytophthora*) damages this species by causing root rot, which eventually kills the tree. 'Oregon Blue' might be more resistant. Plant only in very well-drained soil. Mites can infest and discolor foliage. There is also a branch canker that can cause branch die back.

Chamaecyparis nootkatensis 'Pendula'
Nootka Falsecypress, Alaska-Cedar
Pronunciation: cam-eh-SIP-ah-riss noot-kah-TEN-siss

ZONES: 4A, 4B, 5A, 5B, 6A, 6B, 7A, 7B, 8A

HEIGHT: 35 to 50 feet

WIDTH: 20 feet

FRUIT: round; less than 0.5"; dry; brown; no significant litter

GROWTH RATE: slow-growing; long-lived

HABIT: columnar; pyramidal; weeping; moderate density; irregular silhouette; fine texture

FAMILY: *Cupressaceae*

LEAF COLOR: evergreen; blue-green to green or grayish-green

LEAF DESCRIPTION: scale-like

FLOWER DESCRIPTION: not showy

EXPOSURE/CULTURE:

Light Requirement: part shade to full sun

Soil Tolerances: loam; sand; acidic; drought

PEST PROBLEMS: resistant

BARK/STEMS:

Branches droop as tree grows and will require pruning for vehicular or pedestrian clearance beneath canopy; should be grown with a single leader; no thorns

Pruning Requirement: needs little pruning to develop strong trunk and branch structure

Limb Breakage: very resistant to ice and snow loads

Current Year Twig: brownish-green; slender

LANDSCAPE NOTES:

Origin: northern California to southern Alaska

Usage: bonsai; container or above-ground planter; buffer strip; near a deck or patio; specimen

A graceful, weeping, pyramidal evergreen, Nootka Falsecypress can reach 60 to 90 feet in the wild but stays within a height of 50 feet and a spread of 20 feet in cultivation. The trunk remains straight and dominant throughout the life of the tree. The long, pendulous, flattened branches are clothed with scalelike needles, which give off a rank odor when bruised or crushed. Foliage is sharp to the touch. Nootka Falsecypress is native to coastal Alaska and Washington, performing best in areas with high humidity and moist, but not wet, soil conditions.

It makes a striking specimen. Even one plant will soften any landscape. Use it near water or around a patio or as a lawn specimen in residential or commercial landscapes. Several trees planted together will screen an undesirable view. It would be hard to imagine misusing this tree. The exception would occur if the tree was planted too close to a building which would require the tree to be pruned. Its slow growth rate makes it suited for just about anywhere. The best form is obtained without pruning.

This tree does best with some shade from the afternoon sun if roots are confined to a small soil space or are in competition with an adjacent tree. Some shade may also be beneficial if you try to grow the tree in the southern part of its range. It is probably one of the few small, weeping, evergreen trees that is well adapted to the shaded garden. 'Strict Weeping', with a very narrow canopy, is also a good choice for the shade.

Juniper scale and bagworms occasionally infest foliage. Blight can be a problem on young plants in nurseries or old plants in landscape situations in wet weather. On young plants, branch tips turn brown and die back until the whole branch or young tree is killed. Scorch may look like a disease but is caused by freezing stress, drought, or mites.

Chamaecyparis obtusa Hinoki Falsecypress

Pronunciation: cam-eh-SIP-ah-riss ob-TUSE-ah

ZONES: 5A, 5B, 6A, 6B, 7A, 7B, 8A

HEIGHT: 4 to 75 feet

WIDTH: 3 to 20 feet depending on cultivar

FRUIT: round; less than 0.5"; dry; brown; no significant litter

GROWTH RATE: slow-growing; long-lived

HABIT: pyramidal; weeping; dense; symmetrical; fine texture

FAMILY: *Cupressaceae*

LEAF COLOR: evergreen; dark green

LEAF DESCRIPTION: scale-like

FLOWER DESCRIPTION: yellow; not showy

EXPOSURE/CULTURE:

Light Requirement: grows best in full sun

Soil Tolerances: loam; sand; acidic; moderate drought

PEST PROBLEMS: resistant

BARK/STEMS:

Branches droop as tree grows; showy trunk; tall cultivars should be grown with a single leader; no thorns

Pruning Requirement: needs little or no pruning to develop strong trunk and branch structure

Limb Breakage: resistant

Current Year Twig: brownish-green; slender

CULTIVARS AND SPECIES:

There are many cultivars, some quite dwarf. Other cultivars have excellent foliage coloration or unusual growth habit. Cultivars include: 'Aurea'—golden foliage; 'Caespitosa'—rare, miniature, about six inches tall; 'Compacta'—dwarf, about three feet tall, dense, conical; 'Coralliformis'—branchlets reddish and contorted; 'Crippsii'—broad pyramid with spreading branches and golden foliage; 'Erecta'—columnar habit; 'Ericoides'—low, blue-gray foliage; 'Filicoides'—fern-like; 'Gracilis'—compact growth habit, tips of branchlets pendulous; 'Kosteri'—dwarf, three to four feet tall, branch tips curved; 'Mariesii'—dwarf, foliage variegated with yellowish-white; 'Nana'—very dwarf, height and spread of two feet; 'Pygmaea'—dwarf, two feet tall, wider than tall; 'Stoneham'—slow, dwarf, tiered branching; 'Tetragona'—slow, dwarf, erect.

LANDSCAPE NOTES:

Origin: Japan

Usage: bonsai; specimen; buffer; hedge; screen; container

This broad, sweeping, conical-shaped evergreen has graceful, flattened, fern-like branchlets that gently droop at the tips. The species reaches 50 to 75 feet in height with a spread of 10 to 20 feet, has dark green foliage, and attractive, shredding, reddish-brown bark that peels off in long narrow strips.

Many people select the dwarf types for small residential gardens. Larger-growing forms make good screens or windbreaks. Falsecypress responds well to clipping or shearing to create a tall hedge in a sunny location.

Hinoki Falsecypress should be grown in full sun on moist, well-drained soil, in areas of moderate to high humidity, preferably where it can be protected from harsh winds. A protected garden or cove is well suited for planting any of the numerous cultivars. It is fairly free of pests and diseases and usually lasts for many years in the landscape. There are some minor problems, but none usually disfigure or threaten the life of the tree.

Chamaecyparis pisifera 'Filifera' Sawara Falsecypress

Pronunciation: cam-eh-SIP-ah-riss peh-SIFF-er-ah

ZONES: 5A, 5B, 6A, 6B, 7A, 7B, 8A

HEIGHT: 25 to 35 feet

WIDTH: 20 to 25 feet

FRUIT: round; less than 0.5"; dry; brown; no significant litter

GROWTH RATE: slow-growing; long-lived

HABIT: pyramidal; dense; symmetrical; fine texture

FAMILY: *Cupressaceae*

LEAF COLOR: evergreen; light green

LEAF DESCRIPTION: scale-like on slender, drooping branches

FLOWER DESCRIPTION: not showy

EXPOSURE/CULTURE:

Light Requirement: part shade to full sun

Soil Tolerances: all textures; acidic; salt-sensitive; moderate drought

Aerosol Salt Tolerance: none

PEST PROBLEMS: resistant

BARK/STEMS:

Branches droop as tree grows, and will require pruning for vehicular or pedestrian clearance beneath canopy; showy trunk; should be grown with a single leader; no thorns

Pruning Requirement: needs little or no pruning to develop strong trunk and branch structure

Limb Breakage: resistant

Current Year Twig: brown; green; slender

CULTIVARS AND SPECIES:

Numerous other cultivars are available for different foliage color, size, and shape.

LANDSCAPE NOTES:

Origin: Japan

Usage: bonsai; specimen; a wide screen or windbreak

Sawara, or Japanese, Falsecypress grows slowly to at least 30 feet in height and 20 feet wide at the base of the tree, and has thin, horizontal to pendulous branches of a very fine texture, which form a dense, broad pyramid. Trees in Japan grow to be much larger. The very attractive, reddish-brown, smooth, peeling bark is complemented nicely by the foliage color, but is usually not seen because lower branches are normally left on the tree and thus hide the trunk. This tree is quite popular in oriental and rock gardens, but can grow to be quite wide, so allow plenty of room for best form and unobstructed development.

It is best used as a specimen planting for a large, open area of a commercial or large residential landscape. Although it looks great in the nursery, it often grows too wide for a small residential lot.

Sawara Falsecypress should be grown in full sun to partial shade on moist, well-drained, non-alkaline soil in regions with moderate to high humidity. Although moderately drought-tolerant, it is not especially happy in very hot summers unless provided with some irrigation. This tree transplants reasonably well when root-pruned. It must be given full sun so lower branches remain on the tree, to provide the best appearance. Plant looks sloppy if lower branches die or are removed, and this is not recommended. Locate the plant properly to avoid the need to remove lower branches.

Bagworm can devour some foliage upon occasion. Tip blight can cause die-back of branch tips.

Chilopsis linearis Desert-Willow

Pronunciation: chi-LOP-siss lin-ee-AIR-iss

ZONES: 7B, 8A, 8B, 9A, 9B, 10A, 10B, 11

HEIGHT: 15 to 30 feet

WIDTH: 25 feet

FRUIT: elongated; 3 to 12"; dry; brown or tan; causes some litter; showy; persistent; attracts birds

GROWTH RATE: fast; moderate life span

HABIT: round; spreading; open; irregular silhouette when young; fine texture

FAMILY: *Bignoniaceae*

LEAF COLOR: green in summer; no significant fall color change

LEAF DESCRIPTION: alternate or opposite; simple; entire margin; lanceolate; deciduous; 5 to 12" long

FLOWER DESCRIPTION: white, lavender, or pink; pleasant fragrance; showy; spring and summer

EXPOSURE/CULTURE:

Light Requirement: grows best in full sun

Soil Tolerances: loam; sand; alkaline to acidic; requires well-drained soil; extremely drought-tolerant

PEST PROBLEMS: mostly resistant in dry climates

BARK/STEMS:

Branches droop as tree grows and will require pruning for vehicular or pedestrian clearance beneath canopy; routinely grown with multiple trunks; can be pruned to one, short trunk; no thorns

Pruning Requirement: requires pruning to develop strong structure and a single leader

Limb Breakage: resistant

Current Year Twig: green; slender

CULTIVARS AND SPECIES:

'White Storm' has white flowers with a yellow throat. 'Dark Star' and 'Burgundy Lace' have rich burgundy flowers and may not produce fruit; 'Pink Star' has bright pink flowers. Others are also available. *X Chitalpa tashkentensis* is a hybrid between *C. linearis* and *Catalpa bignonioides* that is gaining popularity in the dry parts of the western United States.

LANDSCAPE NOTES:

Origin: southern California to Texas south into Mexico

Usage: above-ground planter; buffer strip; highway; reclamation; near a patio or deck; specimen; hedge

This tree is well known in hot, dry areas where the soft, willow-like leaves and beautiful blooms are a welcome relief. Desert-Willow slowly reaches mature height and width with fairly loose, open branching.

The blossoms of Desert-Willow help make it so special. Trumpet-shaped blooms appear at the tips of the branches on new growth from late spring to early fall, or only during the summer if rainfall is sparse. The fragrant, orchid-like blooms are most often seen in shades of lavender and pink, but a white variety is occasionally found. Bees find the blossoms irresistible and a delightful honey is produced from the flowers. Flowers are regularly visited by hummingbirds.

Unlike the weak wood of true willows, the wood of Desert-Willow was used by Native Americans to craft hunting bows. The wood has also been used for fence posts, and baskets are often woven from the twigs. Its multi-trunked, well-branched habit and thick growth make Desert-Willow well suited for a wide screen or tall hedge. Groups can be planted in a large-scale landscape for a splash of fine-textured color. The tree has also been popular in residential plantings as a specimen. It creates nice light shade near a patio or deck.

Desert-Willow should be grown in full sun and is extremely drought-tolerant. Though the trees will grow better with adequate moisture, they will not tolerate overwatering. Unless planted in the desert, irrigation is not needed after they are established. Alternaria leaf spot and powdery mildew may limit usage in southwestern United States.

Chionanthus retusus Chinese Fringetree

Pronunciation: kye-o-NAN-thuss ree-TOO-suss

ZONES: 5B, 6A, 6B, 7A, 7B, 8A, 8B, 9A, 9B, (10A)

HEIGHT: 20 feet

WIDTH: 15 feet

FRUIT: oval; 0.5 to 1"; fleshy; blue to purple; attracts birds; might cause some litter; showy

GROWTH RATE: slow-growing; moderate life span

HABIT: oval or rounded vase; moderate density; symmetrical; medium texture

FAMILY: *Oleaceae*

LEAF COLOR: dark green in summer; showy yellow fall color

LEAF DESCRIPTION: opposite/subopposite; simple; entire margin; deciduous; 2 to 4" long

FLOWER DESCRIPTION: white; pleasant fragrance; showy; spring flowering

EXPOSURE/CULTURE:

Light Requirement: part shade to full sun

Soil Tolerances: all textures; alkaline to acidic; occasionally wet soil; some drought

PEST PROBLEMS: mostly pest-free

BARK/STEMS:

Branches droop as tree grows and will require pruning for pedestrian clearance beneath canopy; routinely grown with multiple trunks; can be trained to grow with a short, single trunk; bark peels to expose smooth, bright green on young stems; no thorns

Pruning Requirement: needs little or no pruning to develop strong trunk and branch structure; some pruning required to develop single leader

Limb Breakage: resistant

Current Year Twig: brown to gray; medium; fairly stout

LANDSCAPE NOTES:

Origin: China, Korea, and Japan

Usage: container or above-ground planter; large parking lot island; buffer strip; near a deck or patio; specimen; street tree

It is hard to think of a more beautiful small tree than Chinese Fringetree when it is in full bloom in the spring. The pure white, fragrant flowers, emerging just as dogwood flowers fade, hang in four-inch-long, spectacular terminal panicles that appear to cover the tree with snowy white cotton for two to three weeks. Flowers emerge at the terminal end of the spring shoot growth flush. This differs from the North American native Fringetree, which flowers before leaves emerge.

Chinese Fringetree forms a round, multi-trunked, drooping ball if left unpruned, but can be trained into a small tree with one or several trunks and lower branches removed. It is probably one of the finest accents available in the nursery trade. As with other white-flowered trees, Fringetree looks best when viewed against a dark background of evergreen or a dark building. It makes the perfect patio tree, or can be planted along streets if provided with occasional summer irrigation. Although reportedly difficult to transplant, it can be successfully moved quite easily with regular irrigation following planting. It can be grown and sold in containers or B&B.

Chinese Fringetree looks best in a sunny spot sheltered from wind. The tree appears more attractive in warm, humid climates when grown with several hours of shade, but blooms best in full sun, on moist, acid soil, and will gladly grow in occasionally wet soils. It grows very slowly, usually 4 to 10 inches per year, but can grow a foot per year if given rich, moist soil and periodic fertilization. Fringetree tolerates moderate drought, but looks best if irrigated occasionally during extended summer drought.

Generally pest- and disease-free. Leaf spots and powdery mildew occasionally are seen on the foliage, but are of no consequence.

Chionanthus virginicus Fringetree, Old-Mans-Beard

Pronunciation: kye-o-NAN-thuss vir-GIN-ee-cuss

ZONES: 3A, 3B, 4A, 4B, 5A, 5B, 6A, 6B, 7A, 7B, 8A, 8B, 9A, 9B

HEIGHT: 12 to 15 feet

WIDTH: 10 to 15 feet

FRUIT: on female plants only; oval; round; 0.5"; fleshy; blue to purple; a slight litter problem for a short period; persistent on tree until birds find it; showy

GROWTH RATE: slow-growing; moderate life span

HABIT: oval to rounded vase; moderate density; symmetrical; medium texture

FAMILY: *Oleaceae*

LEAF COLOR: dark glossy green in summer; showy yellow fall color

LEAF DESCRIPTION: opposite/subopposite; simple; entire margin; oblong to obovate; deciduous; 3 to 8" long

FLOWER DESCRIPTION: white; pleasant fragrance; very showy; spring

EXPOSURE/CULTURE:

Light Requirement: part shade to full sun

Soil Tolerances: all textures; acidic; occasionally wet soil; moderate drought

Aerosol Salt Tolerance: none

PEST PROBLEMS: resistant

BARK/STEMS:

Bark is thin and easily damaged by mechanical impact; branches droop as tree grows and will require pruning for pedestrian clearance beneath canopy; routinely grown with multiple trunks; can be trained to a single trunk; no thorns

Pruning Requirement: needs little or no pruning to develop strong structure; some pruning needed to develop single trunk

Limb Breakage: resistant

Current Year Twig: brown; green; gray; moderately stout

LANDSCAPE NOTES:

Origin: southern New Jersey to Florida and Texas

Usage: container or above-ground planter; buffer strip; highway; near a deck or patio; specimen; sidewalk cutout; urban tolerant

It is hard to think of a more beautiful small tree than the Fringetrees when they are in full bloom. The upright oval to rounded form adds dark green color in summer, bright white flowers in spring. The slightly fragrant flowers, emerging just as dogwood flowers fade, hang in long, spectacular panicles that appear to cover the tree with cotton for two weeks. As with other white-flowered trees, they look best when viewed against a dark background. Male plants seem to develop the best flower display.

Leaves emerge later in the spring than those of most plants just as the flowers are at peak bloom. This differs from Chinese Fringetree, which flowers at the terminal end of the spring growth flush. Female plants develop purple-blue fruits which are highly prized by many birds.

The plant eventually grows to 25 feet tall in the woods under ideal conditions, but these trees have thin canopies. It forms as a dense, multi-stemmed round shrub if left unpruned, but can be trained into a small tree with lower branches removed. Transplants readily in the spring with irrigation. Fall transplants do not develop roots until the following summer.

The foliage appears more attractive when grown with several hours of shade, but the tree blooms best in full sun. Probably best overall with some afternoon shade. A North American native commonly found in upland woods and stream banks throughout most of the south, Fringetree prefers moist, acid soil and will gladly grow in even wet soils. It grows very slowly, usually six to eight inches per year, but can grow a foot per year if given rich, moist soil and plenty of fertilizer.

Scales and mites in full sun locations. Stem cankers can girdle stems.

x *Chitalpa tashkentensis* Chitalpa

Pronunciation: chi-TAL-pa tas-ken-TEN-siss

ZONES: 7A, 7B, 8A, 8B, 9A, 9B, 10A, 10B, 11

HEIGHT: 20 to 40 feet

WIDTH: 20 to 40 feet

FRUIT: no fruit

GROWTH RATE: moderately rapid; probably moderate life span

HABIT: round; spreading; open; symmetrical; medium texture

FAMILY: *Bignoniaceae*

LEAF COLOR: green in summer; no fall color

LEAF DESCRIPTION: simple; entire margin; ovate; deciduous; 5 to 12" long

FLOWER DESCRIPTION: lavender, pink, or white; showy; nearly continuous flowering from spring to fall; sterile

EXPOSURE/CULTURE:

Light Requirement: full sun for best flowering

Soil Tolerances: loam; sand; slightly alkaline to acidic; salt-sensitive; drought

PEST PROBLEMS: resistant so far (new tree)

BARK/STEMS:

Branches droop as tree grows and will require pruning for vehicular or pedestrian clearance beneath canopy; routinely grown with multiple trunks; can be pruned to one, short trunk; no thorns

Pruning Requirement: requires pruning to develop strong structure and a single leader; encourage growth in branches with a wide angle to trunk

Limb Breakage: susceptible unless properly pruned

CULTIVARS AND SPECIES:

'Morning Cloud' is a larger tree with a more upright form and white flowers; 'Pink Dawn' has pink flowers and is more available.

LANDSCAPE NOTES:

Origin: hybrid

Usage: above-ground planter; buffer strip; highway; reclamation; near a patio or deck; specimen; possible street tree if properly pruned when young

This tree is becoming known in hot, dry areas where the soft leaves and beautiful blooms are a welcome relief. It is a hybrid between *Chilopsis linearis* and *Catalpa bignonioides,* which should prove to be well suited for planting in residential and commercial landscapes. It takes its name from the city in which it was created, Tashkent in central Asia. Growth is more rapid than *Chilopsis,* reaching perhaps 30 feet in height and width, with fairly loose, open branching.

The blossoms help make it so special. Trumpet-shaped blooms appear at the tips of the branches on new growth from late spring to early fall, or only during the summer if rainfall is sparse. The fragrant, orchid-like flowers are most often seen in shades of lavender and pink. A delightful honey is produced by bees that frequent these flowers.

The multi-trunked, well-branched habit makes Chitalpa well suited for a wide screen. Groups can be planted in a large-scale landscape for a splash of color. The tree will also be popular in residential plantings as a specimen. It creates nice light shade near a patio or deck and allows enough light through the canopy for flowers and other plants to prosper. Slow the growth on aggressive major limbs by heading or thinning to encourage a strong attachment to the trunk. Branches that stay smaller than half the diameter of the trunk are likely to be stronger than those that are allowed to grow larger.

This hybrid, like its parents, should be grown in full sun and is drought-tolerant. Though the trees will grow better with adequate moisture, they will not tolerate overwatering. Unless planted in the desert, irrigation is not needed after the tree is established. It may lose some leaves in extended drought without irrigation, but this will do little to permenantly damage the tree.

Mildew, tent caterpillars, and aphids are known to cause some trouble. A fungus leaf spot has been reported on trees grown in Florida.

Chorisia speciosa Floss-Silk Tree

Pronunciation: ko-REE-see-ah spee-see-O-sah

ZONES: 10A, 10B, 11

HEIGHT: 50 feet

WIDTH: 40 to 50 feet

FRUIT: oval; round; 6 to 12"; dry; brown turning white and silky; no significant litter; silk is showy

GROWTH RATE: rapid-growing; moderate longevity

HABIT: oval; round; pyramidal; upright; spreading; moderate to open canopy; irregular silhouette; coarse texture

FAMILY: *Bombacaceae*

LEAF COLOR: green to grayish-green; no fall color change

LEAF DESCRIPTION: alternate; palmate; serrate margin; oval; deciduous; 3 to 6" long

FLOWER DESCRIPTION: showy; pink; white; fall and winter

EXPOSURE/CULTURE:

Light Requirement: flowers best in full sun

Soil Tolerances: all textures; alkaline to acidic; occasionally wet soil; drought

Aerosol Salt Tolerance: low

PEST PROBLEMS: nearly pest-free

BARK/STEMS:

Bright green with nearly lethal spines; should be grown with a single leader

Pruning Requirement: little, if any, needed if single trunk is allowed to develop; multi-trunked and low-branched specimens require pruning to prevent included bark in major limbs

Limb Breakage: mostly resistant

Current Year Twig: green; medium thickness

CULTIVARS AND SPECIES:

Two grafted selections are available: 'Majestic Beauty' has rich pink flowers and 'Los Angeles Beautiful' has wine-red flowers. The cultivar 'Monsa' has a thornless trunk and pink fall flowers. There are other species of *Chorisia* with even more thorns. Grafted trees are preferred as they bloom earlier and at a smaller size.

LANDSCAPE NOTES:

Origin: Brazil and Argentina

Usage: buffer strip; highway; shade tree; specimen; street tree provided there is plenty of soil space for root and trunk growth

This rounded, deciduous tree eventually has wide-spreading branches that are green when young and covered with spines and often become grayish-green and sometimes lose their coarse, sharp spines. Young, unpruned trees can have a columnar or upright form, often growing from a central leader. Floss-Silk Tree grows rapidly the first few years, then more slowly. The large, showy, pink and white, five-petaled flowers are produced in small clusters in fall and winter (usually October) when the tree is nearly bare. The fruits are large, eight-inch-long, pear-shaped, woody capsules, filled with silky, white, kapok-like floss and pea-like seeds.

An excellent specimen tree for parks, parking lots, and other large landscapes, Floss-Silk Tree is spectacular when in bloom, producing an outstanding show of color in the fall. Prune the tree to be sure that only one central trunk develops when the tree is young. The central leader becomes less vigorous in middle age and allows lateral limbs to dominate the main structure of the tree, producing a spreading form. Although most branches are horizontal and well attached to the tree, upright branches can develop with included bark, which can cause the branch to split from the trunk. Prevent this by pruning the major limbs so they remain less than half the diameter of the trunk.

Buttressed trunk, spines, and large roots near the trunk limit usefulness of the tree in urban landscapes. Do not plant closer than 15 feet to pavement. Huge surface roots appear at the base of the trunk as the tree grows older.

Chrysalidocarpus lutescens
Yellow Butterfly Palm, Areca Palm
Pronunciation: chris-ah-lid-uh-CAR-puss LOO-teh-sens

ZONES: 10A, 10B, 11

HEIGHT: 20 to 25 feet

WIDTH: 8 to 10 feet

FRUIT: oval; 0.5 to 1"; fleshy; dark red to brown; no litter

GROWTH RATE: rapid-growing; moderate life span

HABIT: upright vase; open; irregular silhouette; fine texture

FAMILY: *Arecaceae*

LEAF COLOR: yellow-green with yellow petioles in full sun, green in the shade

LEAF DESCRIPTION: spiral; odd pinnately compound; entire margin; 12 to 36" long leaflets

FLOWER DESCRIPTION: white; not showy; periodically throughout the year

EXPOSURE/CULTURE:

Light Requirement: shade to sun

Soil Tolerances: all textures; slightly alkaline to acidic; occasionally wet soil; moderate drought

Aerosol Salt Tolerance: moderate

PEST PROBLEMS: resistant

BARK/STEMS:

Routinely grown with multiple trunks; trunks grow mostly upright and will not droop; showy, bright green 4 to 6" thick; green crown shaft; no thorns

Pruning Requirement: remove dead and yellow fronds

Limb Breakage: resistant

LANDSCAPE NOTES:

Origin: Madagascar

Usage: container or above-ground planter; screen; specimen; interiorscapes; house plant

This graceful, clump-type palm has gently arching, four- to six-inch-wide, ringed, bamboo-like, green, multiple trunks topped with curved, feathery fronds. Known under a variety of names, this beautiful soft palm is quite valued throughout the tropics and is widely planted in frost-free areas. On older palms, the small, white, inconspicuous flowers are periodically produced all year long on three-foot stalks among the leaves, and the small, oblong, black fruits ripen all year.

Yellow Butterfly Palm makes an attractive specimen, screening, or poolside planting, and is well suited for porches and patios. Planted in a large bed of low, fine-textured ground cover or dwarf shrubs, the bright trunks come to life at night when lit from below. For this reason it is often used as a specimen planting alone in the lawn. Containers filled with Butterfly Palm can be scattered about the landscape and taken in during cold weather.

Growing in full sun, where it makes an excellent specimen or screen (on four-foot-centers), to the rather dense shade of patios, porches, or as house plants, Yellow Butterfly Palm prefers fertile, well-drained, acid soil. Small palms benefit from some shade until they are several feet tall and palms should be watered during periods of drought.

They require regular fertilizer applications to maintain dark green foliage. Young palms in full sun and those in high pH soils develop yellow leaves. Older leaves on plants of any age become chlorotic, frequently from a deficiency of potassium. Affected leaves are often speckled with bronze or yellow.

Scales followed by sooty-mold can be a problem. Ganoderma rots roots on wet sites. Potassium deficiency turns old leaves yellow.

Chrysophyllum oliviforme Satinleaf

Pronunciation: chris-o-FILL-um oh-live-eh-FOR-mee

ZONES: 10B, 11

HEIGHT: 35 to 70 feet

WIDTH: 25 feet

FRUIT: elongated; 0.5 to 1"; fleshy; purple; edible; attracts wildlife; no significant litter

GROWTH RATE: slow-growing; moderate life span

HABIT: oval, upright vase; moderate density; nearly symmetrical with age; medium texture

FAMILY: *Sapotaceae*

LEAF COLOR: broadleaf evergreen; dark green top, bright copper underside

LEAF DESCRIPTION: alternate; simple; entire margin; oval; 4 to 8" long

FLOWER DESCRIPTION: white; not showy; periodic year-round flowering

EXPOSURE/CULTURE:

Light Requirement: part shade to full sun

Soil Tolerances: all textures; alkaline to acidic; occasionally wet soil; drought

Aerosol Salt Tolerance: moderate

PEST PROBLEMS: resistant

BARK/STEMS:

Light gray, showy trunk; bark is thin and easily damaged by mechanical impact; branches droop as tree grows and will require pruning for vehicular or pedestrian clearance beneath canopy; should be grown with a single leader; no thorns

Pruning Requirement: needs pruning to develop a strong, single leader

Limb Breakage: resistant

Current Year Twig: brown; medium thickness

CULTIVARS AND SPECIES:

Star-apple, *Chrysophyllum cainito*, closely related, bares leaves of similar decorative quality and is grown for its larger (up to four inches long), tastier fruits.

LANDSCAPE NOTES:

Origin: south Florida to Cuba, Puerto Rico and Jamaica

Usage: near a deck or patio; shade tree; specimen; sidewalk cutout; street tree; reclamation

At mature height of 70 feet, in an oval form, Satinleaf is a medium-sized tree noted for its unusually attractive foliage. The evergreen, four-inch-long leaves are a glossy, dark green above and a glowing, bright copper color beneath, providing a beautiful, two-toned effect when breezes cause the leaves to flutter. Leaves in some respects resemble those of the brown-back Southern Magnolias. The trunks are rather showy because they are covered with a thin, light reddish-brown or gray, scaly bark. Small, inconspicuous flowers are followed by small, sweet, purple fruits.

This Florida native makes an attractive freestanding lawn specimen or blends well in a shrubbery border or naturalized landscape. It could be tried as a street tree or parking lot tree. It would be nice planted 20 to 30 feet apart along an entrance road to a commercial landscape.

Satinleaf should be grown in full sun or partial shade on fertile, well-drained soils. Plants should be mulched and watered faithfully, although they are able to withstand occasional drought. The tree has not been widely planted but should make a good, durable urban tree. Trunk and branch structure are good, making this a clean, long-lasting tree. Be sure to select well-formed trees with one central leader from the nursery.

Caterpillars and leaf notcher will occasionally chew leaves, and gall mites deform leaves. Leaves can become quite unsightly.

Cinnamomum camphora Camphor-Tree

Pronunciation: sin-nah-MO-mum cam-FORE-ah

ZONES: 9B, 10A, 10B, 11

HEIGHT: 40 to 50 feet

WIDTH: 60 to 80 feet

FRUIT: round; less than 0.5"; fleshy; black; attracts birds; causes significant litter for a while

GROWTH RATE: rapid-growing; long-lived

HABIT: oval; round; spreading; dense; symmetrical; medium texture

FAMILY: *Lauraceae*

LEAF COLOR: broadleaf evergreen; light green

LEAF DESCRIPTION: aromatic; alternate; simple; entire margin; obovate to ovate; 2 to 4" long

FLOWER DESCRIPTION: yellow; not showy; spring

EXPOSURE/CULTURE:

Light Requirement: grows best in full sun

Soil Tolerances: all textures; slightly alkaline to acidic; drought

Aerosol Salt Tolerance: low to moderate

PEST PROBLEMS: resistant

BARK/STEMS:

Showy with large branches low on trunk; branches droop as tree grows and will require pruning for vehicular or pedestrian clearance beneath canopy; should be grown with a single leader; no thorns

Pruning Requirement: requires some early pruning to develop strong structure

Limb Breakage: resistant

Current Year Twig: green; medium to slender

CULTIVARS AND SPECIES:

The cultivar 'Monum' has larger, richer green foliage.

LANDSCAPE NOTES:

Origin: China, Formosa, and Japan

Usage: shade tree; street tree; buffer strip; urban tolerant

This large, round-canopied, evergreen tree has broad, large-diameter, unusually strong branches. It can reach 70 feet in height with a broader spread but is usually smaller. The glossy, thin, leathery leaves give off a camphor aroma when crushed and they create dense shade. The stems and bark on young branches of Camphor-Tree are bright green, tinged with red when young, maturing into a dark gray-brown, rugged-looking trunk that appears almost black when wet. Trunk and branch structure on older trees appear similar to mature live oaks.

Too big for all but the largest spaces, Camphor-Tree is ideal when used as a shade tree for a park or large landscapes. It makes an ideal street tree, providing a complete canopy over a residential street. Use along residential streets should be tempered because messy fruits drop on sidewalks and cars, and seeds germinate readily in nearby landscapes. It might be considered for boulevard planting where cars do not park. Prune to develop major branches; space 18 to 30 inches apart along a central trunk to develop good structure. Do not allow major branches to grow from the same spot on the trunk, and avoid upright, multi-trunked trees. Large-diameter roots can appear at soil surface.

Growing in full sun to partial shade, Camphor-Tree is amenable to a variety of soils. It will grow but often develops minor element deficiencies in alkaline soils. It is adapted to grow along the coast exposed to some sea salt and is often used this way in California.

Since seeds germinate easily in the landscape, it has escaped cultivation in Florida, Louisiana, and parts of coastal Texas. As a result, some communities discourage or ban use of this tree. Camphor-Tree is subject to a root rot, especially in poorly drained soils. Scales and mites are common problems on Camphor-Trees.

Citrus spp. Citrus

Pronunciation: SIT-truss species

ZONES: 9A, 9B, 10A, 10B, 11

HEIGHT: 12 to 30 feet

WIDTH: 8 to 25 feet

FRUIT: round; 2 to 6"; fleshy; green turning orange, red, or yellow; attracts mammals; edible; causes significant litter; persistent; showy

GROWTH RATE: moderate; moderate life span

HABIT: round; dense; symmetrical; medium texture

FAMILY: *Rutaceae*

LEAF COLOR: broadleaf evergreen; glossy dark green

LEAF DESCRIPTION: alternate; trifoliate; crenate margin; obovate to ovate; 2 to 4" long

FLOWER DESCRIPTION: white; very pleasant fragrance; showy; winter and spring

EXPOSURE/CULTURE:

Light Requirement: grows best in full sun

Soil Tolerances: all textures; slightly alkaline to acidic; moderate drought

Aerosol Salt Tolerance: low

PEST PROBLEMS: pest-sensitive

BARK/STEMS:

Bark is thin and easily damaged by mechanical impact; branches droop as tree grows and will require pruning for vehicular or pedestrian clearance beneath canopy; trainable to be grown with multiple trunks; should be trained to grow with a short, single trunk; thorns are present on branches

Pruning Requirement: requires pruning to develop strong structure and prevent formation of included bark

Limb Breakage: susceptible to breakage at crotch if included bark develops in crotch

Current Year Twig: shiny green; medium thickness

CULTIVARS AND SPECIES:

Many species of citrus are available. *C. aurantifolia* gives us key limes; *C. limon*, lemons; *C. paradisi*, grapefruit; *C. reticulata*, tangerines or Mandarin oranges; and *C. sinensis*, sweet orange. Many types of each are available at nurseries.

LANDSCAPE NOTES:

Origin: Asia

Usage: fruit; espalier; impenetrable hedge; buffer strip; near a deck or patio; standard; specimen

The dense, rounded crown of foliage and the pure white, extremely fragrant blossoms make Citrus a popular garden choice for frost-free locations. Citrus varies greatly in height and in width, depending upon species and cultivar. Citrus can be used for shade or to screen unwanted views; smaller varieties can be used in containers. The juicy, fragrant fruits, which vary from yellow to orange or red, ripen in the fall and winter.

The green, thorny branches will only occasionally need pruning after a strong branch structure is developed in the nursery or landscape. Prune so that no bark is included or pinched together in branch crotches. This condition could cause the branch to break from the trunk when fruit loads it down. Thin out enough of the foliage on lateral branches so they remain less than about half the diameter of the trunk.

Citrus should be grown in full sun on well-drained, slightly acid soil, and watered faithfully. Mulch can be applied under citrus, but like any other trees or shrubs, do not allow it to touch the trunk. It is very heat-tolerant and can be effectively used in a tough urban site such as a parking lot island or sidewalk cutout if dropping fruits will not pose a problem.

Nematodes, scales, whiteflies, mites, caterpillars, fruitfly, and others are problems, as are viral and fungal diseases. Twig galls sometimes form in response to a fungus infection.

Cladrastis kentukea (lutea) American Yellowwood

Pronunciation: clah-DRASS-tiss ken-TOO-key-ah

ZONES: 4A, 4B, 5A, 5B, 6A, 6B, 7A, 7B, 8A, 8B

HEIGHT: 30 to 50 feet

WIDTH: 40 to 50 feet

FRUIT: elongated; 2 to 4"; dry; brown; some litter problem; persistent; showy

GROWTH RATE: moderate; moderate life span

HABIT: rounded vase; moderate to dense; symmetrical; medium texture

FAMILY: *Leguminosae*

LEAF COLOR: medium green in summer; copper to yellow fall color

LEAF DESCRIPTION: alternate; odd pinnately compound; entire margin; oval; ovate; deciduous; 2 to 4" long leaflets

FLOWER DESCRIPTION: white; very pleasant fragrance; showy; late spring; attracts bees

EXPOSURE/CULTURE:

Light Requirement: part shade to full sun; flowers best in full sun

Soil Tolerances: all textures; alkaline to acidic; occasionally wet soil; low to moderate salt; moderate drought

Aerosol Salt Tolerance: low to moderate

PEST PROBLEMS: resistant

BARK/STEMS:

Bark is thin and easily damaged by mechanical impact; branches droop as tree grows and will require pruning for vehicular or pedestrian clearance beneath canopy; should be grown with a single leader as far up into the canopy as possible; no thorns

Pruning Requirement: requires unusually frequent, rigorous pruning in first 15 years to develop strong structure; prune in summer to avoid bleeding

Limb Breakage: susceptible to breakage at crotch due to multi-trunked habit, poor collar formation, and included bark

Current Year Twig: brown; slender

LANDSCAPE NOTES:

Origin: southern Appalachian Mountains

Usage: buffer strip; highway; shade tree; specimen; street tree; reclamation

Yellowwood is so named because the freshly cut heartwood is yellow and yields a yellow dye. This seldom-used, native, deciduous tree, which can reach 75 feet in the woods, makes a very striking specimen or shade tree, with a broad, rounded canopy and a vase-shaped, moderately dense silhouette. Smooth, gray to brown bark, bright green leaflets, and a strikingly beautiful display of white, fragrant blossoms make Yellowwood a wonderful choice for multiple landscape uses.

Yellowwood should be pruned when young to develop well-spaced branches having a wide angle with the trunk and a U-shaped crotch. This is challenging! Prune back upright limbs to more horizontal branches, and keep them smaller than half the diameter of the trunk. This will help them develop strong attachments to the trunk and give a longer-lived tree than one that is not pruned in this way. Any pruning should be done in summertime because excess bleeding may occur if done in the winter or spring. It could be used as a street tree if properly pruned to avoid weak crotches.

Yellowwood grows slowly on poor soil but looks fine, responding to better soil with moderate growth. It tolerates alkaline soil (it is native to dry limestone outcroppings and stream banks) or acidic soil and urban conditions. Its roots fix nitrogen.

Unpruned trees often fall apart 30 to 40 years after planting. Extended drought may cause die-back. Verticillium wilt and canker can also be troublesome.

Clusia rosea Pitch-Apple, Cupey

Pronunciation: CLUE-zee-ah row-ZEE-ah

ZONES: 10B, 11

HEIGHT: 25 to 40 feet

WIDTH: 30 to 50 feet

FRUIT: poisonous; round; 2 to 4"; fleshy, then hard; greenish; attracts wildlife; individually showy; fruit causes some litter that persists on the ground; persistent on tree

GROWTH RATE: moderate; moderate life span

HABIT: round; upright then spreading; dense; nearly symmetrical; coarse texture

FAMILY: *Clusiaceae*

LEAF COLOR: broadleaf evergreen; deep green

LEAF DESCRIPTION: opposite; simple; entire margin; obovate; 8 to 12" long; very thick

FLOWER DESCRIPTION: pink to white; showy; summer

EXPOSURE/CULTURE:

Light Requirement: part shade to full sun

Soil Tolerances: all textures; alkaline to acidic; salt; drought

Aerosol Salt Tolerance: high

PEST PROBLEMS: pest-free

BARK/STEMS:

Branches droop as tree grows and will require pruning for vehicular or pedestrian clearance beneath canopy; routinely grown with multiple trunks; can be trained to grow with a single trunk; no thorns

Pruning Requirement: requires pruning to develop strong structure, develop a single trunk, and remove aerial roots

Limb Breakage: resistant

Current Year Twig: green; thick

LANDSCAPE NOTES:

Origin: south Florida, West Indies, South America

Usage: above-ground planter; espalier; hedge; parking lot island; buffer strip; highway; near a deck or patio; reclamation; standard; shade; specimen; sidewalk cutout; street tree; urban tolerant

Pitch-Apple is greatly admired as an ornamental and shade tree. Leaves are often written on with a fingernail and will retain the inscription for many years. In summer, the showy, pink and white, two- to three-inch flowers appear at night and sometimes remain open all morning on overcast days. They appear near the branch tips and are followed by a fleshy, light green, poisonous fruit, three inches in diameter. The black material surrounding the seeds was once used to caulk the seams of boats—hence this tree's common name, 'Pitch-Apple'.

Growing well in full sun to dappled shade, Pitch-Apple tolerates many different soil types but grows most rapidly on moist soils. It is quite tolerant of light open sands and salt spray, making it ideal for seaside locations. Pitch-Apple is often used as a screen, due to its low spreading habit, and is ideal for espalier use to cool building walls in the summer. Some maintenance is required to trim prop roots and aerial roots as they descend from the trunk base and lower branches. Some prop and aerial roots should be allowed to reach the ground and root to provide support to branches on large specimens. Otherwise, it is a low-maintenance tree. With lower branches removed, it can make an attractive, small to moderately sized street tree, although some people object to the falling fruits and thick, slowly decomposing leaves. A patio can be kept cooler with a dense canopy of Pitch-Apple nearby.

Purchase trees that have been trained in the nursery to one central leader for street tree use. Those grown for specimen use with several upright trunks are not suited for streets, as vehicle clearance will be difficult to maintain and trees will be less durable.

Scale may infest foliage and twigs.

Coccoloba diversifolia Pigeon-Plum

Pronunciation: co-co-LOW-bah di-ver-sih-FOE-lee-ah

ZONES: 10B, 11

HEIGHT: 20 to 40 feet

WIDTH: 20 to 30 feet

FRUIT: oval; round; 0.5" fleshy; purple-black; attracts wildlife; edible; causes messy litter for a short period

GROWTH RATE: moderate; moderate life span

HABIT: upright rounded; dense; symmetrical; medium texture

FAMILY: *Polygonaceae*

LEAF COLOR: broadleaf evergreen; dark green

LEAF DESCRIPTION: alternate; simple; entire margin; oblong to ovate; 3 to 6" long

FLOWER DESCRIPTION: white; somewhat showy

EXPOSURE/CULTURE:

Light Requirement: part shade to full sun

Soil Tolerances: all textures; alkaline to acidic; occasionally wet; salt; drought

Aerosol Salt Tolerance: high

PEST PROBLEMS: resistant

BARK/STEMS:

Showy, exfoliating bark; bark is thin; routinely grown with multiple trunks; branches grow mostly upright and will not droop; can be trained to grow with a single trunk; no thorns

Pruning Requirement: needs some pruning to develop strong structure; pruning needed to form a single trunk and prevent formation of included bark

Limb Breakage: resistant

Current Year Twig: brown; medium to thick

LANDSCAPE NOTES:

Origin: south Florida

Usage: container or above-ground planter; hedge; parking lot island; buffer strip; highway; near a deck or patio; reclamation; shade; specimen; street tree; fruit; urban tolerant

Mature trees in hammocks grow to 70 feet tall, but trees in an open landscape are usually less than 40 feet. Young trees appear pyramidal until the branches begin spreading, forming a rounded vase on older specimens. It is a wonderful medium-sized tree, typically sporting showy grayish-brown bark that falls off in plates to reveal dark purplish bark beneath. Unless trained to a single trunk, one- to two-foot-thick trunks often grow parallel to each other, forming included bark. This does not appear to compromise the strength of Pigeon-Plum until the tree gets older. It is best to avoid this condition by pruning early. The four-inch-long, shiny, leathery leaves drop uniformly in March, but new growth quickly emerges as bright red new growth.

Lower branches will have to be removed over time for vehicle clearance along streets, but there is a definite place for the tree along boulevards and in parking lot buffer strips where the fruits cannot drop on cars. Trees trained to a single trunk in the nursery can be very useful for planting along streets where vehicle clearance is needed. Although Pigeon-Plum makes a wonderful shade tree, the fallen fruit may create a litter problem on patios, sidewalks, and along streets. However, the several-week inconvenience of messy fruit may be a small price to pay for the wonderful effect this striking tree creates along streets or in a residential yard.

Growing in full sun or partial shade, Pigeon-Plum does best on moist, well-drained soils. It has good salt and drought tolerance. Be sure to slice and otherwise drastically disturb and pull apart the root ball on pot-bound, container-grown trees. Pot-bound trees have a reputation for rooting out poorly into landscape soil.

Chewing insects will occasionally riddle new growth, but control is not usually required. Scales and mites also cause some problems. Chlorosis occurs on highly alkaline soil. Susceptible to ganoderma rot.

Coccoloba uvifera Seagrape

Pronunciation: co-co-LOW-bah u-VIFF-er-ah

ZONES: 10A, 10B, 11

HEIGHT: 30 to 40 feet

WIDTH: 25 to 40 feet

FRUIT: on female trees only; round; 0.5 to 1"; fleshy, grape-like; blue or purple; attracts birds and mammals; edible; causes significant litter; showy

GROWTH RATE: moderate; moderate life span

HABIT: rounded, upright or spreading vase; moderate; nearly symmetrical; coarse texture

FAMILY: *Polygonaceae*

LEAF COLOR: broadleaf evergreen; emerging red, turning green

LEAF DESCRIPTION: alternate; simple; entire margin; orbiculate; 6 to 8" long

FLOWER DESCRIPTION: white; not especially showy; winter

EXPOSURE/CULTURE:

Light Requirement: part shade to full sun

Soil Tolerances: all textures; alkaline to acidic; salt; drought

Aerosol Salt Tolerance: high

PEST PROBLEMS: resistant to lethal pests

BARK/STEMS:

Exfoliating bark; branches droop as tree grows and will require pruning for vehicular or pedestrian clearance beneath canopy; routinely grown with multiple trunks; can be trained to grow with a single trunk; no thorns

Pruning Requirement: requires pruning to prevent formation of weak, included bark in branch crotches and to form a single leader

Limb Breakage: susceptible to breakage at crotch due to poor collar formation

Current Year Twig: brown; thick

LANDSCAPE NOTES:

Origin: south Florida into the Caribbean Islands

Usage: hedge; parking lot island; buffer strip; highway; reclamation; shade tree; specimen; street tree; urban tolerant

Reaching a height of 30 to 40 feet, Seagrape typically forms a multi-stemmed vase shape if left unpruned. Leaves frequently turn completely red before they fall in winter. The new, young foliage is a beautiful bronze color, which is set off nicely by the older, dark green, shiny leaves. Dense clusters of 3/4-inch diameter green grapes form on female trees only, ripening to a luscious deep purple in late summer. Males do not produce fruit.

The contorted, twisting trunk (which can grow to two feet in diameter) and upright branching habit make Seagrape an interesting, picturesque shade tree or specimen plant. It can be maintained as a dense hedge, screen, or windbreak. Pruning is required 2 or 3 times during the first 10 years after planting to train the tree to a strong structure. Be sure branches do not develop included bark, which indicates a poor attachment to the trunk. Large branches with included bark can split from the trunk. The wood and the tree are generally very strong and durable following this developmental and corrective pruning. The tree will then perform well with little care, except for occasional pruning of lower branches to create clearance for vehicles. Some people object to the litter created by the large, slowly decomposing leaves which fall from the tree throughout the year.

Requiring full sun and sandy, well-drained soils, Seagrape is excellent for seaside locations because it is highly salt- and drought-tolerant. Plants should not require irrigation after they are well established in the landscape.

Stems are subject to seagrape borer, which can kill branches. A harmless, red nipple gall forms on the upper leaf surface. Leaf-eating insects disfigure foliage.

Coccothrinax argentata Silverpalm, Thatchpalm

Pronunciation: co-co-THRY-nax are-gen-TAY-tah

ZONES: 10B, 11

HEIGHT: 6 to 10 feet

WIDTH: 6 feet

FRUIT: round; less than 0.5"; fleshy; purple to brown; no significant litter

GROWTH RATE: desperately slow-growing; moderate life span

HABIT: open; symmetrical; coarse texture

FAMILY: *Arecaceae*

LEAF COLOR: blue-green to green with silver undersides

LEAF DESCRIPTION: simple; serrate margins; star-shaped; 24 to 40" wide

FLOWER DESCRIPTION: white; not showy; summer flowering

EXPOSURE/CULTURE:

Light Requirement: shaded to full sun

Soil Tolerances: all textures; alkaline to acidic; salt; drought

Aerosol Salt Tolerance: high

PEST PROBLEMS: resistant

BARK/STEMS:

Single trunk covered with interwoven fibers; 8" thick; no crown shaft; no thorns

Pruning Requirement: remove dead and dying fronds

Trunk Breakage: resistant

LANDSCAPE NOTES:

Origin: south Florida

Usage: container or above-ground planter; near a deck or patio; specimen

This slow-growing, small, native Florida palm can reach 20 feet in height, but is usually seen at 6 to 10 feet. The slender Silverpalm has distinctive, drooping, delicate, deeply divided palmate leaves with a beautiful silver color beneath, providing a bright glint in the landscape when the leaves sway in the wind. The six- to eight-inch-wide trunk is either smooth and gray or is sometimes covered with woven, burlap-like fiber. The small, white flowers are borne in profusion on two-foot-long stalks, hidden among the leaves during the summer. The flowers become prominent if you duck inside the canopy. The small, round, purple fruits ripen in late summer and fall.

This palm is most suited for residential and commercial landscapes where the unusual blue foliage can be appreciated. It makes a nice accent in a shrub border, and can be massed together to create a dramatic colorful and textural impact. Place it in a low-growing groundcover to provide an exclamation point in the landscape.

Growing in full sun or mostly shaded, Silverpalm will tolerate any well-drained soil. The palm will grow straight up and provide a beautiful accent, even in areas receiving only two or three hours of sun. It is highly salt-tolerant and is especially useful for coastal locations and for soils with a high pH.

No limitations of major concern.

Cocos nucifera 'Malayan Dwarf'
'Malayan Dwarf' Coconut Palm
Pronunciation: KOE-kuss new-SIFF-er-ah

ZONES: 10B, 11

HEIGHT: 30 to 50 feet

WIDTH: 15 to 25 feet

FRUIT: oval; round; 6 to 8"; hard; green or yellow to brown; edible; causes significant litter; persistent; showy

GROWTH RATE: moderate; moderate life span

HABIT: palm; open; irregular; coarse texture

FAMILY: *Arecaceae*

LEAF COLOR: green to greenish yellow

LEAF DESCRIPTION: odd pinnately compound; entire margin; 18 to 36" long leaflets

FLOWER DESCRIPTION: white; not showy; periodic year-round flowering

EXPOSURE/CULTURE:

Light Requirement: grows best in full sun

Soil Tolerances: all textures; alkaline to acidic; salt; drought

Aerosol Salt Tolerance: high

PEST PROBLEMS: somewhat resistant

BARK/STEMS:

Single, curved trunk; no crown shaft; 12" thick; no thorns

Pruning Requirement: needs pruning to remove flowers to prevent fruit from forming in urban areas

Trunk Breakage: extremely resistant to hurricanes

CULTIVARS AND SPECIES:

Due to the widespread devastation of lethal yellowing disease, use only the resistant Malayan strains, often called dwarf or pygmy coconuts, and labelled yellow, golden, red, and green, according to the color of their fruits, such as 'Golden Malayan Dwarf'. The Malayan palms are very similar to the 'Jamaican Tall' except for having straight trunks. The red strain is the most rugged of the three, with the most disease resistance, but is considered to have the least attractive, yellow foliage. Unfortunately, the green frond types are less resistant to lethal yellowing disease. The variety 'Maypan', a hybrid of 'Malayan Dwarf' x 'Panama Tall', has the most robust and rapid growth yet retains its resistance to lethal yellowing disease. It also grows well on poor sites.

LANDSCAPE NOTES:

Origin: Pacific Islands

Usage: fruit; highway; reclamation; specimen; sidewalk cutout; street tree; urban tolerant

One of humanity's most useful plants. The heavy crown of long flowing fronds (up to 16 feet long) and gently curved trunks of Coconut Palm lend a tropical effect to any landscape setting. A beautiful street tree, Coconut Palm is also ideal as a background tree, framing tree, or striking free-standing specimen. Coconut Palms located along streets or walkways, or near patios, require pruning to remove the flowers or developing fruit so fruit does not fall and cause injury or property damage.

Coconut Palms grow in full sun on any well-drained soils. No irrigation is needed once established. They make nice street trees if planted when they are tall enough. Be aware that falling fruit can damage vehicles or hit pedestrians; thus the flower stalks (in spring) or developing fruit (summer) may have to be removed.

Nematodes, lethal yellowing disease (spread by the tropical leaf hopper bug), viral diseases, and fungi all affect Coconut Palms. Be sure to plant only selections that are resistant to lethal yellowing disease. Ganoderma butt rot can infect the lower trunk and roots and kill the palm. Avoid injury to the palm in this area. There is no control for butt rot, only prevention. Remove infected palms. 'Malayan Dwarf' selections require more fertilizer and may be more susceptible to palm aphids than 'Jamaican Tall' Coconut Palms.

Conocarpus erectus Buttonwood, Button-Mangrove

Pronunciation: co-no-CAR-puss ee-RECK-tuss

ZONES: 10B, 11

HEIGHT: 30 to 35 feet

WIDTH: 20 to 30 feet

FRUIT: oval; less than 0.5"; dry; fleshy; brownish red; no significant litter; persistent; showy

GROWTH RATE: moderate; moderate life span

HABIT: spreading vase; moderately dense; somewhat symmetrical; fine texture

FAMILY: *Combretaceae*

LEAF COLOR: evergreen; green

LEAF DESCRIPTION: alternate; simple; entire margin; lanceolate to oblong; 2 to 5" long

FLOWER DESCRIPTION: purple; white; not showy; periodic year-round flowering

EXPOSURE/CULTURE:

Light Requirement: full sun

Soil Tolerances: all textures; alkaline to acidic; wet soil; salt; drought

Aerosol Salt Tolerance: high

PEST PROBLEMS: resistant

BARK/STEMS:

Branches droop as tree grows and will require pruning for vehicular or pedestrian clearance beneath canopy; can be grown with a single or multiple trunks; showy trunk; no thorns

Pruning Requirement: needs little or no pruning to develop strong structure; pruning needed to develop single trunk

Limb Breakage: resistant

Current Year Twig: green; slender

LANDSCAPE NOTES:

Origin: south Florida into the Caribbean Islands

Usage: bonsai; hedge; parking lot island; buffer strip; highway; near a deck or patio; reclamation; shade; specimen; sidewalk cutout; street tree; urban tolerant

This low-branching, multi-trunked tree grows larger than the more common variety, Silver Buttonwood. The inconspicuous, small, greenish flowers appear in dense conelike heads in terminal panicles in spring and are followed by ½-inch, conelike, red-brown fruits. The dark brown, attractive bark is ridged and scaly. The tree is clean, having small leaves that fall between the grass blades of the lawn and are easily washed away in the rain.

Planted in the open as a tree, Buttonwood will often take on a picturesque, contorted appearance when exposed to constant seashore winds, creating an attractive specimen. The crown is more symmetrical back from the coast or on the inland side of a tall oceanfront building.

Due to its small size, plant on 15- to 20-foot centers to form a closed canopy over a sidewalk along a street. Purchase single-trunked trees for street and parking lot plantings to provide for pedestrian and vehicle clearance beneath the canopy. Lateral branches should be pruned to keep them from growing to more than half the diameter of the main trunk. Included bark often develops in major branch crotches, but the strong wood appears to compensate for this potential defect. Trees are tough and long-lasting in the landscape. The wood of Buttonwood was formerly used for firewood, cabinetwork, and charcoal making and is very strong. It is an ideal wood for smoking meats and fish.

Sucking insect secretions will result in problems with sooty mold on trees inland from the coast. Drooping branches will have to be removed from multi-trunked trees planted along streets. Prune to a central leader when the tree is young to avoid later removal of low branches.

Conocarpus erectus var. *sericeus* Silver Buttonwood

Pronunciation: co-no-CAR-puss ee-RECK-tuss [variety] sir-EE-see-us

ZONES: 10B, 11

HEIGHT: 15 to 20 feet

WIDTH: 15 to 20 feet

FRUIT: oval; less than 0.5"; dry; fleshy; brown to red; no significant litter; persistent; showy

GROWTH RATE: moderate; moderate life span

HABIT: vase shape; moderately dense; symmetrical; fine texture

FAMILY: *Combretaceae*

LEAF COLOR: blue or blue-green; silver; evergreen

LEAF DESCRIPTION: alternate; simple; entire margin; lanceolate to oblong; hairy; 2 to 4" long

FLOWER DESCRIPTION: purple; white; not showy; periodic year-round flowering

EXPOSURE/CULTURE:

Light Requirement: grows best in full sun

Soil Tolerances: all textures; alkaline to acidic; wet soil; salt; drought

Aerosol Salt Tolerance: high

PEST PROBLEMS: resistant

BARK/STEMS:

Showy trunk; branches droop as tree grows and will require pruning for vehicular or pedestrian clearance beneath canopy; routinely grown with multiple trunks; can be trained to a single trunk; no thorns

Pruning Requirement: needs little pruning to develop strong structure; pruning needed to develop single leader

Limb Breakage: resistant

Current Year Twig: greenish-gray; slender

LANDSCAPE NOTES:

Origin: south Florida into Caribbean Islands

Usage: container or above-ground planter; espalier; hedge; parking lot island; buffer strip; highway; near a deck or patio; reclamation; shade; specimen; sidewalk cutout; street tree; urban tolerant

This low-branching, multi-trunked, shrubby, evergreen tree has beautiful, hairy, silvery leaves. The inconspicuous, small, greenish flowers appear in dense conelike heads in terminal panicles in spring and are followed by ½-inch, conelike, red-brown fruits. The dark brown, attractive bark is ridged and scaly. The tree is clean, having small leaves that fall between the grass blades of the lawn and are easily washed away in the rain.

Due to its small size, plant on 10- to 15-foot centers to form a closed canopy over a sidewalk along a street. Purchase single-trunked trees for street and parking lot plantings to provide for pedestrian and vehicle clearance beneath the canopy. Multi-trunked specimens make nice accents in a landscape full of green-foliaged plants. The dark trunks are set off best when the tree is planted in a low, uniform ground cover. Due to the attractive bark and soft foliage, a multi-stemmed specimen can make a nice patio tree. Trees respond well to clipping and make excellent tall hedges.

Silver Buttonwood tolerates brackish areas and alkaline soils, thriving in the broken shade and wet soils of hammocks. This is a tough tree! It withstands the rigors of urban conditions very well and makes a durable street or parking lot tree.

Sooty mold often grows on sucking insect secretions on trees inland from the coast. Refer to species for pruning guidelines.

Cordia boissieri Wild-Olive, Mexican Olive

Pronunciation: CORE-dee-ah boss-ee-AIR-ee

ZONES: 9A, 9B, 10A, 10B, 11

HEIGHT: 15 to 20 feet

WIDTH: 15 to 20 feet

FRUIT: purple; attracts wildlife; not showy; no significant litter

GROWTH RATE: slow-growing; moderate life span

HABIT: round; vase shape; moderate density; symmetrical; medium texture

FAMILY: *Boraginaceae*

LEAF COLOR: silvery to greenish-blue; evergreen

LEAF DESCRIPTION: alternate; simple; entire margin; ovate; 3 to 5" long

FLOWER DESCRIPTION: white; showy; spring and summer and periodically throughout year

EXPOSURE/CULTURE:

Light Requirement: flowers best in full sun

Soil Tolerances: all textures; alkaline to acidic; wet soil; some salt; drought

Aerosol Salt Tolerance: moderate

PEST PROBLEMS: resistant

BARK/STEMS:

Bark is thin and easily damaged by mechanical impact; branches droop as tree grows and will require pruning for vehicular or pedestrian clearance beneath canopy; routinely grown with multiple trunks; can be trained to grow with a single trunk; no thorns

Pruning Requirement: requires pruning to develop strong structure and a single leader

Limb Breakage: somewhat susceptible

Current Year Twig: brown; medium thickness

LANDSCAPE NOTES:

Origin: southwestern United States into Mexico

Usage: container or above-ground planter; buffer strip; highway; near a deck or patio; standard; parking lot island; specimen; sidewalk cutout

Wild-Olive is a native North American evergreen tree rarely found in the wild, and it is even reportedly close to extinction. The leaves have a velvety texture and the showy, white flowers appear year-round, if enough rainfall or irrigation is available. Otherwise, the three-inch-wide, trumpet-shaped, white blossoms with yellow throats will first appear from late spring to early summer. They produce one of the most brilliant, clear white flowers of any tree. The olive-like, white fruits have a sweet flesh relished by birds and other wildlife and, although edible by humans, should not be eaten in quantities.

This is a versatile plant adapted for use as a specimen tree or an accent in a shrub border. Showy, nearly year-round flowers make it suitable for placing in a lawn area as a free-standing specimen. It can be planted in an above-ground container and kept looking nice for a number of years when it is carefully maintained. Train it to grow with many trunks branched low to the ground for the best flower display in a container. Train it to a central leader for planting along streets or for parking lot planting. It should also grow very nicely in confined soil spaces and would make a nice show planted in a small soil space cut out of the sidewalk. It would be difficult to find a place not to plant this wonderful tree.

Wild-Olive should be grown in full sun or partial shade on well-drained soils and is highly drought-tolerant. Although hardy to about 20°F., Wild-Olive will lose its leaves in a severe frost. This is the cold-hardy relative of *Cordia sebestena,* which is very sensitive to the cold.

Unfortunately, this tree is not generally available.

Cordia sebestena Geiger-Tree

Pronunciation: CORE-dee-ah seh-beh-STEE-nah

ZONES: 10B, 11

HEIGHT: 25 feet

WIDTH: 25 feet

FRUIT: oval; 1 to 2"; greenish may cause significant litter; showy

GROWTH RATE: slow-growing; moderate life span

HABIT: rounded vase; moderate density; nearly symmetrical; coarse texture

FAMILY: *Boraginaceae*

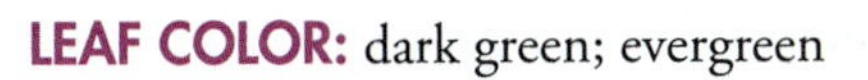

LEAF COLOR: dark green; evergreen

LEAF DESCRIPTION: alternate; simple; rough, similar to sandpaper texture; undulate margin; cordate to ovate; 4 to 8" long

FLOWER DESCRIPTION: orange; very striking; spring and summer

EXPOSURE/CULTURE:

Light Requirement: part shade to full sun

Soil Tolerances: all textures; alkaline to acidic; salt; drought

Aerosol Salt Tolerance: high

PEST PROBLEMS: except for geiger beetle, pest-free

BARK/STEMS:

Bark is thin and easily damaged by mechanical impact; branches droop as tree grows and will require pruning for vehicular or pedestrian clearance beneath canopy; trainable to be grown with a single or multiple trunks; no thorns

Pruning Requirement: requires some early pruning to develop strong structure and a central leader

Limb Breakage: resistant

Current Year Twig: brown, medium thickness

CULTIVARS AND SPECIES:

Cordia boissieri is frost-resistant (tolerating temperatures in the 20s) and has stunning white flowers with yellow centers.

LANDSCAPE NOTES:

Origin: south Florida, West Indies into northern South America

Usage: container or above-ground planter; buffer strip; highway; near a deck or patio; shade; parking lot island; specimen; sidewalk cutout; street tree

This dense, rounded, evergreen native tree grows slowly but can develop a trunk 12 inches thick. The large, seven-inch-long, stiff leaves are rough and hairy, feeling much like sandpaper. Throughout the year, but especially in spring and summer, dark orange, two-inch-wide flowers appear in clusters at branch tips. The splendid flowers are followed by one- to two-inch-long, pear-shaped fruits, which have a pleasant fragrance but are not particularly tasty. They are hard and litter the ground when they fall. There is nothing quite like a Geiger-Tree in full bloom. According to legend, the common name was bestowed by Audubon in commemoration of John Geiger, a Key West pilot and wrecker of the 19th century; it is now used quite universally as the common name for this excellent Florida native tree.

Geiger-Tree works nicely as a free-standing specimen or patio tree. Trees are normally seen as multi-trunked and low-branching, but nurseries can produce single-trunked trees suitable for along streets and parking lots. It has been used as a street tree in some communities, but drops leaves as a drought-avoidance strategy in prolonged dry spells. This makes the tree a very low-maintenance plant suited for planting in medians and buffer strips without irrigation.

Growing in full sun to partial shade, Geiger-Tree is tolerant of light, sandy, alkaline soils and salt spray. It is highly recommended for seaside plantings.

The geiger beetle defoliates the tree upon occasion, but the trees generally grow out of it and do well. The problem can be locally troublesome. Falling fruits and leaves may be considered a nusiance by some.

Cornus alternifolia Pagoda Dogwood

Pronunciation: KOR-nus all-ter-nih-FO-lee-uh

ZONES: 3B, 4A, 4B, 5A, 5B, 6A, 6B, 7A

HEIGHT: 15 to 20 feet

WIDTH: 15 to 20 feet

FRUIT: oval; 0.5"; fleshy; bluish-black; attracts birds; edible; no significant litter

GROWTH RATE: slow to moderate; moderate life span

HABIT: round; dense; layered; irregular silhouette; medium texture; often wider than tall at maturity

FAMILY: *Cornaceae*

LEAF COLOR: dark green in summer; purple to red fall color

LEAF DESCRIPTION: opposite; simple; entire margin; ovate; deciduous; 2 to 4" long

FLOWER DESCRIPTION: creamy white; very showy; spring

EXPOSURE/CULTURE:

Light Requirement: part shade to mostly sunny; some protection from afternoon sun for best form

Soil Tolerances: all textures; slightly alkaline to acidic; some salt; moderate drought in partial shade

Aerosol Salt Tolerance: some

PEST PROBLEMS: resistant

BARK/STEMS:

Branches droop as tree grows and will require pruning for pedestrian clearance beneath canopy; routinely grown with multiple trunks; no thorns

Pruning Requirement: needs little pruning to develop strong structure

Limb Breakage: resistant

Current Year Twig: greenish-red; medium thickness

CULTIVARS AND SPECIES:

'Argentea' is smaller, with variegated foliage that looks pretty good for a variegated tree. It would look nice planted in a partially shaded garden displayed against dark green foliage.

LANDSCAPE NOTES:

Origin: Alabama to Minnesota and New Brunswick

Usage: container or above-ground planter; near a deck or patio; specimen

Pagoda Dogwood has showy, creamy-white flower clusters comprised of small individual flowers, good fall color, and attractive fruit. It is compact and makes a nice complement to the Flowering Dogwood. Branches grow horizontally, then droop to touch the ground, even on young trees. The horizontal layered effect of the branches is even more pronounced than on Flowering Dogwood. The fully opened flowers last for about 10 days followed by the darkened fruit that is quickly devoured by birds. The fruit is borne on bright red stalks and can be showy for a period of time if birds do not find it.

Pagoda Dogwood needs open soil space for root expansion to look its best. Some shade will improve performance in restricted soil spaces, and in the south. To reduce stress, some horticulturists locate the tree so it receives sun only in the morning hours. Growth is best on moist, loamy, well-drained soil (not heavy clay) with mulch or leaf litter accumulated over the soil. Pagoda Dogwood is not particularly drought- or heat-tolerant, requiring irrigation during drought periods in summer.

This dogwood is perfectly suited for planting on the north or east side of a building where direct sun hits it for only part of the day. Use it to complement the horizontal lines or to soften vertical lines of the building. It responds well to planting in the shade of larger trees, where it can brighten the spring and fall landscape with seasonal color. The canopy is also nicely displayed in a soft texture, making Pagoda Dogwood a good choice for near the patio or other quiet area of the yard.

Borers are one of the biggest problems for dogwoods. There are a number of other problems, including anthracnose, cankers, leaf spots, powdery mildew, galls, and scales.

Cornus florida Flowering Dogwood

Pronunciation: KOR-nus FLOOR-eh-dah

ZONES: 5A, 5B, 6A, 6B, 7A, 7B, 8A, 8B, 9A

HEIGHT: 20 to 30 feet

WIDTH: 20 to 30 feet

FRUIT: oval; round; less than 0.5"; fleshy; red; attracts birds; no significant litter; persistent; showy

GROWTH RATE: moderate; moderate life span

HABIT: oval; round; moderate density; nearly symmetrical; medium texture

FAMILY: *Cornaceae*

LEAF COLOR: medium green in summer; red fall color

LEAF DESCRIPTION: opposite; simple; entire margin; ovate; deciduous; 3 to 5" long

FLOWER DESCRIPTION: white; showy; spring

EXPOSURE/CULTURE:

Light Requirement: mostly shaded to partial sun; full sun if roots can grow unimpeded by urban structures

Soil Tolerances: well-drained clay; loam; sand; slightly alkaline to acidic; salt-sensitive; moderate drought in shade, drought-sensitive in other locations

Aerosol Salt Tolerance: low

PEST PROBLEMS: except for anthracnose, resistant to most lethal pests

BARK/STEMS:

Showy on older trees; branches droop as tree grows and will require pruning for vehicular or pedestrian clearance beneath canopy; trainable to be grown with multiple trunks; normally trained to grow with a single trunk; no thorns

Pruning Requirement: needs little pruning to develop strong trunk and branch structure; remove upright branches with included bark

Limb Breakage: resistant

Current Year Twig: green; medium thickness

CULTIVARS AND SPECIES:

Several of the cultivars listed are not readily available. Pink-flowering cultivars grow poorly in zones 8 and 9. 'Apple Blossom'—pink bracts; 'Cherokee Chief—red bracts; 'Cherokee Princess'—white bracts; 'Cloud 9'—white bracts, many blooms, flowers at early age; 'Fastigiata'—upright growth while young, spreading with age; 'First Lady'—leaves variegated with yellow, turning red and maroon in the fall; 'Gigantea'—bracts six inches from tip of one bract to tip of opposite bract; 'Magnifica'—bracts rounded, four-inch-diameter

Continued

pairs of bracts; 'Multibracteata'—double flowers; 'New Hampshire'—flower buds cold-hardy; 'Pendula'—weeping or drooping branches; 'Plena'—double flowers; var. *rubra*—pink bracts; 'Spring Song'—bracts rose-red; 'Springtime'—bracts white, large, blooms at an early age; 'Sunset'—supposedly resistant to anthracnose; 'Sweetwater Red'—bracts red; 'Weaver's White'—large white flowers, adapted to zones 8 and 9; 'Welchii'—leaves variegated with yellow and red; 'White Cloud'—flowers more numerous, bracts white; 'Xanthocarpa'—fruit yellow.

There are many pests and diseases of dogwood.

Anthracnose, caused by *Discula* fungus, has received the most attention lately and is a serious disease (especially in shaded locations) capable of killing large numbers of trees. Most *Cornus* species can become infected with anthracnose. Powdery mildew also disfigures the foliage and distorts new growth. Crosses (hybrids) between *C. florida* and *C. kousa* are more resistant to anthracnose.

C. nuttallii, Western Dogwood, has similar cultural requirements and can grow 50 to 80 feet tall in western landscapes. It flowers in spring and fall, but is also susceptible to anthracnose.

LANDSCAPE NOTES:

Origin: eastern North America

Usage: near a deck or patio; small shade tree; specimen

The state tree of Virginia, Flowering Dogwood grows up to 35 feet tall in the wild and spreads to 30 feet. It can be trained with one central trunk or as a picturesque multi-trunked tree. A central leader usually forms on trees grown in the shade. The flowers consist of four showy bracts which subtend the small head of yellow flowers. The bracts may be white, pink, or red depending on cultivar. The fall color depends on site and seed source but on most sun-grown plants will be red to maroon. Fall color is more vivid in zones 5 to 8A.

Branches on the lower half of the crown grow horizontally; those in the upper half are more upright. In time, this can lend a strikingly horizontal impact to the landscape, particularly if some branches are thinned to open up the crown. Lower branches left on the trunk will droop to the ground, creating a wonderful landscape feature.

Not suited for parking lot planting, but can be grown in a wide street median if planted in well-drained heavy soil or provided with less than full-day sun and irrigation. Dogwood is a standard tree in many gardens where it is used by the patio for light shade, in the shrub border to add spring and fall color, or as a specimen in the lawn or groundcover bed. It can be grown in sun or shade, but shaded trees will be less dense, grow more quickly and taller, and have poor fall color and fewer flowers. Trees prefer part shade (preferably in the afternoon), especially in the southern end of the range. Many nurseries grow the trees in full sun but irrigate regularly in sandy soil. They transplant from field nurseries in the spring.

Flowering Dogwood prefers a deep, rich, well-drained, sandy or loamy soil and has a moderately long life. It is not recommended for heavy, wet soils unless it is grown on a raised bed to keep roots on the dry side. The roots rot in soils without adequate drainage.

Borers pose a big threat to survival, especially when the trunk is regularly damaged by lawn mowing equipment. Locate this tree in a landscape bed to keep mowers away.

Cornus kousa

Kousa Dogwood, Chinese Dogwood, Japanese Dogwood

Pronunciation: COR-nus COO-sah

ZONES: 5A, 5B, 6A, 6B, 7A, 7B, 8A, 8B

HEIGHT: 15 to 20 feet

WIDTH: 15 to 20 feet

FRUIT: oval; round; 0.5 to 1"; fleshy; red; attracts birds; edible; could cause some litter; showy; looks like a raspberry

GROWTH RATE: slow to moderate; moderate life span

HABIT: round; dense; nearly symmetrical; medium texture; often wider than tall at maturity

FAMILY: *Cornaceae*

LEAF COLOR: medium green in summer; purple to red fall color

LEAF DESCRIPTION: opposite; simple; entire margin; ovate; deciduous; 2 to 4" long

FLOWER DESCRIPTION: white; very showy; spring

EXPOSURE/CULTURE:

Light Requirement: part shade to mostly sunny

Soil Tolerances: all textures; slightly alkaline to acidic; some salt; moderate drought; needs good drainage

Aerosol Salt Tolerance: some

PEST PROBLEMS: mostly resistant, except for anthracnose

BARK/STEMS:

Very striking lower trunk; branches droop as tree grows and will require pruning for pedestrian clearance beneath canopy; routinely grown with, or trainable to be grown with, multiple trunks; no thorns

Pruning Requirement: needs little pruning to develop a strong structure

Limb Breakage: resistant

Current Year Twig: brown to green; medium thickness

CULTIVARS AND SPECIES:

'Chinensis'—larger bracts; 'Milky Way'—produces more flowers; the variety *angustata* is evergreen as far north as Philadelphia. *C. florida* x *kousa* hybrids in the Stellar Series are new, becoming available, and have wonderful flowers. Preliminary study shows they have good mildew and *Discula* anthracnose resistance.

LANDSCAPE NOTES:

Origin: Japan and Korea

Usage: container or above-ground planter; near a deck or patio; specimen

Kousa Dogwood has beautiful exfoliating bark, long-lasting flowers, good fall color, and attractive fruit. In most ways it is superior to the native Dogwood. Branches grow upright when the tree is young, but appear in horizontal layers on older trees. The white, pointed bracts are produced several weeks later than Flowering Dogwood and are effective for about a month, sometimes longer.

The bark is so attractive on older trees that lower branches are often selectively thinned to show it off. The tree also makes a great silhouette as a specimen planting and should be allowed to branch close to the ground to show its full character. The strong horizontal branching habit on older plants is difficult to find in other trees, and it looks great when lit at night from beneath the canopy. A Kousa Dogwood can extend the spring flowering season several weeks, as it begins flowering just after Flowering Dogwood.

Kousa Dogwood needs open soil space for root expansion to look its best. Some shade will improve performance in restricted soil spaces and in the south. Growth is best on moist, loamy, well-drained soil (not heavy clay) with mulch or leaf litter accumulated over the roots. Kousa Dogwood is not particularly drought- or heat-tolerant, requiring irrigation during drought periods in summer to look its best.

Not immune to borers, but more resistant than Flowering Dogwood. Susceptible to dogwood anthracnose. Fruit can cause a mushy litter and are more or less resistant to mildew.

Cornus mas Cornelian-Cherry

Pronunciation: KOR-nus mas

ZONES: 5A, 5B, 6A, 6B, 7A, 7B, 8A

HEIGHT: 15 to 20 feet

WIDTH: 12 to 18 feet

FRUIT: oval; 0.5"; fleshy; red; attracts birds; edible; could be a litter problem on walks; showy

GROWTH RATE: slow to moderate; moderate life span

HABIT: round; moderate density; symmetrical; medium texture

FAMILY: *Cornaceae*

LEAF COLOR: medium green in summer; dull green to red fall color

LEAF DESCRIPTION: opposite; simple; entire margin; ovate; deciduous; 2 to 4" long

FLOWER DESCRIPTION: yellow; showy; long flower display; late winter and spring

EXPOSURE/CULTURE:

Light Requirement: part shade to full sun

Soil Tolerances: all textures; alkaline to acidic; salt-sensitive; moderate drought

Aerosol Salt Tolerance: sensitive

PEST PROBLEMS: mostly pest-free

BARK/STEMS:

Showy trunk; branches droop as tree grows and will require pruning for pedestrian clearance beneath canopy; routinely grown with multiple trunks; can be trained to a short, single trunk; no thorns

Pruning Requirement: needs little pruning to develop strong structure

Limb Breakage: resistant

Current Year Twig: green; medium thickness

CULTIVARS AND SPECIES:

Some may not be readily available: 'Alba'—white fruit; 'Aureo-Elegantissima'—leaves bordered in yellow, green, and pink; 'Elegantissima'—leaves yellow, green, and pink; 'Flava'—yellow fruits; 'Fructu Violaceo'—purple fruits; 'Golden Glory'—dark green foliage with a dense, pyramidal habit; 'Macrocarpa'—larger fruits; 'Nana'—dwarf, three feet tall, 'Variegata'—leaves with white variegations; 'Xanthocarpa'—yellow fruit. One of the best for the south is 'Spring Glow,' which has a low chilling requirement, beautiful flowers and handsome, leathery foliage that looks good until frost. *C. officinalis* would be a better choice for the south if you can find it.

LANDSCAPE NOTES:

Origin: central and southern Europe to Asia Minor

Usage: container or above-ground planter; screen; near a deck or patio; specimen

Cornelian-Cherry is a slow-growing, small tree or large shrub preferring sun or partial shade and well-drained soil. Flowers are produced in northern areas but most of the southeastern United States lacks the chilling hours required to set flower buds. 'Spring Glow' is the one cultivar that will flower in the south. Bark is very showy and is often displayed by removing lower foliage. The yellow flowers produced in very early spring are similar to Forsythia and are followed by red fruit, which is edible but partially hidden by the foliage in the summer. Falling fruit is messy and can temporarily stain walks and concrete. Cornelian-Cherry responds well to pruning and may be used as a hedge plant.

The once-popular species has fallen out of the trade recently, but deserves a comeback. It is pest-free and grows well on a variety of soil, including clay. Soil should be kept moist with good drainage. Althought mulching encourages better root growth and moderate drought tolerance, it is not considered tolerant of extended drought. Use it as a specimen or in a group planting or shrub border. The fruit makes an excellent tart jelly and attracts birds. Makes an excellent patio tree in the yard and should be planted more.

May sucker from the base of the trunk and roots. Well adapted to urban landscapes. Shows good resistance to powdery mildew and *Discula* anthracnose in some tests.

Cornus walteri Walter Dogwood

Pronunciation: KOR-nus WALL-terr-eye

ZONES: 6A, 6B, 7A, 7B, 8A

HEIGHT: 30 to 40 feet

WIDTH: 30 to 40 feet

FRUIT: round; less than 0.5"; fleshy; black; attracts wildlife; no significant litter

GROWTH RATE: moderate; moderate life span

HABIT: upright vase; moderate density; nearly symmetrical; medium texture

FAMILY: *Cornaceae*

LEAF COLOR: dark green in summer; red fall color

LEAF DESCRIPTION: opposite; simple; serrate to serrulate margin; ovate; deciduous; 2 to 4" long

FLOWER DESCRIPTION: white; showy; spring

EXPOSURE/CULTURE:

Light Requirement: part shade to full sun

Soil Tolerances: all textures; slightly alkaline to acidic; moderate drought

PEST PROBLEMS: unknown

BARK/STEMS:

Showy trunk; branches grow mostly upright and will not droop; grows with many trunks but can be trained to a single leader; no thorns

Pruning Requirement: needs little pruning to develop strong trunk and branch structure; early pruning needed to develop and maintain a single trunk

Limb Breakage: resistant

Current Year Twig: green; medium thickness

CULTIVARS AND SPECIES:

C. coreana is closely related, with showy white flowers borne in clusters in late spring on upright branches.

LANDSCAPE NOTES:

Origin: Central China

Usage: near a deck or patio; shade tree; residential street tree

Walter Dogwood is a medium-sized deciduous tree. The two- to five-inch-long leaves are joined in June by delicate white flowers arranged in very showy, two- to three-inch diameter cymes. They are among the last dogwoods to bloom. The blossoms are followed by the production of small black fruits that are popular with birds and other wildlife. Walter Dogwood is probably best known for the alligator-like bark on older specimens, even more so than Flowering Dogwood.

Young specimens have an upright branching habit that gives way to an open spreading habit. The branches probably droop less than most other dogwoods, making it a possible candidate for street tree use, although they have not been tested for this in the United States. Most certainly a good patio tree, providing shade quickly.

Grown in the full sun, it develops a dense canopy, but is more open and perhaps more attractive in partial shade. It will flower well in partial shade and maintains the upright growth habit, where most plants might lean toward the light. The plant is now rare in the industry but deserves to be tried. It can be seen at a number of arboreta across the United States. It is drought-tolerant and grows well in clay soil. It is the fastest growing dogwood in North Carolina and is probably well adapted for wider planting when it becomes available.

May suffer leaf spot, twig blight, or canker. See Flowering Dogwood.

Corylus colurna Turkish Filbert, Turkish Hazel

Pronunciation: CORE-e-luss co-LUR-nah

ZONES: 5A, 5B, 6A, 6B, 7A, 7B

HEIGHT: 40 to 50 feet

WIDTH: 25 to 35 feet

FRUIT: elongated; oval; 0.5 to 1"; dry; brown; attracts wildlife; edible; could cause significant litter; showy

GROWTH RATE: slow to moderate; long-lived

HABIT: oval; pyramidal; dense; unusually symmetrical; medium texture

FAMILY: *Betulaceae*

LEAF COLOR: green in summer; muted yellow fall color

LEAF DESCRIPTION: alternate; simple; double serrate margin; obovate to ovate; deciduous; 3 to 5" long; persistent into the fall

FLOWER DESCRIPTION: spring; catkins

EXPOSURE/CULTURE:

Light Requirement: grows best in full sun, tolerates some shade

Soil Tolerances: all textures; alkaline to acidic; salt-sensitive; occasionally wet; drought

Aerosol Salt Tolerance: sensitive

PEST PROBLEMS: except for Japanese beetles, pest-free

BARK/STEMS:

Showy; small-diameter branches grow mostly upright and horizontal, and will not droop; often grown with multiple trunks, but should be grown with a single leader for street planting; no thorns

Pruning Requirement: needs little or no pruning to develop strong trunk and branch structure, early pruning needed to form single trunk

Limb Breakage: resistant

Current Year Twig: brown; gray; thick

LANDSCAPE NOTES:

Origin: southeastern Europe to Asia Minor

Usage: parking lot island; buffer strip; highway; shade; specimen; sidewalk cutout; street tree; urban tolerant

Like many trees with horizontal branches, the main limbs are quite numerous and small in diameter in relationship to the typically straight trunk, and arise at almost a 90° angle. This helps make the tree quite durable in urban areas and helps maintain the symmetrical crown so prized by architects. Trees with multiple trunks are more suited for planting in open spaces such as parks and golf courses. On some older trees, the bark becomes ridged and corky, peeling off in sections to expose the orange/brown bark beneath.

Turkish Hazel makes a wonderful urban shade or street tree due to the dense canopy and its ability to withstand air pollution. It makes a rather formal statement in the landscape due to the tight, consistently shaped, narrow crown. It is well suited for areas with restricted overhead horizontal space for lateral branch development. Once the central leader has been properly trained in the nursery and the tree is established, it should require little, if any, pruning or other maintenance. The nut may be a slight litter problem for some, but is considered more unsightly than a hazard to pedestrians.

Turkish Hazel is susceptible to Japanese beetles, which can defoliate portions of the tree rather quickly, but this should not be cause to eliminate this very adaptable tree from your recommended tree list. Other trees scorch in summer drought, whereas this one will stay green. This tree should enjoy much wider use than it does now, but may have to be grown in containers, or under field conditions that encourage root development inside the root ball, to overcome poor transplant survival and slow recovery. Limited availability.

Cotinus coggygria Smoketree, Wig-Tree, Smokebush

Pronunciation: ko-TIE-nuss co-GUY-gree-ah

ZONES: 5B, 6A, 6B, 7A, 7B, 8A, 8B, 9A, 9B, (10)

HEIGHT: 10 to 15 feet

WIDTH: 10 to 15 feet

FRUIT: irregular; oval; 0.5"; dry; brown; showy; no significant litter

GROWTH RATE: slow-growing; short to moderate life span

HABIT: oval; round; pyramidal when young; dense; symmetrical; medium texture

FAMILY: *Anacardiaceae*

LEAF COLOR: blue-green to green in summer; outstanding orange, purple, red, or yellow fall color

LEAF DESCRIPTION: alternate; simple; entire margin; oval to obovate; deciduous; 2 to 4" long

FLOWER DESCRIPTION: pink to yellowish; showy; spring to summer

EXPOSURE/CULTURE:

Light Requirement: grows and flowers best in full sun

Soil Tolerances: all textures; alkaline to acidic; some salt; drought

Aerosol Salt Tolerance: moderate

PEST PROBLEMS: fairly resistant

BARK/STEMS:

Bark is thin and easily damaged by mechanical impact; branches droop as tree grows and will require pruning for vehicular or pedestrian clearance beneath canopy; routinely grown with multiple trunks; no thorns

Pruning Requirement: could use pruning to space multiple trunks along a short central trunk

Limb Breakage: old trees split at the crotch of multiple trunks

Current Year Twig: brown; medium to thick

CULTIVARS AND SPECIES:

'Royal Purple'—leaves are darker purple than 'Notcutt's Variety', rich red-purple fall color, purplish-red inflorescences; 'Pendulus'—pendulous branches; 'Velvet Cloak'—dark purple leaf form, spectacular fall color of reddish-purple.

LANDSCAPE NOTES:

Origin: southern Europe to central China

Usage: container or above-ground planter; buffer strip; highway; near a deck or patio; reclamation; standard; parking lot island; specimen; small street tree; urban tolerant

The large panicles of wispy cream-colored flowers produced in spring and early summer give the effect of a cloud of smoke. Smoketree makes a wonderful accent in a shrub border and can be planted as a specimen or as a patio tree where the black, showy, multiple trunk can be displayed. Planting Smoketree is a good way to extend the spring flowering-tree season into the summer. Many people grow it simply to enjoy the vivid fall color.

The tree is tough and adapts to restricted soil spaces. It could be used along a street under power lines and would not require pruning for many years, if ever. It is well adapted to urban areas with almost year-round interest and should be used more in our landscapes.

Smoketree grows best in a sunny location with excellent drainage. It will grow asymmetrically and lean toward the light in a partially sunny area, so it is best to locate it in full day sun where the crown will develop symmetrically. Smoketree is useful in dry, rocky soil where there is little or no irrigation. Probably short-lived (30 years, maybe more) in most situations, but who cares—the tree is great while it's around!

Oblique-banded leaf roller causes cosmetic damage by mining and rolling the leaves, usually in June. Leaf spots can be caused by various genera of fungi but are usually not serious. Smoketree is susceptible to verticillium wilt. A stem canker can be a problem in the eastern United States.

Cotinus obovatus American Smoketree, Chittamwood

Pronunciation: ko-TIE-nuss ob-o-VAY-tuss

ZONES: 5A, 5B, 6A, 6B, 7A, 7B, 8A, 8B

HEIGHT: 20 to 30 feet

WIDTH: 15 to 25 feet

FRUIT: oval; less than 0.5"; fleshy; showy like a smoke bomb

GROWTH RATE: very slow-growing; moderate life span

HABIT: upright to oval; open when young, dense later; symmetrical; medium texture

FAMILY: *Anacardiaceae*

LEAF COLOR: light green in summer; striking orange, purple, red, yellow fall color, amazing in fall color even in shade

LEAF DESCRIPTION: alternate; simple; entire margin; oval to obovate; deciduous; 2 to 6" long

FLOWER DESCRIPTION: pinkish-yellow; slightly showy and variable; spring

EXPOSURE/CULTURE:

Light Requirement: mostly shaded to full sun, denser in sun

Soil Tolerances: all textures; alkaline to acidic; excellent drainage is essential; salt-sensitive; drought

Aerosol Salt Tolerance: low to moderate

PEST PROBLEMS: sensitive in poorly drained soil

BARK/STEMS:

Showy form and bark; bark is thin and easily damaged by mechanical impact; branches droop somewhat as tree grows and will eventually require some pruning for vehicular or pedestrian clearance beneath canopy; trainable to be grown with single or multiple trunks; no thorns

Pruning Requirement: requires early pruning to develop upright structure

Limb Breakage: resistant

Current Year Twig: brown to gray; medium thickness

CULTIVARS AND SPECIES:

'Red Leaf' is popular for its especially brilliant fall color. Other cultivars selected for consistent and superior fall color are likely to be made available in the near future. Hybrids between *C. obovatus* x *C. coggygria* including 'Grace' will be available in the near future.

LANDSCAPE NOTES:

Origin: Tennessee to Alabama and Texas

Usage: container or above-ground planter; parking lot island; buffer strip; highway; near a deck or patio; reclamation; standard; specimen; sidewalk cutout; street tree; urban tolerant

The bark on older specimens flakes off at the base of several twisted and gnarled trunks, showing a dark gray to black exfoliating character. The 7- to 12-inch, fairly sparse, fuzzy panicles of springtime flowers give the plant an unusual, somewhat smoky effect, but American Smoketree is most outstanding for its beautiful foliage. The two- to six-inch-long leaves are pinkish-bronze when young, mature to a lush, dark blue/green, and then, in autumn, change to astonishing shades of yellow, red, orange, or purple. It is very striking in the fall—one of the best for fall color—and lasts three to four weeks.

Smoketree is best used in a shrub border or trained as a multiple-trunked patio or accent tree. It can easily be pruned to a single trunk for street tree use. This will provide clearance beneath the canopy for pedestrians and vehicles.

A North American native, American Smoketree is tolerant of a wide range of adverse urban conditions, including wind, drought, and compacted soil. Plants grow well and are native to alkaline soil and should be located in full sun or partial shade. Best flowering, form, and overall attractiveness are achieved in full sun. May need occasional irrigation on sandy soil. It is reportedly difficult to transplant, but this is easily overcome by growing in containers.

Crataegus laevigata English Hawthorn

Pronunciation: crah-TEE-gus lee-vih-GAY-tah

ZONES: 4B, 5A, 5B, 6A, 6B, 7A, 7B, 8A, 8B, (9)

HEIGHT: 15 to 25 feet

WIDTH: 15 to 25 feet

FRUIT: round; less than 0.5"; fleshy; red; attracts birds; no significant litter; persistent; somewhat showy

GROWTH RATE: moderate to rapid; moderate to long life span

HABIT: oval; vase; moderate; irregular silhouette; fine texture

FAMILY: *Rosaceae*

LEAF COLOR: dark green in summer; no fall color change

LEAF DESCRIPTION: alternate; simple; lobed; serrate margin; obovate to ovate; deciduous; less than 2" long

FLOWER DESCRIPTION: lavender, pink, white depending on cultivar; showy; spring

EXPOSURE/CULTURE:

Light Requirement: grows best in full sun

Soil Tolerances: all textures; alkaline to acidic; occasionally wet; salt-sensitive; drought

Aerosol Salt Tolerance: sensitive

PEST PROBLEMS: resistant to lethal pests; sensitive to foliage diseases

BARK/STEMS:

Branches droop as tree grows and will require pruning for vehicular or pedestrian clearance beneath canopy; tree normally grows with several trunks but can be trained to grow with a single trunk; some thorns are present on branches

Pruning Requirement: needs pruning to develop uniform habit and single trunk

Limb Breakage: resistant

Current Year Twig: brown; medium thickness

CULTIVARS AND SPECIES:

A number of improved cultivars are offered in garden centers, including 'Crimson Cloud' with red flowers and more disease resistance. This cultivar is nearly thornless and is preferred over the species. Other cultivars include: 'Flore-Plena—double, white flowers; 'Gireoudii'—foliage variegated, white flowers, large red fruit; 'Masekii'—pale rose, double flowers; 'Paulii' ('Paul's Scarlet')—deep pink, double flowers; 'Pendula'—weeping growth habit; 'Punica'—single, pink flowers; 'Rosea'—pink flowers; 'Stricta'—upright growth habit.

LANDSCAPE NOTES:

Origin: Europe to North Africa

Usage: bonsai; espalier; parking lot island; buffer strip; highway; reclamation; sidewalk cutout; street tree; urban tolerant

English Hawthorn grows rapidly in an open irregular pyramid to about 12 feet, then the crown expands to become an irregular vase. The tree tolerates most soils, growing well in clay, but prefers heavy, dry loam. The main ornamental features are white or pink flowers, borne in spring, and good fall color. Some types produce bright scarlet fruits. Though quite ornamental, Hawthorns is severely affected by insect and disease problems, but is worthy of use nonetheless.

The tree casts fairly dense shade, and turf cannot be grown underneath if lower branches are left on the trunk. Persistent lower branches add to the ornamental characteristic of the tree and it makes a nice specimen in a lawn for all seasons when left unpruned. When lower branches are removed, this tough tree can be used as a street or parking lot tree where overhead space is limited by power lines or other features. Be sure to purchase trees that do not require a stake to hold them erect. Branches bear sharp thorns that can inflict pain if they meet flesh, so select cultivars without these. Grows well in tree pits and other confined soil spaces. Troublesome pests include aphids, lacebug, fireblight, and rust. Plant in the open with good air circulation to help reduce leaf diseases.

Crataegus phaenopyrum Washington Hawthorn

Pronunciation: crah-TEE-gus fee-no-PIE-rum

ZONES: 4A, 4B, 5A, 5B, 6A, 6B, 7A, 7B, 8A

HEIGHT: 20 to 35 feet

WIDTH: 20 to 25 feet

FRUIT: very showy; round; 0.5"; fleshy; orange or red; attracts birds; no significant litter; persistent

GROWTH RATE: moderate; moderate to long life span

HABIT: oval; pyramidal; moderate density; nearly symmetrical; soft, medium texture

FAMILY: *Rosaceae*

LEAF COLOR: dark green in summer; showy copper fall color

LEAF DESCRIPTION: alternate; simple; lobed; serrate margin; ovate; deciduous; 1 to 2" long

FLOWER DESCRIPTION: white; showy; late spring

EXPOSURE/CULTURE:

Light Requirement: full sun is essential to minimize disease incidence

Soil Tolerances: all textures; alkaline to acidic; occasionally wet soil; salt-sensitive; drought

Aerosol Salt Tolerance: low

PEST PROBLEMS: resistant to lethal pests; sensitive to foliage diseases

BARK/STEMS:

Branches droop as tree grows and will require early pruning for vehicular or pedestrian clearance beneath canopy; can be grown with a single leader or trained to several trunks; painfully sharp thorns are present on trunk or branches

Pruning Requirement: needs little or no pruning to develop strong structure; pruning needed to develop one trunk

Limb Breakage: susceptible to ice damage; major trunks with included bark split from tree, causing severe damage; otherwise resistant

Current Year Twig: brown; slender

CULTIVARS AND SPECIES:

The cultivar 'Fastigiata' has an upright growth habit with a 12-foot-wide canopy. *C. marshalli* may be better suited to central and south Texas.

LANDSCAPE NOTES:

Origin: Virginia to Alabama and Missouri

Usage: bonsai; above-ground planter; buffer strips; barriers; highway; specimen; street tree if no sidewalk is present; urban tolerant

Washington Hawthorn grows to mature height in a wide pyramidal shape. The tree has a rapid growth rate when young, slowing with age. The small, white, abundant flowers, produced in clusters in late spring, are followed by showy orange to red fruits that persist into winter if not eaten by birds. The tree looks best in the spring as the new growth and flowers emerge, before the foliage becomes disfigured by diseases. The tree comes to life again in the fall with a brilliant display of red berries.

With lower branches removed, this Hawthorn is quite useful as a street tree where there will not be heavy pedestrian traffic. It has also been successfully used in very small parking lot islands to brighten up a normally dank expanse of pavement. With the lowest branches seven to eight feet off the ground on a single trunk, the thorns stay clear of pedestrians. The thorns are about three inches long and contact with them is unforgettable. For this reason, it is well suited for creating a showy yet impenetrable barrier. A person will attempt to walk through a row of Washington Hawthorn only once.

Left unpruned, it creates a nice specimen in a lawn, with lower branches persisting all the way to the ground. This characteristic also makes it quite suitable as a screen. The bright fruit makes a show in the fall and winter that draws many comments. Like other hawthorns, the major problem with the tree is sensitivity to a large variety of insects and diseases, but it most certainly has its place in the urban landscape.

Affected by the usual array of hawthorn pests, including rust, leaf blight, scab, aphids, borers, tent caterpillars, lacebugs, and others.

Crataegus viridis 'Winter King' 'Winter King' Hawthorn

Pronunciation: crah-TEE-gus VERE-eh-diss

ZONES: 4A, 4B, 5A, 5B, 6A, 6B, 7A, 7B

HEIGHT: 20 to 30 feet

WIDTH: 20 to 30 feet

FRUIT: round; less than 0.5"; fleshy; orange to red; attracts birds; no significant litter; persistent; outstanding color

GROWTH RATE: moderate; moderate to long life span

HABIT: rounded, upright vase; moderate density; symmetrical; fine textures

FAMILY: *Rosaceae*

LEAF COLOR: dark green in summer; showy red to gold fall color

LEAF DESCRIPTION: alternate; simple; lobed; serrate margin; ovate; deciduous; less than 2" long

FLOWER DESCRIPTION: white; showy; spring

EXPOSURE/CULTURE:

Light Requirement: full sun is best to minimize disease infections

Soil Tolerances: all textures; alkaline to acidic; occasionally wet soil; salt-sensitive; drought

Aerosol Salt Tolerance: low

PEST PROBLEMS: resists lethal pests, sensitive to foliage pests and diseases

BARK/STEMS:

Very showy; following early training, branches grow mostly upright and do not droop; trainable to be grown with multiple trunks; can be trained fairly easily to grow with a single trunk; thorns on branches

Pruning Requirement: requires little pruning to develop strong structure; early pruning and staking needed to develop and maintain single trunk

Limb Breakage: resistant

Current Year Twig: brownish-red; medium thickness

LANDSCAPE NOTES:

Origin: eastern North America south of Pennsylvania

Usage: container or above-ground planter; street tree; parking lot island; hedge; reclamation; specimen; urban tolerant

'Winter King' Hawthorn is a North American native tree typically grafted onto Washington Hawthorn root. It is very dense with inch-long thorns, which makes it a popular choice for use as a hedge or screen. Unlike other hawthorns, the thorns are small and inconspicuous. The handsome, silver-gray bark peels off in sections to reveal the inner orange bark, making 'Winter King' Hawthorn a striking specimen in all seasons, including winter. The white blooms are followed by large, orange/red fruits that persist on the naked tree throughout the winter, adding to its landscape interest. The original tree from which the cultivar was selected is about 100 years old and 25 feet tall, with outstanding winter berry color.

The multiple trunks and wide pyramidal to vase shape make this adaptable tree well suited for the low-maintenance landscape as a specimen. If lower branches are removed from the trunks, the vase shape can be maintained. Direct growth to the upper branches when the tree is young to develop a tree suited for planting along streets and in parking lot islands; it is adaptable to this hot environment. Locate the lowest branch six to eight feet from the ground to keep thorns away from passersby. It has become quite popular and is available in many areas.

'Winter King' Hawthorn should be grown in full sun on well-drained soil. It is a very adaptable tree well suited for the urban landscape. It is among the best of the many hawthorns available.

Fruits often become infected with rust, but the foliage is relatively clean for a hawthorn. Occasionally suckers originate from the root stock. Aphids, borers, lacebugs, caterpillars, and leaf miners can infest foliage. Transplants best in the spring.

Crescentia cujete Calabash-Tree

Pronunciation: creh-SEN-she-ah coo-JET-tee

ZONES: 10B, 11

HEIGHT: 20 to 30 feet

WIDTH: 25 to 30 feet

FRUIT: borne on trunk and branches; oval; round; 5 to 10"; fleshy; green; no significant litter; showy

GROWTH RATE: moderate; moderate life span

HABIT: rounded and spreading; unusually open canopy; irregular silhouette; medium texture

FAMILY: *Bignoniaceae*

Photo courtesy of G. Joyner

LEAF COLOR: evergreen; bright green

LEAF DESCRIPTION: alternate; simple; entire margin; obovate; 3 to 6" long

FLOWER DESCRIPTION: green; yellow; showy; periodic year-round

EXPOSURE/CULTURE:

Light Requirement: full sun for fullest canopy

Soil Tolerances: all textures; alkaline to acidic; moderate drought

Aerosol Salt Tolerance: none

PEST PROBLEMS: resistant

BARK/STEMS:

Branches droop as tree grows and will require pruning for vehicular or pedestrian clearance beneath canopy; routinely grown with multiple trunks or low branches; no thorns

Pruning Requirement: requires pruning to develop sturdy trunk and well-formed canopy

Limb Breakage: resistant

Current Year Twig: medium to stout

LANDSCAPE NOTES:

Origin: Puerto Rico, Virgin Islands, other Caribbean Islands

Usage: buffer strip; highway; near a deck or patio; specimen; street tree

Calabash is an evergreen tree with a broad, irregular crown composed of long, spreading branches clothed in two- to six-inch-long leaves, which create light shade beneath the tree. Calabash is most outstanding in the landscape for its year-round production of flowers and fruit, both of which are unusual. The two-inch-wide flowers are yellow/green with red or purple veins, cup-shaped, and emerge directly from the larger branches. These are followed by the emergence of large, round fruits with a smooth, hard shell, which hang from main branches.

Calabash is a popular street tree in some tropical cities, providing filtered shade along the hottest areas of the city. Specimens are planted near patios and decks for the novelty of the fruit borne directly on the trunk. They contrast well with the dark bark and are irresistible to touch. Provide a path to the tree and people are sure to follow. The light green foliage provides refreshing relief from the familiar darker greens in the landscape.

The main limbs on Calabash originate close to the ground, forming a low-branched tree without training and pruning. These main limbs branch infrequently, forming a lanky, awkward-looking canopy. This is suitable for planting in an open area where there is plenty of space. Prune the tips of the branches regularly when the tree is young to develop more secondary branches close to the trunk. This will help build the diameter of the main limbs and thicken the canopy. The trunk can be trained straight by staking, and lateral branches directed to grow upright.

Calabash-Tree should be grown in full sun on any well-drained soil. Trees in the shade become open and branch poorly.

No limitations of major concern, but occasionally bothered by Chinese rose beetles and a leaf-webbing caterpillar. Fruit is poisonous and the flowers smell terrible.

Cryptomeria japonica Japanese-Cedar

Pronunciation: crip-toe-ME-ree-ah jah-PON-eh-cah

ZONES: 6A, 6B, 7A, 7B, 8A, 8B, 9A, (9B), (10A), (10B)

HEIGHT: 50 to 60 feet

WIDTH: 15 to 20 feet

FRUIT: round; 0.5 to 1"; dry; brown; no significant litter

GROWTH RATE: slow-growing; long-lived

HABIT: pyramidal when young, oval with age; dense; symmetrical; fine texture

FAMILY: *Taxodiaceae*

LEAF COLOR: evergreen; copper in winter

LEAF DESCRIPTION: spirally arranged; awl-like; 0.5" long

FLOWER DESCRIPTION: not showy

EXPOSURE/CULTURE:

Light Requirement: grows best in full sun

Soil Tolerances: all textures; acidic; drought

PEST PROBLEMS: pest sensitive in humid climates

BARK/STEMS:

Very showy trunk form and bark; normally grows with a single leader; no thorns

Pruning Requirement: needs little or no pruning to develop strong structure

Limb Breakage: resistant

Current Year Twig: green; medium thickness

CULTIVARS AND SPECIES:

'Benjamin Franklin' makes a good screen. 'Sekan Sugi' has cream-colored branch tips. 'Yoshino' holds green foliage color in the winter and appears to be more disease resistant. 'Elegans' grows to 15 feet tall. 'Globosa Nana' is rounded and compact. There are several dwarf cultivars.

LANDSCAPE NOTES:

Origin: Japan

Usage: screen; large parking lot island; buffer strip; specimen; possible street tree; urban tolerant

The tree keeps a billowy pyramidal form on one central trunk until close to maturity when the crown opens up into a gracefully irregular, narrow oval. Old specimens can develop trunks to three feet in diameter. The reddish-brown bark is ornamental, peeling off in long strips, and is the most pronounced characteristic on old trees. The foliage will become bronzed during the winter but greens up again in spring. Branches usually persist on the tree, with old specimens branched to the ground. Japanese-Cedar is truly a wonderful tree as it grows older. A number of cultivars are available that vary in growth habit and ability to hold green foliage color in the winter.

Provide an acid soil and protection from winter winds. Locate the tree so air circulation is good, particularly during summer, to help prevent leaf blight. Best with afternoon shade in warmer climates. *Cryptomeria* is tolerant of compacted soil and performs well in parking lots and other tough, urban sites with some irrigation in drought. It makes a wonderful accent, screen, or border tree for larger properties. As it grows large on a residential landscape, remove some of the lower branches to enjoy the exquisite trunk and bark. They can be planted as street trees in residential areas to provide an elegant flavor to the neighborhood. It is well suited for small lots due to the narrow canopy and relatively slow growth rate.

Mites can infest the foliage. Leaf blight often causes much of the interior foliage to brown, creating an unsightly specimen. Fungicide sprays help prevent the disease, as does placing the tree so it receives early morning sun to dry the foliage. Keep the needles as dry as possible. Planting in a protected area in the northern part of its range can help prevent foliage burn in winter. Maskell scale can also cause a slow decline in the eastern part of the United States.

Cunninghamia lanceolata China-Fir

Pronunciation: kun-ing-HAM-ee-ah lance-ee-o-LAY-tah

ZONES: 6B, 7A, 7B, 8A, 8B, 9A, (9B), (10A), (10B)

HEIGHT: 50 to 70 feet

WIDTH: 15 to 25 feet

FRUIT: round; 1 to 2"; dry; brown; not showy; no significant litter

GROWTH RATE: slow to moderate; long-lived

HABIT: pyramidal, pendulous; moderate density when young, open later, variable; nearly symmetrical but variable; coarse texture

FAMILY: *Pinaceae*

LEAF COLOR: needle-leaf evergreen; green or blue; turns brown and persists

LEAF DESCRIPTION: spiral; lanceolate; sharp tip; 1 to 2" long

FLOWER DESCRIPTION: not showy

EXPOSURE/CULTURE:

Light Requirement: part shade to full sun

Soil Tolerances: all textures; acidic; occasionally wet; drought

PEST PROBLEMS: pest-free

BARK/STEMS:

Somewhat showy; branches droop as tree grows and will require pruning for vehicular or pedestrian clearance beneath canopy; should be grown with a single leader; no thorns

Pruning Requirement: needs little pruning to develop strong structure

Limb Breakage: resistant

Current Year Twig: green; thick

CULTIVARS AND SPECIES:

'Chason's Gift' (zones 7A, 7B, 8A, 8B, 9A, 9B) has great potential with a dense, conical habit and upright growth even on young trees. 'Glauca' has blue-green foliage. There is great variation among trees in form and color, which shows the possibilities to select more cultivars.

LANDSCAPE NOTES:

Origin: central and southern China

Usage: buffer strip; highway; specimen; urban tolerant

The strong, heavy trunk stands straight up through the coarse, irregular, pyramidal form, which is composed of somewhat pendulous branches densely clothed in 1.5- to 2.5-inch-long, pointed needles. Occasionally the tree will be grown with several trunks arising from the soil, but above that point the trunk almost never branches into a double or multiple leader. The center of the tree grows up like an arrow and lends a strong vertical accent to the landscape. The crown is fairly open, allowing the trunk and large branches to show prominently. Like other tall-growing trees, it is best to maintain only one central trunk.

The leaves take on a bronze hue in autumn and, when they die, persist on the tree on interior branches sometimes for a year or more, giving it a somewhat unkempt appearance according to some horticulturists. The attractive, brown, peeling bark strips away to reveal the inner reddish bark, adding to the tree's overall textural interest.

China-Fir is best used as a specimen at the corner of a large building or as a large-scale screen planted 15 to 20 feet apart. Young specimens maintain a fairly tight crown, but they open up with age, becoming asymmetrical with large pieces of the crown missing. Not a tree you would use everywhere, China-Fir has its place as an occasional accent and conversation piece, particularly in a park, on a golf course, or on campuses and other large-scale landscapes.

Brown foliage persists on interior branches, but this is a normal condition, usually not caused by insects or disease.

Cupaniopsis anacardiopsis Carrotwood

Pronunciation: koo-pan-ee-OP-siss an-ah-car-dee-OP-siss

ZONES: 10A, 10B, 11

HEIGHT: 25 to 35 feet

WIDTH: 25 to 35 feet

FRUIT: irregular; round; 0.5 to 1"; dry; green turning orange and red; attracts birds; showy and abundant; fruit causes significant litter and is very messy

GROWTH RATE: moderate to rapid; moderate life span

HABIT: oval; round; dense; nearly symmetrical; medium texture

FAMILY: *Sapindaceae*

LEAF COLOR: dark glossy green; evergreen

LEAF DESCRIPTION: alternate; odd pinnately compound; entire, undulating margins; oval to oblong; 3 to 5" long leaflets

FLOWER DESCRIPTION: green; not showy; summer

EXPOSURE/CULTURE:

Light Requirement: grows best in full sun

Soil Tolerances: all textures; slightly alkaline to acidic; wet soil; salt; drought

Aerosol Salt Tolerance: moderate to high

PEST PROBLEMS: pest-free

BARK/STEMS:

Smooth and showy; bark is thin and easily damaged by mechanical impact; major branches droop; included bark forms in crotches; should be grown with a single leader, but is often trained to several codominant trunks; no thorns

Pruning Requirement: encourage growth in well-spaced branches in upper canopy by thinning aggressive, low branches

Limb Breakage: somewhat susceptible at weak crotches

Current Year Twig: brown; thick

LANDSCAPE NOTES:

Origin: Australia

Usage: parking lot island; street tree; buffer strip; specimen; shade; urban tolerant

This compact evergreen tree has four-inch-long, divided leaflets and has been used as a shade, specimen, patio, or poolside tree. It is popular in many yards and is used as a small to medium-sized street tree spaced about 20 feet apart. The insignificant, small lime-green flowers are followed by ½-inch diameter fruits that split open to reveal copious amounts of red seeds. They germinate in the landscape and may be a significant litter problem. In addition, the seeds are disseminated by birds, which makes it easy for the tree to spread.

Carrotwood tolerates poor, wet soils, full sun, and hot, salty winds. It is truly a durable, urban tolerant tree, able to grow even in confined planting pits in downtown sidewalks. Perhaps it is best used in these areas. It is very popular in the areas where it can be grown. It is deep-rooting on well-drained soils and will tolerate drought.

Selected, upright branches in the crown can be removed to allow for more light penetration and better turf growth under the crown. If not, the dense canopy will shade out all but the most shade-tolerant plants. The wood is bright apricot-colored in cross-section, and resists breakage because it is hard. If you cut one down, save the wood for a woodworker friend.

Use with caution—this tree has become invasive in Florida and is or will be banned in most communities. Commonly used, and very adaptable along streets and parking lots in southern California and tropical areas, but the fruit is a nuisance.

x *Cupressocyparis leylandii* Leyland Cypress

Pronunciation: x koo-press-o-sye-PAIR-iss lay-LAN-dee-eye

ZONES: 6A, 6B, 7A, 7B, 8A, 8B, 9A, 9B, 10A

HEIGHT: 35 to 50 feet

WIDTH: 15 to 25 feet

FRUIT: round; less than 0.5"; dry or hard; brown; no litter

GROWTH RATE: rapid-growing; moderate life span

HABIT: columnar; oval; pyramidal; dense; symmetrical; fine texture

FAMILY: *Cupressaceae*

LEAF COLOR: blue-green to green; evergreen

LEAF DESCRIPTION: scale-like; ⅛" long

FLOWER DESCRIPTION: not showy

EXPOSURE/CULTURE:

Light Requirement: part shade to full sun

Soil Tolerances: all textures; alkaline to acidic; salt; drought

Aerosol Salt Tolerance: moderate to high

PEST PROBLEMS: becoming sensitive to canker

BARK/STEMS:

Branches grow mostly upright and will not droop; should be grown with a single leader; no thorns

Pruning Requirement: prune to one leader to prevent trunk splitting; California recommends that fungicide be applied to pruning wounds to slow the spread of canker

Limb Breakage: double leaders split in heavy snow

Current Year Twig: green; slender

CULTIVARS AND SPECIES:

'Castlewellan'—compact, gold-tipped leaves; 'Leighton Green'—columnar form; 'Haggerston Gray'—sage-green color; 'Naylor's Blue'—blue-gray foliage, columnar form; 'Silver Dust'—wide-spreading form with blue-green foliage marked with white variegations.

LANDSCAPE NOTES:

Origin: hybrid

Usage: clipped hedge; screen; buffer strip; Christmas tree; highway

A rapid growing evergreen when young, Leyland Cypress will easily grow three to four feet per year, ultimately attaining a majestic height of 50 feet or more in the west (perhaps somewhat shorter in the east). Universally popular in all parts of its hardiness range, Leyland Cypress forms a dense, oval, or pyramidal outline when left unpruned, but will tolerate severe trimming to create a formal hedge, screen, or windbreak. The fine, feathery foliage, which is composed of soft, pointed leaves on flattened branchlets, is dark blue-green when mature, soft green when young.

Leyland Cypress quickly outgrows its space in small landscapes and is too big for most residential landscapes unless it will be regularly trimmed. Although it can be sheared into a tall screen on small lots, Leyland Cypress should probably be saved for large-scale landscapes where it can be allowed to develop into its natural shape. It is surprisingly tolerant of severe pruning, recovering nicely from even severe topping (although this is not recommended), even when half the top is removed. It grows larger than most people want it to. Plant dwarf cultivars of Arborvitae or Falsecypress instead of Leyland Cypress in small landscapes. Substitute *Thuja plicata* 'Hogan', *Juniperus viginiana* or *J. silicicola* or *J. chinensis* 'Spartan', *Calocedrus decurrens,* or *Thuja occidentalis* if concerned about canker disease infection.

It grows well in clay soil and tolerates poor drainage for a short period of time. It also is very tolerant of salt spray and is suited for coastal landscapes.

A canker infects branches and trunk, severely disfiguring or killing the tree. This and *Cupressus macrocarpa* are not recommended for California due to this disease. Prune in the driest time of the year to help prevent canker infection. Bagworm can defoliate a tree in a week or two and can be quite serious. Snow can split older trees apart, especially those with double trunks.

Cupressus glabra (arizonica)
Smooth-Barked Arizona Cypress
Pronunciation: koo-PRESS-us GLAY-brah

ZONES: 7A, 7B, 8A, 8B, 9A, 9B, (10)

HEIGHT: 30 to 70 feet

WIDTH: 10 to 15 feet

FRUIT: round; less than 0.5"; dry; brown; no significant litter

GROWTH RATE: very fast; long-lived in dry climates, moderate life span in many landscapes

HABIT: columnar; oval; pyramidal; dense; symmetrical; fine texture

FAMILY: *Cupressaceae*

LEAF COLOR: evergreen; blue or blue-green

LEAF DESCRIPTION: opposite/subopposite; scale-like; 1/16" long

FLOWER DESCRIPTION: not showy

EXPOSURE/CULTURE:

Light Requirement: all-day sun only; no shade is tolerated

Soil Tolerances: all textures; alkaline to slightly acidic; good drainage is essential; drought

PEST PROBLEMS: sensitive

BARK/STEMS:

Showy; branches grow mostly upright and will not droop; should be grown with a single leader; no thorns

Pruning Requirement: prune to one central leader

Limb Breakage: resistant

Current Year Twig: green; slender

CULTIVARS AND SPECIES:

'Blue Ice' has powdery gray-blue foliage with mahogany-red tips, and a columnar habit; 'Blue Pyramid' is 25 feet tall by 12 feet wide with blue foliage; 'Carolina Safire' has brilliant blue foliage.

LANDSCAPE NOTES:

Origin: central to southern Arizona

Usage: bonsai; hedge; screen; specimen; windbreak

The species, native to the American southwest, is densely clothed in blue-gray leaves and becomes a narrow pyramid or column. *C. arizonica* and this tree are closely related and difficult to tell apart without the cone fruit; many botanists group them together. Both are exceptionally tolerant of hot, dry conditions. The outer red-brown bark breaks off every year, revealing the fresh, new, bright red, smooth inner bark. Foliage of 'Carolina Safire' is one of the brightest blues you will ever see. Jump out of the way after planting this cultivar, as it springs out of the ground, with young trees growing about six feet each year with irrigation. It grows faster than Leyland Cypress.

Smooth-Barked Arizona Cypress should be grown in full sun on well-drained, slightly acid or alkaline soil. If grown with too much moisture or on a poorly drained clay, a shallow root system will develop, making the tree susceptible to wind damage. The species was once widely grown in the southeastern United States, but may perform best in drier climates. It is ideal for use as a screen, clipped hedge, or windbreak because it is strong, durable, and wind-tolerant. Plant individuals far enough apart to keep the canopies away from each other. This will help air to circulate through the trees and may help minimize disease. Plant on a well-drained site in the eastern United States. This plant may be a good substitute for the disease-prone *Juniperus scopulorum.*

Trees can be sheared into formal shapes and maintained that way for years. Cypress trees in general are well suited for this purpose. They are best located in the most exposed part of the landscape so that sun and air keep the tree dry.

A stem canker has devastated large numbers of Arizona Cypress in various parts of the country. Mites and bagworms are potential problems. Resists *Phomopsis* and *Kabatina* diseases.

Cupressus sempervirens Italian Cypress

Pronunciation: koo-PRESS-us sem-per-VIE-wrens

ZONE: 7B, 8A, 8B, 9A, 9B, 10A, 10B, 11

HEIGHT: 40 to 60 feet

WIDTH: 3 to 12 feet

FRUIT: oval; 0.5 to 1"; dry; brown; no significant litter

GROWTH RATE: moderate; moderate life span

HABIT: columnar; dense; symmetrical; fine texture

FAMILY: *Cupressaceae*

LEAF COLOR: evergreen; green

LEAF DESCRIPTION: whorled; scale-like; tiny; sharp to the touch

FLOWER DESCRIPTION: not showy

EXPOSURE/CULTURE:

Light Requirement: grows best in full sun

Soil Tolerances: all textures; alkaline to acidic; salt; drought

Aerosol Salt Tolerance: moderate

PEST PROBLEMS: sensitive

BARK/STEMS:

Branches grow upright and will not droop; grows with a single leader; no thorns

Pruning Requirement: needs no pruning to develop strong trunk and branch structure

Limb Breakage: resistant

Current Year Twig: brown to gray; slender

CULTIVARS AND SPECIES:

'Glauca' has blue-green foliage and tight columnar form; 'Stricta' is very popular; 'Horizontalis' has short, horizontally spreading branches and may be better suited for cooler landscapes.

LANDSCAPE NOTES:

Origin: southern Europe, Iran

Usage: screen, formal specimen; windbreak

With its narrow columnar habit of growth, this evergreen forms a tall, dark green column 40 to 60 feet in height in the western United States, but is often much shorter in the east. Trees are normally no more than 3 to 10 feet wide. Narrow, columnar cultivars are commonly available and planted, and they grow no more than three to four feet wide. The scale-like leaves lend a very fine texture to any setting.

The tree has a very special form and is therefore suited only for certain landscapes. It is most common in a formal landscape and is widely planted in California. The tree is certain to attract attention to an area, and often looks nice planted in groups of three to five. Planted three feet apart, they make a dense, tall screen. Italian Cypress is often used for framing, as a strong accent around large buildings, or in the formal landscape, but does not lend itself well to many home landscapes. It quickly grows much too tall for most residential landscapes, looking much like a green telephone pole.

Growing in full sun on various well-drained soils, Italian Cypress should be planted in a well-prepared site and watered periodically until well established. Italian Cypress should not be pruned. It is very susceptible to mites and trees are often infested.

Bagworms are occasionally a problem for Italian Cypress. Root rot can infect Italian Cypress in poorly drained soil. A twig-boring beetle causes tip die-back. Canker is a devastating disease that kills trees.

Cydonia sinensis Chinese Quince

Pronunciation: sih-DON-yah sigh-NEN-sis

ZONES: (5B), 6A, 6B, 7A, 7B, 8A, 8B

HEIGHT: 15 to 25 feet

WIDTH: 20 feet

FRUIT: round; 3 to 6"; fleshy; yellow; causes a very mushy litter; very fragrant

GROWTH RATE: slow-growing; long-lived

HABIT: upright vase to oval; open to moderate density; nearly symmetrical; medium to fine texture

FAMILY: *Rosaceae*

LEAF COLOR: dark green in summer

LEAF DESCRIPTION: alternate; simple; finely serrate margin; ovate; deciduous; 2 to 4" long

FLOWER DESCRIPTION: pink; showy; spring; fragrant

EXPOSURE/CULTURE:

Light Requirement: part shade to full sun

Soil Tolerances: all textures; slightly alkaline to acidic; at least moderate drought

PEST PROBLEMS: resistant

BARK/STEMS:

Very showy with smooth, exfoliating, green bark marked with yellow, gray, tan, and brown; bark is thin and easily damaged by mechanical impact; branches grow mostly upright and will not droop; routinely grown with multiple trunks; can be trained to grow with a short, single trunk; no thorns

Pruning Requirement: thin the most aggressive upright branches to help prevent formation of included bark

Limb Breakage: resistant

Current Year Twig: brownish-green; slender

CULTIVARS AND SPECIES:

C. oblonga is a similar tree from central and southwestern Asia with multiple trunks and pear-like fruit. It tolerates alkaline soil.

LANDSCAPE NOTES:

Origin: China

Usage: container or above-ground planter; buffer strip; deck or patio; specimen

This tree is now properly named *Pseudocydonia*, but many still refer to it as *Cydonia*. An upright habit, attractive, fragrant flowers, and excellent bark characteristics combine to make Chinese Quince a small tree with good potential for planting in most landscapes. Bark is attractive all year long, flaking off in small patches to show various colors through the medium green background. Another striking attribute is the fruit, which is unusually fragrant and often displayed in a dish to scent a room. It, however, makes an incredible mess on a walk or in the lawn as it is produced in abundance.

The upright, vase-shaped crown makes the tree well suited for street tree planting. After irrigation is supplied to establish the tree, Quince can survive on rainfall alone. Lower branches are often thinned to show off the trunk form and color. As the tree grows older, remove small-diameter low branches to display the true beauty of the trunk and bark structure. Severe pruning or topping can stimulate basal sprouting which can become a nuisance, requiring regular removal.

It should grow well in spaces with plenty of soil for root expansion such as along boulevards and in residential and commercial landscapes. Although it has a small size, it has not been tested enough to recommend it for small soil spaces and urban sites with reflected light and hot temperatures. They appear to tolerate clay so long as it drains well.

Fireblight is the only reported problem.

Dalbergia sissoo Indian Rosewood

Pronunciation: dal-BURR-gee-ah SISS-oo

ZONES: 10A, 10B, 11

HEIGHT: 45 to 60 feet

WIDTH: 40 feet

FRUIT: elongated; pod; 2 to 4"; dry; brown; no significant litter; persistent

GROWTH RATE: rapid-growing; short to moderate life span

HABIT: oval; open; irregular silhouette; medium texture

FAMILY: *Leguminosae*

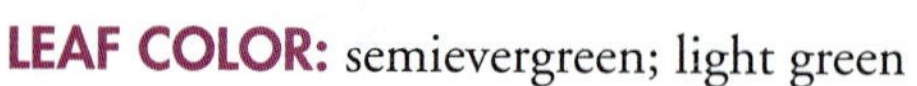

LEAF COLOR: semievergreen; light green

LEAF DESCRIPTION: alternate; odd pinnately compound; entire margin; oval to orbiculate; 2 to 3" long leaflets

FLOWER DESCRIPTION: white; pleasant fragrance; spring

EXPOSURE/CULTURE:

Light Requirement: part shade to full sun

Soil Tolerances: all texture; alkaline to acidic; occasionally wet soil; moderate drought

Aerosol Salt Tolerance: low

PEST PROBLEMS: resistant

BARK/STEMS:

Bark is thin and easily damaged by mechanical impact; branches droop as tree grows and will require pruning for vehicular or pedestrian clearance beneath canopy; should be grown with a single leader; no thorns

Pruning Requirement: requires pruning to develop strong structure

Limb Breakage: susceptible to breakage either at crotch due to poor collar formation, or wood itself is weak and tends to break

Current Year Twig: brown; green; slender

LANDSCAPE NOTES:

Origin: Asia

Usage: parking lot island; shade; urban tolerant

A handsome specimen or shade tree, easily grown Indian Rosewood has delicate, oval pointed leaflets and can quickly reach mature height and spread. The inconspicuous, very fragrant, white flowers are followed by slender, flat, brown, one- to four-seeded pods. The trunks yield a prized cabinet wood for fine furniture and the Rosewood genus is an important timber tree in India. Many *Dalbergia* species grown in the tropical regions of the world are harvested for veneer and lumber.

Though the wood is beautiful, the tree has a reputation for being brittle. Some of this may be due to improper pruning practices or inadequate training when the tree is young. Be sure that lateral branches remain smaller than two-thirds the trunk diameter to help ensure good tree structure. Remove, shorten, or thin branches with included bark in favor of those with strong, U-shaped crotches. This could help keep the tree together in windstorms.

Growing quickly in full sun or high shifting shade, Indian Rosewood will thrive on a variety of soil types, from dry to wet, but is not particularly salt-tolerant. Young plants should be watered until well established. Plants train easily into a well-formed, single-leader tree, which is desirable in urban landscapes. Sprouts often develop from the roots and become a maintenance problem, and roots often lift sidewalks if planted too close. Surface roots often grow large in diameter and can become a nuisance. A number of horticulturists consider this to be a nuisance tree. The tree casts light shade due to the open canopy.

Weedy habit, invasive potential, and large surface roots. Magnesium deficiency is common in alkaline soil.

Delonix regia Royal Poinciana

Pronunciation: deh-LON-icks REE-gee-ah

ZONES: 10B, 11

HEIGHT: 25 to 40 feet

WIDTH: 50 to 60 feet

FRUIT: elongated; pod; 18" or more; dry and hard; brown; fruit and twigs cause significant litter; persistent; showy

GROWTH RATE: rapid; moderate life span

HABIT: flat, rounded, spreading vase; moderate density; symmetrical; fine texture

FAMILY: *Leguminosae*

LEAF COLOR: semievergreen; light green

LEAF DESCRIPTION: alternate; bipinnately compound; tiny oblong leaflets

FLOWER DESCRIPTION: orange or red; astonishingly showy; early summer

EXPOSURE/CULTURE:

Light Requirement: flowers best in full sun

Soil Tolerances: all textures; alkaline to acidic; drought

Aerosol Salt Tolerance: low

PEST PROBLEMS: resistant

BARK/STEMS:

Branches droop as tree grows and will require pruning for vehicular or pedestrian clearance beneath canopy; routinely grown with multiple trunks, but should be grown with one trunk; no thorns

Pruning Requirement: prevent formation of included bark in major branch crotches; space branches along trunk

Limb Breakage: small twigs are shed from tree

Current Year Twig: brownish-green; medium to thick

LANDSCAPE NOTES:

Origin: Madagascar

Usage: reclamation; shade; specimen; street tree; urban tolerant

This many-branched, broad, spreading, flat-crowned deciduous tree is well known for its brilliant display of red-orange bloom, literally covering the tree tops from May to July. There is nothing like a Royal Poinciana in full bloom. The fine, soft, delicate leaflets afford dappled shade during the remainder of the growing season, making Royal Poinciana a favorite shade tree or freestanding specimen in large, open lawns. The tree is often broader than tall, and trunks can become as large as 50 inches or more in diameter. Eighteen-inch-long, dark brown seed pods hang on the tree throughout the winter, then fall on the ground in spring, creating a nuisance. Some nurseries are growing yellow-flowered selections.

Royal Poinciana will grow and flower best when planted in full sun. Tolerant of a wide variety of soils and conditions, Royal Poinciana needs to be well watered until established, then only during the severest droughts. Grass grows poorly beneath Poinciana. Do not plant closer than about 10 feet from pavement or sidewalks, because large surface roots often grow beneath them and can be destructive.

Early pruning is required to encourage development of branches that are well attached to the trunk. Train the tree so the major limbs are spaced along the short trunk. Locate the first branch 8 to 10 feet from the ground to allow for adequate clearance beneath the tree. To develop a strong, durable tree, thin major limbs to prevent them from growing to more than half the diameter of the trunk.

Caterpillars can eat some foliage. There is a root fungus that can kill a weakened tree. Fruit pods profusely litter the ground.

Diospyros kaki Japanese Persimmon

Pronunciation: die-OS-purr-os KAK-eye

ZONES: 7A, 7B, 8A, 8B, 9A, 9B

HEIGHT: 20 to 30 feet

WIDTH: 20 feet

FRUIT: showy; round; 3 to 6"; fleshy; orange; attracts squirrels and other mammals; edible; causes messy litter

GROWTH RATE: moderate; moderate life span

HABIT: round; upright; dense; nearly symmetrical; coarse texture

FAMILY: *Ebenaceae*

LEAF COLOR: showy orange, red or yellow fall color; green in summer

LEAF DESCRIPTION: alternate; simple; serrate or entire; oval to obovate; deciduous; 3 to 7" long

FLOWER DESCRIPTION: white; not showy; spring

EXPOSURE/CULTURE:

Light Requirement: best fruit in full sun

Soil Tolerances: all textures; alkaline to acidic; moderate drought

PEST PROBLEMS: resistant

BARK/STEMS:

Branches droop as tree grows and will require pruning for vehicular or pedestrian clearance beneath canopy; can be trained to grow with a single trunk; no thorns

Pruning Requirement: thin branches to keep them less than half the trunk diameter and to reduce fruit load

Limb Breakage: limbs with included bark split from tree under heavy fruit load

Current Year Twig: brown or gray; medium thickness

CULTIVARS AND SPECIES:

Some nonastringent cultivars have been selected and could be available locally. There are a number to choose from.

LANDSCAPE NOTES:

Origin: Japan and China

Usage: fruit; specimen

Japanese Persimmon is related to Common Persimmon (*D. virginiana*), but is native to Asia. It can grow to about 30 feet when mature. This is an excellent fruit tree for ornamental use and makes an excellent specimen. The tree is a sight to behold when leaves have fallen in autumn, displaying the bright yellow-orange fruits throughout the canopy. Similar to Common Persimmon, its preference is for moist, well-drained soil in full sun locations. The tree has good drought tolerance. Japanese Persimmon develops an attractive red fall color, but the two- to four-inch-diameter fruits can be a big mess when they fall from the tree.

Certainly not a street or parking lot tree, Japanese Persimmon is probably best located as an occasional specimen where it can be viewed from a distance, away from walks. This will ensure that the fruit will drop on the lawn, not on a walk. Better yet, plant the tree in a loose, low-growing groundcover so dropping fruit will be hidden from view in the foliage of the groundcover.

The tree is easy to grow and the fruit is delicious. Plant it in the garden or landscape to provide fruit during the fall and winter.

A trouble-free tree, but messy when fruit drops. Do not grow it unless you are prepared to deal with the fruit.

Diospyros texana Texas Persimmon

Pronunciation: die-OS-purr-os tex-A-nah

ZONES: 7A, 7B, 8A, 8B, 9A, 9B

HEIGHT: 20 to 30 feet

WIDTH: 15 to 25 feet

FRUIT: round; 0.5 to 1"; fleshy; green turning black; attracts wildlife; edible; no significant litter

GROWTH RATE: slow-growing; long-lived

HABIT: upright vase; open; symmetrical; fine texture

FAMILY: *Ebenaceae*

LEAF COLOR: no fall color change; dark green; semi-evergreen

LEAF DESCRIPTION: alternate; simple; entire margin; oblong to obovate; deciduous; 1 to 2" long

FLOWER DESCRIPTION: white; not showy but very fragrant

EXPOSURE/CULTURE:

Light Requirement: grows best in full sun

Soil Tolerances: all textures; alkaline to slightly acidic; drought

PEST PROBLEMS: pest-free

BARK/STEMS:

Bark is thin and easily damaged by mechanical impact; routinely grown with multiple trunks; branches grow mostly upright and will not droop; very showy trunk; no thorns

Pruning Requirement: requires little pruning to develop strong structure

Limb Breakage: resistant

Current Year Twig: gray; slender

LANDSCAPE NOTES:

Origin: Texas into northern Mexico

Usage: bonsai; container or above-ground planter; parking lot island; buffer strip; highway; near a deck or patio; specimen; sidewalk cutout; street tree

This slow-growing, native North American tree reaches a height of about 20 feet with an equal width (usually smaller) and is deciduous from zone 8 northward, remaining evergreen in its southern range. The one- to two-inch-long, leathery leaves are slightly pubescent underneath. The bark of Texas Persimmon is particularly striking, the smooth outside layers of gray, white, and pink peeling off in beautiful layers. Branches ascend into the crown in a twisted fashion unlike most other trees. The inconspicuous, green/white flowers are followed by the production of small, one-inch black fruits which, although edible by humans, contain an unappealing number of seeds. However, these fruits are quite popular with birds and other wildlife who relish the sweet, juicy flesh. In Mexico, the fruits are used to make a black dye. Do not get the fruit on your clothes.

Multiple trunks ascend into the vase-shaped crown forming a tree with a shape and structure similar to Crapemyrtle and American Hornbeam. Use it as an accent planted in a low groundcover to display the muscular-looking bark. It is well suited for planting in a highway median or along a street with overhead power lines, due to its small stature.

Except for the small, black fruits that drop for a short period of time, this is a clean tree that could be planted more often in the urban landscape. It is very well adapted for residential landscapes, having tolerated extended periods of drought and neglect. Plant it in a prominent location to display the striking habit.

Texas Persimmon should be grown in full sun on well-drained soils and is often found on alkaline sites, but tolerates slightly acid soil. It is especially tolerant of drought and neglect and should need only occasional fertilization every year or two. It can be skillfully thinned into a striking specimen to display the beautiful trunk form.

Few pests or diseases, none normally serious.

Diospyros virginiana Common Persimmon

Pronunciation: die-OS-purr-os vir-gin-ee-A-nah

ZONES: 4B, 5A, 5B, 6A, 6B, 7A, 7B, 8A, 8B, 9A, 9B

HEIGHT: 30 to 60 feet

WIDTH: 20 to 35 feet

FRUIT: on female trees only; round; 1 to 2"; fleshy; orange; attracts mammals; edible after first fall freeze; causes messy litter; persistent; showy

GROWTH RATE: moderate; long-lived in the north

HABIT: oval; upright; open to moderate density; pleasing, irregular silhouette; medium texture

FAMILY: *Ebenaceae*

LEAF COLOR: showy red to yellow fall color; dark green in summer

LEAF DESCRIPTION: alternate; simple; serrate margin; oval to ovate; deciduous; 3 to 5" long

FLOWER DESCRIPTION: white; not showy; spring

EXPOSURE/CULTURE:

Light Requirement: part to full sun

Soil Tolerances: all textures; slightly alkaline to acidic; occasionally wet soil; some salt; drought

Aerosol Salt Tolerance: moderate to good

PEST PROBLEMS: resistant to susceptible

BARK/STEMS:

Nearly black and blocky, with beautiful form; branches mostly horizontal to upright; usually grown with a single leader; no thorns

Pruning Requirement: needs little or no pruning to develop strong trunk and branch structure

Limb Breakage: resistant

Current Year Twig: dark brown to gray; reddish; slender

LANDSCAPE NOTES:

Origin: eastern North America

Usage: fruit; buffer strip; highway; reclamation; specimen; urban tolerant

An excellent medium-sized tree, Common Persimmon is an interesting, somewhat irregularly shaped native tree, for possible naturalizing in yards or parks. Bark is gray or black and distinctly blocky. Fall color can be a spectacular red in zones 4 through 8A. It is well adapted to cities, but presents a problem with fruit litter, attracting flies and scavengers such as opossums and other mammals. Its mature height can reach 60 feet, a trunk 2 feet thick, but it is commonly much shorter in landscapes. The trunk typically ascends up through the crown in a curved but very dominant fashion, rarely producing double or multiple leaders. Lateral branches are typically much smaller in diameter than the trunk.

Common Persimmon prefers moist, well-drained bottomland or sandy soils but is also very drought- and urban tolerant—truly an amazing tree in its adaptability to any site conditions, including alkaline soil. It is seen colonizing old fields as a volunteer tree but grows slowly on dry sites. Its fruit is an edible berry that usually ripens after frost, although some cultivars do not require the frost treatment to ripen. Before ripening, however, the fruit is decidedly astringent and not edible. Most American cultivars require both male and female trees for proper fruiting.

Except for cleaning up the messy fruit, Common Persimmon maintenance is quite easy, and the tree could be planted more, especially male clones. Locate it where the slimy fruit will not fall on sidewalks and cause people to slip and fall. Because transplantation is difficult due to a coarsely branched root system, Persimmon trees should be balled and burlapped when young or planted from containers. The wood, sometimes used for golf club heads, is very hard and almost black.

Leaf-spot disease may limit its use in the southern United States. Black spots appear on the leaves and premature defoliation occurs. It is also susceptible to a vascular wilt in the southern United States which can be devastating to established trees. Tent caterpillars defoliate trees. Suckers appear from the roots of some trees.

Elaeagnus angustifolia Russian-Olive

Pronunciation: ee-lee-AG-nuss ann-guss-tih-FOE-lee-ah

ZONES: 3A, 3B, 4A, 4B, 5A, 5B, 6A, 6B, 7A, 7B, 8A, 8B

HEIGHT: 15 to 20 feet

WIDTH: 15 to 20 feet

FRUIT: round; 0.5"; fleshy; yellow; attracts wildlife; not showy; edible; no significant litter

GROWTH RATE: rapid-growing; short-lived

HABIT: round; open; irregular silhouette; fine texture

FAMILY: *Elaeagnaceae*

LEAF COLOR: silver in summer; no fall color change

LEAF DESCRIPTION: alternate; simple; entire; lanceolate to oblong; deciduous; 1 to 3" long

FLOWER DESCRIPTION: white; pleasant fragrance; not showy; spring

EXPOSURE/CULTURE:

Light Requirement: must be grown in full sun

Soil Tolerances: loam; sand; alkaline to acidic; salt; drought

Aerosol Salt Tolerance: very high

PEST PROBLEMS: sensitive

BARK/STEMS:

Branches droop as tree grows and will require pruning for vehicular or pedestrian clearance beneath canopy; routinely grown with multiple trunks; thorns are present on branches

Pruning Requirement: requires early pruning to develop upright structure

Limb Breakage: mostly resistant

Current Year Twig: gray; slender

LANDSCAPE NOTES:

Origin: central Europe to central Asia

Usage: parking lot island; buffer strip; highway; reclamation; specimen

Russian-Olive grows in an open, somewhat irregular globe shape 15 to 20 feet tall and wide (occasionally larger) and has striking, silvery-gray foliage. It has a rapid growth rate when young, becoming moderate with age. Shaping and training the leader and major branches is needed to develop a well-formed tree. Otherwise, an open messy canopy develops.

This thorny tree transplants well but susceptibility to canker diseases and verticillium wilt make it undesirable for large-scale landscape plantings, particularly in the moist, eastern climate. Used as an occasional accent, however, the tree attracts attention due to the color of the foliage. Commonly planted in dry climates where it performs well due to drought tolerance and poor, drier soil. Plant in the area of the landscape with the best drainage. It has a use along highways in more humid climates.

This tough tree fixes its own nitrogen and needs a sunny location and is tolerant of alkaline soil, drought, and coastal conditions, including salty soil. It probably is better suited for dry climates of the plains states than the moist climate typical of the eastern United States.

Large branches, or the entire tree, are often killed by verticillium wilt in the east in poorly drained sites. It is better adapted to drier climates. The most likely insect problem may be scale. It has escaped cultivation in many parts of the western United States along stream banks and moist areas. It is considered an invasive weed in some western regions, but is well-liked elsewhere.

Eriobotrya deflexa Bronze Loquat

Pronunciation: ear-ee-o-BOT-ree-ah dee-FLEX-ah

ZONES: 9B, 10A, 10B, 11

HEIGHT: 15 to 20 feet

WIDTH: 15 to 25 feet

FRUIT: oval; round; less than 0.5"; fleshy; orange to yellow; attracts wildlife; slight litter problem

GROWTH RATE: moderate; short to moderate life span

HABIT: oval; round; vase shape; dense; nearly symmetrical; coarse texture

FAMILY: *Rosaceae*

LEAF COLOR: broadleaf evergreen; emerges red turning shiny dark green

LEAF DESCRIPTION: alternate; simple; dentate to serrate margin; 8 to 10" long; oval to oblong

FLOWER DESCRIPTION: white; pleasant fragrance; showy; spring

EXPOSURE/CULTURE:

Light Requirement: part shade to full sun

Soil Tolerances: all textures; alkaline to acidic; drought

PEST PROBLEMS: somewhat susceptible

BARK/STEMS:

Branches droop as tree grows and will require pruning for vehicular or pedestrian clearance beneath canopy; should be grown with a short, single leader; no thorns

Pruning Requirement: requires pruning to develop strong structure; remove root suckers and branches with included bark

Limb Breakage: resistant

Current Year Twig: green; stout

CULTIVARS AND SPECIES:

'Bronze Improved' has bronze-colored new growth.

LANDSCAPE NOTES:

Origin: Japan

Usage: container; espalier; hedge; buffer strip; highway; near a deck or patio; standard; parking lot island; specimen; street tree

The leaves on this adaptable, small tree emerge bright red-bronze or coppery-colored, but eventually turn dark green. The small white, fragrant flowers are produced on terminal panicles in spring and are followed by small, inedible fruits.

Its neat habit and compact growth make Loquat an ideal specimen or patio shade tree, and it can be used as a street tree or median strip tree in areas where overhead space is limited. Branches will have to be trained to grow up, as they tend to droop with time under the weight of the developing branch. It is not suited for planting next to the street if trucks pass close to the tree, as adequate clearance is not possible, but it is successful in wide median strips. It espaliers well against a sunny wall and makes a good screen due to its dense canopy. Sprouts along the trunk can be a maintenance nuisance.

Providing best fruit and form when grown in full sun, Loquat can tolerate partial shade and a variety of well-drained soils. It grows well on soils with a high pH and maintains the characteristic dark green foliage. Loquat should be well watered until established, but can then survive periodic droughts. Do not overfertilize, as this could increase sensitivity to fire blight disease. It performs well along the coast with some protection from salty air.

Fireblight kills entire branches but trees often survive. Roots rot on wet soils. To reduce fireblight problems, provide good air circulation and keep away from other fireblight hosts, such as Pyracantha and pears. If leaves and stems blacken from the top downward, prune back one foot or more into healthy wood. Sterilize shears between cuts so as not to spread the disease.

Eriobotrya japonica Loquat

Pronunciation: ear-ee-o-BOT-ree-ah jah-PON-eh-cah

ZONES: (7B), 8A, 8B, 9A, 9B, 10A, 10B, 11

HEIGHT: 20 to 30 feet

WIDTH: 30 to 35 feet

FRUIT: oval; round; 1 to 2"; fleshy; orange or yellow; attracts wildlife; edible; causes significant litter; showy

GROWTH RATE: moderate; short to moderate life span

HABIT: round; vase shape; dense; symmetrical; coarse texture

FAMILY: *Rosaceae*

LEAF COLOR: dark green with rusty undersides; broadleaf evergreen

LEAF DESCRIPTION: alternate; simple; serrate margin; oval to oblong; 8 to 10"

FLOWER DESCRIPTION: creamy white; fall and winter; pleasant fragrance; showy clusters

EXPOSURE/CULTURE:

Light Requirement: part shade to full sun

Soil Tolerances: all textures; alkaline to acidic; moderate salt; drought

Aerosol Salt Tolerance: moderate

PEST PROBLEMS: other than fire blight, resistant

BARK/STEMS:

Bark is thin and easily damaged by mechanical impact; branches droop as tree grows and will require pruning for vehicular or pedestrian clearance beneath canopy; should be grown with a single, short leader; no thorns

Pruning Requirement: requires pruning to develop strong structure; remove branches with included bark

Limb Breakage: resistant

Current Year Twig: gray; stout

LANDSCAPE NOTES:

Origin: China

Usage: container or above-ground planter; espalier; fruit; hedge; buffer strip; highway; near a deck or patio; standard; specimen; street tree

Loquat can reach 25 to 30 feet in height in the shade but is frequently seen 15 feet tall with a 15- to 25-foot-spread in a sunny location. Fragrant clusters of creamy white flowers are produced in fall, followed by the delicious, brightly colored, winter fruit. Fruit rarely sets in zones 7 and 8A.

Its neat habit and compact growth make Loquat an ideal specimen or patio shade tree, and it can be used as a street tree or median strip tree in areas where overhead space is limited. An adequate clear trunk with upright branches must be developed early in the life of the tree to provide for vehicle clearance, though. It is not suited for planting next to the street if trucks would pass close to the tree, as adequate clearance is not possible, but it is successful in wide median strips. It also blends well into informal shrubbery borders, and the fruit is attractive to wildlife. Sprouts along the trunk can be a maintenance nuisance.

Providing best fruit and form when grown in full sun, Loquat can tolerate partial shade and a variety of well-drained soils. Clay soil is acceptable as long as there is sufficient slope to allow surface water to run away from the root system. It often looks best in the southern portion of its range when given some shade in the afternoon, especially if it is not irrigated. Do not overfertilize, as this could increase sensitivity to fire blight disease.

To reduce fireblight problems, provide good air circulation and keep away from other fireblight hosts, such as Pyracantha and pears. Root rot occurs on wet soils. Young seedlings can be seen sprouting beneath the canopy.

Erythrina crista-galli Cockspur Coral Tree

Pronunciation: err-eh-THRY-nah kris-tah-GALL-ee

ZONES: 10A, 10B, 11

HEIGHT: 15 to 20 feet

WIDTH: 15 to 20 feet

FRUIT: no significant litter

GROWTH RATE: moderate; long-lived

HABIT: round and spreading; dense; irregular, billowing canopy; medium texture

FAMILY: *Leguminosae*

LEAF COLOR: dark green in summer

LEAF DESCRIPTION: alternate; trifoliate; entire margin; deltoid to ovate leaflets; briefly deciduous; 2 to 6" long leaflets

FLOWER DESCRIPTION: orange-red; showy; spring, periodically through fall

EXPOSURE/CULTURE:

Light Requirement: grows best in full sun

Soil Tolerances: all textures; alkaline to acidic; drought tolerant

PEST PROBLEMS: resistant

BARK/STEMS:

Bark is thin and easily damaged by mechanical impact; branches grow horizontally and droop on older trees; routinely grown with multiple trunks, or can be trained to a short, single trunk

Pruning Requirement: requires pruning to develop an upright or strong structure; remove, shorten, thin, or cable branches with included bark; slow growth rate on aggressive, young main branches by thinning or heading back; dead flower stalks and small twigs can be pruned on canopy margins

Limb Breakage: somewhat susceptible at crotches

Current Year Twig: brown to gray; thick

CULTIVARS AND SPECIES:

E. caffre is also a popular Coral Tree often used in California with similar cultural requirements. It is often used as *E. crista-galla* is used.

LANDSCAPE NOTES:

Origin: Argentinia, Brazil, Paraguay

Usage: specimen; highway median; buffer strip; near a deck or patio

This slow-growing, deciduous tree with up to six-inch-long leaves forms a low, broad canopy. Most older specimens are wider than tall. In spring, before the leaves appear, Cockspur Coral Tree is decorated with showy red blossoms, each two inches long and arranged in dense, six-inch-long racemes. They literally cover the canopy with color periodically during the growing season. The real beauty of the tree shows during the brief deciduous period in winter. The sinuous branches ascend into the canopy creating a coarse silhouette that looks mystical when lit from below the canopy at night.

The small size of Cockspur Coral Tree makes it suited for planting in small residential lots as well as in parks, golf courses, and other large-scale landscapes. Multiple trunks ascend from the lower portion of the main trunk, giving rise to a wide-spreading canopy that casts dense shade. Lower branches eventually droop to the ground if they are allowed to remain on the tree.

Imagine a roadway lined with these trees spaced 20 to 30 feet apart. They would have to be set back from the edge of the pavement because of the low canopy. Planted 10 to 15 feet away from the curb, they would help create a corridor, inviting drivers through the neighborhood. They will not form a canopy over the street.

Borers may infest weakened trees. Caterpillars can eat significant amounts of foliage. Included bark can form in the crotch of main branches, weakening the tree.

Erythrina variegata var. *orientalis* Coral Tree

Pronunciation: err-eh-THRY-nah vair-ee-ah-GAY-tah [variety] or-ee-en-TAY-liss

ZONES: 10A, 10B, 11

HEIGHT: 50 to 60 feet

WIDTH: 40 to 50 feet

FRUIT: pod, 12" long, no significant litter

GROWTH RATE: moderately fast; long-lived

HABIT: rounded vase; dense; symmetrical; coarse texture

FAMILY: *Leguminosae*

LEAF COLOR: green and yellow variegated; no fall color change

LEAF DESCRIPTION: alternate; trifoliate; entire margin; deltoid to ovate leaflets; deciduous; 2 to 6" long leaflets

FLOWER DESCRIPTION: red; showy; spring

EXPOSURE/CULTURE:

Light Requirement: grows best in full sun

Soil Tolerances: all textures; alkaline to acidic; drought

Aerosol Salt Tolerance: moderate

PEST PROBLEMS: resistant

BARK/STEMS:

Very striking; bark is thin and easily damaged by mechanical impact; branches grow mostly upright after some initial training, lower branches droop; routinely grown with multiple trunks; thorns on branches

Pruning Requirement: requires pruning to develop upright or strong structure; remove, thin, drop-crotch, or cable branches with included bark

Limb Breakage: susceptible

Current Year Twig: brown to gray; thick

LANDSCAPE NOTES:

Origin: Philippine Islands, Indonesia

Usage: shade tree; specimen

This fast-growing, deciduous tree with variegated six-inch-long leaves creates a broad canopy. In spring, before the leaves appear, Coral Tree is decorated with showy red blossoms, each flower 2.5 inches long and arranged in dense, 6-inch-long racemes. These blooms are followed by 12-inch-long, red-brown seedpods that contain poisonous seeds. The real beauty of the tree shows during the brief deciduous period in winter. The sinuous branches ascend into the canopy, creating one of the most graceful silhouettes of any tropical tree.

The large size of Coral Tree makes it suited for planting in parks, golf courses, and other large-scale landscapes. Multiple trunks ascend from the lower portion of the main trunk, giving rise to a wide-spreading canopy that casts dense shade. Lower branches eventually droop to the ground if they are allowed to remain on the tree. Give this tree plenty of room to develop, as the canopy is large and the tree looks wonderful if the canopy shape does not have to be altered. Since the trunk often flares at the base, plant it at least 10 feet from a sidewalk or driveway. Large buttress roots form at the base of the trunk.

Coral Tree should be grown in full sun on well-drained soil. Trees are highly drought-tolerant and moderately salt-tolerant.

Borers may infest weakened trees. Caterpillars can eat significant amounts of foliage. Included bark can form in the crotch of main branches, weakening the tree.

Eucalyptus citriodora Lemon-Scented Gum

Pronunciation: you-cah-LIP-tuss sih-tree-oh-DOOR-ah

ZONES: 9B, 10A, 10B, 11

HEIGHT: 75 to 100 feet

WIDTH: 25 to 30 feet

FRUIT: round; 0.5"; dry and hard; brown to green; no significant litter; persistent; somewhat showy

GROWTH RATE: rapid-growing; long-lived

HABIT: oval; open; irregular silhouette; medium to fine texture

FAMILY: *Myrtaceae*

LEAF COLOR: green; evergreen

LEAF DESCRIPTION: alternate; simple; entire margin; lanceolate; 4 to 7" long; lemon-scented

FLOWER DESCRIPTION: white; not showy; winter

EXPOSURE/CULTURE:

Light Requirement: best in full sun

Soil Tolerances: all textures; alkaline to acidic; moderate drought

Aerosol Salt Tolerance: moderate

PEST PROBLEMS: resistant

BARK/STEMS:

Smooth, whitish-yellow bark; branches grow upright and will not droop; should be grown with a single leader; trunks often curve up into canopy; no thorns

Pruning Requirement: requires some early pruning to develop one leader and a strong trunk; trunk needs staking early, or prune branches back to lighten canopy and strengthen trunk; choose branches without included bark for major, structural limbs; prune only in winter or early spring to minimize borer infestation

Limb Breakage: mostly resistant

Current Year Twig: brownish-red; slender

LANDSCAPE NOTES:

Origin: Australia

Usage: buffer strip; parking lot island; street tree; highway; specimen

A native of Australia, Gum grows best on the western coast of the United States and is seldom successful in the interior, primarily due to cold sensitivity. Foliage of Eucalyptus is aromatic, with few distinguishable differences between juvenile and mature leaves. The bark on young and old trees is extremely smooth and soft to the touch. This is a popular tree grown almost everywhere in parts of southern California.

As this is one of the more slender-trunked Eucalyptus trees, it can be used where larger ones cannot be planted. Unlike other Gum trees, it does not have a reputation for displacing pavement and sidewalks, as do the thick-trunked Gum trees. Therefore, it is probably one of the best Gum trees for planting in parking lot islands or along streets.

One common problem with this tree is the tendency to overprune it. Although thinning this tall, slender tree reduces the chance that wind will break it apart in a open landscape, removing all interior branches makes the tree look like several lions' tails have been inserted into the trunk. This malpractice should not be performed on the tree because it actually weakens the branches. The canopy can be properly thinned by removing branches from all along the main limbs, especially at the tips of the limbs.

Long-horned borers and crown gall are Eucalyptus's major problems, but this tree is resistant to both. Other pests of Eucalyptus include leaf spots, psyllids, aphids, mealybugs, scales, mites, and caterpillars. Psyllids disfigure the leaves and can be quite a problem. Eucalyptus are resistant to armillaria root rot and to verticillium wilt. They are susceptible to powdery mildew and to root rots.

Eucalyptus ficifolia Red-Flowering Gum

Pronunciation: you-cah-LIP-tuss fiss-eh-FOE-lee-ah

ZONES: 10A, 10B, 11

HEIGHT: 30 to 40 feet

WIDTH: 20 to 25 feet

FRUIT: round; 0.5"; dry and hard; brown to green; no significant litter; persistent; somewhat showy

GROWTH RATE: moderate; long-lived

HABIT: rounded to oval; moderate density; irregular silhouette; medium texture

FAMILY: *Myrtaceae*

LEAF COLOR: evergreen; dark green

LEAF DESCRIPTION: alternate; simple; entire margin; lanceolate; 2 to 4" long

FLOWER DESCRIPTION: pink to red, variable; very showy and long-lasting

EXPOSURE/CULTURE:

Light Requirement: best in full sun

Soil Tolerances: all textures; alkaline to acidic; moderate drought

PEST PROBLEMS: resistant

BARK/STEMS:

Branches grow mostly upright and will not droop; showy trunk; should be grown with a single leader; no thorns

Pruning Requirement: requires some early pruning to develop one leader; prune only in winter or early spring to minimize borer attraction

Limb Breakage: somewhat susceptible

Current Year Twig: brownish-red; slender

LANDSCAPE NOTES:

Origin: western Australia

Usage: buffer strip; parking lot islands; street tree; highway; specimen; shade

A native of Australia, Red-Flowering Gum grows best on the western coast of the United States and is seldom successful in the interior, primarily due to cold sensitivity. Its flowers are spectacular. Foliage of Eucalyptus is aromatic, with few distinguishing differences between juvenile and mature leaves. The leaves bear some resemblance to several ficus species. The bark on older stems becomes fibrous and furrowed and is quite ornamental.

Because this is one of the smaller Eucalyptus trees, it can be used where others cannot. Place it in the front lawn area or in a landscape bed along with low-growing groundcovers or shrubs. The flower display shows off nicely along the highways and at interchanges. It can also be used along residential streets as a medium-sized shade tree. It adapts well to the dry heat of parking lots and stays fairly small. This makes it well suited for planting near curbs and sidewalks, as roots are not as aggressive as on large Eucalyptus trees. It could be tried as a container plant in cooler regions, if wintered indoors.

See *E. citriodora* for limitations and possible problems. *E. torelliana* is used in Florida as a shade tree.

Eucommia ulmoides Hardy Rubber Tree

Pronunciation: you-COMB-ee-ah ul-MOY-dees

ZONES: 4B, 5A, 5B, 6A, 6B, 7A, 7B

HEIGHT: 40 to 60 feet

WIDTH: 50 to 70 feet

FRUIT: typically absent; oval; 1"; fleshy; no significant litter

GROWTH RATE: slow-growing; moderate life span

HABIT: spreading; round; dense; symmetrical; medium texture

FAMILY: *Eucommiaceae*

LEAF COLOR: shiny dark green in summer; little to no fall color change

LEAF DESCRIPTION: alternate; simple; serrate margin; oval to oblong or ovate; deciduous; 3 to 8" long

FLOWER DESCRIPTION: brown; not showy

EXPOSURE/CULTURE:

Light Requirement: partial to full sun

Soil Tolerances: all textures; alkaline to acidic; moderate salt; drought; good drainage is essential

Aerosol Salt Tolerance: moderate

PEST PROBLEMS: pest-free

BARK/STEMS:

Branches grow mostly upright and will not droop; showy trunk; should be grown with a single leader; no thorns

Pruning Requirement: needs early pruning to develop strong structure; space branches along trunk

Limb Breakage: resistant

Current Year Twig: brown; medium thickness

LANDSCAPE NOTES:

Origin: The former Soviet Union to south China

Usage: container or above-ground planter; parking lot island; buffer strip; highway; shade; specimen; sidewalk cutout; street tree; urban tolerant

If you want green leaves on a tree during a severe drought, this is your tree! This little-known but urban-tough, very attractive, slow-growing, broad deciduous tree has a dense, symmetrical oval to rounded crown and low-branched silhouette, making it ideal for use as a specimen, shade, or street tree. The thin, glossy leaves are almost totally resistant to pests and disease and remain an attractive dark green throughout the summer, changing only to a paler green before dropping in early fall. The foliage is quite striking and appears to glimmer in the moonlight or when lit from above. Branches ascend, forming an upright silhouette in winter. The inconspicuous blooms are followed by the production of small, 1.5-inch-long, flat, winged seeds. Only one or two corrective prunings at an early age is all that is normally needed to develop good structure in the crown.

The Chinese have used the Hardy Rubber Tree for more than 2,000 years for its medicinal value. Trees there rarely reach mature size, as they are harvested regularly and stripped of their ridged or furrowed, gray-brown bark. Hardy Rubber Tree may be the only tree that grows in cold climates from which a rubber product can be obtained. This rubbery substance may be visible as thin strands bridging two sections of a torn leaf.

Hardy Rubber Tree should be grown in full sun on moist soil, but when well established tolerates extensive drought. Trees have been growing in many regions for years without irrigation and have survived extreme drought in very poor, clay soil in the full sun—but they grow slowly. They should be grown and tried more often in urban areas such as in highway medians, along streets, and as medium-sized shade trees. Growth rate appears to be quite slow, but could probably be improved with adequate irrigation. The tree is adapted to high soil pH.

Amazingly free of problems, but do not plant it in poorly drained soil.

Eugenia spp. Stopper

Pronunciation: you-GEE-nee-ah species

ZONES: 10B, 11

HEIGHT: 15 to 30 feet

WIDTH: 15 to 25 feet

FRUIT: round; 0.5 to 1"; fleshy; black or red; attracts wildlife; edible; causes some litter; showy

GROWTH RATE: slow to moderate; moderate life span

HABIT: oval to rounded, upright vase; dense; symmetrical; medium textures

FAMILY: *Myrtaceae*

LEAF COLOR: medium to dark green; evergreen

LEAF DESCRIPTION: opposite; simple; entire margin; ovate; 2 to 4" long

FLOWER DESCRIPTION: white; some are showy; spring and summer

EXPOSURE/CULTURE:

Light Requirement: mostly shade to full sun

Soil Tolerances: all textures; alkaline to acidic; moderate salt; drought

Aerosol Salt Tolerance: moderate

PEST PROBLEMS: resistant to lethal pests

BARK/STEMS:

Smooth, showy trunk; branches grow mostly upright and will not droop; routinely grown with multiple trunks, but can be trained to grow with a single trunk; no thorns

Pruning Requirement: requires pruning to develop single trunk

Limb Breakage: resistant

Current Year Twig: brown; slender

CULTIVARS AND SPECIES:

There are many species. *E. malaccensis* has beautiful pink flowers and is used as a street and ornamental tree in Puerto Rico and the Virgin Islands.

LANDSCAPE NOTES:

Origin: southern Florida to South America

Usage: container or above-ground planter; hedge; screen; parking lot island; buffer strip; highway; near a deck or patio; reclamation; standard; shade; specimen; sidewalk cutout; street tree

Eugenia is a large group of plants, including evergreen trees and shrubs, some of which have been reclassified to the genus *Syzygium*. The evergreen leaves are firm and glossy, and the flowers white and somewhat showy. The dried buds of *E. aromatica* (*Syzygium aromaticum*) become the fragrant "herb" cloves. The flowers are followed by the production of berries, some types of which are edible. All these traits—the attractive foliage, flowers, and berries—help make Eugenia a popular landscape choice in warm climates, such as California, Florida, and Hawaii. *E. confusa* (Ironwood, Red Stopper) is native to Florida, grows to about 35 feet and is well suited for street tree and parking lot planting. *E. foetida* (Spanish Stopper) is also native and grows to about 15 feet tall.

The smooth, brown to gray, mottled bark and tight canopy of fine-textured leaves make Eugenia well suited for planting as a specimen in any yard. Trees can be trained in the nursery to one central trunk or encouraged to develop multiple trunks. They create shade for a patio or deck, but will not grow to the large, often overpowering size of a Ficus tree. They are often used along streets, in highway medians, and in parking lots because they adapt to small soil spaces. Street and parking lot trees are often specified to have one trunk to allow vehicle clearance beneath the crown. Multiple-trunked trees are often specified for specimen planting so the beautiful bark can be displayed.

Eugenia should be grown in full sun or part shade on well-drained soil. Once established in the landscape, they are drought-tolerant, requiring little, if any, irrigation.

Foliage insects, especially psyllids, can disfigure leaves. Psyllids limit the tree's usefulness in parts of California, however, a wasp has been introduced that appears to provide a biological control.

Evodia danielii Korean Evodia, Bebe Tree

Pronunciation: ee-VOE-dee-ah dan-ee-ELL-ee-eye

ZONES: 4A, 4B, 5A, 5B, 6A, 6B, 7A, 7B, 8A

HEIGHT: 25 to 30 feet

WIDTH: 25 to 30 feet

FRUIT: round; less than 0.5"; dry; red to black; persistent; very showy

GROWTH RATE: slow-growing; moderate life span

HABIT: rounded vase; open; moderate density; nearly symmetrical; medium texture

FAMILY: *Rutaceae*

LEAF COLOR: dark green in summer; bland yellow fall color

LEAF DESCRIPTION: opposite; odd pinnately compound; crenate margin; oblong to ovate; deciduous; 2 to 5" long leaflets

FLOWER DESCRIPTION: white; pleasant fragrance; very showy; summer; attracts bees

EXPOSURE/CULTURE:

Light Requirement: part to full sun

Soil Tolerances: all textures; alkaline to acidic; drought

Aerosol Salt Tolerance: moderate

PEST PROBLEMS: pest-free

BARK/STEMS:

Branches droop as tree grows and will require pruning for vehicular or pedestrian clearance beneath canopy; grown with multiple trunks, but can be trained to grow with a short, single trunk; no thorns

Pruning Requirement: requires pruning to develop strong structure and a single leader

Limb Breakage: mostly resistant; some breakage at crotches could occur if included bark develops

Current Year Twig: brown to gray; medium to thick

LANDSCAPE NOTES:

Origin: north China and Korea

Usage: large parking lot island; buffer strip; near a deck or patio; small shade tree; specimen; street tree

Korean Evodia is a little-known but highly desirable small, deciduous tree. Several large-diameter limbs eventually dominate the canopy. The pinnately compound leaves cast a light shade below the tree and remain attractive and disease-free throughout the summer. The leaves often drop in autumn while still green, though some trees have been known to provide a display of clear yellow fall foliage. In early summer, many showy, flat-topped flower clusters appear; the white, fragrant blossoms attract a multitude of bees. *E. hupehensis* is similar but less cold-hardy.

Evodia should be grown and used more often as a small tree, with its mound-shaped crown providing shade for small areas. Well-suited for patios and other small areas, including residences, Evodia deserves to be tried more as an urban tree. It could be suited for planting close to power and telephone lines, as height increase slows down with age, giving way to more horizontal growth. Trunks are covered with smooth gray bark and originate close to the ground (three to five feet) on unpruned trees. This can form a beautiful multi-trunked effect, if desired.

According to one report, the wood is brittle and subject to storm damage, and the trees may be relatively short-lived (15 to 40 years). Others have observed many trees throughout the south that appear to be doing fine. Prune so that main branches remain less than half the diameter of the trunk, to increase longevity by minimizing formation of included bark and reducing branch breakage.

Korean Evodia should be grown in full sun on moist but well-drained, fertile soil. It tolerates drought once established and will grow nicely in poor soil, including moderately drained clay.

Fagus grandifolia American Beech

Pronunciation: FAY-gus gran-deh-FOE-lee-ah

ZONES: 3A, 3B, 4A, 4B, 5A, 5B, 6A, 6B, 7A, 7B, 8A, 8B

HEIGHT: 50 to 75 feet

WIDTH: 40 to 80 feet

FRUIT: oval; 0.5 to 1"; dry or hard; brown; attracts wildlife; edible; no litter

GROWTH RATE: slow-growing; long-lived

HABIT: oval; pyramidal; dense; symmetrical; medium texture

FAMILY: *Fagaceae*

LEAF COLOR: glossy green in summer; showy copper fall color

LEAF DESCRIPTION: alternate; simple; serrate margin; oblong to ovate; deciduous; 2 to 5" long

FLOWER DESCRIPTION: not showy; spring

EXPOSURE/CULTURE:

Light Requirement: part shade to full sun, tolerates nearly full shade

Soil Tolerances: loam; sand; acidic; salt- and drought-sensitive

Aerosol Salt Tolerance: low

PEST PROBLEMS: resistant

BARK/STEMS:

Very striking, smooth gray bark; branches droop as tree grows and will require pruning for vehicular or pedestrian clearance beneath canopy; should be grown with a single leader; no thorns

Pruning Requirement: needs little pruning to develop strong trunk and branch structure; remove upright branches with included bark; plants pruned in spring bleed sap

Limb Breakage: resistant

Current Year Twig: brown to gray; slender

LANDSCAPE NOTES:

Origin: East of the Mississippi River

Usage: clipped hedge; shade tree; specimen

This massive tree will slowly reach a height and spread of 50 or more feet. Forest-grown trees can reach up to 120 feet. The tree is naturally low-branched with attractive leaves providing deep, inviting shade. Little grows in the dense shade of a Beech tree, but if low branches are left on the tree no groundcover or grass is needed. Some brown leaves not blown off by the wind are held late into the winter. The thin, smooth, gray bark looks like elephant skin on older specimens. The four tiny nuts in each spiny bur of this American native are much prized by birds and various mammals, including humans. The tree is very resistant to decay under water, so it was used to make water wheels in colonial times. The wood is also used for tool handles, chairs, cutting boards, and charcoal.

Branch structure is typically very sturdy, with one central trunk growing straight up through the canopy. Many branches are often horizontally oriented, small in diameter, and well secured to the trunk.

American Beech needs a loose, acid soil that is well drained yet capable of retaining enough moisture for its shallow root system. The root zone should be free of compaction and grass competition. Plant it on large estates, along entry drives to large commercial properties, or as a specimen in a park or campus setting. Young trees collected from the woods are difficult to transplant and are best simply preserved where they naturally occur. Plant from nursery-bought stock in the spring. Give this tree plenty of above-ground and below-ground rooting space.

Drought-sensitive. Needs open soil space to expand its root system. Many insects and diseases are known. Nectria canker and beech scale are probably the most devastating, especially in the northeastern United States. Sprouts often emerge from roots.

Fagus sylvatica European Beech
Pronunciation: FAY-gus sill-VAH-tee-cah

ZONES: 4A, 4B, 5A, 5B, 6A, 6B, 7A, 7B

HEIGHT: 50 to 75 feet

WIDTH: 40 to 60 feet

FRUIT: oval; 0.5 to 1"; dry; brown; attracts wildlife; edible; no significant litter

GROWTH RATE: slow growing; long-lived

HABIT: oval; pyramidal; dense; symmetrical; fine to medium texture

FAMILY: *Fagaceae*

***Fagus sylvatica* 'Atropunicea'**
Purple European Beach
foliage emerges as a light bronze and darkens within two weeks to a very dark purple.

***Fagus sylvatica* 'Pendula'**
Weeping European Beach
leave plenty of room for this tree to expand. Its shape would be spoiled by extensive pruning.

LEAF COLOR: light green in summer; showy copper fall color

LEAF DESCRIPTION: alternate; simple; entire, undulating margin; oval to ovate; deciduous; 3 to 5" long

FLOWER DESCRIPTION: not showy; spring

EXPOSURE/CULTURE:

Light Requirement: part shade to full sun

Soil Tolerances: loam; sand; slightly alkaline to acidic; salt-sensitive, moderate drought

Aerosol Salt Tolerance: low

PEST PROBLEMS: except for cankers, resistant

BARK/STEMS:

Smooth, gray bark; bark is thin and easily damaged by mechanical impact; branches droop as tree grows and will require pruning for vehicular or pedestrian clearance beneath canopy; should be grown with a single leader; keep low branches to shade trunk; no thorns

Pruning Requirement: needs little pruning to develop strong structure; remove upright branches with included bark; plants pruned in spring bleed sap

Limb Breakage: resistant

Current Year Twig: brown to gray; slender

CULTIVARS AND SPECIES:

There are many cultivars displaying different foilage colors, shapes, and canopy forms.

LANDSCAPE NOTES:

Origin: Europe

Usage: hedge; shade; specimen

The bright gray trunk ascends straight up into the canopy of light-textured foliage. Branches originate low on the trunk and normally sweep the ground in a graceful fashion. Lower branches on some forms grow upright and compete with the leader. These branches should be removed when the tree is young to increase the tree's longevity. Included bark often forms in the crotches of

Continued

these aggressive branches. Many of the lateral branches are horizontally oriented, small in diameter, and well secured to the trunk.

Though not a street tree, European Beech makes one of the finest specimens of all those available in North America for large-scale landscapes with plenty of soil space for root expansion. Except during a summer drought, it is hard to find a time of year when this tree is not attractive. Allow plenty of room for expansion of the canopy so it can develop the natural shape. The tree grows slowly, recovers slowly from transplanting, and prefers a sunny location in a moist, light soil. European Beech is somewhat tolerant of heat and dry soil, but it is best to locate it where it will receive adequate moisture. Transplant in spring for best survival.

European Beech needs a loose, acid soil that is well drained yet capable of retaining enough moisture for its shallow root system. The root zone should be free of compaction and grass competition. Plant it on large estates along entry drives to large commercial properties or as a specimen in a park or campus setting.

Leaves scorch in dry summers. There are many possible pests, including aphids, borers, and trunk diseases.

Ficus aurea Strangler Fig

Pronunciation: FYE-kus AH-ree-ah

ZONES: 10B, 11

HEIGHT: 50 to 60 feet

WIDTH: 50 to 90 feet

FRUIT: oval; round; less than 0.5"; fleshy; green to yellow; causes a mushy litter

GROWTH RATE: rapid-growing; long-lived

HABIT: round; upright; dense; irregular silhouette; coarse texture

FAMILY: *Moraceae*

LEAF COLOR: dark green; broadleaf evergreen

LEAF DESCRIPTION: alternate; simple; entire, undulating margin; 4 to 8" long; oval to ovate

FLOWER DESCRIPTION: white; not showy

EXPOSURE/CULTURE:

Light Requirement: shade to full sun

Soil Tolerances: all textures; alkaline to acidic; wet soil; salt; drought

Aerosol Salt Tolerance: moderate

PEST PROBLEMS: resistant

BARK/STEMS:

Bark is thin and easily damaged by mechanical impact; branches droop as tree grows and will require pruning for vehicular or pedestrian clearance beneath canopy; showy trunk; should be grown with a single leader; no thorns

Pruning Requirement: requires pruning to develop strong structure; prevent formation of included bark by thinning major branches; remove, prune, or cable branches with included bark; retain some aerial roots so they root in the ground providing support for horizontal branches

Limb Breakage: resistant

Current Year Twig: green; medium thickness

LANDSCAPE NOTES:

Origin: south Florida

Usage: reclamation; shade; buffer strip

Often starting out in nature as an epiphyte nestled in the limbs of another tree, the native Strangler Fig is vine-like while young, later strangling its host with heavy aerial roots and eventually becoming a self-supporting, independent tree. The gray trunk can become massive, forming a thickened base (buttress) capable of wreacking havoc on pavement. Older trees appear to eat pavement. Not recommended for small landscapes, Strangler Fig grows quickly and can reach 60 feet in height with an equal or greater spread.

The broad, spreading, lower limbs are festooned with aerial roots, which create many slim but rigid trunks once they reach the ground and take hold. They help to hold up major lateral limbs, but become a maintenance headache, as these roots are often removed to keep a neat-looking landscape. If these descending roots are clipped from the tree, limbs should be regularly thinned with drop-crotch or thinning cuts to reduce the weight on the lateral branch. The shiny, thick leaves create dense shade and the surface roots add to the problem of maintaining a lawn beneath this massive tree. The fruit drops and makes a mess beneath the canopy.

Easily grown in full sun or partial shade, Strangler Fig can literally be planted, watered a few times, and forgotten. A variety of soils, including wet, will do, and Strangler Fig is moderately salt-tolerant. More often than not, large Strangler Figs were existing trees, not planted. Seeds germinate easily in the landscape, thus allowing the tree to invade nearby land.

Sooty mold follows aphid and scale infestations. Seedlings are often found near large trees.

Ficus benjamina Weeping Fig

Pronunciation: FYE-kus ben-ja-MINE-ah

ZONES: 10B, 11

HEIGHT: 45 to 60 feet

WIDTH: 60 to 100 feet

FRUIT: round; less than 0.5"; fleshy; red; not showy; causes significant litter

GROWTH RATE: rapid-growing; long-lived

HABIT: round; weeping; spreading; dense; symmetrical; fine texture

FAMILY: *Moraceae*

LEAF COLOR: evergreen; medium green in summer

LEAF DESCRIPTION: alternate; simple; entire, undulating margins; oval to ovate; 2 to 4" long

FLOWER DESCRIPTION: white; not showy

EXPOSURE/CULTURE:

Light Requirement: part shade to full sun

Soil Tolerances: all textures; alkaline to acidic; occasionally wet; drought

Aerosol Salt Tolerance: moderate

PEST PROBLEMS: resistant

BARK/STEMS:

Interesting trunk; bark is thin and easily damaged by mechanical impact; branches droop as tree grows and will require pruning for vehicular or pedestrian clearance beneath canopy; should be grown with a single leader; no thorns

Pruning Requirement: requires pruning to control growth and thin the canopy; keep branches less than half the trunk diameter; prune, remove, or cable branches with included bark

Limb Breakage: resistant

Current Year Twig: gray; slender

CULTIVARS AND SPECIES:

'Exotica' has wavy-edged leaves with long, twisted tips. Variegated cultivars are popular indoors. There are other Ficus, such as *F. rubiginosa,* which do not produce aerial roots and are much better suited as landscape trees for shade, because they will not take over the landscape as will Weeping Fig.

LANDSCAPE NOTES:

Origin: southeast Asia

Usage: bonsai; container; hedge; standard; shade; house plant; interiorscape

The dense, rounded canopy and gracefully drooping branches of Weeping Fig made it quite popular as a landscape tree. The thick, shiny, two- to five-inch-long leaves generously clothe the long branches, and the tiny figs eventually turn a deep red. Branches will weep toward the ground, forming a canopy so dense that nothing grows beneath it.

Fruit can stain cars and sidewalks, and it makes quite a mess around the tree as it falls to the ground. The tree is much too large for residential planting unless it is used as a hedge or clipped screen, but can be seen growing into massive trees in parks and other large-scale areas. Aerial roots descend from the branches, touch the ground and take root, eventually forming numerous sturdy trunks that can clog a landscape. Trees can grow to be quite large and spreading in this fashion. Roots grow rapidly, invading gardens and growing under and lifting sidewalks, streets, patios, and driveways.

There have been reports of fertile fruit germinating in some landscapes in south Florida. This is of concern because it could allow the tree to spread and become a pesty weed, which is definitely not needed. Able to tolerate severe pruning, Weeping Fig can be successfully used as a clipped hedge or screen and is probably best used in this fashion, or can be trained into an espalier or topiary. Young trees are often grown in containers, appearing on patios, at entranceways, or indoors.

Massive size and potentially invasive habit. Temperatures below freezing can cause leaf drop up to several years later.

Ficus elastica Rubber Tree, India-Rubber Fig

Pronunciation: FYE-kus ee-LAS-tee-kah

ZONES: 10B, 11

HEIGHT: 25 to 40 feet

WIDTH: 35 to 50 feet

FRUIT: round; less than 0.5"; fleshy; green; causes messy litter in early summer

GROWTH RATE: moderate; moderate to long life

HABIT: oval; round; moderate; irregular silhouette when young; coarse texture

FAMILY: *Moraceae*

LEAF COLOR: broadleaf evergreen; dark green in summer

LEAF DESCRIPTION: alternate; simple; entire margins; oval; 5 to 12" long

FLOWER DESCRIPTION: white; not showy

EXPOSURE/CULTURE:

Light Requirement: part shade to full sun

Soil Tolerances: all textures; alkaline to acidic; occasional wet; drought

Aerosol Salt Tolerance: moderate

PEST PROBLEMS: resistant

BARK/STEMS:

Branches droop as tree grows and will require pruning for vehicular or pedestrian clearance beneath canopy; should be grown with a single leader; no thorns

Pruning Requirement: keep branches less than half the trunk diameter

Limb Breakage: susceptible to breakage at crotch due to poor collar formation

Current Year Twig: green; thick

CULTIVARS AND SPECIES:

'Doescheri' has yellow-variegated leaves; 'Decora' produces broad, reddish-green leaves with ivory-colored veins running down the center of the leaf; 'Variegata' has light green leaves with white or yellow margins. *F. nekbuda*, the African Cloth-Bark Tree, is a related tree grown in the Virgin Islands and Puerto Rico as a street and shade tree.

LANDSCAPE NOTES:

Origin: tropical East Africa, India

Usage: container; espalier; buffer strip; highway; shade; street tree; specimen

Often seen as an interior container plant, Rubber Tree has large, thick, glossy evergreen leaves, multiple trunks, and a spreading, irregular canopy. Able to reach 100 feet in height in its native jungle habitat, but most often seen at about 25 to 40 feet in the landscape, Rubber Tree is useful as a screen, shade, patio, or specimen tree. Its coarse texture makes a strong statement in the landscape. The dense shade under the canopy is ill suited for growing anything but the most shade-tolerant plants. For this reason, Rubber Tree makes a wonderful shade tree.

Use as a street tree is limited by the tree's tendency to break apart in strong winds. The tree could be made stronger by removing branches with weak, tight-angle crotches and spacing major lateral branches along one central trunk. Leave secondary branches along the main limbs to help make them stronger. Eliminate multiple trunks early in the life of the tree and prune lateral branches so they remain smaller than half the diameter of the trunk, to increase longevity in the landscape. These techniques should help develop stronger trees that are suited for planting along streets and in parking lots.

Large, aggressive surface roots invade everything. Give tree plenty of soil space. Scales and mealy bugs can infest the twigs and foliage.

Rubber Tree will grow quickly in sun or partial shade on almost any well-drained soil. The soil should be allowed to become fairly dry between waterings, especially in containers. Rubber Tree makes a nice house plant if it is not overwatered. Provide it with bright light and bring it outside as often as possible.

Ficus lyrata Fiddleleaf Fig

Pronunciation: FYE-kus lye-RAY-tah

ZONES: 10B, 11

HEIGHT: 20 to 40 feet

WIDTH: 25 to 35 feet

FRUIT: paired; round; less than 0.5"; fleshy; green

GROWTH RATE: moderate; moderate life span

HABIT: rounded vase; moderate density; irregular silhouette when young; coarse texture

FAMILY: *Moraceae*

LEAF COLOR: broadleaf evergreen; glossy green

LEAF DESCRIPTION: alternate; simple; entire, undulating margin; obovate; 8 to 12" long; large foliage decomposes slowly and causes litter

FLOWER DESCRIPTION: green; not showy; periodic year-round

EXPOSURE/CULTURE:

Light Requirement: part shade to full sun

Soil Tolerances: all textures; alkaline to acidic; occasionally wet; drought

Aerosol Salt Tolerance: moderate

PEST PROBLEMS: resistant

BARK/STEMS:

Branches droop on young trees and will require pruning to develop clearance for vehicles or pedestrians beneath canopy; should be grown with a single leader; no thorns

Pruning Requirement: requires pruning to develop strong structure

Limb Breakage: resistant

Current Year Twig: brown; thick

LANDSCAPE NOTES:

Origin: western Africa

Usage: container or above-ground planter; espalier; buffer strip; highway; near a deck or patio; street tree; parking lot island; shade; specimen; house plant

An evergreen tree of upright-spreading, irregular growth, Fiddleleaf Fig produces 8- to 15-inch-long and 10-inch-wide, thick, fiddle-shaped leaves that are quite attractive. The trunk can grow to several feet thick. Trees become more symmetrical with age, developing a nice, almost vase shape. Most trees in the landscape are 25 to 30 feet tall. Larger ones sometimes break apart in strong winds due to tight branch crotches and included bark.

Fiddleleaf Fig can be used in containers when young or can be planted to make a striking specimen tree. They create quite an accent by a patio or in a shrub bed because of the coarse leaf texture. Due to their large size, the leaves can be a nuisance to some people when they fall, but there are never too many of them.

Corrective pruning early in the life of the tree can help prevent branch breakage. Spacing major limbs one to two feet along one trunk allows the branch collars to develop properly. These well-formed trees would be suited for planting in parking lot islands and along residential streets. They might even be tried in small sidewalk cutouts for shade in the downtown area. Unless pruned to correct structural weaknesses in the branch crotches, plant Fiddleleaf Fig in a place protected from the wind, such as a courtyard, to increase longevity in the landscape.

Fiddleleaf Fig will grow moderately fast in full sun or partial shade on any well-drained soil and should receive regular watering until established. Be sure to cut roots circling the container before planting, as these can cause the tree to become unstable as it grows older.

Some aerial roots are produced from the branches, but not as many as on some other Ficus, such as *F. benjamina*. Occasionally scales infest the foliage and twigs.

Ficus retusa (microcarpa) Cuban-Laurel

Pronunciation: FYE-kus ree-TOO-sah

ZONES: 10A, 10B, 11

HEIGHT: 60 to 80 feet

WIDTH: 60 to 80 feet

FRUIT: round; less than 0.5"; fleshy; red; causes messy litter

GROWTH RATE: rapid-growing; moderate life span

HABIT: rounded vase; dense; symmetrical; medium texture

FAMILY: *Moraceae*

LEAF COLOR: evergreen; dark green

LEAF DESCRIPTION: alternate; simple; entire margin; oval; 2 to 4" long

FLOWER DESCRIPTION: white; not showy

EXPOSURE/CULTURE:

Light Requirement: part shade to full sun

Soil Tolerances: all textures; alkaline to acidic; occasionally wet; drought

Aerosol Salt Tolerance: moderate

PEST PROBLEMS: resistant to lethal pests

BARK/STEMS:

Showy; bark is thin and easily damaged by mechanical impact; branches grow mostly upright and horizontal; should be grown with a single leader; no thorns

Pruning Requirement: develop well-spaced branches; thin canopy to slow growth on aggressive branches

Limb Breakage: branches with included bark can split at crotch

Current Year Twig: gray; slender

CULTIVARS AND SPECIES:

'Green Gem' is resistant to thrips. The variety *nitida* has more erect branches and is better suited than the species for urban landscapes.

LANDSCAPE NOTES:

Origin: Ceylon, India, southern China

Usage: container or above-ground planter; hedge; buffer strip; highway; standard; shade; urban tolerant

This rapidly growing, rounded, broad-headed, evergreen tree can reach 50 feet or more in height with an equal spread. The glossy, leathery leaves densely clothe large, somewhat weeping branches and are usually infested with thrips. New growth, produced all year long, is a light rose to chartreuse color, giving the tree a lovely two-toned effect. The smooth, light gray trunk is quite striking, can grow to four feet in diameter, and firmly supports the massively spreading canopy. Branches trained to remain less than half the diameter of the trunk are well secured to the trunk.

Used as a park tree, Cuban-Laurel tolerates trimming well and can be shaped and sheared into a hedge, screen topiary, or barrier. It also makes a wonderful shade tree on large properties. Plant at least 10 feet from the curb or sidewalk so large surface roots will not cause damage. It is not suited for parking lots because the fruit stains cars and sidewalks, and it can generally be messy on paved and other hard surfaces. Importation of the wasp that pollinates the flowers allows production of fertile fruit, and has allowed the tree to become a pesty weed in Florida.

Growing easily in full sun or partial shade, Cuban-Laurel thrives on various well-drained soils and is moderately salt-tolerant. It can be planted near the coast, but not directly in the way of salt spray.

Scales and thrips often infest the foliage but do not affect plant health. Limit irrigation and fertilizer applications to reduce thrip infestations. Trees can break apart in strong winds.

Ficus rubiginosa Rusty Fig

Pronunciation: FYE-kus roo-beh-gin-O-sah

ZONES: 10B, 11

HEIGHT: 50 feet

WIDTH: 60 feet

FRUIT: round; 0.5"; fleshy; brown; not showy; no significant litter

GROWTH RATE: moderate; long-lived

HABIT: rounded; dense; symmetrical; medium texture

FAMILY: *Moraceae*

LEAF COLOR: broadleaf evergreen; dark green with rusty brown, pubescent underside

LEAF DESCRIPTION: alternate; simple; entire margin; oval; 3 to 6" long

FLOWER DESCRIPTION: white; not showy

EXPOSURE/CULTURE:

Light Requirement: part shade to full sun

Soil Tolerances: all textures; alkaline to acidic; drought

Aerosol Salt Tolerance: moderate

PEST PROBLEMS: resistant

BARK/STEMS:

Bark is thin and easily damaged by mechanical impact; branches grow mostly upright and horizontally; should be grown with a single leader; no thorns

Pruning Requirement: thin canopy to reduce wind resistance and slow growth on aggressive branches

Limb Breakage: resistant

Current Year Twig: brownish-green, pubescent; medium thickness

CULTIVARS AND SPECIES:

'Variegata' has leaves variegated with cream-yellow.

LANDSCAPE NOTES:

Origin: Australia

Usage: container; hedge; parking lot island; buffer strip; highway; shade; specimen; street tree; urban tolerant

This broad, spreading evergreen tree is densely covered with oval, blunt-tipped, smooth leaves, the undersides of which are brown and hairy. One of the hardiest of the rubber trees, Rusty Fig makes an attractive, durable specimen tree, especially when only a few major branches are allowed to develop and the canopy is thinned to create a more open form. Trees with brown, pubescent leaves develop more aerial roots than green-leaved forms. It grows to about 35 feet in 30 years.

Rusty Fig's moderate growth rate make it better suited for smaller landscapes than most other Ficus trees, though surface roots can grow large. It is a nice tree for creating a medium-sized canopy along a residential street or commercial entry way. Plant on 30- to 40-foot centers to form a closed canopy effect. It is among the best Ficus trees for frost-free climates due to its moderate growth rate and lack of numerous aerial roots. It will not take over a landscape as other Ficus can. It is well suited as a shade or street tree and should require little maintenance once initial pruning creates a good structural habit. Space major branches along the trunk and keep them thinned so they remain less than half the diameter of the trunk. This encourages development of a strong structure.

Easily grown in full sun or partial shade, Rusty Fig will thrive on a variety of well-drained soils. Once established, it can withstand periods of drought and 30°F. for a short time. Mites and scales infest the foliage but do not kill the tree. Rusty Fig is subject to root rot on poorly drained soils.

Firmiana simplex Chinese Parasoltree, Varnish-tree

Pronunciation: fir-mee-A-nah SIM-plecks

ZONES: 7A, 7B, 8A, 8B, 9A, 9B

HEIGHT: 30 to 50 feet

WIDTH: 15 to 20 feet

FRUIT: elongated; 1.5" long; dry; yellowish-green; causes significant litter; showy

GROWTH RATE: rapid-growing; moderate life span

HABIT: oval; upright; moderate density; irregular silhouette; coarse texture

FAMILY: *Sterculiaceae*

LEAF COLOR: bright green in summer; showy yellow fall color

LEAF DESCRIPTION: alternate; simple; lobed; undulate margin; deciduous; 6 to 12" long

FLOWER DESCRIPTION: yellow; showy; summer

EXPOSURE/CULTURE:

Light Requirement: part shade to full sun

Soil Tolerances: all textures; alkaline to acidic; drought

PEST PROBLEMS: resistant

BARK/STEMS:

Bright green trunk; bark is very thin and easily damaged by mechanical impact; branches grow mostly upright and will not droop; should be grown with a single leader; no thorns

Pruning Requirement: heading young branches increases canopy uniformity on young trees

Limb Breakage: susceptible to breakage at crotch, if included bark develops in the crotches

Current Year Twig: green to gray; stout

LANDSCAPE NOTES:

Origin: China and Japan

Usage: buffer strip; highway; shade; specimen

Chinese Parasoltree has a very unusual appearance, having green stems and bark and extremely large, three- to five-lobed leaves. In June or July, 10- to 20-inch-long, upright, loose, terminal panicles of yellow-green blooms appear and are followed by the production of peculiar pods that split open to reveal small, round seeds. Pods are harvested for use in winter decorations. The foliage of Chinese Parasoltree can turn brilliant yellow before dropping in fall to reveal an interesting branching structure of green stems.

The tree appears out of place to some horticulturists, but many people enjoy the dramatic impact one or several of these trees can have on a landscape. The tree lends a tropical effect and is probably best used occasionally as a specimen. It could be tried as a street tree on a small scale but may be objectionable due to its messy nature. Leaves are large, decompose slowly, and blow around after they fall. Falling fruits also contribute to the mess, although they are dry. The tree looks a bit scraggly in winter, with old flower stalks persisting on the branch tips.

Branches can be poorly attached to the trunk, so be sure that branches grow no larger than about half the diameter of the trunk. This will help ensure a stronger attachment to the tree. Roots often grow close to the surface in clay soil, especially near the trunk. Seeds germinate in the landscape and can become weeds under certain circumstances.

Chinese Parasoltree should be grown in a full-sun, wind-protected location. Trees will grow in shade with an upright, almost columnar form as they reach for the sunlight. Trees should be regularly watered when young but become drought-tolerant once established. They tolerate clay soil but often develop root rot if the soil is not well drained.

A trunk scale reportedly kills trees if not controlled. Branches often break out from the top of the tree, leading to trunk rot.

Franklinia alatamaha Franklin-Tree, Franklinia

Pronunciation: frank-LIN-ee-ah ah-lah-tah-MAH-hah

ZONES: 5A, 5B, 6A, 6B, 7A, 7B, 8A, 8B

HEIGHT: 15 to 25 feet

WIDTH: 10 to 15 feet

FRUIT: round; 0.5 to 1"; dry; no significant litter

GROWTH RATE: slow-growing; moderate life span

HABIT: round; pyramidal; upright; dense; symmetrical; medium texture

FAMILY: *Theaceae*

LEAF COLOR: bright green in summer; showy orange to red fall color

LEAF DESCRIPTION: alternate; simple; serrate margin; oblong to oblanceolate; deciduous; 4 to 8" long

FLOWER DESCRIPTION: white; summer; pleasant fragrance; showy

EXPOSURE/CULTURE:

Light Requirement: part shade to full sun

Soil Tolerances: loam to sand; slightly alkaline to acidic; moderate drought

PEST PROBLEMS: sensitive

BARK/STEMS:

Branches droop as tree grows and will require pruning for vehicular or pedestrian clearance beneath canopy; can be trained to a single leader or with multiple trunks; showy white strips along trunk; no thorns

Pruning Requirement: needs little pruning to develop strong trunk and branch structure; early pruning eliminates multiple trunks

Limb Breakage: resistant

Current Year Twig: brown or green; medium thickness

LANDSCAPE NOTES:

Origin: eastern North America in Georgia (possibly extinct)

Usage: container or planter; near a deck or patio; specimen

This excellent small tree, originally native to Georgia, reaches a maximum of 30 feet but is usually smaller in a sunny landscape. It is somewhat pyramidal when young, becoming more rounded with age, with many thin stems and trunks. It is best used as a specimen or in borders, to show off its fragrant, white, camellia-like flowers, three inches across, that bloom from July to late summer when few other trees bloom. The foliage turns a vivid orange-red in the fall while some flowers are still in bloom. Ridged gray bark with prominent vertical white striations adds winter interest. Franklin-Tree typically grows with numerous vertical stems or trunks originating at or near ground level.

Franklin-Tree does best in well-drained, rich, acid soil, with ample water, and partial sun in the southern part of its range. Drainage must be excellent. The tree does not tolerate clay soil and is only slightly or moderately drought-tolerant. It is best to provide a permanent irrigation system for Franklin-Tree, or to irrigate in drought until well established for several years.

The most serious problem of Franklin-Tree is a root-rot disease. The best protection is to plant in a soil where the disease has not been active. The tree has been difficult to establish in some areas, perhaps because of disease problems, but the beauty of this tree makes extra effort worthwhile.

Fraxinus americana White Ash

Pronunciation: FRACK-suh-nuss ah-mare-eh-CANE-ah

ZONES: 3A, 3B, 4A, 4B, 5A, 5B, 6A, 6B, 7A, 7B, 8A, 8B, 9A

HEIGHT: 60 to 80 feet

WIDTH: 40 to 70 feet

FRUIT: 1 to 2"; dry; tan; attracts birds; abundant, causes significant litter

GROWTH RATE: fast-growing; long-lived on moist sites

HABIT: oval to pyramidal; moderate density; symmetrical; medium texture

FAMILY: *Oleaceae*

LEAF COLOR: stunning yellow or purple fall color

LEAF DESCRIPTION: opposite; odd pinnately compound; entire margin; lanceolate to ovate; deciduous; 2 to 4" long

FLOWER DESCRIPTION: green; not showy; spring

EXPOSURE/CULTURE:

Light Requirement: part shade to full sun

Soil Tolerances: all textures; alkaline to acidic; occasionally wet soil; salt; moderate drought

Aerosol Salt Tolerance: good

PEST PROBLEMS: sensitive unless in a moist site

BARK/STEMS:

Branches grow mostly upright and will not droop; should be grown with a single leader; no thorns

Pruning Requirement: eliminate upright trunks and branches and thin one of two aggressive limbs growing opposite each other

Limb Breakage: somewhat susceptible

Current Year Twig: brown to gray; thick

CULTIVARS AND SPECIES:

'Autumn Applause' (purple fall color) and 'Autumn Purple' (yellow, purple, and orange fall color) are seedless. 'Rosehill' is reported to have better branch arrangement. *F. texensis* (zone 7) may be the best Ash for Texas due to drought tolerance. *F. nigra* (zone 2B) tolerates moderate drought.

LANDSCAPE NOTES:

Origin: eastern North America

Usage: wide tree lawns; buffer strips; shade tree; residential street tree

White Ash resists heat, although it is native to moist locations, including flood plains and well-drained upland sites. The trees produce countless seeds every two to three years, and these germinate in the landscape, creating a nuisance and perhaps look a bit messy. The tree grows rapidly and is almost pyramidal when young, but gradually slows down and develops a more spreading round or oval shape.

White Ash is a good tree for large open areas, but is much too big for a home landscape unless you want lots of shade. The potential disease and insect problems limit use to parks and other areas where the tree will not be missed if it dies prematurely. Extensive use as a street tree in some areas could be risky because of these problems, especially borers, and a sensitivity to drought in confined soil spaces.

White Ash that have not been properly pruned can break apart in wind storms. Be sure to space branches along the trunk and remove any that are vigorously growing upright with narrow branch crotches. Major branches often develop opposite each other, and one should be removed or thinned to develop good structure.

Select male trees to avoid fruit litter. Most cultivars are males. Borers are a problem for newly planted trees. Ash decline, a major problem, is believed to be caused by a complex of conditions, including ash yellows caused by phytoplasma. Large surface roots can form.

Fraxinus excelsior Common or European Ash

Pronunciation: FRAK-suh-nuss eck-SELL-see-or

ZONES: 5A, 5B, 6A, 6B, 7A, 7B, 8A

HEIGHT: 60 to 80 feet

WIDTH: 60 to 90 feet

FRUIT: elongated; 1 to 2"; dry; tan; abundant, causes significant litter

GROWTH RATE: rapid growth; well-formed trees can be long-lived

HABIT: round; dense; symmetrical; medium texture

FAMILY: *Oleaceae*

LEAF COLOR: dark green in summer; yellow fall color

LEAF DESCRIPTION: opposite; odd pinnately compound; serrate margin; lanceolate to ovate; deciduous; 2 to 4" long

FLOWER DESCRIPTION: green; not showy; spring

EXPOSURE/CULTURE:

Light Requirement: grows best in full sun

Soil Tolerances: all textures; alkaline to acidic; salt and drought; occasionally wet

Aerosol Salt Tolerance: high

PEST PROBLEMS: susceptible

BARK/STEMS:

Branches droop as tree grows and will require pruning for vehicular or pedestrian clearance beneath canopy; often seen with no leader, but should be grown with a central leader; no thorns

Pruning Requirement: challenging, but trees are stronger if branches are spaced along trunk

Limb Breakage: branches clustered together often split from trunk

Current Year Twig: brown or gray; stout

CULTIVARS AND SPECIES:

'Aurea'—slow-growing, perhaps to 50 feet tall and wide, deep yellow fall color, yellow twigs; 'Aurea Pendula'—yellow, pendulous branchlets; 'Aureovariegata'—leaves variegated or edged with yellow; 'Hessei'—60 feet tall, single leaves, very disease-resistant, seedless, limited yellow fall color; 'Nana'—compact, slow-growing dwarf form with small leaflets; 'Pendula'—weeping form; 'Rancho' (also known as 'Kimberly')—30 feet tall, round canopy, yellow fall foliage; 'Spectabillis'—pyramidal shape. *F. quadrangulata* is a beautiful tree with interesting form native to limestone soil. *F. latifolia* is native from central California to Washington.

LANDSCAPE NOTES:

Origin: Europe and Asia Minor

Usage: specimen

Common Ash is a broad, spreading, deciduous tree, capable of reaching 100 feet or more in height in the woods. The multi-divided leaves usually drop off in autumn while still green, but some cultivars turn a brilliant yellow. The inconspicuous springtime flowers are followed by abundant clusters of 1.5-inch-long, winged fruits that turn brown and remain on the trees well after the leaves have fallen. The low-branched, rounded silhouette of naked branches on top of the short trunk and the black, dormant leaf buds help to make Common Ash an attractive winter landscape element.

Like many Ashes, the tree requires careful training and pruning to develop a central leader with strong branch structure. Without pruning, many branches originate at the same position on the trunk, which makes them prone to breakage and shortens their life. Select and develop up to a dozen main branches well spaced along the trunk as far up in the canopy as practical.

Common Ash prefers deep, moist, rich soil, with adequate moisture, although trees are drought-tolerant. It tolerates poorly drained, low-quality, and alkaline soil. Its use may be limited by borers, which often infest the trunk of young trees and cause the trees to decline or die. Probably best suited for landscapes with plenty of soil for root expansion, not in a confined urban space.

Fraxinus ornus Flowering Ash

Pronunciation: FRAK-suh-nuss OR-nuss

ZONES: 5A, 5B, 6A, 6B, 7A, 7B, 8A, 8B

HEIGHT: 40 to 50 feet

WIDTH: 25 to 35 feet

FRUIT: elongated; 1 to 2"; dry; green to tan; abundant, causes some litter; persistent

GROWTH RATE: moderate; moderate life span

HABIT: round; dense; symmetrical; fine to medium texture

FAMILY: *Oleaceae*

F. ornus 'Rotundafolia'

LEAF COLOR: dark green in summer; showy, yellow fall color

LEAF DESCRIPTION: opposite; odd pinnately compound; serrate margin; lanceolate to ovate; deciduous; 3 to 4" long

FLOWER DESCRIPTION: white; showy; spring

EXPOSURE/CULTURE:

Light Requirement: grows best in full sun

Soil Tolerances: all textures; alkaline to acidic; drought; good drainage essential

PEST PROBLEMS: susceptible when stressed by wet soil

BARK/STEMS:

Branches grow mostly upright; often seen with multiple trunks and no leader; no thorns

Pruning Requirement: reduce number of branches and space them out along trunk as much as possible; requires constant vigilance

Limb Breakage: branches clustered together can split from trunk

Current Year Twig: brown or gray; stout

CULTIVARS AND SPECIES:

'Rotundafolia' has slightly smaller leaves and the tree is smaller than the species.

LANDSCAPE NOTES:

Origin: Europe and Asia Minor

Usage: above-ground planter; specimen

Flowering Ash is a small to moderate-sized, deciduous tree, capable of reaching 50 feet or more in height in the woods, but more often smaller in the landscape. It is not common in the United States, but has been grown in Europe for a long time. The wonderful flowers covering the canopy in late spring make it worthy of consideration for the American landscape. The multi-divided leaves usually drop off in autumn after putting on a spectacular fall color show. The showy flowers are followed by clusters of 1.5-inch-long, winged fruits that turn brown and remain on the trees well after the leaves have fallen. The low-branched, rounded silhouette of naked branches on top of the short trunk and the black, dormant leaf buds help to make Flowering Ash an attractive winter landscape element.

Like many Ashes, the tree requires careful training and pruning to develop a central leader with strong branch structure. Without pruning, many branches originate at the same position on the trunk, which makes them prone to breakage and shortens their life. Select and develop up to a dozen main branches well spaced along the trunk as far up in the canopy as practical.

Flowering Ash prefers deep, moist, rich soil, with adequate moisture, although trees are drought-tolerant. Its use may be limited by borers, which often infest the trunk of young ashes and cause the trees to decline or die. Probably best suited for landscapes with plenty of soil for root expansion, not in a confined urban space. Some communities recommend planting Flowering Ash along the streets in the strip of soil between the curb and the sidewalk. The small size of the trunk at maturity allows this ash to be planted here, even if the strip is only four feet wide, whereas the trunks on other ashes grow too large for this space.

Leaf spots and aphids may appear, but usually are not a major problem unless the tree is stressed by growing in soil that is too wet.

Fraxinus pennsylvanica Green Ash

Pronunciation: FRAK-suh-nuss pen-sill-VAN-ee-cah

ZONES: 2B, 3A, 3B, 4A, 4B, 5A, 5B, 6A, 6B, 7A, 7B, 8A, 8B, 9A

HEIGHT: 60 to 70 feet

WIDTH: 45 feet

FRUIT: elongated; 1 to 2"; dry; tan; attracts birds; abundant, causes significant litter

GROWTH RATE: rapid; long-lived

HABIT: oval; upright; moderate density; symmetrical; medium texture

FAMILY: *Oleaceae*

LEAF COLOR: medium green in summer; showy yellow fall color

LEAF DESCRIPTION: opposite; odd pinnately compound; entire or serrate margin; lanceolate; deciduous; 2 to 5" long

FLOWER DESCRIPTION: green; not showy; spring

EXPOSURE/CULTURE:

Light Requirement: full sun

Soil Tolerances: all textures; alkaline to acidic; wet soil; salt and drought

Aerosol Salt Tolerance: moderate to high

PEST PROBLEMS: susceptible when young

BARK/STEMS:

Branches grow mostly upright and will not droop; should be grown with a single leader, but often grows with weak, double leaders; no thorns

Pruning Requirement: remove upright, aggressive branches; thin major limbs early to slow their growth to form a strong crotch; remove branches broken in storms

Limb Breakage: susceptible, especially in ice storms

Current Year Twig: brown to gray; thick

CULTIVARS AND SPECIES:

'Marshall Seedless', 'Newport', and 'Patmore' have good fall color and are mostly seedless. 'Wahpeton' with a central leader, 'Rugby' with a narrow canopy, and 'Leeds' with a small, globose canopy are adapted to the northern plains. 'Summit' walls off decay well at wound sites. 'Johnson' is compact with a globose canopy expected to be 15 to 20 feet tall. Cultivars developed in the north could do poorly in zone 9.

LANDSCAPE NOTES:

Origin: Montana to Texas to the eastern seaboard; the most widely distributed of North American ashes

Usage: large parking lot island; buffer strip; highway; reclamation; shade; street tree; urban tolerant

Upright growth habit on young trees gives way to an oval shape with age. The annual abundant seed-set on female trees is used by many birds, but the seeds sprout everywhere. Seedless cultivars are available and are recommended instead of the species.

Once established, Green Ash will grow in wet or dry sites, whether acid or alkaline. It adapts quite well to confined soil spaces, probably due to its tolerance for flooded and wet sites. The tree adapts to city street tree planting pits and tree lawns, and has been widely planted throughout the country. Like some other rapidly growing trees, large surface roots can develop and become a nuisance, as they lift sidewalks and curbs or make mowing difficult. Planting only in well-drained soil may keep surface rooting in check. Using root barriers around the edge of planting pits and along sidewalks could deflect roots away from pavement. Extensive use as a street tree is tempting but risky, because of the potential insect and disease problems (especially borers when the trees are young).

Green Ash requires regular pruning when it is young to develop a nice central trunk. It tends to develop a number of dominant upright trunks or multiple leaders if it is pruned improperly or left unpruned. Be sure the trees have branches that are well spaced along one central trunk. Prune in the fall.

Borers are common on Ashes and can kill trees, especially new plantings. Plant male trees to avoid abundant seeds. Branches break in storms. Leaf anthracnose can be severe. Intermediate susceptibility to ash yellows.

Fraxinus uhdei Evergreen, or Shamel Ash

Pronunciation: FRAK-suh-nuss YOU-dee-eye

ZONES: 7A, 7B, 8A, 8B, 9A, 9B, 10A, 10B

HEIGHT: 60 to 80 feet

WIDTH: 50 to 70 feet

FRUIT: elongated; 0.5 to 1"; dry; green turning tan; causes significant litter

GROWTH RATE: rapid; moderate life span, unless properly pruned

HABIT: rounded vase; moderate density; irregular silhouette; medium texture

FAMILY: *Oleaceae*

LEAF COLOR: medium green in summer; showy, yellow fall color

LEAF DESCRIPTION: opposite; odd pinnately compound; serrate margin; oval to lanceolate; 2" semi-evergreen leaflets; 2" long

FLOWER DESCRIPTION: green; not showy; spring

EXPOSURE/CULTURE:

Light Requirement: grows best in full sun

Soil Tolerances: all textures; alkaline to acidic; wet soil; moderate drought

PEST PROBLEMS: sensitive

BARK/STEMS:

Branches grow mostly upright and will not droop; usually grown with codominant leaders, but should be grown with a single leader; no thorns

Pruning Requirement: requires regular pruning when tree is young to space branches along trunk

Limb Breakage: susceptible to breakage at crotch, due to poor collar formation, unless branches are spaced along trunk and kept small

Current Year Twig: gray; medium to stout

CULTIVARS AND SPECIES:

'Sexton' and 'Tomlinson' are more compact and slower growing; 'Majestic Beauty' has a narrower canopy. They are suited for planting provided they are trained to a single leader.

LANDSCAPE NOTES:

Origin: central Mexico

Usage: large parking lot island; reclamation; shade tree; street tree

This fast-growing, deciduous, native Mexican tree has four-inch-long leaflets that turn a brilliant yellow in fall before dropping. The inconspicuous, green, springtime flowers are followed by the production of showy, persistent fruits. The tree has been widely planted in California and in parts of the west, including Texas, for its rapid growth, but it has gained a reputation for splitting apart due to improper or no pruning.

The tree has been traditionally difficult to maintain, due to the development of many upright trunks originating from the same position on the main trunk. This condition leads to the creation of weak trees that often break apart at the base of the multiple trunks. Careful pruning and branch selection is required during the first 10 years after planting to ensure good, strong trunk and branch development. It is important to purchase good-quality planting stock that has a central leader and no upright multiple trunks. This will make the tree much easier to train and maintain in the landscape. Some horticulturists do not recommend planting this tree, because of its high pruning requirement and susceptibility to breakage.

Shamel Ash should be grown in full sun on any soil and will tolerate both alkaline and rocky soils. It is tolerant of wet soil and has been extensively used along streets in areas with poor drainage. Roots can grow close to the soil surface, causing a nuisance by breaking sidewalks and curbs. Locate them six feet or more from pavement.

Very susceptible to borers. Limbs can fall from the tree unless it was properly pruned when young. Trees have a moderately short life unless they are properly pruned. Foliage scorches in strong winds, but resists anthracnose. When seeds fall, they germinate everywhere creating a terrible weed problem.

Fraxinus velutina Velvet Ash, Modesto Ash, Arizona Ash

Pronunciation: FRAK-suh-nuss vell-u-TEE-nah

ZONES: 7A, 7B, 8A, 8B, 9A, 9B, 10A, 10B

HEIGHT: 40 to 50 feet

WIDTH: 40 to 50 feet

FRUIT: elongated; 0.5 to 1"; dry; green turning tan; causes significant litter

GROWTH RATE: rapid; moderate life span

HABIT: round; variable; moderate density; irregular silhouette; medium texture

FAMILY: *Oleaceae*

LEAF COLOR: dark green in summer; showy yellow fall color

LEAF DESCRIPTION: opposite; odd pinnately compound; serrate margin; oval to lanceolate; deciduous; leaflets 1 to 2" long

FLOWER DESCRIPTION: green; not showy; spring

EXPOSURE/CULTURE:

Light Requirement: grows best in full sun

Soil Tolerances: all textures; alkaline to acidic; wet soil; moderate drought

PEST PROBLEMS: very sensitive

BARK/STEMS:

Branches grow mostly upright and will not droop; usually grown with codominant leaders, but should be grown with a single leader; no thorns

Pruning Requirement: requires regular pruning when young to space branches along trunk

Limb Breakage: susceptible to breakage at crotch, due to poor collar formation, unless branches are spaced along trunk and kept small

Current Year Twig: gray; medium to stout

CULTIVARS AND SPECIES:

The variety *glabra* is fruitless and is a cleaner tree than the species, but it is disease-sensitive and susceptible to branch breakage unless properly pruned.

LANDSCAPE NOTES:

Origin: southwestern United States into Mexico

Usage: large parking lot island; reclamation; shade; street tree

This fast-growing, deciduous, native North American tree reaches a height of 40 to 50 feet, depending upon cultural conditions. The multiple leaflets turn a brilliant yellow in fall before dropping. The inconspicuous, green, springtime flowers are followed by the production of showy, persistent fruits. The tree has been widely grown in California and in parts of the west, including Texas.

The tree has traditionally been difficult to maintain due to the development of many upright trunks originating from the same position on the main trunk. This condition leads to the creation of weak trees that often break apart at the base of the multiple trunks. Careful pruning and branch selection are required during the first 15 years after planting to ensure good, strong trunk and branch development. It is important to purchase good-quality planting stock that has a central leader and no upright multiple trunks. This will make the tree much easier to maintain in the landscape. Some horticulturists do not recommend planting this tree, because of its high pruning requirement and susceptibility to breakage.

Velvet Ash should be grown in full sun on any soil and will tolerate both alkaline and rocky soils. It is tolerant of wet soil and has been extensively used along streets in areas with poor drainage. Roots can grow close to the soil surface, causing a nuisance by breaking sidewalks and curbs.

Very susceptible to borers and leaf blight. Limbs fall from the tree. Trees have a moderately short life. Irrigation and fertilization help trees recover from verticillium wilt.

Geijera parviflora Australian-Willow

Pronunciation: gay-GEE-ruh parv-eh-FLOOR-ah

ZONES: 9A, 9B, 10A, 10B, 11

HEIGHT: 30 to 35 feet

WIDTH: 20 to 25 feet

FRUIT: bright orange; round; 0.5"; fleshy; some litter problem

GROWTH RATE: rapid-growing; moderate life span

HABIT: round; weeping; moderate density; symmetrical; fine texture

FAMILY: *Rutaceae*

LEAF COLOR: evergreen; olive green

LEAF DESCRIPTION: alternate; simple; entire margin; lanceolate; 3 to 6" long

FLOWER DESCRIPTION: white; fall and spring; showy

EXPOSURE/CULTURE:

Light Requirement: grows best in full sun

Soil Tolerances: loam; sand; slightly alkaline to acidic; moderate drought

Aerosol Salt Tolerance: moderate

PEST PROBLEMS: resistant

BARK/STEMS:

Branches droop as tree grows and will require pruning for vehicular or pedestrian clearance beneath canopy; should be grown with a single leader; no thorns

Pruning Requirement: locate the lowest permanent branch high enough for adequate clearance beneath canopy

Limb Breakage: resistant

Current Year Twig: green; slender

LANDSCAPE NOTES:

Origin: Australia

Usage: parking lot island; near a deck or patio; shade; specimen; street tree

Australian-Willow is an attractive evergreen with a weeping, rounded silhouette. The main inner branches are composed of strong, wind-resistant wood and are directed upward; the outer, smaller branches are pendulous. This gives a decided weeping habit to older trees. Younger trees are more oval-shaped. This characteristic, combined with the thin, narrow leaves that droop from the branches, gives the tree much the same effect as a young Weeping Willow. Short panicles of small, creamy white, showy flowers appear in early spring and early fall.

Australian-Willow casts light shade and is ideal for use around a patio or as a lawn specimen, but may be too weeping for use as a street tree unless it is properly trained and regularly trimmed when young. If the lowest main limbs are located high enough above the ground and are oriented in an upright-spreading fashion, then there will be clearance even for large trunks. Only small branches may have to be removed later as they droop from the main limbs toward the ground. Otherwise the tree has very good cultural adaptability to urban spaces. It is well suited for planting along wide medians and highways.

Australian-Willow will give best growth in full sun on moist but well-drained soil, although plants can tolerate light shade and very dry conditions. Trees grown in the open rarely require any pruning if located where the drooping branches will not interfere with traffic below. A popular tree for the dry southwest, not really known or planted in humid climates. California has many fine examples of Australian-Willow used as street trees.

No limitations of major concern, except perhaps root rot on sites without excellent drainage.

Ginkgo biloba Maidenhair Tree, Ginkgo

Pronunciation: GINK-go bye-LOW-bah

ZONES: 3A, 3B, 4A, 4B, 5A, 5B, 6A, 6B, 7A, 7B, 8A, 8B, (9)

HEIGHT: 50 to 75 feet

WIDTH: 50 to 60 feet

FRUIT: oval; round; 1"; fleshy; green; not showy; female trees cause messy, smelly litter; can cause skin irritation

GROWTH RATE: slow at first, then moderate; long-lived

HABIT: oval; pyramidal; open when young, denser later; irregular silhouette; medium texture

FAMILY: *Ginkgoaceae*

LEAF COLOR: medium green in summer; traffic-stopping yellow fall color

LEAF DESCRIPTION: alternate; simple; lobed; fan-shaped; deciduous; 2 to 4" long

FLOWER DESCRIPTION: pleasant fragrance; not showy; spring

EXPOSURE/CULTURE:

Light Requirement: part shade to full sun

Soil Tolerances: all textures; alkaline to acidic; occasionally wet soil, once established; some salt; drought

Aerosol Salt Tolerance: moderate

PEST PROBLEMS: pest-free

BARK/STEMS:

Branches grow mostly upright and do not droop; showy trunk; should be grown with a single leader, but often grows with several trunks; no thorns

Pruning Requirement: needs some pruning to develop a central leader and strong structure

Limb Breakage: resistant

Current Year Twig: brownish-gray; medium thickness

CULTIVARS AND SPECIES:

Fruitless male cultivars include: 'Autumn Gold', 'Fairmont', and 'Fastigiata'—narrow canopy, 'Lakeview' and 'Princeton Sentry'—very narrow canopy.

LANDSCAPE NOTES:

Origin: originally native to Asia, Australia, and North America; still native in China

Usage: buffer strip; highway; specimen; sidewalk cutout; street tree; urban tolerant

Ginkgo is practically pest-free, resistant to storm damage, and casts light to moderate shade. Young trees are often very open, but they fill in to form a denser canopy. It makes a durable street tree where there is enough overhead space to accommodate the large size. The shape is often irregular, with a large branch or two seemingly forming its own tree on the trunk. If overhead space is narrow, select from the narrow upright cultivars such as 'Princeton Sentry' and 'Fairmont'. The tree is easily transplanted and has a vivid yellow fall color which is second to none in brilliance, even in hot climates. However, leaves fall quickly and the fall color show is short, sometimes only two or three days.

Ginkgo may grow extremely slowly for several years after planting, but will then pick up and grow at a moderate rate, particularly if it receives an adequate supply of water and some fertilizer. Do not overwater or plant in a poorly drained area. Be sure to keep turf several feet away from the trunk to help trees become established. It is very tolerant of urban soils and pollution. It is adapted for use as a street tree, even in confined soil spaces, as surface roots do not seem to become very large. Some early pruning to form one central leader is essential.

Pest-free and considered resistant to gypsy moth. Only male plants should be used, as the female produces foul-smelling fruit in late autumn. Most cultivars are males. The only way to select a male plant is to purchase a named cultivar, because there is no reliable way to select a male plant from a seedling until it fruits. It could take as long as 20 years or more for Ginkgo to fruit. Recovery from transplanting is very slow. The tree may grow only a few inches in the first year or two after planting. The random branching habit on the species makes it less desirable than the more uniformly branched cultivars.

Gleditsia triacanthos var. *inermis* Thornless Honeylocust

Pronunciation: gleh-DIT-see-uh tri-ah-CAN-thos [variety] in-ER-miss

ZONES: 3A, 3B, 4A, 4B, 5A, 5B, 6A, 6B, 7A, 7B, 8A, 8B, (9)

HEIGHT: 50 to 75 feet

WIDTH: 35 to 50 feet

FRUIT: pod; about 12"; dry; brownish-purple; causes significant litter; persistent; mostly lacking on many cultivars

GROWTH RATE: rapid-growing; moderate life span

HABIT: upright spreading; open; irregular silhouette; fine texture

FAMILY: *Leguminosae*

LEAF COLOR: light green in summer; showy copper to yellow fall color

LEAF DESCRIPTION: alternate; bipinnately or pinnately compound; deciduous; leaflets up to 1" long

FLOWER DESCRIPTION: yellow; pleasant fragrance; spring

EXPOSURE/CULTURE:

Light Requirement: part shade to full sun

Soil Tolerances: all textures; alkaline to acidic; wet soil; salt and drought; compaction

Aerosol Salt Tolerance: good

PEST PROBLEMS: mostly resistant

BARK/STEMS:

Branches grow mostly upright and will not droop; should be grown with a single leader; no thorns

Pruning Requirement: space limbs along trunk and eliminate crossed and touching branches

Limb Breakage: susceptible to ice storms

Current Year Twig: brown; moderate thickness

CULTIVARS AND SPECIES:

'Imperial'—upright-spreading, seedless, and thornless until 10 to 15 years old, when some seeds develop; 'Majestic'—upright, seedless, thornless; 'Moraine'—spreading, usually seedless, thornless; 'Shademaster'—upright, spreading, usually seedless and thornless until 10 to 15 years old, when some seeds develop (perhaps the best cultivar); 'Skyline'—central leader, pyramidal, generally seedless, thornless; 'Sunburst'—spring foliage is golden yellow, turning light green, 50 to 60 feet tall, seedless, thornless, favored by plant bugs and leafhoppers.

LANDSCAPE NOTES:

Origin: Iowa to New Jersey and Texas

Usage: parking lot island; buffer strip; highway; reclamation; light shade; sidewalk cutout; street tree; urban tolerant

The species has undesirable thorns on the trunk and main branches and large seed pods. Most nurseries sell cultivars of the variety *inermis,* which are thornless and some are also nearly seedless. The seed pods look rather unsightly hanging on the tree into the fall and make quite a mess as they litter the ground below the canopy. The tree casts light shade, making it easy to maintain a lawn beneath the canopy. Trees often defoliate early in the south and are bare by October.

Some cultivars reportedly grow a central leader (don't count on it) and would require only minor pruning, but others grow many upright codominant trunks. Strive to develop one central trunk with upright spreading branches spaced several feet apart along the trunk. Purchase good-quality trees with one leader to reduce the pruning requirement. The tree has no particular soil preferences. Unfortunately, it has been overplanted in some areas and insect problems are beginning to catch up with Honeylocust, including the cultivars.

Seed pods, cankers (including Nectria), borers, pod gall midge, leaf spots, rusts, and other potential problems. Mimosa webworm devours foliage in summer. Leaf drop is an unattractive feature in the deep south due primarily to poor heat tolerance. Plant bugs defoliate trees in spring.

Gleditsia triacanthos var. *inermis*
'Shademaster' Thornless Honeylocust
Pronunciation: gleh-DIT-see-uh tri-ah-CAN-thos [variety] in-ER-miss

ZONES: 3A, 3B, 4A, 4B, 5A, 5B, 6A, 6B, 7A, 7B, 8A, 8B, (9)

HEIGHT: 50 to 75 feet

WIDTH: 35 to 50 feet

FRUIT: pod; about 12"; dry; brownish-purple; causes significant litter if produced; persistent; mostly lacking

GROWTH RATE: rapid-growing; moderate life span

HABIT: upright spreading; open; irregular silhouette; fine texture

FAMILY: *Leguminosae*

LEAF COLOR: light green in summer; showy copper to yellow fall color

LEAF DESCRIPTION: alternate; bipinnately compound; deciduous; leaflets up to 1" long

FLOWER DESCRIPTION: yellow; pleasant fragrance; spring

EXPOSURE/CULTURE:

Light Requirement: part shade to full sun

Soil Tolerances: all textures; alkaline to acidic; wet soil; salt and drought; compaction

Aerosol Salt Tolerance: good

PEST PROBLEMS: mostly resistant

BARK/STEMS:

Branches grow mostly upright and will not droop; should be grown with a single leader; no thorns

Pruning Requirement: space limbs along trunk and eliminate crossed and touching branches

Limb Breakage: susceptible to ice storms

Current Year Twig: brown; moderate thickness

LANDSCAPE NOTES:

Origin: Iowa to New Jersey and Texas

Usage: parking lot island; buffer strip; highway; reclamation; light shade; sidewalk cutout; street tree; urban tolerant

Most trees are nearly seedless, making Shademaster a much cleaner tree than the species, but occasionally some trees will flower and produce red-brown, 10-inch-long seed pods. Some pods may begin appearing 10 to 15 years after planting, especially on trees that are under environmental stress. Although a legume, Honeylocust does not fix nitrogen in the roots. On mature trees, the bark becomes very dark, which makes the bare trees in winter especially striking against glistening white snow or blue sky.

Some trees grow a central leader and would require little pruning, but others grow many upright, codominant branches. These trees must be trained when they are young with two or three prunings spaced several years apart. Strive to develop one central trunk with upright, spreading branches spaced several feet apart along the trunk. Street trees planted on 25- to 30-foot-centers form a canopy over a residential street. It makes a nice medium-sized tree, but it has been overused in some areas.

Gordonia lasianthus Loblolly-Bay

Pronunciation: gore-DOAN-ee-ah laz-ee-ANN-thuss

ZONES: 7A, 7B, 8A, 8B, 9A, 9B

HEIGHT: 50 to 60 feet

WIDTH: 10 to 15 feet

FRUIT: round; less than 0.5"; fleshy; green turning brown; no significant litter

GROWTH RATE: slow-growing; moderate life span

HABIT: columnar; pyramidal; open; symmetrical; medium texture

FAMILY: *Theaceae*

LEAF COLOR: broadleaf evergreen; dark green

LEAF DESCRIPTION: alternate; simple; serrulate margin; oblong to oblanceolate; 2 to 7" long broadleaf

FLOWER DESCRIPTION: white; showy; spring and summer

EXPOSURE/CULTURE:

Light Requirement: shade to part shade; full sun if irrigated

Soil Tolerances: all textures; alkaline to acidic; wet soil; drought-sensitive

Aerosol Salt Tolerance: none

PEST PROBLEMS: resistant on a wet or irrigated site

BARK/STEMS:

Bark is thin and easily damaged by mechanical impact; branches grow mostly upright and horizontal and will not droop; normally grows with a single leader; no thorns

Pruning Requirement: needs little or no pruning to develop strong structure

Limb Breakage: resistant

Current Year Twig: brown; medium thickness

CULTIVARS AND SPECIES:

The obscure 'Variegata' selection has white and green foliage.

LANDSCAPE NOTES:

Origin: coastal southeast North America

Usage: espalier; wetland reclamation; specimen

A native American, usually single-trunked, evergreen tree, Loblolly-Bay reaches mature height with a columnar or pyramidal, open growth habit. The canopy thickens in the full sun. The glossy leaves are light gray on the underside, giving a two-toned effect in the wind. Although evergreen, several individual leaves at a time will turn a brilliant scarlet color in the fall, adding to the tree's attractiveness. The white, two- to three-inch-wide, five-petalled, cup-shaped flowers open from late spring through summer and are very attractive but sparsely produced throughout the canopy.

Loblolly-Bay makes an attractive specimen planting, and its light, airy growth habit lends itself well to smaller, partially enclosed locations. In moist soils, Loblolly-Bay naturalizes well and is well suited to the low-maintenance landscape. Because of the shape of the crown, it makes a suitable tree in urban areas with restricted, narrow overhead space, and might do well as a street tree if located in moist soil. It is nicely suited for ditch banks and near retention ponds. Mulch the area beneath the canopy to reduce turf competition.

Preferring partial shade and moist to wet soil, Loblolly-Bay can tolerate full sun only with sufficient moisture. Loblolly-Bay has a shallow root system and will die if not watered during periods of drought. It is found in the wild most often growing in wet sites in the shade of maple, cypress, and pine. It is well suited for planting in boggy and other poorly drained soils.

Borers, aphids, and caterpillars may attack weakened trees. Plant on a moist or wet site only, unless irrigation will be provided. Root knot nematodes can infect roots causing decline.

Grevillea robusta Silk-Oak

Pronunciation: grah-VILL-ee-ah row-BUS-tah

ZONES: 9B, 10A, 10B, 11

HEIGHT: 80 to 120 feet

WIDTH: 25 to 30 feet

FRUIT: dry; black

GROWTH RATE: rapid-growing; long-lived in dry climates

HABIT: columnar; pyramidal; moderate density; symmetrical; fine texture

FAMILY: *Proteaceae*

LEAF COLOR: evergreen; gray-green above, silvery beneath

LEAF DESCRIPTION: alternate; odd pinnately compound; parted; 2 to 4" long leaflets; twigs and foliage decompose slowly after they fall

FLOWER DESCRIPTION: orange and yellow; showy; flowers in spring and periodically throughout year

EXPOSURE/CULTURE:

Light Requirement: grows best in full sun, bends toward light in part sun

Soil Tolerances: all textures; slightly alkaline to acidic; occasionally wet; drought

Aerosol Salt Tolerance: low

PEST PROBLEMS: resistant

BARK/STEMS:

Branches grow mostly upright and will not droop; should be grown with a single leader; no thorns

Pruning Requirement: needs little pruning to develop a central leader; occasionally, an aggressive lateral branch may need pruning to slow growth if it will compete with leader; thin canopy to reduce branch breakage

Limb Breakage: susceptible to breakage, especially following topping; topping sets the course for continued branch breakage

Current Year Twig: brownish-gray; medium thickness

LANDSCAPE NOTES:

Origin: Australia

Usage: specimen

Reaching a height of 100 feet or more, Silk-Oak is pyramidal to oval in shape, eventually developing a few heavy horizontal limbs and a trunk up to five feet in diameter. The light, ferny leaves are accented by large clusters of bright yellow-orange flowers in spring. Black, leathery seed capsules follow the flowers. A great quantity of leaves fall in the spring immediately preceding the emergence of new growth, and leaves also fall sporadically throughout the year, creating quite a litter problem to some people.

Silk-Oak works as a specimen in large, open landscapes but probably should not be located near houses, due to its large size, messy habit, and the brittleness of the wood as it ages. Tops have been known to snap out of the trees in strong winds. Thinning, not topping, helps to prevent this. It is a valuable timber tree in its native Australia, growing to more than 125 feet tall.

Quick-growing Silk-Oak requires full sun and well-drained soils to perform its best, as roots rot in poorly drained, wet soils. Silk-Oak thrives in heat and is quite tolerant of drought. It grows extremely well in southern California, where it easily reaches 100 feet tall. Tall trees are often hit by lightning in Florida.

Malpracticing arborists top the tree. Oleander pit scale limits growth in Puerto Rico. Mushroom root rot can damage the root system on poorly drained soils. Tops die back in drought. The tree has naturalized in most Hawaiian Islands, as it is a prolific reproducer. Forms a large, invasive root system.

Guaiacum sanctum Lignumvitae

Pronunciation: goo-AYE-uh-kum SANK-tum

ZONES: 10B, 11

HEIGHT: 8 to 15 feet

WIDTH: 8 to 10 feet

FRUIT: round; less than 0.5"; fleshy; yellow; no significant litter; showy

GROWTH RATE: slow-growing; long-lived

HABIT: rounded vase; dense; symmetrical; fine texture

FAMILY: *Zygophyllaceae*

LEAF COLOR: evergreen; dark green

LEAF DESCRIPTION: opposite; even pinnately compound; entire margin; oval to obovate; leaflets less than 2" long

FLOWER DESCRIPTION: blue or purple; showy; periodic year-round flowering, mainly blooms in fall and spring or when dry

EXPOSURE/CULTURE:

Light Requirement: flowers best in full sun

Soil Tolerances: all textures; alkaline to acidic; wet soil; salt and drought

Aerosol Salt Tolerance: high

PEST PROBLEMS: pest-free

BARK/STEMS:

Very showy, exfoliating bark; branches droop as tree grows and will require pruning for vehicular or pedestrian clearance beneath canopy; routinely grown with multiple trunks; no thorns

Pruning Requirement: requires little pruning to develop strong structure; thin canopy to display showy trunk and branch form

Limb Breakage: resistant

Current Year Twig: gray; slender

CULTIVARS AND SPECIES:

G. officinale grows 15 to 30 feet tall and has purplish-blue flowers. *G. angustifolium* is native to Mexico and grows to 10 feet tall.

LANDSCAPE NOTES:

Origin: south Florida into Caribbean islands

Usage: bonsai; container or above-ground planter; parking lot island; buffer strip; near a deck or patio; specimen; sidewalk cutout

Lignumvitae is an extremely slow-growing evergreen that ultimately reaches 8 to 20 feet in height and casts light shade, but few people have seen plants of this size because it is not extensively available in nurseries. Most are seen 8 to 10 feet tall with a beautiful array of twisted, multiple trunks and a rounded canopy much like that of a mature Crape-Myrtle. The one- to two-inch-long, leathery leaves are joined several times throughout the year by large clusters of deep blue flowers, the old flowers fading to a light silvery-blue and creating a shimmering haze over the rounded canopy. These flowers are followed by small, heart-shaped, yellow berries, appearing on the tree at the same time as the blue flowers and creating a lovely sight.

Underneath the smooth, beige/gray bark of Lignumvitae is some of the heaviest of all wood (specific gravity of 1.09), which sinks instead of floating in water. This dense wood was once popular for use in the manufacture of bowling balls, and has also been used for steamship propeller shafts, gears, and mallets. The picturesque crooked trunks, evergreen leaves, and beautiful flowers and fruit would all combine to make Lignumvitae a popular choice for use as a container, patio, or specimen planting if it were widely available in a range of sizes. Unfortunately, like many other slow-growing trees, this one is not often grown in nurseries. One must travel to arboreta or private gardens to view nice specimens of this tree.

Gymnocladus dioicus Kentucky Coffeetree

Pronunciation: jim-no-CLAY-duss die-o-EYE-cuss

ZONES: 3B, 4A, 4B, 5A, 5B, 6A, 6B, 7A, 7B, 8

HEIGHT: 65 to 70 feet

WIDTH: 50 to 70 feet

FRUIT: pod; 5 to 10"; dry and hard; green turning brown; female trees cause significant litter; persistent; showy

GROWTH RATE: slow to moderate; long-lived

HABIT: oval; round; vase shape; open; irregular silhouette; medium texture

FAMILY: *Leguminosae*

LEAF COLOR: light green in summer; showy yellow fall color

LEAF DESCRIPTION: alternate; bipinnately compound; entire margins; oval to ovate; deciduous; leaflets 2 to 4" long, leaves up to 18" long

FLOWER DESCRIPTION: greenish-white; pleasant fragrance; somewhat showy up close; spring

EXPOSURE/CULTURE:

Light Requirement: grows best in full sun

Soil Tolerances: all textures; alkaline to acidic; occasionally wet soil; some salt; extreme drought tolerance

Aerosol Salt Tolerance: some

PEST PROBLEMS: nearly pest-free

BARK/STEMS:

Very showy; branches droop as tree grows and will require pruning for vehicular or pedestrian clearance beneath canopy; should be grown with a single leader; no thorns

Pruning Requirement: needs little pruning following some initial training to develop strong trunk and branch structure; space major limbs along one central trunk; sap bleeds from spring pruning cuts

Limb Breakage: resistant to ice storms

Current Year Twig: brownish-green; stout

CULTIVARS AND SPECIES:

Availability of male cultivars in the near future will increase Coffeetree's popularity, especially as a street tree, due to lack of fruit and more upright canopy.

LANDSCAPE NOTES:

Origin: northeast-central United States

Usage: buffer strip; highway; reclamation; shade tree; specimen; urban tolerant

This medium-growing tree will reach about equal height and spread. The state tree of Kentucky should be used more often because it is adaptable to many soils (including alkaline), has interesting bark, and grows with an open canopy that allows light to penetrate to the ground. The coarse branch texture in the winter is also unique, forming an interesting silhouette of several large branches.

Large seed pods hang on the tree in the winter but can be a litter problem when they fall in the spring. They are very hard and can be "shot" from a lawnmower that runs over the fruit. Male trees that do not set fruit are sometimes available. The seeds and leaves may be poisonous to humans and cattle.

The trunk normally ascends straight up through the crown and is very strong. Branches grow at wide angles to the trunk and are usually well spaced along the trunk. This configuration adds to the durability of the tree. Some people object to the sparse branching when this tree is young, but some pruning to create more branches can help. Any shortcomings of the tree are made up for by the almost total lack of insect or disease problems. Lawns grow well beneath the tree due to the light shade cast by the thin, open canopy.

Large seed pod limits usage in some areas. The tree recovers slowly from transplanting. Using root-pruning containers in the nursery should help produce a high-quality, transplantable tree.

Halesia carolina Carolina Silverbell

Pronunciation: ha-LEE-she-ah care-o-LYE-nah

ZONES: 5A, 5B, 6A, 6B, 7A, 7B, 8A, 8B

HEIGHT: 20 to 40 feet

WIDTH: 15 to 30 feet

FRUIT: oval; 1 to 2"; fleshy, then dry; yellowish-green; no significant litter; persistent; very showy

GROWTH RATE: moderate; short to moderate life span

HABIT: rounded vase; dense in sun; nearly symmetrical; fine texture

FAMILY: *Styracaceae*

LEAF COLOR: medium green in summer; yellow fall color

LEAF DESCRIPTION: alternate; simple; serrulate margin; oval, oblong, or ovate; deciduous; 2 to 5" long

FLOWER DESCRIPTION: white; very showy and delicate; spring

EXPOSURE/CULTURE:

Light Requirement: part shade to full sun; open canopy, fewer flowers in shade

Soil Tolerances: all textures; acidic; occasionally wet; drought-sensitive in full sun

Aerosol Salt Tolerance: some

PEST PROBLEMS: pest-free

BARK/STEMS:

Showy white lines on rapid-growing stems; branches droop as tree grows and will require some pruning for vehicular or pedestrian clearance beneath canopy; can be trained to grow with a single or multiple trunk; no thorns

Pruning Requirement: once a central leader is developed, needs little pruning to maintain strong trunk and branch structure

Limb Breakage: resistant

Current Year Twig: brown; slender

CULTIVARS AND SPECIES:

H. monticola grows taller and makes a nice shade tree. The variety *magniflora* has beautiful large flowers borne on a symmetrical canopy.

LANDSCAPE NOTES:

Origin: scattered throughout southeastern United States

Usage: container or above-ground planter; near a deck or patio; specimen; street tree

A North American native tree, Carolina Silverbell grows into a pyramidal, vase, or rounded silhouette. The two- to five-inch-long leaves turn yellow in fall and are among the first to drop. An understory tree naturally growing in a partially shaded or shaded location, Silverbell prefers moist, fertile soil with an accumulation of leaf litter and/or mulch. But it adapts to sun very well if soil is not allowed to completely dry out and roots are not confined.

The white, bell-shaped, showy blossoms are borne in two- to five-inch-long clusters, even on young, shorter trees. Flowering occurs along last year's branches in mid-May. Because the flowers point downward, they are partially hidden by the foliage and best viewed from under the canopy. Other ornamental features are the yellow fall color and the bark, which peels off in large, flat scales. The pale yellow fruits are quite attractive as they hang down from last year's branches.

This tree is interesting all year long, with attractive foliage, pretty flowers, showy fruits, and exfoliating bark. It branches low to the ground, making a nice lawn or specimen tree, and when pruned to one central leader can be used as a street tree in residential areas. The bark shows off nicely with foliage removed from the lower branches, and multistemmed specimens come to life when lit from below at night. It is a splendid small tree to locate near a patio or deck. Branches pruned in the spring often bleed sap.

Trees and seedlings transplant best in the spring. Water the tree during a drought and avoid compacted soil.

Hamamelis mollis Chinese Witch-Hazel

Pronunciation: ham-ah-MEE-liss MOLL-lis

ZONES: 6A, 6B, 7A, 7B, 8A

HEIGHT: 10 to 20 feet

WIDTH: 12 to 18 feet

FRUIT: 0.5 to 1"; dry; black; no significant litter

GROWTH RATE: slow-growing; short-lived

HABIT: rounded vase; open; irregular silhouette; medium texture

FAMILY: *Hamamelidaceae*

LEAF COLOR: dull gray-green in summer; yellow fall color

LEAF DESCRIPTION: alternate; simple; sinuate margin; obovate to orbiculate; deciduous; 3 to 6" long

FLOWER DESCRIPTION: yellow; very pleasant fragrance; showy; winter; long-lasting

EXPOSURE/CULTURE:

Light Requirement: part shade to full sun

Soil Tolerances: all textures; acidic; moderate drought

PEST PROBLEMS: resistant

BARK/STEMS:

Bark is thin and easily damaged by mechanical impact; branches droop as tree grows and will require pruning for vehicular or pedestrian clearance beneath canopy; routinely grown with multiple trunks; can be trained to a short, single trunk; no thorns

Pruning Requirement: requires early pruning to develop strong structure

Limb Breakage: resistant

Current Year Twig: gray; slender

LANDSCAPE NOTES:

Origin: central China

Usage: container; buffer strip; highway; specimen; deck or patio

Chinese Witch-Hazel is a small, slow-growing, deciduous tree capable of reaching 20 feet high and wide, but is more often seen at 10 to 15 feet. The leaves usually put on a showy display in fall, as the dying leaves change to shades of yellow and orange before dropping. The long-lasting, showy, yellow flowers appear in late winter or early spring and are quite fragrant. However, they may occasionally be injured by cold temperatures (–10°F.). These blooms are followed by the production of inconspicuous, green fruits that turn black and persist on the tree.

Plant Chinese Witch-Hazel near a deck or patio or in the lawn or low groundcover as a specimen. Space branches apart on a short trunk so they can properly develop their horizontal, layered habit. The lowest branch can be located one or two feet from the ground to form a thick canopy all the way to the ground, or if planted close to a walk or patio, five to seven feet up to allow for pedestrian clearance beneath the crown. Trees can also be purchased and trained with multiple trunks for planting in open areas as specimens. This is an attractive, versatile small tree that could be used more in the urban landscape for its compact size and ornamental habit.

Chinese Witch-Hazel should be grown in full sun or partial shade on well-drained, moist, acid soils. Trees grown in the partial shade are very nice, developing an open crown, but do not become leggy and unkempt like some other trees in partial shade. Nice specimens can be found in clay soils, even those that dry out for a period of time in the summer.

No limitations appear to be serious, although the plant has not been widely planted or tested.

Hibiscus syriacus Rose-of-Sharon

Pronunciation: hi-BISS-cuss sear-ee-A-cuss

ZONES: 5B, 6A, 6B, 7A, 7B, 8A, 8B, 9A, 9B

HEIGHT: 8 to 12 feet

WIDTH: 4 to 10 feet

FRUIT: oval; 0.5 to 1"; dry; green then brown; persistent; no litter

GROWTH RATE: slow-growing; short-lived

HABIT: upright vase; open; symmetrical; fine texture

FAMILY: *Malvaceae*

LEAF COLOR: light green in summer; no fall color

LEAF DESCRIPTION: alternate; simple; dentate margin; ovate to rhomboid; deciduous; 2 to 4" long

FLOWER DESCRIPTION: blue, lavender, pink, purple, red, white; showy; summer

EXPOSURE/CULTURE:

Light Requirement: part shade to full sun

Soil Tolerances: all textures; acidic; occasionally wet soil; some drought

Aerosol Salt Tolerance: moderate to good

PEST PROBLEMS: resistant

BARK/STEMS:

Bark is thin and easily damaged by mechanical impact; routinely grown with multiple trunks; branches grow mostly upright and will not droop, except when in flower

Pruning Requirement: needs little pruning to maintain upright growth habit

Limb Breakage: branches with included bark split at crotch, and wood itself is weak

Current Year Twig: gray; slender

CULTIVARS AND SPECIES:

'Admiral Dewey'—single, white flowers; 'Ardens'—purple, semi-double flowers; 'Bluebird'—single, bluish-purple flowers; 'Coelestis'—single violet-blue with reddish-purple throat; 'Mauve Queen'—mauve flowers; 'Red Heart'—single pure white flowers with deep red center; 'Woodbridge'—single flowers, reddish-purple, darker at the base. Some of the newer triploid hybrids including 'Aphrodite', 'Diane', 'Helen', and 'Minerva' are denser with large flowers.

LANDSCAPE NOTES:

Origin: China and India

Usage: container; deck or patio; standard; specimen

Rose-of-Sharon is valued for its large flowers produced in summer when few other shrubs bloom. Individual flowers stay open for one day and close at night. It is useful as a garden accent due to its strict, upright habit. The open, loose branches and light green leaves make Rose-of-Sharon ideally suited to formal or informal plantings, and with a little pruning it makes an attractive, small specimen tree. The plant grows in sun or partial shade and in any soil. Several roots are usually located just beneath the soil surface.

Shaping or pruning can be done at any time but late winter or early spring pruning minimizes loss of emerging flower buds on new growth. Pruning is usually not required, as the plant grows slowly and keeps a tight upright form. Frequent severe pruning gives fewer but larger flowers.

Although tolerant of poor soils and drought in sun or light shade, this upright, deciduous shrub requires ample moisture to flower its best and to avoid leaf drop. Some protection from midday or afternoon sun is beneficial for optimum plant appearance. Tolerance to aerosol salt and wet soils, combined with drought tolerance, make this a fine plant for many landscapes.

Although usually easy to grow, Hibiscus can be bothered by aphids that accumulate at the tips of stems, causing new growth to be misshapen. Japanese beetles are particularly fond of the flowers.

Hovenia dulcis Japanese Raisintree

Pronunciation: hoe-VEE-knee-ah DULL-sis

ZONES: 6A, 6B, 7A, 7B, 8A, 8B, 9A, 9B, 10A

HEIGHT: 30 to 35 feet

WIDTH: 15 to 25 feet

FRUIT: round; less than 0.5"; fleshy; red; attracts wildlife; edible; causes messy litter

GROWTH RATE: moderate; moderate life span

HABIT: oval; pyramidal; open; symmetrical; medium texture

FAMILY: *Rhamnaceae*

LEAF COLOR: glossy green in summer; no fall color change

LEAF DESCRIPTION: alternate; simple; serrate margin; oval to ovate; deciduous; 4 to 6" long

FLOWER DESCRIPTION: white; pleasant fragrance; very showy; summer

EXPOSURE/CULTURE:

Light Requirement: part shade to full sun

Soil Tolerances: all textures; alkaline to acidic; occasionally wet; moderate drought

PEST PROBLEMS: resistant

BARK/STEMS:

Branches grow mostly upright or horizontal and will not droop; showy trunk; should be grown with a single leader; no thorns

Pruning Requirement: needs little pruning to maintain a central leader and well-attached branches

Limb Breakage: resistant

Current Year Twig: brown; slender

LANDSCAPE NOTES:

Origin: China

Usage: container; parking lot island; buffer strip; shade tree; specimen; street tree

Japanese Raisintree can reach 40 to 50 feet in height, but is most often seen smaller with an open, upright, oval silhouette. Unfortunately, it is a little too big for planting beneath most power lines. The tree usually maintains a fairly good central leader with small-diameter main branches.

The long, glossy leaves are particularly striking and create light shade below the trees, but they show no appreciable color change in autumn, dropping while they are still green. In early summer, the branch tips are festooned with small, two- to three-inch-long cymes of sweetly fragrant, greenish-white flowers that are quite attractive to insects. These blooms are followed by the production of small, fleshy, brown drupes that ripen to bright red and have a flavor similar to a sweet raisin, giving the tree its common name.

The bark of Japanese Raisintree is ridged and an attractive light gray with brown valleys. Lower branches can be removed or trimmed to display the lovely bark. The trunk often grows straight up the center of the tree without training.

The moderate growth rate, upright habit, and striking flowers and fruit make Japanese Raisintree useful as a featured accent in the landscape. It is a candidate for a street tree, because of its small to medium size and ascending branches, but it has not yet been extensively used for this purpose.

Trees will grow in almost any well-drained soil in full sun or partial shade. Not for poorly drained areas or compacted soil, Japanese Raisintree prefers adequate soil space for root exploration. No limitations of major concern.

Hydrangea paniculata Panicle Hydrangea

Pronunciation: hi-DRAN-gee-ah pah-nick-you-LAY-tah

ZONES: 4A, 4B, 5A, 5B, 6A, 6B, 7A, 7B, 8A

HEIGHT: 15 to 25 feet

WIDTH: 10 to 20 feet

FRUIT: oval; 0.5 to 1"; dry; pink turning brown; no significant litter

GROWTH RATE: moderate; short-lived

HABIT: upright vase shape; moderate density; irregular silhouette; coarse texture

FAMILY: *Hydrangeaceae*

LEAF COLOR: dark green in summer

LEAF DESCRIPTION: opposite or whorled; simple; serrate margin; oval to ovate; deciduous; 3 to 6" long

FLOWER DESCRIPTION: white; very showy; summer

EXPOSURE/CULTURE:

Light Requirement: part shade to full sun

Soil Tolerances: all textures; alkaline to acidic; moderate drought

PEST PROBLEMS: susceptible to numerous nonlethal problems

BARK/STEMS:

Branches droop as tree grows and will require pruning for vehicular or pedestrian clearance beneath canopy; routinely grown with multiple trunks; no thorns

Pruning Requirement: requires pruning to develop balanced canopy, to remove drooping branches, and to remove suckers

Limb Breakage: susceptible

Current Year Twig: brownish-gray; thick

LANDSCAPE NOTES:

Origin: China, Japan, Sachalin

Usage: container; hedge; near a deck or patio; standard; specimen; urban tolerant

This low-branching, multiple-trunked tree or large shrub is fast-growing and is often seen at about 10 feet tall. The deciduous leaves are 3 to 6 inches long and 1.5 to 3 inches wide, and fade only to a sickly yellow in fall before dropping. The spectacular summertime blooms appear in six- to eight-inch-long panicles, the cream-colored flowers gradually fading over time to purplish-pink. The upright, spreading branches often bend down under the weight of the blooms.

Panicle Hydrangea can be used in a shrub border as a large-sized accent shrub. It may be best to locate it away from the house or patio, because of its large size and spreading habit. Lower branches can be pruned to clean up the bottom of the plant and make it grow into a multi-stemmed tree. Early training can create a single-stemmed small tree that would be well suited for planting as a specimen in a low groundcover or lawn area. They are also suited for planting in above-ground containers to display the nice flowers that develop in the summer. The plant will gracefully droop over a wall if left unpruned.

The faded blooms should be removed by late September to keep the tree looking neat. Without pruning, some people object to the brown fruits and old flowers that hang onto the tree during the winter. Pruning also is suggested to keep the plant to a consistent, neat shape.

Panicle Hydrangea should be grown in full sun or partial shade on well-drained, moist, loamy soil. Plants are fuller in the sun.

Usually no lethal limitations. Aphids, rose chafer, oystershell scale, two-spotted mites, nematodes, bacterial wilt, bud blight, leaf spot, powdery mildew, and rust can be found.

Ilex x attenuata 'East Palatka' 'East Palatka' Holly

Pronunciation: EYE-lecks x ah-ten-u-A-tah

ZONES: 7A, 7B, 8A, 8B, 9A, 9B

HEIGHT: 30 to 45 feet

WIDTH: 10 to 15 feet

FRUIT: round; 0.25"; fleshy; red; attracts wildlife; causes minimum litter; persistent; showy

GROWTH RATE: moderate; long-lived

HABIT: columnar; pyramidal; open when young, moderate density later; symmetrical; medium texture

FAMILY: *Aquifoliaceae*

LEAF COLOR: evergreen; shiny green

LEAF DESCRIPTION: alternate; simple; entire margin; oval to oblong; 2 to 3" long

FLOWER DESCRIPTION: white; not showy; spring

EXPOSURE/CULTURE:

Light Requirement: part shade to full sun

Soil Tolerances: all textures; acidic; salt and drought

Aerosol Salt Tolerance: moderate

PEST PROBLEMS: mostly resistant

BARK/STEMS:

Bark is thin and easily damaged by mechanical impact; branches droop as tree grows and will require pruning for vehicular or pedestrian clearance beneath canopy; should be grown with a single leader; no thorns

Pruning Requirement: if not sheared in nursery, needs little pruning to develop a strong central leader and good branch structure; tip branches to create a denser canopy on young trees; do not top

Limb Breakage: resistant

Current Year Twig: green; medium thickness

LANDSCAPE NOTES:

Origin: a natural hybrid found in Florida

Usage: container; hedge; parking lot island; buffer strip; highways; standard; specimen; sidewalk cutout; residential street tree; urban tolerant

Discovered in 1927 growing near East Palatka, Florida, this female Holly is one of a group of hybrids between *Ilex cassine* x *Ilex opaca*. The shiny, rounded leaves have one spine at the tip and few, if any, along the blade edge. The trees eventually take on a moderately tight, pyramidal shape, although many young trees have an open canopy. A row of East Palatka Hollies looks quite uniform, a trait that adds to the popularity of the tree among landscape architects and designers. Foster's Holly forms a more uniform canopy without pruning.

East Palatka Holly makes a durable street tree throughout its range and is quite drought-tolerant once it becomes well established. The tree grows well in small soil spaces carved out of downtown sidewalks.

Unfortunately, most trees are sheared in the nursery, and this practice is often repeated in the landscape after planting. The natural shape of the tree is rarely seen, but is a graceful pyramid of drooping branches growing from a strong central trunk, laden with bright red berries that remain on the trees until eaten by birds. When grown with one central leader, the crown of East Palatka Holly is narrow, making it well suited for urban areas with restricted vertical space and limited soil space. Multi stemmed trees form a wider canopy after planting in the landscape because new branches droop to the horizontal, then weep, creating an unkempt, asymmetrical mess.

Occasional and sometimes serious problems due to gall, witches' broom, and die-back caused by a fungus (*Sphaeropsis*) in parts of Florida.

Ilex x attenuata 'Fosteri' Foster's Holly

Pronunciation: EYE-lecks x ah-ten-you-A-tah

ZONES: 6A, 6B, 7A, 7B, 8A, 8B, 9A, 9B

HEIGHT: 15 to 25 feet

WIDTH: 8 to 12 feet

FRUIT: round; 0.25"; fleshy; red; attracts birds; no litter; persistent; showy

GROWTH RATE: slow-growing; moderate life span

HABIT: columnar; pyramidal; dense; symmetrical; fine texture

FAMILY: *Aquifoliaceae*

LEAF COLOR: evergreen; dark green

LEAF DESCRIPTION: alternate; simple; spiny margin; oval to ovate; 1 to 3" long

FLOWER DESCRIPTION: white; showy; spring

EXPOSURE/CULTURE:

Light Requirement: part shade to full sun

Soil Tolerances: all textures; acidic; occasionally wet; drought

Aerosol Salt Tolerance: low

PEST PROBLEMS: resistant

BARK/STEMS:

Bark is thin and easily damaged by mechanical impact; branches grow upright then droop forming a tight canopy; should be grown with a single leader; no thorns

Pruning Requirement: needs little pruning to develop strong structure and neat habit

Limb Breakage: resistant

Current Year Twig: green to gray; medium thickness

LANDSCAPE NOTES:

Origin: hybrid

Usage: container; hedge; parking lot island; buffer strip; standard; specimen; sidewalk cutout; street tree; urban tolerant

Foster's Holly #2 is one of the better cultivars of *Ilex x attenuata*, part of a group of hybrids between *Ilex cassine x Ilex opaca*. Foster's Holly reaches mature height with a dense, pyramidal silhouette without pruning. The trunk usually grows straight up through the crown, unless the tree is topped. The small, glossy, almost black-green, linear leaves have spiny margins, and are joined in spring by showy, small, white flowers. The blooms are followed by the heavy production of brilliant red berries, which persist on trees from fall through winter.

With its dense, compact, upright growth and neat habit, Foster's Holly is ideal for use as a clipped or unclipped screen or hedge, or as a specimen, foundation, or container plant. A dense, unclipped screen can be formed by planting trees five to seven feet apart. It is also suited for planting in a small soil space or in a tall, narrow overhead space. It would probably make a suitable street tree but has not been extensively tried.

Foster's Holly should be grown in full sun or partial shade on well-drained, slightly acid, moist soil. It is very drought-tolerant once established.

Scale and leaf miners occasionally infest the foliage, but are usually not serious. *Sphaeropsis* fungus, a serious problem on other *I. x attenuata* cultivars, probably infects 'Fosters' as well.

There are other Foster's Hollies—#1 and #4—but these are less available and perhaps not as showy.

Ilex x attenuata 'Savannah' Savannah Holly

Pronunciation: EYE-lecks x ah-ten-you-A-tah

ZONES: 6A, 6B, 7A, 7B, 8A, 8B, 9A, 9B

HEIGHT: 30 to 45 feet

WIDTH: 6 to 10 feet

FRUIT: round; 0.25"; fleshy; red; attracts birds; no litter; persistent on the tree; showy

GROWTH RATE: moderate; moderate life span

HABIT: columnar; pyramidal; medium density; symmetrical; medium texture

FAMILY: *Aquifoliaceae*

LEAF COLOR: evergreen; medium green

LEAF DESCRIPTION: alternate; simple; spiny; oval to ovate; 2 to 4" long

FLOWER DESCRIPTION: white; not showy; spring

EXPOSURE/CULTURE:

Light Requirement: part shade to full sun

Soil Tolerances: all textures; acidic; drought

Aerosol Salt Tolerance: moderate

PEST PROBLEMS: resistant

BARK/STEMS:

Bark is thin and easily damaged by mechanical impact; branches grow mostly upright then droop; should be grown with a single leader; no thorns

Pruning Requirement: needs little pruning to develop strong structure; remove upright, competing branches early to develop one trunk

Limb Breakage: resistant

Current Year Twig: green; medium thickness

LANDSCAPE NOTES:

Origin: eastern United States

Usage: hedge; screen; parking lot island; buffer strip; specimen; sidewalk cutout; street tree

Savannah Holly is a beautifully shaped tree, with a narrow, open pyramidal to columnar form. A 35-foot-tall tree can be 8 feet wide in 40 years, indicating a moderate growth rate. The spiny, dull, medium green leaves have wavy or spiny margins and are accented in fall with heavy clusters of red berries, which persist throughout the fall and winter. Male and female flowers appear on separate trees and must be located in the same neighborhood to ensure production of berries. Many nurseries propagate from female trees, so most nursery trees have berries. Many trees are grown with a central trunk and skinny lateral branches, although some nurseries offer ones with several upright trunks growing straight up through the crown. Many trees are sheared in the nursery to create more branches and a fuller canopy. The canopy often opens up when the tree is planted and shearing ceases.

Savannah Holly is ideal for use as a street tree, framing tree, specimen, or barrier planting. A row of 'Savannah' Holly planted 8 feet apart makes a nice screen. Roots are rarely invasive, because of their great number and relatively small diameter. This native tree is ideal for naturalizing on moist, slightly acid soils, and the fruit is very attractive to wildlife, serving as an excellent food source. Berry production is best, and the canopy is most dense, in full sun. Savannah Holly foliage thins slightly during drought, but insect infestations are usually minimal. Scales and spider mites infest foliage. Leaf miner occasionally makes tracks in the leaves. The most severe disease is caused by *Sphaeropsis,* a fungus that causes severe die-back and can kill trees.

A popular landscape plant, this broad-leafed evergreen has served a variety of uses through the years. Native Americans used preserved Holly berries as decorative buttons and as barter items. The wood has been used for making canes, scroll work, and furniture, and has even been substituted for ebony in inlay work when stained black.

Ilex cassine Dahoon Holly

Pronunciation: EYE-lecks cass-SEEN

ZONES: 7A, 7B, 8A, 8B, 9A, 9B, 10A, 10B

HEIGHT: 20 to 30 feet

WIDTH: 8 to 12 feet

FRUIT: round; 0.25"; fleshy; red; attracts wildlife; abundant, some litter when berries drop; persistent; showy

GROWTH RATE: slow-growing; moderate life span

HABIT: oval; pyramidal; open; irregular silhouette; medium texture

FAMILY: *Aquifoliaceae*

LEAF COLOR: light green; evergreen

LEAF DESCRIPTION: alternate; simple; entire or spiny margin; oval to oblong; 1 to 3" long

FLOWER DESCRIPTION: white; showy; spring

EXPOSURE/CULTURE:

Light Requirement: mostly shaded to full sun

Soil Tolerances: all textures; slightly alkaline to acidic; wet soil; moderate drought

Aerosol Salt Tolerance: moderate

PEST PROBLEMS: resistant

BARK/STEMS:

Bark is thin and easily damaged by mechanical impact; branches droop as tree grows and will require pruning for vehicular or pedestrian clearance beneath canopy; trainable to be grown with single or multiple trunks; no thorns

Pruning Requirement: needs some training and pruning to develop neat habit and strong structure

Limb Breakage: multi-trunked trees with included bark are susceptible to breakage; prune to a central leader to avoid breakage

Current Year Twig: green; medium thickness

LANDSCAPE NOTES:

Origin: southeastern coastal plain of North America

Usage: container; wetland reclamation; buffer strip; highway; near a deck or patio; specimen; street tree; urban tolerant

Attractive when tightly clipped into a tall screen or allowed to grow naturally into its single-trunked, small tree form, Dahoon Holly is ideal for a variety of landscape settings. Capable of reaching 40 feet in height, Dahoon Holly is usually smaller in the landscape. Trunks on old specimens can be 24 inches in diameter. At least two Hollies (male and female) must be planted in the landscape to ensure production of the brilliant red berries in fall and winter. The berries are an excellent food source for wildlife.

The tree has been largely used to reclaim moist areas, but could be used more in the urban and suburban landscape, especially for its spectacular fruit display. It would be nicely suited for informal plantings around water retention ponds. A row of them planted on 15-foot centers creates an open, colorful screen. It lends itself well to use as a specimen or street tree. Little pruning is needed to create a loose, open habit. It appears to adapt well to the confined spaces of urban and downtown landscapes.

Dahoon grows well in full sun to partial shade, but does best on moist soils because it is native to wet, boggy soils near swamps. Dahoon Holly can tolerate drier locations with some watering, but often has a thin crown in this environment. It is not recommended in the southern part of its range in a dry, exposed site unless irrigation is provided.

Mites can infest foliage on trees planted on dry sites. Occasional and sometimes serious problems due to galls, witches' broom, and die-back caused by a fungus.

Ilex cornuta 'Burfordii' Burford Holly

Pronunciation: EYE-lecks core-NEW-tah

ZONES: 7A, 7B, 8A, 8B, 9A, 9B

HEIGHT: 15 to 25 feet

WIDTH: 15 to 25 feet

FRUIT: round; 0.25"; fleshy; red; attracts birds; no significant litter; persistent; showy

GROWTH RATE: slow-growing; moderate life span

HABIT: round; vase shape; dense; symmetrical; medium texture

FAMILY: *Aquifoliaceae*

LEAF COLOR: evergreen; glossy dark green

LEAF DESCRIPTION: alternate; simple; entire margin; oval to oblong; 1 to 3" long

FLOWER DESCRIPTION: white; spring

EXPOSURE/CULTURE:

Light Requirement: part shade to full sun

Soil Tolerances: all textures; alkaline to acidic; wet; moderate salt; drought

Aerosol Salt Tolerance: moderate

PEST PROBLEMS: resistant

BARK/STEMS:

Bark is thin and easily damaged by mechanical impact; branches droop as tree grows and will require pruning for vehicular or pedestrian clearance beneath canopy; routinely grown with multiple trunks; no thorns

Pruning Requirement: requires little pruning to develop strong structure and uniform shape

Limb Breakage: resistant

Current Year Twig: green turning gray; medium thickness

LANDSCAPE NOTES:

Origin: eastern China and Korea

Usage: container; hedge; street tree; parking lot island; buffer strip; highway; deck or patio; standard; sidewalk cutout; urban tolerant

This dense shrub or small tree has glossy leaves, each with a single terminal spine. Leaves appear varnished and are among the darkest green of any tree. The somewhat-showy clusters of fragrant, springtime, white flowers attract many bees. The large, bright red, long-lasting berries during the fall and winter provide a nice contrast with leaves. The plants are self-fertile and do not need a male plant located nearby for pollination. It is one of the most popular shrubs in some areas of the country and could be used more as a small tree.

Although typically pruned for formal hedges, the large form and gracefully drooping branches of Burford Holly make it ideal for unpruned natural plantings or as a specimen for spacious areas and large buildings. There are much better, finer-textured plants for pruning into formal hedges. Burford Holly can also be trained as an attractive vase-shaped multi-stemmed small tree. Trees trained in this fashion often have a thick canopy comprised of many branches and small-diameter trunks. The cultivar 'Sizzler' can also be used as a small tree.

Growing best in rich, well-drained, slightly acid soil, Burford Holly does well in full sun or part shade. However, flowering and subsequent fruiting are reduced in shady locations. Burford Holly is drought-tolerant and easy to grow once established. In many ways this holly defines drought-tolerant. It is well suited for low-maintenance landscapes that receive little or no irrigation or fertilizer after trees are established. Once the tree reaches 10 or 15 feet tall, growth rate slows. 'Burfordii Nana' reaches 12 to 18 feet tall and makes a better shrub than 'Burfordii', due to its slower growth rate and smaller leaves.

Burford Holly can be plagued with severe infestations of tea scale.

Ilex decidua Possumhaw

Pronunciation: EYE-lecks dih-SID-jew-ah

ZONES: 5A, 5B, 6A, 6B, 7A, 7B, 8A, 8B, 9A

HEIGHT: 10 to 15 feet

WIDTH: 10 to 15 feet

FRUIT: round; 0.25"; fleshy; orange to red; attracts birds; no litter; persistent; showy

GROWTH RATE: slow-growing; moderate life span

HABIT: rounded vase; dense; symmetrical; fine texture

FAMILY: *Aquifoliaceae*

LEAF COLOR: dark green in summer; no fall color

LEAF DESCRIPTION: alternate; simple; crenate margins; oval to oblong; deciduous; 1 to 3" long

FLOWER DESCRIPTION: white; spring

EXPOSURE/CULTURE:

Light Requirement: mostly shaded to full sun

Soil Tolerances: all textures; alkaline to acidic; wet soil; drought

PEST PROBLEMS: resistant

BARK/STEMS:

Bark is thin and easily damaged by mechanical impact; branches droop as tree grows and will require pruning for pedestrian clearance beneath canopy; routinely grown with multiple trunks; no thorns

Pruning Requirement: needs little pruning except to direct growth and remove drooping branches

Limb Breakage: ice can bend and break trunks

Current Year Twig: brown or gray; slender

CULTIVARS AND SPECIES:

Cultivars with nice fruit include: 'Byers Golden'—yellow fruit; 'Council Fire'—persistent orange-red fruit well into the winter; 'Sentry'—cold-hardy only to zone 6, but columnar habit makes it potentially suited for planting in highway medians. There are many other cultivars.

LANDSCAPE NOTES:

Origin: Virginia to Florida and west to Texas

Usage: bonsai; container; buffer strip; highway; near a deck or patio; wetland reclamation; specimen

This native North American tree is often seen as a spreading, 8- to 10-foot-high, multi-stemmed thicket, but can also become a 20-foot-tall tree when planted in partial shade or with proper training and pruning. Although the leaves are deciduous, they do not present any appreciable fall color change. From March to May, small white flowers appear among the leaves. These tiny blooms are followed by the production of small fruits that become brilliant orange/red when they ripen in early autumn. As male trees will not fruit, be sure to purchase females so you will not miss the abundant fruit production. These fruits persist on the tree throughout the winter and are quite showy against the bare branches. After the fruits have been exposed to freezing and thawing, they become a favorite food source of many birds and mammals.

Lower branches are often removed to form a small tree with a tight head of foliage along the outer portion of the crown. Interior leaves are often shaded out and drop from the tree. If lower branches are not removed, the plant develops into a large, spreading mound of foliage reaching to the ground. Often found along stream banks in the wild, Possumhaw tolerates wet soil and can be used to stabilize stream banks. It can also be utilized as a large accent shrub or small tree planted in a lawn area as a specimen or near a patio. Allow for plenty of room for this plant to spread; it looks its best when it develops a symmetrical canopy. It would make a good plant near water retention ponds and other areas that regularly accumulate water.

There do not appear to be many serious problems affecting this small tree.

Ilex latifolia Lusterleaf Holly

Pronunciation: EYE-lecks lat-eh-FOE-lee-ah

ZONES: 7A, 7B, 8A, 8B, 9A, 9B, 10A, 10B

HEIGHT: 20 to 25 feet

WIDTH: 15 to 25 feet

FRUIT: round; 0.25"; fleshy; red; attracts birds; no litter; persistent; showy

GROWTH RATE: slow to moderate; moderate life span

HABIT: round; pyramidal; dense; nearly symmetrical; coarse texture

FAMILY: *Aquifoliaceae*

LEAF COLOR: broadleaf evergreen; dark green

LEAF DESCRIPTION: alternate; simple; serrate margin; oval to oblong; 4 to 8" long

FLOWER DESCRIPTION: creamy yellow; pleasant fragrance; spring

EXPOSURE/CULTURE:

Light Requirement: mostly shaded to full sun

Soil Tolerances: all textures; slightly alkaline to acidic; moderate drought

Aerosol Salt Tolerance: none

PEST PROBLEMS: resistant

BARK/STEMS:

Bark is thin and easily damaged by mechanical impact; branches droop as tree grows and will require pruning for vehicular or pedestrian clearance beneath canopy; trainable to be grown with a single or multiple trunks; no thorns

Pruning Requirement: needs little pruning to develop strong structure; leader remains dominant, which makes training into tree-form fairly simple

Limb Breakage: resistant

Current Year Twig: green; stout

CULTIVARS AND SPECIES:

'Wirt L. Winn' is known for its excellent form and foliage color.

LANDSCAPE NOTES:

Origin: Japan

Usage: container; espalier; hedge; parking lot island; street tree; buffer strip; deck or patio; specimen

This broad-leaved tree can reach 40 feet in height with a 20- to 25-foot spread, but is usually seen at half that size. The glossy, leathery leaves have serrate margins and are unusually coarse-textured and large for a Holly. The thick, newly emerging green shoots add to the coarseness of this attractive tree. New shoots droop under the weight of the emerging leaves, creating a full-crowned, round, or slightly pyramidal canopy. The inconspicuous, yellowish-white spring flowers are followed by a profusion of small, brick-red berries (on female plants), which appear in dense clusters and persist on the plant throughout the winter.

Lusterleaf Holly can be clipped or can grow on its own into a dense screen when located in bright light, but has a more open crown in a mostly shaded garden. Plants will require only occasional pruning to maintain a neat form, but they can be espaliered on a wall or fence quite easily. It makes a nice background plant in a shrub border due to the dark foliage color. Other plants contrast well when planted in front of it. It can be trained into a multi-stemmed specimen or pruned to develop a single trunk for patio or street tree use.

Lusterleaf Holly looks its best if planted in an area receiving less than all-day sun. Some shading in the afternoon would be fine. Not for exposed, windy, dry, full-sun areas, this Holly does well with some shade. Provide some irrigation in a prolonged drought in an exposed site. The bold foliage of Lusterleaf Holly contrasts well with other plants and it is also a striking specimen planting on its own. Thin the canopy and light the tree at night from below to create a nice night-time specimen.

A caterpillar occasionally chews holes in the leaves.

Ilex x *'Nellie R. Stevens'* 'Nellie R. Stevens' Holly

Pronunciation: EYE-lecks x 'Nellie R. Stevens'

ZONES: 6A, 6B, 7A, 7B, 8A, 8B, 9A, 9B

HEIGHT: 20 to 30 feet

WIDTH: 10 to 15 feet

FRUIT: round; 0.25"; fleshy; red; no significant litter; showy

GROWTH RATE: moderate; moderate life span

HABIT: oval; pyramidal; very dense; symmetrical; medium texture

FAMILY: *Aquifoliaceae*

LEAF COLOR: evergreen; dark green

LEAF DESCRIPTION: alternate; simple; spiny margins; oblong; 2 to 3" long

FLOWER DESCRIPTION: white; not showy; spring

EXPOSURE/CULTURE:

Light Requirement: part shade to full sun

Soil Tolerances: all textures; slightly alkaline to acidic; occasionally wet; drought

PEST PROBLEMS: resistant

BARK/STEMS:

Bark is thin and easily damaged by mechanical impact; routinely grown with multiple trunks; can be trained to grow with a single trunk; branches droop

Pruning Requirement: needs little or no pruning to develop strong structure

Limb Breakage: resistant

Current Year Twig: green; medium thickness

LANDSCAPE NOTES:

Origin: hybrid

Usage: hedge or screen; parking lot island; buffer strip; highway; sidewalk cutout; street tree

A hybrid between *Ilex aquifolium* and *Ilex cornuta*, Nellie R. Stevens Holly has kept the best traits of both parents, with lustrous leaves and abundant fruit production. Leaves are among the darkest of any plant. Vigorous and fast-growing, this Holly quickly grows into an attractive, broad pyramidally-shaped evergreen. It will need a male Holly nearby to ensure pollination and production of the vivid red berries. Chinese Holly, *Ilex cornuta,* will flower at the proper time and may be used for this purpose.

Nellie R. Stevens Holly is one of the best Hollies for the warmer regions of the country, and is ideally suited for use as a screen or border due to its very dense, symmetrical habit. It maintains a nice, uniform shape without pruning. It is also now becoming widely available. Locate it where it will have enough space to spread, as trees become wide at the base.

Lower branches should be removed to create a clear trunk for planting along a walk or near a patio, but the tree really shines (literally) as a specimen or screen allowed to develop with all branches intact to the ground. Allow plenty of room for lower branch growth since plants quickly spread 15 feet wide or more. Nurseries grow the tree either as a multi-stemmed clump or with one central leader. Multi-stemmed trees may not hold up in ice storms as well as those with a central leader. Main branches on single-leadered trees are usually well secured to the trunk, making the tree sturdy and a permanent fixture for almost any landscape.

Nellie R. Stevens Holly should be grown in full sun or partial shade on well-drained, slightly acid soil. Plants are very drought-resistant once established. No limitations of major concern; perhaps scale on occasion.

Ilex opaca American Holly

Pronunciation: EYE-lecks oh-PACK-ah

ZONES: 6A, 6B, 7A, 7B, 8A, 8B, 9A, 9B

HEIGHT: 40 to 50 feet

WIDTH: 15 to 25 feet

FRUIT: round; 0.33"; fleshy; red; attracts birds; no litter; persistent; showy

GROWTH RATE: slow-growing; long-lived

HABIT: pyramidal; dense; symmetrical; medium texture

FAMILY: *Aquifoliaceae*

LEAF COLOR: evergreen; dull green

LEAF DESCRIPTION: alternate; simple; spiny margin; oval; 2 to 3" long

FLOWER DESCRIPTION: whitish-green; pleasant fragrance; attracts bees

EXPOSURE/CULTURE:

Light Requirement: mostly shaded to full sun

Soil Tolerances: all textures; slightly alkaline to acidic; occasionally wet soil; salt and drought

Aerosol Salt Tolerance: extremely high

PEST PROBLEMS: resistant

BARK/STEMS:

Bark is thin and easily damaged by mechanical impact; branches droop as tree grows and will require pruning for vehicular or pedestrian clearance beneath canopy; should be grown with a single leader; no thorns

Pruning Requirement: needs little pruning to develop strong structure because leader remains dominant

Limb Breakage: resistant

Current Year Twig: brownish-green; medium thickness

LANDSCAPE NOTES:

Origin: New Jersey to Texas and Florida

Usage: bonsai; hedge; parking lot island; street tree; buffer strip; highway; reclamation plant; standard; specimen; sidewalk cutout; urban tolerant

A popular landscape plant since the beginning of American history, this broad-leafed evergreen has served a variety of uses through the years. The Native Americans used preserved Holly berries as decorative buttons, which were much sought after as barter items. The wood has been used for making canes, scroll work, and furniture, and has even been substituted for ebony in inlay work when stained black.

American Holly is a beautifully shaped tree, with a symmetrical, dense, wide pyramidal form. The spiny leaves are accented with clusters of red berries that persist throughout the fall and winter. Male and female flowers appear on separate trees, and trees of both sexes must be located in the same neighborhood to ensure production of berries on female plants.

American Holly is ideal for use as a street or courtyard tree (with lower branches removed), framing tree, specimen, barrier planting, or screen. Roots are shallow and finely branched, and rarely invasive due to their great number and relatively small diameter. This native tree is ideal for naturalizing on moist, slightly acid soils, and the fruit is very attractive to wildlife, serving as an excellent food source. A 35-foot-tall tree can be 20 feet wide in 40 years. They are common along the coastal dunes in the mid-Atlantic states, but can also be found in Florida.

Berry production is highest in full sun. American Holly foliage thins during drought but this usually does not permanently injure the tree. Hundreds of cultivars of American Holly, too numerous to list, have been developed and hybridized over the years, providing variety of form, leaf characteristics, and fruit color.

Many insects and diseases are listed for Holly. Scale may be the most prevalent and can make trees become unthrifty.

Ilex rotunda Lords Holly, Round-leaf Holly

Pronunciation: EYE-lecks row-TUN-dah

ZONES: 8B, 9A, 9B, 10A, 10B

HEIGHT: 25 to 35 feet

WIDTH: 25 to 35 feet

FRUIT: round; 0.25"; fleshy; red; attracts birds; no litter; persistent; showy

GROWTH RATE: slow to moderate; moderate life span

HABIT: spreading and rounded; dense; nearly symmetrical; medium texture

FAMILY: *Aquifoliaceae*

LEAF COLOR: broadleaf evergreen; dark green

LEAF DESCRIPTION: alternate; simple; entire margin; oval to oblong; 1.5 to 3" long

FLOWER DESCRIPTION: yellow; spring

EXPOSURE/CULTURE:

Light Requirement: part shade to full sun

Soil Tolerances: all textures; slightly alkaline to acidic; occasionally wet; moderate drought

Aerosol Salt Tolerance: none

PEST PROBLEMS: resistant

BARK/STEMS:

Light gray; bark is thin and easily damaged by mechanical impact; branches droop as tree grows and will require some pruning for vehicular or pedestrian clearance beneath canopy; trainable to be grown with a single trunk or multiple trunks; no thorns

Pruning Requirement: needs initial pruning to develop strong structure; stake trunk in nursery to create an upright, tree form, then space main branches along a short, single trunk

Limb Breakage: resistant

Current Year Twig: green; medium thickness

LANDSCAPE NOTES:

Origin: Japan, Korea

Usage: container or above-ground planter; espalier; hedge; parking lot island; street tree; buffer strip; near a deck or patio; specimen

This small-leaved, evergreen tree can eventually reach 40 feet in height with a 20- to 25-foot spread, but is usually smaller. The glossy, leathery leaves have entire margins and are similar to the *Ilex x attenuata* selections. New shoots droop under the weight of the emerging leaves, creating a full, rounded to spreading canopy. The inconspicuous, yellowish-white spring flowers are followed by a profusion of small, brick-red berries (on female plants), which appear in dense clusters and persist on the plant throughout the winter. The berries are unusually abundant and persistent, making this a very showy tree during the winter months.

Lords Holly can be clipped or can grow on its own into a dense-canopied, medium-sized shade tree. It has not been extensively used in zones 9 and 10, so it is not well known there, but past experience in zone 8B shows that this can be a very nice shade tree useful for many landscapes. Unfortunately, the unusually cold weather in the mid 1980s killed trees to the ground in the southern part of zone 8B. The tree deserves to be tried in zones 9 and 10.

Plants will require only occasional pruning to maintain a neat form, but they can be espaliered on a wall or fence quite easily. Lords Holly makes a nice background plant in a shrub border due to the foliage color. Other plants contrast well when planted in front of it. It can be trained into a multi-stemmed specimen or pruned to develop a single trunk for patio or street tree use. The spreading canopy makes it well suited for planting in a parking lot to provide shade for cars.

Provide some irrigation in a prolonged drought in an exposed site. Thin the canopy and light the tree at night from below to create a nice night-time specimen.

A caterpillar occasionally chews holes in the leaves.

Ilex serrata Finetooth Holly

Pronunciation: EYE-lecks sir-A-tah

ZONES: 5A, 5B, 6A, 6B, 7A, 7B, 8A, 8B

HEIGHT: 12 to 16 feet

WIDTH: 12 to 16 feet

FRUIT: round; 0.33"; fleshy; red; attracts birds; no significant litter; persistent; showy

GROWTH RATE: very slow-growing; long-lived

HABIT: rounded vase; moderate density; somewhat irregular; fine texture

FAMILY: *Aquifoliaceae*

LEAF COLOR: dark green in summer; no fall color

LEAF DESCRIPTION: alternate; simple; serrate margin; oblanceolate; deciduous; 1 to 3" long

FLOWER DESCRIPTION: white; not showy; spring

EXPOSURE/CULTURE:

Light Requirement: part shade to full sun

Soil Tolerances: all textures; acidic; occasionally wet; drought

PEST PROBLEMS: resistant

BARK/STEMS:

Bark is thin and easily damaged by mechanical impact; branches on young trees grow mostly upright, then droop on older trees; small, lower branches may need pruning for pedestrian clearance beneath canopy; routinely grown with multiple trunks; no thorns

Pruning Requirement: needs little pruning to develop strong structure; drooping branches can be removed to create a more upright habit

Limb Breakage: resistant

Current Year Twig: brown; slender

CULTIVARS AND SPECIES:

'Sundrops' grows to about 10 feet tall by 15 feet wide with yellow fruit.

LANDSCAPE NOTES:

Origin: Japan

Usage: bonsai; container or above-ground planter; buffer strip; highway; reclamation; specimen; patio; parking lot island

Finetooth Holly may take as long as 35 years to reach mature height. The thin, multi-stemmed branches rise from the ground forming a fountain of branches and foliage similar to Winterberry Holly. Branches and stems weep under the weight of the foliage and berries, creating a graceful, vase-shaped, somewhat irregular canopy. The bright red, persistent berries provide vibrant color in early winter after the leaves have fallen off. Plants with a full load of fruit appear to glow red. At least one white and one yellow cultivar are available at selected nurseries. Young plants are somewhat irregularly shaped, but fill out to display a wonderful, upright and spreading canopy.

Use it in a shrub border or in the landscape as a specimen, or in any other area to attract birds. When planted in turf as a specimen, be sure to keep the soil under the canopy mulched so weeping stems and branches can droop to display the nice form. If turf is allowed to grow beneath the crown, it will thin due to the dense shade, and low branches will interfere with mowing equipment. Little pruning is needed if the plant is properly located to allow for its spread. The small tree is well adapted to the rigors experienced in a parking lot buffer strip, provided the soil does not become too compacted.

A neater canopy form can easily be developed by selective clipping of the branches. The tree should require little, if any, maintenance once it is well established in the landscape. The slow growth ensures that pruning needs will be minimal. The plant grows in sun or partial shade in a rich, well-drained, acid soil. No serious limitations.

Ilex verticillata Winterberry

Pronunciation: EYE-lecks ver-tiss-ih-LAY-tah

ZONES: 4A, 4B, 5A, 5B, 6A, 6B, 7A, 7B, 8A, 8B

HEIGHT: 6 to 10 feet

WIDTH: 6 to 10 feet

FRUIT: round; 0.33"; fleshy; bright red; attracts birds; no significant litter; persistent; very showy

GROWTH RATE: slow-growing; moderate life span

HABIT: rounded vase; moderate density; symmetrical; fine texture

FAMILY: *Aquifoliaceae*

LEAF COLOR: dark green in summer

LEAF DESCRIPTION: alternate; simple; serrate margin; oblanceolate; deciduous; 1 to 3" long; showy yellow fall color

FLOWER DESCRIPTION: white; not showy; spring

EXPOSURE/CULTURE:

Light Requirement: part shade to full sun

Soil Tolerances: all textures; acidic; wet soil; drought-sensitive

PEST PROBLEMS: resistant

BARK/STEMS:

Bark is thin and easily damaged by mechanical impact; branches droop as tree grows and will require pruning for pedestrian clearance beneath canopy; routinely grown with multiple trunks; no thorns

Pruning Requirement: needs little pruning to develop good structure; drooping branches can be removed to create a more upright habit

Limb Breakage: resistant

Current Year Twig: brown; slender

CULTIVARS AND SPECIES:

'Afterglow'—bright red berries and a heavy fruit set; 'Aurantiaca'—bright orange fruit; 'Chrysocarpa'—yellow fruits; 'Fastigiata'—narrow, upright; 'Nana'—3.5 feet tall, large fruits; 'Red Sprite'—large, bright red fruit, grows to about five feet tall; 'Winter Red'—dense branching, dark foliage, heavy fruit production. There are many more. *I. serrata* is a similar plant that slowly reaches about 16 feet tall and is well suited for planting in confined soil spaces.

LANDSCAPE NOTES:

Origin: eastern North America

Usage: bonsai; container or above-ground planter; buffer strip; highway; reclamation; specimen; patio; wetlands

Winterberry's thin, multi-stemmed branches grow slowly. Branches and stems weep under the weight of the foliage and berries, creating a graceful, vase-shaped, symmetrical canopy. Winterberry's bright red, persistent berries provide vibrant color in early winter after the leaves have fallen off. Plants with a full load of fruit appear to glow red. Young plants are somewhat irregularly shaped, but fill out to become a wonderful, drooping mass. Winterberry is dioecious, so both male and female plants are needed for fruit production. The fruits of this native are relished by birds.

Use it in a shrub border or in a landscape as a specimen, or in any other area to attract birds. When planted in turf as a specimen, be sure to keep the soil under the canopy clear of turf by mulching so weeping stems and branches can droop to display the nice form. If turf is allowed to grow beneath the crown, it will thin due to the dense shade, and low branches will interfere with mowing equipment. Little pruning is needed if the plant is properly located to allow for its spread.

The plant grows in sun or partial shade in a rich, well-drained, acid soil, though it thrives in swampy areas. Its tolerance to wet soil makes this a useful plant in poorly drained landscapes. It is native to swampy areas with standing water for long periods. There are no serious limitations.

Ilex vomitoria Yaupon Holly

Pronunciation: EYE-lecks vom-ih-TOUR-ee-ah

ZONES: 7A, 7B, 8A, 8B, 9A, 9B

HEIGHT: 15 to 20 feet

WIDTH: 15 to 20 feet

FRUIT: round; 0.25"; fleshy; red; attracts wildlife; no significant litter; persistent; showy

GROWTH RATE: moderate; moderate life span

HABIT: rounded vase; open; nearly symmetrical; fine texture

FAMILY: *Aquifoliaceae*

LEAF COLOR: evergreen; gray-green

LEAF DESCRIPTION: alternate; simple; crenate margins; ovate; 1 to 1.5" long

FLOWER DESCRIPTION: white; spring; not showy; attracts bees

EXPOSURE/CULTURE:

Light Requirement: shade to full sun

Soil Tolerances: all textures; alkaline to acidic; occasionally wet; salt and drought

Aerosol Salt Tolerance: high

PEST PROBLEMS: resistant, except for possible *Sphaeropsis* infection which can be lethal

BARK/STEMS:

Bark is thin and easily damaged by mechanical impact; branches droop as tree grows and will require pruning for vehicular or pedestrian clearance beneath canopy; routinely grown with multiple trunks; no thorns

Pruning Requirement: needs pruning to remove lower branches and basal sprouts

Limb Breakage: resistant

Current Year Twig: gray; slender

CULTIVARS AND SPECIES:

'Pride of Houston'—shrub to small tree with heavy fruit production; 'Nana'—dwarf, compact shrub form, male plant, no berries; 'Schelling's'—supposedly more compact than 'Nana'.

LANDSCAPE NOTES:

Origin: New York to Florida and Texas

Usage: bonsai; container; hedge; parking lot island; buffer strip; highway; deck or patio; standard; sidewalk cutout; urban tolerant

Yaupon Holly has small, leathery leaves densely arranged along smooth, stiff, light gray branches. Young plants can appear awkward and require five to eight years to develop a distinct vase shape. The flowers attract bees for several weeks. Female plants produce brilliant red berries (yellow on some cultivars) that are quite attractive to wildlife. If you want a berry-producing plant, purchase plants with berries on them or ones propagated from cuttings of female plants.

Yaupon Holly is ideal for training into a small tree, with lower branches removed to reveal the interestingly contorted, multiple trunks. It can also be used for topiaries, espaliers, specimens, screens, or barriers. It sprouts readily from the roots, forming clumps of upright shoots beneath the canopy. Sprouting is most troublesome if the soil beneath the canopy is disturbed (as when shrubs or flowers are planted under the tree).

A tough native of the southern United States, Yaupon Holly grows in a variety of locations, in full sun or shade, seaside or swamps, sand or clay. Crowns will be thin in the shade. It will grow in soil with a pH in the 7s, is very tolerant of drought, and adapts to a small soil space. Yaupon Holly is well suited for hot, dry locations, such as exposed parking lots, along highways, and other stressful, urban sites. It can also be planted on 15- to 20-foot centers along a walkway to create a canopy. The crown is spreading with branches drooping toward the ground. This makes use as a street tree difficult, requiring regular pruning for vehicular clearance.

Ilex vomitoria 'Pendula' Weeping Yaupon Holly

Pronunciation: EYE-lecks vom-ih-TORE-ee-ah PEN-dyoo-lah

ZONES: 7A, 7B, 8A, 8B, 9A, 9B

HEIGHT: 20 to 30 feet

WIDTH: 6 to 12 feet

FRUIT: round; 0.25"; fleshy; red; attracts birds; no significant litter; persistent; showy

GROWTH RATE: moderate; moderate life span

HABIT: upright, weeping vase; open; irregular silhouette; fine texture

FAMILY: *Aquifoliaceae*

LEAF COLOR: evergreen; gray-green

LEAF DESCRIPTION: alternate; simple; crenate margin; ovate; 1 to 1.5 " long

FLOWER DESCRIPTION: white; spring; attracts bees

EXPOSURE/CULTURE:

Light Requirement: part shade to full sun

Soil Tolerances: all textures; alkaline to acidic; occasionally wet; salt and drought

Aerosol Salt Tolerance: high

PEST PROBLEMS: free of most pests

BARK/STEMS:

Bark is thin and easily damaged by mechanical impact; branches droop as tree grows and will require pruning for vehicular or pedestrian clearance beneath canopy; routinely grown with multiple trunks; can be trained to a single leader; no thorns

Pruning Requirement: needs some early pruning to develop upright tree form; remove lower branches to create tree form

Limb Breakage: resistant

Current Year Twig: gray; slender

LANDSCAPE NOTES:

Origin: eastern United States

Usage: bonsai; container; parking lot island; buffer strip; highway; deck or patio; specimen; screen

Weeping Yaupon Holly makes a very distinct, irregular, weeping form with its upright, crooked trunks and slender, curved, pendulous branches clothed with small, oval leaves. Many nursery operators produce this tree with several trunks in a clump. Capable of reaching 30 feet or more in height, Weeping Yaupon Holly is most often much smaller. Old plants will spread to 25 feet if left unpruned. The inconspicuous male and female flowers appear on separate plants and are followed in fall and winter by a spectacular display of translucent red berries, which attract wildlife. The flowers attract bees for several weeks.

The tree is best used as an accent or specimen due to its unusual form, but planted about 8 to 10 feet apart makes a nice, wide screen in the full sun. On older trees next to a sidewalk or patio, lower branches can be removed to allow for pedestrians to pass beneath. It is well suited for planting close to buildings because the crown is fairly narrow and adapts to obstacles well. Because new, upright stems emerge from the upper sides of weeping branches, a contorted or twisted trunk is formed on older trees. Young, establishing trees can add two to three feet of height each year.

A sturdy North American native, Weeping Yaupon Holly is adaptable to a wide range of cultural conditions, from well-drained to wet, acid to alkaline, and sun to part shade. It is very tolerant of drought and sea salt, and is one of the most durable and adaptable of the small-leaved evergreen Hollies for use in southern landscapes. Light pruning may be necessary to maintain a uniform shape, but unlike the species it requires less maintenance because it does not sprout from the roots.

Scale, leaf miners, mites, and aphids form part of a long list of potential problems, but except for *Sphaeropsis* none are normally serious.

Jacaranda mimosifolia (acutifolia) Jacaranda

Pronunciation: jah-cah-RAN-dah mih-moe-sih-FOE-lee-ah

ZONES: 10A, 10B, 11

HEIGHT: 35 to 45 feet

WIDTH: 45 to 60 feet

FRUIT: irregular pod; 1 to 3"; dry; brown; causes some litter; persistent

GROWTH RATE: rapid-growing; short to moderate life span

HABIT: vase shape; open; nearly symmetrical; fine texture

FAMILY: *Bignoniaceae*

LEAF COLOR: light green

LEAF DESCRIPTION: alternate; bipinnately compound; entire margin; obovate to rhomboid leaflets; deciduous; 0.33" long

FLOWER DESCRIPTION: lavender or purple; pleasant fragrance; showy; spring and summer

EXPOSURE/CULTURE:

Light Requirement: flowers best in full sun

Soil Tolerances: loam; sand; slightly alkaline to acidic; moderate drought

Aerosol Salt Tolerance: none

PEST PROBLEMS: resistant

BARK/STEMS:

Bark is thin and easily damaged by mechanical impact; branches droop as tree grows and will require pruning for vehicular or pedestrian clearance beneath canopy; should be grown with a single leader; watch for included bark; no thorns

Pruning Requirement: remove or shorten aggressive lower limbs when tree is young; branches regularly die in canopy and must be removed to maintain neat apearance

Limb Breakage: susceptible to breakage at crotch due to poor collar formation

Current Year Twig: brown or gray; thick

CULTIVARS AND SPECIES:

'Alba' is a white-flowered cultivar with a longer blooming period but sparser blooms.

LANDSCAPE NOTES:

Origin: Brazil

Usage: large parking lot island; shade tree; specimen; street tree

Soft, delicate, fernlike, deciduous foliage and dense terminal clusters of lavender-blue, lightly fragrant, trumpet-shaped flowers make this large, spreading tree an outstanding specimen plant. Flowering is reportedly best in poor soil and following a winter with several nights in the upper 30s. Jacarandas can reach 50 to 60 feet in height with an equal spread, with arching trunks covered with light gray bark. Unpruned trees are often shorter than this and are wider than tall.

Jacaranda can make an ideal street tree, creating a spectacular sight when in full bloom, but trees are rarely pruned appropriately. The arching branch habit is ideal for creating a canopy over a street or boulevard, but lower branches are usually left on the tree too long. They eventually droop too low for vehicular clearance and must be removed. The 6- to 10-inch diameter wound left by this pruning is very damaging to the tree. Remove or shorten these lower limbs much earlier to force growth in the upper canopy. Be sure to plant only trees with one central trunk and major limbs spaced well apart for street tree and other high-use areas. Prune branches so they remain less than half the diameter of the trunk to help keep the plant intact and increase durability.

Heaviest-flowering when grown in full sun, small trees of Jacaranda can tolerate light shade and will grow quickly. They thrive in any well-drained soils, but should be watered during dry periods. Grafted trees flower when young.

Mushroom root rot on poorly drained soil. Trunks decay after removal of large branches. Seedlings can appear near the tree and become weeds.

Jatropha integerrima Peregrina, Fire-Cracker

Pronunciation: juh-TRO-fah in-tuh-GAIR-ruh-muh

ZONES: 10B, 11

HEIGHT: 10 to 15 feet

WIDTH: 10 to 15 feet

FRUIT: oval; 0.5 to 1"; no litter

GROWTH RATE: slow-growing; short life

HABIT: weeping, rounded vase; open; symmetrical; medium texture

FAMILY: *Euphorbiaceae*

LEAF COLOR: dark green; evergreen

LEAF DESCRIPTION: alternate; simple; lobed, variable; oblong to obovate; 3 to 7" long

FLOWER DESCRIPTION: red; showy; year-round

EXPOSURE/CULTURE:

Light Requirement: flowers in part shade to full sun

Soil Tolerances: all textures; alkaline to acidic; moderate drought

Aerosol Salt Tolerance: moderate

PEST PROBLEMS: resistant

BARK/STEMS:

Bark is thin and easily damaged by mechanical impact; branches droop as tree grows; routinely grown with multiple trunks; can be trained to grow with a single trunk; no thorns

Pruning Requirement: head back and thin branches to create a more upright canopy

Limb Breakage: resistant

Current Year Twig: brown; medium thickness

CULTIVARS AND SPECIES:

A pink-flowered form is available at some nurseries.

LANDSCAPE NOTES:

Origin: Cuba

Usage: container or above-ground planter; buffer strip; highway; deck or patio; standard; specimen

This slender-stemmed, multi-trunked tropical evergreen tree or large shrub, a native of Cuba, has unusual, long leaves varying in shape from oblong to fiddle-shaped or even-lobed. The one-inch-wide red flowers are produced year-round in beautiful clusters held upright above the foliage and help make Fire-Cracker an interesting specimen plant. The seed capsules that follow hold several smooth, speckled, toxic seeds, a fact to be considered when placing this plant in the landscape; it should be kept out of the reach of children.

Jatropha makes a delightful red-flowered accent in a shrub border planted to attract attention to an area. It is quite popular as a patio tree or garden accent, and looks great in a container. The tree can be staked and trained to grow with a short trunk for three or four feet. This is a nice way to display the plant as an accent or specimen. Do not expect this small tree to provide shade, but it will attract hummingbirds and butterflies.

Peregrina should be grown in full sun or partial shade on any well-drained soil. Plants in full sun flower best. No limitations of major concern, but occasionally bothered by mites, scales, and superficial leaf miner. Seeds are toxic.

Juglans nigra Black Walnut

Pronunciation: JUG-lanz NYE-grah

ZONES: 5A, 5B, 6A, 6B, 7A, 7B, 8A, 8B, 9A, (9B), (10A), (10B)

HEIGHT: 60 to 80 feet

WIDTH: 60 to 80 feet

FRUIT: round; 1 to 2"; dry and hard; green; attracts mammals; causes significant, golfball-like litter

GROWTH RATE: moderate; long-lived

HABIT: rounded; very open; irregular silhouette; coarse texture

FAMILY: *Juglandaceae*

LEAF COLOR: medium green in summer; usually no fall color change, occasionally turns strikingly yellow

LEAF DESCRIPTION: alternate; odd pinnately compound; serrate margin; lanceolate to ovate; deciduous; leaflets 2 to 5" long

FLOWER DESCRIPTION: not showy; spring

EXPOSURE/CULTURE:

Light Requirement: part to full sun

Soil Tolerances: all textures; slightly alkaline to acidic; occasionally wet soil; salt and drought

Aerosol Salt Tolerance: good

PEST PROBLEMS: susceptible to fungus diseases in humid climates

BARK/STEMS:

Large trunk; branches grow mostly upright and horizontal and will not droop; should be grown with a single leader; no thorns

Pruning Requirement: needs little pruning to develop strong structure; space main branches along trunk; to prevent bleeding prune in summer or fall, not spring

Limb Breakage: resistant

Current Year Twig: brownish-gray; thick

CULTIVARS AND SPECIES:

'Laciniata' is a beautiful tree with cut leaflets. Other cultivars exist for nut quality. *Juglans microcarpa, regia, californica,* and *cineria* are similar in many ways. *J. cinerea* is very sensitive to a deadly fungus pathogen. *J. microcarpa* (zone 4B) tolerates high pH soils. *J. regia* is smaller than the rest.

LANDSCAPE NOTES:

Origin: South Dakota to New Jersey south to Florida and Texas

Usage: nuts; specimen; highway; shade

Black Walnut grows with a rounded crown to about 70 feet (it can reach 100 to 150 feet in the woods) and spreads 60 to 80 feet when open-grown. Best growth occurs in a sunny, open location with the moist, rich soil common along stream banks in its native habitat. The tree grows rapidly when young, but slows down with age and develops with a number of massive branches well spaced along the trunk, forming a very strong, durable tree. Though valued as a lumber tree, it may not make the best yard tree. The nuts are a nuisance to clean up and leaves often fall prematurely from some type of leaf disease. Seeds are used for candy-making, cleaning abrasives, and explosives.

Black Walnut roots contain juglone, which inhibits growth of some plants beneath the tree. Plants such as tomato and evergreens are quite sensitive to juglone. Trees produce a strong tap root on well-drained, loose soils and recover poorly after transplanting, unless planted in spring. Trees with trunks to five feet in diameter can be found in the eastern part of the country.

The tree is probably best used in a park, campus, or other open space area. However, the fruit is very hard and can dull lawn mower blades quickly, and a mower can "shoot" the fruit at a high rate of speed, possibly injuring people in the area. Place the tree so it will receive an adequate supply of water. It drops leaves in dry spells and is poorly adapted for compacted, urban soils. It is really happiest in the loose, gravelly soil of stream banks and other undisturbed areas, but tolerates alkaline and wet soil very well.

Fall webworm and other caterpillers eat foliage. Some leaf scorch occurs in droughts. Leaves usually drop during late summer and litter the ground along with the fruit.

Juniperus ashei Ashe Juniper, Mountain-Cedar

Pronunciation: jew-NIP-ur-us ASH-ee-eye

ZONES: 7A, 7B, 8A, 8B, 9A

HEIGHT: 35 to 40 feet

WIDTH: 15 to 25 feet

FRUIT: oval; round; 0.25"; fleshy; blue; attracts birds; no significant litter; persistent; showy

GROWTH RATE: moderate; long-lived in dry climates

HABIT: columnar to oval; pyramidal when young; open; irregular silhouette; fine texture

FAMILY: *Cupressaceae*

LEAF COLOR: evergreen; green to blue-green

LEAF DESCRIPTION: scale-like

FLOWER DESCRIPTION: spring

EXPOSURE/CULTURE:

Light Requirement: full sun

Soil Tolerances: all textures; alkaline to slightly acidic; extreme drought tolerance

PEST PROBLEMS: resistant

BARK/STEMS:

Showy; branches grow mostly upright and will not droop; should be trained to grow with a single trunk; no thorns

Pruning Requirement: needs some pruning to develop a strong central leader; remove aggressive, upright branches

Limb Breakage: resistant

Current Year Twig: grayish to red; slender

LANDSCAPE NOTES:

Origin: Missouri south into Mexico

Usage: bonsai; buffer strip; highway; reclamation; street tree

This multi-stemmed, native North American tree has fragrant, resinous leaves. Although the flowers are small and insignificant, the fruits that follow ripen in August and September into large, blue, berry-like cones which are relished by wildlife. The twigs and foliage are occasionally browsed by mammals. The bark of Ashe Juniper peels off in long, thin strips, which are used by birds for nesting material.

Ashe Juniper is best suited for dry landscapes. The buffer strip around an exposed parking lot would be a perfect place to plant it, provided it was not irrigated much (if at all). Be sure to provide irrigation until it is well established, though. A row of Junipers makes a nice, fine-textured screen or windbreak. Birds use the tree for nesting and food.

The wood of Ashe Juniper is strong enough to use for posts and poles. It can be found growing with a single trunk or in clumps. It is probably best used in unirrigated landscapes, as it is susceptible to Juniper blight, a foliage disease that thins the canopy. Many locations in the eastern part of the country are probably too humid to successfully grow this tree. It is most often found in the west and in the drier parts of Texas, Oklahoma, and Mexico.

Ashe Juniper should be grown in full sun on well-drained soil and easily tolerates alkaline soil. Overirrigated trees often grow poorly due to root diseases. Locate the tree in the driest part of the landscape.

Juniper blight can cause leaf discoloration and defoliation. Ashe Juniper is reportedly resistant to Cedarapple rust, a serious disease of other Junipers.

Juniperus chinensis 'Torulosa' 'Torulosa' Juniper

Pronunciation: jew-NIP-ur-us chi-NEN-sis tore-oo-LOW-sah

ZONES: 5A, 5B, 6A, 6B, 7A, 7B, 8A, 8B, 9A, 9B, 10A, 10B, 11

HEIGHT: 10 to 15 feet

WIDTH: 6 to 10 feet

FRUIT: oval; less than 0.5"; fleshy to dry; blue; no litter

GROWTH RATE: slow growth; moderate life span

HABIT: columnar; pyramidal; open; irregular silhouette; fine texture

FAMILY: *Cupressaceae*

LEAF COLOR: evergreen; emerald green

LEAF DESCRIPTION: scale-like; ⅛" long

FLOWER DESCRIPTION: not showy

EXPOSURE/CULTURE:

Light Requirement: part shade to full sun

Soil Tolerances: all textures; alkaline to acidic; salt and drought

Aerosol Salt Tolerance: moderate

PEST PROBLEMS: resistant

BARK/STEMS:

Routinely grown with multiple trunks; branches grow mostly upright and will not droop until tree is older; no thorns

Pruning Requirement: needs little pruning to develop strong structure

Limb Breakage: resistant

Current Year Twig: green; slender

LANDSCAPE NOTES:

Origin: China, Mongolia, and Japan

Usage: bonsai; container; espalier; buffer strip; highway; specimen; screen

Torulosa Juniper grows into a narrow cone shape when young, then opens up as the plant ages. Although the main body of its foliage is as thick as if it had been sheared, delicately twisted, upright branches emerge gracefully all around the plant, in almost a flame-like manner. It may grow to 15 feet tall in 15 years, perhaps leaning to one side in a picturesque manner. Do not expect a row of them to provide a uniform shape, as crown form varies. The varying crown form is just the characteristic that makes this a popular small tree or shrub.

Torulosa Juniper develops into a showcase specimen without pruning and is probably best used for this purpose. Planted on four- to six-foot centers, it can develop into a thick screen that could be useful along a driveway, where a narrow, bright green screen is often needed to create privacy. It is well suited for planting in a dry soil, such as along a highway or in an unirrigated buffer strip around the edge of a parking lot in a humid, moist climate. Some irrigation is needed in drier climates. Any area with good drainage will support a Chinese Juniper.

Growing best in full sun, more open in partial shade, Torulosa Juniper needs well-drained soil or it will decline from root rot. It tolerates alkaline soil and is quite drought-tolerant but root regeneration is slow after transplanting from a field nursery. For this reason, it is frequently offered in containers from a nursery. It is adapted to all areas within its hardiness range.

Mites and bagworms can infest the foliage. Torulosa Juniper is susceptible to root rot and tip blight.

Juniperus deppeana 'McFetter'
McFetter Alligator Juniper
Pronunciation: jew-NIP-ur-us dep-pee-A-nah

ZONES: 7A, 7B, 8A, 8B, 9A, 9B, (10)

HEIGHT: 50 to 60 feet

WIDTH: 20 to 25 feet

FRUIT: round; 0.25"; fleshy; copper-colored; no litter

GROWTH RATE: moderate; long-lived

HABIT: columnar; oval; pyramidal; dense; symmetrical; fine texture

FAMILY: *Cupressaceae*

LEAF COLOR: evergreen; striking blue or blue-green

LEAF DESCRIPTION: scale-like

FLOWER DESCRIPTION: not showy

EXPOSURE/CULTURE:

Light Requirement: grows best in full sun

Soil Tolerances: all textures; alkaline to slightly acidic; drought

PEST PROBLEMS: resistant

BARK/STEMS:

Branches grow mostly upright and will not droop; showy trunk; should be grown with a single leader; no thorns

Pruning Requirement: needs little pruning to develop strong structure

Limb Breakage: resistant

Current Year Twig: brownish-green; slender

LANDSCAPE NOTES:

Origin: west Texas to Arizona into Mexico

Usage: bonsai; buffer strip; highway; reclamation; specimen; screen

This native North American evergreen tree is found in the southwestern United States and forms a broad, pyramidal or rounded canopy that eventually spreads to about 20 or 25 feet. The trunk is normally rather short and covered with brown, scaly bark. The common name is derived from the bark, which eventually splits into blocks to resemble alligator skin. The rich blue foliage stands out in the landscape against any background, especially dark-colored evergreens.

McFetter Alligator Juniper can be used as Leyland Cypress has been, but probably will not grow as fast. This could be an advantage on a residential landscape where space may be limited and unsuited for a fast-growing tree such as Leyland Cypress. Plant it on 10- to 15-foot centers to form a wide, solid blue screen or windbreak. Growth rate is most rapid, and foliage stays full, in sites with all-day sun. Keep it away from the house and other structures so air circulates freely through the canopy.

McFetter Alligator Juniper should grow in full sun on well-drained, alkaline to slightly acid soil. It appears to perform well on clay soils provided they drain well, and should be suited for many areas of the southern United States.

No limitations of major concern at this time. However, the plant has not been cultivated in the eastern United States for long. Many other Junipers get a foliage blight in overirrigated landscapes.

Juniperus scopulorum

Rocky Mountain Juniper, Colorado Redcedar

Pronunciation: jew-NIP-ur-us scop-you-LORE-um

ZONES: 4A, 4B, 5A, 5B, 6A, 6B, 7A, 7B, 8A, 8B, 9A, 9B, (10)

HEIGHT: 30 to 40 feet

WIDTH: 10 to 15 feet

FRUIT: round; 0.25"; fleshy; blue; attracts birds; no litter; persistent

GROWTH RATE: slow-growing; long-lived in dry sites

HABIT: columnar; oval; pyramidal; moderate density; symmetrical; fine texture

FAMILY: *Cupressaceae*

LEAF COLOR: evergreen; blue or blue-green

LEAF DESCRIPTION: scale-like

FLOWER DESCRIPTION: not showy

EXPOSURE/CULTURE:

Light Requirement: grows best in full sun

Soil Tolerances: all textures; alkaline to slightly acidic; drought

PEST PROBLEMS: susceptible, especially in humid climates

BARK/STEMS:

Can be grown with a single or multiple trunks; branches grow mostly upright and will not droop; no thorns

Pruning Requirement: needs little pruning to develop strong structure

Limb Breakage: resistant

Current Year Twig: brownish-gray; slender

CULTIVARS AND SPECIES:

Many, including: 'Blue Heaven'—blue foliage; 'Gray Gleam'—silvery gray foliage; 'Skyrocket'—very narrow columnar growth, bluish-green foliage; 'Table Top'—semi-upright, flat-topped growth habit, silvery gray foliage, 5 feet high in 10 years; 'Taylor'—nice, tight canopy—'Wichita Blue'—bright blue cast to foliage, pyramidal form.

LANDSCAPE NOTES:

Origin: Rocky Mountains

Usage: bonsai; buffer strip; highway; specimen

Rocky Mountain Juniper can be found in dry, rocky soils at high elevations in the western United States. It outcompetes other trees on poor, rocky soils where it thrives on limited rainfall. In general, this slow-growing evergreen tree has a narrow, pyramidal habit that grows to a mature height up to 40 feet. The form, foliage color, and height vary greatly, as displayed by the large number of cultivars that have been selected from the species.

Rocky Mountain Juniper is useful as a privacy screen or specimen, and is well suited for use as a windbreak in an open, exposed location. The cultivars are well deserving of specimen planting, especially in a small, personal landscape where the fine-textured foliage contrasts with other garden elements.

It is similar to other Junipers in that it requires full sun exposure and thrives in dry, droughty soils. This Juniper is difficult to grow in moist, humid climates due to foliage and root diseases. If you try it in these climates, provide excellent drainage and keep the foliage dry by placing it in an open, breezy location; do not irrigate it once established. It has been overused in some regions.

Trunk, branch, foliage, and root diseases can be serious. Many insects can potentially cause damage. Rocky Mountain and other Junipers transplant poorly, especially large specimens. Many Junipers are sold in containers. *Thuja plicata* and *T. occidentalis* are good substitutes for this plant.

Juniperus scopulorum
'Tolleson's Green Weeping' Juniper
Pronunciation: jew-NIP-ur-us scop-you-LORE-um

ZONES: 4A, 4B, 5A, 5B, 6A, 6B, 7A, 7B, 8A, 8B, 9A, 9B, (10)

HEIGHT: 25 to 30 feet

WIDTH: 25 to 30 feet

FRUIT: round; fleshy; blue; attracts birds; no significant litter

GROWTH RATE: moderate; moderate life span on dry sites

HABIT: weeping; moderate density; irregular silhouette; fine texture

FAMILY: *Cupressaceae*

LEAF COLOR: evergreen; green to greenish-blue
LEAF DESCRIPTION: scale-like
FLOWER DESCRIPTION: not showy
EXPOSURE/CULTURE:

Light Requirement: best growth in full sun

Soil Tolerances: all textures; alkaline to acidic; drought

PEST PROBLEMS: susceptible
BARK/STEMS:

Branches droop as tree grows; should be trained with major limbs spaced along a single trunk; no thorns

Pruning Requirement: prevent formation of included bark in crotches of major branches by thinning some foliage; remove aggressive, young branches that are opposite each other

Limb Breakage: resistant

Current Year Twig: brownish-gray; slender

LANDSCAPE NOTES:

Origin: species is from the Rocky Mountains

Usage: specimen

This cultivar of the native Rocky Mountain Juniper eventually grows to mature height with an equal spread. Arching branches grow up and out from the trunk, bearing foliage that hangs almost like Weeping Willow.

The trunk may have to be staked initially to develop a central leader; otherwise the tree may simply droop over toward the ground. This training begins in the nursery but must be carried over into the landsacpe if the tree is still small at planting. Train lateral branches to grow all along the central leader and keep them cut back in the early years to help build the diameter and strength of the main branches. This will help keep them more upright and create a tree form. A tree not pruned in this way often droops to the ground, becoming a sprawling shrub.

The tree is very striking and will provoke comments from neighbors and passersby. This and other weeping trees look very nice planted close to water, but be sure to keep the root zone on the dry side. It is similar to other Junipers in that it requires full sun exposure and will tolerate dry, droughty soils. It is useful as a privacy screen, and makes a wonderful specimen. This Juniper is difficult to grow in the south due to disease problems, but might be accomplished in a well-drained, dry site that receives little or no irrigation after the tree is well established.

Limitations are the same as for the species.

Juniperus silicicola Southern Redcedar

Pronunciation: jew-NIP-ur-us sih-liss-ih-KO-lah

ZONES: 8A, 8B, 9A, 9B, 10A, 10B

HEIGHT: 30 to 45 feet

WIDTH: 20 to 30 feet

FRUIT: round; 0.25"; fleshy; blue; attracts birds; no significant litter; showy

GROWTH RATE: moderate; long-lived

HABIT: columnar; oval; pyramidal; open to moderate density; nearly symmetrical; fine texture

FAMILY: *Cupressaceae*

LEAF COLOR: evergreen; medium green

LEAF DESCRIPTION: awl- or scale-like

FLOWER DESCRIPTION: not showy

EXPOSURE/CULTURE:

Light Requirement: best in full sun

Soil Tolerances: all textures; alkaline to slightly acidic; salt and drought

Aerosol Salt Tolerance: high

PEST PROBLEMS: resistant to lethal pests

BARK/STEMS:

Showy; branches droop as tree grows and will require pruning for vehicular or pedestrian clearance beneath canopy; should be grown with a single leader; lower limbs become large as tree ages; no thorns

Pruning Requirement: needs little pruning to develop strong structure

Limb Breakage: susceptible to ice storms

Current Year Twig: brownish-green; slender

LANDSCAPE NOTES:

Origin: southeast coastal plain of the United States

Usage: buffer strip; highway; reclamation; street tree; windbreak; urban tolerant

This densely foliated, wide pyramidal, columnar, or oval evergreen grows fairly quickly. Some plants become wider than tall as they grow older. Some botanists do not make a distinction between *J. silicicola* and *J. virginiana.* Its fine-textured leaves and drooping branchlets help to soften the rather symmetrical, oval juvenile form. Mature specimens of Southern Redcedar take on a more open, flat-topped, almost windswept appearance, making them very picturesque. Bark and trunk on older specimens take on a delightful, twisted look.

The dense growth and attractive foliage make Southern Redcedar a favorite for windbreaks, screens, and wildlife cover for large-scale landscapes. Its high salt tolerance makes it ideal for seaside locations. Redcedar can make a nice Christmas tree, and the fragrant wood is popular for repelling insects. Cedar Key, Florida, once had large Redcedar forests before the lumber was extensively harvested and the wood used for chests, paneling, and pencils.

Although not currently used often as a street tree, the foliage is clean and the fruit is small, making it a suitable candidate. There are some nice examples of street tree use in southern cities. The roots will not usually raise sidewalks or curbs, so the tree can be planted in narrow tree lawns and other sites with small soil space.

Trees are difficult to transplant because of a coarse root system, except when quite small. Plants from containers can be planted at any time. Water until well established and then forget about the tree. Redcedar performs admirably with no care, even on alkaline soil and along the coast. Usually insects and diseases are not a problem if the tree is grown in full sun. There may be local restrictions on planting this tree near apple orchards because it is the alternate host for cedar-apple rust.

Juniperus virginiana Eastern Redcedar

Pronunciation: jew-NIP-ur-us vir-gin-ee-A-nah

ZONES: 2A, 2B, 3A, 3B, 4A, 4B, 5A, 5B, 6A, 6B, 7A, 7B, 8A, 8B, 9A, 9B, 10A, 10B

HEIGHT: 40 to 50 feet

WIDTH: 8 to 25 feet

FRUIT: round; 0.25"; fleshy; blue; attracts birds; no significant litter; persistent until eaten by birds; showy

GROWTH RATE: rapid; long-lived

HABIT: columnar; oval; pyramidal; open to dense, variable; symmetrical; fine texture

FAMILY: *Cupressaceae*

LEAF COLOR: evergreen; medium green

LEAF DESCRIPTION: awl- or scale-like

FLOWER DESCRIPTION: greenish-yellow; not showy

EXPOSURE/CULTURE:

Light Requirement: best in full sun

Soil Tolerances: all textures; alkaline to slightly acidic; occasional wetness; salt and drought

Aerosol Salt Tolerance: high

PEST PROBLEMS: mostly resistant to lethal pests

BARK/STEMS:

Showy; branches droop as tree grows and will require pruning for vehicular or pedestrian clearance beneath canopy; should be grown with a single leader; no thorns

Pruning Requirement: needs little or no pruning to develop strong trunk and branch structure

Limb Breakage: susceptible to ice storms

Current Year Twig: brownish-green; slender

CULTIVARS AND SPECIES:

The cultivars have more predictable form and growth rates. They include: 'Burkii' and 'Canaerttii'—20 feet, pyramidal; 'Hillspire'—columnar, dark green foliage; 'Manhattan Blue'—dense blue-green foliage; 'Skyrocket'—narrow column with silver-gray foliage.

LANDSCAPE NOTES:

Origin: eastern United States

Usage: buffer strip; highway; reclamation; street tree; windbreak; urban tolerant

Redcedar is an evergreen growing in an oval, columnar, or pyramidal form (very diverse) and spreading 8 to 25 feet. It develops a brownish tint in winter in the north. The fruit is a blue berry on female trees and is ornamental when produced in quantity. Birds devour the fruit and "plant" it along farm fences and in old abandoned fields. Some botanists do not separate *J. virginiana* from *J. silicicola.*

The dense growth and attractive foliage make Eastern Redcedar a favorite for windbreaks, screens, and wildlife cover for large-scale landscapes. Redcedar can make a nice Christmas tree, and the fragrant wood is popular in homes for repelling insects. Although not currently used often as a street tree, its wood is fairly strong, the foliage is clean, and the fruit is small, making it a suitable candidate. There are some nice examples of street tree use in eastern cities. With proper pruning to remove lower branches, it should adapt well to streetscapes. Roots do not raise sidewalks, even when planted closeby.

Planted in full sun or partial shade, Eastern Redcedar will easily grow on a variety of soils, including clay, but will not do well on soils kept continually moist. Growth may be poor in overirrigated landscapes. Field-grown trees are difficult to transplant, due to a coarse root system, except when quite small. Once established, Redcedar performs admirably with no care, even on alkaline soil and along the coast. Usually insects and diseases are not a problem if the tree is grown in full sun.

There may be local restrictions on planting this tree near apple orchards because it is the alternate host for cedar-apple rust. Bagworms can strip foliage in summer. In some regions, *Kabatina* blight can be devastating on Junipers.

Kalopanax pictus Castor-Aralia

Pronunciation: cal-oh-PAN-ax PICK-tuss

ZONES: 5A, 5B, 6A, 6B, 7A, 7B, 8A

HEIGHT: 40 to 60 feet

WIDTH: 40 to 60 feet

FRUIT: round; 0.25"; fleshy; black; attracts wildlife; not showy; some litter

GROWTH RATE: moderate; long-lived

HABIT: round; dense; nearly symmetrical; coarse texture

FAMILY: *Araliaceae*

LEAF COLOR: dark green in summer; muted red to yellow fall color (occasionally showy)

LEAF DESCRIPTION: alternate; simple; star-shaped; deciduous; 8 to 12"

FLOWER DESCRIPTION: white; showy; summer

EXPOSURE/CULTURE:

Light Requirement: full sun

Soil Tolerances: all textures; alkaline to acidic; drought

PEST PROBLEMS: pest-free

BARK/STEMS:

Showy black trunk; branches grow mostly upright and horizontal and will not droop; should be grown with a single leader; thorns are present on twigs and branches

Pruning Requirement: space major limbs along a central trunk and prevent included bark by thinning major limbs; following this initial training, needs little or no pruning to maintain strong trunk and branch structure

Limb Breakage: resistant

Current Year Twig: brown; stout

LANDSCAPE NOTES:

Origin: Japan, Korea, China

Usage: shade tree; specimen; street tree; buffer strip

With massive, spreading branches and large, 7- to 12-inch-diameter, multi-lobed leaves, Castor-Aralia provides dense shade below its canopy and makes an ideal shade tree. Growing to mature height with an equal spread, Castor-Aralia is deciduous (the leaves turn a faint red or yellow in fall before dropping). Although the young stems are armed with short, yellow prickles, the mature trunk is attractively ridged and blackened and free of spines. The one-inch-diameter, white flowers appear in very showy, dense, 12- to 24-inch-long terminal panicles and attract quite a few bees. This is not a problem on larger specimens because flowers are borne up in the tree away from the ground. The small black fruit, which ripens in early fall, is eagerly consumed by birds.

This tree could be planted more often in landscapes. The large size and coarse texture probably make it best suited for large-scale sites, such as golf courses, parks, business complexes, and campuses, but it could be tried along streets where there is plenty of soil space for root expansion. A row of these along the street or entrance to a commercial landscape could prove quite dramatic. Surface roots do not appear to be a problem, but large trees should be assumed guilty in this respect until proven innocent through experience. This tree simply has not been available for testing in large numbers. Because seeds germinate readily in the landscape, the tree could become a weed.

Castor-Aralia should be grown in full sun on well-drained soil and will tolerate alkaline soil. Though drought-tolerant once established, Castor-Aralia should receive ample moisture until then. Any pruning should be done in late spring.

Koelreuteria bipinnata Bougainvillea Goldenraintree

Pronunciation: coal-rue-TEA-ree-ah bye-pin-A-tah

ZONES: 7B, 8A, 8B, 9A, 9B, 10A, 10B, 11

HEIGHT: 20 to 35 feet

WIDTH: 20 to 30 feet

FRUIT: elongated; 1 to 2"; dry; pink then brown; no significant litter; persistent; showy

GROWTH RATE: rapid-growing; moderate life span

HABIT: round; dense; irregular when young, nearly symmetrical later; fine texture

FAMILY: *Sapindaceae*

LEAF COLOR: medium green in summer; showy yellow fall color

LEAF DESCRIPTION: alternate; bipinnately compound; incised to serrate margin; oblong to ovate; deciduous; 2 to 4" long leaflets

FLOWER DESCRIPTION: yellow; late summer and fall; very showy

EXPOSURE/CULTURE:

Light Requirement: flowers best in full sun

Soil Tolerances: all textures; alkaline to acidic; occasionally wet soil; drought

Aerosol Salt Tolerance: low to moderate

PEST PROBLEMS: resistant

BARK/STEMS:

Bark is thin and easily damaged by mechanical impact; branches droop as tree grows and will require early pruning for vehicular or pedestrian clearance beneath canopy; should be grown with a single leader as far up into canopy as possible; no thorns

Pruning Requirement: high

Limb Breakage: reportedly susceptible to breakage, but most trees appear to stay together

Current Year Twig: brown; stout

CULTIVARS AND SPECIES:

Easily distinguished from *K. paniculata,* because *K. bipinnata* has twice compound leaves, whereas *K. paniculata* has single pinnate compound leaves; very similar to *K. elegans*.

LANDSCAPE NOTES:

Origin: China

Usage: above-ground planter; buffer strip; highway; reclamation plant; shade tree; parking lot island; specimen; street tree; sidewalk cutout; urban tolerant

A yellow carpet of fallen petals, delicate leaflets that cast a mosaic of welcoming shade year-round, and large clusters of persistent, rose-colored, papery capsules all help to make Goldenraintree a nice landscape tree. This broad-spreading tree eventually takes on a flat-topped silhouette. It is used as a patio, shade, street, or specimen tree. The small, fragrant, yellow flowers appear in very showy, dense, terminal panicles in late summer when nothing else is in bloom, and are followed in fall by large clusters of pink, two-inch-long "Chinese lanterns". These papery husks are held above the foliage, retain their pink color after drying, and are very popular for use in everlasting flower arrangements.

Spacing of main branches apart along a central trunk is crucial to developing a durable tree. Remove or head any double or multiple trunks as soon as possible, especially if they have included or pinched bark in the branch crotch. Train major branches so they grow up and out, spreading from the trunk to create the clearance needed for street tree or parking lot planting. This tree takes a great deal of effort to train into an attractive young specimen.

Scale and mushroom root rot may occur on wet soil or on old trees. Goldenraintree bug feeds on the fruit, but can be a nuisance. A canker causes dead and sunken areas on the bark. Verticillium wilt attacks *Koelreuteria*; the disease causes wilting and death of leaves on infected branches. Eventually the entire tree may be killed. Seeds germinate in the landscape and can become weeds.

Koelreuteria elegans Flamegold, Golden Raintree

Pronunciation: coal-rue-TEA-ree-ah ELL-eh-gans

ZONES: 9A, 9B, 10A, 10B, 11

HEIGHT: 35 to 50 feet

WIDTH: 35 to 50 feet

FRUIT: elongated; 1 to 2"; dry; pink then brown; no significant litter; persistent; showy

GROWTH RATE: rapid-growing; moderate life span

HABIT: round; open; irregular silhouette; coarse texture

FAMILY: *Sapindaceae*

LEAF COLOR: medium green in summer; showy yellow fall color

LEAF DESCRIPTION: alternate; bipinnately compound; incised to serrate margin; oblong to ovate; briefly deciduous; 2 to 4" long leaflets

FLOWER DESCRIPTION: yellow; summer and fall; very showy

EXPOSURE/CULTURE:

Light Requirement: flowers best in full sun

Soil Tolerances: all textures; alkaline to acidic; occasionally wet; drought

Aerosol Salt Tolerance: moderate

PEST PROBLEMS: resistant

BARK/STEMS:

Bark is thin and easily damaged by mechanical impact; branches droop as tree grows and will require early pruning for vehicular or pedestrian clearance beneath canopy; should be grown with a single leader; no thorns

Pruning Requirement: high; remove dead branches

Limb Breakage: reportedly susceptible to breakage, but most trees without included bark appear to stay together

Current Year Twig: brown; stout

CULTIVARS AND SPECIES:

Easily distinguished from *K. paniculata,* because *K. elegans* has twice compound leaves, whereas *K. paniculata* has single pinnate compound leaves. Differs from *K. bipinnata* only in that *K. elegans* is nearly evergreen in the extreme southern part of its range.

LANDSCAPE NOTES:

Origin: Formosa

Usage: buffer strip; highway; reclamation plant; shade tree; parking lot island; specimen; street tree; urban tolerant

A yellow carpet of fallen petals, delicate leaflets that cast a mosaic of welcoming shade year-round, and large clusters of persistent, rose-colored, papery capsules all help to make Flamegold a very popular landscape tree. This broad-spreading, semi-evergreen tree eventually takes on a flat-topped, somewhat irregular silhouette. It is often used as a patio, shade, street, or specimen tree. The small, fragrant, yellow flowers appear in very showy, dense, terminal panicles in summer, and are followed in fall by large clusters of two-inch-long "Chinese lanterns". These papery husks are held above the evergreen foliage, retain their pink to brownish color after drying, and are very popular for use in everlasting flower arrangements.

Spacing of main branches apart along a central trunk is crucial to developing a durable tree. Remove or shorten any double or multiple trunks as soon as possible, especially if they have included or pinched bark in the branch crotch. Train major branches so they grow up and out, spreading from the trunk to create the clearance needed for street tree or parking lot planting.

Scale and mushroom root rot may occur on wet soil or on old trees. Goldenraintree bug can be a nuisance. A canker causes dead and sunken areas on the bark. Verticillium wilt attacks *Koelreuteria*; the disease causes wilting and death of leaves on infected branches. Eventually the entire tree may be killed. Seeds germinate in the landscape and can become invasive weeds.

Koelreuteria paniculata Goldenraintree

Pronunciation: coal-rue-TEA-ree-ah pah-nick-you-LAY-tah

ZONES: 5B, 6A, 6B, 7A, 7B, 8A, 8B, 9A, 9B

HEIGHT: 30 to 40 feet

WIDTH: 30 to 40 feet

FRUIT: elongated; oval; 1 to 2"; dry; green turning tan then brown; no litter; persistent; showy

GROWTH RATE: moderate; moderate life span

HABIT: rounded vase; open; nearly symmetrical; coarse texture

FAMILY: *Sapindaceae*

LEAF COLOR: medium green in summer; showy, yellow fall color

LEAF DESCRIPTION: alternate; bipinnately compound; incised to serrate margin; deciduous; 2 to 4" long leaflets

FLOWER DESCRIPTION: yellow; very showy; early summer

EXPOSURE/CULTURE:

Light Requirement: flowers best in full sun, will grow in part sun

Soil Tolerances: all textures; alkaline to acidic; occasionally wet soil; well-drained; salt and drought

Aerosol Salt Tolerance: low to moderate

PEST PROBLEMS: resistant

BARK/STEMS:

Bark is thin and easily damaged by mechanical impact; branches droop as tree grows and will require pruning for vehicular or pedestrian clearance beneath canopy; should be grown with a single leader; no thorns

Pruning Requirement: space branches along a short, central trunk; prune in dormant season; remove dead wood and old fruit stalks for neater appearance

Limb Breakage: reportedly susceptible to breakage, but most trees without included bark appear to stay intact

Current Year Twig: brown; stout

LANDSCAPE NOTES:

Origin: China, Japan, and Korea

Usage: above-ground planter; parking lot island; buffer strip; highway; reclamation; small shade tree; specimen; sidewalk cutout; street tree; urban tolerant

Goldenraintree grows in a broad, somewhat irregular globe shape. Some trees appear vase-shaped. Although it has a reputation for being weak-wooded, trees pruned to prevent formation of included bark appear to hold together. It is rarely attacked by pests and grows in a wide range of soils. It makes a good street or parking lot tree, particularly where overhead or soil space is limited, due to its adaptive abilities. The tree grows moderately and bears large panicles of bright yellow flowers in May (zone 9) to July (zone 6) when few other trees bloom. The tree flowers when it is only two to three years old. It is not as showy as *K. bipinnata,* but is much more cold-tolerant and a better-formed tree when young. The seed pods look like brown "Chinese lanterns" and are held on the tree well into the fall.

The root system is coarse, with only a few but large roots, so transplant when young, in the spring, or from containers. It is considered a city-tolerant tree due to its ability to withstand air pollution, drought, heat, and alkaline soils. It also tolerates some salt spray. It would be hard to find a more adaptive yellow-flowering tree for urban planting. It makes a nice patio tree, creating light shade. The tree has only a few branches when it is young and some pruning to increase branchiness helps sell the tree. Prune the tree early to space major branches along the trunk to create a strong branch structure; the tree will be longer-lived and require little maintenance. Only single-stemmed trees trained in the nursery with well-spaced branches should be planted along streets and parking lots.

Goldenraintree bugs feed on the fruit, but can be a nuisance. Occasional attacks by scale may be seen. Wilt can kill trees. Seeds germinate in the landscape. Recovery from transplanting is slow. Do not let these potential problems prevent you from planting this wonderful tree. May flower poorly in zone 9.

Laburnum alpinum Scotch Laburnum, Goldenchain Tree

Pronunciation: lah-BURR-num al-PINE-um

ZONES: 5B, 6A, 6B, (7A)

HEIGHT: 15 to 20 feet

WIDTH: 12 to 18 feet

FRUIT: elongated pod; dry; no significant litter

GROWTH RATE: slow-growing; usually short-lived

HABIT: upright, weeping vase; open; irregular silhouette; medium texture

FAMILY: *Leguminosae*

LEAF COLOR: dark green in summer; no appreciable fall color

LEAF DESCRIPTION: alternate; trifoliate; entire; oval; deciduous; 1 to 3" long

FLOWER DESCRIPTION: beyond beautiful—in a world by itself; spring

EXPOSURE/CULTURE:

Light Requirement: half-day sun, morning or afternoon

Soil Tolerances: loam; sand; alkaline to slightly acidic; exceedingly well-drained soil required; drought- and heat-sensitive

Aerosol Salt Tolerance: none

PEST PROBLEMS: sensitive

BARK/STEMS:

Branches droop as tree grows; should be grown with a single leader to help prevent splitting at crotch; no thorns

Pruning Requirement: requires pruning to develop strong structure; slow growth on aggressive, upright lower limbs by thinning to prevent formation of included bark; cable large limbs with included bark

Limb Breakage: susceptible

Current Year Twig: green; medium thickness

CULTIVARS AND SPECIES:

L. x *waterei* 'Vossii' is a popular hybrid and is nearly identical.

LANDSCAPE NOTES:

Origin: southern Alps

Usage: espalier; container or above-ground container; near a deck or patio; specimen

This is an airy, graceful, tall shrub or small tree, with rich, yellow, wisteria-like, drooping flower panicles, up to 10 inches long, that appear in late spring. The tree is truly incredible in full bloom. Even a good photograph of the tree does not portray the true beauty of the plant. It is difficult to grow, but if tried should be grown in a spot protected from full-day sun, with very well-drained, alkaline pH soil and high organic matter. It reportedly does not grow well in soil below pH 7.0, although this photograph was taken in soil with a pH in the low 6s.

Goldenchain Tree is a good specimen, patio, or border accent tree, and can also be espaliered against a wall. Plant this tree in a prominent location so you can enjoy the blooms. It looks nice planted in a bed of low-growing, green shrubs or groundcover. Blue flowers nicely complement the bright yellow blooms on Goldenchain Tree. Thin or remove upright-growing branches that might form included bark in the crotches to increase the longevity of the tree.

Locate it on a hill or slope so water runs away from this tree. Do not overwater this tree. Apply mulch beneath the canopy to reduce competition from other plants. In zone 7, young plants are extremely attractive but do not usually last long, as the plant does not tolerate heat well. Use as a novelty in a protected spot with good air circulation where a brightly colored accent is needed in the spring.

A soil with exceptional drainage is essential. Do not attempt to grow it if water stands in the planting site for even an hour or two. Tree can be attacked by aphids or mealy bugs. A number of diseases also infect the tree. All parts of the plant are poisonous.

Lagerstroemia fauriei Japanese Crape-Myrtle

Pronunciation: lag-er-STRO-mee-ah FAR-ee-eye

ZONES: 6B, 7A, 7B, 8A, 8B, 9A, 9B, 10A, 10B

HEIGHT: 35 to 50 feet

WIDTH: 25 to 35 feet

FRUIT: oval; round; dry; brown; no litter; persistent

GROWTH RATE: moderate; long-lived

HABIT: oval; upright, vase; dense; symmetrical; medium texture

FAMILY: *Lythraceae*

LEAF COLOR: medium green in summer; no fall color change

LEAF DESCRIPTION: alternate to subopposite; simple; entire margin; oval to oblong; deciduous; 2 to 4" long

FLOWER DESCRIPTION: white; very showy

EXPOSURE/CULTURE:

Light Requirement: full sun to part shade

Soil Tolerances: all textures; slightly alkaline to acidic; drought

Aerosol Salt Tolerance: moderate

PEST PROBLEMS: resistant

BARK/STEMS:

Bark is thin and easily damaged by mechanical impact; grows with multiple trunks but can be trained to a short, single leader; branches grow mostly upright and will not droop; no thorns

Pruning Requirement: requires minimum pruning to develop strong structure; select branches with wide crotches for major limbs

Limb Breakage: branches with included bark can split from trunk

Current Year Twig: brownish-green; slender

CULTIVARS AND SPECIES:

'Fantasy' will become more available. It grows 40 to 50 feet tall; has small white flowers, with an extremely showy flower display even on old trees; may develop 2-foot-diameter trunk 35 years after planting; and is resistant to powdery mildew. 'Townhouse' is smaller with a round canopy and beautiful exfoliating bark.

LANDSCAPE NOTES:

Origin: Japan

Usage: above-ground planter; parking lot island; buffer strip; highway; shade tree; specimen; street tree; urban tolerant

Japanese Crape-Myrtle is very similar to the more common Crape-Myrtle (*L. indica*) except that it has smaller, white blossoms and larger leaves that show no appreciable fall color change. It is also larger in size than the common Crape-Myrtle. The flaking, peeling, red and brown bark is especially attractive in the winter after the leaves have fallen. With lower branches removed, as is often the case with Crape-Myrtle, bark is also very showy during the entire year. The upright, vase-shaped crown makes Japanese Crape-Myrtle well-suited for street tree planting. Hybrids with this parentage grow well in zone 10.

Pruning should be done in late winter or early spring before growth begins, because it is easier to see which branches require pruning. New growth on young trees can be pinched during the growing season, when the tree is very small, to increase branchiness and flower number, but this is not needed to develop good form. The tree flowers very nicely without pruning. Lower branches are often thinned to show off the trunk form and color.

Japanese Crape-Myrtle will make a wonderful street tree for residential or commercial landscapes. Planted on 20-foot-centers, it would form a canopy over the sidewalk, but the crown may not spread enough to canopy the street. The effect would be similar to a smaller version of American Elm. Choose single-trunked trees for street planting, or well-formed, multi-trunked specimens with no included bark. The outstanding bark character makes it one of the best specimen trees for the South. This will become a very popular tree.

No limitations of major concern. Aphids probably infest the new growth to a certain extent, causing a harmless but unsightly sooty mold to grow on the foliage. Japanese Crape-Myrtle appears very resistant to powdery mildew.

Lagerstroemia indica (hybrids and cultivars)
Crape-Myrtle
Pronunciation: lag-er-STRO-mee-ah INN-dih-cah

ZONES: 7A, 7B, 8A, 8B, 9A, 9B, 10A, 10B

HEIGHT: 10 to 30 feet

WIDTH: 15 to 25 feet

FRUIT: round; dry; brown; no litter; persistent

GROWTH RATE: moderate; moderate longevity

HABIT: vase; moderate density; symmetrical; medium texture

FAMILY: *Lythraceae*

LEAF COLOR: showy orange, red, or yellow fall color

LEAF DESCRIPTION: alternate to subopposite; simple; entire margin; oval to oblong; deciduous; 2 to 3" long

FLOWER DESCRIPTION: lavender; pink; purple; red; white; showy; spring and summer

EXPOSURE/CULTURE:

Light Requirement: flowers best in full sun

Soil Tolerances: all textures; slightly alkaline to acidic; drought

Aerosol Salt Tolerance: moderate

PEST PROBLEMS: species is susceptible; National Arboretum selections are resistant

BARK/STEMS:

Very showy with smooth, exfoliating bark; bark is thin and easily damaged by mechanical impact; branches droop as tree grows and will require pruning for vehicular or pedestrian clearance beneath canopy; routinely grown with multiple trunks; can be trained to grow with a short, single trunk; no thorns

Pruning Requirement: head back aggressive limbs to strengthen wood for a more upright habit; remove or shorten aggressive branches with included bark

Limb Breakage: resistant

Current Year Twig: brownish-green; slender

LANDSCAPE NOTES:

Origin: China and Korea

Usage: container; buffer strip; highway; deck or patio; standard; small shade tree; parking lot island; specimen; street tree; urban tolerant

A long period of striking summer flower color, attractive fall foliage, and good drought tolerance all combine to make Crape-Myrtle a favorite small tree for either formal or informal landscapes. Cultivars have been selected for size, flower color, and disease resistance. Flower color, includes all shades of white, pink, red, or lavender, borne on 6- to 12-inch-long clustered blooms appearing on the tips of branches during late spring and summer in zones 8B and 9, and summer in other areas. Most tree types grow to 20 to 25 feet tall, although more dwarf forms are available. The upright, vase-shaped crown makes the tall-growing selections well suited for street tree planting. The national champion *L. indica* is in central Florida.

New growth can be pinched during the growing season to increase branchiness and flower number. Lower branches are often thinned to show off the trunk form and color. Remove spent flower heads to encourage a second flush of flowers and to prevent formation of the brown fruits. As cultivars are now available in a wide range of growth heights, severe pruning should not be necessary to control size. Severe pruning or topping can stimulate basal sprouting, which can become a constant nuisance requiring regular removal. Some trees sprout from the base of the trunk and roots even without severe heading. To help reduce sprouting, pull these out when succulent instead of pruning them.

It grows well in limited soil spaces in urban areas, such as along boulevards, in parking lots, and in small pavement cutouts, if provided with some irrigation until well established. Crape-Myrtle tolerates clay and slightly alkaline soil well. However, the flowers of some selections may stain car paint.

Selections for zone 10 such as 'Natchez', 'Muskogee', and 'Tuskegee' are resistant to powdery mildew and aphids and are the selections of choice. 'Tuscarora' is susceptible to aphids. The species is sensitive to aphids and powdery mildew, especially in the shade. Honeydew drops from infested trees create a sticky mess under the canopy.

Lagerstroemia speciosa Queens Crape-Myrtle

Pronunciation: lag-er-STRO-mee-ah spee-see-O-sah

ZONES: 10B, 11

HEIGHT: 30 to 50 feet

WIDTH: 30 to 40 feet

FRUIT: round; dry; brown; no litter; persistent and unsightly

GROWTH RATE: rapid-growing; moderate life span

HABIT: round; upright vase; moderate density; symmetrical; medium texture

FAMILY: *Lythraceae*

LEAF COLOR: dark green; no fall color change

LEAF DESCRIPTION: alternate; subopposite; simple; entire margin; oval to oblong; deciduous; 2 to 3" long

FLOWER DESCRIPTION: lavender; pink; showy; summer

EXPOSURE/CULTURE:

Light Requirement: best flowering in full sun

Soil Tolerances: all textures; slightly alkaline to acidic; drought

Aerosol Salt Tolerance: moderate

PEST PROBLEMS: resistant

BARK/STEMS:

Bark is thin and easily damaged by mechanical impact; branches droop as tree grows and will require pruning for vehicular or pedestrian clearance beneath canopy; showy trunk; can be trained to grow with a single trunk or multiple trunks; no thorns

Pruning Requirement: requires pruning to develop upright structure; head back aggressive limbs to strengthen wood for a more upright habit; remove aggressive limbs with included bark

Limb Breakage: resistant

Current Year Twig: brownish-green; slender

CULTIVARS AND SPECIES:

There are other species of tropical *Lagerstroemia* (e.g., *hirsuta*), all with a nice flower display, and some are available in selected nurseries.

LANDSCAPE NOTES:

Origin: Asia

Usage: parking lot island; buffer strip; shade tree; street tree

This is one of only a few deciduous trees that grow in tropical and subtropical areas of the country. A profusion of large, three-inch-wide, bright pink to lavender blooms appear in dense, foot-long, terminal panicles from June to July, making Queen's Crape-Myrtle a spectacular specimen or street tree. This large, upright-rounded, deciduous tree is clothed with 12-inch-long, oblong, leathery leaves. The attractive bark is smooth, mottled, and peeling. In India, the wood is used for railroad ties and construction because it is very strong.

Where there are no overhead restrictions, this makes a nice, medium-sized street tree. To train in the nursery for street tree use, secure the main leader to a stake and cut back other upright stems and branches for the first several years when they begin to droop. This will help the tree develop a central trunk that will be tall and strong enough to hold the tree erect later. The first permanent lateral branch can be selected at about eight feet from the ground. Space major branches 8 to 12 inches apart on the central trunk.

Plants should be watered faithfully and planted only in frost-free climates. Queen's Crape-Myrtle leaves remain green with regular fertilization but may drop in drought. Fertilize regularly to prevent chlorosis on alkaline soils. May suffer aphids and scale, followed by sooty mold.

Larix decidua European or Common Larch

Pronunciation: LAIR-icks dee-SID-u-ah

ZONES: 2A, 2B, 3A, 3B, 4A, 4B, 5A, 5B, 6A, 6B

HEIGHT: 60 to 70 feet

WIDTH: 30 to 35 feet

FRUIT: oval; 1.5" long; dry; brown; abundant; showy for a short period; no significant litter

GROWTH RATE: moderate; long-lived

HABIT: pyramidal; very open; irregular; fine texture

FAMILY: *Pinaceae*

LEAF COLOR: light green in summer; showy, yellow fall color

LEAF DESCRIPTION: spiral; simple; linear; needle-like; deciduous; 1.5" long

FLOWER DESCRIPTION: yellow male, red female; showy; spring

EXPOSURE/CULTURE:

Light Requirement: grows best in full sun

Soil Tolerances: all textures; slightly alkaline to acidic; salt; moderate drought

PEST PROBLEMS: resistant

BARK/STEMS:

Showy bark; branches droop as tree grows and will require pruning for vehicular or pedestrian clearance beneath canopy; normally grown with a single leader; no thorns

Pruning Requirement: requires little pruning to develop strong structure

Limb Breakage: resistant

Current Year Twig: tan to brown; slender

CULTIVARS AND SPECIES:

'Pendula' has weeping branches. Form varies coinsiderably from plant to plant. *L. laricina* is native to northern North America, tolerates wet soil well, and naturally grows on alkaline soil. *L. kaempferi* may grow a little wider than *L. decidua*.

LANDSCAPE NOTES:

Origin: Europe

Usage: specimen

This large but graceful, deciduous conifer reaches mature height with a straight trunk and a spread of about 30 feet. Growth is most rapid on a moist site. The silhouette is a rather open, irregular pyramidal shape with pendulous branch tips. Form varies considerably from tree to tree. The fine-textured, needled foliage is 1.5 inches long in dense whorls on side shoots, or singly on long shoots. The foliage turns a brilliant yellow in the fall for a short time—but long enough to make a striking landscape statement. The 1.5-inch-long, upright cones are clustered along the branches, making a showy display throughout the year. The reddish-brown bark is rugged and furrowed, showing nicely in the winter.

The attractive bark, strong pyramidal shape, and widespreading, horizontal branches make Larch particularly attractive in the winter landscape. Its slow growth and attractive form make it a popular choice for large above-ground planters and for use as a bonsai. It can be planted in residential landscapes because it grows so slowly. Plant it in the open so the form can be fully appreciated. Locate it far enough from sidewalks and buildings so lower branches will not have to be pruned. If lower branches reach the walk or building, remove them back to the trunk.

Larch should be grown in full sun on deep, rich, well-drained, moist acid soil where the trees can be somewhat protected from harsh, cold winds. Afternoon shade and regular irrigation help transplanted trees recover. The trees should not be planted in limestone soils, and they are not tolerant of clay unless located on a slope where drainage is excellent. Planted on an appropriate site, Larch will provide light shade and will last for many years. It is best to locate the tree where the roots can expand into the surrounding soil unobstructed by pavement, buildings, or other urban structures.

Cankers formed by various species of fungi are common on Larch. Other occasional problems include leaf casts, needle rusts, aphids, and sawflies.

Latania spp. Latan Palm

Pronunciation: lah-TAY-nee-ah species

ZONES: 10B, 11

HEIGHT: 25 to 35 feet

WIDTH: 10 to 15 feet

FRUIT: oval; 1 to 2" long; fleshy; brown; no significant litter

GROWTH RATE: slow-growing; short life in Florida

HABIT: coarse

FAMILY: *Arecaceae*

LEAF COLOR: blue-green to silver

LEAF DESCRIPTION: enormous palmate fronds 6 to 8 feet across; pubescent petioles

FLOWER DESCRIPTION: white; spring and summer

EXPOSURE/CULTURE:

Light Requirement: part shade to full sun

Soil Tolerances: all textures; slightly alkaline to acidic; drought

Aerosol Salt Tolerance: moderate

PEST PROBLEMS: sensitive to lethal yellowing

BARK/STEMS:

Single, thick trunk; no thorns; no crown shaft

Pruning Requirement: remove old, declining fronds

Limb Beakage: resistant

CULTIVARS AND SPECIES:

The different species of Latan Palm can be told apart by leaf color. Only young leaves that have not yet turned silvery should be used for this determination. *L. loddigesii*, Blue Latan Palm, has blue-gray leaves. *L. lontaroides*, Red Latan Palm, has reddish petioles, leaf margins, and veins. *L. verschaffeltii*, Yellow Latan Palm, has leaf margins, veins, and petioles of a deep orange-yellow.

LANDSCAPE NOTES:

Origin: Mauritius mountains

Usage: container; deck or patio; specimen

A single-trunked palm, Latan Palm is noted for its distinctive, coarse-textured leaves. The large, very thick and stiff leaves, up to eight feet in diameter, are held aloft on five-foot-long petioles. The surface of each leaf is covered with a whitish, waxy or wooly down, providing a silvery appearance to the palm. The three- to six-foot-long flower stalks are present among the leaves in spring and some of the glossy brown, two-inch-wide fruits are always ripening. The 10- to 12-inch-wide trunks have thick, swollen bases.

Latan Palm makes a striking specimen planting and is well suited to seaside locations due to its moderate salt tolerance. Plant it in an area where you would like to attract attention. People's eyes will always be drawn to this plant, no matter where it is planted. The palm looks great combined with any plant groupings, especially those with small leaves. This textural contrast between the large leaves of this palm and fine-textured shrubs provides a very pleasant, formal landscaped feeling.

Preferring full sun, but tolerant of partial shade, slow-growing Latan Palm should be located on fertile, well-drained soil. Regular irrigation following root pruning or transplanting helps plants recover quickly. Unfortunately, it is susceptible to lethal yellowing disease, which kills the tree, and so it should be used sparingly in the landscape.

Leptospermum laevigatum Australian Tea Tree

Pronunciation: lep-toe-SPER-mum leh-va-GAY-tum

ZONES: 9A, 9B, 10A, 10B, 11

HEIGHT: 15 to 20 feet

WIDTH: 20 to 35 feet

FRUIT: oval; 0.25"; dry; no significant litter

GROWTH RATE: moderate; moderate life span

HABIT: round and spreading; dense; irregular; fine texture

FAMILY: *Myrtaceae*

LEAF COLOR: gray-green in summer; no fall color change

LEAF DESCRIPTION: opposite; simple; entire margin; ovate; deciduous; 1" long

FLOWER DESCRIPTION: white, pink, or red; showy; resemble pinwheels; spring

EXPOSURE/CULTURE:

Light Requirement: full sun

Soil Tolerances: all textures; acid only; good drainage is essential; salt and drought

PEST PROBLEMS: pest-free in well-drained soil

BARK/STEMS:

Thick bark; twisted branches and trunk; trunk can grow to 2' thick; branches droop as tree grows and will require pruning for vehicular or pedestrian clearance beneath canopy; routinely grown with multiple trunks; can be trained to grow with a short, single trunk; no thorns

Pruning Requirement: requires pruning to develop one trunk; can be sheared into a hedge; can be pruned into an attractive, open specimen with twisting branches

Limb Breakage: resistant

Current Year Twig: brown or gray; slender

CULTIVARS AND SPECIES:

There are many species of *Leptospermum*, most of them ground covers or shrubs. Some grow into small trees, but none reach the size of *L. laevigatum*.

LANDSCAPE NOTES:

Origin: Australia

Usage: bonsai; container or above-ground planter; hedge or screen; near a deck or patio; specimen; buffer strip; highway

Australian Tea Tree is an excellent, low-growing tree for small yards, patios, and other small-scale landscapes most commonly planted in California and the southwestern United States. It can be grown as a multi-stemmed clump or trained into a small tree with a single trunk up to six feet tall. The tree eventually grows to about 20 feet tall and has a spreading, rounded, finely branched growth habit on several thick main branches, which creates dense shade under the crown. Branches droop and spread as they grow older, eventually touching the ground. Tea Tree grows rapidly when young if it receives water and fertilizer, but slows down later as the canopy becomes more spreading.

The main ornamental value of Tea Tree is the unusual, low-branched habit. The twisted trunks and branches eventually grow along the ground unless trained to grow more upright. Branches on the lower trunk can be headed back several times when the tree is young to create a more upright form. The young twigs and foliage on upright branches weep toward the ground, providing a nice, delicate texture. It is often planted near water or on a slope to accent the wonderful form. It makes a good container tree, and is used near decks and patios due to its small size.

The plant tolerates clipping well and so is often used in hedges or screens. It is durable, tolerating poor soil, but will show chlorosis on alkaline sites.

Australian Tea Tree is usually pest-free. The drooping, spreading form needs to be considered when locating the tree close to walks.

Leucaena retusa Goldenball Leadtree

Pronunciation: lew-SEE-nah ree-TOO-sah

ZONES: 7A, 7B, 8A, 8B, 9A, 9B

HEIGHT: 12 to 15 feet

WIDTH: 15 feet

FRUIT: elongated pod; 3 to 10"; dry and hard; brown; causes some litter; showy

GROWTH RATE: moderate; short-lived

HABIT: upright vase; open; irregular silhouette; fine texture

FAMILY: *Leguminosae*

Courtesy B. Simpson

LEAF COLOR: evergreen; blue-green to bright green

LEAF DESCRIPTION: alternate; bipinnately compound; entire margin; oval to oblong; up to 1" long

FLOWER DESCRIPTION: yellow; summer and fall; showy; fragrant

EXPOSURE/CULTURE:

Light Requirement: full sun for best growth

Soil Tolerances: all textures; alkaline to slightly acidic; drought

PEST PROBLEMS: resistant

BARK/STEMS:

Routinely grown with multiple trunks, but can be trained to a single trunk; branches droop and require pruning for clearance beneath canopy; no thorns

Pruning Requirement: requires pruning to develop strong, uniform, upright structure; space branches along trunk to allow for proper attachment to trunk

Limb Breakage: susceptible

Current Year Twig: brown; slender

LANDSCAPE NOTES:

Origin: southeastern United States into Mexico

Usage: above-ground planter; bonsai; buffer strip; highway; near a deck or patio; specimen

Goldenball Leadtree is a small, native North American evergreen tree that can reach 25 feet in height. The showy, yellow to white, rounded blooms appear from April to October and are especially prominent after a heavy rain. The attractive fruits that follow ripen in late summer. The trunk grows to about eight inches in diameter.

The crown can develop a one-sided or asymmetrical habit when it is young, so pruning and training may be needed to create a more uniformly shaped tree. Stake the trunk in the upright position, cut back but keep branches on the lower trunk, and reduce the length of lateral branches located below the lowest permanent branch. When the tree is tall enough to select the first permanent branch, the trunk stake can probably be removed.

The drooping habit of the branches makes this tree difficult to maintain near a street without early training, but unpruned trees make a nice accent for a shrub border or backyard garden. The tree tends to seed itself into the surrounding landscape and spreads rapidly. Some consider it a weed, and the wood is brittle.

Goldenball Leadtree should be grown in full sun on well-drained soil and will tolerate alkaline conditions very well. It is native to soils with a high limestone content.

Ligustrum japonicum Japanese Privet, Wax-Leaf Privet

Pronunciation: lie-GUSS-trum jah-PON-ih-come

ZONES: 7B, 8A, 8B, 9A, 9B, 10A

HEIGHT: 8 to 12 feet

WIDTH: 15 to 25 feet

FRUIT: oval; round; 0.25"; fleshy; blackish-blue; attracts birds; no significant litter

GROWTH RATE: slow-growing; moderate life span

HABIT: rounded vase; moderate density; symmetrical; medium texture

FAMILY: *Oleaceae*

LEAF COLOR: evergreen; dark green

LEAF DESCRIPTION: opposite; simple; entire margin; oblong to ovate; 2 to 4" long

FLOWER DESCRIPTION: white; showy; spring and summer; strong fragrance

EXPOSURE/CULTURE:

Light Requirement: part shade to full sun

Soil Tolerances: all textures; slightly alkaline to acidic; moderate drought

Aerosol Salt Tolerance: moderate

PEST PROBLEMS: resistant

BARK/STEMS:

Bark is thin and easily damaged by mechanical impact; branches droop as tree grows and will require pruning for vehicular or pedestrian clearance beneath canopy; routinely grown with multiple trunks; can be trained to a short trunk; showy trunk structure; no thorns

Pruning Requirement: requires training and pruning to develop upright tree form

Limb Breakage: resistant

Current Year Twig: greenish-gray; medium thickness

LANDSCAPE NOTES:

Origin: Japan and Korea

Usage: container; hedge; buffer strip; highway; deck or patio; standard; specimen; sidewalk cutout; urban tolerant

Although often used as a shrub or hedge, Japanese Privet works well when allowed to grow into a small tree. Its curved multiple trunks and dark green canopy create an interesting architectural focus. Old specimens can grow to 25 feet across. The glossy evergreen leaves are abundantly produced on the upright, spreading branches. The small, white, malodorous flowers appear in terminal panicles during spring in the south and in the summer in northern climes. The blooms are followed by abundant blue-black berries that persist most of the year.

Although tolerant of tight clipping, Japanese Privet is quite attractive when allowed to retain its natural multi-stemmed form, making it ideal for use in shrubbery borders and as a featured accent. It makes a nice specimen in any landscape where a small, dark tree is needed. Planted close together on about 10- to 15-foot centers, *Ligustrum* will form a canopy over a pedestrian walkway. The tree looks best in a landscape setting with a low groundcover planted around its base.

Japanese Privet grows in full sun or partial shade and is tolerant of a wide range of soil types, including calcarious clay, as long as water is not allowed to stand in the root zone. Plants grow quickly while young but slow with age. Although it can withstand drought, Japanese Privet requires protection from direct salt spray. If you decide to use this plant as a clipped hedge, be sure that the top is kept narrower than the bottom to provide light to the lower branches. This will help ensure that the plant will remain full to the ground.

Soil nematodes can cause serious plant decline and they can be prevalent, particularly in sandy soil.

Ligustrum lucidum Glossy Privet, Tree Ligustrum

Pronunciation: lie-GUSS-trum LEW-seh-dum

ZONES: 8A, 8B, 9A, 9B, 10A, 10B

HEIGHT: 25 to 40 feet

WIDTH: 25 to 35 feet

FRUIT: oval; round; 0.25"; fleshy; bluish-purple; causes some messy litter; eaten by birds; showy

GROWTH RATE: moderate; moderate life span

HABIT: rounded vase; dense; symmetrical; medium texture

FAMILY: *Oleaceae*

LEAF COLOR: broadleaf evergreen; dull green

LEAF DESCRIPTION: opposite; simple; entire margin; lanceolate to ovate; 3 to 6" long

FLOWER DESCRIPTION: white; very showy; summer; strong fragrance; attracts bees

EXPOSURE/CULTURE:

Light Requirement: part shade to full sun

Soil Tolerances: all textures; alkaline to acidic; moderate salt; moderate drought

Aerosol Salt Tolerance: moderate

PEST PROBLEMS: resistant

BARK/STEMS:

Bark is thin and easily damaged by mechanical impact; branches droop as tree grows and will require pruning for vehicular or pedestrian clearance beneath canopy; routinely grown with multiple trunks; can be trained to grow with a single, short trunk; no thorns

Pruning Requirement: pruning or staking may be needed to develop single trunk; head back or drop-crotch lower limbs to make them stronger and help prevent them from drooping

Limb Breakage: resistant; crotches with included bark split from old trees

Current Year Twig: greenish-gray; medium thickness

LANDSCAPE NOTES:

Origin: China, Korea, Japan

Usage: container; large parking lot island; buffer strip; highway; deck or patio; small shade tree; specimen; street tree; urban tolerant

This fast-growing evergreen tree has a dense canopy of bending branches composed of large, four- to six-inch-long leaves that have narrow, translucent margins. Terminal, 6- to 10-inch-long, eyecatching panicles of small, white, malodorous flowers are produced in late spring to summer. The berries are popular with birds; the dispersed seeds germinate in the landscape and could become somewhat of a nuisance. Care must be taken in the location of this multi-trunked tree because the profuse berry production can create a litter problem on hard surfaces. The fallen berries may temporarily stain car paint, walks, and patios.

Trees trained with sturdy branches spaced along a single, short trunk are well suited for planting along streets. Those with low-growing branches are not well suited for streets because the low branches will have to be removed as they droop. Most nurseries grow Tree Ligustrum with several trunks originating close to the ground. They spread out from each other as they ascend into the rounded, vase-shaped canopy. These can be planted as specimens in locations where lower branches can droop, forming the classic vase-shaped canopy.

The tree seems to thrive on neglect and is used along highways with little or no irrigation in humid climates. Clayey soil and high pH do not seem to cause any problems as long as water drains away from the roots. Scales, whiteflies, sooty mold, and soil nematodes can be problems, but usually are not serious. Roots rot in wet soil.

Liquidambar styraciflua Sweetgum

Pronunciation: lick-wid-AM-burr sty-rass-uh-FLEW-ah

ZONES: 5B, 6A, 6B, 7A, 7B, 8A, 8B, 9A, 9B, (10)

HEIGHT: 75 feet

WIDTH: 50 feet

FRUIT: round; 1"; dry; brown; attracts wildlife; causes significant litter and easily germinates; showy

GROWTH RATE: moderate to fast; moderate life span

HABIT: pyramidal to rounded; moderate density; symmetrical; coarse texture

FAMILY: *Hamamelidaceae*

LEAF COLOR: medium green in summer; showy orange, purple, red, or yellow fall color

LEAF DESCRIPTION: alternate; simple; serrate margin; star-shaped; deciduous; 4 to 8" long

FLOWER DESCRIPTION: yellowish-green; not showy; spring

EXPOSURE/CULTURE:

Light Requirement: part shade to full sun

Soil Tolerances: all textures; slightly alkaline to acidic; wet soil; moderate drought

Aerosol Salt Tolerance: some

PEST PROBLEMS: resistant to most lethal problems

BARK/STEMS:

Branches droop as tree grows and will require early pruning for vehicular or pedestrian clearance beneath canopy; should be grown with a single leader; no thorns

Pruning Requirement: needs little or no pruning to develop strong trunk and branch structure

Limb Breakage: moderately resistant

Current Year Twig: reddish-brown; medium thickness

CULTIVARS AND SPECIES:

Cultivars have been selected for their fall color, leaf shape, or growth habit. 'Festival' and 'Burgundy' grow well in warmer climates. The cultivar 'Rotundiloba' is fruitless, but leaves do not look like the typical sweetgum, and included bark develops in crotches. *L. formosana* (zone 6B) differs by having a three-lobed leaf.

LANDSCAPE NOTES:

Origin: New Jersey to Texas and Florida

Usage: large parking lot island; buffer strip; reclamation; shade tree; specimen; street tree

On some trees, particularly in the northern part of the range, branches are covered with characteristic corky projections. The trunk is straight and usually does not divide into double or multiple leaders, and side branches are small in diameter on young trees, creating a narrow, pyramidal form. Sweetgum makes a nice conical park, campus, or residential shade tree for large properties when it is young, developing a more oval or rounded canopy as it grows older as several branches become dominant and increase in diameter.

Plant trees 8 to 10 feet or more from walks to prevent root damage; surface roots can be aggressive and large. Much of the root system is shallow (particularly in its native, moist habitat), but there are deep vertical roots directly beneath the trunk in well-drained soils. Cut circling roots at planting to prevent girdling roots from forming. The seeds provide food for wildlife and will readily germinate in shrub and groundcover beds, requiring removal to maintain a neat landscape appearance. The fruit may also be a litter nuisance to some in the fall, but this is usually only noticeable on hard surfaces, such as roads, patios, and sidewalks, where people could slip and fall on the fruit.

The tree should be planted only in soil with a pH of 7.5 or less. Chlorosis occurs on alkaline soils. Tree thickets form in old fields, creating dense monocultures of Sweetgum. Existing trees often die back near the top of the crown, apparently due to extreme sensitivity to soil disturbance or drought injury.

Limitations include trunk cankers, fruit drop, and surface roots. Leaf spots are common in humid climates, but cause no lasting harm to the tree. Sweetgum is a new host for bacterial leaf scorch.

Liriodendron tulipifera

Tuliptree, Tulip-Poplar, Yellow-Poplar

Pronunciation: lear-ee-oh-DEN-dron too-lih-PIFF-err-ah

ZONES: 5A, 5B, 6A, 6B, 7A, 7B, 8A, 8B, 9A, 9B, (10)

HEIGHT: 80 to 120 feet

WIDTH: 25 to 40 feet

FRUIT: elongated; 1 to 2"; dry; brown; attracts birds; causes considerable litter

GROWTH RATE: rapid-growing, long lived

HABIT: pyramidal when young, oval later; open to moderate density; symmetrical when young; coarse texture

FAMILY: *Magnoliaceae*

LEAF COLOR: light green in summer; showy yellow fall color

LEAF DESCRIPTION: alternate; simple; lobed; entire margin; deciduous; 4 to 7" long

FLOWER DESCRIPTION: greenish-yellow; pleasant fragrance; showy; late spring

EXPOSURE/CULTURE:

Light Requirement: grows best in full sun

Soil Tolerances: all textures; acidic; occasionally wet soil; salt-sensitive; withstands moderate drought in humid climates, drought sensitive elsewhere

Aerosol Salt Tolerance: none

PEST PROBLEMS: resistant

BARK/STEMS:

Branches grow mostly upright and will not droop; showy trunk; should be grown with a single leader; bark on young trees is thin and easily damaged; no thorns

Pruning Requirement: needs little pruning to develop strong structure; prune limbs with included bark

Limb Breakage: susceptible to ice damage, otherwise resistant

Current Year Twig: brown; medium thickness

LANDSCAPE NOTES:

Origin: eastern United States

Usage: shade tree; possible street tree for very wide boulevard

Tuliptree maintains a fairly narrow oval crown when young, but spreads to an open oval or rounded canopy with age. Trunks become massive in old age, becoming deeply furrowed with thick bark. The tree maintains a straight trunk and generally does not form double or multiple leaders. Older trees have several large-diameter major limbs in the top half of the crown. The soft wood reportedly is subject to storm damage, but the trees have held up remarkably well in the southern United States during hurricanes. It is probably stronger than given credit for due to the strong branch attachments.

Although a rather large tree, Tulip-Poplar could be used along residential streets with very large lots and plenty of soil for root growth if set back 10 or 15 feet. Trees can be planted from containers at any time, but transplanting from a field nursery should be done in spring, followed by faithful watering. It is usually recommended only for moist sites in many parts of Texas, including Dallas, but has grown in an open area with plenty of soil space for root expansion near Auburn and Charlotte without irrigation, where the trees are vigorous and look nice. The tree also grows nicely with irrigation into southern California, but branches with included bark split from the tree in windy weather.

Aphids leave heavy deposits of sticky honeydew on lower leaves, cars, and other hard surfaces below. A black, sooty mold grows on the honeydew. Although this does little permanent damage to the tree, the honeydew and sooty mold can be annoying. During hot, dry weather, interior leaves turn yellow and fall off, creating a constant mess. 'Arnold' (tall and narrow habit) drops less leaves in summer. This nonlethal condition is due to the weather and is not a disease. Branches occasionally break in storms.

Litchi chinensis Lychee

Pronunciation: LEE-chee chi-NEN-sis

ZONES: 10A, 10B, 11

HEIGHT: 25 to 40 feet

WIDTH: 20 to 40 feet

FRUIT: round; 0.5 to 1"; fleshy; red; attracts wildlife and people because it is delicious; causes significant mess on hard surfaces; showy

GROWTH RATE: moderate; moderate life span

HABIT: rounded vase; dense; symmetrical; medium texture

FAMILY: *Sapindaceae*

LEAF COLOR: broadleaf evergreen; dark green

LEAF DESCRIPTION: alternate; odd pinnately compound; serrate margin; oval to lanceolate; 3 to 6"

FLOWER DESCRIPTION: creamy yellow; showy; spring

EXPOSURE/CULTURE:

Light Requirement: grows best in full sun

Soil Tolerances: all textures; slightly alkaline to acidic; moderate drought

Aerosol Salt Tolerance: none

PEST PROBLEMS: resistant

BARK/STEMS:

Branches droop as tree grows and will require pruning for pedestrian clearance beneath canopy; routinely grown with multiple trunks; no thorns

Pruning Requirement: requires pruning to develop strong structure; thin or remove branches with included bark

Limb Breakage: resistant

Current Year Twig: green; slender

CULTIVARS AND SPECIES:

Several named cultivars are available for best fruit production: 'Brewster', 'Mauritius', 'Sweet Cliff', 'Kate Sessions', and 'Kwai Mi'.

LANDSCAPE NOTES:

Origin: China

Usage: container; fruit; hedge; deck or patio; specimen

This attractive fruit tree has particularly handsome, glossy, evergreen leaves, and forms a compact, round-headed canopy. New leaves emerge an attractive bronze-red. Lychee trees can eventually reach 50 feet in height, but will reach about 30 feet tall 30 years after planting in a landscape, creating a wonderful shade, framing, or specimen tree. Small, yellow flowers appear in drooping, foot-long panicles in early spring and are followed by clusters of delicious fruit in late June and July. When ripe, the warty outer surface of the strawberry-like fruit turns bright red and becomes brittle. Easily peeled, the interior sweet, juicy, white flesh surrounds a single, large, glossy brown seed. The trees are quite decorative when laden with fruit.

The tree may be located near a patio, in a shrub border, or as an accent in the lawn. The thick canopy also makes it well suited as a screen. Spaced 20 to 30 feet apart, they make a nice boulevard tree. Consider locating the tree in the back yard if you are planting on a residential lot. This will prevent passersby from helping themselves to the delectable fruit.

Easily grown in full sun on deep, fertile, well-drained soil, Lychee should be located where it can be protected from strong winds. The dense canopy can catch the wind and the tree can topple over. Proper thinning can help prevent this. Plants should receive regular fertilization, as micronutrient deficiency, especially iron, can show in alkaline soil.

Scales and mushroom root rot may occur on soils where oaks were grown. Cuban May beetle chews foliage.

Lithocarpus densiflorus Tanbark or Tan Oak

Pronunciation: Lith-oh-CAR-puss den-sih-FLOOR-uss

ZONES: 7B, 8A, 8B, 9A, 9B, 10A, 10B

HEIGHT: 30 to 50 feet

WIDTH: 30 to 50 feet

FRUIT: round; 0.75"; dry and hard; pubescent; brown; attracts wildlife; cause some litter; showy; edible

GROWTH RATE: moderate; long-lived

HABIT: rounded and spreading; dense; symmetrical; fine texture

FAMILY: *Fagaceae*

LEAF COLOR: evergreen; dark green above, hairy and gray-green beneath

LEAF DESCRIPTION: alternate; simple; dentate or serrate margin; oblong to obovate; 1.5 to 4" long

FLOWER DESCRIPTION: whitish; showy; rank odor; spring

EXPOSURE/CULTURE:

Light Requirement: part shade to full sun

Soil Tolerances: loam or sand; acidic; drought

PEST PROBLEMS: resistant

BARK/STEMS:

Branches on open-grown trees droop and will need pruning to create clearance beneath canopy for pedestrians and vehicles; in forest, grows with a single leader; no thorns

Pruning Requirement: needs early pruning to develop a single trunk

Limb Breakage: resistant

Current Year Twig: brown; gray; slender

LANDSCAPE NOTES:

Origin: along the coast from southern Oregon to Santa Barbara, California

Usage: parking lot island; buffer strip; highway; reclamation; shade tree; sidewalk cutout; street tree

Tanbark Oak grows slowly, sometimes to more than 80 feet in height in its native, mountainous habitat, but is more often seen 30 to 40 feet high and wide in a landscape, forming a rounded silhouette with a tight canopy, large-diameter branches, and a twisted trunk. The leathery leaves have fuzzy undersides that probably contribute to the drought tolerance of this relative of the oaks. The acorns formed part of the main diet of Native Americans in this tree's native habitat. The acorns are also popular with birds, deer, bear, and raccoons.

Tan Oak should be grown with a single trunk and a few widely spaced branches to mimic its growth habit in the wild. Branches should form a wide angle with the trunk. The first permanent branch can be located three to five feet from the ground if the tree will be planted in an open lawn area and allowed to develop a wide crown. For those planted as street trees or in areas requiring clearance for vehicles or pedestrians, the first permanent branch should be higher on the trunk. The interior portion of the crown often cleans itself of small branches, displaying the nice branch arrangement common on most specimens.

Tan Oak should be grown in full sun or partial shade on almost any soil, as it tolerates dry, sandy sites quite well. Do not be afraid to plant this tree on sites where there is only occasional irrigation after trees are established in the landscape. However, *Arbutus mensiezii, Quercus garryana,* and *Q. kelloggii* are even more tolerant of drought. Tanbark Oak is a popular choice for planting along streets and other urban areas because of its tolerance to soils typical in urban areas. However, there is no evidence that the tree tolerates highly alkaline soils.

No limitations of major concern; probably susceptible to the usual array of Oak problems.

Lithocarpus henryi Henry Tanbark Oak

Pronunciation: lith-oh-CAR-puss HEN-ree-eye

ZONES: 8A, 8B, 9A, 9B, 10A, 10B

HEIGHT: 15 to 20 feet

WIDTH: 15 to 20 feet

FRUIT: round; 0.75"; dry and hard; brown; attracts wildlife; cause some litter; showy

GROWTH RATE: slow-growing; moderate life span

HABIT: round; open when young, dense later; symmetrical; coarse texture

FAMILY: *Fagaceae*

LEAF COLOR: evergreen; light green; sometimes discolors in winter

LEAF DESCRIPTION: alternate; simple; entire margin; oblong to lanceolate; 6 to 10" long

FLOWER DESCRIPTION: creamy white; showy; late summer

EXPOSURE/CULTURE:

Light Requirement: part shade to full sun

Soil Tolerances: loam or sand; acidic; drought

PEST PROBLEMS: resistant

BARK/STEMS:

Branches on open-grown trees droop and will need pruning to create clearance beneath canopy for pedestrians and vehicles

Pruning Requirement: shorten lower branches to encourage growth higher in canopy

Limb Breakage: resistant

Current Year Twig: brown to gray; slender

LANDSCAPE NOTES:

Origin: China

Usage: container or above-ground planter; near a deck or patio; parking lot island; buffer strip; highway; specimen; shrub border; screen

Henry Tanbark Oak grows slowly, occasionally to 50 feet in a landscape, forming a rounded silhouette with a tight canopy by the time it is 12 years old. The pointed leaves make the billowing canopy a handsome sight on this little-known tree. The fruit is popular among birds, deer, bear, and raccoons.

Tanbark Oak should be grown with a short, single trunk. Training main branches to a more upright growth habit, and heading branches to increase density on young trees, might help the tree gain acceptance in the trade. Branches usually droop toward the ground, forming a shrub-like habit unless the tree is staked and pruned to form a central trunk. After a central leader is developed, space main branches along the trunk to create a durable form.

This could be a popular small tree for planting in a container, near a deck or patio, or as an accent in a shrub border. The coarse, glossy foliage makes it suited for planting as a specimen in any landscape. Good tolerance to drought allows it to be incorporated into a landscape that receives little or no irrigation. Use in downtown and other highly urban areas should be tempered because there is no evidence that the tree tolerates alkaline soils.

Possibly susceptible to the usually array of Oak problems, as it is so closely related.

Livistona chinensis Chinese Fan Palm, Fountain Palm

Pronunciation: lih-veh-STOW-nah chi-NEN-sis

ZONES: (9A), 9B, 10A, 10B, 11

HEIGHT: 30 to 50 feet

WIDTH: 10 to 12 feet

FRUIT: round; 1"; fleshy; blackish-blue; no significant litter

GROWTH RATE: moderate; moderate life span

HABIT: palm; open; fine texture

FAMILY: *Arecaceae*

LEAF COLOR: light green

LEAF DESCRIPTION: costapalmate; entire margin; 5 to 6 feet across; sharp teeth on petiole; distinctly drooping tips

FLOWER DESCRIPTION: white; not especially showy; spring and summer

EXPOSURE/CULTURE:

Light Requirement: part shade to full sun

Soil Tolerances: all textures; alkaline to acidic; some drought

Aerosol Salt Tolerance: medium

PEST PROBLEMS: sensitive

BARK/STEMS:

Single trunk, 12" thick; nonpersistent leaf bases; no crown shaft; no thorns

Pruning Requirement: remove dying fronds

Limb Breakage: resistant

CULTIVARS AND SPECIES:

Livistona chinensis subglobosa is a dwarf Chinese Fan Palm. There are a number of other Livistona species with similar habits.

LANDSCAPE NOTES:

Origin: China

Usage: large parking lot island; buffer strip; deck or patio; specimen; sidewalk cutout; street tree

Able to reach 50 feet in height, but usually seen at 30 feet with a 10- to 12-foot spread, Chinese Fan Palm has a single straight trunk and large, 6-foot-long leaves with drooping tips. It is one of about 30 species of *Livistona*. The divided frond sections have long, tapering, ribbon-like segments that gracefully sway beneath the leaves, creating an overall weeping, fountain-like effect. Only a slight breeze is needed to move leaf segments, adding a lovely, fine-textured effect to any landscape. The petioles are armed with sharp teeth. The inconspicuous flowers are hidden among the fronds and are followed by small, blue-black, olive-like fruits.

Although Chinese Fan Palm has long been used as a container palm, its neat leaf habit and interesting form make it ideal for further landscape uses, such as in staggered groupings or as a freestanding specimen or street tree. Extensive use as a street tree is not recommended in areas where lethal yellowing disease is present, as this palm is moderately susceptible. Fan Palms form a closed canopy when planted about eight to ten feet apart along a walk or street, and grow well in confined soil spaces. The palm is self-cleaning of old leaves and will require little or no pruning.

Tolerant of full sun, young specimens of Chinese Fan Palm should be partially shaded. Any reasonably fertile, well-drained soil, including alkaline, is suitable. Chinese Fan Palm should be fertilized two or three times during the year to maintain the green foliage. Plants should be watered during dry spells and will benefit from an organic mulch placed beneath the canopy. Scales may occur.

Lysiloma spp. Wild Tamarind

Pronunciation: lie-sih-LOW-mah species

ZONES: 10B, 11

HEIGHT: 40 to 60 feet

WIDTH: 45 feet

FRUIT: elongated pod; 3 to 6"; dry and hard; brown; no significant litter; showy

GROWTH RATE: rapid-growing; long-lived

HABIT: weeping vase; open; irregular silhouette; fine texture

FAMILY: *Leguminosae*

LEAF COLOR: evergreen; pale green

LEAF DESCRIPTION: alternate; bipinnately compound; entire margin; obovate; leaflets 0.5" long

FLOWER DESCRIPTION: white; pleasant fragrance; not showy; spring and summer

EXPOSURE/CULTURE:

Light Requirement: part shade to full sun

Soil Tolerances: very adaptable; all textures; alkaline to acidic; salt and drought

Aerosol Salt Tolerance: high

PEST PROBLEMS: pest free

BARK/STEMS:

Showy; branches droop as tree grows and will require pruning for vehicular or pedestrian clearance beneath canopy; routinely grown with multiple trunks; should be trained to grow with a short, single trunk with well-spaced branches; no thorns

Pruning Requirement: requires pruning to develop strong structure; thin major branches to slow growth and prevent formation of included bark

Limb Breakage: resistant

Current Year Twig: green; slender

CULTIVARS AND SPECIES:

L. candida and *L. thornberi* (zone 9A) are drought-tolerant, small trees planted in frost-free landscapes in the western United States.

LANDSCAPE NOTES:

Origin: south Florida, Cuba, Caribbean islands

Usage: container; parking lot island; buffer strip; highway; reclamation plant; shade tree; specimen; sidewalk cutout; street tree; urban tolerant

Wild Tamarind is native to South Florida and grows moderately fast. Its slender, short trunk, topped with long, somewhat arching branches, forms an umbrella-like silhouette. The fern-like leaves are a showy red when young and make a striking contrast when new and old growth appear together. Developing into a more open tree with age, Cuban Tamarind makes an ideal street tree, shade, park, or seaside planting.

Cities have planted Cuban (*L. bahamensis*) and Wild Tamarind (*L. latisiliqua*) along streets with good success. Codominant stems form very low on the trunk without proper pruning and training, and branches will droop toward the ground. Specify trees for planting along streets and in parking lots that have a clear trunk to about six to eight feet to help avoid this problem. If large branches are allowed to develop below this point, the tree could become disfigured, as these branches will have to be removed to allow for passage of vehicles and pedestrians under the canopy. Locate the first permanent branch 6 (preferably 10) or more feet from the ground to allow for clearance. Low branches can be left on the tree if it will be planted in a yard, park, or other location where vehicle clearance is not a concern.

Major branches often develop included bark, because they grow at the same rate as the trunk. They often grow to about the same size as the trunk. This does not appear to be a problem on small trees, but could encourage branch breakage as the tree grows older. Try to keep the major branches from growing larger than about one-half the diameter of the trunk by thinning.

Stem galls and rust disease can be an occasional nuisance. Caterpillars eat foliage.

Maackia amurensis Amur Maackia

Pronunciation: MACK-ee-uh a-more-EN-siss

ZONES: 3A, 3B, 4A, 4B, 5A, 5B, 6A, 6B, 7A, 7B

HEIGHT: 20 to 35 feet

WIDTH: 25 to 50 feet

FRUIT: pod; 2 to 3"; dry and hard; causes some litter

GROWTH RATE: slow-growing; moderate life span

HABIT: rounded vase; moderate density; symmetrical; medium texture

FAMILY: *Leguminosae*

LEAF COLOR: new leaves emerge silvery gray; mature leaves dark green; yellow fall color

LEAF DESCRIPTION: alternate; odd pinnately compound; entire margin; oblong to ovate; deciduous; leaflets 1 to 3" long

FLOWER DESCRIPTION: white; interesting but not especially showy except in droughty years; summer

EXPOSURE/CULTURE:

Light Requirement: full sun

Soil Tolerances: all textures; alkaline to acidic; extreme drought

PEST PROBLEMS: pest-free

BARK/STEMS:

Shiny, brown bark; branches mostly upright and do not droop; routinely grown with multiple trunks; forms undesirable included bark; no thorns

Pruning Requirement: requires pruning to develop strong structure; space branches along a short, central trunk

Limb Breakage: resistant

Current Year Twig: brown; medium to slender

LANDSCAPE NOTES:

Origin: China, Manchuria

Usage: container; buffer strip; highway; deck or patio; shade tree; parking lot island; specimen; street tree

A very hardy and adaptable tree, Amur Maackia is slow-growing and deciduous, reaching 45 feet tall in the wild but most often seen at 20 to 35 feet in landscapes. The multi-divided leaflets are bright gray/green when young but mature to dark green, and drop in fall with or without significantly changing color. In summer, dense, erect, eight-inch-long racemes of bloom appear, each small, off-white flower tinged with very pale, dark blue. These blooms are followed by the abundant appearance of flat fruit pods.

Amur Maackia may be well suited to use as a street tree (or in other confined soil spaces in urban areas), where the peeling, orange-brown, shiny bark is especially noticeable on young trees. It is usually seen with numerous small-diameter trunks originating from the lower trunk. Thirty-five-year-old, single-trunked trees have a diameter of about one foot.

Lower branches droop slowly toward the horizontal as the tree grows, forming an attractive outline—almost the perfect small tree form. If grown near a walk or along the street, be sure to prune early in the life of the tree to locate the lower branches far enough up on the trunk so that branches will not interfere with traffic below. This will eliminate the need to make large pruning cuts on older trees, which reportedly close over pruning wounds very slowly. Space major branches apart on the trunk to form a sturdy tree for urban planting sites.

Amur Maackia should be grown in full sun on well-drained soil, either acid or alkaline. Amur Maackia has nitrogen-fixing bacteria associated with the root system and should require little maintenance other than some corrective pruning early in the life of the tree to space main limbs apart and to prevent included bark in the crotches of major limbs. It should be grown and planted more often. It held up better than *any* other tree in the drought of 1988 in the Midwest.

Maclura pomifera Osage-Orange, Bois-D'Arc

Pronunciation: mah-CLUE-rah pom-IF-err-ah

ZONES: 5A, 5B, 6A, 6B, 7A, 7B, 8A, 8B, 9A, 9B

HEIGHT: 35 to 45 feet

WIDTH: 20 to 40 feet

FRUIT: round; 4 to 6"; fleshy; green; causes significant litter; attracts mammals; showy

GROWTH RATE: rapid-growing; long-lived

HABIT: rounded vase; dense; irregular silhouette; symmetrical; coarse texture

FAMILY: *Moraceae*

LEAF COLOR: dark green; showy yellow fall color

LEAF DESCRIPTION: alternate; simple; entire, undulating margin; lanceolate to oblong or ovate; deciduous; 3 to 6" long

FLOWER DESCRIPTION: green; somewhat showy; spring

EXPOSURE/CULTURE:

Light Requirement: grows best in full sun

Soil Tolerances: all textures; alkaline to acidic; wet soil; some salt; drought

Aerosol Salt Tolerance: some

PEST PROBLEMS: resistant

BARK/STEMS:

Branches droop as tree grows and will require pruning for vehicular or pedestrian clearance beneath canopy; routinely grown with multiple trunks, but can be trained with a central leader; showy; thorns are present on trunk or branches

Pruning Requirement: requires some early pruning to develop a dominant trunk

Limb Breakage: resistant

Current Year Twig: brown; thick

CULTIVARS AND SPECIES:

Thornless, fruitless trees are being developed and should prove very adapted to the urban landscape. These include 'Park', 'Witchita', and 'White Shield'.

LANDSCAPE NOTES:

Origin: Arkansas, Oklahoma, Texas

Usage: reclamation plant; windbreak; specimen; urban tolerant

This deciduous North American native tree rapidly creates a dense canopy, making it useful as a windbreak. The large, two to three-inch-wide, shiny leaves turn bright yellow in fall before dropping, although this color change is not quite as noticeable on trees grown in the southeastern United States. The bark is deeply furrowed and the strong, durable wood is bright orange in color.

It is reported that the Osage Indians made their hunting bows from this beautiful and hard wood, and it is also used to make furniture. From April to June, Osage-Orange puts out its inconspicuous green flowers, but these are followed by the very conspicuous fruits. The fruits are four- to five-inch-diameter, rough-textured, heavy green balls that ripen to yellow-green and fall to the ground in October and November. These fruits are inedible, but squirrels relish the small seeds buried inside the pulp. When the fruits drop, they can be very messy; for this reason, male, fruitless trees should be selected if you plant this tree. Osage-Orange is thorny, just like true citrus trees, and forms thickets if left to grow on its own.

Osage-Orange should be grown in full sun on well-drained soil. This tough, native plant can withstand almost anything once established—heat, cold, wind, drought, poor soil, ice storms, vandalism—but appreciates regular watering when young. Roots can be aggressive and raise walks if planted too close.

Magnolia acuminata

Cucumbertree, Cucumber Magnolia

Pronunciation: mag-NO-lee-ah ah-cue-muh-NAY-tah

ZONES: 4A, 4B, 5A, 5B, 6A, 7A, 7B, 8A, 8B

HEIGHT: 60 to 80 feet

WIDTH: 50 to 60 feet

FRUIT: elongated; 2 to 3"; dry; red; attracts wildlife; no significant litter; persistent and showy

GROWTH RATE: rapid-growing; long-lived

HABIT: narrow pyramidal when young; dense; irregular silhouette; coarse texture

FAMILY: *Magnoliaceae*

LEAF COLOR: yellow-green in summer; striking yellow to bronze fall color in some years

LEAF DESCRIPTION: alternate; simple; entire margin; oval to ovate; deciduous; 6 to 12" long

FLOWER DESCRIPTION: yellow; pleasant fragrance; hidden by foliage; spring

EXPOSURE/CULTURE:

Light Requirement: part shade to full sun

Soil Tolerances: all textures; alkaline to acidic; occasionally wet soil; moderate salt and drought

Aerosol Salt Tolerance: some

PEST PROBLEMS: resistant

BARK/STEMS:

Bark is thin and easily damaged by mechanical impact; branches do not droop; should be grown with a single leader; no thorns

Pruning Requirement: needs little pruning to develop strong structure

Limb Breakage: resistant

Current Year Twig: brown; medium thickness

LANDSCAPE NOTES:

Origin: Appalachian Mountains south to Louisiana

Usage: shade tree; specimen; potential street tree

One of the fastest-growing Magnolias, Cucumbertree is pyramidal when young but becomes broad and oval with age. Older trees have a stately silhouette, particularly in the winter with branches bare, sporting a number of large-diameter branches growing from a dominant central trunk. The trunk can grow to be five feet thick and the wood has been used, along with Tuliptree, for "poor man's walnut."

The large, deciduous leaves are lighter and fuzzy underneath and cast very dense shade below, making Cucumbertree ideal as a shade or specimen tree. The slightly fragrant, three-inch-wide flowers appear in May or early June, but their greenish-yellow to yellow coloring causes them to become lost among the foliage.

Many Magnolias have a root system that spreads more than those of other trees, and there are only a few fine roots. This is thought to contribute to the poor growth that follows transplanting Magnolia from a field nursery. There is no problem planting from containers, provided adequate irrigation is available until established. Transplant field material in the spring.

This tree is best on large estates and open-soil areas such as parks and golf courses, or along either side of an entrance road with plenty of soil space for root expansion. It does not tolerate the compacted, disturbed soils of urban areas unless it is extensively mulched. Be sure young trees receive adequate irrigation until the root system is well established in loose, open soil.

No limitations of major concern, but occasionally bothered by scale, as are many other Magnolias.

Magnolia cordata Yellow Magnolia

Pronunciation: mag-NO-lee-ah core-DAH-tah

ZONES: 5A, 5B, 6A, 6B, 7A, 7B, 8A, 8B, 9A, 9B, (10)

HEIGHT: 25 feet

WIDTH: 20 to 30 feet

FRUIT: elongated and irregular; 1 to 3"; dry; red seeds; attracts birds; no significant litter

GROWTH RATE: slow-growing; moderate life span

HABIT: pyramidal when young; open; irregular silhouette; coarse texture

FAMILY: *Magnoliaceae*

LEAF COLOR: green in summer; muted yellow fall color

LEAF DESCRIPTION: alternate; simple; entire; oblong to obovate; deciduous; 3 to 6" long

FLOWER DESCRIPTION: yellow; very showy; spring

EXPOSURE/CULTURE:

Light Requirement: part shade to full sun

Soil Tolerances: all textures; slightly alkaline to acidic; occasionally wet; moderate drought

Aerosol Salt Tolerance: none

PEST PROBLEMS: resistant

BARK/STEMS:

Bright gray bark; bark is thin and easily damaged by mechanical impact; branches grow mostly upright, especially on young trees; grows with a single, dominant leader, unless pruned differently; no thorns

Pruning Requirement: needs little pruning to develop strong structure; if pruning is needed, do it after flowering to enjoy the flower display

Limb Breakage: somewhat susceptible

Current Year Twig: brown; medium

LANDSCAPE NOTES:

Origin: eastern North America

Usage: espalier; near a deck or patio; specimen; potential street tree

Young Yellow Magnolia is distinctly upright and pyramidal, becoming more oval, then round about 30 years later. Blooms open in late spring before the leaves, producing large, yellow flowers and creating a spectacular display. Few other yellow-flowering trees put on a spring display like Yellow Magnolia. Recent taxonomic changes list this as a variety of *M. acuminata*.

The tree is best used as a specimen in a sunny spot where it can develop a uniform crown. Although lower limbs can be removed to allow for pedestrian clearance if planted close to a walk or patio, Magnolia looks its best when branches are left to droop to the ground. The light gray bark shows off nicely, particularly during the winter when the tree is bare.

Transplant in the spring, just before growth begins, and use balled-in-burlap or containerized plants. Bareroot material may take a long time to recover from transplanting. Pruning wounds may not close well, so train plants early in their life to develop the desired form so that large pruning cuts will not be necessary. This will help ensure that pruning wounds will be small. If you have to remove a large branch on an older tree, decay may begin in the trunk. This could be the beginning of a slow decline in the tree.

This tree is not common in the nursery trade. Those at arboreta are often grafted onto another magnolia understock, such as *M. stellata*. Generally pest-free, but scales of various types may infest twigs and foliage. Magnolia may be subject to leaf spots. Canker diseases will kill entire branches.

Magnolia denudata (heptapeta) Yulan Magnolia

Pronunciation: mag-NO-lee-ah den-you-DAY-tah

ZONES: 5B, 6A, 6B, 7A, 7B, 8A, 8B, 9A, 9B, (10A)

HEIGHT: 30 to 40 feet tall

WIDTH: 30 to 40 feet

FRUIT: elongated; 1 to 3"; dry; brown with red seeds; attracts birds; no significant litter

GROWTH RATE: moderate; moderate life span

HABIT: upright rounded; open; irregular silhouette; coarse texture

FAMILY: *Magnoliaceae*

LEAF COLOR: dark green in summer; no fall color

LEAF DESCRIPTION: alternate; simple; entire margin; oblong to obovate; deciduous; 4 to 6" long

FLOWER DESCRIPTION: white; pleasant fragrance; extremely showy, one of the best; spring

EXPOSURE/CULTURE:

Light Requirement: part shade to full sun

Soil Tolerances: all textures; slightly alkaline to acidic; some drought

PEST PROBLEMS: resistant

BARK/STEMS:

Bark is thin and easily damaged by mechanical impact; branches droop as tree grows and will require early pruning for vehicular or pedestrian clearance beneath canopy; routinely grown with multiple trunks, but can be trained to a short, single trunk; no thorns

Pruning Requirement: requires pruning to develop strong structure; remove lower branches early if they will be in the way later

Limb Breakage: resistant

Current Year Twig: greenish-brown; medium thickness

CULTIVARS AND SPECIES:

'Japanese Clone'—larger flowers; 'Lacey'—flowers up to eight inches across. Yulan Magnolia was used as a parent plant along with *Magnolia acuminata* to produce the hybrid x 'Elizabeth', which has a pyramidal shape and clear yellow, fragrant blooms.

LANDSCAPE NOTES:

Origin: central China

Usage: container; espalier; deck or patio; specimen

This broad, spreading, deciduous tree has a relatively fast growth rate and eventually reaches mature height with an equal spread. The crown is open and forms a rounded outline several years after planting in full sun. The off-white, saucer-shaped, six-inch-diameter, fragrant blooms appear on the trees before the emergence of leaves. The blooms are followed by brown fruits, which ripen in early fall to reveal bright red, inner seeds. Although there is no appreciable fall color change, the multi-stemmed, irregular form of Yulan Magnolia makes it quite striking in the winter garden after the leaves have fallen.

This small tree is best used as a patio tree for shade and accent, due to the low branching habit, attractive foliage, and striking gray bark. It would look very nice lining an entrance walk to a commercial buiding or in a double or single row set back from an entrance roadway or long driveway. Grows well in urban areas but avoid poor, compacted soil. Nice for an urban garden.

Yulan Magnolia should be grown in full sun or partial shade on rich, moisture-retentive soil in an area protected from strong, drying winds. In the warmer climates, it performs best in partial sun locations. Plants should not be exposed to overly wet or dry conditions. No limitations of major concern, but occasionally bothered by Magnolia scale, as are many other Magnolias. Frost often injures flowers in spring.

Magnolia grandiflora Southern Magnolia

Pronunciation: man-NO-lee-ah gran-deh-FLOOR-ah

'Hasse'

ZONES: 6B, 7A, 7B, 8A, 8B, 9A, 9B, 10A, (10B)

HEIGHT: 60 to 80 feet

WIDTH: 30 to 40 feet

FRUIT: elongated; 3 to 8"; dry; red seeds; attracts birds; causes some litter

GROWTH RATE: moderate; long-lived

HABIT: oval; pyramidal; dense, variable; nearly symmetrical when young, irregular later; coarse texture

FAMILY: *Magnoliaceae*

LEAF COLOR: broadleaf evergreen; dark green

LEAF DESCRIPTION: alternate; simple; entire margin; oval to ovate; 5 to 10" long

FLOWER DESCRIPTION: white; showy; spring and summer

EXPOSURE/CULTURE:

Light Requirement: mostly shaded to full sun

Soil Tolerances: all textures; slightly alkaline to acidic; moderate salt; occasionally wet; some are quite drought-tolerant

Aerosol Salt Tolerance: moderate

PEST PROBLEMS: resistant to most lethal pests

BARK/STEMS:

Bark is thin and easily damaged by mechanical impact; branches droop as tree grows and will require pruning for vehicular or pedestrian clearance beneath canopy; should be grown with a single leader; no thorns

Pruning Requirement: remove any competing leaders; needs little pruning to develop strong structure

Limb Breakage: resistant

Current Year Twig: green; thick

CULTIVARS AND SPECIES:

Trees germinated from seed are quite variable in growth rate and form. 'Bracken's Brown Beauty' (hardy to about −10°F) has an unusually dark brown lower leaf surface and is considered one of the best selections; 'Cairo' has an early and long flowering period; 'Charles Dickens' has broad, nearly blunt leaves, large flowers, and large red fruit; 'Claudia Wannamaker' has brown-backed leaves and a tight habit with a dense root system; 'Edith Bogue' is a hardy cultivar and will bloom when only two to three years old; 'Glen St. Mary' has a compact form, will bloom when young, is slow-growing, and has leaves with a bronze underside; 'Gloriosa' has large flowers and leaves; 'Goliath' has flowers up to 12 inches across, a long

Continued

Magnolia grandiflora *Continued*

blooming period, and a bushy habit; 'Hasse' is upright, columnar shaped, and can be used for a compact, dense hedge or screen or street tree, but has been tough for growers to transplant due to a coarse root system; 'Lanceolata' has a narrow pyramidal form and narrower leaves with rusty undersides; 'Little Gem' has a compact, upright form, reaches probably 50 feet tall, has small leaves and flowers, is very slow-growing, flowers heavily at an early age and for a long time during the summer (five months), has bronze leaf undersides. It will bloom when only three to four feet tall and is excellent as a pruned evergreen hedge, a small street tree, or an espalier; 'Majestic Beauty' (patented) has large, dark green leaves, a pyramidal shape, and profuse flowering; 'Migtig' (Greenback™) has beautiful green-backed foliage, and a tight pyramidal canopy; 'Praecox Fastigiata' has an upright, narrow growth habit; 'Samuel Sommer' has an upright, rapid growth habit and flowers up to 14 inches across; 'Victoria' is very hardy, has small flowers and rust-red leaf-undersides. It is often difficult to see significant differences among a number of cultivars from a plant usage standpoint.

LANDSCAPE NOTES:

Origin: east Texas to Florida and North Carolina

Usage: espalier; buffer strip; shade tree; specimen; street tree; median; highway

This large, stately, native North American evergreen tree, with its large, beautiful, saucer-shaped, fragrant flowers, is almost a Southern landscape tradition. It is the state tree of Mississippi. Southern Magnolia forms a dense (more open in the shade), pyramidal shape, with lower branches often bending to the ground. However, form and growth rate on seedlings are extremely variable. Some are dense and make great screens; others are very open, with large spaces between branches; some have a narrow, almost columnar form; others are as wide as they are tall. Select from the many available cultivars to ensure the desired shape and density.

The trunk can grow to more than three feet in diameter, and frequently grows straight up through the center of the crown. Branches are typically numerous and small in diameter. Remove those few branches that occasionally form weak, tight crotches.

The leathery, oblong, shiny leaves are shed as new foliage emerges, but the debris is hidden by the dense foliage of the lower limbs, if they are left on the tree. Some consider this a litter nuisance when the large, slow-to-decompose leaves drop on the sidewalk, lawn, or patio. The underside of the leaves is covered with a fine, red-brown fuzz (more prominent on some selections than others). In late spring and sporadically throughout the summer, huge, eight-inch-diameter, waxy, fragrant, white blossoms open to perfume the garden. Fuzzy brown cones follow these blooms, ripening in fall and winter to reveal bright red seeds that are used by a variety of wildlife.

Long used as a striking garden specimen, Southern Magnolia can also serve as a dense screen (select one of the dense cultivars), windbreak, or street tree (with lower limbs removed). The only objection to this tree as a street tree might be the falling leaves and fruit. Its ease of growth and carefree nature make Southern Magnolia ideal for the low-maintenance landscape, provided the area beneath the canopy is mulched. With proper pruning, Southern Magnolia trees can also be used as an interesting espalier. Early summer branch tipping can extend the summer flowering period. Some nurseries remove flower buds to improve growth rate.

If moist, peaty soils or irrigation is available, Southern Magnolia will thrive in full sun and hot conditions once established. If irrigation cannot be provided periodically, plants located in partial shade for several years after planting seem to grow better. Drought-tolerant when grown in areas with plenty of soil and moisture for root expansion, it is only moderately drought-tolerant in restricted-soil areas or in poor, dry soil. Tops die back in severe drought. Leaves remain green with regular nitrogen applications.

Southern Magnolia prefers acid soil but will tolerate a slightly basic, even occasionally wet or clay soil. The root system is wider spreading than most other trees, extending from the trunk a distance equal to nearly four times the canopy spread. This makes it very difficult to save existing Magnolia trees on construction sites. Be sure that there are no roots circling close to the trunk, as Magnolia is prone to girdling roots. Cut any circling roots, especially those toward the top of the root ball, prior to planting. Field-grown trees recover slowly from transplanting due to the wide-spreading root system, and trees often transplant best in winter and spring, not in the fall.

Scales infest twigs and leaves. Magnolia stem borer is a problem on young nursery stock. Algal leaf spots and lichens are common and highly visible on leaves in humid climates.

Magnolia kobus Kobus Magnolia

Pronunciation: mag-NO-lee-ah KO-bus

ZONES: 5A, 5B, 6A, 6B, 7A, 7B, 8A, 8B

HEIGHT: 25 feet

WIDTH: 35 feet

FRUIT: elongated; irregular; 1 to 3"; dry; very showy; pinkish-red; attracts birds; no significant litter

GROWTH RATE: slow-growing; moderate life span

HABIT: round and spreading with age; dense; nearly symmetrical; coarse texture

FAMILY: *Magnoliaceae*

LEAF COLOR: dark green in summer; showy, yellow fall color

LEAF DESCRIPTION: alternate; simple; entire margin; obovate; deciduous; 3 to 6" long

FLOWER DESCRIPTION: pink or white; pleasant fragrance; showy; spring

EXPOSURE/CULTURE:

Light Requirement: part shade to full sun

Soil Tolerances: all textures; slightly alkaline to acidic; some drought

Aerosol Salt Tolerance: some

PEST PROBLEMS: resistant

BARK/STEMS:

Bark is thin and easily damaged by mechanical impact; branches droop as tree grows and will require pruning for vehicular or pedestrian clearance beneath canopy; routinely grown with multiple trunks; no thorns

Pruning Requirement: needs little pruning to develop strong structure

Limb Breakage: resistant

Current Year Twig: brownish-green; medium to thick

CULTIVARS AND SPECIES:

The cultivar 'Wada's Memory' has dark-green leaves, large, five- to six-inch blooms, and an upright or columnar growth habit (at least in youth), making it suited for street tree planting.

LANDSCAPE NOTES:

Origin: Japan

Usage: container; specimen; deck or patio

A striking tree in summer or winter. Dropping its large leaves in fall with a moderate to good display of color, Kobus Magnolia forms an attractive winter specimen with its rounded silhouette and multiple trunks originating close to the ground. Grows 30 to 40 feet tall but most often is 25 feet or less in an open, sunny landscape site. It is capable of reaching 75 feet in in its native forest habitat. In an open site, spread is often greater than height, with 25-foot-tall trees 35 feet wide if given the room to grow unobstructed. Branches gracefully touch the ground on older specimens as the tree spreads, in a manner not unlike open-grown Live Oaks. Allow plenty of room for proper development.

The lightly fragrant blooms, which appear all over the tree in spring before the new leaves unfold, are ivory-colored to pale pink and four inches in diameter. Young trees flower poorly. The pink fruits split open to reveal bright red seeds, which sway from slender threads before dropping to the ground.

Trees grow wide, but they do so slowly. This makes them suited for planting next to the deck or patio as a shade tree, although you may not benefit much from the shade the tree will eventually provide. This is a fine Magnolia much deserving of specimen planting as a feature in the landscape. Designers enjoy lighting the tree from beneath because of the nice, twisting trunk structure that develops on older trees.

Kobus Magnolia should be grown in full sun or partial shade on any well-drained soil. It is supposedly tolerant of soil with an alkaline pH. Occasionally bothered by scale, as are other Magnolias.

Magnolia macrophylla Bigleaf Magnolia

Pronunciation: mag-NO-lee-ah mack-row-FYE-lah

ZONES: 5B, 6A, 6B, 7A, 7B, 8A, 8B

HEIGHT: 30 to 40 feet

WIDTH: 20 to 25 feet

FRUIT: elongated; 1 to 3"; dry; red seeds; attracts birds; causes some litter; persistent; showy

GROWTH RATE: moderate; moderate life span

HABIT: oval; pyramidal; upright; moderate density; irregular silhouette; coarse texture

FAMILY: *Magnoliaceae*

LEAF COLOR: bright green above, silver-gray beneath in summer; showy yellow fall color

LEAF DESCRIPTION: alternate; simple; entire margin; oblong to obovate; deciduous; 18 to 36" long

FLOWER DESCRIPTION: white; pleasant fragrance; showy to 12" across; summer

EXPOSURE/CULTURE:

Light Requirement: part shade to full sun

Soil Tolerances: loam; sand; slightly alkaline to acidic; moderate drought

PEST PROBLEMS: resistant

BARK/STEMS:

Branches droop as tree grows and will require pruning for vehicular or pedestrian clearance beneath canopy; not particularly showy; should be grown with a single leader; no thorns

Pruning Requirement: needs little or no pruning to develop strong trunk and branch structure

Limb Breakage: resistant

Current Year Twig: brown; green; thick

CULTIVARS AND SPECIES:

'Palmberg' has very large flowers; 'Purple Spotted' has flowers with purple markings in the center; var. *ashei* is found only in the Florida panhandle.

LANDSCAPE NOTES:

Origin: eastern United States

Usage: shade tree; specimen

This North American native tree is deciduous in most areas but semi-evergreen in the deep South. Bigleaf Magnolia grows slowly and forms a rounded, broad canopy. The leaves of Bigleaf Magnolia are truly large, 18 to 36 inches long and 7 to 12 inches wide, when found in the wild, though somewhat smaller when grown in landscapes. Leaves are bright green above with a fuzzy, silver-gray underside, creating a beautiful, two-toned effect with each passing breeze. From May to July, the showy, fragrant blossoms appear, each 8- to 12-inch-wide, ivory-colored bloom having a slight rose tint at its base. These blooms are followed by the production of 2.5- to 3-inch-long, hairy, red, egg-shaped fruits. Bigleaf Magnolia trees must be 12 to 15 years of age before they begin to bloom.

The tree may be rather short-lived in many landscape sites unless its cultural requirements are met fairly closely. Branches break easily in wind storms and ice-laden branches snap off. The large leaves are easily damaged by wind and decompose slowly after they fall, blowing around and creating litter that some people will find objectionable. It may be best to locate this tree in a groundcover bed where leaves can drop and filter down beneath the low-growing plants unseen.

Bigleaf Magnolia should be grown in a sheltered location, such as a valley, in full sun or partial shade on well-drained soil. It does not tolerate wet soil or extended drought. It appears to be somewhat picky in its requirements. In its native habitat, it is found on low, rich, moist soils.

Magnolia x soulangiana Saucer Magnolia
Pronunciation: mag-NO-lee-ah x sue-lan-gee-A-nah

ZONES: 5A, 5B, 6A, 6B, 7A, 7B, 8A, 8B, 9A

HEIGHT: 25 to 30 feet

WIDTH: 20 to 30 feet

FRUIT: elongated; irregular; 1 to 2"; dry; red seeds; attracts birds; no significant litter

GROWTH RATE: moderate; moderate life span

HABIT: rounded or vase; open when young, dense later; irregular silhouette, becoming symmetrical with age; coarse texture

FAMILY: *Magnoliaceae*

LEAF COLOR: medium green in summer; yellow fall color

LEAF DESCRIPTION: alternate; simple; entire margin; oblong to obovate; deciduous; 3 to 6" long

FLOWER DESCRIPTION: pink and white; showy; spring and winter

EXPOSURE/CULTURE:

Light Requirement: part shade to full sun

Soil Tolerances: all textures; acidic; occasionally wet soil; moderate drought

Aerosol Salt Tolerance: none

PEST PROBLEMS: resistant

BARK/STEMS:

Bright gray bark; bark is thin and easily damaged by mechanical impact; branches droop as tree grows and will require some pruning for vehicular or pedestrian clearance beneath canopy; routinely grown with multiple trunks; no thorns

Pruning Requirement: needs pruning to develop strong structure; head back aggressive branches after flowering to increase canopy density and flower number

Limb Breakage: somewhat susceptible

Current Year Twig: brown; medium

CULTIVARS AND SPECIES:

'Alba'—flowers almost white; 'Burgundy'—deep purple flowers, blooms earlier; 'Lennei'—one of the best, flowers rosy-purple outside, white flushed with purple inside, flowers large, blooms later; 'Lilliputian'—slow grower to 10 to 15 feet tall; 'Lombardy Rose'—similar to Lennei, except flowers continue to open for several weeks; 'Verbanica'—flowers clear rose-pink outside, late-blooming, slow-growing to 10 feet tall.

LANDSCAPE NOTES:

Origin: hybrid

Usage: container; espalier; near a deck or patio; specimen

Young Saucer Magnolias are distinctly upright, becoming more oval and then round by 10 years of age. Blooms open in late winter to early spring before the leaves, producing large, white flowers shaded in pink or purple, creating a spectacular flower display. However, a late frost can ruin the flowers. In warmer climates, late-flowering cultivars avoid frost damage, but some are less showy than the early-flowered forms, which blossom when little else is in flower.

The tree is best used as a specimen in a sunny spot where it can develop a symmetrical crown. It can be pruned up if planted close to a walk or patio to allow for pedestrian clearance, but probably looks its best when branches are left to droop to the ground. The light gray bark shows off nicely, particularly during the winter when the tree is bare.

Transplant in spring, just before growth begins, and use balled-in-burlap or containerized plants. Pruning wounds may not close well, so train plants early in their life to develop the desired form to avoid large pruning wounds. Generally pest-free, but scales of various types may infest twigs and foliage. Magnolia may be subject to leaf spots. Canker diseases will kill entire branches.

Magnolia stellata Star Magnolia

Pronunciation: mag-NO-lee-ah stell-LAY-tah

ZONES: 5A, 5B, 6A, 6B, 7A, 7B, 8A, 8B

HEIGHT: 15 to 20 feet

WIDTH: 10 to 15 feet

FRUIT: elongated; irregular; 1 to 3"; dry; brown; attracts birds; no significant litter

GROWTH RATE: slow-growing; moderate life span

HABIT: oval; round; moderate density; symmetrical; medium texture

FAMILY: *Magnoliaceae*

LEAF COLOR: dark green; showy, copper to yellow fall color

LEAF DESCRIPTION: alternate; simple; entire margin; oblong to obovate; deciduous; 2 to 4" long

FLOWER DESCRIPTION: white or pink; showy; spring

EXPOSURE/CULTURE:

Light Requirement: part shade to full sun

Soil Tolerances: all textures; acidic; moderate drought

Aerosol Salt Tolerance: none

PEST PROBLEMS: resistant

BARK/STEMS:

Branches droop as tree grows and will require pruning for pedestrian clearance beneath canopy; routinely grown with multiple trunks; no thorns

Pruning Requirement: needs little or no pruning to develop good structure and uniform habit; prune before flower buds form

Limb Breakage: resistant

Current Year Twig: brown; medium; slender

CULTIVARS AND SPECIES:

'Jane Platt'—new, superior type with many pink petals; 'Keiskei'—flowers purplish on the outside; 'Rosea' (Pink Star Magnolia)—pale pink flowers; 'Rubra' (Red Star Magnolia)—purplish flowers, darker than 'Rosea'. The 'Little Girl' hybrids flower later than the species, thus avoiding frost injury in most years. They include 'Ann', 'Betty', 'Jane', 'Judy', 'Randy', 'Ricki', and 'Susan'.

LANDSCAPE NOTES:

Origin: central Japan

Usage: container; near a deck or patio; specimen; possible street tree

Star Magnolia is the hardiest of the Magnolias. Recent taxonomic changes list this as *M. kobus* var. *stellata*. Typically branching close to the ground and flowering at an early age, the multi-stemmed form develops with a dense head of foliage. Lower foliage can be removed to show off the trunk and create more of a tree form. Otherwise, the persistent lower branches and oval-to-round form lend a "large bush" look to the plant. When planted against a dark background, the branching pattern and light gray trunk will show off nicely, particularly when lit up at night. The leafless winter silhouette looks great shadowed on a wall by a spotlight at night. The white flowers are produced in spring before the leaves appear, even on young plants. Flowers are usually not as sensitive to cold as Saucer Magnolia, but they can still be injured if freezing temperatures arrive during flowering.

Star Magnolia is intolerant of root competition or dryness, and plants grow slowly, about one foot each year. Plant in the full sun or partial shade in a rich, porous, and slightly acid soil. It is hard to transplant from the field successfully, and in the north this plant is often moved balled and burlapped in the summer. In zones 7 and 8, transplant in late winter through the spring, or plant from containers at any time.

Basically trouble-free, although scales of various types may infest twigs and leaves. Magnolia may be subject to leaf spots, blights, scabs, and black mildews caused by a large number of fungi or bacteria.

Magnolia virginiana Sweetbay Magnolia

Pronunciation: mag-NO-lee-ah vir-gin-ee-A-nah

ZONES: 5A, 5B, 6A, 6B, 7A, 7B, 8A, 8B, 9A, 9B, 10A

HEIGHT: 40 to 50 feet

WIDTH: 15 to 25 feet

FRUIT: elongated; 1 to 3"; dry; fleshy; green with red seeds; attracts wildlife; no significant litter; showy

GROWTH RATE: moderate; moderate life span

HABIT: columnar or rounded vase, variable; moderate density; symmetrical; medium texture

FAMILY: *Magnoliaceae*

LEAF COLOR: green, whitish-green undersides; no fall color change

LEAF DESCRIPTION: alternate; simple; entire margin; oval to oblong; deciduous or semievergreen; 2 to 4"

FLOWER DESCRIPTION: white; pleasant fragrance; showy; summer

EXPOSURE/CULTURE:

Light Requirement: part shade to full sun

Soil Tolerances: all textures; acidic; wet soil; moderate drought

Aerosol Salt Tolerance: low to moderate

PEST PROBLEMS: resistant

BARK/STEMS:

Bark is thin and easily damaged by mechanical impact; routinely grown with, or trainable to be grown with, multiple trunks; can be trained to grow with a single trunk; branches grow mostly upright and will not droop; no thorns

Pruning Requirement: needs little or no pruning to develop strong structure

Limb Breakage: resistant

Current Year Twig: green; slender

LANDSCAPE NOTES:

Origin: Massachusetts to south Florida and Texas

Usage: espalier; buffer strips; highway; wetland reclamation; near a deck or patio; specimen; possilble street tree

Sweetbay Magnolia is a graceful Southern, wide columnar tree, ideal for use as a patio tree or specimen. It can grow to a mature height of 40 feet in the north or to 60 feet in the south in its native habitat. Trees glimmer in the wind due to the whitish-green undersides of the leaves. They are very noticeable in water-logged woodlands. The tree provides excellent vertical definition in a shrub border or as a free-standing specimen, and flourishes in moist, acid soil such as the swamps in the eastern United States and along stream banks. The creamy white, lemon-scented flowers appear from June through September, and are followed by small red seeds that are used by a variety of wildlife.

Sweetbay Magnolia makes an excellent tree for planting next to buildings, in narrow alleys or corridors, or in other urban areas with limited space for horizontal crown expansion. It has not been planted extensively in downtown urban areas, but its flood and drought tolerance and narrow crown combine to make it a good candidate. It usually maintains a good, straight central leader, although occasionally the trunk branches low to the ground, forming a round, multi-stemmed, spreading tree.

Sweetbay Magnolia grows freely near coastal areas, and is happiest in Southern climates. It is thriving in the Auburn Shade Tree Evaluation trials in Alabama without irrigation. However, in the confined soil spaces typical of some urban areas, occasional irrigation may be needed.

The species is deciduous in zones 5, 6, 7, and 8 (evergreen farther south) but the variety *australis* and cultivar 'Henry Hicks' are evergreen; 'Havener' has larger flower petals.

Scales sometimes infest foliage and twigs, particularly on dry sites where the tree is under stress.

Malpighia glabra Barbados Cherry

Pronunciation: Mal-PIG-ee-ah GLAY-brah

ZONES: 10A, 10B, 11

HEIGHT: 8 to 12 feet

WIDTH: 8 to 12 feet

FRUIT: round; 1"; fleshy; red; attracts birds; no significant litter; edible

GROWTH RATE: moderate; moderate life span

HABIT: rounded vase; moderate density; symmetrical; medium texture

FAMILY: *Oleaceae*

LEAF COLOR: evergreen; dark green

LEAF DESCRIPTION: alternate; simple; entire margin; oblong to ovate; 1 to 2" long

FLOWER DESCRIPTION: pink and white; somewhat showy; year-round

EXPOSURE/CULTURE:

Light Requirement: part shade to full sun

Soil Tolerances: all textures; alkaline to acidic; salt and drought

Aerosol Salt Tolerance: moderate to high

PEST PROBLEMS: somewhat sensitive

BARK/STEMS:

Bark is thin and easily damaged by mechanical impact; branches droop as tree grows and will require pruning for vehicular or pedestrian clearance beneath canopy; routinely grown with multiple trunks; no thorns

Pruning Requirement: requires training and pruning to develop into an upright tree form

Limb Breakage: mostly resistant

Current Year Twig: brown; medium thickness

CULTIVARS AND SPECIES:

M. emarginata is a similar species. *M. coccigera* is low-growing and is often used as a groundcover or low shrub.

LANDSCAPE NOTES:

Origin: tropical America

Usage: container; fruit tree; hedge; buffer strip; highway; deck or patio; standard; specimen; buffer strip

Although often used as a shrub or hedge, Barbados Cherry works well when allowed to grow into a small tree, its curved multiple trunks and dark green canopy creating an interesting architectural focus. Old specimens can grow to 20 feet across. The dull, evergreen leaves are abundantly produced on the upright, spreading branches. The small, pink and white flowers appear scattered throughout the canopy periodically all year long. The blooms are followed by abundant bright red, tasty berries which persist most of the year. The fruit is rich in vitamin C.

Although tolerant of clipping, Barbados Cherry is quite attractive when allowed to retain its natural multi-stemmed form, making it ideal for use in shrubbery borders and as a featured accent. It makes a nice specimen in any landscape where a small tree is needed. Planted close together on about 10- to 15-foot centers, a row of them will form a canopy over a pedestrian walkway. The tree looks best in a landscape setting with a low groundcover planted around its base. Many homeowners plant the tree for its edible fruit.

Barbados Cherry grows in full sun or partial shade and is tolerant of a wide range of soil types, including calcarious clay, as long as water is not allowed to stand in the root zone. Plants grow quickly while young but slow with age. Although it can withstand drought, Barbados Cherry will require protection from direct salt spray. If you decide to use this plant as a clipped hedge, be sure that the top is kept narrower than the bottom to provide light to the lower branches. This will help ensure that the plant will remain full to the ground.

Soil nematodes can cause serious plant decline, and they can be prevalent, particularly in sandy soil. Scales and whiteflies can also infest the foliage and twigs.

Malus spp. Crabapple

Pronunciation: MAY-luss species

Malus 'Adams'

Malus baccata 'Jackii'

Malus 'Donald Wyman'

Malus 'Brandywiner'

Malus floribunda

Malus 'Professor Sprenger'

Malus hupehensis

Malus x 'Zumi Calocarpa'

Continued

Malus spp. *Continued*

ZONES: 4A, 4B, 5A, 5B, 6A, 6B, 7A, 7B, (8A)

HEIGHT: 10 to 25 feet

WIDTH: 10 to 25 feet

FRUIT: round; 0.25 to 2"; fleshy; green, orange, red, yellow; attracts wildlife; edible; fruit, twigs, and foliage cause significant litter; most are persistent; showy

GROWTH RATE: slow to moderate; moderate life span

HABIT: rounded; upright; vase; weeping, columnar; moderate density; symmetrical; medium texture

FAMILY: *Rosaceae*

LEAF COLOR: medium green; showy yellow fall color

LEAF DESCRIPTION: alternate; simple; oval; deciduous; 1 to 4" long

FLOWER DESCRIPTION: pink; red; white; pleasant fragrance; showy; spring

EXPOSURE/CULTURE:

Light Requirement: full sun

Soil Tolerances: all textures; alkaline or acidic; occasionally wet soil; salt-tolerant to sensitive, depending on cultivar; drought

Aerosol Salt Tolerance: moderate, varies with cultivar

PEST PROBLEMS: resistant to very susceptible, depending on cultivar

BARK/STEMS:

Branches droop on many cultivars as tree grows and will require pruning for vehicular or pedestrian clearance beneath canopy; others are upright; routinely grown with, or trainable to be grown with, multiple trunks; some cultivars can be trained to grow with a single trunk; no thorns

Pruning Requirement: needs little pruning to develop strong structure; pruning is required to remove some suckers and sprouts along major branches and to thin canopy for disease suppression

Limb Breakage: resistant

Current Year Twig: brown; reddish; medium; slender

CULTIVARS AND SPECIES:

There are hundreds of Crabapple cultivars with single or double, red, pink, or white flowers, and varying fruit size. Disease resistance can vary depending on where a particular cultivar is grown, so be sure to choose one that has been shown to be resistant to disease in your area. Your urban forestry program could suffer if you plant the wrong cultivar, but it could blossom if the correct ones are installed. One of the best Crabapples for the southern United States is *Malus* x 'Callaway'. 'Louisa' has also performed well in zone 7B.

Disease-resistant cultivars include 'Adams', 'Bobwhite', 'David', 'Dolga', 'Donald Wyman', 'Ellwangeriana', *floribunda*, 'Inglis', 'Jackii', 'Jewelberry', 'Margaret', 'Mount Arbor Special', 'Prairifire', 'Professor Sprenger', *sargentii*, 'Selkirk', 'Sentinel', 'Sugar Tyme', and 'Tomiko' and several *M. baccata* cultivars, including 'Jackii'. Contact the Ornamental Crabapple Society for up-to-date information on Crabapples.

Japanese-beetle-resistant Crabapples: most resistant are *baccata*, 'Brandywine', *hupehensis*, 'Jewelberry', 'Harvest Gold', 'Mary Potter', 'Molten Lava', *sargentii*, 'Silver Moon', and 'Strawberry Parfait'; resistant are 'Candy Mint', 'David', *floribunda*, 'Molten Lava', 'Red Jade', 'Red Jewel', 'Silver Moon', and 'Zumi Calocarpa'. Susceptible cultivars include 'Adams', 'Radiant', 'Red Splendor', 'Sentinel', 'Sinai Fire', 'Snowdrift', and 'White Angel'.

LANDSCAPE NOTES:

Origin: variable

Usage: container or above-ground planter; espalier; parking lot island; buffer strip; highway; near a deck or patio; standard; specimen; street tree; urban tolerant

Tree size, flower color, fruit color, growth, and branching habit vary considerably with the cultivar grown. Crabapples are grown for their adaptability, showy flowers, and attractive, brightly colored fruit. A few Crabapples, including 'Indian Magic', 'Indian Summer', 'Red Barron', and *tschonoskii,* have exceptional fall color.

Continued

***Malus* spp.** *Continued*

Double-flowered types like 'Brandywine' hold blossoms longer than single-flowered cultivars. Some Crabapples, including *hupehensis*, 'Mary Potter', *sargentii*, and 'Selkirk', are alternate bearers, blooming heavily only every other year.

Plants are used for specimens, patios (small-fruiting types), and along streets to create a warm glow of color each spring. Most disease-resistant types are attractive during the summer, bearing glossy green foliage. Popular around overhead powerlines due to their small stature, a row of Crabapples along each side of the street or median strip can really enhance a neighborhood. Select plants that have been grafted onto EMLA 106, 111, or Melling rootstock, or grown on their own roots, to reduce root suckering. In the coldest climates, seedlings of 'Dolgo' or 'Red Splendor' are often used.

Crabapples are best grown in a sunny location with good air circulation, but have no particular soil preferences. Some cultivars withstand periodic flooding quite well. Some tolerate salt spray.

Crabapple is well adapted to compacted urban soil and tolerates drought. It is well adapted to all areas within its hardiness zone range, including very dry climates. Do not overfertilize, as this could increase the incidence of disease.

Large-fruited types, such as 'Callaway', 'Bastulong', 'Brandywine', 'Dolga' (zone 2), 'Selkirk' (zone 2), and *tschonoskii,* can create a maintenance problem, because rotting fruits attract insects and rodents and are quite messy. Some Crabapples sprout vigorously from the roots, and these will require regular pruning to maintain their attractiveness. Trees used as street trees will require regular pruning early in life to train lower branches for pedestrian and vehicle clearance. The upright cultivars, including 'Centurion', 'Indian Summer', 'Prairifire', 'Professor Sprenger', 'Red Barron', 'Red Splendor' (zone 3), *Malus transitoria* 'Schmidtcut-leaf', 'Silver Moon', and 'Sentinal', are some of the best adapted to streets because the canopy stays mostly upright, with only minimum pruning needed to remove drooping branches. Some other *Malus* adapted for street tree and urban use include 'Adams', 'Bob White', 'David', 'Donald Wyman', 'Profusion', 'Red Jewel', and *M. floribunda*. Disease resistance is good, fruits are ½" or less and branches are not spreading. Be sure to specify tree form plants for street tree use, as branching may be too low on multi-stemmed trees grown for specimen use.

Aphids infest branch tips and suck plant juices, and are quite common. Fall webworm makes nests on the branches and feeds on foliage inside the nest. Borers can be a problem on stressed trees. Eastern tent caterpillar feeds on foliage outside the nest and can defoliate trees.

Scab infection takes place early in the season (May) and dark olive green spots appear on the leaves. By late summer, the infected leaves fall off when they turn yellow, with black spots. Infected fruits have black, slightly raised spots. Use resistant varieties to help avoid this potentially severe problem.

Fire-blight-susceptible trees get blighted branch tips, particularly when the tree is growing rapidly. Leaves on infected branch tips turn brown or black, droop, and hang on the branches. The leaves look scorched as if by fire. Use resistant cultivars; severe infections on susceptible trees can kill the tree. Fire blight is a major problem in regions receiving regular rain during bloom time.

Cedar apple rust causes brown to rusty-orange spots on the leaves and fruit. Redcedars (*Juniperus virginiana*) are the alternate host. Badly spotted leaves fall prematurely, and defoliation can be heavy.

Crabapples are subject to several canker diseases. Avoid unnecessary wounding, keep trees healthy, and prune out infected branches when seen.

Mangifera indica Mango

Pronunciation: man-JIFF-err-ah INN-dih-cah

ZONES: 10B, 11

HEIGHT: 30 to 45 feet

WIDTH: 50 feet

FRUIT: oval; 3 to 6"; fleshy; green and red or yellow; edible; causes messy litter; persistent; showy

GROWTH RATE: fast-growing; moderate life span

HABIT: round; dense; symmetrical; coarse texture

FAMILY: *Anacardiaceae*

LEAF COLOR: broadleaf evergreen; dark green

LEAF DESCRIPTION: alternate; simple; entire, undulating margin; lanceolate; 12 to 18" long

FLOWER DESCRIPTION: white; showy; winter and spring

EXPOSURE/CULTURE:

Light Requirement: full sun

Soil Tolerances: all textures; alkaline to acidic; moderate drought

Aerosol Salt Tolerance: moderate

PEST PROBLEMS: somewhat susceptible

BARK/STEMS:

Branches droop as tree grows and will require pruning for vehicular or pedestrian clearance beneath canopy; should be grown with a single leader to maintain structural integrity in canopy; no thorns

Pruning Requirement: requires pruning to develop strong structure; space major limbs along trunk

Limb Breakage: susceptible to breakage at crotch due to poor collar formation on major branches with included bark

Current Year Twig: brown; gray; medium thickness

CULTIVARS AND SPECIES:

Cultivars have been selected for fruit quality: 'Keitt', 'Hent', 'Edward', 'Glenn', 'Haden' and others are best for Florida; 'Alolia', 'Edgehill', 'Haden', 'Manila', and others are recommended for California. Those resistant to disease include 'Carrie', 'Early-Gold', 'Florigon', and 'Saigon'.

LANDSCAPE NOTES:

Origin: Asia

Usage: fruit; hedge; shade tree

An abundant harvest of juicy, red-gold fruit and attractive dark green, tropical foliage make Mango a popular home landscape item for large yards in warm climates. The tree grows tall and wide, so allow plenty of room for canopy development. The tree is covered with very showy, white flower spikes in March and early April, which extend well beyond the long, glossy leaves. New foliage emerges brillant purple-red following the showy flower display.

Mango trees grow quickly into round, multi-branched, dense, spreading shade trees, but placement is crucial because of the falling fruit. Some people are allergic to the pollen, the sap, and even the fruit, and can suffer an allergic reaction simply from walking past the tree.

Mango trees grow best in full sun on fertile, well-drained soils and should have ample moisture. Leaf, flower, twig, and fruit litter are a constant nuisance, and branches are subject to breakage during severe wind storms. It seems like something is always falling from a Mango tree to litter the lawn. Place it in a bed with other plants to hide the litter.

Scales followed by sooty mold and the Mediterranean fruit fly can be problems. Anthracnose on fruit and leaves is a serious problem for Mango.

Manilkara zapota Sapodilla

Pronunciation: man-ill-CAR-ah zah-POE-tah

ZONES: 10B, 11

HEIGHT: 40 to 45 feet

WIDTH: 40 feet

FRUIT: round; 3 to 4"; fleshy; brown; attracts bats, which stain nearby walks and buildings; edible and very tasty; causes some litter

GROWTH RATE: moderate; long-lived

HABIT: rounded; dense; symmetrical; billowing, fine texture

FAMILY: *Sapotaceae*

LEAF COLOR: dark green; broadleaf evergreen

LEAF DESCRIPTION: alternate; simple; entire margin; oval; 4 to 8" long

FLOWER DESCRIPTION: white; not showy; year-round

EXPOSURE/CULTURE:

Light Requirement: full sun

Soil Tolerances: all textures; alkaline to acidic; salt and drought

Aerosol Salt Tolerance: high

PEST PROBLEMS: resistant

BARK/STEMS:

Showy trunk; usually grows with a single short leader; no thorns

Pruning Requirement: requires little pruning to develop strong structure; if the tree was not topped in the nursery, lateral branches are small and usually well attached to central trunk

Limb Breakage: resistant

Current Year Twig: brown; green; medium thickness

CULTIVARS AND SPECIES:

Superior fruit cultivars are available: 'Brown Sugar', 'Modello', 'Prolific', and 'Russel'. *Manilkara bahamensis*, the Wild Dilly, which is native to the Florida Keys, makes a nice street tree but has less desirable fruit.

LANDSCAPE NOTES:

Origin: southern Mexico and Central America

Usage: fruit; hedge or screen; buffer strips; highway; shade tree; specimen; street tree; urban tolerant

A superb shade, street (where falling fruit will not be a problem), or fruit tree, Sapodilla has smooth, dark, and glossy, evergreen leaves clustered at the tips of twigs. The small, cream-colored solitary flowers appear in the leaf axils throughout the year. The brown fruits have a juicy, sweet, yellow-brown flesh and ripen to softness in spring and summer. The flower-to-fruit period is about 10 months. The bark and branches, when injured, bleed a white latex which is the source of chicle, the original base for chewing gum. The trunk on older specimens is flaky and quite attractive, and flares at the base into numerous surface roots. The tree requires only minimal structural pruning because the trunk usually grows straight up the center of the canopy. Remove competing leaders as they develop.

Requiring full sun for best growth and form, Sapodilla is a tough tree tolerating a variety of poor soils, but will grow better on well-drained sites. It has good salt tolerance and is very drought- and wind-resistant, enduring storms very well. These traits make it ideal for seaside locations. Thinning the very dense crown will help to increase grass and other plant growth under the tree. Thinning also increases the tree's capability to withstand wind storms. The trunk and roots grow quite large on older specimens, so locate no closer than about eight feet from sidewalks and curbs. It makes a superb specimen tree for a large residential or commercial landscape.

Scales and fruit flies occasionally cause problems. Seedlings develop in the landscape where they could become a slight weed problem. Bats stain nearby walks and buildings with their urine when they visit the tree.

Melaleuca quinquinervia Melaleuca

Pronunciation: Mell-ah-LEW-kah kwin-kwi-NERVE-ee-ah

ZONES: 9B, 10A, 10B, 11

HEIGHT: 40 to 60 feet

WIDTH: 20 to 30 feet

FRUIT: round; less than 0.25"; hard; no significant litter; easily dispersed

GROWTH RATE: rapid-growing; moderate life span

HABIT: columnar; oval; open; somewhat irregular; fine texture

FAMILY: *Mrytaceae*

LEAF DESCRIPTION: simple; entire margin; oblong to lanceolate; 1 to 3" long

FLOWER DESCRIPTION: white; showy; smelly; spring and summer

EXPOSURE/CULTURE:

Light Requirement: shade to part shade

Soil Tolerances: all textures; alkaline to acidic; wet soil; drought

Aerosol Salt Tolerance: low

PEST PROBLEMS: resistant

BARK/STEMS:

Bark is soft and thick and resists mechanical impact; normally grows with a single leader; no thorns

Pruning Requirement: needs some pruning to develop strong structure; prevent included bark from forming by thinning major branches to slow growth

Limb Breakage: mostly resistant

Current Year Twig: brown; medium thickness

CULTIVARS AND SPECIES:

Many other Melaleucas are grown in the warmest regions of the world. Most have thick, showy bark similar to *M. quinquinervia. M. decora* reportedly is noninvasive and might have potential in Florida, but more likely no tree bearing the *Melaleuca* name will ever be recommended for, or planted in, Florida. *M. linarifolia* is a small tree planted in California.

LANDSCAPE NOTES:

Origin: Australia

Usage: live trees are cut down and used as mulch in Florida; specimen, buffer strip, parking lot island, and shade tree in California and warm climates

A pretty tree, but unfortunately Melaleuca has become an unimaginable weed in the Florida everglades. Ironically, it is planted frequently in parts of California along streets, in parking lot islands and buffer strips, and in other adverse sites. It is usually single-trunked, reaching mature height with a columnar or oval, open growth habit. The canopy thickens in full sun. Melaleuca forms a very attractive specimen when it is thinned to display the showy, whitish bark on the trunk and major branches. The small, white flowers open in summer to perfume nearby neighborhoods with the smell of boiling potatoes. The flowers are actually quite attractive.

In California, Melaleuca makes an attractive specimen planting and its light, airy growth habit lends itself well to smaller, partially enclosed locations. Melaleuca is well suited to a low-maintenance landscape due to its rapid growth and carefree nature. Its crown shape makes it suitable for urban areas with restricted, narrow overhead space, and it does well as a street tree if located in moist soil. Mulch the area beneath the canopy to reduce turf competition.

Melaleuca prefers full sun in any soil and has a shallow root system. It grows well in confined soil spaces, but roots will lift pavement if planted too close. The tree is notoriously weedy in Florida where it has taken over tens of thousands of acres of wetlands. It cannot be planted in Florida, but is common in the warmest regions of California.

Melia azedarach Chinaberry

Pronunciation: MEE-lee-ah as-eh-DARE-etch

ZONES: 7A, 7B, 8A, 8B, 9A, 9B, 10A, 10B

HEIGHT: 30 to 40 feet

WIDTH: 20 to 25 feet

FRUIT: round; 0.3"; hard; green turning yellow; attracts birds; causes significant litter; persistent; showy

GROWTH RATE: rapid-growing; moderate life span

HABIT: rounded vase, variable; open or dense; irregular or symmetrical silhouette; fine texture

FAMILY: *Meliaceae*

LEAF COLOR: medium green in summer; yellow fall color

LEAF DESCRIPTION: alternate; bipinnately compound; incised or serrate margin; oval to ovate; deciduous; less than 2" long

FLOWER DESCRIPTION: lavender; pleasant fragrance; not showy; spring

EXPOSURE/CULTURE:

Light Requirement: part shade to full sun

Soil Tolerances: all textures; alkaline to acidic; drought

Aerosol Salt Tolerance: moderate

PEST PROBLEMS: none of any consequence

BARK/STEMS:

Branches droop somewhat as tree grows and will require pruning for vehicular or pedestrian clearance beneath canopy; should be grown with a single leader; no thorns

Pruning Requirement: requires pruning to develop strong structure; head back aggressive branches to prevent formation of included bark

Limb Breakage: susceptible to breakage at crotch, due to poor collar formation; wood itself is weak and tends to break

Current Year Twig: brown; stout

CULTIVARS AND SPECIES:

'Umbracultiformis' has a dome-like form and could be the plant seen commonly in some wild stands. It is often sold as Texas Umbrella-Tree. It would be nice to find a fruitless selection.

LANDSCAPE NOTES:

Origin: India and China

Usage: urban tolerant

Chinaberry is a round, deciduous, shade tree that grows 5 to 10 feet during the first and second years after seed germination. Growth slows as the tree reaches 15 or 20 feet tall. It is successfully grown in a wide variety of situations, including alkaline soil, where other trees might fail. Truly an urban survivor, Chinaberry has become naturalized in much of the southern United States.

The clusters of lilac flowers are fragrant in the evening but are often hidden by emerging foliage. The leaves turn a vivid yellow for a short time in the fall. The golden yellow fruit is quite attractive as it persists on the tree during the fall and winter. When eaten in quantities, the fruit is poisonous to people, but not to the birds that happily distribute the seeds. The wood is very brittle, but it has been used in cabinet making.

Chinaberry is considered a "weed" tree in the southeastern United States, so it is not usually available from nurseries. Many people despise the tree because it has taken over waste areas and other disturbed soil areas, and has naturalized over large areas of the South. It grows anywhere in any soil except wet. With proper pruning to create a well-formed trunk and branch structure, the plant could improve its reputation. If you have one and would like to increase its life span, prune to open up the crown to encourage development of a few well-spaced major limbs. You will not find anyone recommending that you plant this tree, but fine examples can be found growing in the worst soil.

Scale, whitefly, and sooty mold infest Chinaberry. Leaf spot causes premature defoliation.

Metasequoia glyptostroboides Dawn Redwood

Pronunciation: met-ah-see-KWOY-yah glip-toe-stro-BOY-dees

ZONES: 5A, 5B, 6A, 6B, 7A, 7B, 8A, 8B

HEIGHT: 70 to 90 feet

WIDTH: 25 to 35 feet

FRUIT: elongated; round; 0.5 to 1"; dry; brown; no significant litter

GROWTH RATE: rapid-growing; long-lived

HABIT: pyramidal; open; symmetrical; fine texture

FAMILY: *Taxodiaceae*

LEAF COLOR: light green in summer; orange to yellow showy fall color

LEAF DESCRIPTION: opposite; linear; deciduous; 0.5"

FLOWER DESCRIPTION: not showy

EXPOSURE/CULTURE:

Light Requirement: full sun to part shade

Soil Tolerances: all textures; acidic; wet soil; salt-sensitive; moderate drought

PEST PROBLEMS: resistant

BARK/STEMS:

Striking bark and lower trunk structure; some trees flare out at base; branches droop as tree grows and will require pruning for vehicular or pedestrian clearance beneath canopy; normally grows with a single leader; no thorns

Pruning Requirement: needs little or no pruning to develop strong structure

Limb Breakage: resistant

Current Year Twig: brown; green; slender

CULTIVARS AND SPECIES:

'National' and 'Sheridan Spire' have been selected for their narrow growth habit.

LANDSCAPE NOTES:

Origin: China

Usage: large parking lot island; buffer strip; highway; specimen; street tree

Dawn Redwood grows in a perfect pyramid and was known from fossils before living plants were discovered in China. It is similar in appearance to Baldcypress. Although it looks like an evergreen, the needles are deciduous. The orange-red to brown trunk base is the most outstanding part of the tree. It tapers and thickens quickly with numerous, buttress-like root flares extending several feet up the tree in a manner unlike any other tree except some tropical trees. As with any other tree, butt flare (increased caliper at the base of the trunk) can be reduced somewhat by removing the lower branches at an early age. The small, upright-spreading branches are well attached to the typically straight trunk and make for excellent climbing. The tree requires little if any pruning to maintain the pyramidal form. Lightning protection is recommended for older trees, as they usually grow taller than many trees.

The tree has been used primarily as a specimen, but there are reports of it being very tolerant of air pollution, and it has done well as a street tree with lower branches removed. Lower branches should be left on the tree for most other uses to enjoy the graceful form and delightful foliage. Do not plant in soil with a high pH.

The tree grows rapidly, but late-season growth may be injured by early frosts. The preferred soil is moist and moderately fertile, but *Metasequoia* does not appear to be damaged by drought, having survived without irrigation in urban landscapes during recent summer droughts. It will also tolerate wet and slightly alkaline soil. Appears to be adapted to clay soil and grows best when located on continually moist sites with slightly acidic pH.

Relatively free of pests, except for a canker that has been reported recently. Japanese beetles consume some foliage.

Morus alba (fruitless types) White Mulberry

Pronunciation: MOE-russ AL-buh

ZONES: 3B, 4A, 4B, 5A, 5B, 6A, 6B, 7A, 7B, 8A, 8B, 9A, 9B

HEIGHT: 30 feet

WIDTH: 35 to 45 feet

FRUIT: no fruit; species bears tasty fruit, but creates a mess beneath canopy

GROWTH RATE: astoundingly fast; moderate life span

HABIT: round and spreading; dense; irregular silhouette when young, more uniform later; coarse texture

FAMILY: *Moraceae*

LEAF COLOR: dark green in summer; yellow fall color
LEAF DESCRIPTION: alternate; simple; lobed or dentate margin; ovate; deciduous; 2 to 8" long; variably showy
FLOWER DESCRIPTION: green; not showy; spring
EXPOSURE/CULTURE:

Light Requirement: part shade to full sun

Soil Tolerances: all textures; alkaline to acidic; wet soil; salt; drought

Aerosol Salt Tolerance: good

PEST PROBLEMS: somewhat susceptible

BARK/STEMS:

Branches droop as tree grows and will require pruning for vehicular or pedestrian clearance beneath canopy; can be grown with a single or multiple leader; no thorns

Pruning Requirement: requires pruning to develop strong structure

Limb Breakage: susceptible to breakage either at crotch, due to poor collar formation, or wood itself is weak and tends to break; breaks in ice storms

Current Year Twig: gray; slender

CULTIVARS AND SPECIES:

'Chaparral' White Mulberry is completely different from what most people expect when they hear the name Mulberry. This is a weeping form composed of deeply cut, dark green leaves that clothe the gently draping, thin branches all the way to the ground. Usually grafted onto a standard, Chaparral White Mulberry forms a broad silhouette, 8 to 20 feet high by 15 feet wide, and is quite striking in the winter garden. It may be listed as 'Pendula' at nurseries.

LANDSCAPE NOTES:

Origin: China

Usage: shade tree; specimen; urban tolerant

This group of Mulberries is fruitless, a definite plus when compared to the mess created by the abundant fruits of the common White Mulberry. The plant quickly forms a dark mass of foliage from a short trunk or group of trunks. This gives many people reason to plant the tree. However, it is quite sensitive to ice damage, has fast-growing and invasive surface roots, and drops leaves in summer without regular irrigation.

The fruitless Mulberries may be well suited for planting as a quick shade tree in a park or other open area. It is most popular in the drier parts of the country. They can be used in a residential area for quick shade near a patio or other area where dense shade is desired. However, there are so many other more superior trees available that planting large quantities of this one seems pointless.

White Mulberry should be grown in full sun or partial shade on any soil. It adapts to any soil and has naturalized in many parts of the world. Although it is tolerant of air pollution and dry conditions, the tree will perform its best on moist soils. Scale and mites, leaf spot, bacterial leaf scorch, powdery mildew, and cankers are among the many potential pests.

Murraya paniculata

Orange Jessamine, Orange Jasmine

Pronunciation: muir-AA-yah pan-ick-you-LA-tah

ZONES: 9B, 10A, 10B, 11

HEIGHT: 10 to 15 feet

WIDTH: 10 to 15 feet

FRUIT: ¼"; dry; red turning brown; no litter

GROWTH RATE: slow to moderate; moderate life span

HABIT: vase; moderate density; nearly symmetrical; fine texture

FAMILY: *Rutaceae*

LEAF COLOR: medium green; evergreen

LEAF DESCRIPTION: alternate; pinnately compound; entire margin; oval to oblong; ½ to 1" long leaflets

FLOWER DESCRIPTION: white; extremely fragrant; showy; spring and summer

EXPOSURE/CULTURE:

Light Requirement: grows and flowers best in full sun

Soil Tolerances: all textures; slightly alkaline to acidic; drought

Aerosol Salt Tolerance: moderate

PEST PROBLEMS: somewhat sensitive

BARK/STEMS:

Very showy with longitudinally, exfoliating bark; bark is thick and not easily damaged; branches grow mostly upright and will require little pruning for vehicular or pedestrian clearance beneath canopy; routinely grown with multiple trunks; can be trained to grow with a short, single trunk; no thorns

Pruning Requirement: stake to develop a central trunk, then space branches along trunk to create a nice, small specimen tree

Limb Breakage: quite resistant

Current Year Twig: green; slender

LANDSCAPE NOTES:

Origin: Australia, southern Asia, Pacific islands

Usage: espalier or topiary; container or above-ground container; buffer strip; highway; deck or patio; standard; parking lot island; specimen; street tree; urban tolerant

A long period of fragrant, summer flowers, attractive bark and trunk form, and good drought tolerance all combine to make Orange Jasmine a potentially popular small tree for either formal or informal landscapes. The fine-textured foliage moves in the slightest breezes, creating a light, airy effect in any landscape. Most plants are installed and maintained as clipped hedges, but the true beauty of the plant is displayed when it is allowed to grow into a small tree. The upright, vase-shaped crown makes the tree well suited for street tree planting, especially where utility lines are directly overhead.

New growth can be pinched during the growing season to increase branchiness and flower number. Because the tree responds so well to shearing and shaping, it makes a nice espalier or specimen pruned to any number of shapes. Lower branches are often removed or thinned to show off the trunk form and color as the tree becomes larger. A row of trees pruned this way, planted 10 to 15 feet apart along a sidewalk or roadway, can make a striking statement.

It grows well in limited soil spaces in urban areas, such as along boulevards, in parking lot islands, and in small pavement cutouts, if provided with some irrigation until well established. The small size and tolerance to drought make Jasmine a good candidate for planting in a raised planter or container near the patio or on the deck, or even along a street. It tolerates slightly alkaline soil well, although chlorosis often develops in poorly drained soil.

Soil nematodes, scales, and whitefly can become problems for Orange Jasmine.

Musa spp. Banana

Pronunciation: MEW-sah species

ZONES: 9A, 9B, 10A, 10B, 11

HEIGHT: 6 to 30 feet

WIDTH: 5 to 15 feet

FRUIT: elongated; 2 to 10" long; fleshy; green to yellow; edible; causes significant litter; persistent; showy

GROWTH RATE: rapid-growing; short life span

HABIT: palm-like; upright; open; irregular silhouette; coarse texture

FAMILY: *Musaceae*

LEAF COLOR: broadleaf evergreen; medium green

LEAF DESCRIPTION: simple; entire; oblong; 3 to 8' long; tears along veins in the wind

FLOWER DESCRIPTION: orange, purple, or red, depending on species; showy; spring

EXPOSURE/CULTURE:

Light Requirement: part shade to full sun

Soil Tolerances: all textures; alkaline to acidic; wet soil; drought-sensitive

Aerosol Salt Tolerance: none

PEST PROBLEMS: resistant

BARK/STEMS:

Bright green and showy; clump-forming; no thorns

Pruning Requirement: remove old leaves; cut tall stems to stimulate sprouting; remove sprouts as they form for a more open, neater form

Limb Breakage: susceptible to wind

CULTIVARS AND SPECIES:

Many different species of Banana are available. Some ornamental types are grown for foliage or flowers. *M. coccinea* has brilliant red bracts; *M. rosea* has pink bracts. Both hold up very well as cut flowers. *M. acuminata*, 'Dwarf Cavendish', is one of the best fruit cultivars. It has large bunches with large fruit and the plant's small size makes it easier to protect from wind. The tall-growing 'Ladyfinger' has small bunches of small Bananas, but they are very thin-skinned and delicious. *M. velutina* grows three to four feet tall with three-foot leaves that are green above and bronzy beneath. The upright pink bracts have orange flowers and yield velvety pink fruit.

LANDSCAPE NOTES:

Origin: Asia, Indochina

Usage: fruit; specimen

Large, fleshy, upright stalks topped with soft, smooth, arching leaves signify the Banana plant. Ranging from 6 feet for dwarf species to over 30 feet for the largest types, Banana trees are guaranteed to lend a tropical flavor to any landscape setting. The broad, tender leaves are easily torn by winds, and plants should be located in a sheltered area to prevent this. The easily grown Banana tree is ideal for planters near the pool, around garden ponds, or as a cluster for an exotic effect. The unusual reddish-purple flowers are followed by clusters of upwardly-pointing green fruit that matures to either a beautiful yellow or green.

Growing best on fertile, moist soil, Bananas will thrive in full sun or partial shade and should be protected from both wind and cold. Plants respond well to regular fertilization. Do not allow too many suckers to develop, as this will decrease the ability of any one plant to produce a good bunch of fruit. By allowing suckers to develop only at periodic intervals, a succession of fruiting can be obtained. Banana bunches should be harvested when the fruit is still green and allowed to ripen in a cool, dark place. It produces fruit in zones 8B and 9A only when winter temperatures stay above freezing. Plants killed to the ground which sprout from the soil in the spring will not produce fruit until the following year.

Scales and nematodes, sigatoka leaf-spot, cercospora leaf-spot, and Panama disease are all potential problems.

Myrica cerifera Southern Waxmyrtle

Pronunciation: MY-reh-cah suh-RIFF-err-ah

ZONES: 7B, 8A, 8B, 9A, 9B, 10A, 10B, 11

HEIGHT: 15 to 20 feet

WIDTH: 20 to 25 feet

FRUIT: round; less than 0.5"; fleshy to hard; blue; attracts wildlife; inconspicuous; no significant litter

GROWTH RATE: rapid-growing; short-lived

HABIT: rounded vase; open to moderate density; irregular silhouette; fine texture

FAMILY: *Myricaceae*

LEAF COLOR: evergreen; olive green

LEAF DESCRIPTION: alternate; simple; entire to serrate margin; oval to oblanceolate or spatulate; 2 to 5" long

FLOWER DESCRIPTION: green; not showy; spring

EXPOSURE/CULTURE:

Light Requirement: part shade to full sun; canopy is thin in shade

Soil Tolerances: all textures; alkaline to acidic; wet soil; drought and salt

Aerosol Salt Tolerance: high

PEST PROBLEMS: susceptible to a vascular wilt

BARK/STEMS:

Bark is thin and easily damaged by mechanical impact; branches droop as tree grows and will require pruning for vehicular or pedestrian clearance beneath canopy; routinely grown with, or trainable to be grown with, multiple trunks; no thorns

Pruning Requirement: requires pruning to develop strong structure; begin pruning when young to create desired canopy form and avoid need to remove large branches later; rot often begins at large pruning wounds

Limb Breakage: susceptible to breakage at crotch, due to poor collar formation; wood itself is weak and tends to break

Current Year Twig: brown; gray; slender

CULTIVARS AND SPECIES:

'Compacta' and 'Emperor PAT' are listed as cultivars growing to about half the size of the species. *M. cerifera var. pussila* also stays small. *M. californica* is similar to *C. cerifera* and native to California.

LANDSCAPE NOTES:

Origin: Maryland along the coast to Texas on wet sites

Usage: container or above-ground planter; hedge; parking lot island; buffer strip; highway; near a deck or patio; reclamation; standard; specimen; sidewalk cutout; urban tolerant

Multiple, twisted trunks with smooth, light gray bark, aromatic leaves, and clusters of gray-blue, waxy berries (on female plants) that attract wildlife are just some of the reasons Southern Waxmyrtle is such a popular landscape plant. Most specimens form a multi-stemmed, open, rounded canopy of weak trunks and branches. This rapid-growing, small, evergreen native tree is capable of reaching a height of 25 feet with an equal spread, but is usually seen in the 10- to 20-foot range.

Very tough and easily grown, Southern Waxmyrtle can tolerate a variety of landscape settings from full sun to partial shade, wet swamplands or high, dry, and alkaline areas. Trees propagated from dry sites are most drought tolerant. Growth is thin in total shade. It is adapted to parking lot and street tree planting, especially beneath power lines, but branches tend to droop, hindering the flow of vehicular traffic if planted too close.

Removing excess shoot growth two times each year eliminates the tall, lanky branches and reduces the tendency for branches to droop. Some landscape managers hedge the crown into a multi-stemmed dome-shaped topiary. Plants spaced 10 feet apart, maintained in this manner, can create a nice canopy of shade for pedestrians.

Cankers may form on old branches and trunks and kill them. Also, a lethal wilt disease caused by the fungus *Fusarium oxysporum* has been noted attacking Waxmyrtle plants in central and south Florida. Root injury and nitrogen fertilization encourage the disease. Sprouts from the lower trunk and from the roots are a maintenance nuisance. Roots fix nitrogen.

Neodypsis decaryi Triangle Palm

Pronunciation: knee-oh-DIP-sis dee-CAR-ee-eye

ZONES: 10B, 11

HEIGHT: 20 to 25 feet

WIDTH: 15 feet

FRUIT: red-brown; 0.75"; no significant litter

GROWTH RATE: slow-growing; long-lived

HABIT: palm; open; symmetrical; coarse texture; three-sided

FAMILY: *Arecaceae*

LEAF COLOR: bluish-green

LEAF DESCRIPTION: pinnately compound; entire margin; 18 to 36" long leaflets; 6 to 12' long fronds

FLOWER DESCRIPTION: white; not showy; spring

EXPOSURE/CULTURE:

Light Requirement: part shade to full sun

Soil Tolerances: all textures; alkaline to acidic; salt-sensitive; drought

Aerosol Salt Tolerance: low

PEST PROBLEMS: pest resistant

BARK/STEMS:

Single, 12"-diameter trunk; 3-sided; no thorns

Pruning Requirement: remove dead and dying fronds

Limb Breakage: resistant

LANDSCAPE NOTES:

Origin: Madagascar

Usage: buffer strip; highway; near a deck or patio; specimen; street tree

This single-trunked palm is easily recognized by its canopy of graceful fronds held in three planes. Frond color varies from green through bluish green. Leaflets are displayed in an uneven fashion, giving an informal, ragged appearance to the palm. The heavy, stocky trunks are clean of leaf bases, forming a fairly smooth surface. This slow- to moderate-growing palm eventually will reach 20 feet tall and is attractive as a freestanding specimen or grouped with other palms. Most are seen smaller than this in landscapes.

Planted 10 feet apart as a street tree, they create a wall of bluish-green foliage. They can be planted beneath power lines due to slow growth and small size. Locate one or a group in a low-growing groundcover to set it off from the rest of the landscape and to protect the trunk from damage. The bluish foliage stands out in a shaded garden, adding a certain warmth. This is an unusual palm with a striking habit of growth. There is no mistaking this palm for any other due to the three-sided frond display.

Growing in full sun or part shade on a wide variety of soils, including alkaline, Triangle Palm has little salt tolerance. It survives hot, windy conditions, asphalt, and concrete areas, but looks better in good soil with adequate moisture.

Few nurseries have large numbers for sale. Potassium deficiencies often show up as speckled foliage. A regular, preventive fertilization program will help the palm look its best at all times.

Nerium oleander Oleander

Pronunciation: NEE-ree-um OH-lee-ann-der

ZONES: 9A, 9B, 10A, 10B, 11

HEIGHT: 15 feet

WIDTH: 10 to 12 feet

FRUIT: elongated; 4 to 6"; dry; no significant litter

GROWTH RATE: rapid; short to moderate life span

HABIT: round; moderate density; symmetrical; medium texture

FAMILY: *Apocynaceae*

LEAF COLOR: evergreen; dark green

LEAF DESCRIPTION: opposite/subopposite or whorled; simple; entire margin; lanceolate; 3 to 6" long

FLOWER DESCRIPTION: pink, red, yellow, or white; pleasant fragrance; showy; year-round

EXPOSURE/CULTURE:

Light Requirement: part sun to full sun

Soil Tolerances: all textures; alkaline to acidic; drought and salt

Aerosol Salt Tolerance: moderate

PEST PROBLEMS: resists most lethal pests

BARK/STEMS:

Routinely grown with multiple trunks, but can be trained with one short trunk; branches grow mostly upright; no thorns

Pruning Requirement: requires pruning to develop tree form; trunk sprouts require regular removal to maintain tree form

Limb Breakage: susceptible to breakage in exposed location at crotch, due to poor collar formation; wood itself is weak and tends to break

Current Year Twig: green; thick

LANDSCAPE NOTES:

Origin: Asia and the Mediterranean

Usage: container or above-ground planter; hedge; parking lot islands; buffer strip; highway; near a deck or patio; reclamation; standard; specimen; sidewalk cutout; street tree

Oleander is a wonderful, easy-care, rounded shrub or small tree, with long leaves and an abundance of single or double, sometimes fragrant, flowers, in shades of white, pink, red, or yellow depending on the cultivar. Often trained into an attractive small tree, Oleander also does well as a quick-growing screen or specimen tree. Planted on five- to seven-foot centers, a row of Oleander makes a nice screen for a large residence or other large-scale landscape. A dwarf shrub selection, 'Petite', is most suited for residential landscapes.

Growing well with only one yearly fertilization and springtime pruning, Oleander is one of the easiest plants to care for. The plant can be trained with a short central leader in the nursery and is often sold as a "standard" Oleander. Suckers produced at the base of the plant must be removed to maintain one trunk. It grows into a round-headed ball, flowering year-round in zones 9B through 11. Flowering is reduced in winter in zone 9A. Every few years, tops of trees in 9A are injured by cold.

All parts of the plant are extremely poisonous, so care must be taken when locating Oleander near areas frequented by small children; burning of the trimmings will produce toxic fumes. Even chewing once or twice on a leaf or twig can send a person to the hospital.

Oleander survives drought well and is well suited to growing on soil too poor for most other plants, even tolerating salt spray, brackish water, and alkaline soil. It is commonly planted in highway medians as a no-maintenance plant. It grows following wet weather, slowing down in drought, but always looks good even in powder-dry soil.

Pest problems include scale and Oleander caterpillar, which can do quite a bit of damage to the foliage if left unchecked. Oleander caterpillar, common in Florida, can defoliate a plant in a week or two, and bacterial scorch in the southwest is also a pest problem.

Noronhia emarginata Madagascar Olive

Pronunciation: no-ROW-nee-ah ee-mar-gin-A-tah

ZONES: 10B, 11

HEIGHT: 20 feet

WIDTH: 20 to 35 feet

FRUIT: round; 1"; fleshy; green to yellow; no significant litter; showy

GROWTH RATE: slow-growing; moderate life span

HABIT: oval or rounded; dense; irregular silhouette; coarse texture

FAMILY: *Oleaceae*

LEAF COLOR: broadleaf evergreen; olive green

LEAF DESCRIPTION: opposite; simple; entire margin; oval to obovate; 4 to 7" long

FLOWER DESCRIPTION: yellow; slight pleasant fragrance; not showy; produced year-round

EXPOSURE/CULTURE:

Light Requirement: part shade to full sun

Soil Tolerances: all textures; alkaline to acidic; drought and salt

Aerosol Salt Tolerance: high

PEST PROBLEMS: resistant

BARK/STEMS:

Branches droop as tree grows and will require some pruning for vehicular or pedestrian clearance beneath canopy; trainable to be grown with a single or multiple trunk; no thorns

Pruning Requirement: requires early pruning to develop upright structure

Limb Breakage: resistant

Current Year Twig: brown; green; thick

LANDSCAPE NOTES:

Origin: Madagascar

Usage: container or above-ground planter; espalier; fruit; parking lot island; buffer strip; highway; near a deck or patio; reclamation; specimen; sidewalk cutout; street tree; urban tolerant

A very tough tree well suited for coastal and seaside locations, Madagascar Olive is an attractive, upright oval, evergreen with long, leathery leaves and a high tolerance for salt and wind. The crown appears to stay well-formed even right up on the coast where it is exposed to direct salt spray. The inconspicuous but fragrant, small yellow blooms are followed by one-inch-diameter, bright yellow fruits, which turn to dark purple with a sweet, edible, cream-colored flesh and add to the tree's ornamental value. The dropping fruit may be undesirable to some people when this tree is planted near walks or pavement, but located in a bed or lawn is of little concern.

Growing easily in full sun or partial shade, Madagascar Olive is quite adaptable to a wide range of soils. It works well as a dense screen or windbreak, and can form an ideal specimen or framing tree for landscapes with limited room after the lower branches are removed. A cluster of trees creates a handsome, informal grouping. This is a tough, small tree suitable for many locations. Use it to create shade on a patio or grow it in a large, heavy container on a deck. The small size makes it ideal for planting along streets where there are overhead power lines.

The fruit causes some litter along streets and sidewalks, but it is usually not abundant enough to cause a great deal of concern. No limitations of major concern.

Nyssa ogeche Ogeechee Tupelo

Pronunciation: NISS-ah o-GEE-chee

ZONES: 7A, 7B, 8A, 8B, 9A, 9B

HEIGHT: 40 feet

WIDTH: 30 to 35 feet

FRUIT: oval; 1 to 1.5" long; fleshy; red; attracts wildlife; edible; causes some messy litter; showy

GROWTH RATE: moderate; moderate life span

HABIT: round; moderate density; symmetrical; medium texture

FAMILY: *Nyssaceae*

LEAF COLOR: dark green in summer; orange, purple, red, or yellow fall color

LEAF DESCRIPTION: alternate; simple; entire or sinuate margin; oval to obovate; deciduous; 4 to 6" long

FLOWER DESCRIPTION: white or yellow; not showy; spring

EXPOSURE/CULTURE:

Light Requirement: part shade to full sun

Soil Tolerances: all textures; acidic; wet soil; moderate drought

PEST PROBLEMS: unknown

BARK/STEMS:

Trunk is thick at base; branches droop as tree grows and will require pruning for vehicular or pedestrian clearance beneath canopy; should be grown with a single leader; no thorns

Pruning Requirement: needs little pruning to develop strong structure; young trees usually produce a central leader

Limb Breakage: resistant

Current Year Twig: green; medium thickness

LANDSCAPE NOTES:

Origin: southern Georgia and north Florida

Usage: parking lot island; buffer strip; highway; wetland reclamation; shade tree; specimen; street tree; sidewalk cutout

First discovered by William Bartram along the Ogeechee River in Georgia, Ogeechee-Lime is a lovely native tree that is pyramidal when young and matures to a spreading, flat-topped crown. The multiple, irregular branches emerge from a trunk covered with dark brown or gray, ridged bark. The base of an older tree often develops swollen buttress-type roots. The leaves are joined in early spring by dense, hanging clusters of small, white blooms. The 1.5-inch-long, showy red fruits on female trees are produced in abundance and ripen in autumn. The juice can be used as a substitute for lime juice; hence this tree's common name. Fruits can make a slight mess on a sidewalk, driveway, or patio. In autumn, the trees put on a brilliant display of colorful foliage ranging from vivid yellow to deep purple, which would make Ogeechee-Lime a popular landscape choice. Unfortunately, it is not grown by many nurseries.

The sex on trees can be determined at an early age in the nursery, so it should be easy to select male trees. These might be preferrable in some urban landscapes because the fruits on female trees can be somewhat messy.

Most often found along streams and in low-lying areas that are regularly flooded in spring and winter, Ogeechee-Lime will prefer a moist site on acid soil. It is well adapted to sites that are wet for prolonged periods, once the tree becomes established. Located in full sun or partial shade, Ogeechee-Lime will easily adapt to drier locations having been grown on dry, poor, clayey soils, but should be protected from harsh winds.

No known limitations, but other Tupelos are occasionally bothered by leaf miner, scale, rust, and leaf spot.

Nyssa sylvatica Blackgum, Sourgum, Black Tupelo

Pronunciation: NISS-ah sill-VAH-tee-cah

ZONES: 4B, 5A, 5B, 6A, 6B, 7A, 7B, 8A, 8B, 9A, 9B, (10A)

HEIGHT: 65 to 75 feet

WIDTH: 25 to 35 feet

FRUIT: oval to round; less than 0.5"; fleshy; blue; attracts wildlife; not showy; causes some litter

GROWTH RATE: slow-growing; long-lived

HABIT: pyramidal when young, oval later; moderate density; symmetrical; medium texture

FAMILY: *Nyssaceae*

LEAF COLOR: dark green in summer; orange or bright red, showy fall color

LEAF DESCRIPTION: alternate; simple; entire margin; oblong to obovate or ovate; deciduous; 2 to 6" long

FLOWER DESCRIPTION: white; not showy; spring

EXPOSURE/CULTURE:

Light Requirement: part shade to full sun

Soil Tolerances: all textures; acidic, below 6 is best; wet soil; some salt; drought

Aerosol Salt Tolerance: moderate

PEST PROBLEMS: resistant

BARK/STEMS:

Branches grow horizontal, then droop as tree ages, so will require some early pruning for vehicular or pedestrian clearance beneath canopy; normally grows with a single leader; no thorns

Pruning Requirement: needs little or no pruning to develop strong structure

Limb Breakage: resistant

Current Year Twig: brown; gray; medium thickness

LANDSCAPE NOTES:

Origin: eastern North America

Usage: parking lot island; buffer strip; highway; reclamation; shade tree; specimen; sidewalk cutout; street tree

Two forms of the tree are commonly recognized. *Nyssa sylvatica* var. *sylvatica* grows on moist and mesic sites; var. *biflora* has a buttressed and swollen trunk base and grows in swamps. Sourgum forms horizontal branches from a straight trunk, making it well suited for many urban uses. Crown form varies considerably from tree to tree. After 10 to 15 years, the canopy becomes more uniform among trees. The brilliant display of red to deep purple foliage in the fall surprises most people, as the tree does not particularly stand out in the landscape until then. The small, blue fruits may be considered a litter nuisance, but are quite popular with many birds and mammals and wash away quickly from a sidewalk or street.

Sourgum prefers a moist, slightly acid soil, but tolerates wet or dry soil and low-oxygen soil. Little pruning is required to form a well-structured tree, because the trunk stays straight and branches usually grow at wide angles to the trunk. Sourgum makes a good street or parking lot tree for suburban neighborhoods and should be used more often for this purpose.

Sourgum planted in dry sites may benefit from occasional irrigation during drought, especially in the southern part of its range, although it is very drought-tolerant. It does amazingly well in the wet soils typical of many urban areas. It is also somewhat salt-tolerant for planting along the shore and salted streets.

Sourgum is rarely attacked by pests, and when it is they are rarely serious enough to warrant control. A large number of trees have been discovered dying in the mountains of North Carolina, Tennessee, Virginia, and Georgia, but the causal agent is unknown. Larger specimens may be difficult to transplant from deep, well-drained field soil because of deep roots, and should be transplanted from the field only in the spring. Sourgum is offered by some nurseries in containers.

Ochrosia elliptica Ochrosia

Pronunciation: o-CROW-she-ah ee-LIP-tee-cah

ZONES: 10B, 11

HEIGHT: 20 to 25 feet

WIDTH: 15 to 25 feet

FRUIT: oval; 1 to 2"; fleshy; bright red; no significant litter; persistent and very showy; poisonous

GROWTH RATE: moderate; moderate life span

HABIT: upright oval; open; symmetrical; coarse texture

FAMILY: *Apocynaceae*

LEAF COLOR: broadleaf evergreen; medium green

LEAF DESCRIPTION: opposite or whorled; simple; entire or undulate margin; oval to oblong or obovate; 4 to 8" long

FLOWER DESCRIPTION: yellow; fall, winter or summer; pleasant fragrance; showy

EXPOSURE/CULTURE:

Light Requirement: part shade to full sun

Soil Tolerances: loam; sand; alkaline to acidic; salt and drought

Aerosol Salt Tolerance: moderate to good

PEST PROBLEMS: resistant

BARK/STEMS:

Routinely grown with multiple trunks; can be trained to grow with a single trunk; branches grow mostly upright and horizontal; no thorns

Pruning Requirement: requires pruning to develop strong structure

Limb Breakage: resistant

Current Year Twig: green; thick

LANDSCAPE NOTES:

Origin: Australia

Usage: container or above-ground planter; espalier; buffer strip; near a deck or patio; specimen

This large, upright, evergreen shrub or small tree has glossy, leathery leaves and clusters of fragrant, yellow/white flowers from late summer into winter, followed by bright red, two-inch-long, poisonous fruit borne in pairs. It will make a nice tree for a patio area, providing shade with the lower branches removed, or a visual screen of coarse, dense foliage without growing too tall. Suited for planting beneath power lines because of low maximum height. Seedlings germinate in the landscape.

Salt-tolerant Ochrosia grows well close to the ocean in full sun or partial shade on a wide range of soils, including alkaline. It is very drought-tolerant but responds well to irrigation and fertilizer. Its dark green, dense foliage makes it ideal for tall screens or at the rear of a shrub border.

Scale and occasionally mites can be locally troublesome.

Olea europaea European Olive

Pronunciation: OH-lee-ah your-OH-pea-ah

ZONES: 9A, 9B, 10A, 10B, 11

HEIGHT: 30 to 40 feet

WIDTH: 30 to 40 feet

FRUIT: oval; 0.75"; fleshy; black; attracts wildlife; causes significant, messy litter; showy; edible

GROWTH RATE: slow to moderate; moderate life span

HABIT: rounded; dense; irregular silhouette; fine texture

FAMILY: *Oleaceae*

LEAF COLOR: gray-green above, silvery below; evergreen

LEAF DESCRIPTION: opposite; simple; entire margin; ovate to lanceolate; 1.5 to 3" long

FLOWER DESCRIPTION: white; showy; spring

EXPOSURE/CULTURE:

Light Requirement: full sun

Soil Tolerances: all textures; alkaline to acidic; good drainage is essential; salt and drought

Aerosol Salt Tolerance: high

PEST PROBLEMS: mostly resistant

BARK/STEMS:

Smooth on young trees, twisted later; bark is thick; branches droop as tree grows and will require early pruning for vehicular or pedestrian clearance beneath canopy; trainable to one trunk or several trunks; no thorns

Pruning Requirement: space major limbs along trunk and prevent formation of included bark by thinning and drop-crotching; prune only in winter to prevent spread of olive knot bacteria

Limb Breakage: resistant

Current Year Twig: smooth, gray; slender

LANDSCAPE NOTES:

Origin: Mediterranean region

Usage: bonsai; hedge; buffer strip; highway; windbreak; urban tolerant; fruitless types are useful as street trees and in parking lots

Olive is considered one of the ten most useful trees in the world. The fruit is well known for use as a fresh fruit and for production of oil. It also makes a durable urban tree for a variety of landscapes. The fruitless types are especially useful for planting in tough sites, such as along streets and in parking lot islands, in environments that are unsuited for many other trees.

The small, leathery leaves are densely arranged along smooth, stiff, light gray branches. Young plants can appear awkward and irregularlly shaped, requiring up to 10 years to develop a distinct, rounded canopy. The trunk is usually smooth and gray, becoming gnarled and twisted on older specimens.

European Olive is ideal for training into a medium-sized tree, with lower branches removed to reveal the interestingly contorted, multiple trunks. Trees planted along streets should be trained to grow with one central leader, and major branches should be spaced along the trunk to ensure durability. Thin the foliage on the largest limbs so that diameter remains less than about half that of the trunk directly above the branch. This will help create a stronger tree and allow other plants to grow beneath the dense canopy.

European Olive is well suited for hot, dry locations, such as exposed parking lots, along highways and other stressful urban sites. It can also be planted on 20- to 30-foot centers along a walkway or street to create a dense canopy of shade.

The species produces abundant fruit that litters the ground. Fruitless cultivars, including 'Swan Hill' and 'Wilson', should be better suited for many landscape sites. It would be quite appropriate to plant fruitless Olive along streets for a durable shade tree.

Osmanthus spp. Osmanthus

Pronunciation: oz-MAN-thuss species

ZONES: 7A, 7B, 8A, 8B, 9A, 9B

HEIGHT: 15 to 20 feet

WIDTH: 6 to 10 feet

FRUIT: black; no significant litter

GROWTH RATE: slow-growing; moderate longevity

HABIT: columnar; oval; upright; dense; symmetrical; medium texture

FAMILY: *Oleaceae*

LEAF COLOR: evergreen; medium to dark green

LEAF DESCRIPTION: opposite; simple; entire or serrate margin; oval to ovate; 2 to 4" long; spiny

FLOWER DESCRIPTION: white; fall and spring; strong, pleasant fragrance; not especially showy

EXPOSURE/CULTURE:

Light Requirement: part shade to full sun

Soil Tolerances: all textures; acidic; drought

PEST PROBLEMS: resistant

BARK/STEMS:

Routinely grown with, or trainable to be grown with, multiple trunks; branches grow mostly upright on young plants, but droop later; no thorns

Pruning Requirement: needs little pruning to develop strong trunk and branch structure; lower branches can be removed to create clear trunk and produce more of a tree form

Limb Breakage: resistant

Current Year Twig: gray; medium thickness

CULTIVARS AND SPECIES:

The cultivar 'San Jose' has cream to orange flowers; 'Variegatus' has white margins and a nice appearance for a variegated plant. *O. fragans* is a beautiful, upright, columnar tree that spreads with age with very fragrant, small white blooms produced in the fall and winter months.

LANDSCAPE NOTES:

Origin: one native; others hybrids

Usage: container or above-ground planter; hedge; specimen; screen; windbreak

This large, vigorous, evergreen shrub or small tree forms a dense, round or oval silhouette. Very old specimens form a spreading vase. A hybrid of Holly Osmanthus (*O. heterophyllus*) and Fragrant Tea Olive (*O. fragrans*), Fortune's Osmanthus has the spiny, holly-like, dark green, leathery foliage of one parent plant and the extremely fragrant white flowers of the other. Frequently trimmed into a hedge or screen, Fortune's Osmanthus could be used often as a specimen or container planting, and the barbed leaves make it suitable as a barrier planting. *O. americanus* is native to moist sites in eastern North America but makes a nice small tree for dry sites as well.

The narrow, upright canopy on young plants makes Osmanthus well suited for a living screen or tall clipped hedge. Plant individuals on six- to eight-foot centers and provide irrigation until they are well established. Tolerance to extended drought and neglect makes them adaptable to planting along highways and other areas that receive little, if any, care.

Fortune's Osmanthus should be grown in full sun or partial shade on any well-drained soil, including clay. Drought tolerance is good; established specimens apparently do fine without irrigation.

Scales are one of the few problems. They can cause some defoliation if infestation is serious. Mushroom root rot may occur when the tree is grown on wet soils. *O. americanus* grows fine on moist to wet sites.

Ostrya virginiana American Hophornbeam, Ironwood

Pronunciation: os-TRY-yah vir-gin-ee-A-nah

ZONES: 3A, 3B, 4A, 4B, 5A, 5B, 6A 6B, 7A, 7B, 8A, 8B, 9A

HEIGHT: 30 to 40 feet

WIDTH: 25 to 35 feet

FRUIT: elongated; 1 to 2"; dry; green turning tan; attracts birds; no significant litter; persistent; showy

GROWTH RATE: slow to moderate; moderate life span

HABIT: oval to rounded; moderate density; symmetrical; fine texture

FAMILY: *Betulaceae*

LEAF COLOR: light green; showy, yellow or tan fall color

LEAF DESCRIPTION: alternate; simple; double serrate margin; oblong to ovate; deciduous, some leaves persist into winter; 3 to 5" long

FLOWER DESCRIPTION: brown; green; somewhat showy; summer

EXPOSURE/CULTURE:

Light Requirement: part shade to full sun

Soil Tolerances: all textures; alkaline to acidic; salt-sensitive; drought

PEST PROBLEMS: except for trunk rot on old trees, nearly pest-free

BARK/STEMS:

Branches droop slightly as tree grows and will require some pruning when tree is young for vehicular or pedestrian clearance beneath canopy; trainable to be grown with a single trunk or multiple trunks; showy, orange-brown trunk; no thorns

Pruning Requirement: needs little pruning to develop single trunk; prune in winter or early spring

Limb Breakage: very resistant

Current Year Twig: brown; slender

LANDSCAPE NOTES:

Origin: eastern North America

Usage: above-ground planter; parking lot island; buffer strip; highway; near a deck or patio; reclamation; shade tree; specimen; sidewalk cutout; street tree; urban tolerant

This shade-tolerant tree may slowly grow to 50 feet in height, but is often 25 to 40 feet tall, forming an oval or round canopy. Hophornbeam has a lovely yellow fall color, and the small nutlets, which ripen in summer and fall, are used by birds and mammals during the winter. Bark varies but is usually an attractive orange or grayish-brown, peeling off in longitudinal strips. Older trunks become darkened and lose the showy, orange coloration. The finely textured crown casts a medium or dense shade in full sun, but is more open in the shade, casting a light shadow. Variety *lasia* replaces the typical species in the southern United States.

This rugged tree is tolerant of poor soil conditions found in urban areas and should be grown and planted more. Can be purchased as a single or multi-trunked specimen from nurseries. Multi-stemmed trees have a dramatic impact in the landscape with bright bark and wonderful form. Great for climbing.

Hophornbeam has a shallow root system and will grow in most soils except those that are wet. It is well adapted to downtown city plantings provided soil drainage is good. Often found on dry, rocky slopes with little soil, Hophornbeam is quite tolerant of drought and needs little care once established. Locate it close to people so they can enjoy the wonderful bark and foliage. Transplanting field-grown trees in spring is best. Containerized trees can be planted anytime.

Two-lined chestnut borer may infest Hophornbeam. Keep trees healthy by regular fertilization, and irrigate soil around tree during extended drought periods to reduce susceptibility to borer attack. Basswood leafminer may cause damage to foliage. Orange-humped mapleworm may be found feeding on *Ostrya* north of Pennsylvania. Leaf spots are common but generally not serious. Recovers slowly after transplanting.

Oxydendrum arboreum Sourwood

Pronunciation: oks-ee-DEN-drum are-BORE-ee-um

ZONES: 5A, 5B, 6A, 6B, 7A, 7B, 8A, 8B

HEIGHT: 40 to 60 feet

WIDTH: 30 to 35 feet

FRUIT: oval; less than 0.5"; dry; brown; no significant litter; persistent

GROWTH RATE: slow-growing; moderate life span

HABIT: pyramidal when young, oval later; dense; nearly symmetrical; medium texture

FAMILY: *Ericaceae*

LEAF COLOR: dark green in summer; brilliant red fall color

LEAF DESCRIPTION: alternate; simple; entire or serrulate margin; lanceolate to oblong; deciduous; 4 to 8" long

FLOWER DESCRIPTION: white; showy; summer and fall

EXPOSURE/CULTURE:

Light Requirement: part shade to full sun

Soil Tolerances: all textures; acidic; moderate drought

Aerosol Salt Tolerance: some

PEST PROBLEMS: resistant

BARK/STEMS:

Branches droop as tree grows and will require pruning for vehicular or pedestrian clearance beneath canopy; normally grown with a single leader; no thorns

Pruning Requirement: needs little pruning to develop strong structure

Limb Breakage: resistant

Current Year Twig: green; reddish; medium thickness

LANDSCAPE NOTES:

Origin: Pennsylvania to Louisiana and northern Florida

Usage: buffer strip; highway; shade tree; specimen

Sourwood grows as a pyramid or narrow oval, with a more or less straight trunk, usually to a height of 25 to 35 feet, though it can reach 50 to 60 feet tall. Occasionally young specimens have a more open spreading habit reminiscent of Redbud, especially on drier sites. Leaves are lustrous green and appear to weep or hang from the twigs. Branches droop toward the ground, forming a graceful outline when the tree is planted as a single specimen. The branching pattern and persistent fruit make the tree interesting in winter. The mid- to late-summer flowers are borne in terminal clusters of racemes that droop downward, creating a graceful, fine-textured effect at flowering time. The fall color is a striking red and orange which is rivaled by only a few other trees, such as Blackgum, Chinese Pistache, the pears, and Chinese Tallowtree. Few sights are as striking as a row of Sourwood in fall color.

The tree grows slowly, adapts to sun or shade, and prefers a slightly acid, peaty loam. The tree transplants easily when young and from containers of any size. Cut any roots that circle the container at planting. Sourwood grows well in confined soil spaces with good drainage, making it a candidate for urban planting, but it is largely untried as a street tree. It is reportedly sensitive to air pollution injury. Irrigation is required during hot, dry weather in the southern part of its range to keep leaves on the tree. There are beautiful specimens in zones 5, 6, and 7 growing in the open sun in poor clay with no irrigation.

Pests are usually not a problem for Sourwood. Fall webworm can defoliate portions of the tree in summer and fall, but control usually is not needed. Twig blight kills leaves at the branch tips. Trees in poor health seem to be more susceptible. Prune out infected branch tips and fertilize. Leaf spots can discolor some leaves, but are not serious other than causing premature defoliation.

Pandanus utilis Screw-pine

Pronunciation: pan-DAY-nuss YOU-tih-liss

ZONES: 10B, 11

HEIGHT: 25 feet

WIDTH: 15 feet

FRUIT: oval; round; 8 to 9"; fleshy and hard; orange-yellow; attracts mammals; edible; causes significant litter; showy

GROWTH RATE: slow-growing; long-lived

HABIT: pyramidal; open; symmetrical; coarse texture

FAMILY: *Pandanaceae*

LEAF COLOR: evergreen; blue or blue-green

LEAF DESCRIPTION: spiral, strap-like; simple; serrate margin; 18 to 36" long; drops periodically throughout summer

FLOWER DESCRIPTION: white; pleasant fragrance; winter

EXPOSURE/CULTURE:

Light Requirement: part shade to full sun

Soil Tolerances: all textures; alkaline to acidic; wet soil; drought

Aerosol Salt Tolerance: high

PEST PROBLEMS: mostly resistant

BARK/STEMS:

Branches grow mostly upright or horizontal and will not droop; showy trunk; should be grown with a single leader; no thorns

Pruning Requirement: needs little pruning to develop good structure

Limb Breakage: mostly resistant

Current Year Twig: brown; unusually stout

CULTIVARS AND SPECIES:

Veitch Screw-Pine or Ribbon-Plant (*P. veitchii*) has white-banded, spiny leaves, does not fruit, and is often used as a potted plant. Spineless varieties exist. Sander Screw-Pine (*P. sanderi*) has denser, more tufted foliage with golden yellow bands from center of leaf to margin.

LANDSCAPE NOTES:

Origin: Madagascar

Usage: container or above-ground planter; near a deck or patio; specimen; street tree

Creating a striking landscape effect wherever it is used, Screw-Pine has a pyramidal, sometimes irregular, open, but much-branched silhouette. The smooth, stout trunks are topped with full, graceful heads of thin leaves, three feet long and three inches wide, emerging spirally from stubby branches. It is not a true pine tree. The blue-green foliage color adds to the striking nature of this exotic tree. The leaves are edged with small red spines and are used to make mats and baskets in the tropics. Branches have prominent leaf scars which encircle the stems. Large brace-roots emerge from the trunk several feet above the ground, helping to support the plant. Screw-Pine is capable of reaching 60 feet in height, but is not usually seen over 25 feet in zones 10 and 11, with a spread of 15 feet. Growth rate is slow to moderate, depending upon fertilization and watering schedules, and Screw-Pine is very popular for use as a specimen or container planting.

Although the male plants have conspicuous, fragrant flowers, the female plant is preferred for landscape use because of the large, globular fruits that hang from thick cords. The fruits are made up of 100 to 200 tightly compressed drupes, similar to those of a pineapple, and change from green to yellow when ripe. Each contains only a small amount of edible pulp, but the fruits are quite showy.

Screw-Pine produces fruit when grown in full sun, but young plants may be kept in the shade. Soil should be well-drained while plants are establishing and plants kept well-watered. Screw-Pine may be considered messy due to constant leaf drop throughout the year. No limitations of major concern except scales.

Parkinsonia aculeata
Jerusalem-Thorn, Palo Verde, Retama
Pronunciation: par-kin-SO-nee-ah ah-coo-lee-A-tah

ZONES: 8B, 9A, 9B, 10A, 10B, 11

HEIGHT: 15 to 20 feet

WIDTH: 20 to 25 feet

FRUIT: pod; 3 to 6"; dry; brown; some litter; persistent; showy

GROWTH RATE: rapid-growing; short-lived

HABIT: spreading vase; open in humid climates, denser in dry climates; irregular when young, symmetrical later; very fine texture

FAMILY: *Leguminosae*

LEAF COLOR: light green

LEAF DESCRIPTION: alternate; pinnately compound; entire margins; oblanceolate; deciduous; leaflets less than 2" long

FLOWER DESCRIPTION: yellow; slight pleasant fragrance; very showy; spring and summer

EXPOSURE/CULTURE:

Light Requirement: grows best in full sun

Soil Tolerances: all textures; alkaline to acidic; salt and drought

Aerosol Salt Tolerance: high

PEST PROBLEMS: resistant

BARK/STEMS:

Bark is thin and easily damaged by mechanical impact; branches droop as tree grows and will require pruning for vehicular or pedestrian clearance beneath canopy; routinely grown with, or trainable to be grown with, multiple trunks; should be trained to a short central leader for planting close to walks and pavement; showy trunk; thorns are present on trunk or branches

Pruning Requirement: requires regular pruning to develop upright, strong structure, direct growth on low branches, and remove dead wood

Limb Breakage: resistant

Current Year Twig: green; medium; slender

LANDSCAPE NOTES:

Origin: Mexico and southwest United States

Usage: parking lot island; buffer strip; highway; specimen; street tree (with regular pruning); urban tolerant

Pendulous, delicate leaflets, a low-branching growth habit, and a profusion of small, but very striking, slightly fragrant, yellow blooms combine to make this a popular, small landscape tree. *P. texana* (zone 8B) is smaller and more drought tolerant. Open-grown trees are beautiful if left unpruned, forming a fountain of fine texture. Branch bark often remains bright green even on several-year-old limbs. The stems are armed with short, sharp spines and the trees should be located where they will not injure passersby.

Trees have a short life of approximately 15 to 40 years. Poor drainage may account for short life on many sites, so locate it in areas where water does not pond after rain. The bark is thin and easily injured and the tree appears to compartmentalize decay poorly. Locate the tree properly and design and maintain the site to minimize trunk injury.

Adapted to arid regions, Jerusalem-Thorn is one of the best choices for hot, dry locations and its salt tolerance makes it ideal for seaside plantings. The light shade afforded by the fine-textured foliage allows lawns to thrive beneath this tree. Be sure to purchase trees with a six- to eight-foot clear trunk and upright branches for street and parking lot plantings to allow passage of vehicles. This can be a tough chore because branches weep toward the ground as they spread from the tree.

Limitations include scales and thorn bugs, but these are not serious. Root rot may occur on wet soils. Witches' broom occasionally causes a proliferation of branches forming tight heads of foliage. Sharp spines. Seeds germinate in the landscape. Foliage drops in the western United States in dry weather unless irrigation is provided.

Parrotia persica Persian Parrotia

Pronunciation: parr-ROW-tee-ah PER-seh-cuh

ZONES: 5A, 5B, 6A, 6B, 7A, 7B, 8A, (8B)

HEIGHT: 20 to 40 feet

WIDTH: 20 to 40 feet

FRUIT: irregular; less than 0.5"; dry; brown; not abundant; no significant litter

GROWTH RATE: slow-growing; moderately long life span

HABIT: upright oval, eventually rounded; moderate density; symmetrical; medium texture

FAMILY: *Hamamelidaceae*

LEAF COLOR: reddish-purple when young, dark green later; brilliant orange, red, or yellow fall color

LEAF DESCRIPTION: alternate; simple; crenate, dentate or serrate margins; oblong to obovate; deciduous; 2 to 5" long

FLOWER DESCRIPTION: red; somewhat showy; spring

EXPOSURE/CULTURE:

Light Requirement: part shade to full sun

Soil Tolerances: all textures; slightly alkaline to acidic; drought

PEST PROBLEMS: except for Japanese beetles, nearly pest-free

BARK/STEMS:

Bark is thin and easily damaged by mechanical impact; routinely grown with, or trainable to be grown with, multiple trunks; can be trained to one central trunk; branches grow mostly upright and will not droop; showy trunk; no thorns

Pruning Requirement: needs little or no pruning to develop strong trunk and branch structure

Limb Breakage: resistant

Current Year Twig: brown; slender

CULTIVARS AND SPECIES:

'Pendula' reportedly forms a rounded, weeping silhouette, 5 to 6 feet high by 10 feet wide, but is rare in the trade.

LANDSCAPE NOTES:

Origin: Iran

Usage: container or above-ground planter; buffer strip; highway; near a deck or patio; shade tree; specimen; street tree; urban tolerant

This deciduous tree forms a low-branched, oval to rounded silhouette, usually 15 to 30 feet wide, and often has multiple trunks, although it can be trained to a single trunk. The flowers, which appear before the leaves in spring, are somewhat interesting, showing no petals but a profusion of relatively inconspicuous deep crimson stamens. Fruits are not set in abundance and are of little consequence. The foliage of Persian Parrotia attracts the most attention, unfolding as reddish-purple young leaves, maturing to a lustrous, dark green through the summer, and then finally putting on a brilliant fall display of various hues of vivid yellow, burnt orange, and deep, pure scarlet. Even in winter Persian Parrotia is a striking landscape element, as the much-branched canopy and multiple trunks will finally be able to clearly display their attractive peeling bark and spectacular form. Trunk and bark character can be displayed year-round by removing or thinning lower branches and foliage.

These ornamental characteristics and a pest-free nature make Persian Parrotia ideal for use as a specimen or street tree. Accent the tree in a landscape by setting it off by itself in a lawn or in a bed of low groundcover. Space 20 to 30 feet apart along a street or walk to create a canopy over the walk. It will not canopy over the street, but will form a wall of wonderful foliage along the sides of a residential street. This tree should be grown and planted more.

Persian Parrotia should be grown in full sun or partial shade on well-drained, slightly acid soil, and will adapt to alkaline soil provided other cultural requirements are met. Trees will not tolerate wet soil conditions, but should show considerable drought tolerance once established.

Paulownia tomentosa Princess-Tree, Royal Paulownia

Pronunciation: paw-LOW-nee-ah toe-men-TOE-sah

ZONES: (5B), 6A, 6B, 7A, 7B, 8A, 8B, 9A, 9B

HEIGHT: 50 feet

WIDTH: 50 feet

FRUIT: oval; 1 to 2"; dry; brown; causes some litter; persistent; showy

GROWTH RATE: rapid-growing; moderate life span

HABIT: round; moderate density; irregular silhouette; coarse texture

FAMILY: *Scrophulariaceae*

LEAF COLOR: medium green; no fall color change

LEAF DESCRIPTION: opposite; simple; entire margin; ovate; deciduous; 5 to 10" long; persistent and messy

FLOWER DESCRIPTION: lavender; very showy; spring

EXPOSURE/CULTURE:

Light Requirement: grows best in full sun

Soil Tolerances: all textures; slightly alkaline to acidic; wet soil; salt; drought

Aerosol Salt Tolerance: moderate

PEST PROBLEMS: nearly pest-free

BARK/STEMS:

Bark is thin and easily damaged by mechanical impact; branches on young trees droop as tree grows and will require pruning for vehicular or pedestrian clearance beneath canopy; should be grown with a single leader; no thorns

Pruning Requirement: requires pruning to develop strong structure; prune branches when small to avoid trunk decay from large pruning wounds

Limb Breakage: susceptible to breakage at crotches with included bark

Current Year Twig: brown; thick

LANDSCAPE NOTES:

Origin: Asia

Usage: reclamation; highly prized lumber tree; urban tolerant

This native of China gives a most dramatic, coarse-textured appearance, with its huge heart-shaped leaves and large clusters of lavender flowers in the spring. Flowers are borne before leaf emergence, so they stand out nicely, especially against an evergreen background. With a rapid growth rate, Princess-Tree may reach 50 feet in height, with an equal spread, in an open landscape, though most trees are seen 30 to 40 feet tall and wide. Fuzzy, brown flower buds form in early autumn, persist over the winter, and bloom in early spring. Buds may freeze in very cold weather and drop off. Woody seed capsules, containing up to 2,000 seeds, form in autumn and persist on the tree through the winter. Seeds germinate readily in the landscape and wherever they are carried. Leaves drop periodically throughout autumn and within one week following the first frost in autumn.

Close-grained Paulownia wood has become extremely valuable during the last 15 years. It may be the highest priced saw timber in the United States at this time. It is exported to Japan where it is milled into furniture and jewelry boxes. The wood is very lightweight and makes good crating material. It has become naturalized in many parts of the South.

Princess-Tree should be planted where falling flowers and leaves are not objectionable. In some areas, the tree is considered a "weed" tree, and has naturalized in the edge of woodlands. If it is planted, consider placing it in a park or other open-space area.

Aggressive and invasive habit. There have been occasional reports of problems with mildew, leaf spot, and twig canker.

Peltophorum pterocarpum (inerme)

Yellow Poinciana

Pronunciation: pell-to-FOR-um tair-o-CAR-pum

ZONES: (9B), 10A, 10B, 11

HEIGHT: 45 to 50 feet

WIDTH: 35 to 45 feet

FRUIT: elongated pod; 3 to 4"; dry; purple to red; causes significant litter; showy

GROWTH RATE: rapid-growing; moderate life span

HABIT: rounded vase; open; irregular silhouette; symmetrical; fine texture

FAMILY: *Leguminosae*

LEAF COLOR: dark green; semievergreen

LEAF DESCRIPTION: alternate; bipinnately compound; entire margin; oblong; leaflets less than 2" long

FLOWER DESCRIPTION: yellow; pleasant fragrance; showy; summer

EXPOSURE/CULTURE:

Light Requirement: grows and flowers best in full sun

Soil Tolerances: all textures; alkaline to acidic; occasionally wet; moderate salt; drought

Aerosol Salt Tolerance: moderate

PEST PROBLEMS: resistant

BARK/STEMS:

Usually grown with multiple trunks, but should be trained to one central leader as far up into canopy as possible; main branches grow mostly upright and will not droop; no thorns

Pruning Requirement: requires pruning to develop strong structure and a central leader

Limb Breakage: susceptible to breakage at crotch due to included bark; wood itself is weak and tends to break

Current Year Twig: brown; medium thickness

CULTIVARS AND SPECIES:

Peltophorum dubium is cold-hardy to Central Florida (zone 9B) and has slightly smaller leaflets and fruit.

LANDSCAPE NOTES:

Origin: Australia to India

Usage: buffer strip; highway; reclamation; shade tree; specimen

The form can be quite variable from tree to tree, unfortunately eliminating this plant from the palette of many architects. With proper training in the nursery and in the landscape, a more uniform crown will develop. The delicate, feathery leaflets provide a softening effect for the tree's large size and create a welcoming, dappled shade. From May through September, the entire canopy is smothered with a yellow blanket of flowers, appearing in showy, terminal panicles and exuding a delicious, grape-like perfume. These flower clusters are followed by long seed pods that ripen to a brilliant, dark, wine-red.

Yellow Poinciana is a wonderful shade or specimen tree for a large landscape, especially when in full bloom, and it can make a street tree as long as it receives regular pruning to maintain a leader, minimize formation of included bark, and control its weedy, somewhat unkempt habit. Its large size makes it a natural for the wide-open spaces of large lawns or city parks. Not much litter is produced from its tiny leaflets when they fall in winter.

Trunks or branches of multi-trunked trees should be well spaced along a central stem and not allowed to grow larger than half the diameter of the main stem. This will increase wind hardiness. Thinning can also help prevent wind damage to the canopy. Locate the tree about 10 feet from sidewalks or pavement so the large surface roots do not cause damage. Plant only single-trunked trees along streets and other public areas to ensure formation of a durable plant.

A fast-growing tree, Yellow Poinciana grows best in full sun on any well-drained soil. Temperatures in the high 20s cause the leaves to drop, but these are quickly replaced. Even though Yellow Poinciana will develop a very large trunk, it is susceptible in shallow soils to being blown over during severe wind storms.

Persea americana Avocado

Pronunciation: PER-see-ah ah-mare-eh-CAY-nah

ZONES: 9B, 10A, 10B, 11

HEIGHT: 35 to 40 feet

WIDTH: 25 to 40 feet

FRUIT: oval; 3 to 6"; fleshy; green to purple; edible; causes significant litter; persistent; showy

GROWTH RATE: moderate to rapid; moderate life span

HABIT: round; dense; symmetrical; coarse texture

FAMILY: *Lauraceae*

LEAF COLOR: broadleaf evergreen; dark green

LEAF DESCRIPTION: alternate; simple; entire margin; oval; 4 to 8" long

FLOWER DESCRIPTION: green; somewhat showy; winter and spring

EXPOSURE/CULTURE:

Light Requirement: part shade to full sun

Soil Tolerances: all textures; alkaline to acidic; moderate drought

Aerosol Salt Tolerance: none

PEST PROBLEMS: somewhat susceptible

BARK/STEMS:

Branches droop; should be grown with a short, single leader; no thorns

Pruning Requirement: requires pruning to develop strong structure; prevent formation of included bark by thinning major branches to keep them less than half the diameter of trunk

Limb Breakage: susceptible to breakage at crotch due to poor collar formation; wood itself is brittle

Current Year Twig: green; moderate thickness

CULTIVARS AND SPECIES:

Some of the many cultivars available for variety of fruit production and season are: 'Lula', 'Tonnage', 'Taylor', 'Booth 7', 'Booth 8', 'Pollack', 'Trapp', 'Walden', 'Linda', and 'Itzamna'.

LANDSCAPE NOTES:

Origin: tropical America

Usage: fruit tree; shade tree; specimen

The large, lustrous, evergreen leaves and low-branching canopy of Avocado make it a wonderful shade tree, but it is most often grown for the abundant production of its well-known, delicious, buttery fruits. Depending on cultivar and variety, the fruits may vary from smooth-skinned to rough, and yellow-green or green to purple. Commonly seen at 35 to 40 feet in height, but capable of growing much larger, Avocado fits well into large residential landscapes in frost-protected locations. It can be pruned to an open spreading form or left to grow tall forming a rather narrow oval. The somewhat showy, greenish flowers appear on terminal panicles in late winter to early spring and are followed by the large, pendulous, pear-shaped fruits, which ripen late summer to early spring, depending upon cultivar.

Avocado trees grow quickly in either full sun or light shade on any well-drained soil. Trees should be watered regularly until established and later during droughts. A forest tree in its native habitat, Avocado responds well to a thick mulch and periodic fertilization. Lawn grasses should be kept away from the trunk. The brittle wood of Avocado trees is subject to storm damage when trees grow taller than 40 feet in the open.

Mites and scale infestations can become quite serious in local areas. Roots rot on poorly drained soils. Leaf-spotting diseases can be troublesome.

Persea borbonia Redbay

Pronunciation: PER-see-ah bore-BONE-ee-ah

ZONES: 7B, 8A, 8B, 9A, 9B, 10A, 10B, 11

HEIGHT: 30 to 50 feet

WIDTH: 30 to 50 feet

FRUIT: oval; round; less than 0.5"; fleshy; blue; attracts wildlife; causes some messy litter; persistent; somewhat showy

GROWTH RATE: moderate; moderate life span

HABIT: oval; round; dense; nearly symmetrical; medium texture

FAMILY: *Lauraceae*

LEAF COLOR: broadleaf evergreen; medium green

LEAF DESCRIPTION: alternate; simple; entire; oblong; 2 to 6" long; browsed by deer and bear

FLOWER DESCRIPTION: green; not showy; spring

EXPOSURE/CULTURE:

Light Requirement: part shade to full sun

Soil Tolerances: all textures; alkaline to acidic; occasionally wet soil; salt and drought

Aerosol Salt Tolerance: very high

PEST PROBLEMS: resistant

BARK/STEMS:

Branches droop as tree grows and will require pruning for vehicular or pedestrian clearance beneath canopy; showy trunk; should be grown with a single leader as far up as possible; specimen trees can be trained to a low-branching, multi-trunked habit; no thorns

Pruning Requirement: requires pruning to develop strong structure; prune and train when young to avoid large pruning wounds later, as decay appears to spread quickly in trunk

Limb Breakage: limb wood decays and becomes weak on older trees possibly from old wounds

Current Year Twig: green; thick

LANDSCAPE NOTES:

Origin: Virginia, through Florida along the coast to Texas

Usage: buffer strips; highway; near a deck or patio; reclamation; shade tree; specimen; street tree

The glossy, leathery leaves emit a spicy fragrance when crushed. Inconspicuous, springtime flower clusters are followed by small, dark blue fruits that ripen in fall. The trunk bears very showy, ridged, red-brown bark and frequently branches low to the ground, forming a multi-stemmed habit. It can be pruned to make a single, central leader, which would be most suitable for many urban planting sites.

Thriving on little care in full sun or partial shade, Redbay can tolerate a wide range of soils, from hot and dry to wet and swampy. *P. palustris* tolerates wet soil very well. Pruning to keep lateral branches less than half the diameter of the trunk will increase the tree's longevity and help prevent branches from separating from the trunk.

The densely foliated, spreading branches create a lush, billowy, rounded canopy, making Redbay a wonderful shade tree. It can make a nice street tree planted on 20- to 25-foot centers. Birds love the fruit and often visit the tree, leaving their droppings on cars. Its ease of growth and neat, dense crown habit also make Redbay ideal for the low-maintenance and naturalized landscape. The dark brown, furrowed bark is particularly attractive on older specimens. Seeds germinate readily in the landscape.

Removing large-diameter branches could initiate decay in the trunk. Prune early and at regular intervals to avoid large pruning wounds. Occasionally twig die-back is caused by a boring insect that enters small twigs. This is only a nuisance on large trees and control is not needed, but infested small trees in a nursery can become unsightly. Galls formed by the magnolia psyllid regularly form on the leaves, but are merely a cosmetic problem.

Phellodendron amurense Amur Corktree

Pronunciation: fell-o-DEN-dron ah-more-EN-see

ZONES: 3B, 4A, 4B, 5A, 5B, 6A, 6B, 7A, 7B, 8A, 8B

HEIGHT: 30 to 40 feet

WIDTH: 40 to 60 feet

FRUIT: on females only; oval; 0.5"; fleshy; green turning black; attracts wildlife; causes messy litter; showy

GROWTH RATE: moderate; long-lived

HABIT: rounded, spreading vase (males are more upright); open to moderately dense; irregular when young, symmetrical later; medium texture

FAMILY: *Rutaceae*

LEAF COLOR: dark green; copper or yellow showy fall color

LEAF DESCRIPTION: opposite; odd pinnately compound; entire or undulate margins; ovate; deciduous; 2 to 4" long leaflets

FLOWER DESCRIPTION: greenish-white; not showy; spring

EXPOSURE/CULTURE:

Light Requirement: grows best in full sun

Soil Tolerances: all textures; alkaline to acidic; occasionally wet soil; drought

Aerosol Salt Tolerance: moderate

PEST PROBLEMS: pest-free

BARK/STEMS:

Showy, ridged bark; branches droop on young trees and will require pruning for vehicular or pedestrian clearance beneath canopy; routinely grown with multiple trunks and clustered branches, but should be grown with branches spaced along a single trunk; no thorns

Pruning Requirement: requires pruning to develop strong structure and space major limbs along trunk; prevent formation of included bark

Limb Breakage: resistant, provided branches are spaced along trunk

Current Year Twig: brown; gray; thick

CULTIVARS AND SPECIES:

The fruitless (male) cultivar 'Macho' should be planted instead of the species due to its less spreading and less drooping habit. This makes it more suited for street tree planting. Little pruning is needed after major branches are spaced along a central trunk so they become well secured to the trunk. Allow for adequate soil space for root development so that roots do not lift sidewalks and pavement. 'Shademaster' is also fruitless and available in the trade.

LANDSCAPE NOTES:

Origin: China, Japan, Manchuria

Usage: buffer strip; highway; reclamation; shade tree; specimen; residential street tree

Growing to mature height with an equal or greater spread, Amur Corktree has a short trunk and an open, rounded, spreading canopy, which makes it ideal as a durable shade tree. To avoid the potentially messy fruit, male clones can be used for street and parking lot planting. The deciduous, leaflets change to bronze and yellow in the fall before dropping. The black fruits give off a strong odor when crushed and create a mess on a sidewalk, so some people may find it more desirable to plant only male trees ('Macho' or 'Shademaster') that will not produce fruit. The species could naturalize, as seeds readily germinate in the landscape.

The attractive gray-brown bark is deeply ridged and furrowed, and on mature trees it takes on a corky texture. Branches are usually borne low on the trunk, droop, and grow horizontally, forming a spreading habit.

Amur Corktree tolerates wet soil and drought. It is most drought-tolerant on sites where roots are allowed to expand freely. Once highly recommended as a street and urban-tolerant tree, it does not appear to hold up to the rigors of city life under certain conditions, particularly restricted soil spaces.

Virtually pest-free, but recovers slowly from transplanting. Seeds itself into the landscape. Pruning is needed to space limbs along the trunk.

Phoenix canariensis Canary Island Date Palm

Pronunciation: FEE-nix cah-nair-ee-EN-sis

ZONES: 9A, 9B, 10A, 10B, 11

HEIGHT: 40 to 60 feet

WIDTH: 15 to 25 feet

FRUIT: round; 0.5 to 1"; fleshy; orange to yellow; showy on female trees

GROWTH RATE: slow-growing; long-lived

HABIT: palm; open; symmetrical; coarse texture

FAMILY: *Arecaceae*

LEAF COLOR: bright orange petioles; evergreen

LEAF DESCRIPTION: odd pinnately compound; entire margin; 12 to 18" long leaflets; fronds 8 to 12' long

FLOWER DESCRIPTION: white; not showy; spring and winter

EXPOSURE/CULTURE:

Light Requirement: grows best in full sun

Soil Tolerances: all textures; slightly alkaline to acidic; some salt; drought

Aerosol Salt Tolerance: moderate to high

PEST PROBLEMS: moderately susceptible to lethal yellowing

BARK/STEMS:

Showy with persistent frond bases; single, 3 to 4' thick trunk; long spines are present on petioles; no crown shaft

Pruning Requirement: remove dying and drooping older fronds; clean pruning tools with bleach after pruning to prevent spreading Fusarium wilt

Limb Breakage: very resistant

LANDSCAPE NOTES:

Origin: Canary Islands

Usage: parking lot island; buffer strip; specimen; sidewalk cutout; street tree

This large, stately palm often becomes too massive for most residential landscapes but, fortunately it is very slow-growing and will take a considerable amount of time to reach mature height. Canary Island Date Palm is most impressive, with its single, upright, thick trunk topped with a crown of 8- to 12-foot-long, stiff leaves with extremely sharp spines at their bases. The stalks of inconspicuous flowers are replaced with clusters of one-inch-diameter, orange-yellow, date-like, ornamental fruits that ripen in early summer. The trunk can reach a diameter of four feet and is covered with an attractive, diamond-shaped pattern from old leaf scars.

Canary Island Date Palm should be located in full sun on fertile, moist soil for best growth, but is tolerant of any well-drained soil. It can be planted on the inland side of coastal condominiums and large homes due to moderately high salt tolerance. It does well as a street or avenue tree, even in confined soil spaces. Canary Island Date Palm will require pruning to remove old fronds. Older leaves frequently become chlorotic from magnesium or potassium deficiency in soils with a pH above 7. Preventive applications of appropriate fertilizer helps avoid this. Avoid damage to the trunk by locating it properly in the landscape and keeping landscape maintenance equipment away. Only prune fronds that hang below the horizontal. Do not remove those growing upright, as this may slow the growth and reduce vigor.

Giant palm weevil can kill recently transplanted palms, especially those with small root balls or those which are injured. Preventing injury is the best way to avoid the weevil since control is not practical. Some landscape managers conduct a preventive spray program following transplanting on these highly valued palms until they are well established in the landscape. Palm leaf skeletonizer devours leaves. Mildly susceptible to lethal yellowing disease and leaf spot. Stressed and damaged trees often are infected with the Ganoderma fungus (butt rot). Remove the conk that forms and remove the tree to help control the spread of disease. Be sure sprinklers do not irrigate the trunk so it remains wet; a wet trunk and wet soil encourage this disease. There is no control for butt rot, only prevention. Fusarium wilt can kill this palm; it is spread by pruning tools.

Phoenix dactylifera Date Palm

Pronunciation: FEE-nix dac-tah-LIFF-er-ah

ZONES: (8B), 9A, 9B, 10A, 10B, 11

HEIGHT: 50 to 80 feet

WIDTH: 15 feet

FRUIT: oval; 0.5 to 1"; fleshy; orange to yellow; causes significant litter beneath female trees; showy; rarely fruits in Florida

GROWTH RATE: slow-growing; long-lived

HABIT: palm; open; symmetrical; coarse texture

FAMILY: *Arecaceae*

LEAF COLOR: evergreen; blue-green

LEAF DESCRIPTION: pinnately compound; entire margin; 12 to 18" long leaflets; 6 to 10' long fronds; spiny petioles

FLOWER DESCRIPTION: white; not showy; spring and winter

EXPOSURE/CULTURE:

Light Requirement: grows best in full sun

Soil Tolerances: all textures; slightly alkaline to acidic; drought

Aerosol Salt Tolerance: moderate to high

PEST PROBLEMS: slightly susceptible to lethal yellowing

BARK/STEMS:

Showy with persistent frond bases; usually develops several trunks, but often pruned to a single trunk by removing suckers; long spines are present on petioles; no crown shaft

Pruning Requirement: remove dying and drooping older fronds; clean pruning tools following pruning

Limb Breakage: mostly resistant

CULTIVARS AND SPECIES:

'Deglet Noor', 'Medjool', and 'Zahedi' are the most common cultivars.

LANDSCAPE NOTES:

Origin: north Africa

Usage: parking lot island; buffer strip; specimen; sidewalk cutout; street tree; interiorscapes

This large, stately palm often becomes too massive for most residential landscapes. It is very slow-growing and will take a considerable amount of time to reach mature height. Date Palm is most impressive with its single, upright, moderately thick trunk topped with a crown of 8- to 10-foot-long, stiff leaves with extremely sharp spines at their bases. The stalks of inconspicuous flowers are replaced with clusters of one-inch-diameter, edible fruits, which ripen in early summer. The trunk can reach a diameter of three feet and is covered with an attractive, diamond-shaped pattern from old leaf scars.

Date Palm should be grown in full sun on fertile, moist soil for best growth, but is tolerant of any well-drained soil. It can be planted on the inland side of coastal condominiums and large homes due to moderate salt tolerance. It does well as a street or avenue tree, even in confined soil spaces, provided the soil pH is less than 8. Older leaves can become chlorotic from magnesium or potassium deficiency in soils with a pH above 7. Preventive applications of appropriate fertilizer help avoid this.

Avoid damage to the trunk by locating the tree properly in the landscape and by keeping landscape maintenance equipment away. Trees with trunk damage are susceptible to Ganoderma rot. Date Palm will require pruning to remove old fronds. Remove only those that droop; do not remove those growing upright, as this may slow growth and reduce vigor. Cleaning pruning tools in a solution of bleach and water may help prevent spread of Fusarium wilt, a deadly vascular disease.

See *P. canariensis* for limitations.

Phoenix reclinata Senegal Date Palm

Pronunciation: FEE-nix rek-linn-A-tah

ZONES: 9B, 10A, 10B, 11

HEIGHT: 25 to 35 feet

WIDTH: 12 to 20 feet

FRUIT: round; 0.5 to 1"; fleshy; orange; no significant litter; showy

GROWTH RATE: moderate; moderate life span

HABIT: palm; open; irregular silhouette; medium texture

FAMILY: *Arecaceae*

LEAF DESCRIPTION: odd pinnately compound; entire margin; 12 to 18" long leaflets

FLOWER DESCRIPTION: white; showy; spring and summer

EXPOSURE/CULTURE:

Light Requirement: part shade to full sun

Soil Tolerances: all textures; slightly alkaline to acidic; moderate drought

Aerosol Salt Tolerance: moderate

PEST PROBLEMS: resistant

BARK/STEMS:

Multiple trunks grow mostly upright and will not droop; showy trunk with persistent leaf bases; thorns are present on base of petioles; no crown shaft

Pruning Requirement: remove dead and dying fronds; remove suckers from base of palm if desired

Limb Breakage: resistant

LANDSCAPE NOTES:

Origin: Africa

Usage: container or above-ground planter; specimen

This striking palm creates an interesting silhouette with its multiple, gracefully curved, often reclining, slender brown trunks, and dense crowns of stiff but feathery leaf fronds. Old frond bases are medium brown and remain on the trunk, forming a showy trunk that is attractive all year long. A mature specimen of Senegal Date Palm is a striking tree and casts a light shade. The palm is elegant when lighted from below at night. The somewhat showy flower stalks, often lost within the thick foliage, are followed by one-inch-long, bright orange dates that are incredibly showy. These can be very attractive, particularly when viewed from a balcony above the tree.

The multiple trunks lose older fronds as the palm grows, clearing lower trunks of all foliage. This characteristic makes Senegal Date Palm a wonderful tree for accenting in a bed of groundcover or a grouping of low shrubs. Trunks that bend as the palm ages may need to be supported with a brace or cable. This palm is best used as an accent for large landscapes and parks.

Growing easily in full sun or partial shade, Senegal Date Palm will thrive on any well-drained soil, but shows nutrient deficiency symptoms in high pH soil. Plants should receive adequate moisture during periods of drought and a regular fertilization program should be established. This palm is too large for all but the largest residential landscapes. They are very costly to purchase, due to the slow growth rate—large specimens command a high price.

A variety of scales infest this palm. Somewhat susceptible to lethal yellowing disease and leaf spot. Stressed and damaged trees often are infected with the Ganoderma fungus (butt rot). A conk is formed at the base of the tree which appears as a varnished shelf or mushroom. Remove the conk and the tree to help control the spread of the disease to other plants. Prevent injury to the trunk and roots and plant in well-drained soil. Be sure sprinklers do not irrigate the trunk so that it remains wet. A wet trunk and wet soil encourage this disease. There is no control for butt rot, only prevention. Seedlings have been spotted growing in woodlands in Florida.

Phoenix roebelenii Pygmy Date Palm

Pronunciation: FEE-nix row-beh-LEE-nee-eye

ZONES: 10A, 10B, 11

HEIGHT: 6 to 12 feet

WIDTH: 6 to 8 feet

FRUIT: round; 0.5 to 1"; fleshy; red to black; no significant litter

GROWTH RATE: slow-growing; moderate life span

HABIT: palm; open; irregular; fine texture

FAMILY: *Arecaceae*

LEAF COLOR: evergreen; glossy green

LEAF DESCRIPTION: odd pinnately compound; entire margin; 8 to 12" long leaflets

FLOWER DESCRIPTION: white; not showy; spring

EXPOSURE/CULTURE:

Light Requirement: grows best in part shade, but tolerates full sun

Soil Tolerances: all textures; slightly alkaline to acidic; well-drained; moderate drought

Aerosol Salt Tolerance: low to moderate

PEST PROBLEMS: somewhat susceptible

BARK/STEMS:

Showy trunk; forms one trunk; spines are present on petioles; no crown shaft

Pruning Requirement: remove dead fronds

Limb Breakage: resistant

LANDSCAPE NOTES:

Origin: southeast Asia

Usage: container or above-ground planter; near a deck or patio; specimen; interiorscape

One of the finest of the dwarf palms, Pygmy Date Palm slowly reaches mature height with an upright or curving, single trunk topped with a dense, full crown of gracefully arching, three- to four-foot-long leaves. The insignificant flower clusters, hidden by the foliage, are present periodically throughout the year and produce small, jet-black dates that ripen to a deep red. Pygmy Date Palm is quite popular as a specimen planting or in containers, especially attractive at poolside. It is often used as a single specimen, although it is also effective in groups of three or more.

Many nurseries grow the palm with several trunks of differing heights planted in the same container. This form makes a nice specimen, accenting a particular area of the landscape. Only low-growing groundcovers and shrubs should be planted at the base of this palm. Those growing more than two feet tall may all but hide this small, slow-growing palm.

Pygmy Date Palm should be grown only in frost-free areas, in sun or shade, on well-drained soils and regularly watered. Magnesium or potassium deficiency symptoms (chlorotic and spotted fronds) often develop on older leaves when grown in soils with a pH above 7. It has some salt tolerance, surviving on the inland side of coastal condominiums.

Palm leaf skeletonizer and a large number of scales can infest this palm. In soils with a pH above 7, small, deformed leaves develop due to manganese deficiency. Leaf spot and bud rot can also disfigure the plam.

Photinia x fraseri Fraser Photinia, Redtip

Pronunciation: foe-TIN-ee-ah x FRAY-zur-eye

ZONES: 7B, 8A, 8B, 9A, 9B

HEIGHT: 12 to 18 feet

WIDTH: 8 to 12 feet

FRUIT: round; less than 0.5"; fleshy; red; not especially showy; no significant litter

GROWTH RATE: moderate; short-lived

HABIT: oval; upright; dense; symmetrical; medium texture

FAMILY: *Rosaceae*

LEAF COLOR: emerging purple or red, turning green later; broadleaf evergreen

LEAF DESCRIPTION: alternate; simple; serrate margins; oval to oblong or obovate; 3 to 8" long

FLOWER DESCRIPTION: white; very showy; late spring to summer

EXPOSURE/CULTURE:

Light Requirement: grows best in full sun

Soil Tolerances: all textures; slightly alkaline to acidic; well-drained soil is essential; salt-sensitive; drought

PEST PROBLEMS: susceptible

BARK/STEMS:

Bark is thin and easily damaged by mechanical impact; routinely grown with, or trainable to be grown with, multiple trunks; can be trained to grow with a short, single trunk; no thorns; branches grow mostly upright and will not droop

Pruning Requirement: requires pruning to develop tree form

Limb Breakage: somewhat susceptible

Current Year Twig: reddish; medium; slender

LANDSCAPE NOTES:

Origin: hybrid

Usage: container or above-ground planter; hedge; buffer strip; highway; near a deck or patio; standard; urban tolerant

Fraser Photinia forms an upright silhouette of glossy, evergreen leaves. One of its most striking features is the burgundy-red new foliage, which contrasts nicely against the dark green mature foliage. Most people do not see the showy white flower clusters borne at the ends of the branches in the summer because they regularly prune the new growth. Too bad, because it is really quite attractive in flower, as the entire canopy fills with white for two or three weeks.

Ideal for use as a clipped hedge or screen (although the bottom may be on the thin side), Fraser Photinia produces some attractive new red growth all throughout the growing season in zones 8 and 9. Unfortunately, everyone has one or more of the disease-prone plants trained as a hedge. Try it as a tree for something different. Plants trained as trees have less leaf spot disease. *P. serrulata* makes an even better tree form. It is best used as a tall screen in the full sun, but is often misused as a foundation plant in front of the house. It grows fine with some shade, but leaf spot is sure to follow, causing defoliation.

Fraser Photinia needs well-drained soil and a full-sun location. Leaves often become infected with leaf spot fungi when grown in shade or when the leaves remain too moist. Plants grow at a moderate rate and tolerate pruning very well, although the bottom of the plant often thins when clipped into a hedge. Micronutrient problems occur on alkaline soil, although plants continue to grow. They tolerate heat well and are suited for exposed sites like parking lots and median strips in highways.

Caterpillars, mites, scales, and European fruit-tip moth can be found on Photinia, but are often of little consequence. Fire blight can kill the plant rather quickly, and leaf-spot diseases are very serious and common. Root rot can kill plants, particularly those in wet soils. This is not a pest-free plant. It is probably best suited for dry climates, or for highway plantings and other areas that receive little irrigation and bright sun all day.

Photinia serrulata Chinese Photinia

Pronunciation: foe-TIN-ee-ah sair-u-LAY-tah

ZONES: 7A, 7B, 8A, 8B, 9A, 9B

HEIGHT: 15 to 25 feet

WIDTH: 15 to 25 feet

FRUIT: round; 0.25"; fleshy; red; no significant litter; persistent; showy

GROWTH RATE: moderate; moderate life span

HABIT: rounded vase; dense if not pruned; symmetrical; medium texture

FAMILY: *Rosaceae*

LEAF COLOR: broadleaf evergreen; dark green

LEAF DESCRIPTION: alternate; simple; serrate margin; lanceolate to oblong; 4 to 6" long

FLOWER DESCRIPTION: white; showy; late spring to summer; unpleasant fragrance

EXPOSURE/CULTURE:

Light Requirement: grows best in full sun

Soil Tolerances: all textures; alkaline to acidic; drought

Aerosol Salt Tolerance: none

PEST PROBLEMS: somewhat susceptible

BARK/STEMS:

Bark is thin and easily damaged by mechanical impact; routinely grown with, or trainable to be grown with, multiple trunks; can be trained to a short, single trunk; branches grow mostly upright and will not droop; no thorns

Pruning Requirement: needs little pruning to develop strong structure

Limb Breakage: resistant

Current Year Twig: brownish-red; green; thick

CULTIVARS AND SPECIES:

'Aculeata' and 'Nova' reach about 10 feet in height. There is a hybrid between red-leaved Photinia (*P. glabra*) and Chinese Photinia (*P. serrulata*) which perhaps displays the best characteristics of both parents, called *P.* x *fraseri*. It is the most popular Photinia. *P. villosa* (zone 5B) has an upright form with beautiful orange fall foliage.

LANDSCAPE NOTES:

Origin: China

Usage: container or above-ground planter; tall screen; parking lot island; buffer strip; highway; near a deck or patio; specimen; sidewalk cutout; street tree; urban tolerant

Small, white flowers arranged in dense, showy, eight-inch-diameter clusters at branch tips and large, shiny, dark leaves combine to make Chinese Photinia an attractive evergreen for multiple landscape applications. An underused landscape tree with very showy flowers and bright red fruit. Often kept trimmed into a hedge shape, multi-trunked Chinese Photinia creates an ideal small tree with the proper training. When the lower branches are removed, the natural rounded canopy creates an attractive silhouette. New growth is tinged pink, though not as much as *Photinia glabra*. Plants may reach 40 feet in height under ideal conditions but are more often seen at 15 to 25 feet.

Nurseries occasionally produce trees with a single leader and market them as street trees. Chinese Photinia stays fairly small and grows fast when young, but slows down later as the crown broadens. Great for beneath power lines. Prune to distribute main branches along a central trunk to increase longevity.

Chinese Photinia needs well-drained soil and full sun. Leaves often become infected with leaf spot fungi when grown in shade or when the leaves remain too moist. Plants grow at a moderate rate and tolerate pruning very well, although the bottom of the plant often thins when clipped into a hedge. Photinia could be used more as a multi-stemmed specimen, street, or patio tree. It is available in the nursery trade and is used as a hedge or foundation plant—a usage that is quite inappropriate due to its large size and rapid growth rate.

Caterpillars, mites, scales, and European fruit-tip moth may attack. Usually disease-free except for leaf spot, which can be devastating. Fire blight, and mildew in shady locations, can be troublesome.

Picea abies Norway Spruce

Pronunciation: PIE-see-ah A-bees

ZONES: 2B, 3A, 3B, 4A, 4B, 5A, 5B, 6A, 6B, 7A

HEIGHT: 80 to 100 feet

WIDTH: 25 to 40 feet

FRUIT: elongated; oval; 3 to 5"; dry; brown; no significant litter; persistent; showy

GROWTH RATE: slow-growing; long-lived

HABIT: pyramidal; moderate density; symmetrical; fine texture

FAMILY: *Pinaceae*

LEAF COLOR: dark green; needle-leaf evergreen

LEAF DESCRIPTION: spiral; simple; up to 1" long

FLOWER DESCRIPTION: pink; not showy

EXPOSURE/CULTURE:

Light Requirement: grows best in full sun

Soil Tolerances: all textures; slightly alkaline to acidic; salt-sensitive; moderate drought

Aerosol Salt Tolerance: poor

PEST PROBLEMS: resistant

BARK/STEMS:

Branches droop as tree grows and will require pruning for vehicular or pedestrian clearance beneath canopy; normally grows with a single leader; no thorns

Pruning Requirement: needs little or no pruning to develop a straight trunk and good structure

Limb Breakage: resistant

Current Year Twig: brown; medium thickness

CULTIVARS AND SPECIES:

Some cultivars are dwarf and shrublike; others are trees. Not all will be available in nurseries. Cultivars include: 'Clanbrasiliana'—dwarf, about 4 feet tall and twice as wide; 'Columnaris'—narrow, columnar; 'Gregoryana'—rounded, broad, about 3 feet tall but much wider, slow-growing; 'Humulis'—about 2 feet tall; 'Inversa'— drooping habit; 'Maxwelli'—4 feet tall and 10 feet wide, slow-growing, dense; 'Nidiformis'—dwarf, very dense mound; 'Pendula'—weeping; 'Procumbens'—flat, dense, can be 3 feet tall; 'Pumila'—spreading, about 4 feet tall; 'Pygmea'—conical, slow-growing; 'Reflexa'—branchlets pendulous, 1 foot high but 10 feet wide; 'Repens'—flat and prostrate, less than 3 feet tall but quite wide; 'Stricta'—slender, spirelike, 40 to 50 feet tall, 8 feet wide.

LANDSCAPE NOTES:

Origin: Europe

Usage: specimen; screen

Norway Spruce can grow very tall, though some listed cultivars are shrublike. Small-diameter branches sweep horizontally from the straight trunk, which can grow to four feet thick. Crown geometry and texture often vary among the numerous forms of this tree. Branchlets droop from the branches toward the ground in a graceful, weeping fashion, forming a delicate pyramid. On very old specimens, the lower branches increase to 12 inches or more in diameter and the top becomes open. Many small-diameter roots originate from the base of the trunk, and they are often found fairly close to the surface of the soil. The root system is shallow and often dense, particularly close to the trunk, which makes growing grass difficult.

Norway Spruce is best used as a specimen in a lawn area or as a windbreak or screen, planted on 20-foot-centers. (Rockefeller Center in New York City erects a Norway Spruce each Christmas next to the skating rink and decorates it for the holiday season.)

Norway Spruce tolerates most soils if moist and transplants easily if balled-in-burlap or potted. Trees are much happier if they receive periodic irrigation, although they tolerate drought fairly well.

Mites often speckle the foliage with yellow in dry summers. Bagworms can defoliate portions of trees. Foliage diseases can be nuisance at times.

Picea abies 'Inversa' 'Inversa' Norway Spruce

Pronunciation: PIE-see-ah A-bees

ZONES: 2B, 3A, 3B, 4A, 4B, 5A, 5B, 6A, 6B, 7A

HEIGHT: depends on training

WIDTH: 10 to 40 feet

FRUIT: elongated; oval; 3 to 5" long; dry; brown; no significant litter; persistent; showy

GROWTH RATE: slow-growing; long-lived

HABIT: pyramidal; weeping; moderate density; irregular silhouette; fine texture

FAMILY: *Pinaceae*

LEAF COLOR: dark green; needle-leaf evergreen

LEAF DESCRIPTION: spiral; simple; about 1" long

FLOWER DESCRIPTION: pink; not showy

EXPOSURE/CULTURE:

Light Requirement: grows best in full sun

Soil Tolerances: all textures; slightly alkaline to acidic; moderate drought

PEST PROBLEMS: resistant

BARK/STEMS:

Branches droop onto ground as tree grows; usually trained to a single leader; no thorns

Pruning Requirement: requires staking and pruning to develop upright structure

Limb Breakage: resistant

Current Year Twig: brown; medium thickness

LANDSCAPE NOTES:

Origin: Europe

Usage: specimen

This cultivar of Norway Spruce can grow 30 to 50 feet tall and spread 25 to 40 feet, depending on how it is trained. Small-diameter branches sweep downward touching and growing along the ground as a groundcover if not trained into a tree. Branchlets droop toward the ground in a graceful, weeping fashion, forming a delicate pyramid. On very old specimens that have been trained with a central trunk, large, picturesque gaps form in the canopy. Many small-diameter roots originate from the base of the trunk and are often found fairly close to the surface of the soil. The root system is shallow and often dense, particularly close to the trunk, which makes growing grass difficult.

Weeping Norway Spruce is best used as a specimen in a lawn area or as a windbreak or screen, planted on 20-foot-centers. To display this tree properly, plan on leaving the area beneath the canopy free of turf and other plants to allow the lower branches to touch the ground.

The tree is often staked in the nursery to create a central trunk. Staking must be continued in the landscape if you want to grow a tall, tree-form specimen. A weeping, sprawling shrub forms without staking.

Norway Spruce tolerates most soils if moist and transplants easily if balled-in-burlap or potted. Trees are much happier if they receive periodic irrigation, although they tolerate drought well. Mites often infest the needles in prolonged dry weather, and in hot weather they can build to populations that require control. They can be a major problem in summer after hot, dry weather, especially near concrete, buildings, and other urban surfaces that reflect heat. The small insects cannot be readily seen with the naked eye. The first noticeable symptom is yellowing at the base of the oldest needles on infested branches. Close inspection with a magnifying glass will confirm the presence of mites.

Picea glauca White Spruce

Pronunciation: PIE-see-ah GLAW-cah

ZONES: 2A, 2B, 3A, 3B, 4A, 4B, 5A, 5B, 6A, 6B

HEIGHT: 40 to 60 feet

WIDTH: 10 to 20 feet

FRUIT: elongated; 1 to 3"; dry; brown; no significant litter; persistent

GROWTH RATE: moderate; long-lived

HABIT: columnar; pyramidal; dense; symmetrical; fine texture

FAMILY: *Pinaceae*

LEAF COLOR: needle-leaf evergreen; blue-green; silver-green; green

LEAF DESCRIPTION: spiral; simple; 0.75 to 1" long; blunt tips

FLOWER DESCRIPTION: reddish-yellow; not showy; spring

EXPOSURE/CULTURE:

Light Requirement: full sun

Soil Tolerances: all textures; acidic; some salt; occasional wetness; drought

Aerosol Salt Tolerance: low to moderate

PEST PROBLEMS: resistant

BARK/STEMS:

Branches droop as tree grows and will require pruning for vehicular or pedestrian clearance beneath canopy; normally grows with a single leader; no thorns

Pruning Requirement: needs little or no pruning to develop strong structure

Limb Breakage: resistant

Current Year Twig: brown; thick

CULTIVARS AND SPECIES:

The cultivar 'Conica' is a 10- to 15-foot-high dwarf form with soft, blue-green needles and is ideal for planting in a small, residential landscape. 'Pendula' is a pyramidal conifer with a weeping habit that resists snow loads. Staking young plants encourages upright growth, but in time, the plant stiffens. The subspecies or variety *densata* (*densa*) is the state tree of South Dakota. *P. englemannii* is native to the Rocky Mountains, has beautiful blue-green needles, and grows nicely on dry, high elevation sites.

LANDSCAPE NOTES:

Origin: Alaska to Labrador, South Dakota to Maine

Usage: specimen; screen; windbreak; buffer strip

This North American native tree has a pyramidal silhouette when young, but matures into a dense, tall column, 10 to 20 feet wide. The short, silver-green, or green needles densely clothe the horizontal branches, making White Spruce ideally suited for use as a windbreak. Small, 1- to 2.5-inch-long, light brown, pendulous cones decorate the branches throughout the year. The new layers of purplish-gray bark have a soft, silvery sheen that adds to the tree's attractiveness as a specimen planting.

Found on stream banks and lake shores in its native habitat, White Spruce should be grown in the landscape in full sun. Well-established trees are quite tolerant of wind, heat, cold, and drought, but they also tolerate wet soil. Trees develop an adventitious root layer close to the surface if subjected to wet soil after they are well established. Native Americans used the roots of White Spruce to lace canoes and baskets. Recent transplants do not withstand wet soil and often develop root rot.

Keep turf away from the base of the tree to reduce root competition for moisture, especially during the establishment period.

Mites, aphids, and bagworms are common. Roots rot on constantly wet soils.

Picea omorika Serbian Spruce

Pronunciation: PIE-see-ah o-more-EE-cah

ZONES: 4A, 4B, 5A, 5B, 6A, 6B, 7A, 7B

HEIGHT: 50 feet

WIDTH: 20 feet

FRUIT: elongated; oval; 1 to 2"; dry; brown; no significant litter; persistent

GROWTH RATE: slow-growing; moderate life span

HABIT: columnar; pyramidal; gracefully narrow at top; moderate density; symmetrical; fine texture

FAMILY: *Pinaceae*

LEAF COLOR: light green or blue-green; needle-leaf evergreen

LEAF DESCRIPTION: spiral; simple; 0.5 to 1" long

FLOWER DESCRIPTION: not showy

EXPOSURE/CULTURE:

Light Requirement: part shade to full sun

Soil Tolerances: all textures; alkaline to acidic; moderate drought

Aerosol Salt Tolerance: low to moderate

PEST PROBLEMS: resistant

BARK/STEMS:

Branches droop to ground as tree grows and will require pruning for pedestrian clearance beneath canopy; should be grown with a single leader; no thorns

Pruning Requirement: needs little or no pruning to develop strong structure

Limb Breakage: resistant

Current Year Twig: brown; medium thickness

CULTIVARS AND SPECIES:

'Pendula' has an elegant, weeping form and makes a beautiful specimen tree.

LANDSCAPE NOTES:

Origin: Europe

Usage: specimen; screen; protect from direct wind in northern sites

Serbian Spruce develops a narrow, concave-sided pyramidal silhouette with thin, gracefully arching branches. The slender trunk grows straight and the tree requires no pruning to keep the nearly symmetrical form. The upper surface of the needles is glossy; the lower surface is marked with two white stomata lines. The plant is one of the most elegant Spruces. Nice specimens can be found at botanical gardens and arboreta, but the tree may be difficult to locate at a local garden center. A search for the tree is worth the extra effort. Like Oriental Spruce, this is one of a few Spruces that stand out in a crowd.

This tree is best used as a specimen in a landscape bed or large, open lawn area, or as a screen. It can probably be planted as a windbreak in the middle and southern end of its hardiness range where winters are not so cold. Windy weather in the north appears to dry the needles in the winter. Some horticulturists have even suggested it as an evergreen street tree for narrow overhead spaces, but Serbian Spruce has not been tested as a street tree. It is apparently tolerant of urban conditions, probably more so than other Spruces.

Serbian Spruce tolerates full-day sun if the roots are allowed to extend into open soil. It also tolerates soils with a high pH.

Bagworms make a sack by webbing needles and debris together but are usually not serious. In northern areas, Spruce budworm larvae feed on developing buds and young needles on a number of Spruces. The yellowish-brown caterpillars are difficult to see. The Spruce needle miner makes a small hole in the base of a needle, then mines out the center. Dead needles are webbed together and can be found on infested twigs. Spider mites can be a problem in summer after hot dry weather. Spruce may be attacked by needle casts, causing needles to turn yellow or brown and drop off.

Picea orientalis Oriental Spruce

Pronunciation: PIE-see-ah or-ee-en-TAY-luss

ZONES: 5A, 5B, 6A, 6B, 7A, 7B

HEIGHT: 25 to 60 feet

WIDTH: 15 to 25 feet

FRUIT: elongated; oval; 3 to 5"; dry; brown; no significant litter; persistent; showy

GROWTH RATE: slow to moderate; long-lived

HABIT: pyramidal; dense; symmetrical; fine texture

FAMILY: *Pinaceae*

LEAF COLOR: evergreen; dark green

LEAF DESCRIPTION: spiral; simple; needles 0.5" long

FLOWER DESCRIPTION: red; not showy

EXPOSURE/CULTURE:

Light Requirement: part shade to full sun

Soil Tolerances: all textures; slightly alkaline to acidic; drought

PEST PROBLEMS: resistant

BARK/STEMS:

Branches droop as tree grows and will require pruning for pedestrian clearance beneath canopy; should be grown with a single leader; no thorns

Pruning Requirement: needs little or no pruning to develop strong structure

Limb Breakage: resistant

Current Year Twig: brown; medium thickness

CULTIVARS AND SPECIES:

'Aurea', 'Aurea-spicata', and 'Skylands'—new growth is yellow, gradually changes to green; 'Gowdy'—narrow columnar form, small green leaves; 'Gracilis'—small conical form, 15 to 20 feet tall, bright green needles; 'Pendula' ('Weeping Dwarf')—compact, pyramidal weeping form.

LANDSCAPE NOTES:

Origin: Asia Minor

Usage: specimen; screen

Soaring to 120 feet in its native habitat, Oriental Spruce is more often 25 to 40 feet tall in the landscape, growing slowly into a pyramidal silhouette that casts dense shade beneath. There are some specimens 50 to 60 feet tall in open landscapes, but these are rare. The horizontal branches on some trees bend downward slightly at the tips, and are generously clothed with short needles. Both male and female flowers are considered insignificant, although the male flowers resemble small red strawberries. The flowers are followed by two- to four-inch-long and one-inch-wide, reddish-purple cones that mature to a shiny brown. Unfortunately, the tree is not common in the industry, but it should be.

Best used as a specimen in protected landscapes, Oriental Spruce is a graceful addition to any yard. Leave plenty of room for lateral branch growth near the base of the tree, as it looks odd when lower branches are removed. The plant is a real standout among the Spruces. It can be planted in small landscapes and can look nice for a long time due to its slow growth rate. Watch for mite infestations during hot weather.

Oriental Spruce should be grown in full sun or partial shade on well-drained soil, and will tolerate infertile, even rocky soils. However, Oriental Spruce should only be used where winters are not extremely dry, and the plants should be located where they will not be exposed to harsh winter winds. Excessively dry, windy, winter weather can brown the foliage. Generous irrigation in the fall will help the tree pull through the winter. Mites, aphids, and bagworms are probably the most common pests, although this Spruce is reportedly less susceptible to mites than Norway or White Spruce.

Picea pungens Colorado Spruce, Blue Spruce

Pronunciation: PIE-see-ah PUN-jens

ZONES: 3A, 3B, 4A, 4B, 5A, 5B, 6A, 6B, 7A, 7B

HEIGHT: 30 to 50 feet

WIDTH: 10 to 20 feet

FRUIT: elongated; oval; 3 to 4"; dry; brown; not showy; no significant litter

GROWTH RATE: slow-growing; long-lived

HABIT: columnar; pyramidal; dense; symmetrical; fine texture

FAMILY: *Pinaceae*

LEAF COLOR: blue or blue-green; needle-leaf evergreen

LEAF DESCRIPTION: spiral; simple; 1 to 1.5" long

FLOWER DESCRIPTION: green; orange; purple; not showy; spring

EXPOSURE/CULTURE:

Light Requirement: part shade to full sun

Soil Tolerances: all textures; alkaline to acidic; occasionally wet; moderate salt and drought

Aerosol Salt Tolerance: moderate to high

PEST PROBLEMS: susceptible

BARK/STEMS:

Branches droop as tree grows and will require pruning for vehicular or pedestrian clearance beneath canopy; normally grown with a single leader; no thorns

Pruning Requirement: needs little or no pruning to develop strong structure; remove lower branches as they die from disease

Limb Breakage: resistant

Current Year Twig: brown; thick

CULTIVARS AND SPECIES:

For reliable blue color, purchase a grafted, named cultivar selected for its blue color. 'Argentea' (Silver Colorado Spruce)—silvery foliage color; 'Glauca' (Blue Colorado Spruce)—bluish foliage; 'Hoopsii'—bluish foliage, probably the best Blue Spruce; 'Foxtail'—a new introduction, could be the most heat-adapted Blue Spruce—further testing will tell; 'Pendens' (Koster's Blue Spruce)—weeping habit, must be staked to make tree form, blue foliage; 'Thompsoni'—bluish foliage.

LANDSCAPE NOTES:

Origin: Rocky Mountains

Usage: specimen; windbreak; screen; buffer strip

Colorado Spruce has a horizontal branching habit and grows taller than 75 feet in its native habitat, but is normally seen at 30 to 50 feet in landscapes. The tree grows about 12 inches per year once established, but may grow slower for several years following transplanting. Needles emerge as a soft clump, changing to a stiff, pointed needle sharp to the touch.

Colorado Spruce casts dense shade when branched to the ground, so no grass grows beneath it. It lends a formal effect to any landscape due to the stiff, horizontal branches, and blue to bluish-green foliage. It is often used as a specimen or as a screen planted 10 to 15 feet apart.

The tree prefers a rich, moist soil, and benefits from irrigation in dry weather. Trees benefit from a layer of mulch extending beyond the edge of the branches. This keeps roots cool and reduces moisture loss from the soil. It grows in full sun or the shade on the north side of a building. Susceptibility to canker makes this tree a questionable choice for large-scale planting in the eastern states. Substitute a blue form of White Fir for the same blue foliage effect without the disease problem. The wax coating on the needles, which gives the blue color, can be washed off by some pesticides. Check the label and test the spray on a small scale before spraying.

Cytospora canker is common. It infects a branch, then eventually kills it. The lower branches are attacked first, then progressively higher branches die. Two gall-forming insects commonly attack Spruce. Eastern Spruce gall adelgid forms pineapple-like galls at the base of twigs. Galls caused by Cooley's Spruce gall adelgid look like miniature cones at the branch tips. The gall adelgids do not kill trees unless the infestation is heavy. A few galls on a large tree are not serious.

Pinckneya pubens Pinckneya, Fevertree

Pronunciation: PINK-nee-ah PEW-bens

ZONES: 7B, 8A, 8B, 9A, 9B

HEIGHT: 15 to 20 feet

WIDTH: 15 feet

FRUIT: round; 0.5 to 1"; dry; brown; no significant litter; persistent; showy

GROWTH RATE: moderate; short-lived

HABIT: columnar; oval; open; irregular silhouette; coarse texture

FAMILY: *Rubiaceae*

LEAF COLOR: dark green; no fall color change

LEAF DESCRIPTION: opposite; simple; entire margin; oval to oblong; deciduous; 5 to 8" long

FLOWER DESCRIPTION: showy red to pinkish sepals; summer

EXPOSURE/CULTURE:

Light Requirement: part shade to full sun

Soil Tolerances: all textures; acidic; wet soil; some drought

PEST PROBLEMS: probably none serious

BARK/STEMS:

Branches droop as tree grows and will require pruning for vehicular or pedestrian clearance beneath canopy; routinely grown with, or trainable to be grown with, multiple trunks; can be trained to grow with a single trunk; no thorns

Pruning Requirement: requires pruning to develop strong structure

Limb Breakage: resistant

Current Year Twig: brown; gray; medium thickness

LANDSCAPE NOTES:

Origin: South Carolina and Florida

Usage: container or above-ground planter; near a deck or patio; specimen

In earlier days, malaria and other fevers were treated at home with a medicine made from the inner bark of this deciduous, 30-foot-tall, North American native tree, giving it the common name "Fevertree." The tree is most commonly seen at 15 to 20 feet tall by 15 feet wide as an open-grown landscape tree. The large leaves, five to eight inches long and three to four inches wide, have a lighter underside and are covered with a light fuzz when young. The clusters of yellowish-green, 2.5-inch-long tubular flowers are made more conspicuous by the cream- to bright rose-colored, petal-like sepals, making Fevertree quite spectacular when it is in full bloom in early summer. The brown, spotted, round seed capsules that follow stay on the tree throughout the winter.

The unusual, open, coarse habit of growth and showy flowers make this a conversation piece in any yard. Use it as an accent in a sunny or shaded shrub border or as a specimen near the patio or deck. This is certainly a tree to consider when planning a yard, due to its small size and pest resistance. It is especially adapted to planting in subdivisions built in existing woods, because it flowers in partial shade provided it receives at least five hours of direct sun.

Found in its native habitat on poorly drained land or along swamp margins, Fevertree should only be planted on moist soils or areas that are flooded with rains periodically throughout the year. Although best flowering appears to be in full sun, trees grown in partial shade will grow but flower poorly. Not for a landscape unless it can be occasionally irrigated during dry summers.

Pinus bungeana Lacebark Pine

Pronunciation: PIE-nuss bun-gee-A-nah

ZONES: 4A, 4B, 5A, 5B, 6A, 6B, 7A, 7B, 8A

HEIGHT: 30 to 40 feet

WIDTH: 20 to 35 feet

FRUIT: oval; 2"; dry; brown; no significant litter; persistent; showy

GROWTH RATE: slow-growing; long-lived

HABIT: pyramidal in youth, oval to round later; moderately dense; symmetrical; fine texture

FAMILY: *Pinaceae*

LEAF COLOR: medium green; needle-leaf evergreen

LEAF DESCRIPTION: spiral; simple; entire; three needles; 2 to 4" long

FLOWER DESCRIPTION: yellow; not showy; spring

EXPOSURE/CULTURE:

Light Requirement: part sun to full sun

Soil Tolerances: all textures; slightly alkaline to acidic; moderate drought

PEST PROBLEMS: mostly resistant

BARK/STEMS:

Showy, exfoliating, brown, red, and green bark; routinely grown with multiple trunks; can be trained to grow with a single trunk; branches grow mostly upright and will not droop; no thorns

Pruning Requirement: encourage multi-trunked development to show off bark; remove developing multiple trunks early for a single-trunked specimen

Limb Breakage: susceptible to breakage at crotch due to poor collar formation; wood itself is weak and tends to break; susceptible to ice and heavy snow storms especially if included bark is present

Current Year Twig: green; gray; medium thickness

LANDSCAPE NOTES:

Origin: China

Usage: bonsai; buffer strip; highway; near a deck or patio; shade tree; specimen

Lacebark Pine is usually grown with a very picturesque multi-trunked and upright, oval habit. Lower branches borne on a central trunk on young trees begin to grow upright forming a multiple-trunked tree. The tree then begins spreading into a flattened canopy, allowing filtered sunlight beneath. Its beautiful gray-green, mottled, exfoliating bark make it one of the best Pines for showy bark character. Cultivated in the Orient for its striking bark, Lacebark Pine can be often seen on the grounds of Buddhist temples.

Capable of reaching 75 feet in height, it is more often seen at 30 to 40 feet with a 20- to 35-foot spread, and has a rounded to pyramidal outline when young. The stiff, two- to four-inch-long needles are sharp to the touch, as are the light brown, two- to three-inch-wide, rounded cones. Locate the tree where passersby can view the trunk character, such as along a walk or near a patio or deck. Those planting the tree will wait a long time to see the tree display its true virtue, for growth is slow at best.

Trees have a reputation for splitting apart in snow and ice storms. This can probably be reduced by training newly planted trees to one central leader and eliminating or pruning weakly attached upright trunks and branches that have included bark. Slow the growth on the developing, aggressive lower branches by thinning to remove the upright portion of the branch. This will modify the striking multi-trunked habit so commonly associated with this tree. It could be worth trying as a single-trunked tree with small, horizontal branches that would most likely be more tolerant of ice and snow loads.

Growing best in full sun on well-drained, acid soil, Lacebark Pine may tolerate soil with a higher pH but should not be exposed to soils that are excessively wet or dry. The trees are also reportedly sensitive to air pollution.

A large number of pests and diseases are reported for Pine.

Pinus canariensis Canary Island Pine

Pronunciation: PIE-nuss ka-nair-ee-EN-sis

ZONES: 9A, 9B, 10A, 10B, 11

HEIGHT: 70 to 80 feet

WIDTH: 20 feet

FRUIT: elongated oval cone; 4 to 8"; dry; brown; attracts mammals; causes some litter; persistent; showy

GROWTH RATE: rapid-growing; long-lived

HABIT: columnar to narrow pyramidal; moderate density; irregular and open when young, more symmetrical and denser later; fine texture

FAMILY: *Pinaceae*

LEAF COLOR: evergreen; dark green
LEAF DESCRIPTION: spiral; simple; three needles; 8 to 10" long
FLOWER DESCRIPTION: yellow; showy; spring
EXPOSURE/CULTURE:

Light Requirement: full sun

Soil Tolerances: all textures; acidic; moderate drought

PEST PROBLEMS: resistant
BARK/STEMS:

To 3 feet in diameter; showy, reddish-brown blocky bark; branches grow mostly upright and do not droop; should be grown with a single leader; no thorns

Pruning Requirement: needs little or no pruning to develop strong structure; remove double leaders as they form, and pinch emerging candles to thicken the open canopy on young trees; tolerates shearing; pruning January through October increases susceptibility to bark beetles, gall rust, and pitch canker

Limb Breakage: resistant

Current Year Twig: brown; thick

LANDSCAPE NOTES:

Origin: Canary Islands

Usage: buffer strip; highway; street tree; windbreak; mass planting; clipped hedge or screen

Canary Island Pine is a large, stately, heavily branched, long-needled conifer with a moderate growth rate. At mature height it may have a three- to four-foot-diameter trunk. Lower branches usually remain on the tree to form a column of dark green foliage. The long cones contain seeds favored by wildlife. Enough sun filters through the moderately open canopy to sustain a lawn beneath the tree. Pines have deep roots beneath the trunk, except in poorly drained soil where all roots are shallow. The tap root is prominent in well-drained soil and can make this Pine difficult to transplant from the field.

This is one of only a few Pines that make good street trees. Pruning requirement is minimal because branches grow slowly and they do not droop into traffic lanes. It is fairly clean, grows fast, and retains lower branches, forming an attractive canopy. Branches will not grow over the street to form a canopy of shade, but the colonnade effect of a row of trees on either side of the street is unsurpassed. It creates a nice buffer between the street and a row of homes or a parking lot.

Canary Island Pine grows well on a variety of acidic soils in full sun. It does poorly in alkaline soil or where irrigation water has a high pH. It is not highly drought-tolerant and needs irrigation in desert climates. Pinyon, Limber, and Coulter Pines are more suited for rocky or dry sites with little or no irrigation.

This tree is awkward when young, but fills in nicely as it reaches 10 years old. It is very common in southern California as a landscape plant and is often planted in groups to create a natural-like setting. The city of Whittier, California, planted it as a street tree 70 years ago, and these trees look great today.

Some may be tempted to pinch the emerging candles on young trees to increase the canopy density on this open Pine. This is not really necessary because trees fill in as they grow older.

Pinus cembra Swiss Stone Pine

Pronunciation: PIE-nuss SEM-bra

ZONES: 3A, 3B, 4A, 4B, 5A, 5B, 6A, 6B, 7A, 7B

HEIGHT: 25 to 40 feet

WIDTH: 10 to 20 feet

FRUIT: elongated; oval; 3"; dry; brown; no significant litter; persistent

GROWTH RATE: slow-growing; long-lived

HABIT: columnar to narrow pyramidal; dense; symmetrical; fine texture .

FAMILY: *Pinaceae*

LEAF COLOR: blue-green; light green; evergreen

LEAF DESCRIPTION: dense; spiral; simple; five needles; 3.5" long

FLOWER DESCRIPTION: males purple to yellow, females red; not showy

EXPOSURE/CULTURE:

Light Requirement: full sun

Soil Tolerances: all textures; slightly alkaline to acidic; some salt; drought

PEST PROBLEMS: resistant

BARK/STEMS:

Branches droop as tree grows and will require some pruning for pedestrian clearance beneath canopy; should be grown with a single leader; no thorns

Pruning Requirement: needs little or no pruning to develop strong structure

Limb Breakage: resistant

Current Year Twig: covered with brownish hairs; medium thickness

CULTIVARS AND SPECIES:

'Columnaris' has a narrower canopy and presents a formal appearance; 'Nana' is smaller. *P. koraiensis*, from Japan and Korea, is closely related, with a more spreading canopy and a moderate growth rate.

LANDSCAPE NOTES:

Origin: Europe, Asia

Usage: specimen; screen, windbreak; possible street tree

Growing to more than 50 feet in its native habitat in the mountains, Swiss Stone Pine is more often 25 to 40 feet tall in the landscape, growing slowly into a dense, narrow pyramidal silhouette that casts dense shade beneath. Some specimens in landscapes are 50 or 60 feet tall, but these are rare. The horizontal branches are held firm on thick branches, with lower limbs reaching the ground if not pruned from the trunk. The canopy remains dense from youth into middle age. The flowers are followed by the production of three-inch-long and one-inch-wide, bluish-purple cones. Unfortunately the tree is not common in the industry.

Useful as a specimen in small landscapes, Swiss Stone Pine is a graceful addition to any yard. Because trees are dense even when young, they would sell well at nurseries if they were made more available. The plant is a real standout among Pines. It can be planted in small landscapes and can look nice for a long time due to its slow growth rate. The narrow canopy and drought tolerance make it a good candidate for street tree planting.

A row of trees planted 10 to 15 feet apart forms a beautiful, durable windbreak and screen that will not have to be pruned. Growth rate might be increased if trees are provided with irrigation and fertilizer.

Swiss Stone Pine should be grown in full sun on well-drained soil, though it will tolerate infertile, even rocky soils. The tree is extremely cold-hardy and is useful in windy climates.

This Pine resists white pine blister rust, a disease common to other five-needled pines.

Pinus cembra 'Nana' Dwarf Stone Pine

Pronunciation: PIE-nuss SEM-bra

ZONES: 3A, 3B, 4A, 4B, 5A, 5B, 6A, 6B, 7A, 7B

HEIGHT: 15 to 25 feet

WIDTH: 5 to 10 feet

FRUIT: elongated; oval; 3"; dry; brown; no significant litter; persistent

GROWTH RATE: very slow-growing; long-lived

HABIT: fat column or pyramid; very dense; perfectly symmetrical; fine texture

FAMILY: *Pinaceae*

LEAF COLOR: light green; evergreen

LEAF DESCRIPTION: dense; spiral; simple; five needles; 3.5" long

FLOWER DESCRIPTION: males purple to yellow, females red; not showy

EXPOSURE/CULTURE:

Light Requirement: full sun

Soil Tolerances: all textures; slightly alkaline to acidic; some salt; drought

PEST PROBLEMS: resistant

BARK/STEMS:

Should be grown with a single leader; tree looks best with branches to the ground; no thorns

Pruning Requirement: needs little or no pruning to develop a nice tree

Limb Breakage: mostly resistant

Current Year Twig: medium thickness

LANDSCAPE NOTES:

Origin: species is from Europe and Asia

Usage: specimen; screen

Probably growing to about 25 feet in 40 or 50 years, Dwarf Swiss Stone Pine will be shorter in most gardens. It grows slowly into a dense pyramidal silhouette. The dense canopy will be responsible for the popularity of this tree once it moves into production and is available at nurseries. The flowers are followed by the production of three-inch-long and one-inch-wide, bluish-purple cones. Unfortunately the tree is rare in the industry.

Useful as a specimen in small landscapes and gardens, the Dwarf Swiss Stone Pine would be the perfect small tree for planting near a water feature or along a walk in a small-scale landscape. Its slow growth and compact habit should make it popular with homeowners looking for a low-maintenance, perfectly symmetrical specimen plant. Because trees are dense even when young, they would sell well at nurseries if they were made more available. The plant is a real standout among the Pines, creating a superfine foliage texture. It can look nice for a long time due to its slow growth rate.

A row of trees planted 8 to 10 feet apart forms a beautiful, durable visual screen that will not have to be pruned, because of the slow growth rate. It will provide a clipped-hedge look without clipping.

Swiss Stone Pine should be grown in full sun on well-drained soil, and will tolerate infertile, even rocky soils. The tree is extremely cold-hardy and is useful in windy climates. The species usually does not discolor in the winter as other Pines can.

The species resists white pine blister rust, a disease common to other five-needled Pines.

Pinus cembroides Mexican Pinyon

Pronunciation: PIE-nuss sem-BROY-dees

ZONES: 5B, 6A, 6B, 7A, 7B, 8A, 8B

HEIGHT: 20 to 40 feet

WIDTH: 15 to 20 feet

FRUIT: irregular oval; 1 to 3"; dry; brown; attracts wildlife; edible; not especially showy; causes some litter

GROWTH RATE: slow-growing; long-lived

HABIT: pyramidal to oval; moderate density; symmetrical; fine texture

FAMILY: *Pinaceae*

LEAF COLOR: blue-green; evergreen

LEAF DESCRIPTION: spiral; simple; entire; two or three needles; about 1.5" long

FLOWER DESCRIPTION: red or yellow; not showy; spring

EXPOSURE/CULTURE:

Light Requirement: grows best in full sun

Soil Tolerances: loam; sand; alkaline to acidic; drought

PEST PROBLEMS: resistant

BARK/STEMS:

Branches droop as tree grows and will require pruning for vehicular or pedestrian clearance beneath canopy; should be grown with a single leader; no thorns

Pruning Requirement: needs little or no pruning to develop strong structure

Limb Breakage: mostly resistant

Current Year Twig: gray; thick

CULTIVARS AND SPECIES:

P. edulis (*P. cembroides* var. *edulis*) is a similar but smaller, long-lived, very drought-tolerant tree that is found naturally at elevations above 6,000 feet. It is generally a two-needled Pine. *P. monophylla* has one needle.

LANDSCAPE NOTES:

Origin: west Texas into Mexico

Usage: bonsai; buffer strip; highway; specimen; reclamation; possible street tree

A native North American tree, Mexican Pinyon may sometimes be larger that its usual 20 to 40 feet. It forms a compact, conical silhouette, producing a rounded canopy with age, and the lower branches are maintained on the tree, providing dense cover to the ground unless shaded out by other trees. The stiff needles are 1 to 1.5 inches long and the small, one- to two-inch, yellow or red-brown cones mature and drop in autumn. The seeds found within the cones are quite popular with many birds and mammals, including humans. Eaten either raw or toasted, Pinyon Pine 'nuts' have a wonderful flavor.

This small to medium-sized Pine is nicely suited for reclamation projects, along highways, unirrigated buffer strips around urban spaces, parks, and other low-maintenance areas, because of its tolerance to drought.

It would also grow well in a small residential garden where space is limited, due to its small size and fine texture. Its compact habit makes it ideal for bonsai. Even unpruned trees bring an 'old-tree' look to the landscape. It can easily be shaped into any form, or maintained at any height, by removing selected branches and twigs. This opens up the canopy, creating a windblown look.

Mexican Pinyon should be grown in full sun on well-drained soil, tolerating dry, poor, alkaline soils but not wet sites. It is probably best not to irrigate this tree much after it is well established in the landscape, unless rainfall is less than 20 inches per year.

Pine tip moth may kill some new growth, but the result is simply to make the tree more compact, as it has a pruning effect.

Pinus coulteri Coulter, or Big-cone Pine

Pronunciation: PIE-nuss cull-TEAR-ee

ZONES: 8A, 8B, 9A, 9B, 10A, 10B

HEIGHT: 30 to 70 feet

WIDTH: 40 feet

FRUIT: oval; enormous, 10 to 12" long; dry; brown; attracts wildlife; causes significant litter; persistent; showy

GROWTH RATE: rapid-growing; long-lived

HABIT: oval; open; irregular when young, more symmetrical later; fine texture

FAMILY: *Pinaceae*

LEAF COLOR: dark gray-green; evergreen

LEAF DESCRIPTION: spiral; simple; three needles; 8 to 10" long

FLOWER DESCRIPTION: yellow; not showy; spring

EXPOSURE/CULTURE:

Light Requirement: full sun

Soil Tolerances: all textures; acidic; drought

PEST PROBLEMS: resistant

BARK/STEMS:

To three feet in diameter; gray; branches droop as tree grows and will require pruning for pedestrian or vehicular clearance beneath canopy; should be grown with a single leader; no thorns

Pruning Requirement: needs little or no pruning to develop strong structure; remove double leaders as they form, and pinch emerging candles to thicken open canopy on young trees; pruning January through October can increase disease and insect problems

Limb Breakage: susceptible to breakage at junction of a double leader

Current Year Twig: tan; stout

LANDSCAPE NOTES:

Origin: California

Usage: buffer strip; highway; windbreak; mass planting; shade tree; specimen

Coulter Pine is a large, heavily branched, long-needled conifer with a moderate to rapid growth rate, and is capable of reaching 80 feet in height with a 3- to 4-foot-diameter trunk. Lower branches usually remain on the tree and droop to the ground unless removed from the trunk. These branches are aggressive and can grow to a large diameter, becoming trunk-like. The form of the canopy becomes unpredictable when trees from the mountains are brought to lower elevations. The huge cones appear among the 10- to 12-inch-long needles, and seeds are favored by wildlife. The tree litters the ground with spiny cones that can inflict pain when they meet flesh. Enough sun filters through the moderately open canopy to sustain a lawn beneath the tree.

This pine makes a great tree for parks and other low-maintenance areas due to its tolerance to drought and neglect. Big-cone Pine is sometimes used along highways and at interchanges where it receives little maintenance other than occasional irrigation. Once it is well established, this pine often grows fine with no irrigation if roots are allowed to expand into open soil.

It is often grouped together in a mass planting to create shade for a large, open area, but the canopy spreads too much for planting in a small residential landscape. The spreading habit is unique among the Pines, making this tree a good candidate for planting as a specimen in a lawn or landscape bed. Open-grown trees form an attractive, low-branched habit that likens it to Deodar Cedar.

Coulter Pine grows well on a variety of acidic soils in full sun. It does poorly in alkaline soil or where irrigation water has a high pH. It is highly drought-tolerant, requiring irrigation only until it is well established, even in high desert climates in California. Pinyon and Limber Pines are also well suited for rocky or dry sites with little or no irrigation.

Locate the tree away from pavement because the large cones create quite a mess as they litter the ground beneath the canopy. Remove *Madia* spp. within 1,000 feet to eliminate the alternate host for needle rust. Pitch canker can kill the tree.

Pinus densiflora Japanese Red Pine

Pronunciation: PIE-nuss den-sih-FLOOR-ah

ZONES: 3B, 4A, 4B, 5A, 5B, 6A, 6B, 7A

HEIGHT: 30 to 50 feet

WIDTH: 30 to 50 feet

FRUIT: oval; 1.5 to 2"; dry; tan; not showy; causes some litter; persistent

GROWTH RATE: slow-growing; long-lived

HABIT: oval; open; irregular silhouette; fine texture

FAMILY: *Pinaceae*

LEAF COLOR: light green; needle-leaf evergreen

LEAF DESCRIPTION: spiral; two needles; 3 to 5" long

FLOWER DESCRIPTION: yellow; not showy

EXPOSURE/CULTURE:

Light Requirement: grows best in full sun; tolerates some shade

Soil Tolerances: loam; sand; well-drained clay; acidic; moderate drought

Aerosol Salt Tolerance: low to none

PEST PROBLEMS: somewhat susceptible

BARK/STEMS:

Showy, orange, flaking bark; trainable to be grown with single or multiple trunks; branches grow mostly upright or horizontal and will not droop; no thorns

Pruning Requirement: requires little pruning; well suited for bonsai

Limb Breakage: susceptible to breakage in ice and snow storms

Current Year Twig: green; medium

CULTIVARS AND SPECIES:

'Alboterminata'—yellowish needle tips; 'Aurea'—yellow needles; 'Oculis-draconis' (Dragon's Eye Pine)—two yellow lines on needles; 'Pendula'—weeping, spreading habit; 'Umbraculifera' (Tanyosho Pine)—20 feet tall, multi-trunked.

LANDSCAPE NOTES:

Origin: China, Japan, and Korea

Usage: bonsai; specimen; near a deck or patio; small shade tree

Japanese Red Pine reaches a height and spread of 30 to 50 feet in the landscape, growing much taller in the woods. Needles are arranged in pairs and remain on the tree for about three years. A distinguishing feature of this tree is the often crooked or sweeping trunk, which shows reddish-orange peeling bark. Because lower branches are held nearly horizontal on the trunk, forming a picturesque silhouette in the landscape, it is used best as a specimen. Needles may turn yellowish during winter on some soils.

Japanese Red Pine is a favorite for sculpting into a bonsai of various sizes in Japanese gardens. The canopy is thinned to show off the attractive bark. Branches are often clipped to increase density and to manipulate and direct growth. This pruning complements and displays the curved trunk to generate a tree worthy of featuring in any garden or landscape. A grouping of trees massed together forms a nice, informal display of light green foliage held on bright orange trunks. The trunks appear to light up in the morning and evening hours when the sun shines directly on the bark.

The tree prefers a site with full sun and well-drained, slightly acid soil. Clayey soil is usually not suitable unless it is very well drained.

Usually pest-free (occasionally scale), but the list of potential problems is long, including needle blight and rusts. Canker diseases may cause die-back of landscape Pines. Keep trees healthy and prune out infected branches. Needle cast is common on small trees and plantation or forest trees. Infected needles yellow and fall off.

Pinus densiflora 'Pendula'
Weeping Japanese Red Pine
Pronunciation: PIE-nus den-sih-FLOOR-ah

ZONES: 3B, 4A, 4B, 5A, 5B, 6A, 6B, 7A

HEIGHT: 6 to 10 feet

WIDTH: 15 feet

FRUIT: oval; 1.5 to 2"; dry; tan; not showy; causes some litter; persistent

GROWTH RATE: slow-growing; moderate life span

HABIT: weeping, spreading, and irregular; fine texture

FAMILY: *Pinaceae*

LEAF COLOR: light-green; needle-leaf evergreen
LEAF DESCRIPTION: spiral; two needles; 3 to 5" long
FLOWER DESCRIPTION: yellow; not showy
EXPOSURE/CULTURE:

Light Requirement: grows best in full sun; tolerates some shade

Soil Tolerances: loam; sand; well-drained clay; acidic; moderate drought

Aerosol Salt Tolerance: low to none

PEST PROBLEMS: somewhat susceptible

BARK/STEMS:

Showy, orange, flaking bark; trainable to be grown with single or multiple trunks; branches grow mostly upright or horizontal and will not droop; no thorns

Pruning Requirement: requires little pruning; well suited for bonsai

Limb Breakage: susceptible to breakage in ice and snow storms

Current Year Twig: green; medium

LANDSCAPE NOTES:

Usage: bonsai; specimen

This cultivar needs to be staked to create a plant tall enough to be called a tree. With the trunk secured in the upright position, the branches descend from the tree, forming a traditional weeping habit of growth.

The fine foliage and delicate habit make this a perfect specimen tree. It is best to feature the plant by placing it in a prominent location. Mulch the area beneath the canopy to reduce turf competition from other plants and to keep mowing equipment away from the trunk and foliage. The tree can also be located in a low groundcover bed to display it prominently.

Pinus densiflora 'Umbraculifera'

'Umbraculifera' Japanese Red Pine

Pronunciation: PIE-nus den-sih-FLOOR-ah

ZONES: 3B, 4A, 4B, 5A, 5B, 6A, 6B, 7A

HEIGHT: 20 to 30 feet

WIDTH: 20 to 30 feet

GROWTH RATE: slow-growing; moderate life span

HABIT: oval; open; symmetrical; fine texture

FAMILY: *Pinaceae*

LEAF COLOR: light green; needle-leaf evergreen

LEAF DESCRIPTION: spiral; two needles; 3 to 5" long

FLOWER DESCRIPTION: yellow; not showy

EXPOSURE/CULTURE:

Light Requirement: grows best in full sun; tolerates some shade

Soil Tolerances: loam; sand; well-drained clay; acidic; moderate drought

Aerosol Salt Tolerance: low to none

PEST PROBLEMS: somewhat susceptible

BARK/STEMS:

Showy, orange, flaking bark; trainable to be grown with single or multiple trunks; branches grow mostly upright or horizontal and will not droop; no thorns

Pruning Requirement: requires little pruning; well suited for bonsai

Limb Breakage: susceptible to breakage in ice and snow storms

Current Year Twig: green; medium

LANDSCAPE NOTES:

Usage: bonsai; specimen; buffer strip; possible street or parking lot tree

This cultivar of Japanese Red Pine can reach a height of 20 to 30 feet, but grows very slowly and is often much smaller. A distinguishing feature of this tree is the upright, spreading branching habit, which is uncommon in the Pine genus. The bark is unusually striking, showing reddish-orange as it exfoliates. The form is compact and the tree stays small, making it ideally suited for the residential yard. It can be used as a screen planted in mass or in a row, or alone as a specimen. Needles may turn yellowish during winter on some soils.

The tree is perfectly suited for planting as a specimen to attract attention to an area. People will not miss or forget this tree once they see it. The trunk and habit of the plant are unmistakable. The open canopy lets light pass through, allowing good turf growth beneath the canopy. The tree is featured nicely by planting it in a lawn area to display it like a statue. The only drawback to using this tree as a street tree would be the occasional severe damage from heavy ice and snow loads. This might be partially overcome by cabling the main trunks together.

Pinus echinata Shortleaf Pine

Pronunciation: PIE-nuss eck-i-NAY-tah

ZONES: 6A, 6B, 7A, 7B, 8A, 8B

HEIGHT: 50 feet

WIDTH: 35 feet

FRUIT: oval to nearly round; 2"; dry; brown; attracts wildlife; causes some litter; persistent; showy

GROWTH RATE: moderate, once established; long-lived

HABIT: oval; pyramidal; open; irregular silhouette; fine texture

FAMILY: *Pinaceae*

LEAF COLOR: evergreen; light green

LEAF DESCRIPTION: spiral; simple; two or three needles; 3 to 5" long

FLOWER DESCRIPTION: yellow; not showy; spring

EXPOSURE/CULTURE:

Light Requirement: full sun

Soil Tolerances: all textures; acid only; requires well-drained soil; moderate drought

Aerosol Salt Tolerance: moderate

PEST PROBLEMS: somewhat susceptible

BARK/STEMS:

Branches droop as tree grows; should be grown with a single leader; no thorns

Pruning Requirement: needs little or no pruning to develop strong structure; remove double leaders as they form

Limb Breakage: susceptible to breakage at junction of a double leader; wood itself is weak and tends to break, especially in ice storms

Current Year Twig: brown; thick

CULTIVARS AND SPECIES:

Foresters have crossed Shortleaf Pine with a number of other Pines to increase disease resistance and improve growth rate. Natural crosses exist with Loblolly Pine west of the Mississippi River.

LANDSCAPE NOTES:

Origin: New Jersey to Oklahoma and north Florida

Usage: buffer strip; highway; reclamation; shade tree; mass planting

Shortleaf Pine is one of four major timber species in the southern United States. The tree develops an open canopy and is capable of reaching 100 feet in height on a good site, but is usually smaller. Trunks on old trees can be up to four feet in diameter. The small cones that appear among the four-inch-long needles are favored by wildlife. Enough sun filters through the rounded, open canopy to sustain a lawn beneath the tree. This Pine can develop a tap root beneath the trunk in the forest, except in poorly drained or hard soil, where all roots are shallow. Tap roots are not common on trees planted in urban and suburban landscapes.

Shortleaf Pine grows well on a variety of acidic soils in full sun only. It does poorly in alkaline or poorly drained soil. The horizontal branches break easily. Because shaded lower branches die and drop as the tree grows taller, do not plant them too close to high traffic areas unless there is a regular pruning and maintenance plan to remove them. Open-grown trees keep more lower branches, probably due to greater sun exposure.

This native plant is not popular as a landscape plant, but could be planted along with Loblolly and Slash Pines in groups to create a natural-like setting. It is reported to be more tolerant of drought than Loblolly Pine. *P. clausa* and *P. palustris* are more suited for very sandy, dry sites.

Needles and cones fall from the tree during the year, creating slippery walks. Chlorosis develops on sites with high-pH irrigation water. There are many potential insect and disease problems. Littleleaf disease, Nantucket pine tip moth, and southern pine beetle can be troublesome, especially if the tree is under stress from some other problem such as root loss. Shortleaf Pine is a preferred host for southern pine beetles.

Pinus elliottii Slash Pine

Pronunciation: PIE-nuss ell-lee-OTT-tee-eye

ZONES: 7A, 7B, 8A, 8B, 9A, 9B, 10A, 10B, 11

HEIGHT: 75 to 100 feet

WIDTH: 35 to 50 feet

FRUIT: elongated; oval; 2 to 6"; dry; brown; attracts mammals; causes significant litter; persistent; showy

GROWTH RATE: rapid-growing; long-lived

HABIT: oval; pyramidal; moderately open; irregular silhouette; fine texture

FAMILY: *Pinaceae*

LEAF COLOR: needle-leaf evergreen; dark green

LEAF DESCRIPTION: spiral; two or three needles; 6 to 12" long

FLOWER DESCRIPTION: yellow; not showy; spring; induces allergic reaction in some people

EXPOSURE/CULTURE:

Light Requirement: part shade to full sun

Soil Tolerances: all textures; acidic to neutral; occasionally wet; drought

Aerosol Salt Tolerance: moderate

PEST PROBLEMS: susceptible

BARK/STEMS:

Branches droop as tree grows; should be grown with a single leader; no thorns

Pruning Requirement: needs little pruning to develop strong structure; remove double leaders as they form

Limb Breakage: susceptible to breakage at junction of a double leader; wood itself is weak and tends to break, especially in ice and snow storms

Current Year Twig: brown; thick

CULTIVARS AND SPECIES:

P. taeda (Loblolly Pine) has smaller cones and often has a shorter trunk.

LANDSCAPE NOTES:

Origin: southern United States

Usage: buffer strip; highway; reclamation; shade tree; specimen or mass planting

The species *elliottii* is a large, stately, heavily branched, long-needled conifer with a rapid growth rate, and is capable of reaching 100 feet in height with a 3- to 4-foot-diameter trunk. The six-inch-long cones that appear among the needles are favored by wildlife. Enough sun filters through the rounded, open canopy to sustain a lawn beneath the tree. Pines have deep roots beneath the trunk, except in poorly drained soil where all roots are shallow. The tap root is prominent in well-drained soil and can make this tree difficult to transplant from the wild. The variety *densa* is very similar and should be planted south of Orlando, Florida.

Slash Pine grows well on a variety of acidic soils in full sun or partial shade. It does poorly in alkaline soil or where irrigation water has a high pH. Once established, it is more tolerant of wet sites than most other Pines and is moderately salt-tolerant. It is not highly drought-tolerant, but is more so than many other eastern Pines. Sand Pine and Longleaf Pines are more suited for very sandy, dry sites. The horizontal branches break easily. Because shaded lower branches die and drop as the tree grows taller, do not plant this tree too close to high traffic areas unless there is a regular pruning and maintenance plan to remove branches. Open-grown trees keep more lower branches, probably due to greater sun exposure.

This native plant is becoming more popular as a landscape plant and is often planted in groups to create a natural-like setting. Needles and cones fall from the tree during the year, creating slippery walks.

Chlorosis may occur on sites with high-pH irrigation water. There are many potential insect and disease problems. Pitch canker may be problematic in intensively managed, highly fertile landscapes. Bark beetles often kill stressed trees, but can infest and kill healthy trees under certain circumstances. Slash and Longleaf Pines are more resistant to southern pine beetle (SPB) than Loblolly and Shortleaf. In severe outbreaks, SPB can devastate hundreds of thousands of trees.

Pinus flexilis Limber Pine

Pronunciation: PIE-nuss FLEX-ill-iss

ZONES: 4A, 4B, 5A, 5B, 6A, 6B, 7A

HEIGHT: 30 to 50 feet

WIDTH: 15 to 35 feet

FRUIT: elongated; 3 to 6"; dry; brown; causes some litter; persistent; showy

GROWTH RATE: slow-growing; long-lived

HABIT: narrow oval; pyramidal; moderately dense; symmetrical or irregular; fine texture

FAMILY: *Pinaceae*

LEAF COLOR: dark blue-green; needle-leaf evergreen
LEAF DESCRIPTION: spiral; five needles; 2 to 4" long
FLOWER DESCRIPTION: yellow; not showy; spring
EXPOSURE/CULTURE:

Light Requirement: part shade to full sun

Soil Tolerances: all textures; slightly alkaline to acidic; occasionally wet soil; extended drought

PEST PROBLEMS: somewhat susceptible to lethal diseases

BARK/STEMS:

Bark is thin and easily damaged by mechanical impact; branches droop as tree grows, and will require pruning for vehicular or pedestrian clearance beneath canopy; routinely grown with, or trainable to be grown with, multiple trunks; can be trained to grow with a single trunk; no thorns; twigs very flexible, and can be tied in a knot

Pruning Requirement: requires some pruning to develop strong structure; eliminate double leaders

Limb Breakage: resistant

Current Year Twig: green; medium

CULTIVARS AND SPECIES:

'Columnaris'—very upright form, 10 feet wide; 'Glauca'—foliage is a deeper blue-green than the species; 'Glauca Pendula'—irregular, wide-spreading shrub with blue-green needles; 'Glenmore Dwarf'—small, upright, pyramidal tree with blue-gray foliage; 'Nana'—dwarf bushy type; 'Pendula'—wide, weeping silhouette; 'Vanderwolf's Pyramid'—forms a tight pyramid and grows vigorously.

LANDSCAPE NOTES:

Origin: western North America

Usage: specimen; windbreak; Christmas tree

This North American native tree has a fairly narrow, pyramidal silhouette when young, but slowly matures into a broad tree with a flattened canopy. It almost looks square as it reaches 20 years old. The 2.5- to 3.5-inch-long needles are joined by light brown, 1.5-inch-wide, hanging cones that add to the tree's overall attractiveness. Limber Pine is so named due to the flexible nature of the branches. Young branches can literally be tied into a knot. The tree has the overall look of a White Pine when it is young.

Limber Pine is a tough tree that maintains its foliage color throughout even the coldest winter. You can rely on it being in the landscape for a long time if it is properly situated. Plant it 20 feet apart as a fine-textured windbreak in the central and western parts of the United States. Use it as an ornamental specimen in the middle of the lawn, but supply plenty of mulch under the canopy of the tree to reduce root competition.

Limber Pine grows best in full sun or partial shade on moist, well-drained soil, but will easily adapt to harsher sites. Some horticulturists say they grow better under poor cultural conditions of drought and compacted clay soil. Its natural range extends into regions receiving less than 20" annual rainfall. It adapts to wet soil by growing a shallow root system. One of the most tolerant of cold, windy weather in the winter, Limber Pine is not damaged when most other Pines show winter needle browning.

There are a large number of pests and diseases on Pine. Armillaria root rot and white pine blister rust can kill trees. Borers can finish off weakened trees.

Pinus glabra Spruce Pine

Pronunciation: PIE-nuss GLAY-brah

ZONES: 8A, 8B, 9A, 9B

HEIGHT: 30 to 60 feet

WIDTH: 25 to 40 feet

FRUIT: dry; brown; attracts mammals; causes significant litter; persistent for several years; showy

GROWTH RATE: slow-growing; moderate life span

HABIT: oval; open; irregular silhouette; fine texture

FAMILY: *Pinaceae*

LEAF COLOR: evergreen; medium green

LEAF DESCRIPTION: spiral; two or three needles; 2 to 4" long

FLOWER DESCRIPTION: yellow; not showy

EXPOSURE/CULTURE:

Light Requirement: grows surprisingly well in part shade and full sun

Soil Tolerances: all textures; acidic to slightly alkaline; wet soil; moderate drought

Aerosol Salt Tolerance: low

PEST PROBLEMS: resistant when grown in its native range

BARK/STEMS:

Branches grow mostly horizontal and droop toward ground; should be grown with a single leader; no thorns

Pruning Requirement: needs little pruning to develop strong structure

Limb Breakage: somewhat susceptible

Current Year Twig: brown; slender

LANDSCAPE NOTES:

Origin: southern Alabama to South Carolina and north Florida

Usage: bonsai; parking lot island; buffer strip; highway; reclamation; shade tree; specimen; Christmas tree; possible street tree

This heavily foliated, much-branched evergreen has a bushy, irregular canopy, when grown in the open, of soft, two- to three-inch, twisted needles and a trunk that often becomes twisted and curved with age. Do not expect a row of Spruce Pine to form a uniformly shaped canopy. The persistent lower branches on Spruce Pine make it ideal for use as a windbreak, large-scale screen, or specimen but create only light shade. Although capable of reaching 80 feet in height with a thick, straight trunk in the woods, Spruce Pine is usually seen at 30 to 50 feet when grown in the open. The reddish gray-brown bark has shallow ridges and furrows. The 2.5-inch-diameter cones remain on the branches for three to four years and are a source of food for wildlife.

Growing in full sun on moist, fertile soils, this North American native will also tolerate poor, dry soils, as well as wet sites, better than other Pines. Many people forget how picturesque this Pine can become as it grows older. It should be used more as a specimen tree. Pines are deep-rooted except on poorly drained sites, where there will be only shallow roots. The tap root can make them difficult to transplant from well-drained sites in the wild.

Pines grow best on acid soil and are usually not recommended for planting in soil with a high pH; growth is best without grass competition. Spruce Pine is unusual among the Pines in that it will grow in partial shade.

Some adelgids will appear as white cottony growths on the bark. All types produce honeydew, which may support sooty mold. Resistant to fusiform rust. Canker diseases may occasionally cause die-back of landscape Pines. Keep trees healthy and prune out infected branches.

Pinus halepensis Aleppo Pine

Pronunciation: PIE-nuss hal-ah-PEN-sis

ZONES: 8A, 8B, 9A, 9B, 10A, 10B, 11

HEIGHT: 30 to 50 feet

WIDTH: 25 to 40 feet

FRUIT: elongated cone; 3"; dry; reddish-brown; abundant; causes some litter; persistent on tree for several years; showy

GROWTH RATE: moderate; long-lived

HABIT: irregular silhouette; moderately dense; fine texture

FAMILY: *Pinaceae*

LEAF COLOR: light green; evergreen

LEAF DESCRIPTION: spiral; simple; two needles; 2 to 4" long

FLOWER DESCRIPTION: yellow; not showy; spring

EXPOSURE/CULTURE:

Light Requirement: full sun

Soil Tolerances: all textures; acidic; salt and drought

PEST PROBLEMS: mostly resistant

BARK/STEMS:

Bark is thin and easily damaged by mechanical impact; branches droop as tree grows and will require pruning for vehicular or pedestrian clearance beneath canopy; should be trained to grow with a single trunk; no thorns; twigs are very flexible and twisted

Pruning Requirement: requires pruning and staking when young to develop an upright form; pruning January through October increases susceptibility to bark beetles, gall rust, and pitch canker

Limb Breakage: susceptible

Current Year Twig: greenish-tan; medium thickness

LANDSCAPE NOTES:

Origin: Mediterranean region

Usage: specimen; windbreak; mass planting

This drought-tolerant pine has a distinctly irregular growth habit when young, becoming somewhat pyramidal as it ages. Most trees grow no more than 50 feet high with a 30-foot spread, although larger specimens exist. Young branches grow in seemingly random directions, sometimes twisting away from the rest of the canopy. Young trees are often staked in the nursery to develop a straight central leader.

Aleppo Pine is a tough tree that maintains its foliage color throughout even the driest summer, as long as some irrigation is supplied periodically. You can rely on it being in the landscape for a long time if it is properly situated. There are many fine examples of this tree through the southwestern United States because it is readily available in nurseries. Plant it 20 feet apart as a fine-textured windbreak, or massed together in a grouping. Use it as an ornamental specimen in the middle of the lawn, but supply plenty of mulch under the canopy of the tree to reduce turf root competition.

The unusual, irregular growth habit is a welcome relief from the mostly uniform canopies so common to other trees planted in modern landscapes. Although medium-sized at maturity, the pine grows nicely in residential lots, providing an upright accent or open windbreak.

Aleppo Pine grows best in full sun on moist, well-drained soil, but will easily adapt to harsher sites. Some horticulturists say they grow better under poor cultural conditions of drought and compacted clay soil. Others attribute a blight to dry soil, and they recommend regular irrigation, especially in winter.

There are a large number of pests and diseases on Pine. Western gall rust can deform trees. Mites are an occasional problem in the summer. Canker can be serious.

Pinus mugo Mugo Pine, Swiss Mountain Pine

Pronunciation: PIE-nuss MEW-go

ZONES: 2A, 2B, 3A, 3B, 4A, 4B, 5A, 5B, 6A, 6B, 7A, 7B

HEIGHT: 15 to 30 feet

WIDTH: 15 to 25 feet

FRUIT: oval; 1 to 2"; dry; brown; cause some litter; persistent

GROWTH RATE: slow-growing; long-lived

HABIT: pyramidal when young, oval later; dense; symmetrical; fine texture

FAMILY: *Pinaceae*

LEAF COLOR: dark green; needle-leaf evergreen

LEAF DESCRIPTION: spiral; two needles; 1 to 2" long

FLOWER DESCRIPTION: yellow; not showy; spring

EXPOSURE/CULTURE:

Light Requirement: part shade to full sun

Soil Tolerances: all textures; alkaline to acidic; occasionally wet; salt; moderate drought

Aerosol Salt Tolerance: moderate to good

PEST PROBLEMS: sensitive

BARK/STEMS:

Main branches grow mostly upright and will not droop; can be grown with a single trunk or multiple trunks; no thorns

Pruning Requirement: needs little pruning to develop strong structure

Limb Breakage: resistant

Current Year Twig: brown; green; medium thickness

CULTIVARS AND SPECIES:

'Compacta'—rounded, 3 feet tall; 'Gnome'—about 12 feet tall; 'Gold Spire'—has bright yellow new growth; 'Hesse'—dwarf; var. *mugo*—2.5 to 6 feet tall but very broad; var. *pumilo*—prostrate.

LANDSCAPE NOTES:

Origin: Spain to the Balkans

Usage: bonsai; container or above-ground planter; buffer strip; highway; specimen

Mugo Pine is a shrub or small, round to broad pyramidal plant 4 to 30 feet tall, which grows best in sun or partial shade in moist loam. Needles of this two-needled Pine are held on the tree for more than four years, making this one of the more dense Pines suitable for a screen planting. Most other Pines are not suited for screens because they lose their inner needles and lower branches as they grow older. There seems to be great variability in height among individual trees, so select nursery plants with the form which you desire or choose one of the cultivars with known form. When selecting a Mugo Pine to grow into a tree, choose one with a central leader; if looking for a more dwarf type Mugo Pine, choose among the many compact selections.

Trees recover best from transplanting when moved from containers or balled-in-burlap, not bare-root. It performs remarkably well on soils with a high pH and is fairly well adapted to urban sites. Plant size and density can be controlled by pinching the elongating candles just before or as the needles begin emerging, but this is usually not needed on Mugo Pine as growth is normally very dense. Pines are deep rooted except on shallow, poorly drained soil, where there will be only shallow roots.

Mugo Pine is a favored host for pine sawfly and pine needle scale. Some adelgids will appear as white cottony growths on the bark. All types produce honeydew, which may support sooty mold. European pine shoot beetle can cause severe problems. European pine shoot moth causes young shoots to fall over. Infested shoots may exude resin. The insects can be found in the shoots during May. Various foliage rust diseases and rots affect the tree.

Pinus nigra Austrian Pine, European Black Pine

Pronunciation: PIE-nuss NYE-grah

ZONES: 5A, 5B, 6A, 6B, 7A, 7B, 8A

HEIGHT: 40 to 60 feet

WIDTH: 25 to 35 feet

FRUIT: oval; 1 to 3"; dry; brown; causes some litter; persistent; showy

GROWTH RATE: moderate; short-lived where disease and pine wilt nematode are prevalent

HABIT: oval; pyramidal; moderate density; symmetrical; fine texture

FAMILY: *Pinaceae*

LEAF COLOR: needle-leaf evergreen; dark green

LEAF DESCRIPTION: spiral; simple; two needles; 3 to 6" long

FLOWER DESCRIPTION: yellow; not showy

EXPOSURE/CULTURE:

Light Requirement: grows best in full sun

Soil Tolerances: all textures; slightly alkaline to acidic; salt; drought

Aerosol Salt Tolerance: high

PEST PROBLEMS: sensitive

BARK/STEMS:

Branches droop as tree grows; should be grown with a single leader; no thorns

Pruning Requirement: needs little or no pruning to develop strong structure

Limb Breakage: resistant

Current Year Twig: brown; stout; thick

CULTIVARS AND SPECIES:

'Austriaca'—stout, broadly ovate; 'Pyramidalis'—tight, pyramidal form.

LANDSCAPE NOTES:

Origin: Europe from Austria to the Balkans

Usage: bonsai; buffer strip; highway; specimen; urban tolerant

Austrian Pine is medium- to fast-growing, reaching 50 to 60 feet in the landscape and even taller on very old specimens. Combined with the dark green needles, the dense habit makes for an outstanding specimen tree. The pyramidal crown, which becomes flat-topped and somewhat irregular on older specimens, is comprised of thick horizontal branches sweeping horizontally and up as they spread from the trunk. Lower branches are held on the trunk, making this a nice, short-trunked specimen or screen plant. Plant it so the lower branches can be left on the trunk to show the true beauty of this Pine. Dark furrowed bark is very attractive, particularly on older trees.

This tough Pine is well adapted to planting along highways and other areas that receive very little care. It is probable best saved for areas such as this, as the potential exists for devastating losses from disease and nematodes. In areas with these problems, choose another Pine for the home or commercial landscape. Some communities use them as street trees.

It will stand dryness and exposure, is well adapted to urban conditions (including alkaline and clay soil), and tolerates road and seaside salt well. Some members of the population may not be adapted to alkaline soil. Unfortunately, Austrian Pine is very susceptible to tip blight in the east, and for this reason should be used sparingly. There are also reports of pine wilt nematode infesting trees and killing them in one season. But it is one of the best Pines for Texas and in other areas and is still highly recommended there. Be aware that these two problems may spread to other areas. Austrian Pine is difficult to transplant, so it should be planted from containers or moved balled-in-burlap after being root-pruned.

Tip blight (*Sphaeropsis*) kills major branches and disfigures trees as they reach 25 to 30 years of age. The new growth turns brown and the branch is stunted. Pine wilt nematode can kill the tree in one growing season. Pine sawflies can quickly eat needles to the fascicle. European pine shoot moth and beetle can kill branches, disfiguring the tree.

Pinus palustris Longleaf Pine

Pronunciation: PIE-nuss pah-LUSS-triss

ZONES: 7A, 7B, 8A, 8B, 9A, 9B, 10A

HEIGHT: 60 to 80 feet

WIDTH: 30 to 40 feet

FRUIT: elongated; 6 to 12"; dry; brown; attracts mammals; causes some litter; persistent; showy

GROWTH RATE: rapid-growing; long-lived

HABIT: oval; pyramidal; open; irregular silhouette; fine texture

FAMILY: *Pinaceae*

LEAF COLOR: needle-leaf evergreen; bright green

LEAF DESCRIPTION: spiral; two or three needles; 8 to 14" long

FLOWER DESCRIPTION: yellow; not showy; spring

EXPOSURE/CULTURE:

Light Requirement: grows best in full sun

Soil Tolerances: all textures; slightly alkaline to acidic; drought

PEST PROBLEMS: resistant

BARK/STEMS:

Very showy trunk; branches grow mostly upright and will not droop; should be grown with a single leader; no thorns

Pruning Requirement: needs little pruning to develop strong structure

Limb Breakage: wood is weak and tends to break

Current Year Twig: brown; stout

LANDSCAPE NOTES:

Origin: southeastern United States

Usage: reclamation; specimen; shade tree; possible street tree where soil space permits

This beautiful, native North American Pine is capable of reaching 80 to 125 feet in height, but is usually smaller. It is usually saved on a construction site for use as a specimen in the landscape or for providing dappled shade. Be sure to protect the area beneath the dripline from heavy equipment during construction. Roots are very sensitive to disturbance.

Longleaf Pine stays in its tufted, grass-like stage for five to seven years after seed germination, growing very slowly while it develops a root system. Following this grass stage, it grows at a moderate to rapid rate. The needles are up to 14 inches long and very flexible, giving an almost weeping effect to the tree. A distinctive characteristic of Longleaf Pine is the new growth clusters, or buds, which are an inch long and silvery white during the winter. The inconspicuous spring flowers are followed by large, spiny cones, 6 to 10 inches long, which persist on the tree for a couple of years. These are often used for decorating at Christmas time.

Longleaf Pine is not usually planted in landscapes, but could be used due to its beautiful bark and nice, open habit. It would be suited for planting in large landscapes, such as golf courses and parks, and in other areas with plenty of overhead space. It would probably adapt to the hot conditions created near concrete and asphalt, but dropping needles and cones often discourage people from planting Pines near streets or other pavement. This may be a small price to pay for having this tree in the landscape. If people would start planting this tree, it might catch on as Slash Pine has in parts of the South.

Longleaf Pine should be grown in full sun or partial shade on well-drained, acidic soil. Once established, trees are very drought-tolerant and require no irrigation for survival.

Limitations include borers, sawflies, pine-shoot moth, and pine weevils. Pine bark beetles will attack old trees or those which are stressed. Longleaf Pine is resistant to fusiform rust. Irrigating nearby landscapes with alkaline water causes chlorosis.

Pinus parviflora Japanese White Pine

Pronunciation: PIE-nuss parv-eh-FLOOR-ah

ZONES: 4B, 5A, 5B, 6A, 6B, 7A

HEIGHT: 25 to 50 feet

WIDTH: 25 to 50 feet

FRUIT: oval; 2 to 4"; dry; brownish-red; causes some litter; persistent; showy

GROWTH RATE: slow-growing; moderate life span

HABIT: oval; pyramidal; dense; symmetrical when young, irregular later; fine texture

FAMILY: *Pinaceae*

LEAF COLOR: blue or blue-green; needle-leaf evergreen

LEAF DESCRIPTION: spiral; five needles; 1 to 2.5" long

FLOWER DESCRIPTION: yellow; not showy; spring

EXPOSURE/CULTURE:

Light Requirement: grows best in full sun

Soil Tolerances: all textures; acidic; salt and drought

Aerosol Salt Tolerance: good

PEST PROBLEMS: somewhat susceptible

BARK/STEMS:

Bark is thin and easily damaged by mechanical impact; branches droop as tree grows and will require pruning for vehicular or pedestrian clearance beneath canopy; should be grown with a single leader; no thorns

Pruning Requirement: needs little pruning to develop strong structure

Limb Breakage: susceptible to breakage at the crotch, due to poor collar formation; wood itself is weak and tends to break

Current Year Twig: brown; green; medium thickness

CULTIVARS AND SPECIES:

'Brevifolia'—upright, narrow tree, sparsely branched, blue-green foliage in tight bundles; 'Glauca'—available in nurseries, greenish foliage with a touch of silver, wide-spreading tree, 45 feet high or more.

LANDSCAPE NOTES:

Origin: Japan

Usage: bonsai; specimen; coastal planting

Japanese White Pine creates a striking landscape element wherever it is used. Often seen as a dense, conical form when young, Japanese White Pine develops into a graceful, irregularly shaped tree with a broad, flattened canopy. The 1- to 2.5-inch-long needles are stiff and twisted, forming blue-green tufts of foliage at branch tips and creating an overall fine texture to the tree's silhouette. The brownish-red cones persist on the tree for six to seven years.

When looking for a small, picturesque specimen Pine for a coastal landscape, search no more. One of the best specimens in any landscape, Japanese White Pine is a pleasure to behold, with attractive foliage in all seasons. Set it off in the landscape with a low groundcover beneath or locate it in the lawn, but keep the grass cleared away from the thin-barked trunk. Mulch the area beneath the canopy to reduce turf competition.

Japanese White Pine should be grown in full sun on well-drained soil with adequate moisture. The trees are salt-tolerant, and tolerate moderate drought and moist, clay soil. They can be planted in dry climates with adequate irrigation.

There are a large number of potential pests and diseases on Pine.

Pinus pinea, Stone Pine, Umbrella Pine

Pronunciation: PIE-nuss pie-NEE-aye

ZONES: 7A, 7B, 8A, 8B, 9A, 9B, 10A, 10B

HEIGHT: 35 to 65 feet

WIDTH: 35 to 45 feet

FRUIT: oval; dry; brown; edible; causes some litter; persistent; showy

GROWTH RATE: moderate; long-lived

HABIT: rounded, spreading; dense; symmetrical; fine texture

FAMILY: *Pinaceae*

LEAF COLOR: evergreen; bright green

LEAF DESCRIPTION: spiral; two needles; 2 to 4" long

FLOWER DESCRIPTION: yellow; not showy; spring

EXPOSURE/CULTURE:

Light Requirement: grows best in full sun

Soil Tolerances: all textures; slightly alkaline to acidic; salt and drought

Aerosol Salt Tolerance: good

PEST PROBLEMS: resistant

BARK/STEMS:

Showy bark; branches grow mostly horizontal and do not droop significantly; should be grown with a single leader; no thorns

Pruning Requirement: needs some pruning to develop strong structure with one leader; prevent included bark from forming by eliminating or pruning young branches with narrow crotches; pruning January through October increases susceptibility to bark beetle, gall rust, and pitch canker.

Limb Breakage: susceptible to breakage from ice or snow loads

Current Year Twig: brown; thick

LANDSCAPE NOTES:

Origin: Asia Minor, Europe

Usage: bonsai; fruit tree; parking lot island; street tree; buffer strip; highway; shade tree; specimen; shade tree

Just as its name implies, Umbrella Pine has a broad, somewhat flattened round canopy. The tree can ultimately reach 80 to 100 feet in height, though it is more often seen at 35 to 45 feet tall and wide. The stiff, six-inch-long needles are arranged in slightly twisted bundles, and are joined by the heavy, five- to six-inch-long cones, which remain tightly closed on the tree for three years. The trunk is showy with narrow, foot-long orange plates set off nicely by darker fissures.

The seeds inside the cones are of particular interest (they are called pignolia or pine nuts in the stores). The closed cones are gathered, then laid out in direct sun, which forces them to open so the seeds can be harvested. Eaten raw or roasted like peanuts, pignolia nuts are a traditional ingredient of certain Italian dishes. In fact, remains of the cones have been discovered in Roman camps in Britain, attesting to their long-time popularity.

Umbrella Pine has not been extensively tried as a street tree except in parts of California, where it has done well, but it has the attributes of an urban-tolerant tree. Use as a street tree in the cooler part of its range could be hindered by susceptibility to ice and snow damage. It certainly makes a nice shade tree for parks and golf courses, due to the spreading canopy. Drought and salt tolerance are high, making Stone Pine suited for the dry western states. It grows well in soil with a pH in the mid 7s or lower, but may do poorly with higher pH. Stone Pine should be grown in full sun on well-drained soil. Winter winds and low temperature in zone 7A cause needles to brown, and they remain unsightly all winter.

There are a large number of potential pests on Pine, though Stone Pine is usually disease-free. The tree has not been extensively tested in the eastern United States. As with many trees, bark beetles often finish off stressed pines.

Pinus ponderosa Ponderosa or Western Yellow Pine

Pronunciation: PIE-nuss pon-derr-OH-sah

ZONES: 3A, 3B, 4A, 4B, 5A, 5B, 6A, 6B, 7A, 7B, 8A, 8B

HEIGHT: 50 to 80 feet

WIDTH: 30 feet

FRUIT: oval; 4"; dry; reddish-brown; attracts wildlife; causes some litter; persistent; showy

GROWTH RATE: rapid-growing; long-lived

HABIT: upright oval; open; irregular silhouette; fine texture

FAMILY: *Pinaceae*

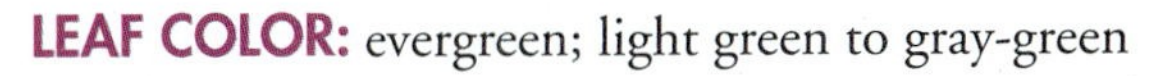

LEAF COLOR: evergreen; light green to gray-green

LEAF DESCRIPTION: spiral; simple; three needles; 8 to 10" long

FLOWER DESCRIPTION: males purple, females red; not showy; spring

EXPOSURE/CULTURE:

Light Requirement: grows best in full sun

Soil Tolerances: all textures; slightly alkaline to acidic; drought

PEST PROBLEMS: somewhat susceptible

BARK/STEMS:

Thick, fire-resistant bark; branches grow mostly horizontal or upright and will not droop; should be grown with a single leader; no thorns

Pruning Requirement: needs little pruning to develop strong structure; remove lower branches as they are shaded out and die

Limb Breakage: susceptible to breakage in heavy snow or ice storms

Current Year Twig: yellow to reddish-brown; medium thickness

CULTIVARS AND SPECIES:

The varieties *ponderosa* and *scopulorum* are recognized, each consisting of two or three races. The species varies widely in growth rate, survival, and root regeneration. Plant trees that were grown from seeds collected locally for the best survival and growth.

LANDSCAPE NOTES:

Origin: western North America

Usage: reclamation; lumber; windbreak; specimen; buffer strip; highway

Ponderosa Pine is a North American native which is usually seen from 50 to 80 feet tall with a 25- to 30-foot spread, though it is capable of reaching more than 150 feet in height. It is the most widely distributed Pine in western North America. This moderately fast-growing pine is pyramidal when young and loses its lower limbs as it grows older, unless grown in full sun on an exposed site. The canopy remains rather open throughout the life of the tree.

The 10-inch-long evergreen needles emerge a light green and turn to a gray-green during the growing season. The often-paired cones are four or more inches long, red-brown, and have very sharp spines. The bark of Ponderosa Pine is very thick, which helps make this tree very resistant to fire in the wild and also helps it maintain park-like stands in its native habitat.

Pines are often grouped together in a landscape in a mass planting, perhaps mimicking nature. They are becoming more popular for planting in parks and in commercial landscapes. They create a light shade that allows grass and other plants to grow easily beneath the canopy. People often complain about the dropping needles and cones.

Ponderosa Pine should be grown in full sun on well-drained, acid or slightly alkaline soil. It is highly drought tolerant once established and can live without irrigation in its native range once it is well established. Some horticulturists routinely incorporate native peat into the backfill soil to help ensure mycorrhizal inoculation.

Limitations include needle cast, bark beetles, air pollution sensitivity, pitch canker, and root diseases. European pine shoot moth can cause twig deformation and die-back.

Pinus radiata Monterey Pine

Pronunciation: PIE-nuss ray-dea-AH-tah

ZONES: 10A, 10B, 11

HEIGHT: 30 to 50 feet

WIDTH: 25 to 35 feet

FRUIT: elongated cone; 5"; dry; brown; causes some litter; persistent on tree for several years; showy

GROWTH RATE: rapid-growing; long-lived

HABIT: pyramidal when young, then irregular and sometimes layered; open to moderate density; fine texture

FAMILY: *Pinaceae*

LEAF COLOR: bright green; needle-leaf evergreen

LEAF DESCRIPTION: spiral; simple; five needles; 4 to 6" long

FLOWER DESCRIPTION: males are yellowish-brown, females red-purple; not showy; spring

EXPOSURE/CULTURE:

Light Requirement: part shade to full sun when young; full sun is needed for older trees

Soil Tolerances: clay or loam; acidic; occasionally wet soil; salt; drought-sensitive

Aerosol Salt Tolerance: good

PEST PROBLEMS: somewhat susceptible

BARK/STEMS:

Bark is thick and resists impact; lower branches droop as tree grows and will require pruning for vehicular or pedestrian clearance beneath canopy; should be grown with a single leader; no thorns

Pruning Requirement: needs little pruning to develop strong structure; pruning January through October can encourage disease and insect infestation

Limb Breakage: mostly resistant, although newly planted and fast-growing specimens are susceptible to windthrow

Current Year Twig: brown; green; slender

LANDSCAPE NOTES:

Origin: California, near San Francisco

Usage: screen; shade tree; specimen; Christmas tree

Despite its tiny native range, Monterey Pine is the most widely planted Pine in the world. Popularity is owed to a rapid growth rate and desirable lumber and pulp characteristics. It has gained popularity in the landscape trade and is planted in many areas within its hardiness range. Several branches begin to develop in random directions on open-grown trees, eventually forming an open, irregular canopy. It is widely planted on golf courses, parks, and landscapes of any size in California.

The lower branches are retained on open-grown trees in full sun, making Monterey Pine an excellent candidate for specimen use, although group plantings with trees spaced 15 to 25 feet apart add a soft accent to any landscape. Pinch or clip the candles as the needles begin to emerge to slow the growth rate and to create a denser canopy on young trees, if you wish. It is best to leave the leader intact, as the tree should be grown with one central trunk.

Root systems are usually shallow and highly branched, with many fine roots close to the surface of the soil. Roots in this Pine's native habitat penetrate into a clay layer that supplies the tree with water during the dry season. Trees transplant well balled-in-burlap or from containers.

Young Monterey Pines are quite tolerant of half-day shade, but mature trees prefer a sunny location and loamy, moist, well-drained soils. They do not grow well and often die on dry, sandy soil unless roots reach a water source. Trees appear to have little tolerance for drought and soil compaction. Trees are sensitive to air pollution, especially ozone.

Needle blight, pitch canker, western gall rust, dwarf mistletoe, bark and pine cone beetles are all problems. Declines as it ages in warm climates. California five-spindle engraver beetle is a serious pest. Needles brown in smoggy urban areas.

Pinus strobus Eastern White Pine

Pronunciation: PIE-nuss STROH-bus

ZONES: 3B, 4A, 4B, 5A, 5B, 6A, 6B, 7A, 7B

HEIGHT: 50 to 80 feet

WIDTH: 25 to 35 feet

FRUIT: elongated; 6 to 8"; dry; brown; causes some litter; persistent; showy

GROWTH RATE: rapid; long-lived

HABIT: pyramidal when young, then oval and layered; open to moderately dense; symmetrical; fine texture

FAMILY: *Pinaceae*

LEAF COLOR: blue-green to green; needle-leaf evergreen

LEAF DESCRIPTION: spiral; five needles; 3 to 5" long

FLOWER DESCRIPTION: pink; yellow; not showy; spring

EXPOSURE/CULTURE:

Light Requirement: part shade to full sun

Soil Tolerances: well-drained clay; loam; sand; acidic; occasionally wet sandy, organic, or loamy soil; salt- and drought-sensitive

Aerosol Salt Tolerance: none

PEST PROBLEMS: sensitive

BARK/STEMS:

Bark is thin and easily damaged by mechanical impact; branches droop as tree grows and will require pruning for vehicular or pedestrian clearance beneath canopy; should be grown with a single leader; no thorns

Pruning Requirement: needs little pruning to develop strong structure; tolerates shearing

Limb Breakage: susceptible, because branch wood is weak

Current Year Twig: brown; green; slender

LANDSCAPE NOTES:

Origin: northeastern North America into the Appalachians; *P. monticola* is native to the west

Usage: screen; hedge; shade tree; specimen; Christmas tree

Eastern White Pine has needles borne in groups of five; foliage color varies greatly from one tree to the next. It is the state tree of Maine and Missouri. Some specimens keep their bluish color throughout the winter, others lose it. Older yellow needles fall in September. Several branches on young trees normally originate from the same point on the trunk, forming a tree appearing to be built of layers of foliage. Although young trees are pyramidal and usually grow with one central leader, the layers (or whorls) of horizontal branches give White Pine a distinctive appearance in middle and old age.

The lower branches are retained, making White Pine an excellent candidate for specimen use. Group plantings with trees spaced 15 to 25 feet apart add a soft accent to any landscape. Planted 8 to 15 feet apart, they are one of only a few Pines that make a nice clipped hedge or screen of soft foliage. Pinch or clip the candles as the needles begin to emerge.

Root systems are usually shallow and highly branched, with many fine roots close to the surface of the soil. Trees transplant well balled-in-burlap or from containers.

Young White Pines are quite tolerant of half-day shade, but mature White Pines prefer a sunny location and tolerate loamy, moist, well-drained soils. They do not grow well and often die on clayey soil or soil with a pH above 7. Trees appear to have little tolerance for drought, soil compaction, and heat and should be used only in the cooler climates. Eastern White Pine is susceptible to salt injury from roads or drain fields and is sensitive to air pollution (particularly ozone and sulfur dioxide).

White Pine weevil bores into the terminal growth and disfigures the tree. Be sure to select White Pine trees certified to be blister-rust-resistant. White Pine decline is used to describe the slow decline of trees planted in dry, clay soils low in organic matter especially those planted in full sun. Plants with this disorder have only a small cluster of needles at the end of each branch. European pine shoot beetle and European pine shoot moth are the latest problems to have a dramtic impact on White Pine, but there are other diseases and pests as well.

Pinus strobus 'Nana' Dwarf White Pine

Pronunciation: PIE-nuss STROH-bus

HEIGHT: 15 to 20 feet

WIDTH: 10 feet

GROWTH RATE: very slow-growing; long-lived

HABIT: pyramidal; dense; symmetrical; fine, billowing texture

LANDSCAPE NOTES:

Usage: screen; specimen; small garden

Dwarf Eastern White Pine has soft blue-green needles borne in groups of five in a billowing canopy. The tight foliage and slow growth make this small tree well suited for planting in small gardens and most residential landscapes.

Pinus strobus 'Pendula' Weeping White Pine

Pronunciation: PIE-nuss STROH-bus

HEIGHT: 15 to 20 feet

WIDTH: 25 to 30 feet

HABIT: weeping and spreading

EXPOSURE/CULTURE:

Light Requirement: full sun is needed for best development

CULTIVARS AND SPECIES:

'Alba'—creamy white new growth lasting nearly two months; 'Hillside Winter Gold'—maintains light yellow needles into the winter.

LANDSCAPE NOTES:

Usage: specimen; screen

Weeping White Pine is a beautiful example of horticultural selection and know how. This special habit is developed in the nursery by careful training and pruning. The tree needs to be trained when it is young to develop a leader that stands erect. Otherwise, a spreading shrub forms, not a tree.

Pinus sylvestris Scotch or Scots Pine

Pronunciation: PIE-nuss sill-VESS-triss

ZONES: 3A, 3B, 4A, 4B, 5A, 5B, 6A, 6B, 7A, 7B, 8A

HEIGHT: 40 to 50 feet

WIDTH: 30 feet

FRUIT: oval; 1 to 3"; dry; brown; causes some litter; persistent

GROWTH RATE: moderate; moderate life span

HABIT: pyramidal when young, oval later; open; irregular silhouette; fine texture

FAMILY: *Pinaceae*

LEAF COLOR: green to blue-green; needle-leaf evergreen

LEAF DESCRIPTION: spiral; two needles; 2 to 4" long

FLOWER DESCRIPTION: yellow; spring; not showy

EXPOSURE/CULTURE:

Light Requirement: grows best in full sun

Soil Tolerances: loam; sand; slightly alkaline to acidic; moderate salt; moderate drought

Aerosol Salt Tolerance: low to moderate

PEST PROBLEMS: susceptible

BARK/STEMS:

Showy, orange bark; branches grow mostly horizontal and upright and will not droop; should be grown with a single leader; often forms several trunks; no thorns

Pruning Requirement: needs little pruning, except to eliminate double leaders, to develop strong structure

Limb Breakage: susceptible to breakage either at crotch, due to poor collar formation, or wood itself is weak and tends to break

Current Year Twig: green; medium thickness

CULTIVARS AND SPECIES:

'Aurea' is one of the oldest cultivars, with light-green summer foliage and bright yellow winter color; 'Fastigiata' grows to about 15 or 20 feet tall with a 4-foot-wide canopy; 'French Blue' has bluish, dense foliage. Ice and snow storms usually bend upright branches and trunks to the ground, disfiguring the tree and shortening its life. Other cultivars have been selected for their dwarf habit and variegated foliage.

LANDSCAPE NOTES:

Origin: Europe and Asia

Usage: bonsai; reclamation; Christmas tree; specimen

In recent years this tree has been bothered in the eastern United States with fatal attacks of Pine wilt nematode; therefore, its use in landscapes is not recommended in many areas. A widely planted evergreen in the past, Scotch Pine has bluish-green to green foliage that usually turns yellowish-green in winter. Orange bark on the trunk and major limbs peels in papery flakes and is visible through the canopy, especially early or late in the day when the sun is low in the sky. The plant will tolerate dry soil and exposed sites, forming an open, picturesque, asymmetrical canopy. It grows best with a soil pH below 7.5.

It is a good tree for reclamation sites because it often seeds itself into the site. It is tough and durable, and is very popular as a Christmas tree. (It is often sprayed with green dye to give it a desirable green color for the holiday season.) It is widely planted for its showy bark and adaptability to urban soils.

Although it is considered tolerant of dry soil, the Pinyon and other western Pines are better suited for dry climates. Scotch Pine benefits from some irrigation in dry times in the western United States.

Pine wilt nematode and tip blight are the most significant pest problems in the eastern and central United States. European pine shoot beetle is the latest recognized problem beginning to attack this tree. A host of other potential problems can also plague Scots Pine.

Pinus sylvestris 'Watereri' 'Watereri' Scotch Pine

Pronunciation: PIE-nuss sill-VESS-triss

ZONES: 3A, 3B, 4A, 4B, 5A, 5B, 6A, 6B, 7A, 7B, 8A

HEIGHT: 12 to 20 feet

WIDTH: 10 feet

FRUIT: oval; 1 to 3"; dry; brown; no litter; persistent

GROWTH RATE: very slow-growing; long-lived

HABIT: pyramidal if unpruned, or rounded vase with lower branches removed; dense; symmetrical; fine texture

FAMILY: *Pinaceae*

LEAF COLOR: evergreen; bluish-green

LEAF DESCRIPTION: spiral; simple; two needles; 2" long

FLOWER DESCRIPTION: yellow; not showy

EXPOSURE/CULTURE:

Light Requirement: grows best in full sun

Soil Tolerances: loam; sand; slightly alkaline to acidic; moderate salt; moderate drought

Aerosol Salt Tolerance: low to moderate

PEST PROBLEMS: susceptible

BARK/STEMS:

Showy, orange bark; branches grow mostly horizontal and upright and will not droop; strongest structure is with a short, single leader; no thorns

Pruning Requirement: needs little pruning to develop strong structure; removing lower branches shows off attractive bark

Limb Breakage: mostly resistant

Current Year Twig: green; medium thickness

LANDSCAPE NOTES:

Origin: Europe and Asia

Usage: bonsai; specimen

In recent years the species has been bothered in the eastern United States with fatal attacks of Pine wilt nematode; therefore, its use in landscapes is not recommended in many areas. There is no reason to believe that this cultivar will be more resistant to the nematode than the species.

'Watereri' Scotch Pine's foliage is held in a compact canopy. Orange bark on the trunk and major limbs peels in papery flakes, and is visible through the canopy if it is thinned or if the small-diameter lower branches are removed from the trunk. The bark is especially pretty early or late in the day when the sun strikes the trunk directly. The plant forms a tight canopy on upright branches originating from the lower trunk. This cultivar is often grafted onto a pine understock.

This small tree is perfectly sized for small landscapes and gardens. Its slow growth combines with a compact habit to fit into tight, urban spaces. Tolerance to urban soil conditions might make it a candidate for planting along streets in areas where the Pine wilt nematode is not a serious problem. It is one of the few pines that would be suited for planting in parking lot islands. However, it is often used more for the residential or commercial landscape where this beautiful tree can be displayed in a lawn or low-growing groundcover.

Scotch Pine is planted for its showy bark and adaptability to urban soils. It is somewhat tolerant of soil pH above 7.5, however, best growth is below 7.5 pH, and grows well in dry soil, but the Pinyon and other western Pines are better suited for dry western sites. 'Watereri' benefits from irrigation in dry times in the western United States.

Pine wilt nematode and tip blight are the most significant pest problems on Scots Pine in the eastern and central United States. This tree is not recommended for regions where pine wilt nematode is prevalent. A host of other potential problems can also plague Scotch Pine.

Pinus taeda Loblolly Pine

Pronunciation: PIE-nuss TEE-dah

ZONES: 6B, 7A, 7B, 8A, 8B, 9A, 9B

HEIGHT: 50 to 80 feet

WIDTH: 30 feet

FRUIT: oval; 3 to 6"; dry; brown; attracts mammals; causes significant litter; persistent; showy

GROWTH RATE: rapid; long-lived

HABIT: oval; open; irregular silhouette; fine texture

FAMILY: *Pinaceae*

LEAF COLOR: needle-leaf evergreen; light green to brown

LEAF DESCRIPTION: spiral; three needles, 6 to 9" long

FLOWER DESCRIPTION: yellow; not showy; spring

EXPOSURE/CULTURE:

Light Requirement: grows best in full sun

Soil Tolerances: all textures; acidic; occasionally wet soil; moderate drought

Aerosol Salt Tolerance: low

PEST PROBLEMS: susceptible

BARK/STEMS:

Branches grow mostly horizontal or upright and will not droop; should be grown with a single leader; no thorns

Pruning Requirement: needs little pruning to develop strong structure; remove lower branches as they are shaded out and die

Limb Breakage: susceptible

Current Year Twig: brown; medium thickness

CULTIVARS AND SPECIES:

'Nana' reportedly reaches only 8 to 16 feet in height, making it ideal for use as a specimen or screen. It has a dense, rounded silhouette and may become popular, especially for small-scale landscapes, once people discover it and it becomes available.

LANDSCAPE NOTES:

Origin: New Jersey to Florida and Texas

Usage: reclamation; windbreak; specimen; lumber

Loblolly Pine is a North American native which is capable of reaching more than 150 feet in height in its natural habitat. This extremely fast-growing pine is pyramidal when young, making it ideal for screening, but loses its lower limbs as it grows older, becoming a tall, stately specimen, windbreak, or shade tree casting dappled shade. The six- to nine-inch-long evergreen needles turn light green to brown during the winter. The often-paired cones are three to six inches long, red-brown, and have very sharp spines. They persist on the tree for several years and mature in the fall. The bark of Loblolly Pine is very thick, which helps make this tree very resistant to fire in the wild. Trees with crooked trunks are more common than on Slash Pine.

Pines are often grouped together in a landscape. They are becoming more popular for planting in parks and in commercial landscapes. They create a light shade that allows grass and other plants to grow easily beneath the canopy. People sometimes complain about the dropping needles on sidewalks and pavement.

Loblolly Pine should be grown in full sun on well-drained, acid soil. It is drought-tolerant once established. Pine bark beetle, borers, pine tip moth, sawflies, fusiform rust, and heart rot can limit growth on Loblolly Pine. Loblolly Pine is the preferred host of the southern pine beetle.

Pinus thunbergiana Japanese Black Pine

Pronunciation: PIE-nuss thun-burr-gee-A-nah

ZONES: 6A, 6B, 7A, 7B, 8A, 8B, 9A, (9B), (10A), (10B)

HEIGHT: 25 to 30 feet

WIDTH: 20 to 35 feet

FRUIT: oval; 1 to 3" long; dry; brown; causes some litter; persistent; showy

GROWTH RATE: moderate; moderate life span

HABIT: pyramid; open to moderately dense; irregular silhouette; medium texture

FAMILY: *Pinaceae*

LEAF COLOR: dark green; needle-leaf evergreen

LEAF DESCRIPTION: spiral; two needles; 5 to 7" long

FLOWER DESCRIPTION: yellow; not showy

EXPOSURE/CULTURE:

Light Requirement: grows best in full sun

Soil Tolerances: all textures; alkaline to acidic; salt and drought

Aerosol Salt Tolerance: high

PEST PROBLEMS: sensitive in parts of eastern United States

BARK/STEMS:

Can be trained with a single trunk or multiple trunks; branches droop as tree ages; no thorns

Pruning Requirement: needs little pruning to develop strong structure; can be trained and pinched to a more uniform, symmetrical canopy

Limb Breakage: resistant

Current Year Twig: brown; medium thickness

LANDSCAPE NOTES:

Origin: Japan

Usage: bonsai; container or above-ground planter; buffer strip; parking lot island; highway; specimen

An excellent, small, irregularly shaped Pine, the size and shape of Japanese Black Pine are variable. The five- to seven-inch-long twisted needles are borne in groups of two. 'Angelica's Thunderhead' features bright white candles that contrast with the dark-green needles. Although trees can be grown with or without a central leader, prune to develop one for the most durable form. Branches on older trees are held horizontally in a picturesque silhouette, and sometimes can outgrow the central leader, forming an attractive multi-stemmed specimen tree. It is often used in this way.

Trunks are usually not straight but sweep up in a gentle curve. This tree is often used as a specimen on smaller properties, or is planted in groups in larger-scale landscapes. Space 10 to 20 feet apart for mass planting or to create an open screen. Japanese Black Pine can be successfully trained into a bonsai or small specimen plant with selective pruning. It lends itself well to thinning by removing selected branches and small twigs and by pinching new growth, and is often trimmed this way in Japanese gardens. Some nurseries offer the tree pruned into an attractive, dense, pyramidal form for use in landscaping residential and commercial properties.

The plant tolerates dry, sandy soil and is extremely salt-tolerant, being used successfully along beachfront property in full sun. It also grows well in clay soil with a pH of 8.0. Drought tolerance is good. It may grow two to three feet in a year, but this rapid growth under optimum cultural conditions may lead to the asymmetrical, open-growth habit.

Pine wilt nematode in the eastern United States has deformed and killed many trees. Tip moth also affects recently transplanted Pines. Maskell scale has recently devastated large numbers of trees in New Jersey. Some adelgids will appear as white cottony growths on the bark. All types produce honeydew, which may support sooty mold. European pine shoot moth causes young shoots to fall over. Sawfly larvae caterpillars are variously colored and generously feed in groups on the needles. Resistant to tip blight. More pest-resistant in the western half of North America.

Pinus virginiana Virginia Pine, Scrub Pine

Pronunciation: PIE-nuss vir-gin-ee-A-nah

ZONES: 5A, 5B, 6A, 6B, 7A, 7B, 8A, 8B

HEIGHT: 20 to 40 feet

WIDTH: 20 to 35 feet

FRUIT: oval; 1 to 3"; dry; brown; attracts mammals; causes significant litter; persistent; showy

GROWTH RATE: moderate; moderate life span

HABIT: oval or round; open; irregular silhouette; fine texture

FAMILY: *Pinaceae*

LEAF COLOR: needle-leaf evergreen; yellowish-green

LEAF DESCRIPTION: spiral; two needles; 2 to 3" long

FLOWER DESCRIPTION: yellow; not showy; spring

EXPOSURE/CULTURE:

Light Requirement: grows best in full sun

Soil Tolerances: all textures; slightly alkaline to acidic; drought

Aerosol Salt Tolerance: low

PEST PROBLEMS: somewhat susceptible

BARK/STEMS:

Bark is thin and easily damaged by mechanical impact; branches droop as tree ages; should be grown with a single leader; no thorns

Pruning Requirement: needs little pruning to develop strong structure; can be sheared and pinched to increase canopy density

Limb Breakage: susceptible to breakage at crotch, due to poor collar formation; wood itself is weak and tends to break

Current Year Twig: reddish; slender

LANDSCAPE NOTES:

Origin: Long Island to Tennessee and South Carolina

Usage: bonsai; buffer strip; highway; reclamation; Christmas tree

This scrubby North American native tree is most often found growing in the poorest sites and will easily adapt to most soil conditions except alkaline. Capable of reaching up to 70 feet in height with a straight trunk on a good site, Virginia Pine on a poor site forms a twisted trunk from 20 to 40 feet in height. The 1.5 to 3-inch-long, flexible needles are joined by numerous, mature, prickly cones. The thin, orange-brown bark becomes ridged and furrowed on older trees and is easily seen through the open canopy.

Although considered to have an untidy appearance because of an irregular habit, the very low branches that persist on the tree help make Virginia Pine a popular choice in the South for culture as a Christmas tree. Rotations can be as quick as 3 years but average 5 to 10 years. The trees grow moderately fast in cultivation and have a branching structure that tolerates pruning quite well. Pruning or shearing increases branchiness to create a nice Christmas tree. Sheared Christmas trees planted in the landscape may look odd for a couple of years while the natural form of the tree develops. Virginia Pine can also be used for an open-branched specimen in a large-scale landscape.

Tolerant of a wide variety of soil types, Virginia Pine prefers to be grown in full sun on well-drained, loamy soil. It grows on soil too dry, rocky, or clayey for most other plants, particularly Pines, but prefers acidic pH. It is useful as a reclamation tree due to its ability to seed itself in and its tolerance to poor, dry soil. It has been successfully used in strip-mine reclamation sites in the eastern United States without serious pest infestation.

Normally, no serious limitations, although the list of possible problems is long. Pitch canker enters the tree through wounds, enlarges rapidly, and girdles the stem. Heart rot in the trunk is common on trees more than 50 years old. Ips beetles can finish off trees under stress from other problems, such as a reduced root system near a recently constructed building. Pine sawflies can quickly eat significant quantities of foliage.

Pistacia chinensis Chinese Pistache

Pronunciation: piss-TAY-she-ah chi-NEN-sis

ZONES: 6B, 7A, 7B, 8A, 8B, 9A, 9B, (10)

HEIGHT: 25 to 35 feet

WIDTH: 25 to 35 feet

FRUIT: round; 0.33"; fleshy; orange or red; causes some litter; very showy on female trees

GROWTH RATE: moderate; moderate life span

HABIT: rounded vase; moderate density; irregular when young, symmetrical later; medium texture

FAMILY: *Anacardiaceae*

LEAF COLOR: dark green; strikingly beautiful orange or red fall color, especially female trees

LEAF DESCRIPTION: alternate; even pinnately compound; entire margin; lanceolate; deciduous; 2 to 4" long leaflets

FLOWER DESCRIPTION: greenish-white; somewhat showy; spring

EXPOSURE/CULTURE:

Light Requirement: grows best in full sun, tolerates partial shade

Soil Tolerances: all textures; alkaline to acidic; drought

Aerosol Salt Tolerance: none

PEST PROBLEMS: pest-free

BARK/STEMS:

Showy, orange, exfoliating bark; branches droop as tree grows and will require pruning for vehicular or pedestrian clearance beneath canopy; should be grown with a single leader; no thorns

Pruning Requirement: requires some early pruning to develop strong structure and uniform branching in nursery; head back lateral branches on young trees to increase branching

Limb Breakage: resistant

Current Year Twig: brown; thick

LANDSCAPE NOTES:

Origin: China and Taiwan

Usage: parking lot island; buffer strip; highway; near a deck or patio; reclamation; shade tree; specimen; sidewalk cutout; street tree; urban tolerant

Finely divided, lustrous foliage, bright red fruit (on female trees) that ripens to dark blue, peeling, attractive bark, and wonderful fall colors combine to make Chinese Pistache an outstanding specimen, shade, or street tree. Young trees are asymmetrical and a bit awkward-looking, requiring regular pruning in the nursery to produce a well-formed canopy. Lower branches often droop to the ground with time, forming a wonderfully spreading crown. Older, more mature trees become more dense and uniformly shaped. This is one of the last trees to color in the fall, extending the fall color show in many parts of the south into November. Leaves drop over a short span of time.

The tree needs special pruning and training in the early years to create branches in desirable places along the trunk. To train an unbranched young sapling, prune the top to force development of several branches. Pick one to be the trunk, another to be a branch, and remove or head back the rest. Allow the tree to grow taller, and again top the unbranched trunk 12 to 18 inches above the first pruning cut to force branch development there. Build the tree in this fashion until a desirable structure with well-spaced branches is achieved.

This is certainly an underutilized urban tree. It has merit for wider use in street tree plantings and in other adverse sites due to its drought tolerance, adaptability, moderate size, and wonderful form and fall color. It is hard to go wrong with Chinese Pistache in all areas within its hardiness range. Chinese Pistache grows quickly in full sun to partial shade on moderately fertile, well-drained soils and will withstand heat and drought extremely well.

Chinese Pistache is used as the understock on which the commercial pistachio nut (*Pistacia vera*) is grafted. The cultivar 'Keith Davey' has outstanding autumn color. Verticillium wilt and oak root fungus occasionally affect Chinese Pistache.

Pithecellobium flexicaule
Ebony Blackbead, Texas Ebony
Pronunciation: pith-eh-sell-LOW-bee-um flex-eh-CALL-ee

ZONES: (9A), 9B, 10A, 10B, 11

HEIGHT: 35 to 50 feet

WIDTH: 20 to 30 feet

FRUIT: elongated; pod; 4 to 6"; dry; black to brown; edible; no significant litter; persistent; showy

GROWTH RATE: slow to moderate; long-lived

HABIT: rounded to vase; spreading; moderate; irregular when young, more symmetrical later; fine texture

FAMILY: *Leguminosae*

LEAF COLOR: evergreen; dark green

LEAF DESCRIPTION: alternate; bipinnately compound; entire margin; oval to obovate leaflets; less than 0.5" long

FLOWER DESCRIPTION: whitish-yellow; pleasant fragrance; showy; summer; attracts bees

EXPOSURE/CULTURE:

Light Requirement: grows best in full sun

Soil Tolerances: all textures; alkaline to slightly acidic; very drought-tolerant

Aerosol Salt Tolerance: moderate

PEST PROBLEMS: resistant

BARK/STEMS:

Routinely grown with, or trainable to be grown with, multiple trunks; branches grow somewhat upright and horizontal, but droop on older trees; short thorns are present on branches

Pruning Requirement: requires pruning to develop strong structure; space branches along a short, central trunk

Limb Breakage: resistant

Current Year Twig: brown; green; thick

CULTIVARS AND SPECIES:

P. pallens is a Texas native growing to 10 or 15 feet tall. *P. mexicanum* (zone 8B) grows 25 to 30 feet tall.

LANDSCAPE NOTES:

Origin: south Texas into Mexico

Usage: raised planter or container; parking lot island; buffer strip; highway; reclamation; small shade tree; specimen

This tree is native to Texas and Mexico and is ideal for use in dry, desert landscapes. The short branches are clothed in very small leaflets and make up a rounded canopy that casts medium to dense shade below. Short thorns are interspersed among the branches, but this should not present a problem for most landscape uses. From June to August, Texas Ebony is decorated with dense, plume-like spikes of very fragrant, light yellow to white blossoms. The dark brown to black, woody seed capsules that follow are four to six inches long and persist on the tree. In Mexico, the seeds from these pods are eaten, and the black woody shells have been roasted as a coffee substitute in times past. The attractive, short trunk of Texas Ebony is covered with smooth, gray bark.

It makes a nice small to medium-sized shade tree for yards, parks, and courtyards. Texas Ebony is well adapted to grow in a raised planter or container along a street. Because of the slightly thorny branches, pruning and early training will be required to raise the crown high enough for locating near a sidewalk. The tree seems to thrive on neglect along highways and in other low-maintenance landscapes. Preferring well-drained, alkaline soils, Texas Ebony will thrive in full sun with little water once established. Trees are tolerant of wind and compacted soil, helping to make them a popular choice for arid landscapes. Although a Texas native, there are some nice trees in landscapes throughout its hardiness range.

Be careful not to overirrigate the tree once it becomes established.

Pittosporum spp. Pittosporum

Pronunciation: pit-uh-SPORE-um species

ZONES: 8A, 8B, 9A, 9B, 10A, 10B, 11

HEIGHT: varies by species; 12 to 40 feet

WIDTH: 10 to 35 feet

FRUIT: round; 0.25"; hard; green; some messy litter

GROWTH RATE: moderate; moderate life span

HABIT: rounded vase; dense; symmetrical; fine to medium texture

FAMILY: *Oleaceae*

LEAF COLOR: evergreen; dark green

LEAF DESCRIPTION: alternate; simple; entire or undulating margin; generally obovate; 2 to 4" long

FLOWER DESCRIPTION: white; showy in clusters; spring and summer; some with strong fragrance

EXPOSURE/CULTURE:

Light Requirement: part shade to full sun

Soil Tolerances: all textures; slightly alkaline to acidic; moderate drought

Aerosol Salt Tolerance: moderate to high

PEST PROBLEMS: resistant

BARK/STEMS:

Bark is thin and easily damaged by mechanical impact; branches droop as tree grows and will require pruning for vehicular or pedestrian clearance beneath canopy; routinely grown with multiple trunks; showy trunk structure; no thorns

Pruning Requirement: many species require training and pruning to develop into an upright tree form and to prevent formation of included bark

Limb Breakage: mostly resistant

Current Year Twig: greenish; medium thickness

CULTIVARS AND SPECIES:

P. undulatum can grow 40 to 60 feet tall (though usually seen smaller) and makes a good street tree. *P. tobira* grows to about 15 feet tall and is often planted as a hedge, but can be trained and pruned into a small tree. *P. rhombifolium* grows to 35 to 50 feet tall and is occasionally planted as a street tree. *P. eugenioides* grows to 40 feet tall.

LANDSCAPE NOTES:

Origin: many are from Australia

Usage: container or above-ground planter; clipped hedge; buffer strip; highway; near a deck or patio; standard; specimen; sidewalk cutout; street tree; windbreak; urban tolerant

Although often used as a shrub or hedge, Pittosporum works well when allowed to grow into a small tree, as its curved multiple trunks and tight canopy create an interesting architectural focus. The glossy or dull leaves are abundantly produced in tight clusters on the upright, spreading branches. The canopy on most species stays tight, displaying a billowing form and casting dense shade. Growth rate is moderate to slow, making this a good plant for low-maintenance landscapes. The small, white, fragrant flowers appear in terminal clusters during spring and summer. The blooms are followed by abundant, green fruit that persists on the tree.

Although tolerant of tight clipping, Pittosporum is quite attractive when allowed to retain its natural multistemmed form, making it ideal for use in shrubbery borders and as a featured accent. It makes a nice specimen in any landscape where a small, dark-green tree is needed. Planted close together on about 10- to 20-foot centers, the larger species of Pittosporum will form a canopy over a pedestrian walkway. The tree looks best in a landscape setting with a low groundcover planted around its base.

Pittosporum grows in full sun or partial shade and is tolerant of a wide range of soil types, including calcarious clay, as long as water is not allowed to stand in the root zone. Plants grow quickly while young but slow with age.

Soil nematodes can cause serious plant decline and they can be prevalent, particularly in sandy soil. Aphids and scale insects often are found on Pittosporum.

Platanus x acerifolia 'Bloodgood'

'Bloodgood' London Planetree

Pronunciation: PLAT-ah-nuss a-sir-ee-FOE-lee-ah

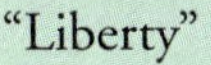

"Liberty"

"Bloodgood"

"Bloodgood" bark

ZONES: 5B, 6A, 6B, 7A, 7B, 8A, 8B, 9A

HEIGHT: 70 to 85 feet

WIDTH: 70 feet

FRUIT: round; 1" in pairs; dry and soft; brown; causes some litter; persistent; showy

GROWTH RATE: rapid; moderate life span

HABIT: pyramidal when young, rounded later; dense; symmetrical; coarse texture

FAMILY: *Platanaceae*

LEAF COLOR: medium green; showy yellow fall color

LEAF DESCRIPTION: alternate; simple; lobed with incised margins; ovate; deciduous; 6 to 10" long; decomposes slowly and litters ground

FLOWER DESCRIPTION: deep red; not showy; spring

EXPOSURE/CULTURE:

Light Requirement: grows best in full sun

Soil Tolerances: all textures; alkaline to acidic; wet soil; salt; moderate drought

Aerosol Salt Tolerance: moderate

PEST PROBLEMS: susceptible

BARK/STEMS:

Showy, exfoliating bark flakes off and can be messy; young branches droop and will require pruning for vehicular or pedestrian clearance beneath canopy; should be grown with a single leader; no thorns

Pruning Requirement: needs little pruning to develop strong structure; prune in dormant season to avoid bleeding; responses well to pollarding

Limb Breakage: resistant, except in ice storms

Current Year Twig: brown; medium thickness

CULTIVARS AND SPECIES:

'Columbia' (easy to propagate) and 'Liberty' are reportedly more resistant to powdery mildew and eastern strains of anthracnose, though not immune. 'Yarwood' is resistant to mildew but susceptible to anthracnose. 'Bloodgood' resists anthracnose but is highly susceptible to mildew.

LANDSCAPE NOTES:

Origin: hybrid

Continued

Platanus x acerifolia *Continued*

Usage: parking lot island; buffer strip; highway; shade tree; sidewalk cutout; street tree; urban tolerant

Pyramidal in youth, with age Planetree develops a spreading rounded crown supported by a few, large-diameter branches. These branches should be spaced two to four feet apart along the trunk to develop a strong structure. The dominant central leader that typically develops on London Planetree usually assures that the structure and arrangement of major limbs is desirable. Little corrective pruning is required other than removing occasionally occurring upright branches with tight crotches. It is also helpful to remove some of the many branches that develop on the central trunk.

The plant tolerates dry soil (but scorches in dry weather) and city conditions well, adapts to most soils (including alkaline), and is more resistant (though not immune) to the anthracnose that afflicts *Platanus occidentalis*. However, it is susceptible to canker stain and leaf scorch, diseases that have caused its demise in some areas, and is often seen infested with lace bugs, which will not kill the tree but cause premature defoliation in late summer. Some tree managers limit use as a street tree due to its large size and pest problems.

Canker stain is very serious on London Planetree and can kill the tree. Bacterial leaf scorch can devastate London Planetree. Aggressive surface roots.

Platanus occidentalis Sycamore

Pronunciation: PLAT-ah-nuss ox-see-den-TAY-liss

ZONES: 4B, 5A, 5B, 6A, 6B, 7A, 7B, 8A, 8B, 9A

HEIGHT: 75 to 90 feet

WIDTH: 60 to 70 feet

FRUIT: abundant; round, 1", borne singly; dry and soft; brown; causes some litter; persistent; showy

GROWTH RATE: rapid; long-lived

HABIT: pyramidal when young, rounded later; dense; symmetrical; coarse texture

FAMILY: *Platanaceae*

LEAF COLOR: medium green; showy yellow fall color

LEAF DESCRIPTION: alternate; simple; lobed; ovate; deciduous; 6 to 10" long

FLOWER DESCRIPTION: deep red; not showy; spring

EXPOSURE/CULTURE:

Light Requirement: grows best in full sun

Soil Tolerances: all textures; alkaline to acidic; wet soil; moderate salt and moderate drought

Aerosol Salt Tolerance: moderate

PEST PROBLEMS: susceptible

BARK/STEMS:

Showy, exfoliating bark; young branches droop and will require pruning for vehicular or pedestrian clearance beneath canopy; should be grown with a single leader; no thorns

Pruning Requirement: needs little pruning to develop strong structure except to eliminate an occasional double leader

Limb Breakage: resistant, except in ice storms

Current Year Twig: brown; medium thickness

LANDSCAPE NOTES:

Origin: Maine to Iowa, Texas and Florida

Usage: large parking lot island; wide buffer strip; highway; shade tree; properly designed sidewalk cutout; street tree; reclamation; urban tolerant

Sycamore is a massive, tall tree, with a rapid growth rate, that tolerates wet and compacted soil. Pyramidal in youth, it develops a spreading, rounded, and irregular crown with age, supported by a few very large-diameter branches. The dominant central leader that typically develops on Sycamore usually assures that the structure of major limbs is desirable. Little corrective pruning is required other than removing occasional upright, aggressive branches with tight crotches.

It is best suited for soils that are continuously moist, as dry soil can lead to short life. The tree grows in places that appear most unsuitable to plant growth, such as in areas with low soil oxygen and high pH. Unfortunately, aggressive roots often raise and destroy nearby sidewalks when planted too close. The dense shade created by the tree's canopy may interfere with growth of lawn grasses beneath it. In addition, the leaves that fall in autumn reportedly release a substance that can kill newly planted grass. It is suited for the toughest sites, not residential landscapes, and should be supplied with some irrigation in drought. Plant at least six feet from the curb or sidewalk. The variety *glabrata* is best suited for conditions and climates similar to the eastern two thirds of Texas.

Bacterial leaf scorch can kill the tree in several growing seasons, and can cause significant losses. Branch and stem cankers form on limbs of trees stressed by drought. Powdery mildew causes a white fuzz on the tops of leaves and distorts leaves but will not kill the tree. Sycamore is subject to anthracnose disease in wet, cool springs. The disease causes moderate to severe leaf drop and many trees are removed with this disease each year in major cities. Many trees also defoliate early in the fall due to lace bug infestation. Aggressive surface roots often form.

Platanus racemosa Western, or California, Sycamore

Pronunciation: PLAT-ah-nuss race-MOE-sah

ZONES: 7A, 7B, 8A, 8B, 9A, 9B, 10A, 10B, 11

HEIGHT: 75 to 90 feet

WIDTH: 40 to 70 feet

FRUIT: abundant; round; 1" borne in groups of 3 to 7; dry and soft; brown; causes some litter; persistent; showy

GROWTH RATE: rapid-growing; moderate life span

HABIT: pyramidal or upright when young, rounded later; dense; nearly symmetrical; coarse texture

FAMILY: *Platanaceae*

LEAF COLOR: green; some muted yellow fall color

LEAF DESCRIPTION: alternate; simple; lobed; ovate; deciduous; 10 to 15" long

FLOWER DESCRIPTION: red; not showy; spring

EXPOSURE/CULTURE:

Light Requirement: grows best in full sun

Soil Tolerances: all textures; alkaline to acidic; wet soil; moderate salt and drought

Aerosol Salt Tolerance: moderate

PEST PROBLEMS: susceptible

BARK/STEMS:

Showy, exfoliating bark; young branches droop and will require pruning for vehicular or pedestrian clearance beneath canopy; should be grown with a single leader; no thorns

Pruning Requirement: needs little pruning to develop strong structure

Limb Breakage: resistant, except in ice storms

Current Year Twig: brown; medium thickness

LANDSCAPE NOTES:

Origin: along streams in central California to Mexico

Usage: susceptibility to anthracnose limits usefulness; wide buffer strip; highway; reclamation; urban tolerant

Sycamore is a massive, tall tree, with a rapid growth rate, that tolerates wet and compacted soil. It is a huge tree, suited only for the largest landscapes. Upright or pyramidal in youth, it develops a spreading, oval crown with age, supported by a few very large-diameter branches. The dominant central leader that typically develops on Sycamore usually assures that the structure of major limbs is desirable, with little corrective pruning required other than removing occasional upright, aggressive branches with tight crotches.

Although considered drought-tolerant, Sycamore is best suited for soils that are continually moist, as dry soil can lead to short life. Unfortunately, aggressive roots often raise and destroy nearby sidewalks when planted too close. Allow at least 12 feet (preferably more) of soil between the sidewalk and curb when planting as a street tree. The dense shade created by the tree's canopy may interfere with the growth of lawn grasses beneath it. Best not planted in yards, due to a messy habit, Sycamore should be saved for the toughest sites and supplied with some irrigation in drought.

Sycamore is subject to attacks of anthracnose, which causes moderate to severe leaf drop and can kill trees. Leaves often fall periodically during the growing season from this disease and from dry soil. For these reasons, it is often best to save this potentially enormous tree for planting in parks and other large-scale landscapes. Many horticulturists recommended against planting this tree due to the disease problems. It is too messy and large for most home landscapes.

Spider mites are especially troublesome in dry weather. Dead, brown leaves often persist on the tree through the winter in warm climates.

Plumeria alba White Frangipani

Pronunciation: ploo-ME-ree-ah AL-bah

ZONES: 10B, 11

HEIGHT: 20 feet

WIDTH: 20 feet

FRUIT: elongated; 6 to 12"; dry; brown; not showy; no significant litter; persistent

GROWTH RATE: slow-growing; short to moderate life span

HABIT: rounded vase; open; nearly symmetrical; coarse texture

FAMILY: *Apocynaceae*

LEAF COLOR: green

LEAF DESCRIPTION: alternate; simple; entire margin; oval to obovate; deciduous; 10 to 20" long

FLOWER DESCRIPTION: white; pleasant fragrance; showy; spring and summer

EXPOSURE/CULTURE:

Light Requirement: grows best in full sun

Soil Tolerances: all textures; alkaline to acidic; salt and drought

Aerosol Salt Tolerance: moderate to good

PEST PROBLEMS: resistant

BARK/STEMS:

Bark is thin and easily damaged by mechanical impact; branches are low and droop as tree grows and will require pruning for vehicular or pedestrian clearance beneath canopy; routinely grown with multiple trunks; can be trained to grow with a short, single trunk; no thorns

Pruning Requirement: requires pruning to develop strong structure; remove branches that are forming included bark

Limb Breakage: susceptible to breakage at crotch, due to poor collar formation; wood itself is weak and tends to break

Current Year Twig: brown; stout

CULTIVARS AND SPECIES:

P. rubra has reddish-pink flowers. *P. obtusa* has white blooms centered in yellow and is variable in form and color. There are many other selections of Frangipani which display a variety of flower colors. They are widely grown in Hawaii where they are used in flower arrangements and in leis.

LANDSCAPE NOTES:

Origin: west Indies

Usage: container or above-ground planter; buffer strip; parking lot island; highway; near a deck or patio; specimen; sidewalk cutout; street tree

Frangipani is well known for its intensely fragrant, lovely, spiral-shaped blooms, which appear at branch tips June through November. The tree itself is rather unusual in appearance, the 20-inch-long, coarse, deciduous leaves clustered only at the tips of the rough, blunt, sausage-like, thick, gray-green branches. A milky sap exudes from the branches when they are bruised or punctured. Branches are upright and rather crowded on the trunk, forming a vase or umbrella shape with age. They are rather soft and brittle and can break, but are usually sturdy against the wind unless they are mechanically hit or disturbed.

Frangipani is very susceptible to freezing temperatures and should be adequately protected, or planted only in areas that do not freeze in the winter. Plants will grow quickly in full sun on a variety of well-drained soils and are fairly drought- and salt-tolerant. Reaching a height of 20 to 25 feet with an equal spread, Frangipani works well as a freestanding specimen, as a patio tree, or as part of a shrubbery border. It displays well in a front yard as an attention-grabber by the entrance. It can be grown with a single trunk or branched low to the ground into a multi-trunked specimen. Single-trunk specimens could be planted as median or street trees on 15- to 20-foot centers.

Limitations include scales, frangipani caterpillar, and nematodes. Root rot can infect plants planted in soils with poor drainage. Rust disease occasionally infects foliage.

Plumeria rubra Frangipani

Pronunciation: ploo-ME-ree-ah ROO-brah

ZONES: 10B, 11

HEIGHT: 20 feet

WIDTH: 20 feet

FRUIT: elongated; 6 to 12"; dry; brown; not showy; no significant litter; persistent

GROWTH RATE: slow-growing; short to moderate longevity

HABIT: rounded vase; open; nearly symmetrical; coarse textures

FAMILY: *Apocynaceae*

LEAF COLOR: green

LEAF DESCRIPTION: alternate; simple; entire, undulate margins; oval to obovate; deciduous; 12 to 20" long

FLOWER DESCRIPTION: pinkish-red; pleasant fragrance; showy; spring and summer

EXPOSURE/CULTURE:

Light Requirement: grows best in full sun

Soil Tolerances: all textures; alkaline to acidic; salt and drought

Aerosol Salt Tolerance: high

PEST PROBLEMS: resistant

BARK/STEMS:

Bark is thin and easily damaged by mechanical impact; low branches persist and droop as tree grows and will require pruning for vehicular or pedestrian clearance beneath canopy; routinely grown with multiple trunks; can be trained to grow with a short, single trunk; no thorns

Pruning Requirement: requires pruning to develop strong structure; remove branches developing included bark

Limb Breakage: susceptible to breakage at crotch, due to poor collar formation; wood itself is weak and tends to break

Current Year Twig: brown; stout

CULTIVARS AND SPECIES:

P. rubra produces red-toned flowers; other species offer a variety of colors. *P. alba* has white flowers. *P. obtusa* has white blooms centered in yellow and is variable in form and color. There are other species and many cultivars.

LANDSCAPE NOTES:

Origin: Mexico and the Caribbean Islands

Usage: container or above-ground planter; buffer strip; parking lot island; highway; near a deck or patio; specimen; sidewalk cutout; street tree

Frangipani is well known for its intensely fragrant, lovely, spiral-shaped, reddish blooms, which appear at branch tips June through November. The tree itself is rather unusual in appearance, the 12- to 20-inch-long, coarse, deciduous leaves clustered only at the tips of the rough, blunt, sausage-like, thick, gray-green branches. A milky sap exudes from the branches when they are bruised or punctured. Branches are upright and rather crowded on the trunk, forming a vase or umbrella shape with age. They are rather soft and brittle and can break, but are usually sturdy against the wind unless they are mechanically hit or disturbed. The crown loses its leaves for a short time during the winter, displaying the coarse-textured, stubby branches.

Frangipani is very susceptible to freezing temperatures and should be adequately protected, or planted only in areas that do not freeze in the winter. Plants will grow quickly in full sun on a variety of well-drained soils and are fairly drought- and salt-tolerant. Reaching a height of 20 to 25 feet with an equal spread, Frangipani works well as a freestanding specimen, as a patio tree, or as part of a shrubbery border. It displays well in a front yard as an attention-grabber by the entrance. It can be grown with a single trunk or branched low to the ground into a multi-trunked specimen. Single-trunk specimens could be planted as median or street trees on 15- to 20-foot centers.

Limitations include scales, frangipani caterpillar, and nematodes. Root rot can infect plants planted in soils with poor drainage. Rust disease can discolor and disfigure foliage.

Podocarpus falcatus Podocarpus

Pronunciation: poe-doe-KAR-puss fowl-CAY-tuss

ZONES: 10A, 10B, 11

HEIGHT: 40 feet

WIDTH: 25 to 35 feet

FRUIT: round; less than 0.5"; fleshy; no significant litter

GROWTH RATE: slow to moderate; moderate life span

HABIT: oval; pyramidal; dense; symmetrical; fine texture

FAMILY: *Podocarpaceae*

LEAF COLOR: blue or blue-green; evergreen

LEAF DESCRIPTION: opposite; simple; linear; 1.5 to 2" long

FLOWER DESCRIPTION: not showy

EXPOSURE/CULTURE:

Light Requirement: part shade to full sun

Soil Tolerances: all textures; alkaline to acidic; moderate drought

PEST PROBLEMS: unknown

BARK/STEMS:

Branches droop as tree grows and will require pruning for pedestrian clearance beneath canopy; should be grown with a single leader; no thorns

Pruning Requirement: needs little pruning to develop strong structure

Limb Breakage: resistant

Current Year Twig: green; medium thickness

CULTIVARS AND SPECIES:

P. latifolius is also moderately drought-tolerant.

LANDSCAPE NOTES:

Origin: South Africa

Usage: buffer strip; highway; specimen; screen; windbreak

Podocarpus falcatus grows very slowly, probably to 40 feet or more in an open landscape, but it can reach 100 feet in its native habitat. The two-inch-long foliage, borne on a rigid pyramidal canopy, would make a striking specimen in any landscape. It casts dense shade when branched to the ground, so no grass grows beneath it. It lends a rigid, formal structure to any landscape, due to the stiff, horizontal branches, but the blue foliage softens this effect. It could be used as a specimen or as a screen planted 10 to 15 feet apart.

The tree prefers a rich, moist soil and benefits from irrigation in dry weather. Trees benefit from a layer of mulch extending beyond the edge of the branches. This keeps roots cool and reduces moisture loss from the soil. Grows in full sun or shade on the north side of a building. This tree has not been grown in nurseries but should be tried.

Because the tree has not been grown much, the pest and disease problems are poorly understood. Most Podocarpus species can become infested with scale insects.

Podocarpus gracilior Weeping, or Fern, Podocarpus

Pronunciation: poe-doe-KAR-puss grah-SILL-ee-or

ZONES: 10A, 10B, 11

HEIGHT: 30 to 50 feet

WIDTH: 25 to 35 feet

FRUIT: irregular; up to 1"; fleshy; red; not showy; no significant litter

GROWTH RATE: slow; long-lived

HABIT: rounded to oval; weeping; dense; symmetrical; fine texture

FAMILY: *Podocarpaceae*

LEAF COLOR: evergreen; bright green to dark green

LEAF DESCRIPTION: opposite; spiral; simple; entire margin; linear; 2 to 4" long

FLOWER DESCRIPTION: yellow; not showy; spring

EXPOSURE/CULTURE:

Light Requirement: part shade to full sun

Soil Tolerances: all textures; alkaline to acidic; moderate drought

Aerosol Salt Tolerance: low to moderate

PEST PROBLEMS: pest-free

BARK/STEMS:

Young branches droop and will require pruning for vehicular or pedestrian clearance beneath canopy; should be grown with a single leader; no thorns

Pruning Requirement: needs some pruning to develop strong structure; space major branches along a dominant trunk

Limb Breakage: resistant

Current Year Twig: brown; green; medium thickness

LANDSCAPE NOTES:

Origin: east Africa

Usage: container or above-ground planter; espalier; buffer strip; highway; standard; shade tree; specimen; sidewalk cutout; street tree; screen; hedge; urban tolerant

This tree has a soft, graceful, billowy canopy of bright green new growth, dark green mature leaves, and weeping branch tips. If left unpruned, the lower branch tips will touch the ground and the tree will appear as a stack of foliage emerging from the earth. With lower limbs removed, a rounded vase shape or oval emerges, creating dense shade beneath. Often used as an espalier, Weeping or Fern Podocarpus is quite striking when used as a specimen, shade tree, or screening plant.

It can make a beautiful street or parking lot tree, but lower branches must be removed in time as the tree grows, as they tend to droop and could hinder traffic visibility. This is a small price to pay for this wonderful tree, though. Podocarpus is a tough plant that can grow very well in urban conditions. The trunk will slowly grow to be two feet in diameter or larger. Very adapted to downtown, restricted-soil planting sites, the roots rarely lift sidewalks or cause other problems.

Used as a large patio tree, the fine-textured foliage is sure to draw compliments from visitors and friends. Appearing almost like a large, soft, green cloud, Fern Podocarpus creates a beautiful effect when placed beside a pond or other water interest point. The tree combines well with Yellow Lantana or other low-growing, yellow-flowering groundcovers planted under it. It complements glass-sided structures well, its soft foliage shimmering and reflecting on a breezy, sunny day.

Growing in full sun or partial shade, it will tolerate a wide range of well-drained soils, but should be protected from frost. It is considerably tolerant of dry soils, requiring little irrigation once established in humid climates. Occasional irrigation is beneficial in dry climates.

No limitations of major concern. Root pruning several months prior to digging field stock is considered essential for survival. A very durable, low-maintenance tree that should be grown more often by nursery operators for use in urban and suburban landscapes.

Podocarpus macrophyllus Yew Podocarpus

Pronunciation: poe-doe-KAR-puss mack-roe-FILL-luss

ZONES: 8B, 9A, 9B, 10A, 10B, 11

HEIGHT: 30 to 40 feet

WIDTH: 20 to 30 feet

FRUIT: irregular; oval; 0.5"; fleshy; purple; attracts birds; edible and tasty; not showy; no significant litter

GROWTH RATE: slow to moderate; long-lived

HABIT: round; pyramidal; fairly open; irregular silhouette; fine texture

FAMILY: *Podocarpaceae*

LEAF COLOR: dark green; evergreen

LEAF DESCRIPTION: spiral; simple; linear; 1 to 3" long

FLOWER DESCRIPTION: creamy yellow; somewhat showy; spring

EXPOSURE/CULTURE:

Light Requirement: mostly shaded to full sun

Soil Tolerances: all textures; alkaline to acidic; moderate salt; drought

Aerosol Salt Tolerance: moderate

PEST PROBLEMS: mostly resistant

BARK/STEMS:

Showy, flaking bark; branches grow mostly upright or horizontal; should be grown with a single leader; no thorns

Pruning Requirement: needs little pruning to develop strong structure; remove competing leaders to develop strong tree form

Limb Breakage: resistant

Current Year Twig: green; medium thickness

CULTIVARS AND SPECIES:

The variety *angustifolius* is a narrow, columnar tree with curved leaves, 2 to 4.5 inches long; variety *maki* has erect branches, columnar form, and 1.5- to 3-inch-long leaves.

LANDSCAPE NOTES:

Origin: Japan

Usage: container or above-ground planter; espalier; clipped hedge or screen; parking lot island; buffer strip; highway; near a deck or patio; reclamation; shade tree; specimen; sidewalk cutout; street tree; urban tolerant

With densely foliated lower limbs that reach the ground, and neat, evergreen leaves, Podocarpus is very popular as a dense screen or hedge. However, Podocarpus can reach 40 feet in height when not sheared and is quite attractive as a tree with the lower branches removed, revealing the light brown, peeling bark. The tree grows in an open manner with large spaces between the branches, creating a pleasing, irregular oval silhouette in middle and old age.

This is a tough tree, adaptable to urban conditions, and should be used much more extensively as a street tree. It could be used more in areas of poor soils and restricted rooting space. Unfortunately, most people choose to trim the tree into a column or hedge, so few have seen the true beauty of the tree. It makes an attractive specimen, street, or parking lot tree, even for the smallest soil space in a downtown planting pit. Roots are not a problem in restricted-soil planting areas and usually do not lift sidewalks.

Showing best growth and form in full sun, Podocarpus will grow more slowly and have a looser appearance when grown in shade. It grows on the north side of a tall building with little or no direct sun. It will tolerate a wide variety of well-drained, acidic soils, but not wet soils.

Root pruning prior to digging field stock is considered essential. Occasionally bothered by scale, mites, and sooty mold, but not seriously. Some magnesium deficiency may occur on sandy soil, which is easily corrected with magnesium sulfate.

Podocarpus nagi Nagi Podocarpus

Pronunciation: poe-doe-KAR-puss NAY-guy

ZONES: (8B), 9A, 9B, 10A, 10B, 11

HEIGHT: 30 to 50 feet

WIDTH: 15 to 25 feet

FRUIT: elongated; 0.5"; fleshy; purple; not showy; no significant litter

GROWTH RATE: moderate; moderate life span

HABIT: pyramidal when young, oval later; dense; mostly symmetrical; medium texture

FAMILY: *Podocarpaceae*

LEAF COLOR: dark green; evergreen

LEAF DESCRIPTION: opposite; simple; oval to ovate; 2 to 3" long

FLOWER DESCRIPTION: yellowish; not showy; spring

EXPOSURE/CULTURE:

Light Requirement: part shade to full sun; tolerates shade, but grows slowly and more open

Soil Tolerances: all textures; alkaline to acidic; salt and drought

Aerosol Salt Tolerance: moderate

PEST PROBLEMS: resistant

BARK/STEMS:

Major branches grow mostly upright and will not droop; should be grown with a single leader; no thorns

Pruning Requirement: needs some initial pruning to develop a strong central leader; remove competing leaders as they form at base of trunk

Limb Breakage: resistant

Current Year Twig: green; medium thickness

CULTIVARS AND SPECIES:

P. henkelii develops a drooping habit with masses of delicate foliage.

LANDSCAPE NOTES:

Origin: Japan

Usage: hedge or screen; parking lot island; buffer strip; highway; shade tree; specimen; sidewalk cutout; street tree; urban tolerant

This upright, dense evergreen has pointed, leathery, flattened leaves arranged on stiff, symmetrical branches, and works very well as a screen, hedge, strong accent plant, or framing tree. The crown forms a somewhat pyramidal to oval outline. Specimens allowed to grow with several trunks grow wider canopies. Able to reach 90 feet in height, Nagi Podocarpus is usually seen at 30 to 40 feet tall due to its moderately slow growth rate. Compact branching habit and very dark foliage make this a dense tree in full sun, more open but still surprisingly dense in shade.

This tough tree is adaptable to urban conditions. It should be used more in areas with poor soils and restricted rooting space. With some pruning to create a more open canopy, with well-spaced ascending branches along the trunk, Nagi Podocarpus could make a good street tree.

Eliminate branches with narrow angles of attachment in favor of those with a wider angle, to enhance structural strength. Thinning the thick canopy periodically will enhance wind-firmness and increase longevity. This could be a popular tree for planting in a variety of landscape situations if it were more avaliable in nurseries.

Growing well in full sun, partial or deep shade, Nagi Podocarpus tolerates a wide range of well-drained soils. Trees grow well in humid climates with no irrigation, once they are established, but will show nutrient deficiencies, especially on alkaline soils. The typical symptom is a wide yellow band or stripe across the leaves, usually attributed to magnesium deficiency.

Scale and sooty mold can be found on Nagi Podocarpus, but are usually not serious. Roots rot on soils with poor drainage.

Pongamia pinnata Pongam

Pronunciation: pon-GAM-ee-ah pin-NAY-tah

ZONES: 10B, 11

HEIGHT: 40 feet

WIDTH: 40 feet

FRUIT: elongated pod; 1 to 3"; dry; brown; causes significant litter; showy; poisonous

GROWTH RATE: rapid; moderate life span

HABIT: spreading and rounded; dense; symmetrical; medium texture

FAMILY: *Leguminosae*

LEAF COLOR: evergreen; glossy green

LEAF DESCRIPTION: alternate; odd pinnately compound; entire margin; oval; 1 to 3" long leaflets

FLOWER DESCRIPTION: pink; pleasant fragrance; very showy; spring

EXPOSURE/CULTURE:

Light Requirement: grows and flowers best in full sun

Soil Tolerances: all textures; slightly alkaline to acidic; drought

Aerosol Salt Tolerance: moderate

PEST PROBLEMS: resistant

BARK/STEMS:

Routinely grown with multiple trunks; branches grow mostly upright and will not droop; can be trained to grow with a short, single trunk; no thorns

Pruning Requirement: requires early pruning to develop strong structure; space major branches along central trunk, and keep them less than half trunk diameter by thinning

Limb Breakage: resistant as long as included bark does not form

Current Year Twig: green; slender

LANDSCAPE NOTES:

Origin: Asia, Africa, and Australia

Usage: parking lot island; buffer strip; highway; near a deck or patio; shade tree; specimen; street tree

Pongam is a fast-growing evergreen tree forming a broad, spreading canopy casting moderate shade. The three-inch-long, pinnately compound leaves drop for just a short period of time in early spring, but are quickly replaced by new growth. Pongam is at its finest in the spring when the showy, hanging clusters of white, pink, or lavender, pea-like, fragrant blossoms appear. The clusters are up to 10 inches long. These beautiful blossoms and the glossy, nearly-evergreen leaves help make Pongam a favorite for use as a specimen or shade tree.

Pongam should be grown in full sun or partial shade on well-drained soil. Pongam is resistant to strong winds provided tree is trained to a dominant leader and branches are kept less than half trunk diameter without included bark in the crotches. It is susceptible to freezing temperatures below 30°F. Pongam will show nutritional deficiencies if grown on alkaline soil with a pH above 7.5. Mulch the area beneath the canopy to reduce turf competition.

It has been planted as a street tree, but dropping pods often litter the ground. The seeds contained within the oval, 1.5-inch-long, brown seedpods are poisonous, a fact that should be considered when placing the tree where children may be present. It is overused in some areas, probably due to its rapid growth rate and showy flowers. It transplants easily and adapts to most soils, so it is much used by homeowners. Many other trees could be planted in place of this overly popular tree. This would increase the health of the urban forest by increasing diversity.

Caterpillars occasionally chew holes in the leaves, but this is not serious enough to warrant control.

Populus alba White Poplar

Pronunciation: POP-you-luss AL-bah

ZONES: 4A, 4B, 5A, 5B, 6A, 6B, 7A, 7B, 8A, 8B, 9A, 9B

HEIGHT: 60 to 100 feet

WIDTH: 40 to 60 feet

FRUIT: elongated; less than 0.5"; dry; brown; causes some litter

GROWTH RATE: rapid; short to moderate life span

HABIT: oval; open; irregular silhouette; coarse texture

FAMILY: *Salicaceae*

LEAF COLOR: dark green above, densely hairy and white below; yellow fall color

LEAF DESCRIPTION: alternate; simple; lobed, sinuate or undulate margin; oval to triangular; deciduous; 2 to 5" long; some foliage drops all during the growing season

FLOWER DESCRIPTION: yellow; not showy; spring

EXPOSURE/CULTURE:

Light Requirement: grows best in full sun

Soil Tolerances: all textures; alkaline to acidic; salt and drought

Aerosol Salt Tolerance: moderate to high

PEST PROBLEMS: susceptible

BARK/STEMS:

Light-colored, showy bark; branches grow mostly upright and will not droop; should be grown with a single leader; no thorns

Pruning Requirement: requires early pruning to develop strong structure

Limb Breakage: wood tends to be weak and break

Current Year Twig: gray; medium thickness

CULTIVARS AND SPECIES:

'Bolleana' has irregularly lobed leaf margins and a narrow, upright growth habit. It could substitute for the disease-prone Lombardy Poplar. 'Pendula' has a weeping form.

LANDSCAPE NOTES:

Origin: Europe to Asia, naturalized in the United States

Usage: reclamation; possible large shade tree

White Poplar is a fast-growing, deciduous tree that makes a nice shade tree quickly. The dark-green, lobed leaves have a fuzzy, white underside that gives the tree a sparkling effect when breezes stir the leaves. Leaves are totally covered with this white fuzz when they are young and first open. The fall color is pale yellow. The flowers appear before the leaves in spring but are not showy, and are followed by tiny, fuzzy seedpods containing numerous seeds.

The white trunk and bark of White Poplar are particularly striking, along with the beautiful, two-toned leaves. The bark stays smooth and white until very old, when it can become ridged and furrowed. The wood of White Poplar is fairly brittle and subject to breakage in storms, and the soft bark on young trees is subject to injury by vandals. Leaves often drop from the tree beginning in summer, and continue dropping through the fall.

The only place for this tree is in a large landscape such as a park or commercial landscape. Keep it away from sidewalks, curbs, drain fields, and sewer lines, because the roots grow fast and can invade everything. Root suckers are common.

White Poplar should be grown in full sun and tolerates almost any soil, wet or dry. Suckering may be a problem on stressed trees, but those growing vigorously are usually not bothered. In areas with much air pollution and soot, the fuzzy white undersides of the leaves act as a filter attracting and holding dirt and dust, making them unattractive.

Crown gall and trunk cankers cause frequent troubles for this tree, among a host of potential problems too long to list. One of the biggest limitations is the large size and brittle wood.

Populus deltoides Eastern Cottonwood

Pronunciation: POP-you-luss dell-TOY-dees

ZONES: 2A, 2B, 3A, 3B, 4A, 4B, 5A, 5B, 6A, 6B, 7A, 7B, 8A, 8B, 9A, 9B

HEIGHT: 80 to 100 feet

WIDTH: 40 to 60 feet

FRUIT: elongated; less than 0.5"; dry; white, like cotton; causes significant litter

GROWTH RATE: rapid; moderate to long life span

HABIT: oval; open; irregular silhouette; coarse texture

FAMILY: *Salicaceae*

LEAF COLOR: green; yellow fall color

LEAF DESCRIPTION: alternate; simple; lobed, sinuate or undulate margin; oval to deltoid; deciduous; 2 to 5" long; some foliage drops all during growing season

FLOWER DESCRIPTION: yellow; not showy; spring

EXPOSURE/CULTURE:

Light Requirement: grows best in full sun

Soil Tolerances: all textures; alkaline to acidic; wet soil; salt and drought

Aerosol Salt Tolerance: good

PEST PROBLEMS: susceptible

BARK/STEMS:

Light brown and furrowed, showy bark; branches grow mostly upright and will not droop; should be grown with a single leader; no thorns

Pruning Requirement: requires early pruning to develop strong structure

Limb Breakage: wood tends to be weak and break, especially branches with included bark

Current Year Twig: gray; medium thickness

CULTIVARS AND SPECIES:

The variety *deltoides* occurs in the eastern part of its natural range, and variety *occidentalis* in the western part. Male cultivars, such as 'Siouxland', have been selected that do not produce the undesirable cottony fruit.

LANDSCAPE NOTES:

Origin: eastern North America

Usage: reclamation; soil stabilization; possible large shade tree for plains states; urban tolerant

Cottonwood is a fast-growing, deciduous tree. The leaves have a flattened petiole, which makes them move in even the lightest breeze. The fall color is pale yellow. The flowers appear before the leaves in spring but are not showy, and are followed by a cottony fruit that litters the ground and surrounding trees, shrubs, groundcovers, and everything else around. The bark becomes deeply furrowed on older trees and is quite striking on large trees. Trunks can grow to more than six feet in diameter. The wood of Cottonwood is fairly brittle and subject to breakage in storms. Leaves often drop from the tree beginning in summer, and continue dropping through the fall.

The best place for this tree is in a large landscape such as a park, away from people, or along stream banks and other reclamation sites. Their rapid growth makes them well suited for quickly stabilizing the soil and reducing run off. Keep it away from sidewalks, curbs, play areas, buildings, drain fields, and sewer lines, because the roots grow fast and can invade everything. Cottonwood is the junkyard dog of the tree world, growing just about anywhere.

Cottonwood should be grown in full sun and tolerates almost any soil, wet or dry. Suckering may be a problem on stressed trees, but those growing vigorously are usually not bothered. The forestry industry uses the tree for making crates, pallets, and similar products.

Several fungi, including *Cytospora*, cause serious cankers. More than 20 insects and diseases can cause serious troubles for Cottonwood, among a host of potential problems too long to list. One of the biggest limitations is the large size and brittle wood.

Populus freemontii

Fremont Poplar, Western Cottonwood

Pronunciation: POP-you-luss free-MONT-tea-eye

ZONES: 6A, 6B, 7A, 7B, 8A, 8B, 9A, 9B

HEIGHT: 60 feet

WIDTH: 50 to 70 feet

FRUIT: dry; white, like cotton; female trees produce significant litter

GROWTH RATE: rapid; moderate to long life span

HABIT: oval; moderate density; irregular silhouette; coarse texture

FAMILY: *Salicaceae*

LEAF COLOR: green; showy, yellow fall color

LEAF DESCRIPTION: alternate; simple; serrate margin; triangular; deciduous; 2 to 3" long; flattened petiole

FLOWER DESCRIPTION: yellow; not showy; spring

EXPOSURE/CULTURE:

Light Requirement: grows best in full sun

Soil Tolerances: all textures; alkaline to acidic; wet soil; salt and drought

Aerosol Salt Tolerance: good

PEST PROBLEMS: trees survive and grow despite susceptibility

BARK/STEMS:

Smooth gray bark on young trees, turning dark brown; furrowed, showy bark; branches grow mostly upright and will not droop; should be grown with a single leader although many unpruned trees have several trunks; no thorns

Pruning Requirement: needs early pruning to develop and maintain strong structure

Limb Breakage: wood tends to be weak and break

Current Year Twig: gray; medium thickness

CULTIVARS AND SPECIES:

Male cultivars have been selected that do not produce the undesirable cottony fruit.

LANDSCAPE NOTES:

Origin: Arizona, California, Colorado, Texas, Utah, and Mexico

Usage: reclamation; soil stabilization; windbreak; large shade tree; urban tolerant

Cottonwood is a fast-growing, deciduous tree. The green leaves have a flattened petiole, which makes them move in even the lightest breeze. The flowers appear before the leaves in spring but are not showy, and on female trees are followed by a cottony fruit that litters the ground and surrounding trees, shrubs, groundcovers, and everything else around. The bark becomes deeply furrowed on older trees and is quite striking on large trees. Trunks can grow to more than six feet in diameter on moist sites. The wood of Cottonwood is fairly brittle and subject to breakage in storms.

The best place for this tree is in a large landscape such as a park, away from people, or along stream banks or other reclamation sites. Their rapid growth makes them well suited for quickly stabilizing the soil, especially in drier climates. This is one of the poplars found along water courses in the desert areas of North America. Keep it away from sidewalks, curbs, play areas, buildings, drain fields, and sewer lines, because the roots grow fast and can invade everything. It is a suitable shade tree for drier climates because this tree can get by with infrequent water where others would require frequent irrigation, once it is well established.

Western Cottonwood should be grown in full sun and tolerates almost any soil, wet or dry, alkaline or acid. Roots grow fast and can quickly reach a water table or other nearby water source. This helps the tree through periods of dry weather. Dead inner branches require regular removal to keep the tree looking clean.

More than 20 insects and diseases can cause serious troubles for Cottonwood, including tent caterpillars and shield borers. Canker disease infects trees with low vigor. One of the biggest limitations is the large size and brittle wood. Branches with mistletoe may have to be removed.

Populus nigra 'Italica' Lombardy Poplar

Pronunciation: POP-you-luss NYE-gruh

ZONES: 4A, 4B, 5A, 5B, 6A, 6B, 7A, 7B, 8A, 8B, 9A

HEIGHT: 30 to 40 feet

WIDTH: 6 to 10 feet

FRUIT: fruitless

GROWTH RATE: rapid; very short life span

HABIT: columnar; moderate density; symmetrical; fine texture

FAMILY: *Salicaceae*

LEAF COLOR: bright green; yellow fall color

LEAF DESCRIPTION: alternate; simple; crenate to serrate margins; deltoid; deciduous; 2 to 4" long

FLOWER DESCRIPTION: red; not showy; spring

EXPOSURE/CULTURE:

Light Requirement: grows best in full sun

Soil Tolerances: all textures; alkaline to acidic; wet soil; moderate salt; drought

Aerosol Salt Tolerance: moderate

PEST PROBLEMS: defines the term *sensitive*

BARK/STEMS:

Bark is thin and easily damaged by mechanical impact; branches grow mostly upright and will not droop; should be grown with a single leader; no thorns

Pruning Requirement: needs only one pruning—at ground level when it dies

Limb Breakage: susceptible

Current Year Twig: brown; slender

CULTIVARS AND SPECIES:

Choose from the many other available columnar or upright screening trees for a more durable planting, including *Acer* x *freemanii* 'Armstrong Two'; *A. platanoides* 'Erectum'; *Acer saccharum* 'Temple Upright'; *Alnus glutinosa* 'Fastigiata'; *Carpinus betulus* 'Fastigiata'; *Populus alba* 'Pyramidalis' (Bolleana); *Quercus robur* 'Attention' 'Fastigiata'; and *Thuja plicata* 'Hogan'. *P. tremuloides* is native to a wide region in the Rocky Mountains. It is especially suited for higher elevations, provides spectacular fall color, but suffers in hot weather from borers and other insects and from tip blight.

LANDSCAPE NOTES:

Origin: Mediterranean region

Usage: not generally recommended for planting due to disease-induced short life

Often planted for its fast growth and usefulness as a short-lived screen or windbreak, Lombardy Poplar forms a slender column of many short, upward-pointing branches. Canker disease almost always infects the tree by the time it is 5 to 15 years old, so except for the Pacific northwest trees are rarely seen larger than about 30 feet tall by 5 feet wide. The triangular to diamond-shaped, 2- to 3.5-inch-long by 1.5- to 3-inch-wide deciduous leaves are bright green on both sides throughout the year, turning a blazing golden yellow in fall before dropping. The small, inconspicuous flowers appear in spring. The bark is gray-green on young trees and on new growth, but becomes black, thickened, and furrowed on older, larger trunks.

If planted, Lombardy Poplar should be grown in full sun. It tolerates wet soil well but also performs in drought, losing leaves early in very dry summers. Multiple suckers often appear at the base of the tree and occasionally on roots far from the tree, and the roots are considered invasive. Plant other narrow-crowned trees instead of this tree.

Stem canker disease is so devastating that this tree should not be included on any recommended tree list, with the possible exception of a reclamation site. Until recently the tree was disease-free in the Pacific northwest.

Prosopis glandulosa Mesquite, Honey Mesquite

Pronunciation: pro-SOP-iss gland-you-LOW-sah

ZONES: 6B, 7A, 7B, 8A, 8B, 9A, 9B, 10A, 10B

HEIGHT: 30 feet

WIDTH: 30 to 40 feet

FRUIT: elongated pod; 4 to 8"; dry; brown; attracts wildlife; causes some litter; showy

GROWTH RATE: moderate; long-lived

HABIT: rounded; open; irregular when young, symmetrical later; fine texture

FAMILY: *Leguminosae*

LEAF COLOR: bright green; no fall color change

LEAF DESCRIPTION: alternate; bipinnately compound; linear; deciduous; 1.5 to 2" long leaflets

FLOWER DESCRIPTION: yellow; somewhat showy; fragrant; spring and summer; attracts bees

EXPOSURE/CULTURE:

Light Requirement: full sun

Soil Tolerances: loam; sand; alkaline to acidic; occasionally wet soil; salt; extremely drought-tolerant

PEST PROBLEMS: resistant

BARK/STEMS:

Showy trunk form; branches droop as tree grows and will require pruning for vehicular or pedestrian clearance beneath canopy; routinely grown with multiple trunks; thorns are very sharp

Pruning Requirement: pruning helps to develop strong, lasting structure; slow growth of aggressive branches by thinning them

Limb Breakage: resistant

Current Year Twig: brown; slender

CULTIVARS AND SPECIES:

There are scores of species. *P. pubescens* makes a nice, floriforous, small tree. *P. chilensis* (Chilean Mesquite) is a thornless, semievergreen with fern-like, soft foliage and grows to a 30-foot-tall ball.

LANDSCAPE NOTES:

Origin: southwestern United States and northern Mexico

Usage: bonsai; reclamation; shade tree; specimen; windbreak; buffer strip

This North American native tree forms a spreading, rounded canopy with many drooping, crooked branches emanating from low on the trunk. These branches are armed with one-inch-long spines that can damage flesh, but thornless selections are available from nurseries. Mesquite has a tendency to form thickets which, along with these thorns, help make the dense growth impenetrable. The reddish-brown bark is rough and fissured. The root system of Mesquite is quite extensive and will spread far and wide to consume whatever moisture is available, sometimes to the detriment of other plantings near the tree. Mesquite is considered a weed along the Rio Grande and in other areas in Texas, where it easily invades adjacent land.

The leaves lend a fine texture to this irregular-shaped tree, which casts light shade. When given adequate moisture and trained and pruned to create a somewhat uniformly shaped crown, Mesquite can be an attractive, somewhat weeping landscape specimen. From May to September, Mesquite is adorned with two-inch-wide, extremely fragrant blooms, which are not especially showy due to their yellow to greenish-white coloration, but are still easily found by bees who love them. The seeds that follow are quite popular with birds and other wildlife and, at times, humans. The southwestern Native Americans used the seed, or beans, as a food source. The beans contain as much as 30 percent sugar and, when fermented, produce an alcoholic beverage. A meal made from the beans was also used by Native Americans to make bread.

Mesquite should be grown in full sun on well-drained soil. The tree is very drought-tolerant. Young plants can be successfully transplanted while small.

Mesquite has become an unimaginable weed problem in parts of Texas. It was naturally suppressed by fire, but now has spread everywhere because of fire control.

Prunus americana American, or Wild, Plum

Pronunciation: PROO-nuss ah-mer-ih-CANE-ah

ZONES: 3A, 3B, 4A, 4B, 5A, 5B, 6A, 6B, 7A, 7B, 8A, 8B

HEIGHT: 12 to 25 feet

WIDTH: 12 to 20 feet

FRUIT: round; 1"; fleshy, but hard; yellow; attracts wildlife; edible; somewhat showy; no significant litter

GROWTH RATE: slow-growing; moderate life span

HABIT: rounded; moderate density; irregular silhouette; fine texture

FAMILY: *Rosaceae*

LEAF COLOR: medium green; somewhat showy, yellow fall color

LEAF DESCRIPTION: alternate; simple; serrate to serrulate margin; oval to ovate; deciduous; 1.5" long

FLOWER DESCRIPTION: white; very showy; early spring

EXPOSURE/CULTURE:

Light Requirement: grows best in full sun, but tolerates part shade

Soil Tolerances: loam; sand; slightly alkaline to acidic; moderate drought

Aerosol Salt Tolerance: none

PEST PROBLEMS: resistant

BARK/STEMS:

Branches droop as tree grows and will require pruning for vehicular or pedestrian clearance beneath canopy; typically grows with many trunks, but can be grown with a single leader; thorns 2 to 3" long grow from branches

Pruning Requirement: requires pruning to develop tree form; remove lower branches, trunks, and suckers early to develop a single leader

Limb Breakage: mostly resistant

Current Year Twig: brown; slender

LANDSCAPE NOTES:

Origin: eastern North America to Utah

Usage: bonsai; container or above-ground planter; parking lot island; buffer strip; highway; near a deck or patio; specimen; reclamation; soil stabilization

A native along the edge of woods, American Plum is a round-topped, deciduous tree that is occasionally planted for its spectacular display of blooms. Because the tree has several forms, ranging from shrub-like to tree form, do not expect a row of them to form a uniform shape. In early spring, before the two-inch-long, finely serrate leaves appear, these small trees take on a white, billowy, almost cloud-like appearance when they are clothed in the profuse, small, white flower clusters. These inch-long blooms are followed by one-inch-long, edible, fruits that vary in flavor from very tart to sweet. These plums are very attractive to various forms of wildlife.

A bit weedy in growth habit, but proper training and pruning can create an attractive specimen or small tree for planting in a highway median or buffer strip around a parking lot. Lower branches must be removed to train this tree for pedestrian clearance beneath the canopy. They are especially suited for planting beneath power lines or in other areas where overhead space is limited. The tree branches low to the ground, making it a nice tall element in a backyard shrub border. Small-diameter, interior branches can be removed in winter to open up the crown for a more formal, attractive shape and habit. The tree may live 30 to 40 years on a good site.

American Plum thrives in full sun or partial shade on a wide variety of soils. When placed in sandy soil, it grows best with irrigation and some shade in the afternoon. Trees grow quickly when young but considerably slower when mature and bearing fruit. They reproduce rapidly and can quickly invade surrounding land.

Limitations include weedy habit and sharp thorns. Tent caterpillars can cause defoliation.

Prunus caroliniana Cherry-Laurel

Pronunciation: PROO-nuss care-o-lin-ee-A-nah

ZONES: 8A, 8B, 9A, 9B, 10A

HEIGHT: 25 to 40 feet

WIDTH: 15 to 25 feet

FRUIT: round; 0.33"; fleshy; black; attracts wildlife; edible; not showy; causes significant messy litter

GROWTH RATE: short to moderate; short to moderate life span

HABIT: oval to round; dense; symmetrical; medium texture

FAMILY: *Rosaceae*

LEAF COLOR: broadleaf evergreen; dark green

LEAF DESCRIPTION: alternate; simple; entire to serrulate margins; lanceolate or oblong; 2 to 4" long

FLOWER DESCRIPTION: white; pleasant fragrance; very showy; spring; attracts bees

EXPOSURE/CULTURE:

Light Requirement: mostly shaded to full sun

Soil Tolerances: all textures; alkaline to acidic; moderate drought; good drainage essential

Aerosol Salt Tolerance: moderate

PEST PROBLEMS: resistant

BARK/STEMS:

Bark is thin and easily damaged by mechanical impact; branches grow mostly upright and will not droop; should be grown with a single leader; can be trained to several trunks; no thorns

Pruning Requirement: needs pruning to prevent formation of included bark; remove upright, competing branches and trunks as they form

Limb Breakage: resistant

Current Year Twig: brown; slender

CULTIVARS AND SPECIES:

'Compacta' and 'Bright n Tight' have a very dense, compact habit of growth.

LANDSCAPE NOTES:

Origin: Virginia to Florida, west to Louisiana

Usage: hedge; screen; buffer strip; highway; near a deck or patio; reclamation; standard; sidewalk cutout; street tree; urban tolerant

Cherry-Laurel is densely foliated with glossy leaves. It can reach 40 feet in height with a 25-foot-spread though is often seen smaller when grown in the open. The tree usually maintains a good central leader and small-diameter, strong lateral branches.

In late winter to spring, tiny, creamy white, showy flowers appear in dense, fragrant clusters for about two weeks. They are followed by small, shiny, black cherries that are quite attractive to wildlife. The great quantity of fruit may create a short-term litter problem if the trees are located near a patio or walkway, but the fruit is small and washes away quickly. Root systems are quite shallow, but usually they are not aggressive and do not cause problems.

Cherry-Laurel will create a very dense screen or hedge with regular pruning, but is also attractive when allowed to grow naturally into its dense, upright oval tree form. Properly trained to a central leader, the plant could make a good, small to medium-sized street tree.

Preferring ample moisture while young, Cherry-Laurel is otherwise well suited to sunny or shady locations on any average, well-drained soil. Once established, it is salt- and drought-tolerant, requiring little or no irrigation. Overirrigating can cause chlorosis. The tree adapts well to soils with high pH but not wet, soggy sites. Clay soil is fine as long as water does not stand after rain.

Cherry-Laurel can be short-lived. Mites, borers, caterpillars, leaf spot, fire blight, and stem canker may occur; borers are troublesome on stressed trees. It is difficult to grow in containers because of root rot from overirrigation. Flowers attract bees. Seedlings beneath the canopy can be numerous and weedy. Plants are poisonous, causing cyanide-like poisoning symptoms.

Prunus cerasifera Cherry Plum

Pronunciation: PROO-nuss sair-ah-SIFF-er-ah

ZONES: 5A, 5B, 6A, 6B, 7A, 7B, 8A, 8B

HEIGHT: 15 to 25 feet

WIDTH: 15 to 25 feet

FRUIT: round; 1"; fleshy; green to purple; attracts wildlife; edible; not showy; causes significant messy litter

GROWTH RATE: moderate; short-lived

HABIT: upright-spreading to round; dense; symmetrical; medium texture

FAMILY: *Rosaceae*

LEAF COLOR: purple to greenish-red summer and fall, depending on cultivar

LEAF DESCRIPTION: alternate; simple; serrate margin; oval to obovate or ovate; deciduous; 1.5 to 3" long

FLOWER DESCRIPTION: pinkish-white; showy; spring

EXPOSURE/CULTURE:

Light Requirement: grows best in full sun

Soil Tolerances: all textures; slightly alkaline to acidic; moderate drought

Aerosol Salt Tolerance: good

PEST PROBLEMS: susceptible

BARK/STEMS:

Bark is thin and easily damaged by mechanical impact; branches grow mostly upright and horizontal and do not droop; should be grown with a short, single leader; no thorns

Pruning Requirement: prune lightly and at regular intervals to eliminate branches with included bark

Limb Breakage: susceptible

Current Year Twig: brown or reddish; slender

CULTIVARS AND SPECIES:

'Atropurpurea' sets abundant fruit; 'Krauter Vesuvius' has deep purple leaves, bears little or no fruit, and is heat tolerant; 'Mt. St. Helens' holds purple color into the fall; 'Newport' (flood tolerant) and *P.* x *blireiana* (heat tolerant, double flower) bear little or no fruit. 'Thundercloud' has purple leaves and fruit.

LANDSCAPE NOTES:

Origin: Asia

Usage: container or above-ground planter; small shade tree; specimen

Purple-leaved Plum has new foliage that unfolds as ruby red, then turns reddish-purple, and finally matures to greenish-bronze by the end of the summer. 'Atropurpurea', 'Newport', and 'Thundercloud' retain purple foliage throughout the growing season. Cherry Plum's fast growth rate and upright to upright-spreading nature allow it to quickly form a dense silhouette. One-inch-diameter purple fruits make a temporary yet considerable mess beneath the tree as they mature and drop during the summer.

Very popular due to the unusual leaf color, many gardeners want one of these things in their yard. But one plant really attracts attention to an area, and it can be overpowering in a small landscape, creating a cramped feeling. Due to the strong effect, this tree is best used in a large-scale landscape as a single specimen, not in a row or mass planting. It makes a nice, small ornamental near the deck or patio, but locate it far enough away so that dropping fruit will fall on the lawn or in a groundcover bed. Do not rely on this tree to be around for a long time, because decline often begins by the time the tree is 20 years old. Enjoy the tree in the meantime.

Purple-leaved Plum should be grown in full sun to bring out the richest color of the leaves. Leaves on the species turn almost green in shade and lose the characteristic purple that people are usually looking for. Well-drained soil with an acid pH is preferred, but it will grow on slightly alkaline soil. Tolerant of moderate heat and drought, it often succumbs to borers on poor, compacted soil.

Short-lived, but who cares! Aphids, borers, scales, mealy bugs, tent caterpillars, and many others may occur—Cherry Plum is not a pest-free plant. Canker and leaf spots can also be troublesome. Trees seed themselves into the landscape.

Prunus 'Hally Jolivette' Hally Jolivette Cherry

Pronunciation: PROO-nuss

ZONES: 6A, 6B, 7A, 7B, (8)

HEIGHT: 15 feet

WIDTH: 15 feet

FRUIT: rare

GROWTH RATE: slow to moderate; moderate life span

HABIT: round; dense; symmetrical; fine texture

FAMILY: *Rosaceae*

LEAF COLOR: dark green; yellow fall color

LEAF DESCRIPTION: alternate; simple; serrate margin; oval to obovate or ovate; deciduous; 2 to 3" long

FLOWER DESCRIPTION: pinkish-white; showy; spring

EXPOSURE/CULTURE:

Light Requirement: grows best in full sun

Soil Tolerances: all textures; alkaline to acidic; drought

PEST PROBLEMS: resistant

BARK/STEMS:

Bark is thin and easily damaged by mechanical impact; branches droop as tree grows and will require pruning for vehicular or pedestrian clearance beneath canopy; routinely grown with multiple trunks; often trained to grow with a short, single trunk; no thorns

Pruning Requirement: needs little pruning to develop well-formed tree

Limb Breakage: resistant

Current Year Twig: brown; slender

LANDSCAPE NOTES:

Origin: hybrid

Usage: bonsai; container or above-ground planter; buffer strip; highway; near a deck or patio; specimen; sidewalk cutout

'Hally Jolivette' Cherry forms a dense-branching, small tree. The main delight of growing this tree is the two-week-long display of blooms in spring. Starting out as pink buds, the flowers unfold to pale pink or pure white double blooms, about an inch in diameter, helping the tree to create a striking specimen in the landscape. The foliage lends a fine texture to the landscape, and it maintains a dark green color through moderate drought. In severe drought, foliage burns a little, becoming brown at the margins, but there appears to be no long-term harm to the tree. This tree will live longer than many of the other cherries.

Prune to open up the canopy to develop more of a tree form; otherwise it grows into a large shrub. Remove interior branches and space main branches along a short trunk. A more upright shape can be created by removing lateral branches, whereas a more spreading shape can be promoted by removing upright branches.

It is a popular tree for planting next to the patio or deck so as to enjoy the flower display, soft texture, and nice fall color. It is also well suited for maintaining in a container for placing anywhere a small tree is needed in the landscape. Use the tree along an entrance road to a commercial development planted on 15- to 20-foot centers. It also makes a nice small-scale courtyard tree. It is probably too shrubby for planting in a tree lawn strip between the street curb and the sidewalk, although properly trained trees could be tried for this use. Set the trees back about 8 feet from the edge of a street on 15-foot centers along a residential or commercial landscape for a nice, uniform, ornamental planting.

'Hally Jolivette' Cherry should be grown in full sun on well-drained soil. It grows well in clay soil that is not too compacted. Occasionally aphids and mites infest the foliage.

Prunus x incamp 'Okame' 'Okame' Cherry

Pronunciation: PROO-nuss x inn-KAMP

ZONES: 7B, 8A, 8B, 9A, 9B

HEIGHT: 15 to 25 feet

WIDTH: 20 feet

FRUIT: attracts birds; not common or especially showy; no significant litter

GROWTH RATE: moderate; short-lived

HABIT: upright when young, oval to round later; moderate density; symmetrical; medium texture

FAMILY: *Rosaceae*

LEAF COLOR: dark green; showy, copper, orange, red, or yellow fall color

LEAF DESCRIPTION: alternate; simple; serrate margin; oval; deciduous; 1 to 2.5" long

FLOWER DESCRIPTION: pink; showy; winter and spring

EXPOSURE/CULTURE:

Light Requirement: part shade to full sun

Soil Tolerances: all textures; slightly alkaline to acidic; some drought

PEST PROBLEMS: somewhat susceptible

BARK/STEMS:

Smooth, showy bark; bark is thin and easily damaged by mechanical impact; branches grow mostly upright and will not droop; should be grown with a single leader as far up as possible; no thorns

Pruning Requirement: requires early pruning to develop strong structure; remove or drop-crotch aggressive, upright branches growing parallel with trunk

Limb Breakage: resistant

Current Year Twig: brown; medium to slender

LANDSCAPE NOTES:

Origin: hybrid

Usage: container or above-ground planter; buffer strip; near a deck or patio; specimen

This hybrid of *Prunus incisa* and *Prunus campanulata* grows quickly when young and forms an upright, oval silhouette. The 1- to 2.5-inch-long leaves cast medium shade beneath the tree, and turn lovely shades of yellow, orange, and red in the fall in the northern part of its range. In late winter to early spring, trees are gaily decorated with a multitude of single pink blooms with red calyces and reddish flower stalks, making them quite outstanding. The trunk is wrapped with prominent lenticels raised above the smooth reddish-bronze bark. 'Okame' Cherry is one of the most delicate and finest flowering trees available and one of the best for the deep south, as it apparently has a low chilling requirement. It has been seen flowering in January and early February well into central Florida.

To produce a nice specimen tree, thin the canopy to open it up to allow for light penetration to the interior foliage. Remove upright-oriented branches in favor of those with a wider angle with the trunk to form a pleasing, upright, spreading habit. Space main branches along the trunk. A more upright shape can be created by removing horizontal branches.

Use the tree along an entrance road to a commercial development, planted on 20-foot centers, or alongside the patio or deck in the back yard. It also makes a nice small-scale courtyard tree or specimen planted in the lawn or in a groundcover bed. It could be planted along a residential street where there is plenty of soil for root expansion. It is generally available in the industry. Seedlings of *P. campanulata* can often be seen growing near established trees.

'Okame' Cherry can be grown in full sun in clay, loam, or sandy soil, but benefits from irrigation in dry weather. The tree prefers some filtered afternoon shade in the southern part of its range.

Occasionally bothered by canker worms, which can almost completely defoliate the tree. The usual assortment of *Prunus* insects and diseases also affect this tree. This should not be cause to avoid planting the tree.

Prunus maackii Amur Chokecherry

Pronunciation: PROO-nuss MACK-ee-eye

ZONES: 2B, 3A, 3B, 4A, 4B, 5A, 5B, 6A, 6B

HEIGHT: 30 to 40 feet

WIDTH: 25 to 35 feet

FRUIT: round; less than 0.5"; fleshy; black; attracts birds; not showy; causes some messy litter

GROWTH RATE: short-lived

HABIT: upright rounded to vase; moderately dense; nearly symmetrical; medium texture

FAMILY: *Rosaceae*

LEAF COLOR: medium green; yellow fall color

LEAF DESCRIPTION: alternate; simple; serrate margin; oval; oblong to ovate; deciduous; 2 to 4" long

FLOWER DESCRIPTION: white; pleasant fragrance; showy; spring

EXPOSURE/CULTURE:

Light Requirement: part shade to full sun

Soil Tolerances: loam; sand; slightly alkaline to acidic; moderate drought

Aerosol Salt Tolerance: moderate

PEST PROBLEMS: fairly resistant

BARK/STEMS:

Extraordinarily showy bark; trainable to be grown with a single or multiple trunks; branches grow mostly upright and will not droop; no thorns

Pruning Requirement: requires pruning to develop strong structure and prevent formation of included bark

Limb Breakage: resistant

Current Year Twig: brown; reddish; slender

LANDSCAPE NOTES:

Origin: Manchuria, Korea

Usage: container or above-ground planter; buffer strip; near a deck or patio; specimen; street tree

Amur Chokecherry is pyramidal when young but ultimately forms a dense, rounded canopy that provides light shade below. The deciduous leaves are joined in early to mid-May by an explosion of white, fragrant flowers in two- to three-inch-long racemes. The multitude of tiny black fruits that follow ripen in August and are quite attractive to birds. The bark is a handsome cinnamon brown, peeling off in shaggy masses on the trunk, or remaining smooth and shiny with minimum exfoliation. The bark is a sight for sore eyes.

Unless pruned to open up and thin the canopy to develop more of a tree form, this plant grows more like a large shrub. Remove some interior branches and space main branches along a short trunk, or prune to a single trunk to create a taller tree. A more upright shape can be created by removing lateral branches; a more spreading shape can be promoted by removing upright branches. However you decide to train this tree, place it in a prominent location so the nice texture and wonderful bark can be enjoyed.

Use the tree along an entrance road to a commercial development, planted on 20- to 25-foot centers, or alongside the patio or deck in the back yard. Mulch beneath the entire canopy to prevent turf grass from competing for moisture and nutrients.

Amur Chokecherry should be grown in full sun on well-drained soil, and performs well only in the north. The trees should be located where the roots can remain moist, but not wet, for best growth. Girdling roots are not uncommon.

Limitations include borers in warm climates, aphids, scale. Leaf spots can disfigure some foliage.

Prunus mexicana Mexican Plum

Pronunciation: PROO-nuss mecks-eh-KAY-nah

ZONES: 6B, 7A, 7B, 8A, 8B

HEIGHT: 15 to 30 feet

WIDTH: 20 to 25 feet

FRUIT: round; 1 to 2"; fleshy; red to purple; attracts birds; edible; causes some messy litter; showy

GROWTH RATE: slow-growing; moderate life span

HABIT: rounded; open; irregular silhouette; medium texture

FAMILY: *Rosaceae*

Courtesy of B. Simpson

LEAF COLOR: yellow-green; occasional good orange or yellow fall color

LEAF DESCRIPTION: alternate; simple; serrate margin; oval to obovate; deciduous; 2 to 4" long

FLOWER DESCRIPTION: white; pleasant fragrance; showy; early spring, along with Redbud

EXPOSURE/CULTURE:

Light Requirement: grows and flowers best in full sun; tolerates and flowers in partial sun

Soil Tolerances: all textures; slightly alkaline to acidic; moderate drought

PEST PROBLEMS: resistant

BARK/STEMS:

Somewhat showy, resembles dark-barked birches; branches grow mostly upright and will not droop; can be trained to grow with a single trunk or multiple trunks; thorns are present on trunk or branches

Pruning Requirement: needs little pruning to develop strong structure

Limb Breakage: resistant

Current Year Twig: brown; gray; slender

LANDSCAPE NOTES:

Origin: Texas and Mexico

Usage: container or above-ground planter; buffer strip; highway; near a deck or patio; specimen; street tree

Mexican Plum is a North American native tree that forms an irregular canopy composed of shiny leaves with fuzzy undersides on thorny branches. In spring, the trees are smothered with fragrant, white blooms, which are followed by purple-red, juicy fruits. These tart fruits can be used to make jams and jellies. The blooms help make Mexican Plum ideal for use as a specimen tree, but it also fits in nicely in the naturalized landscape as an understory tree. The gray-black, ridged, and furrowed bark is quite attractive, and the deciduous leaves often turn a showy orange color in fall before dropping.

This small tree is well suited for residential landscapes. It might be best to locate the tree back from the edge of a patio, deck, or walk, as the fruits can be a little messy for a short period in the summer or early fall. Planted in the lawn or in a bed of low-growing groundcover, fruits drop unnoticed and are of no concern. It complements Redbud well in an understory planting because it flowers at about the same time and tolerates shifting shade. The tree can be a show-stopper when it is in bloom. It makes a nice tree for planting near power lines due to its small size.

Mexican Plum should be grown in full sun or partial shade on well-drained, rich soil, though it will tolerate almost any soil. It is native to river bottoms but is quite drought-tolerant once established. Some irrigation should be provided during extended drought on exposed sites.

Apparently no limitations. It is a wonderful tree that could be planted more often. Prune lower branches away from where pedestrians will walk, to avoid contact with the somewhat thorny branches.

Prunus mume Japanese Apricot

Pronunciation: PROO-nuss MEW-mee

ZONES: 6A, 6B, 7A, 7B, 8A, 8B

HEIGHT: 12 to 20 feet

WIDTH: 15 to 20 feet

FRUIT: round; 1"; fleshy; yellow; attracts birds; no significant litter; showy

GROWTH RATE: moderate; short life span

HABIT: oval or vase; moderate density; irregular silhouette; fine texture

FAMILY: *Rosaceae*

LEAF COLOR: medium green; insignificant fall color change

LEAF DESCRIPTION: alternate; simple; serrate margin; ovate; deciduous; 2 to 4" long

FLOWER DESCRIPTION: pink; pleasant fragrance; showy; winter

EXPOSURE/CULTURE:

Light Requirement: grows and flowers best in full sun

Soil Tolerances: all textures; acidic; some drought

PEST PROBLEMS: fairly resistant

BARK/STEMS:

Bark is thin and easily damaged by mechanical impact; branches droop as tree grows and will require early pruning for vehicular or pedestrian clearance beneath canopy; routinely grown, or trainable to be grown, with multiple trunks; can be trained to grow with a short, single trunk; no thorns

Pruning Requirement: requires pruning to develop strong structure; space main branches 4 to 6" apart to prevent breakage

Limb Breakage: mostly resistant

Current Year Twig: green; slender

CULTIVARS AND SPECIES:

'Bonita'—semidouble rose-red blossoms; 'Dawn'—large ruffled double pink; 'Peggy Clarke'—double deep rose; 'Rosemary Clarke'—double white flowers with red calyces; 'W.B. Clarke'—double pink flowers, weeping form.

LANDSCAPE NOTES:

Origin: Japan and China

Usage: bonsai; buffer strip; highway; near a deck or patio; specimen

Japanese Flowering Apricot may be one of the longest-lived of the showy flowered fruit trees, eventually forming a gnarled, picturesque, 20-foot-tall tree. Appearing during the winter on bare branches are the multitude of small, fragrant, pink flowers that add to the uniqueness of the tree's character. The small yellow fruits that follow the blooms are inedible but attractive.

The tree is well suited for planting near a patio or deck. Locate it where it will receive sun on all sides of the tree to develop a uniform crown, for it becomes one-sided when exposed to sun on only one side. It would add color to a shrub border during the winter when most other plants are dormant. It makes a very nice specimen in a lawn (with a large mulched area under canopy) or planted as a group to accent a building entrance.

Japanese Flowering Apricot should be grown in full sun on well-drained, fertile, acid soils; it is not adapted to poor or dry soils. Plants will require heavy pruning to flower their best. Adhere to cultural requirements for best growth. The species, along with several cultivars, is now being grown by some nurseries.

Aphids cause distortion of new growth, and deposit honeydew that supports the growth of sooty mold. Borers attack stressed trees. Keep trees healthy with regular fertilizer applications and irrigation during drought. Scales of several types infest the cherries.

Prunus persica Peach

Pronunciation: PROO-nuss PURR-see-cah

ZONES: 5B, 6A, 6B, 7A, 7B, 8A, 8B, (9)

HEIGHT: 15 to 25 feet

WIDTH: 15 to 25 feet

FRUIT: round; 3"; fleshy; reddish-yellow; attracts wildlife; edible; causes significant litter; showy

GROWTH RATE: moderate; short-lived

HABIT: rounded vase; dense; irregular silhouette; coarse texture

FAMILY: *Rosaceae*

LEAF COLOR: dark green; showy yellow fall color

LEAF DESCRIPTION: alternate; simple; serrate to serrulate margin; lanceolate to oblanceolate; deciduous; 4 to 8" long

FLOWER DESCRIPTION: pink; red; white; showy; spring

EXPOSURE/CULTURE:

Light Requirement: grows best in full sun

Soil Tolerances: well-drained clay; loam; sand; acidic; some drought

PEST PROBLEMS: susceptible

BARK/STEMS:

Bark is thin and easily damaged by mechanical impact; branches droop as tree grows and will require pruning for vehicular or pedestrian clearance beneath canopy; routinely grown with multiple branches on a short, central trunk; no thorns

Pruning Requirement: requires pruning to develop strong structure and to thin canopy; remove upright water sprouts and suckers

Limb Breakage: susceptible when in full fruit

Current Year Twig: green; reddish; medium thickness

CULTIVARS AND SPECIES:

The double-flowering cultivars are beautiful in flower but are as prone to problems as the species.

LANDSCAPE NOTES:

Origin: China

Usage: bonsai; fruit tree; hedge

Widely popular for their sweet, juicy fruits and beautiful blossoms, Peach trees are actually plagued by so many different pests and diseases that they should probably be planted only by the horticulturally dedicated homeowner. A low, broad tree, Peach forms a rounded crown with upward-reaching branches clothed in four- to eight-inch-long, deciduous leaves. The lovely flowers, which appear in April before the new leaves unfold, are available in single, semi-double, and double forms, in colors ranging from pure white to deep red, and bicolors. The flowers are susceptible to damage by late spring frosts or especially cold winters. The luscious, three-inch-diameter fruits mature in July to August. Bright yellow fall color really stands out in many years.

They have been successfully used in medians of boulevards and around parking lots. They make effective screens for six to seven months, because of their dense, low-branching habit, but are not particularly attractive in winter. Avoid excessive pruning, as this stimulates sprouting on internal branches. Many trees live only 8 to 15 years.

Peach trees should be located in full sun or partial shade on very well-drained, moist, acid soils. The trees often require a regular spray and fertilization schedule to ensure best fruit production, but this is not needed if fruit is not important. Do not allow water to stand around the roots.

Aphids cause distortion of new growth, deposits of honeydew, and sooty mold. Borers attack flowering Peach and can kill the tree. Keep trees healthy with regular fertilizer applications. Various foliage and stem diseases can disfigure the tree.

Prunus sargentii Sargent Cherry

Pronunciation: PROO-nuss sar-JEN-tee-eye

ZONES: 5A, 5B, 6A, 6B, 7A, 7B, 8A

HEIGHT: 25 to 40 feet

WIDTH: 25 to 40 feet

FRUIT: oval; less than 0.5"; fleshy; black to purple or red; attracts birds; not showy; no significant litter

GROWTH RATE: rapid; moderate life span

HABIT: rounded vase to spreading; dense; symmetrical; medium texture

FAMILY: *Rosaceae*

LEAF COLOR: dark green; copper, orange, red, or yellow, showy fall color

LEAF DESCRIPTION: alternate; simple; serrate margin; oval to obovate; deciduous; 3 to 5" long

FLOWER DESCRIPTION: pink; showy; spring

EXPOSURE/CULTURE:

Light Requirement: full sun

Soil Tolerances: all textures; slightly alkaline to acidic; well-drained soil is essential; moderate salt and drought

Aerosol Salt Tolerance: moderate

PEST PROBLEMS: fairly resistant

BARK/STEMS:

Bark is thin and easily damaged by mechanical impact; routinely grown, or trainable to be grown, with multiple trunks; branches grow mostly upright; can be trained to grow with a short, single trunk; no thorns

Pruning Requirement: requires pruning when young to develop strong structure; prune regularly to avoid making large wounds because pruning scars close slowly

Limb Breakage: somewhat susceptible at crotch

Current Year Twig: brown; reddish; medium thickness

CULTIVARS AND SPECIES:

The cultivar 'Columnaris' has a narrow, upright-spreading silhouette 30 to 40 feet tall by 15 feet wide; 'Rancho' has an even narrower canopy. Both are very suitable for tight urban spaces and for along streets.

LANDSCAPE NOTES:

Origin: Japan

Usage: parking lot island; buffer strip; highway; shade tree; specimen; possible street tree if plenty of soil space is available for root growth

With attractive bark, good fall foliage color, and delicate pink blooms, Sargent Cherry is highly recommended for the home and urban landscape. It is often grown with several multiple trunks or upright branches originating from the same position on the trunk and ascending gracefully. This structure could be somewhat of a problem in ice storms. Training to develop well-spaced branches without included bark along the trunk may help reduce this problem. The attractive, cinnamon-brown bark has a shiny, almost polished appearance, with prominent lenticels arranged around the trunk. In late April or early May, the one-inch-wide, pink to deep pink single blooms appear before the emerging red-tinged leaves unfold.

Sargent Cherry works well as a street tree (probably the best of the Cherries for street planting) in areas that can accommodate the spreading canopy. It can be planted along the entry road to a subdivision or commercial landscape on 20-foot centers, or in the tree lawn space between the curb and sidewalk. It is also very effective as a specimen in a lawn or landscape bed.

Sargent Cherry should be grown in full sun on very well-drained, acid soil. Although it grows moderately fast and can reach up to 60 feet tall in the wild, it is relatively short-lived in urban landscapes, with perhaps a 20-year life span. Nevertheless, it provides reliable service during this period. Provide plenty of soil space for root expansion to increase life span. Sargent Cherry requires little maintenance once established with appropriate structure and is quite tolerant of drought and clay soil. Transplant in the spring.

It may be bothered by tent caterpillars, aphids, borers, and scales. Probably less susceptible to disease than most other Cherries, but very susceptible to Japanese beetles.

Prunus serotina Black Cherry

Pronunciation: PROO-nuss sair-o-TEE-nah

ZONES: 3B, 4A, 4B, 5A, 5B, 6A, 6B, 7A, 7B, 8A, 8B, 9A

HEIGHT: 60 to 90 feet

WIDTH: 35 to 50 feet

FRUIT: round; 0.33"; fleshy; purple-red; attracts wildlife; edible; fruit causes messy litter

GROWTH RATE: rapid; long-lived

HABIT: columnar or oval; moderate density; irregular outline or silhouette; medium texture

FAMILY: *Rosaceae*

LEAF COLOR: dark green; yellow to orange fall color

LEAF DESCRIPTION: alternate; simple; serrulate margin; oblong to ovate; deciduous; 3 to 6" long

FLOWER DESCRIPTION: white; pleasant fragrance; showy; spring

EXPOSURE/CULTURE:

Light Requirement: part shade to full sun

Soil Tolerances: all textures; alkaline to acidic; occasionally wet soil; salt and drought

PEST PROBLEMS: resistant to lethal pests

BARK/STEMS:

Bark is thin and easily damaged by mechanical impact; branches grow mostly upright and will not droop; should be grown with a single leader; no thorns

Pruning Requirement: needs little pruning to develop strong trunk and branch structure

Limb Breakage: resistant until middle age

Current Year Twig: brown; slender

LANDSCAPE NOTES:

Origin: Eastern North America into Central America

Usage: reclamation plant; along highways

Black Cherry is a native North American tree with an oval silhouette and low-hanging branches. These are easily removed to create clearance beneath the canopy. The finely toothed, shiny, deciduous leaves change in fall for a short period to lovely shades of yellow, orange, or red, but this varies with weather conditions and among seedlings. The leaves and twigs contain hydrocyanic acid, which could poison livestock or other animals if consumed in large quantities. Wild Cherry cough syrup is made from the reddish-brown, fragrant, and bitter inner bark. The wood is highly prized by woodworkers and has been used since colonial days for fine furniture.

In early spring, as the new leaves are unfolding, Black Cherry produces small, white, fragrant blossoms. These are followed by small, bitter fruits that mature during summer and fall from red to dark purple or black. Sometimes used for jams, jellies, or liqueurs, these fruits are highly prized by birds and other wildlife, which quickly devour them as they ripen. They stain concrete as they fall in summer, and people can roll on the hard seeds. If you plant Black Cherry, it is probably best to locate it away from walks and pavement. The tree appears to be tolerant of drought in its native habitat where roots are allowed to explore a large volume of soil, but growth is often poor in the restricted soil spaces characteristic of urban areas.

Black Cherry should be grown in full sun or partial shade on well-drained, noncompacted soil where it will not receive excessive heat or competition from grasses. Although somewhat tolerant of dry conditions, Black Cherry will respond best to rich, moist soil and a mulch to keep the root zone cool. Plants should not be disturbed after becoming established. They have a fairly shallow root system, making them susceptible to damage from anything stacked, stored, or parked within the dripline.

Limitations include Eastern tent caterpillar, fall webworm, and other chewing insects. These are not usually enough of a problem to warrant control unless trees are repeatedly infested. If eaten, foliage is poisonous causing cyanide-like symptoms.

Prunus serrulata 'Kwanzan' Kwanzan Cherry

Pronunciation: PROO-nuss sair-you-LAY-tah

ZONES: 5A, 5B, 6A, 6B, 7A, 7B, 8A, 8B, 9A

HEIGHT: 15 to 25 feet

WIDTH: 15 to 25 feet

FRUIT: no fruit

GROWTH RATE: moderate; short-lived

HABIT: upright-spreading or vase; moderate density; symmetrical; medium texture

FAMILY: *Rosaceae*

LEAF COLOR: dark green; copper, orange, or yellow fall color

LEAF DESCRIPTION: alternate; simple; serrate margins; lanceolate to ovate; deciduous; 2 to 5" long

FLOWER DESCRIPTION: pink; very showy; spring

EXPOSURE/CULTURE:

Light Requirement: full sun

Soil Tolerances: all textures; slightly alkaline to acidic; occasionally wet soil; some drought

Aerosol Salt Tolerance: low to moderate

PEST PROBLEMS: somewhat susceptible

BARK/STEMS:

Bark is thin and easily damaged by mechanical impact; branches grow mostly upright and will not droop; should be grown with a short, single leader; no thorns

Pruning Requirement: needs little pruning

Limb Breakage: branches can break from tree at top of short trunk

Current Year Twig: brown; medium thickness

CULTIVARS AND SPECIES:

There are many other cultivars of Oriental Cherry, but few are readily available. Some that may be locally available include: 'Amanogawa' ('Erecta')—semi-double, light pink, fragrant flowers, narrow columnar habit, about 20 feet tall; 'Fugenzo' ('James H. Veitch', 'Kofugen')—spreading habit, flowers 2.5 inches across, rose pink fading to light pink, double, 'Kofugen' sometimes described as having deeper color; 'Shirofugen'—rapid growth rate, young foliage bronze, later turning green, flowers double, 2.5 inches across, pink fading to white; 'Shirotae' ('Mt. Fuji', 'Kojima')—flowers double to semi-double, white, ruffled, about 2.5 inches across; 'Shogetsu'—tree 15 feet tall, broad and flat-topped, flowers double, pale pink, center may be white, can be 2 inches across; 'Ukon'—young foliage bronze, flowers pale yellow, semi-double.

LANDSCAPE NOTES:

Origin: China, Japan, Korea

Usage: bonsai; container or above-ground planter; buffer strip; highway; near a deck or patio; standard; specimen; street tree given plenty of soil space

Kwanzan Cherry has double-pink, very attractive flowers and is usually purchased and planted for this reason. The upright-spreading form is quite attractive in many locations, including near a patio or as a specimen away from lawn grass competition. The tree is glorious in flower and has been planted, along with Yoshino Cherry, in Washington, D.C., for the annual Cherry Blossom Festival. Kwanzan Cherry has good yellow fall color and does not bear fruit, but is troubled with pests.

Neither stress-tolerant nor highly drought-tolerant, Kwanzan Cherry should be located on a site with loose soil and plenty of moisture. Not for an urban parking lot island or exposed street tree planting; there borers and other problems normally attack. The trees have some tolerance to salt, and tolerate clay if the soil is well-drained. It prefers full sun, is intolerant of poor drainage, and is easily transplanted. The useful life of the species is limited to about 15 to 25 years for 'Kwanzan', on a good site, but the tree is a joy during this short period and should be planted.

Borers attack flowering cherries. Cankers and other problems also limit life span.

Prunus subhirtella var. autumnalis
Autumnalis Higan Cherry
Pronunciation: PROO-nuss sub-her-TELL-ah

ZONES: 5A, 5B, 6A, 6B, 7A, 7B, 8A, 8B

HEIGHT: 20 to 35 feet

WIDTH: 20 to 35 feet

FRUIT: oval; less than 0.5"; fleshy; black; attracts birds; not showy; no significant litter

GROWTH RATE: moderate; moderate life span

HABIT: rounded vase; dense; symmetrical; medium texture

FAMILY: *Rosaceae*

LEAF COLOR: dark green; yellow fall color

LEAF DESCRIPTION: alternate; simple; serrate margin; oblong to ovate; deciduous; 1.5 to 4" long

FLOWER DESCRIPTION: pink; showy; spring and occasionally fall

EXPOSURE/CULTURE:

Light Requirement: full sun

Soil Tolerances: all textures; acidic; moderate drought

PEST PROBLEMS: somewhat sensitive

BARK/STEMS:

Showy, dark trunk; branches grow mostly upright and horizontal; should be grown with a short, single leader; no thorns

Pruning Requirement: needs initial pruning to develop strong structure; space main branches along trunk

Limb Breakage: resistant

Current Year Twig: brown; slender

LANDSCAPE NOTES:

Origin: Japan

Usage: bonsai; buffer strip; highway; near a deck or patio; specimen

Double-flowered Autumnalis Higan Cherry is usually seen with a multiple trunk, slender, upright branches, and a rounded canopy. The canopy on older trees often grows wider than the tree grows tall. Before the leaves appear in spring, the trees are covered with many semi-double pink flowers; some flowers may also appear in autumn if the weather is sufficiently warm. The one- to four-inch-long leaves are later joined by small, black berries. The leaves cast light shade below the spreading canopy, and turn an attractive yellow or bronze before dropping in the fall.

Use this tree along an entrance road to a commercial development, planted on 20-foot centers, or alongside the patio or deck in the back yard. It also makes a nice small-scale courtyard tree or specimen planted in a lawn or groundcover bed. It could be planted along a residential street where there is plenty of soil for root expansion. It is becoming available in the industry.

Double-flowered Higan Cherry should be grown in full sun or partial shade on well-drained, acid soil with sufficient moisture. It is moderately drought-tolerant and grows well in clay soil, as long as the site drains well. Trees grow quickly when young but slow down with age.

Aphids cause distortion of new growth, deposits of honeydew, and sooty mold. Borers attack trees under stress. Keep trees healthy with regular fertilizer applications and irrigation in extended drought. Scales of several types infest *Prunus*. A bacterium causes leaf spot and twig cankers on cherry. Small, reddish spots dry and drop out, giving a shot-holed appearance. Defoliation can be severe when weather conditions favor disease development. Fertilize infected trees and prune out infected branches.

Prunus subhirtella 'Pendula' Weeping Higan Cherry

Pronunciation: PROO-nuss sub-her-TELL-ah

ZONES: 5A, 5B, 6A, 6B, 7A, 7B, 8A, 8B

HEIGHT: 25 to 30 feet

WIDTH: 15 to 25 feet

FRUIT: oval; 0.33"; fleshy; black; no significant litter

GROWTH RATE: moderate; moderate life span

HABIT: weeping vase; moderate density; irregular silhouette; fine texture

FAMILY: *Rosaceae*

LEAF COLOR: green; yellow fall color

LEAF DESCRIPTION: alternate; simple; serrate margin; oblong to ovate; deciduous; 1.5 to 3" long

FLOWER DESCRIPTION: pink; very showy; spring

EXPOSURE/CULTURE:

Light Requirement: full sun

Soil Tolerances: all textures; acidic; some salt; moderate drought

PEST PROBLEMS: somewhat susceptible

BARK/STEMS:

Branches droop as tree grows and will require pruning for vehicular or pedestrian clearance beneath canopy; should be trained with a single leader as far up into canopy as possible; no thorns

Pruning Requirement: drooping branches can be pruned to encourage more upright growth

Limb Breakage: mostly resistant

Current Year Twig: brown; slender

LANDSCAPE NOTES:

Origin: Japan

Usage: shade tree; specimen

Weeping Higan Cherry spreads in a graceful weeping habit. Leaves stay glossy green throughout the summer and into the fall and then turn a vivid yellow before leaving the tree bare in winter. Even a light breeze will sway the thin, drooping branches. The drooping bare branches lend a soothing grace to the landscape in winter. There is nothing quite like the Weeping Higan Cherry in full bloom in the spring. The light pink (almost white), one-inch-diameter flowers cover the branches before the leaves emerge, giving the appearance that fresh snow has fallen on the tree.

Weeping Higan Cherry makes a striking specimen in a large yard, park, or commercial landscape. Locate it to bring attention to an area, as its form is attractive all year long. Be prepared to provide more maintenance dollars to care for this beautiful tree than others. One of the premier weeping trees for American gardens, Weeping Cherry has a place in any large-scale landscape as a specimen to accent a lawn area. It is best to maintain the tree with mulch out to the edge of the canopy and to allow branches to almost touch the ground. It is very attractive close to water and is often used this way.

Nurseries often graft Weeping Cherry onto a fast-growing rootstock. The rootstock cherry is trained to one straight trunk and small Weeping Cherry branches are grafted to the trunk four to six feet from the ground.

It has a fairly rapid growth rate and prefers an open, sunny location sheltered from wind. The tree grows in almost any soil and transplants easily when young. Clay soil is suitable, as adequate moisture helps keep stress to a minimum. Irrigation is a must in sandy soil. Weeping Higan Cherry is rather prone to problems, including borers, aphids, and various diseases, particularly in dry soil.

Prunus triloba var. *multiplex* Flowering-Almond

Pronunciation: PROO-nuss try-LOW-bah [variety] MULL-tee-plecks

ZONES: 3B, 4A, 4B, 5A, 5B, 6A, 6B

HEIGHT: 10 to 15 feet

WIDTH: 10 to 15 feet

FRUIT: round; 0.5 to 1"; fleshy; brown; attracts birds; not showy; no significant litter

GROWTH RATE: slow to moderate; short-lived

HABIT: rounded, weeping habit; moderate density; symmetrical; medium texture

FAMILY: *Rosaceae*

LEAF COLOR: medium green; copper or yellow fall color

LEAF DESCRIPTION: alternate; simple; double serrate margin; oval to obovate; deciduous; 1.5 to 3" long

FLOWER DESCRIPTION: pink; very showy; spring

EXPOSURE/CULTURE:

Light Requirement: part shade to full sun

Soil Tolerances: all textures; acidic; some drought

Aerosol Salt Tolerance: none

PEST PROBLEMS: susceptible

BARK/STEMS:

Droop to the ground; can be grown with a short, single leader or with several trunks; no thorns

Pruning Requirement: needs little pruning to develop strong structure; training and pruning are required to develop a single trunk and a tree form

Limb Breakage: resistant

Current Year Twig: brown; slender

CULTIVARS AND SPECIES:

'Plena' has double flowers; var. *simplex* has single flowers sometimes followed by red fruit.

LANDSCAPE NOTES:

Origin: China

Usage: bonsai; container or above-ground planter; buffer strip; near a deck or patio; standard; specimen

Reaching approximately equal height and spread, Flowering-Almond has beautiful, double pink flowers that appear in midspring. It provides a wonderful accent in a residential yard or courtyard when in flower. Be sure to locate it in a groundcover or mulched bed, as mechanical injury or stress of any kind hastens the demise of this short-lived tree.

It is very suitable in a shrub border as a tall accent. It can be sculptured nicely into a unique form with proper pruning and training and is well suited for container gardening. Regular pruning is needed for best flowering performance. Branches cut in early spring can be forced into bloom indoors. Plants can be kept small by regular thinning and drop-crotching.

Flowering-Almond grows best in sun or partial shade on rich, moist soil. Keep turf grass cleared away from a four- to six-foot-diameter circle around the tree. It transplants well and grows at a moderate pace.

Aphids cause distortion of new growth, deposits of honeydew, and sooty mold. Borers attack trees under stress. Keep trees healthy with regular fertilizer applications and irrigation in drought. Scales of several types infest *Prunus* species. There are many other potential problems.

Prunus umbellata Flatwoods Plum

Pronunciation: PROO-nuss um-bell-LAY-tah

ZONES: 8A, 8B, 9A, 9B

HEIGHT: 12 to 20 feet

WIDTH: 12 to 20 feet

FRUIT: round; 0.5"; fleshy but hard; purple; attracts wildlife; edible; somewhat showy; no significant litter

GROWTH RATE: slow-growing; moderate life span

HABIT: rounded; moderate density; irregular silhouette; fine texture

FAMILY: *Rosaceae*

LEAF COLOR: medium green; yellow fall color

LEAF DESCRIPTION: alternate; simple; serrate to serrulate margin; oval to ovate; deciduous; 1.5" long

FLOWER DESCRIPTION: white; showy; late winter to early spring

EXPOSURE/CULTURE:

Light Requirement: grows best in full sun, but tolerates part shade

Soil Tolerances: loam; sand; slightly alkaline to acidic; moderate drought

Aerosol Salt Tolerance: none

PEST PROBLEMS: resistant

BARK/STEMS:

Branches droop as tree grows and will require pruning for vehicular or pedestrian clearance beneath canopy; can be grown with a single leader or with several trunks; no thorns

Pruning Requirement: requires little pruning to develop strong structure; shorten then remove lower branches early to develop a single-leader street tree

Limb Breakage: resistant

Current Year Twig: brown; slender

CULTIVARS AND SPECIES:

P. angustifolia, the Chickasaw Plum, is more shrubby, producing many suckers.

LANDSCAPE NOTES:

Origin: Texas to Florida

Usage: bonsai; container or above-ground planter; parking lot island; buffer strip; highway; near a deck or patio; specimen; street tree; reclamation

A native of the woods of the southeastern United States, Flatwoods Plum is a round-topped, deciduous tree that is most often planted for its spectacular display of blooms. It may look a little ragged in winter, but can be pruned to thin and balance the canopy. In late February or March, before the two-inch-long, finely serrate leaves appear, these small trees take on a white, billowy, almost cloud-like appearance when they are clothed in the profuse, small, white flower clusters. These half-inch blooms are followed by one-half- to one-inch-long, edible fruits, which vary in flavor from very tart to sweet. These plums are very attractive to various forms of wildlife.

It is a bit weedy in growth habit, but proper training and pruning can create an attractive specimen or small median strip or street tree. Once adequate trunk diameter develops, lower branches should be removed if the plant is to be trained as a street tree. This tree is especially suited for planting beneath power lines or in other areas where overhead space is limited. Do not expect a row of them to form a uniform shape. The tree branches low to the ground, making it a nice tall element in a backyard shrub border. Some small-diameter, interior branches can be removed in winter to open up the crown for a more formal, attractive shape and habit. The tree may live 30 to 40 years on a good site.

Flatwoods Plum thrives in full sun or partial shade on a wide variety of well-drained soils. When placed in sandy soil, it grows best with irrigation and some shade in the afternoon. Trees grow quickly when young but considerably slower when mature and bearing fruit.

No limitations of major concern. Tent caterpillars occasionally infest Flatwoods Plum.

Prunus x yedoensis Yoshino Cherry

Pronunciation: PROO-nuss x yed-o-EN-sis

ZONES: 5B, 6A, 6B, 7A, 7B, 8A

HEIGHT: 35 to 45 feet

WIDTH: 35 to 45 feet

FRUIT: round; 0.5 to 1"; fleshy; black; attracts birds; no significant litter

GROWTH RATE: moderate; moderate life span

HABIT: upright spreading to vase; moderate density; symmetrical; medium texture

FAMILY: *Rosaceae*

LEAF COLOR: dark green; yellow fall color

LEAF DESCRIPTION: alternate; simple; double serrate to serrate margin; oval to obovate; deciduous; 2 to 4" long

FLOWER DESCRIPTION: pinkish-white; showy; spring

EXPOSURE/CULTURE:

Light Requirement: full sun for best flowers

Soil Tolerances: all textures; acidic; some drought

Aerosol Salt Tolerance: none

PEST PROBLEMS: somewhat susceptible

BARK/STEMS:

Showy bark; bark is thin and easily damaged by mechanical impact; branches droop somewhat as tree grows and will require some pruning for vehicular or pedestrian clearance beneath canopy; should be grown with a single leader; no thorns

Pruning Requirement: requires pruning to develop strong structure; space branches along a short, central trunk

Limb Breakage: resistant

Current Year Twig: brown; slender

CULTIVARS AND SPECIES:

'Akebona' ('Daybreak')—flowers softer pink or white, resists Japanese beetles; 'Perpendens'—irregularly pendulous branches; 'Shidare Yoshino' ('Perpendens')—irregularly pendulous branches trained from a staked central trunk.

LANDSCAPE NOTES:

Origin: Japan

Usage: buffer strip; highway; near a deck or patio; small shade tree; specimen; sidewalk cutout; street tree if irrigated or growing with adequate soil space

Yoshino Cherry grows quickly to 20 feet, and has beautiful bark marked with prominent lenticels, but like most cherries is a relatively short-lived tree. Young trees have mostly upright to horizontal branching, making them ideal for planting along walks and over patios. The white to pink flowers, which bloom in early spring before the leaves develop, are sometimes damaged by late frosts or very windy conditions. This is the tree, along with 'Kwanzan' Cherry, that makes such a show each spring in Washington, D.C.

Best used as a specimen or near the deck or patio for shade, Yoshino Cherry also works nicely along walks or near a water feature. Cherry is not suited for planting in parking lot islands due to drought sensitivity. Large specimens take on a weeping habit, with delicate branchlets arranged on upright-spreading branches affixed to a short, stout trunk. A lovely addition to a sunny spot where a beautiful specimen is needed. Winter form, yellow fall color, and pretty bark make this a year-round favorite.

Provide good drainage in an acidic soil for best growth. Crowns become one-sided unless they receive light from all around the plant, so locate in full sun. Select another tree to plant if soil is poorly drained. Roots should be kept moist and should not be subjected to prolonged drought.

Aphids cause distortion of new growth, deposits of honeydew, and sooty mold. Borers attack flowering Cherries under stress. Scales of several types infest *Prunus.* Spider mites cause yellowing or stippling. Yoshino Cherry may be subject to witches' broom.

Pseudolarix x kaempferi Golden Larch

Pronunciation: sue-doe-LAIR-icks KEM-fair-eye

ZONES: 5A, 5B, 6A, 6B, 7A, 7B

HEIGHT: 40 to 70 feet

WIDTH: 30 to 40 feet

FRUIT: oval; 2 to 3" long; dry; brown; showy for a short period; no significant litter

GROWTH RATE: slow to moderate; long-lived

HABIT: pyramidal; very open; irregular; fine texture

FAMILY: *Pinaceae*

LEAF COLOR: pale green above, blue-green below; showy yellow to copper fall color

LEAF DESCRIPTION: spiral; simple; linear; needle-like; deciduous; 1 to 2" long

FLOWER DESCRIPTION: yellow; not showy; spring

EXPOSURE/CULTURE:

Light Requirement: grows best in full sun

Soil Tolerances: all textures; acidic; moderate drought; good drainage essential

PEST PROBLEMS: resistant

BARK/STEMS:

Showy bark; branches droop as tree grows and will require pruning for vehicular or pedestrian clearance beneath canopy; normally grown with a single leader; no thorns

Pruning Requirement: requires little pruning to develop strong structure

Limb Breakage: resistant

Current Year Twig: brownish-green; slender

CULTIVARS AND SPECIES:

'Nana' is a dwarf form. 'Annesleyana' is a dense, dwarf form with weeping branches.

LANDSCAPE NOTES:

Origin: China

Usage: specimen

Recently reclassified as *P. amabilis*, this large but graceful, uncommon deciduous conifer reaches mature height with a straight trunk and a spread of up to 40 feet. Because it grows slowly, it is usually not seen more than 30 to 40 feet tall unless it is provided with some irrigation or is located on a moist site. The silhouette is a rather open, irregular pyramidal shape with pendulous branch tips. The fine-textured, needled foliage is 1.5 to 2 inches long and pale green above, blue-green below. The foliage turns a brilliant yellow in the fall for only a short time, but long enough to make a very striking landscape statement. The three-inch-long, upright cones are interspersed along the upper branches, and the reddish-brown bark is rugged and furrowed.

The attractive bark, strong pyramidal shape, and wide-spreading branches make Golden Larch particularly attractive in the winter landscape. Its slow growth and attractive form make it a popular choice for containers and for use as a bonsai. It can be used in residential landscapes, because it grows slowly, but is best when located in a large-scale landscape. Plant it in the open so the form can be fully appreciated. Locate it far enough from sidewalks and buildings so that lower branches will not have to be pruned.

Golden Larch should be grown in full sun on deep, rich, well-drained, moist acid soil where it can be protected from harsh, cold winds. The trees should not be planted in limestone soils and they are not tolerant of clay unless located on a slope where drainage will be excellent. No limitations of major concern.

Pseudotsuga menziesii Douglas-fir

Pronunciation: sue-doe-SOO-gah men-ZEE-zee-eye

ZONES: 4A, 4B, 5A, 5B, 6A, 6B

HEIGHT: 40 to 60 feet

WIDTH: 25 to 30 feet

FRUIT: oval; 3 to 4"; dry; brown; no significant litter; showy

GROWTH RATE: moderate; long-lived

HABIT: pyramidal; dense; symmetrical; fine texture

FAMILY: *Pinaceae*

LEAF COLOR: green to blue-green; needle-leaf evergreen

LEAF DESCRIPTION: spiral; simple; needle-like (filiform); 1 to 1.5" long

FLOWER DESCRIPTION: red; not showy

EXPOSURE/CULTURE:

Light Requirement: full sun

Soil Tolerances: loam or sand; alkaline to acidic; salt-sensitive; drought-sensitive; good drainage is essential

Aerosol Salt Tolerance: low

PEST PROBLEMS: susceptible

BARK/STEMS:

Branches droop as tree grows and will require pruning for vehicular or pedestrian clearance beneath canopy; normally grown with a single leader; no thorns

Pruning Requirement: needs little pruning to develop strong structure

Limb Breakage: susceptible when branches are weighted down with rain or snow

Current Year Twig: green; slender

CULTIVARS AND SPECIES:

'Anguina'—long, snake-like branches; 'Brevifolia'—short leaves; 'Compacta'—compact, conical growth; 'Fastigiata'—dense, pyramidal; 'Fretsii'—dense bush, short broad leaves; 'Glauca'—bluish foliage; 'Nana'—dwarf; 'Pendula'—long, drooping branchlets; 'Revoluta'—curled leaves; 'Stairii'—variegated leaves.

LANDSCAPE NOTES:

Origin: western North America into Mexico

Usage: specimen; screen; Christmas tree; lumber

Douglas-Fir grows 40 to 60 feet and spreads 15 to 25 feet in an erect pyramid in the landscape, but grows more than 200 feet tall in its native habitat in the west. Hardiness varies with seed source, so be sure seed was collected from an area with suitable cold-hardiness to the area in which it will be used.

Douglas-Fir is most commonly used as a screen or occasionally a specimen in the landscape. Not suited for a small residential landscape, it is often a fixture in a commercial setting. Allow room for the spread of the tree, as this tree looks terrible with lower limbs removed. It is grown and shipped as a Christmas tree in many parts of the country.

The tree prefers a sunny location with a moist soil and is not considered a good tree for much of the southern United States. In the eastern United States, it grows but struggles in zone 7. Douglas-Fir transplants best when balled-in-burlap. It tolerates pruning and shearing but will not tolerate dry soil for extended periods. Protect from direct wind exposure for best appearance. Some occasional watering during summer dry spells will help the tree stay vigorous, especially in the southern end of its range.

Scale and bark beetles may infest Douglas-Fir, especially those under stress. Root rot can be a serious problem on clay and other wet soils. Needles infected by leaf cast fungi in spring turn brown and fall off. Several fungi cause canker diseases leading to branch die-back. Pitch canker can infect the trunk. Maintain tree health and prune out infected branches.

Psidium littorale Cattley Guava, Strawberry Guava

Pronunciation: SID-ee-um lih-teh-RAH-lee

ZONES: 10A, 10B, 11

HEIGHT: 15 to 25 feet

WIDTH: 12 to 20 feet

FRUIT: round; 1 to 3"; fleshy; red; attracts wildlife; edible; causes significant litter; showy

GROWTH RATE: moderate; moderate life span

HABIT: pyramidal when young, rounded vase later; moderate density; irregular when young, symmetrical later; medium texture

FAMILY: *Myrtaceae*

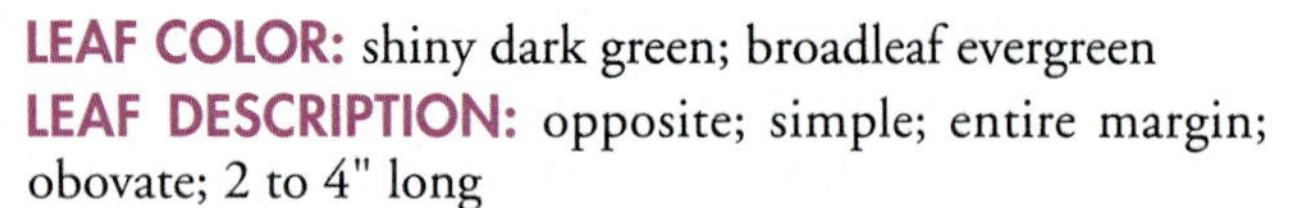

LEAF COLOR: shiny dark green; broadleaf evergreen

LEAF DESCRIPTION: opposite; simple; entire margin; obovate; 2 to 4" long

FLOWER DESCRIPTION: white; somewhat showy; year-round

EXPOSURE/CULTURE:

Light Requirement: part shade to full sun

Soil Tolerances: all textures; alkaline to acidic; occasionally wet soil; drought

Aerosol Salt Tolerance: none

PEST PROBLEMS: resistant

BARK/STEMS:

Showy bark and interesting trunk structure; routinely grown with multiple trunks; branches grow mostly upright and will not droop; no thorns

Pruning Requirement: needs little pruning to develop strong structure

Limb Breakage: resistant

Current Year Twig: brown; green; slender

CULTIVARS AND SPECIES:

The variety *littorale*, Lemon Guava, is slower-growing and has large yellow fruits with delightful, spicy flavor. Variety *longipes* has purple-red fruit.

LANDSCAPE NOTES:

Origin: Brazil

Usage: container or above-ground planter; espalier; fruit tree; buffer strip; near a deck or patio; small shade tree; specimen; standard

This upright, multi-branched, evergreen shrub or small tree has many features that make it a popular landscape choice. Unfortunately, Cattley Guava has escaped cultivation and has infested ditch banks and other wild areas in south Florida. Capable of reaching 25 feet in height, but often seen at 10 to 15 feet, the thick, smooth, 4-inch-long, leathery leaves nicely complement the smooth, gray-brown to golden bark, which attractively peels off in thin sheets. The single, white, one-inch-diameter flowers have many prominent stamens and look like powderpuffs in miniature. The blooms appear heaviest in April, but can open sporadically throughout the year.

The 1.5-inch-diameter, bright red, pear-shaped fruits ripen to a very dark red in July and are a popular treat for people and wildlife. Birds, raccoons, and squirrels all love the delicious, sweet-tart, white flesh and will widely spread the small, grape-like seeds. The abundant, fallen fruits may be messy on hard surfaces such as walks and cars, and trees should probably not be planted along residential streets or over sidewalks or patios. It would make a good small boulevard or median street tree, and accents an area with its well-shaped canopy and interesting trunk form.

Cattley Guava grows well in full sun or partial shade on a wide range of soils, including sand or clay, but fruit quality is improved on rich soils. Plants are drought-tolerant once established, but benefit from a thick, organic mulch. Any necessary pruning should be done after fruiting. Pick the fruit for eating before the fruit fly larvae discover it. Fruit fly larvae devour the fruit but do no harm to the tree.

Pterocarya stenoptera Chinese Wingnut

Pronunciation: tee-row-CARE-ee-yah sten-OP-tur-ah

ZONES: 5A, 5B, 6A, 6B, 7A, 7B, 8A, 8B, 9A, 9B, 10A, 10B, 11

HEIGHT: 60 to 80 feet

WIDTH: 60 to 70 feet

FRUIT: elongated; 1"; dry; green turning brown; abundant; causes some litter; persistent; showy in long, drooping spikes

GROWTH RATE: phenomenal; moderate to long life span

HABIT: pyramidal when young, vase later; open; irregular silhouette; coarse texture

FAMILY: *Juglandaceae*

LEAF COLOR: dark green; no fall color change

LEAF DESCRIPTION: alternate; odd pinnately compound; serrate margin; oval to oblong; deciduous; 2 to 5" long leaflets

FLOWER DESCRIPTION: green; not showy; spring

EXPOSURE/CULTURE:

Light Requirement: grows best in full sun

Soil Tolerances: all textures; acidic; drought

PEST PROBLEMS: resistant

BARK/STEMS:

Branches grow mostly upright; should be grown with a single leader; trunk flares out at base; no thorns

Pruning Requirement: requires pruning to develop strong structure; space main branches along trunk; prune annually during first five years to establish good structure

Limb Breakage: probably resistant on well-formed trees

Current Year Twig: brown; thick

LANDSCAPE NOTES:

Origin: China

Usage: shade tree; possible street tree if planted 12 feet or more from pavement; specimen for a large park or golf course

Chinese Wingnut is a deciduous tree with large, substantial branches that spread as wide as the tree is tall. The 6- to 12-inch-long leaves are composed of many, finely toothed, oval leaflets, and do not display any appreciable fall color. Of particular interest are the 6- to 12-inch-long seed clusters—green strings of winged seeds suspended below the branches, which turn brown and fall in autumn. The tree grows at a phenomenal rate. A 6-year-old tree in Raleigh, N.C., was about 25 feet tall and wide with a 14-inch trunk diameter. The trunk can reportedly grow to at least eight feet in diameter.

Chinese Wingnut may perform well as a street or shade tree, but the large, aggressive roots, which extend a fair distance from the trunk, may make it unsuitable for use in a lawn or garden. It could be considered for broader use as an urban tree, but it is largely untested, so use it with caution. Locate it well away from (12 feet or more) sidewalks and driveways so the large-diameter surface roots will not lift the concrete or asphalt.

Prune early in the life of the tree to form a good, strong structure by spacing major limbs several feet apart along a central trunk. Do not allow these limbs to grow more than about two-thirds the diameter of the trunk, to encourage formation of a strong branch collar.

Chinese Wingnut grows quickly in full sun and moist soil, and is ideally located beside a stream or pond. The trees are tolerant of clay, wind, drought, and compacted soil once they have become well established. Any necessary pruning should be done in the summer to prevent the bleeding that occurs in spring or winter.

Pterostyrax hispida Fragrant Epaulette Tree

Pronunciation: tee-row-STY-racks HISS-puh-dah

ZONES: 5A, 5B, 6A, 6B, 7A, 7B, 8A

HEIGHT: 20 to 30 feet

WIDTH: 20 to 30 feet

FRUIT: oval pod; less than 0.5"; dry; gray, white; no significant litter; persistent; showy

GROWTH RATE: moderate; moderate life span

HABIT: rounded vase; dense; irregular when young, symmetrical later; coarse texture

FAMILY: *Styracaceae*

LEAF COLOR: light green; yellow fall color in the north

LEAF DESCRIPTION: alternate; simple; oblong; 3 to 8" long; deciduous

FLOWER DESCRIPTION: white; pleasant fragrance; showy; summer

EXPOSURE/CULTURE:

Light Requirement: grows best in full sun

Soil Tolerances: all textures; slightly alkaline to acidic; moderate drought

PEST PROBLEMS: resistant

BARK/STEMS:

Branches droop as tree grows and will require pruning for vehicular or pedestrian clearance beneath canopy; can be trained to a single or multiple trunk; showy trunk; no thorns

Pruning Requirement: requires pruning to develop strong structure; head back lower limbs when tree is young for a more upright habit

Limb Breakage: resistant

Current Year Twig: gray; stout

LANDSCAPE NOTES:

Origin: Japan

Usage: near a deck or patio; small shade tree; specimen

Fragrant Epaulette Tree is a deciduous tree that can reach 30 feet in height, with an equal spread, but is often smaller. The three- to eight-inch-long by four-inch-wide, oval leaves have a silvery cast on the undersides, and the leaves turn yellow or yellow-green in fall before dropping.

Of particular interest are the delicate, creamy white, slightly fragrant blooms, which appear in early summer when few other trees are flowering. Hanging in nine-inch-long and four-inch-diameter clusters, these blossoms are quite striking when viewed from below, so the tree should be properly situated to take advantage of this display. Plant it on a bank along a walkway, in a raised bed, or above a bench. This will also allow the attractive bark to be more easily appreciated, with its inner orange-tan color showing through the gray bark at expansion cracks. The seed pods that follow the pendulous blooms are gray, fuzzy, and cling to the branches throughout the winter. They are quite showy and useful in dried flower arrangements.

The upright-spreading branches help to make this a fabulous small shade tree, casting deep shade beneath the tree. Some early pruning may be necessary to remove the lower branches (to allow easier access beneath the tree) or to control the tree's shape. Young, aggressive branches that droop can be headed back to help them grow thicker and remain more upright. Seek out this little-known tree for a specimen or group planting that is not likely to be duplicated due to the rarity of the tree in the nursery trade.

Fragrant Epaulette Tree should be grown in full sun on moist, well-drained soil. It will tolerate both acid and slightly alkaline soils. The trees perform best when they receive no shade, becoming thin and flowering poorly in the shade.

No limitations of major concern, but the tree has not been extensively tested.

Pyrus calleryana 'Aristocrat'
'Aristocrat' Callery Pear
Pronunciation: PIE-russ cal-lurr-ee-A-nah

ZONES: (4B), 5A, 5B, 6A, 6B, 7A, 7B, 8A, 8B, 9A, 9B

HEIGHT: 35 to 40 feet

WIDTH: 25 to 35 feet

FRUIT: abundant; round; 0.33"; hard; brown or tan; attracts birds; not showy; some litter; persistent

GROWTH RATE: rapid; moderate life span

HABIT: oval to pyramidal; moderately dense; symmetrical early, irregular later; medium texture

FAMILY: *Rosaceae*

'Autumn Blaze'

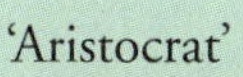

'Aristocrat'

LEAF COLOR: dark green; showy, red fall color

LEAF DESCRIPTION: alternate; simple; crenate to sinuate or undulate margins; ovate; deciduous; 2 to 4" long

FLOWER DESCRIPTION: white; showy; spring

EXPOSURE/CULTURE:

Light Requirement: grows best in full sun

Soil Tolerances: all textures; alkaline to acidic; occasionally wet soil; moderate salt; drought

Aerosol Salt Tolerance: moderate

PEST PROBLEMS: somewhat susceptible

BARK/STEMS:

Bark is thin and easily damaged by mechanical impact; branches droop as tree grows and will require pruning for vehicular or pedestrian clearance beneath canopy; should be grown with a single leader; no thorns

Pruning Requirement: requires some early pruning to develop strong structure; remove any upright branches with narrow crotches

Limb Breakage: more resistant than Bradford Pear

Current Year Twig: brown; thick

LANDSCAPE NOTES:

Origin: China and Korea

Usage: container or above-ground planter; parking lot island; buffer strip; highway; shade tree; specimen; sidewalk cutout; street tree; urban tolerant

The more dominant trunk and open form of 'Aristocrat' Callery Pear helps to make it less susceptible to wind and ice damage than 'Bradford'. Branch angles are wider and lateral branches grow at a slower rate than on 'Bradford'; therefore, the branches are better attached to the trunk. In spring, before the new leaves unfold, the tree puts on a brilliant display of pure white flowers which, unfortunately, do not have a pleasant fragrance. The small, pea-sized, red-brown fruits are quite attractive to birds and other wildlife, and mummify on the tree, persisting for several months to a year. Planting two or more cultivars of Callery Pear together could increase fruit set.

Planted commonly as a street tree or in parking lot islands, it is also quite suited for downtown tree pits, due to its urban tolerance. Like Bradford Pear, it is able to tolerate small soil spaces. 'Aristocrat' appears to be mostly free of the severe branch breakage common to Bradford Pear, but has been shown to be more susceptible to fire blight, particularly in the south. This disease can devastate a planting. Pruning the trees early in their life to space lateral branches along a central trunk should be all that is needed to ensure a strong, well-structured tree. Buy only trees with well-spaced branches.

Callery Pear trees are shallow-rooted and will tolerate most soil types, including alkaline and clay, are pollution-resistant, and tolerate drought and wet soil well. 'Aristocrat' is a very adaptable tree suited for downtown and other restricted soil spaces. 'Redspire' is oval and disease-resistant, producing a small number of fruits. 'Autumn Blaze' is very hardy, shows brilliant red fall color earlier than other cultivars, has wide branch angle and no fruit, but is susceptible to fire blight. 'Capital' is upright and susceptible to disease.

Pyrus calleryana 'Bradford'
'Bradford' Callery Pear
Pronunciation: PIE-russ cal-lurr-ee-A-nah

ZONES: 5A, 5B, 6A, 6B, 7A, 7B, 8A, 8B, 9A, 9B

HEIGHT: 40 feet

WIDTH: 30 to 45 feet

FRUIT: round; 0.33"; hard; brown or tan; no significant litter; persistent; germinates in the landscape

GROWTH RATE: rapid; short to moderate life span

HABIT: pyramidal when young, oval or round later; dense; symmetrical; medium texture

FAMILY: *Rosaceae*

LEAF COLOR: dark green; very showy orange, red, purple fall color

LEAF DESCRIPTION: alternate; simple; crenate or serrate margins; ovate; deciduous; 2 to 4" long

FLOWER DESCRIPTION: white; showy; spring

EXPOSURE/CULTURE:

Light Requirement: grows best in full sun

Soil Tolerances: all textures; alkaline to acidic; occasionally wet soil; moderate salt; drought

Aerosol Salt Tolerance: moderate

PEST PROBLEMS: resistant

BARK/STEMS:

Bark is thin and easily damaged by mechanical impact; branches droop as tree grows and will require pruning for vehicular or pedestrian clearance beneath canopy; should be grown with a single leader as much as possible; no thorns

Pruning Requirement: requires frequent pruning to develop strong structure; good luck!

Limb Breakage: susceptible to breakage at crotch due to poor collar formation

Current Year Twig: brown; thick

LANDSCAPE NOTES:

Origin: China and Korea

Usage: above-ground planter; parking lot island; buffer strip; highway; shade tree; specimen; sidewalk cutout; street tree; urban tolerant

'Bradford' is the original introduction of Callery Pear, but has an inferior branching habit when compared to other cultivars that have since been developed. It forms included bark on many vertical limbs packed closely on the trunk. The crown on young trees is dense and the branches long and not tapered, like an overgrown shrub, making it quite susceptible to wind damage and ice breakage. However, it does put on a gorgeous, early spring display of pure white blossoms, later forming small, brownish fruits. Fall color is incredible, ranging from red and orange to dark maroon. The canopy on older plants (20 years) opens up as branches spread apart.

The major problem with the 'Bradford' Callery Pear has been too many upright branches growing too close together on the trunk. It is nearly impractical to prune the tree to a form that would increase longevity. On young trees, space major branches along the trunk. Do not allow several branches to grow to the same size at one point on the trunk.

Callery Pears are pest- and pollution-resistant, and tolerate soil compaction, drought, and occasionally wet soil. Unfortunately, as trees approach 30 years old, they begin to fall apart in ice and snow storms due to inferior, tight branch structure. They are certainly beautiful and grow extremely well in urban soil until then and probably will continue to be planted because of their urban toughness. As you plan downtown street tree plantings, remember that in downtown sites many other trees succumb before this one, for a variety of reasons.

'Bradford' shows the best resistance to fire blight, in tests conducted in the southeast, of all Callery Pear cultivars tested. Damage is usually noticed only at branch tips. Birds often nest in and visit the tree, dropping their belongings. *Entomosporium* leaf spot causes defoliation in the western United States. *Botryosphaeria* canker has recently been reported.

Pyrus calleryana 'Chanticleer'
'Chanticleer' Callery Pear
Pronunciation: PIE-russ cal-lurr-ee-A-nah

ZONES: 5A, 5B, 6A, 6B, 7A, 7B, 8A, 8B

HEIGHT: 35 to 45 feet

WIDTH: 12 to 15 feet

FRUIT: round; 0.33"; hard; brown or tan; no significant litter; persistent

GROWTH RATE: rapid; moderate life span

HABIT: narrow pyramid, upright-oval; dense; symmetrical; medium texture

FAMILY: *Rosaceae*

LEAF COLOR: dark green; showy, reddish-orange or yellow fall color

LEAF DESCRIPTION: alternate; simple; crenate to sinuate or undulate; ovate; deciduous; 2 to 4" long

FLOWER DESCRIPTION: white; showy; spring

EXPOSURE/CULTURE:

Light Requirement: grows best in full sun

Soil Tolerances: all textures; alkaline to acidic; occasionally wet soil; moderate salt; drought

Aerosol Salt Tolerance: moderate

PEST PROBLEMS: resistant

BARK/STEMS:

Bark is thin and easily damaged by mechanical impact; branches grow mostly upright and will not droop; should be grown with a single leader; no thorns; one of the better formed Callery Pears, with a nice central leader

Pruning Requirement: requires pruning to develop strong structure; eliminate upright branches with tight crotches

Limb Breakage: resistant

Current Year Twig: brown; thick

LANDSCAPE NOTES:

Origin: China and Korea

Usage: container or above-ground planter; parking lot island; buffer strip; highway; shade tree; specimen; sidewalk cutout; street tree; urban tolerant

'Chanticleer' Callery Pear quickly grows to mature height but spreads to only about 15 feet wide, with upright-spreading, thornless branches. The silhouette appears as a fat column, growing wider than 'Whitehouse' and 'Capital' but narrower than 'Bradford' and 'Aristocrat'. In spring, before the new leaves unfold, the tree puts on a nice display of pure white flowers. Leaves turn yellow to orange in fall, putting on an attractive display before dropping. The small, pea-sized, brownish fruits are quite attractive to birds and other wildlife. Planting two or more cultivars of Callery Pear together could increase fruit set.

Planted commonly as a street tree or in parking lot islands, it is also quite suited for downtown tree pits due to its urban tolerance. Like 'Bradford' Pear, it is able to tolerate small soil spaces. It looks great located along a street on 20- to 25-foot-centers and creates a visual corridor for traffic flow.

Its fruit set could create a nuisance for some. Pruning the trees early in their life to space lateral branches along a central trunk should help in developing a strong, well-structured tree. Only buy trees with well-spaced branches. This cultivar has a better form than 'Bradford' and is easier to train to a strong structure.

Callery Pear trees are shallow-rooted and will tolerate most soil types, including alkaline and clay, are pest- and pollution-resistant, and tolerate drought and wet soil well. It is a very adaptable tree suited for downtown and other restricted soil spaces.

Moderately susceptible to fire blight and fungal leaf spot. Branches may fall off due to graft incompatability. 'Cleveland Select' and 'Stonehill' are probably the same trees as 'Chanticleer'.

Quercus acutissima Sawtooth Oak

Pronunciation: KWER-cuss ah-koo-TISS-sih-mah

ZONES: 5B, 6A, 6B, 7A, 7B, 8A, 8B, 9A

HEIGHT: 50 feet

WIDTH: 70 feet

FRUIT: oval; 1"; dry and hard; brown; attracts mammals; some litter; showy

GROWTH RATE: moderate; long-lived

HABIT: pyramidal when young, round and spreading later; open; symmetrical; coarse texture

FAMILY: *Fagaceae*

LEAF COLOR: dark green; bright yellow fall color

LEAF DESCRIPTION: alternate; simple; pectinate to serrate margins; oblong to obovate; deciduous; 4 to 8" long

FLOWER DESCRIPTION: brown; inconspicuous; spring

EXPOSURE/CULTURE:

Light Requirement: grows best in full sun

Soil Tolerances: all textures; slightly alkaline to acidic; occasionally wet soil; drought

Aerosol Salt Tolerance: moderate

PEST PROBLEMS: resistant

BARK/STEMS:

Branches droop as tree grows and will require pruning for vehicular or pedestrian clearance beneath canopy; should be grown with a single leader; no thorns

Pruning Requirement: needs some initial pruning to develop a dominant central leader; space main branches along a central trunk

Limb Breakage: mostly resistant

Current Year Twig: brown; medium thickness

LANDSCAPE NOTES:

Origin: China, Japan, Korea

Usage: large parking lot island; buffer strip; shade tree; specimen; street tree; urban tolerant

Sawtooth Oak is an attractive, large, deciduous tree with a broad, pyramidal or rounded shape. Growth is moderate to rapid; it may add three feet in height each year after it is established. The leaves are similar to Chestnut (*Castanea*) and have small bristles at the edges. Brown leaves hang onto the tree into the winter, which makes the tree unattractive to some people. The trunk flares out at the base, lifting sidewalks and curbing if planted in tree lawns less than eight feet wide or too close to walks.

Lower branches should be shortened, then removed early when planting this as a street or parking lot tree, because they will droop as the tree grows older, requiring removal to allow for vehicular traffic. Leave the more upright, spreading branches to create more clearance beneath. Lower branches can be left on trees grown in more open areas to allow for full development of the picturesque, low-branching, open form. Multi-trunked specimens can break up at about 25 years old due to included bark and poor attachment, so maintain the central leader as far up into the canopy as possible, and keep branches less than half the trunk diameter by pruning.

Following some initial pruning in the first ten years, the trunk remains straight and usually dominant over the lateral branches, forming a pyramidal shape in the early years and making this tree very suitable for urban planting. As it ages, it grows into a broad, rounded canopy that could make it useful as a shade tree, street or parking lot tree, or lawn specimen. The acorns of Sawtooth Oak are large (one inch long) and are produced in large quantities in the fall. They are quite popular with wildlife but may cause a litter problem if the tree is located near walkways, driveways, or house gutters.

Sawtooth Oak prefers a well-drained, acid site but will adapt to most soils except alkaline. Chlorosis due to micronutrient deficiency occurs on high pH soil. Irrigation and transplanting in the spring help trees become established, but once established, they grow very fast without irrigation. Susceptible to defoliation by orange-striped oakwarm.

Quercus agrifolia Coast Live Oak

Pronunciation: KWER-cuss ag-ri-FOLE-ee-ah

ZONES: 8A, 8B, 9A, 9B, 10A, 10B, 11

HEIGHT: 30 to 70 feet

WIDTH: 60 to 100 feet

FRUIT: elongated; oval; 1"; dry and hard; brown; causes some litter

GROWTH RATE: moderately rapid; long-lived

HABIT: spreading and rounded; dense; nearly symmetrical; fine texture

FAMILY: *Fagaceae*

LEAF COLOR: green

LEAF DESCRIPTION: alternate; simple with bristles; holly-like; briefly deciduous; 1 to 3" long

FLOWER DESCRIPTION: brown; not showy; spring

EXPOSURE/CULTURE:

Light Requirement: part shade to full sun

Soil Tolerances: all textures; alkaline to acidic; salt and drought

Aerosol Salt Tolerance: moderate

PEST PROBLEMS: resistant

BARK/STEMS:

Branches droop as tree grows and require pruning for vehicular or pedestrian clearance beneath canopy; should be grown with a single leader as high up into canopy as possible; no thorns

Pruning Requirement: requires some early pruning to develop strong structure; thin aggressive branches to prevent formation of included bark and to encourage growth higher in canopy; can be sheared into topiary; prune November through January to minimize *Diplodia* infection

Limb Breakage: resistant

Current Year Twig: gray; slender

LANDSCAPE NOTES:

Origin: California into Mexico

Usage: large parking lot island; buffer strip; highway; reclamation; shade tree; street tree; urban tolerant

A large, sprawling, picturesque tree, Coast Live Oak is one of the broadest spreading of the Oaks, providing large areas of deep, inviting shade. Aggressive, low branches form a spreading canopy on unpruned trees. An amazingly durable American native, it can measure its lifetime in centuries if properly located and cared for in the landscape. Give it plenty of room, as the trunk can grow to more than four feet in diameter. Construction-impacted trees take a long time to die, giving Live Oak a reputation for being a tough tree. It is usually the last tree to die around a newly constructed building.

Once established, Live Oak will thrive in almost any location and has very good wind resistance. It is a tough, enduring tree that will respond with vigorous growth to plentiful moisture on well-drained soil. Once it is well established, it will survive and grow on rainfall alone, but irrigation during the summer increases vigor. Keep irrigation water away from the base of the trunk to avoid root rot. Like other Oaks, care must be taken to develop a strong branch structure early in the life of the tree. Be sure to eliminate multiple trunks and branches that form a narrow angle or included bark with the trunk, as these can split from the tree when it grows older.

Be sure that adequate soil space is given to Live Oak. Although roots will grow under curbs and sidewalks in sandy soil when planted in confined soil spaces, allowing the tree to thrive in urban sites, in time they lift sidewalks, curbs, and driveways if planted too close. This may be a small price to pay for the bountiful shade cast by a row of healthy trees planted along the street.

Discula and *Cryptocline* fungi cause twig death throughout the canopy, especially when pit scales are present. California oak worm can cause defoliation. Vigorously growing trees can be infected by a powdery mildew fungus that causes witches' broom. Resists orange-striped oakworm.

Quercus alba White Oak

Pronunciation: KWER-cuss AL-bah

ZONES: 3B, 4A, 4B, 5A, 5B, 6A, 6B, 7A, 7B, 8A, 8B

HEIGHT: 60 to 100 feet

WIDTH: 60 to 80 feet

FRUIT: oval; 0.5 to 1"; dry and hard; brown; attracts mammals; not showy; acorns could cause some litter

GROWTH RATE: moderate; long-lived

HABIT: oval; round; moderate density; irregular silhouette; medium texture

FAMILY: *Fagaceae*

LEAF COLOR: green; showy red fall color

LEAF DESCRIPTION: alternate; simple; lobed; oblong to obovate; deciduous; 4 to 8" long

FLOWER DESCRIPTION: brown; not showy; spring

EXPOSURE/CULTURE:

Light Requirement: part shade to full sun

Soil Tolerances: all textures; acidic; salt; occasionally wet; moderate drought

Aerosol Salt Tolerance: good

PEST PROBLEMS: except for gypsy moth, resistant

BARK/STEMS:

Showy, gray bark and trunk structure; branches grow mostly upright and horizontal and will not droop; should be grown with a single leader; no thorns

Pruning Requirement: needs little pruning to develop strong structure

Limb Breakage: resistant

Current Year Twig: brown; medium thickness

CULTIVARS AND SPECIES:

Swamp Chestnut Oak grows with a straight trunk to about 80 feet tall and could be planted more on urban sites if it were available.

LANDSCAPE NOTES:

Origin: eastern North America

Usage: shade tree; specimen

White Oak is a long-lived, slow-growing tree, reaching up to 100 feet in height with a spread of 90 feet in its native upland habitat. Old specimens can become massive and live to be several hundred years old. Because trunks can be six feet in diameter, leave plenty of room for this tree in the landscape. The trunk flares out at the base, lifting sidewalks and curbing if planted in soil spaces less than 10 feet wide. The red fall color is fairly reliable year to year and is outstanding among the oaks in colder climates. Brown leaves may be held on the tree into the early part of the winter.

White Oak has a stately silhouette all year long. It is one of the best-looking Oaks in the winter due to the light gray, platey bark and open crown. The trunk is straight, with main branches well attached to the tree, making this a long-lived, durable tree for large, wide-open landscapes. Leave the area within the drip line *totally* undisturbed if attempting to save an existing tree on a construction site.

Transplant White Oak when the tree is young, because the deep-growing tap root on young trees in well-drained soil can make transplanting very difficult. Production in copper-treated containers could help nurseries overcome this problem. White Oak grows in sun or partial shade and prefers an acid, moist, well-drained soil. Unfortunately, it is not readily available in the nursery trade. Supply new transplants with plenty of water and mulch the area beneath canopy to eliminate grass competition.

Chlorosis due to iron-deficiency occurs on high pH soil. Oaks are often free of lethal pests and diseases if planted on the proper site, but there are many potential problems. White Oak is a host for bacterial leaf scorch. Two-lined chestnut borers infest establishing trees under stress.

Quercus austrina Bluff Oak

Pronunciation: KWER-cuss os-STRY-nah

ZONES: 8A, 8B, 9A, 9B

HEIGHT: 50 to 80 feet

WIDTH: 35 to 50 feet

FRUIT: round; 0.5 to 1"; dry and hard; brown; attracts mammals; causes some litter

GROWTH RATE: moderate; long-lived

HABIT: pyramidal when young, oval later; open; symmetrical; medium texture

FAMILY: *Fagaceae*

LEAF COLOR: green; briefly showy copper or yellow fall color

LEAF DESCRIPTION: alternate; simple; lobed; entire margin; obovate; deciduous; 2 to 4" long

FLOWER DESCRIPTION: brown; not showy; spring

EXPOSURE/CULTURE:

Light Requirement: full sun

Soil Tolerances: loam; sand; acidic; drought; wet soil

PEST PROBLEMS: resistant

BARK/STEMS:

Light gray, showy bark; branches grow mostly upright or horizontal; normally grown with a single leader; no thorns

Pruning Requirement: needs little pruning to develop strong structure; remove upright branches with tight crotches

Limb Breakage: resistant

Current Year Twig: brown; green; medium thickness

LANDSCAPE NOTES:

Origin: southeastern United States

Usage: parking lot island; buffer strip; highway; reclamation; shade tree; specimen; street tree

This North American native oak makes an attractive shade tree, with handsome, scaly gray bark. Mature trees have trunks to four feet in diameter. The green, lobed leaves are deciduous but usually do not change color appreciably before dropping in fall. In cool autumns the fall color can be very nice for a short period. The insignificant, green, spring flowers are followed by small acorns, less than one inch long.

The trunk often grows straight up through the crown with little pruning, and branches are well spaced along the trunk. Numerous, small-diameter branches originate from the central trunk, giving open-grown trees a fine to medium texture. This is one of the Oaks not currently available in most nurseries, but it should be. Urban tree managers will want this oak once they find out about it.

Bluff Oak should be grown in full sun on any acid soil, and has good flood and drought tolerance. Trees can be found growing along streams as well as higher, drier sites. It is well suited for planting in parking lots or along streets and boulevards if there is plenty of space for crown development. A row of Bluff Oaks planted on 30-foot centers lining each side of a street is a wonderful sight. The medium-textured leaves make this Oak stand out from other Oaks in the south. The roots do not appear to break up pavement as much as larger Oaks. Its upright to horizontal branching habit makes this an easy tree to prune for vehicular clearance beneath canopy. Less maintenance should be needed for Bluff Oak than for Live and Laurel Oaks, due to the reduced pruning requirement for this underutilized tree.

Oaks are often free of pests and diseases if planted on the proper site, but there are many potential problems.

Quercus bicolor Swamp White Oak

Pronunciation: KWER-cuss BYE-cull-or

ZONES: 4A, 4B, 5A, 5B, 6A, 6B, 7A, 7B, 8A, 8B

HEIGHT: 50 to 60 feet

WIDTH: 50 to 60 feet

FRUIT: round; 0.5 to 1"; dry and hard; brown; attracts mammals; causes some litter

GROWTH RATE: moderate; long-lived

HABIT: upright then oval to round; moderate density; irregular silhouette; coarse texture

FAMILY: *Fagaceae*

LEAF COLOR: dark green; showy copper or red fall color

LEAF DESCRIPTION: alternate; simple; lobed; oblong to obovate; deciduous; 4 to 8" long

FLOWER DESCRIPTION: brown; not showy; spring

EXPOSURE/CULTURE:

Light Requirement: part shade to full sun

Soil Tolerances: all textures; slightly alkaline to acidic; wet soil; moderate salt and drought; compaction

Aerosol Salt Tolerance: low to moderate

PEST PROBLEMS: resistant

BARK/STEMS:

Branches droop as tree grows and will require pruning for vehicular or pedestrian clearance beneath canopy; should be grown with a single leader; no thorns

Pruning Requirement: requires pruning to develop strong structure; remove double leaders

Limb Breakage: resistant

Current Year Twig: brown; medium thickness

LANDSCAPE NOTES:

Origin: Missouri to Wisconsin, Maine, and North Carolina

Usage: shade tree; specimen; street tree

This deciduous native tree forms a broad, open, rounded canopy and casts dense shade below. The shiny, 5- to 6-inch-long by 2- to 4-inch-wide leaves have fine white hairs on the undersides, and irregular margins. In fall, the leaves turn a showy yellow-brown to red before dropping. The oval, one-inch acorns are usually found in pairs on one- to four-inch-long stems, and are quite attractive to a variety of mammals and birds. Swamp White Oak has deeply ridged and furrowed, dark brown bark, and forms an impressive shade tree. Lower branches are short and persistent, but droop toward the ground on older trees.

As the common name implies, Swamp White Oak is well adapted to low-lying areas that are poorly drained. It is perfectly suited for that wet spot in a park or golf course, or near a stream to help stabilize the soil along the bank. The tolerance to low oxygen should allow it to perform along a street with poorly drained, compacted soil.

Found in the wild along streams and in swampy soils, Swamp White Oak should be grown in full sun to partial shade on acid soils; it shows severe chlorosis on alkaline soils. These trees are very long-lived, surviving for more than 300 years. They tolerate soil compaction, drought, and some salt exposure. They should be tried more often in urban areas. Transplanting from field nurseries is best in the spring, but container-grown material can be planted throughout the year.

Borers, variable oak caterpillar, oak slug caterpillar, and other caterpillars can be troublesome, although natural enemies usually keep infestations under control. Gypsy moth can cause significant damage. There are many other potential pests on Oak, such as anthracnose, canker, powdery mildew, and shoestring root rot. Leaf spot diseases are usually harmless. Leaf blister can cause moderate to severe defoliation. Somewhat resistant to orange-striped oakworm.

Quercus cerris European Turkey Oak

Pronunciation: KWER-cuss SEAR-iss

ZONES: 6A, 6B, 7A, 7B

HEIGHT: 30 to 50 feet

WIDTH: 30 to 60 feet

FRUIT: round; 1"; dry; brown; attracts mammals; causes some litter; showy

GROWTH RATE: moderately rapid; long-lived

HABIT: round; moderate density; symmetrical; medium texture

FAMILY: *Fagaceae*

LEAF COLOR: green

LEAF DESCRIPTION: alternate; simple; lobed; oblong; deciduous; 4 to 8" long

FLOWER DESCRIPTION: brown; not showy; spring

EXPOSURE/CULTURE:

Light Requirement: part shade to full sun

Soil Tolerances: all textures; slightly alkaline to acidic; occasionally wet soil; drought

Aerosol Salt Tolerance: moderate

PEST PROBLEMS: resistant

BARK/STEMS:

Branches droop as tree grows and will require pruning for clearance beneath canopy; should be grown with a single leader; no thorns

Pruning Requirement: needs little pruning to develop strong structure

Limb Breakage: resistant

Current Year Twig: thick; brownish-green

CULTIVARS AND SPECIES:

Q. laevis, Turkey Oak, is native to dry, sandy soils in the southeastern United States, and makes a nice tree for reclaiming dry sites. Acorns serve as food for turkey, bear, and deer, and the wood makes excellent firewood.

LANDSCAPE NOTES:

Origin: Europe and Asia

Usage: parking lot island; buffer strip; shade tree; specimen; street tree

European Turkey Oak is a fast-growing deciduous tree capable of reaching 130 feet in height, but is often only 30 to 50 feet tall and wide. The 2.5- to 5-inch-long leaves are covered with a fine fuzz on both upper and lower surfaces. The one-inch-long acorns are set into big, woolly cups, and ripen in October. They germinate readily in a moist landscape. The attractive, ridged, and furrowed bark reveals an orange color within its fissures. Trunk can grow to at least four or five feet in diameter.

Older specimens often develop a wide-spreading, multi-trunked form, making them great trees for climbing or for use as specimens. Young specimens have been observed growing with a straight central leader with well-spaced major branches. This should be a very durable landscape tree in any large-scale landscape.

European Turkey Oak should be grown in full sun or partial shade, but tolerates almost any soil except wet. It grows well on sandy ridges and is salt-tolerant. Surprisingly, this tree is not grown in the trade and is not available, but its extreme drought tolerance, ability to thrive in poor, clay soil, and attractive habit make it most worthy of use in urban areas. It might make a good street tree for cities.

There may be some limitations, but the tree has not been extensively grown or tested.

Quercus coccinea Scarlet Oak

Pronunciation: KWER-cuss cock-SIN-ee-ah

ZONES: 5A, 5B, 6A, 6B, 7A, 7B, 8A, 8B

HEIGHT: 60 to 75 feet

WIDTH: 45 to 60 feet

FRUIT: round; 1"; dry and hard; brown; attracts mammals; causes some litter

GROWTH RATE: moderate; long-lived

HABIT: round; moderate density; symmetrical; medium texture

FAMILY: *Fagaceae*

LEAF COLOR: medium green; showy red fall color

LEAF DESCRIPTION: alternate; simple; lobed; oval to oblong; deciduous; 3 to 7" long

FLOWER DESCRIPTION: brown; not showy; spring

EXPOSURE/CULTURE:

Light Requirement: full sun

Soil Tolerances: all textures; acidic; moderate drought

Aerosol Salt Tolerance: moderate

PEST PROBLEMS: mostly resistant

BARK/STEMS:

Bark is thin and easily damaged by mechanical impact; branches grow mostly upright and horizontal and do not droop; should be grown with a single leader; no thorns

Pruning Requirement: needs some early pruning to develop a central leader, then little is required to maintain good structure

Limb Breakage: resistant

Current Year Twig: brown or reddish; medium thickness

LANDSCAPE NOTES:

Origin: Alabama to Maine

Usage: parking lot island; buffer strip; shade tree; street tree

Scarlet Oak is so named for its beautiful, red fall color. Fairly smooth gray bark in youth roughens to a dark brown or black on the trunks of older specimens, which can be four feet in diameter. Leaves are more deeply lobed than *Quercus rubra* but otherwise the trees look very much alike. The crown should be built like many other Oaks, with a central trunk and lateral branches well spaced at two- to three-foot intervals. The rounded, spreading canopy makes the tree well suited for planting along streets. The trunk flares out at the base, lifting sidewalks and curbing if planted in soil spaces less than eight feet wide.

Scarlet Oak does well as a street tree if given plenty of soil space to develop. Plant it on 30- to 40-foot centers to form a canopy overhead in residential neighborhoods. Although it appears tolerant of soil compaction, it is not for restricted soil spaces such as downtown planting pits.

This Oak will grow on sandy soils, and is reported to exhibit fewer chlorosis problems than Pin Oak, but should still be grown on acidic soils. Scarlet Oak has a tap root on loose, well-drained soil and a coarse root system. Combined with a slow rate of root regeneration, this makes it difficult to transplant. Spring is the best time to transplant field-grown trees. Scarlet Oak is a good candidate for raising in containers treated with copper to help prevent root circling.

Oak wilt kills trees. Leaf blister symptoms are round raised areas on the upper leaf surfaces, causing depressions of the same shape and size on lower leaf surfaces. Infected areas are yellowish-white to yellowish-brown. The disease is most serious in wet seasons in the spring, but it need not be treated. Chlorosis due to iron deficiency occurs on high-pH soil. Galls cause homeowners much concern, but are mostly harmless, so chemical controls are not suggested. Scales of several types infest branches and are difficult to see. Many caterpillars feed on Oak and orange-striped oakworm can be serious. Where they occur, gypsy moth caterpillars are extremely destructive. Fall cankerworm has been a problem in some years. Anthracnose may be a serious problem in wet weather. Canker diseases occasionally attack the trunk and branches.

Quercus falcata Southern Red Oak

Pronunciation: KWER-cuss fowl-KAY-tah

ZONES: 7A, 7B, 8A, 8B, 9A, 9B

HEIGHT: 60 to 80 feet

WIDTH: 60 to 70 feet

FRUIT: round; 0.5"; dry and hard; brown; attracts mammals; no significant litter

GROWTH RATE: moderate; long-lived

HABIT: oval to round; moderate density; irregular silhouette; coarse texture

FAMILY: *Fagaceae*

LEAF COLOR: shiny green; muted copper or brown fall color over a long period

LEAF DESCRIPTION: alternate; simple; lobed; obovate to ovate; deciduous; 6 to 10" long

FLOWER DESCRIPTION: brown; not showy; spring

EXPOSURE/CULTURE:

Light Requirement: full sun

Soil Tolerances: all textures; not clay unless it is well drained; acidic; drought

Aerosol Salt Tolerance: moderate

PEST PROBLEMS: resistant

BARK/STEMS:

Branches grow mostly upright and will not droop; should be grown with a single leader; no thorns

Pruning Requirement: requires early pruning to develop strong structure

Limb Breakage: resistant

Current Year Twig: brown; medium thickness

CULTIVARS AND SPECIES:

Variety *pagodifolia*, Cherrybark Oak, is adaptable, growing along stream banks and ridge tops throughout its range. Often found alone in nature, possibly due to allelopathy. It may be more commonly available than the species and makes a superior urban tree due to its nicer canopy, tolerance of slightly alkaline soil, and resistance to poorly drained soil. May be called *Q. pagodifolia* in some nurseries.

LANDSCAPE NOTES:

Origin: New Jersey to Texas and Florida

Usage: reclamation; highways; shade tree; specimen; street tree; parking lot island; buffer strip

Southern Red Oak is an excellent, large, durable shade tree with a large, rounded canopy when open grown. The deciduous leaves are five to nine inches long by four to five inches wide, with the terminal lobe much longer and narrower than the others. Leaves brown and fall over an extended period in fall and winter. Some defoliation is noted during the summer in droughty years, but this is probably a drought-avoidance mechanism. No permanent damage appears to come from this. The dark brown to black bark is ridged and furrowed and somewhat resembles Cherry bark. The trunk normally grows straight, with major branches well spaced and strongly attached to the tree. The half-inch-diameter acorns are popular with wildlife.

The drought tolerance of Southern Red Oak makes it well suited for planting in areas such as along roadsides where there is little maintenance after planting. The irregular canopy may not suit it for street tree planting, but tolerance to extended drought allows it to withstand dry soil unsuitable for many other species. The wood is used for furniture, but does not have the quality of *Quercus rubra*.

Naturally found on poor upland soils, Southern Red Oak should be grown in full sun on well-drained soil, whether acid, sandy, or loam (not clay soil unless it is very well drained). It is common on poor-quality, sandy ridges.

Usually no limitations of major concern. Caterpillars can defoliate trees. Fall cankerworm has been a problem in some years. Leaf spots can cause defoliation over an extended period of time during the year. Shows some resistance to orange-striped oakworm.

Quercus glauca Blue Japanese Oak

Pronunciation: KWER-cuss GLAW-cah

ZONES: 8A, 8B, 9A, 9B, (10)

HEIGHT: 25 to 40 feet

WIDTH: 25 to 35 feet

FRUIT: oval; round; 0.5 to 1"; dry and hard; brown; causes some litter

GROWTH RATE: slow-growing; moderate life span

HABIT: oval to round; dense; symmetrical; medium texture

FAMILY: *Fagaceae*

LEAF COLOR: broadleaf evergreen; dark green

LEAF DESCRIPTION: alternate; simple; serrate margin; oblong to obovate; 3 to 5" long

FLOWER DESCRIPTION: brown; not showy; spring

EXPOSURE/CULTURE:

Light Requirement: grows best in full sun

Soil Tolerances: all textures; acidic; occasionally wet; drought

PEST PROBLEMS: resistant

BARK/STEMS:

Interesting, twisted form; branches droop as tree grows and will require pruning for vehicular or pedestrian clearance beneath canopy; can be grown with a single leader or with several trunks; no thorns

Pruning Requirement: requires pruning to develop central leader and to space branches along trunk

Limb Breakage: susceptible to breakage at crotch due to included bark

Current Year Twig: brownish-green; thick

LANDSCAPE NOTES:

Origin: China, Japan

Usage: parking lot island; sidewalk cutout; buffer strip; highway; shade tree; street tree; screen or tall, clipped hedge

At first glance, this plant would hardly be thought of as an Oak. Creating a very formal, dense, round or oval shade, evergreen Blue Japanese Oak grows slowly to mature height and spread. The new growth is often bronze- or purple-tinted, and develops into shiny, dark green, 2.5- to 5.5-inch-long by 1- to 2.5-inch-wide, leathery leaves. It is a very attractive tree suited for many uses, and was widely used in several southern cities until the freeze of 1983–1984 killed most trees. The tree deserves a comeback in the nursery trade, but should be planted only as far north as zone 8, possibly 7B.

Blue Japanese Oak has been used as a street and parking lot tree in parts of the southeast for a number of years. It appears to be tolerant of drought and poor, clay soils, including those that are poorly drained. It grows well in small soil spaces. Its tight crown lends a formal feel to the landscape. It could be the ideal oak for planting near utility lines, because of its slow growth and small size. The dark, rich green foliage, combined with its neat habit, would complement any street in an urban or suburban area.

Many upright trunks and branches originate from one point on the trunk without initial training, and this structure could lead to a weak tree. Following initial training and pruning for approximately ten years to develop a central leader and well-spaced branches, little care should be needed to maintain this tree. It should be grown and planted more in urban areas in the south.

Blue Japanese Oak needs full sun for best growth. It is tolerant of a wide range of soils, including heavy clay soils. Probably no limitations of major concern, though there are many potential problems on Oaks.

Quercus ilex Holly Oak

Pronunciation: KWER-cuss eye-lecks

ZONES: 7A, 7B, 8A, 8B, 9A, 9B, 10A, 10B, 11

HEIGHT: 40 to 70 feet

WIDTH: 60 feet

FRUIT: elongated; oval; 1"; dry and hard; brown; causes some litter

GROWTH RATE: moderately rapid; long-lived

HABIT: rounded; dense; nearly symmetrical; fine texture

FAMILY: *Fagaceae*

LEAF COLOR: gray-green and hairy beneath; evergreen

LEAF DESCRIPTION: alternate; simple; holly-like or entire margin; 1 to 3" long

FLOWER DESCRIPTION: brown; not showy; spring

EXPOSURE/CULTURE:

Light Requirement: part shade to full sun

Soil Tolerances: all textures; slightly alkaline to acidic; salt and drought

Aerosol Salt Tolerance: good; tree is shorter in salty air

PEST PROBLEMS: resistant

BARK/STEMS:

Smooth gray bark; branches do not droop as much as Live Oak; should be grown with a single leader as high up into canopy as possible; no thorns

Pruning Requirement: requires some early pruning to develop strong structure; slow growth rate on aggressive branches to prevent formation of included bark and to generate a dominant trunk; can be sheared; less structural pruning needed than Live Oak

Limb Breakage: very resistant

Current Year Twig: gray; slender

LANDSCAPE NOTES:

Origin: Mediterranean region

Usage: large parking lot island; buffer strip; highway; reclamation; shade tree; street tree; urban tolerant

A moderately large tree, Holly Oak is more upright than Coastal Live Oak. This makes it better suited for planting along streets and in parking lot islands where clearance is needed for vehicle passage beneath canopy. The tree is big enough to provide shade for large areas of the landscape. An amazingly durable tree, it can measure its lifetime in centuries if properly located and cared for in the landscape. Give it plenty of room, as the trunk can grow to more than three feet in diameter.

Once established, Holly Oak will thrive in almost any location and has very good wind resistance. It is a tough, enduring tree that will respond with vigorous growth to plentiful moisture on well-drained soil. Overstimulation with irrigation or heavy fertilizer will invite powdery mildew disease, which deforms foliage on rapid-growing trees. Provide plenty of irrigation until the tree is established. Once it is well established, it will survive and grow on rainfall alone, but irrigation during the summer increases vigor. Keep irrigation away from the base of the trunk to prevent root rot. Like other Oaks, care must be taken to develop strong branch structure early in the life of the tree. Be sure to eliminate multiple trunks and branches that form a narrow angle with the trunk, as these can split from the tree as it grows older. Cut back competing trunks and upright branches to form a dominant trunk.

Be sure that adequate soil space is given to Holly Oak. Although roots will grow under curbs and sidewalks in sandy soil when planted in confined soil spaces, allowing the tree to thrive in urban sites, in time they lift sidewalks, curbs, and driveways if planted too close. This may be a small price to pay for the bountiful shade cast by a row of healthy trees planted along the street.

Quercus imbricaria Shingle Oak

Pronunciation: KWER-cuss im-breh-CARE-ee-ah

ZONES: 5A, 5B, 6A, 6B, 7A, 7B, 8A

HEIGHT: 40 to 60 feet

WIDTH: 40 to 60 feet

FRUIT: oval; 0.5"; dry and hard; brown; attracts mammals; not showy; causes some litter

GROWTH RATE: slow to moderate; long-lived

HABIT: pyramidal when young, oval to round later; dense; symmetrical; fine to medium texture

FAMILY: *Fagaceae*

LEAF COLOR: green; showy red to yellow fall color

LEAF DESCRIPTION: alternate; simple; sinuate to undulate margin; lanceolate to oblong; deciduous; dried leaves persist into winter; 3 to 6" long

FLOWER DESCRIPTION: brown; not showy; spring

EXPOSURE/CULTURE:

Light Requirement: part to full sun

Soil Tolerances: all textures; slightly alkaline to acidic; occasionally wet soil; drought

Aerosol Salt Tolerance: moderate to good

PEST PROBLEMS: resistant

BARK/STEMS:

Branches droop as tree grows and will require pruning for vehicular or pedestrian clearance beneath canopy; should be grown with a single leader; no thorns

Pruning Requirement: needs little pruning to develop strong structure; central leader is easily trained

Limb Breakage: resistant

Current Year Twig: brownish-green; slender

LANDSCAPE NOTES:

Origin: Pennsylvania to Nebraska, Arkansas to Georgia

Usage: parking lot island; shade tree; specimen; street tree; screen

This stately, deciduous, native tree occasionally grows to 100 feet in the forest but is usually seen at 40 to 60 feet with an equal or greater spread. Its broad, strong branches cast medium to deep shade below the rounded canopy. The smooth, three- to six-inch-long by one- to two-inch-wide leaves start out life with a red to yellow cast, deepen to a rich green through the summer, then turn shades of yellow and rust again in the fall before dropping. Some leaves will persist on the tree throughout the winter. In May the flowers appear as drooping yellowish-green catkins; they are followed by the production of one-half- to one-inch-long, dark brown acorns.

The wood of Shingle Oak is extremely durable and was used in pioneer days for split shingles. This strong, resilient nature of Shingle Oak and its pyramidal shape when young help make it suited to use as a screen, or as a street or specimen tree that is unlikely to be damaged by harsh winds or snow loads. It can be planted along residential streets with plenty of soil space, but there are more adaptable trees for downtown areas.

Naturally found along streams or river banks, Shingle Oak should be grown in full sun and prefers moist but well-drained, acid soils, though it will adapt to moderately drier conditions and slightly alkaline soil. It is tolerant of wet sites once established. Field-grown trees should be transplanted in the spring for best survival. Those in containers can be planted any time.

Borer, variable oak caterpillar, and oak slug caterpillar are possible pests, although natural enemies usually provide control. Twig gall, leaf miner, and powdery mildew are common.

Quercus laurifolia Laurel Oak, Darlington Oak

Pronunciation: KWER-cuss lore-eh-FOE-lee-ah

ZONES: 6B, 7A, 7B, 8A, 8B, 9A, 9B, 10A

HEIGHT: 60 to 70 feet

WIDTH: 50 feet

FRUIT: oval; 0.5"; dry and hard; brown; attracts mammals; not showy; causes some litter

GROWTH RATE: rapid; moderate life span

HABIT: oval to round; dense; symmetrical; fine texture

FAMILY: *Fagaceae*

LEAF COLOR: medium green; some yellow fall and winter color

LEAF DESCRIPTION: alternate; simple; entire margin; obovate to spatulate; deciduous in the north; semi-evergreen in the south; 2 to 4" long

FLOWER DESCRIPTION: brown; not showy; spring

EXPOSURE/CULTURE:

Light Requirement: part shade to full sun

Soil Tolerances: all textures; slightly alkaline to acidic; occasionally wet soil; drought

Aerosol Salt Tolerance: low

PEST PROBLEMS: resistant

BARK/STEMS:

Branches droop as tree grows and will require pruning for vehicular or pedestrian clearance beneath canopy; should be grown with a single leader; trunk is often decayed at 50 years old; no thorns

Pruning Requirement: requires pruning to develop strong structure; included bark often forms in crotches; remove upright branches that compete with leader

Limb Breakage: susceptible to breakage at crotch due to poor collar formation

Current Year Twig: brown; gray; slender

LANDSCAPE NOTES:

Origin: Virginia to Texas to south Florida

Usage: parking lot island; buffer strip; highway; reclamation; shade tree

A large, fast-growing, shade tree, Laurel Oak is noted for its dense, oval canopy. *Q. laurifolia* tolerates wet soil; *Q. hemisphaerica* is salt-tolerant and native to upland sites, but is not tolerate of wet soil. Laurel Oaks are taller than they are broad, eventually reaching 60 feet or more in height with a 50- to 70-foot spread. The light gray trunk can be up to four feet in diameter and flares out at the base, lifting sidewalks and curbing if planted in soil areas less than eight feet wide.

Laurel Oaks have a life span of 50 to 70 years. Tree trunks and large branches often become hollow from old injuries and improper pruning cuts. Like other Oaks, care must be taken early in the life of the tree to develop a strong branch structure, with major branches well spaced along the trunk. There are more durable street and parking lot trees, including Live, Southern Red, Northern Red, Swamp Chestnut, and Bluff Oak.

Trees should be pruned to one central trunk, with major branches spaced two to three feet apart. Branch diameter should be less than half the trunk diameter. Prune regularly so you will not have to remove large-diameter branches. This strategy may increase the life span of Laurel Oak.

Laurel Oaks will grow easily in full sun or partial shade and are quite tolerant of a wide range of soils, from moist and rich to dry and sandy. However, they are not adapted to constantly wet or alkaline soil. Trees growing under drier conditions will grow more slowly and, it is thought, will have stronger wood that is less susceptible to breakage.

Wasps can cover branches with galls. Serious infestations slow growth and cause die-back. Mites occasionally infest and discolor foliage, but control is usually not needed. Tubackia leaf spot can defoliate Laurel Oak and slow growth. There are many other potential problems. Oak leaf blister causes premature leaf drop, but is usually nonthreatening (just messy). Chlorosis due to iron deficiency occurs on high-pH soil.

Quercus lobata Valley, or California, White Oak

Pronunciation: KWER-cuss low-BAH-ta

ZONES: 8B, 9A, 9B, 10A, 10B

HEIGHT: 40 to 70 feet

WIDTH: 60 to 80 feet

FRUIT: elongated; oval; 1 to 2"; dry and hard; brown; causes litter

GROWTH RATE: moderate; long-lived

HABIT: spreading and rounded; dense; nearly symmetrical; fine texture

FAMILY: *Fagaceae*

LEAF COLOR: glossy, medium green above, pubescent below

LEAF DESCRIPTION: alternate; simple; deeply lobed; briefly deciduous; 3 to 6" long

FLOWER DESCRIPTION: brown; not showy; spring

EXPOSURE/CULTURE:

Light Requirement: part shade to full sun

Soil Tolerances: all textures; slightly alkaline to acidic; occasionally wet soil, but keep irrigation away from lower trunk; moderate drought

Aerosol Salt Tolerance: none; do not plant along coast

PEST PROBLEMS: resistant

BARK/STEMS:

Checkered bark; branches droop and twist as tree grows and require pruning for vehicular or pedestrian clearance beneath canopy; should be grown with a single leader as high up into canopy as possible; no thorns

Pruning Requirement: requires some early pruning to develop strong structure; slow growth rate on aggressive branches to prevent formation of included bark; prune only November through January to minimize fungi infection.

Limb Breakage: resistant

Current Year Twig: gray; slender

CULTIVARS AND SPECIES:

Q. arizonica and *Q. emoryi* (zone 6B) are evergreens about 35 feet tall and used in the southwest.

LANDSCAPE NOTES:

Origin: California

Usage: large parking lot island; buffer strip; highway; reclamation; shade tree; street tree; urban tolerant

A large, sprawling, picturesque tree, Valley Oak is one of the broadest-spreading of the Oaks, providing large areas of deep, inviting shade. Aggressive, low branches form a spreading canopy on unpruned trees. An amazingly durable American native, it can measure its lifetime in centuries if properly located and cared for in the landscape. Give it plenty of room, as the trunk can grow to more than four feet in diameter. Construction-impacted trees take a long time to die, giving Valley Oak a reputation for being a tough tree.

Once established, Valley Oak will thrive in almost any location. Although native to moist bottomland soils and slopes in California, where it can grow to more than 100 feet tall, it survives drought very nicely. It is a tough, enduring tree that will respond with vigorous growth to plentiful moisture on well-drained soil. Keep irrigation away from trunk to help prevent rot.

Like other Oaks, care must be taken to develop a strong branch structure early in the life of the tree. Be sure to eliminate multiple trunks and branches that form a narrow angle with the trunk, as these can split from the tree as it grows older.

Although roots will grow under curbs and sidewalks in sandy soil when planted in confined soil spaces, allowing the tree to thrive in urban sites, in time they lift sidewalks, curbs, and driveways if planted too close. This may be a small price to pay for the bountiful shade cast by a row of healthy trees planted along the street.

Large galls drop from the tree, but they are harmless and do not affect tree health.

Quercus lyrata Overcup Oak

Pronunciation: KWER-cuss lie-RAY-tah

ZONES: 6A, 6B, 7A, 7B, 8A, 8B, 9A

HEIGHT: 35 to 50 feet

WIDTH: 30 to 50 feet

FRUIT: round; 0.75"; dry and hard; brown; attracts mammals; causes significant litter; showy

GROWTH RATE: moderate; long-lived

HABIT: oval to round; moderate density; symmetrical; medium texture

FAMILY: *Fagaceae*

LEAF COLOR: dark green above, fuzzy white beneath; copper fall color

LEAF DESCRIPTION: alternate; simple; lobed; entire margin; oblong to obovate; deciduous; 6 to 8" long

FLOWER DESCRIPTION: brown; not showy; spring

EXPOSURE/CULTURE:

Light Requirement: part shade to full sun

Soil Tolerances: all textures; acidic; wet soil; drought

PEST PROBLEMS: resistant

BARK/STEMS:

Young branches grow mostly upright or horizontal drooping as the tree ages; should be grown with a single leader; no thorns; showy

Pruning Requirement: needs little pruning to develop strong structure; trunk more or less grows straight up into canopy

Limb Breakage: resistant

Current Year Twig: brown; gray; thick

LANDSCAPE NOTES:

Origin: Maryland to Florida, Texas to Illinois

Usage: parking lot island; buffer strip; highway; reclamation; shade tree; sidewalk cutout; street tree

Overcup Oak grows slowly, sometimes to more than 100 feet in height in its native habitat, but is more often seen 30 to 50 feet high and wide in a landscape, forming a rounded silhouette with an open crown, large-diameter branches. The leathery, lobed leaves have fuzzy, white undersides and turn a rich brown color before dropping in fall. The nuts or acorns are quite popular with squirrels, turkeys, wild hogs, and deer. The rough, reddish or gray-brown bark is attractive and is worthy of display with nighttime lighting.

This Oak should be grown with a single trunk and a few widely spaced branches to mimic its growth habit in the wild. The first permanent branch can be located three to five feet from the ground if the tree will be planted in an open lawn area and allowed to develop a wide crown. For those planted as street trees or in areas requiring clearance for vehicles or pedestrians, the first permanent branch should be much higher on the trunk. The interior portion of the crown often cleans itself of small branches, displaying the nice branch arrangement common on most specimens.

Overcup Oak should be grown in full sun or partial shade on almost any soil; it tolerates wet, poorly drained sites or acid, sandy soils. It would be well suited for planting in poorly drained urban sites, but is not normally available at landscape nurseries. Like Baldcypress and some other wet-site-tolerant trees, best growth is often obtained on sites that are better drained than those in their natural habitat. Despite tolerance to wet sites, do not be afraid of planting this tree on sites where there will be little or no irrigation after trees are established in the landscape. It would be a popular choice for planting along streets and in other urban areas because of its tolerance to soils typical in urban areas. This tree deserves to be widely planted when nurseries begin growing it.

No limitations of major concern. Probably susceptible to the usual array of Oak problems. This Oak roots well from cuttings and transplants, which should make it more available at nurseries.

Quercus macrocarpa Bur Oak

Pronunciation: KWER-cuss mak-row-KAR-pah

ZONES: 3A, 3B, 4A, 4B, 5A, 5B, 6A, 6B, 7A, 7B, 8A, 8B

HEIGHT: 70 to 90 feet

WIDTH: 60 to 80 feet

FRUIT: oval; 1"; dry and hard; brown; nearly enclosed in fringe; attracts mammals; causes significant litter in some years; showy

GROWTH RATE: moderate to rapid; long-lived

HABIT: round; dense; nearly symmetrical; coarse texture

FAMILY: *Fagaceae*

LEAF COLOR: dark green; copper fall color

LEAF DESCRIPTION: alternate; simple; lobed; undulate margin; oblong to obovate; deciduous; 6 to 12" long

FLOWER DESCRIPTION: brown; not showy; spring

EXPOSURE/CULTURE:

Light Requirement: grows best in full sun

Soil Tolerances: all textures; alkaline to acidic; wet soil; moderate salt; drought

Aerosol Salt Tolerance: moderate to good

PEST PROBLEMS: resistant

BARK/STEMS:

Large, showy trunk; branches droop as tree grows and will require pruning for vehicular or pedestrian clearance beneath canopy; should be grown with a single leader; no thorns

Pruning Requirement: needs little pruning to develop strong structure

Limb Breakage: mostly resistant, but can break in ice storms

Current Year Twig: brown; very stout and corky

LANDSCAPE NOTES:

Origin: Maine to Manitoba and Texas

Origin: parking lot island; buffer strip; highway; shade tree; specimen; street tree; urban tolerant

Bur Oak is a huge variable tree with an impressive crown, massive trunk, and stout branches. A specimen in Philadelphia is about 7 feet in diameter and over 90 feet wide. Bark is an unusually light brown to gray, depending on the specimen, and is deeply furrowed on older trees. Young stems have corky ridges, making them appear stout. Acorns are almost completely covered with a furry, bur-like cap and are large, creating a sizeable cleanup job in a maintained landscape. Young trees have an attractive, symmetrical, dense crown borne on a central trunk, and they are well suited for street tree planting. Architects like the tree due to its uniformity in crown shape.

Trees are well suited for street, park, and parking lot planting, but enough soil space should be available to accommodate trunk and root growth. This large Oak should not be planted in tree lawn strips less than 10 feet wide. If soil is not compacted in this wide space, roots usually stay below the soil surface and do not lift sidewalks or curbs for quite some time.

This Oak will adapt to various soils where other Oaks sometimes fail, but is difficult to transplant from well-drained soil due to the tap root. Transplant from a field nursery in the spring for best survival. Bur Oak is a good candidate for raising in a container treated with copper, to help prevent root circling and to promote branching of the dominant tap root, before planting to the landscape. As with most trees grown in urban areas, the tap root becomes much less prominent as the tree grows older, giving way to a more shallow, horizontal root system. Bur Oak is well adapted to alkaline soils, poor drainage, and high clay content in the soil. It is also very drought-tolerant.

Two-lined chestnut borer causes damage in localized areas. Less susceptible to Oak wilt than trees in the Red Oak group. Oakleaf caterpillar and leaf skeletonizer and lace bug can defoliate trees. Bur Oak is susceptible to bacterial leaf scorch. Shows some resistance to orange-striped oakworm.

Quercus michauxii Swamp Chestnut Oak

Pronunciation: KWER-cuss me-SHOW-ee-ii

ZONES: 6A, 6B, 7A, 7B, 8A, 8B, 9A

HEIGHT: 60 to 70 feet

WIDTH: 30 to 50 feet

FRUIT: round; 1"; dry and hard; brown; attracts mammals; causes significant litter

GROWTH RATE: moderate; long-lived

HABIT: oval to round; moderate density; symmetrical; coarse texture

FAMILY: *Fagaceae*

LEAF COLOR: dark green; showy, red to copper fall color

LEAF DESCRIPTION: alternate; simple; coarsely toothed; oblong to obovate; deciduous; 4 to 7" long; litters the ground because of slow decomposition

FLOWER DESCRIPTION: brown; not showy; spring

EXPOSURE/CULTURE:

Light Requirement: part shade to full sun

Soil Tolerances: all textures; acidic; occasionally wet; drought

PEST PROBLEMS: resistant

BARK/STEMS:

Branches grow mostly upright or horizontal and will not droop; should be grown with a single leader; no thorns

Pruning Requirement: needs some early pruning to develop strong structure; eliminate double leaders as they develop when tree is young

Limb Breakage: resistant

Current Year Twig: brown; gray; stout

CULTIVARS AND SPECIES:

Q. prinus is a similar tree that is native to drier soils on ridge tops and slopes. It is probably better adapted to dry sites than Swamp Chestnut Oak, and it transplants well. It is somewhat susceptible to orange-striped oakworm.

LANDSCAPE NOTES:

Origin: New Jersey to Florida, Texas to Illinois

Usage: parking lot island; buffer strip; highway; reclamation; shade tree; street tree

Swamp Chestnut Oak grows to more than 100 feet in height on a good site in its native habitat, but is more often seen 60 to 70 feet high in a landscape, forming a rounded or oval silhouette. Numerous, small-diameter branches arise from the straight trunk when the tree is young. As the tree matures, several large branches dominate the open canopy. The leathery, lobed leaves decompose very slowly after they fall to the ground; some people complain about the leaf litter because of its persistence. The large acorns are quite popular with squirrels, turkeys, wild hogs, and deer. The rough, light gray bark is attractive and is worthy of display with nighttime lighting.

This Oak should be grown with a single trunk and a few widely spaced branches to mimic its growth habit in the wild. The first permanent branch can be located five feet or more from the ground, if the tree will be planted in an open lawn area, and allowed to develop a wide crown. For those planted as street trees or in areas requiring clearance for vehicles or pedestrians, the first permanent branch should be higher on the trunk. The interior portion of the crown often cleans itself of small branches, displaying the nice branch arrangement common on most specimens.

Swamp Chestnut Oak should be grown in full sun on almost any soil. It tolerates poorly drained sites for a short period. Although native to bottomlands and slopes, do not be afraid to plant this tree in humid climates on sites where there will be little or no irrigation after trees are established in the landscape.

Several borers can infest the trunk. Oak leak blister can deform leaves and cause some defoliation, but is usually harmless. Acorns are large and make a mess of streets and walks below the canopy.

Quercus muehlenbergii Chinkapin Oak

Pronunciation: KWER-cuss mew-len-BURR-gee-eye

ZONES: 3A, 3B, 4A, 4B, 5A, 5B, 6A, 6B, 7A, 7B, 8A, 8B, 9A

HEIGHT: 40 to 50 feet

WIDTH: 50 to 60 feet

FRUIT: round; 1"; dry and hard; brown; attracts mammals; edible; causes significant litter

GROWTH RATE: moderate; long-lived

HABIT: round; moderate density; nearly symmetrical; medium texture

FAMILY: *Fagaceae*

LEAF COLOR: yellow-green; showy yellow fall color

LEAF DESCRIPTION: alternate; simple; dentate to undulate margin; lanceolate to oblong; deciduous; brown leaves occasionally persist into winter; 4 to 6" long

FLOWER DESCRIPTION: brown; not showy; spring

EXPOSURE/CULTURE:

Light Requirement: grows best in full sun

Soil Tolerances: all textures; alkaline to slightly acidic; occasionally wet soil; some salt; drought

Aerosol Salt Tolerance: moderate

PEST PROBLEMS: resistant

BARK/STEMS:

Branches grow mostly upright or horizontal and will not droop; should be grown with a single leader; no thorns

Pruning Requirement: needs little pruning to develop strong structure

Limb Breakage: resistant

Current Year Twig: brown; medium; slender

LANDSCAPE NOTES:

Origin: New York to Iowa, Texas and Florida

Usage: parking lot island; buffer strip; highway; reclamation; shade tree; street tree; urban tolerant

Chinkapin Oak is seen at 70 to 90 feet in height in the wild, but is more often 40 to 50 feet in height, with an equal or greater spread, when grown in cultivation. It grows at a moderate rate when young but slows considerably with age, eventually developing a broad, rounded canopy with strong branches. Young trees often exhibit a straight central leader with numerous branches originating at the same node. The deciduous, lobed leaves turn shades of red, yellow, orange, and brown before dropping in fall. Veins are distinctly prominent on the undersides of the coarse-textured leaves. The acorns are edible.

Small specimens are often grown with an upright, oval habit. Older trees develop a more open, rounded form. This Oak should be grown with a single trunk and widely spaced branches to mimic its growth habit in the wild. The first permanent branch can be located three to five feet from the ground, if the tree will be planted in an open lawn area, and allowed to develop a wide crown. For those planted as street trees or in areas requiring clearance for vehicles or pedestrians, the first permanent branch should be much higher on the trunk.

Chinkapin Oak should be grown in full sun on well-drained soil. It reaches its greatest size on loose, bottomland soils and is well adapted to alkaline soils. This adaptable Oak has been planted often in the central part of the country where soils are often clayey and alkaline. It will grow quite nicely in other areas of the country as well.

Propagation is by seed, but plants have been considered difficult to transplant. There are a number of root-promoting techniques developed for nursery production which should improve the branching of the root system, and this should improve the transplantability of this oak. Transplant field-grown trees in the spring for best recovery.

No limitations of major concern. This will soon be a popular street and shade tree.

Quercus nigra Water Oak

Pronunciation: KWER-cuss NYE-grah

ZONES: 6A, 6B, 7A, 7B, 8A, 8B, 9A, 9B, 10A

HEIGHT: 50 to 60 feet

WIDTH: 60 to 70 feet

FRUIT: abundant; round; 0.5"; dry; brown; attracts mammals; causes significant litter

GROWTH RATE: rapid; moderate longevity

HABIT: round; moderate density; symmetrical; fine to medium texture

FAMILY: *Fagaceae*

LEAF COLOR: green; yellow fall color, occasionally and briefly showy

LEAF DESCRIPTION: alternate; simple; entire margin; obovate to spatulate; deciduous to semi-evergreen; 2 to 4" long

FLOWER DESCRIPTION: brown; not showy; spring

EXPOSURE/CULTURE:

Light Requirement: part shade to full sun

Soil Tolerances: all textures; slightly alkaline to acidic; occasionally wet soil; moderate drought

Aerosol Salt Tolerance: low

PEST PROBLEMS: somewhat susceptible

BARK/STEMS:

Branches droop as tree grows and will require pruning for vehicular or pedestrian clearance beneath canopy; should be grown with a single leader; no thorns

Pruning Requirement: requires pruning to develop strong structure; train to a central leader with main branches spaced two feet or more apart; prune regularly to avoid making large pruning wounds

Limb Breakage: susceptible to breakage either at crotch, due to poor collar formation, or wood itself is weak and tends to break

Current Year Twig: reddish-brown; slender

LANDSCAPE NOTES:

Origin: New Jersey to Texas and Florida

Usage: buffer strip; highway; reclamation; shade tree; urban tolerant

Water Oak acorns are abundant and badly stain asphalt and concrete for several months in fall and winter. The leaves vary tremendously, from rounded and entire to three-lobed with several bristle tips, but are most frequently spatulate. Water Oak is deciduous in the north and semi-evergreen in the deep south. Trees can reach 80 feet in height but are shorter when grown in the open. Some trees put on a wonderful yellow fall color show for about a week.

Easily transplanted, young trees should be trained to develop one central trunk, and branches spread two to three feet apart. Naturalized trees often develop with several upright multiple trunks that are poorly attached to the tree. Horizontal branches droop toward the ground as additional growth adds to their weight. They can split from the tree in wind storms, deforming the plant and beginning the process of decay and decline. They appear to be poor compartmentalizers of decay; many are hollow at 50 years old.

The tree often begins to break apart just as it grows to a desirable size. For this reason, Live, Bur, Shumard, Red, White, Swamp, and White Oak, among others, are much better choices. Like other Oaks, care must be taken to develop a strong branch structure early in the life of the tree. This might increase the life span by eliminating the need to remove large-diameter branches. Pruning large branches from the trunk often initiates decay in the trunk.

A North American native, Water Oak is adapted to wet, swampy areas, such as along ponds and stream banks, but can also tolerate well-drained sites and even heavy, compacted clay soils. Not adapted to highly alkaline soil, but will grow well in clay if it stays moist.

Trunk rots by the time the tree is 50 years old. Boring insects are most likely to attack weakened or stressed trees. Mushroom root rot may occur. Leaf blister can cause defoliation. Chlorosis due to iron deficiency occurs on very high-pH soil. Resistant to orange-striped oakworm.

Quercus palustris Pin Oak

Pronunciation: KWER-cuss pah-LUSS-triss

ZONES: 4A, 4B, 5A, 5B, 6A, 6B, 7A, 7B, 8A

HEIGHT: 50 to 75 feet

WIDTH: 35 to 40 feet

FRUIT: round; 0.5"; dry and hard; brown; attracts mammals; causes some litter

GROWTH RATE: rapid; long-lived on acid soil

HABIT: pyramidal; moderate density; symmetrical; medium texture

FAMILY: *Fagaceae*

LEAF COLOR: medium green; showy, copper or red fall color

LEAF DESCRIPTION: alternate; simple; lobed and parted; oval; deciduous; 3 to 6" long

FLOWER DESCRIPTION: brown; not showy; spring

EXPOSURE/CULTURE:

Light Requirement: grows best in full sun

Soil Tolerances: all textures; acid only; wet soil; salt-sensitive; moderate drought

Aerosol Salt Tolerance: low to none

PEST PROBLEMS: somewhat sensitive

BARK/STEMS:

Bark is thin and easily damaged by mechanical impact; branches droop as tree grows and will require pruning for vehicular or pedestrian clearance beneath canopy; should be grown with a single leader; no thorns

Pruning Requirement: needs little pruning to develop strong structure; remove upright branches and double leaders early; branches continue to droop, even on older trees requiring regular removal

Limb Breakage: susceptible to ice storms, but otherwise resistant

Current Year Twig: brown; green; slender

CULTIVARS AND SPECIES:

'Crown Right' and 'Sovereign' showed early promise, but are no longer sold because the trunk eventually snaps off due to graft incompatability.

LANDSCAPE NOTES:

Origin: Massachusetts to Virginia, west to Oklahoma

Usage: parking lot island; buffer strip; highway; reclamation; shade tree; specimen; sidewalk cutout; street tree; urban tolerant in friable acid soil

One of the most utilized (perhaps overused) Oaks in the midwest and eastern United States, Pin Oak's popularity is due to the attractive pyramidal shape and straight, dominant trunk, even on older specimens. Chlorosis develops on compacted and alkaline soils because of iron deficiency, so this Oak is only recommended for friable acid sites. Green, glossy leaves borne on relatively small-diameter branches give way to brilliant red to bronze fall color, attracting attention in the landscape. Some brown leaves persist on the tree into the winter, providing interest to some people. Others do not care to use Pin Oak because of the leaf persistence characteristic.

When grown on a moist, acid soil, this can be a handsome specimen tree. The lower branches tend to droop, middle branches are horizontal, and branches in the upper part of the crown grow upright. Lower branches will require removal when Pin Oak is used as a street or parking lot tree, but the persistent lower branches can be an asset in an open lawn setting, due to the picturesque habit displayed by an open-grown Pin Oak. The trunk is typically straight up through the crown, only occasionally developing a double leader. Prune any double or multiple leaders out as soon as they are recognized, with several prunings in the first 15 to 20 years after planting. The straight trunk and small, well-attached branches make Pin Oak an extremely safe tree to plant in urban areas. Roots are fibrous making transplanting easy.

Iron chlorosis is the most common problem for Pin Oak growing in soils with a pH greater than 7. Chlorosis may not show up until 10 years after planting the tree. Galls can be numerous in some areas but usually do not cause a serious problem. Gypsy moth, oak wilt, and cankers can be troublesome. Susceptible to bacterial leaf scorch and the two-lined chestnut borer. Very susceptible to orange-striped oakworm.

Quercus phellos Willow Oak

Pronunciation: KWER-cuss FELL-us

ZONES: 6A, 6B, 7A, 7B, 8A, 8B, 9A, 9B

HEIGHT: 60 to 75 feet

WIDTH: 40 to 60 feet

FRUIT: round; 0.5"; dry and hard; brown; attracts mammals; not showy; causes some litter

GROWTH RATE: rapid; long-lived

HABIT: pyramidal when young, rounded later; dense; symmetrical; fine texture

FAMILY: *Fagaceae*

LEAF COLOR: light green; showy yellow fall color

LEAF DESCRIPTION: alternate; simple; entire margin; lanceolate; deciduous; 2 to 5" long

FLOWER DESCRIPTION: brown; inconspicuous

EXPOSURE/CULTURE:

Light Requirement: grows best in full sun

Soil Tolerances: all textures; acidic; occasionally wet soil; drought

Aerosol Salt Tolerance: moderate

PEST PROBLEMS: resistant

BARK/STEMS:

Branches grow mostly upright or horizontal, lower branches droop; should be grown with a single leader; branches tend to clump together at one point on trunk; no thorns

Pruning Requirement: requires pruning to develop strong structure; space main branches along a central trunk

Limb Breakage: resistant

Current Year Twig: brown; slender

LANDSCAPE NOTES:

Origin: New Jersey to Texas and Florida

Usage: parking lot island; buffer strip; highway; reclamation; shade tree; sidewalk cutout; street tree; urban tolerant

Widely used as a shade tree, in parks, and to line streets and boulevards, the fast-growing Willow Oak can reach over 70 feet in height. The pyramidal shape in youth gives way to a rounded canopy in middle and old age. The long, willow-like leaves create dense shade and a graceful effect, turning bright yellow before they fall. The tree is easy to transplant but reportedly transplants poorly in the fall.

The city of Charlotte, N.C., and other southern cities planted Willow Oak extensively during this century, and it has proven to be an excellent street tree. Willow Oak forms a tight, compact head with most of the foliage on the outside of the crown. Most trees form a strong central leader. Open-grown trees would benefit from some pruning to thin the crown, to develop more secondary branches along the major limbs, and to increase the taper on major limbs. This would improve the already good wind-hardiness of Willow Oak. Pruning is difficult, as there are many short, spine-like twigs, which are a nuisance to tree climbers. Be sure to prune out double leaders as they form, early in the life of the tree, to help create a strong form. Prune to space large branches along the trunk so they are well attached to the tree.

In its native habitat, Willow Oak grows in low, bottomland wet sites of flood plains, yet is drought-tolerant. It thrives in constantly wet to moist soil, although it has been known to adapt to seemingly impossible habitats. Willow Oak is a tough tree well adapted to urban conditions, but can develop chlorosis on high-pH soils, so be sure to plant on acid or neutral soil only. Willow Oak has few major problems and tolerates clay, salt, poor drainage, and compacted soil.

Root rot can be a problem in confined urban soil spaces. Borers and trunk canker can cause tree loss. Chlorosis due to iron deficiency occurs on high-pH soil. Possible susceptibility to gypsy moth puts fear in the hearts of tree managers with many Willow Oaks in their communities. Susceptible to orange-striped oakworm.

Quercus robur English Oak

Pronunciation: KWER-cuss ROW-burr

ZONES: 5A, 5B, 6A, 6B, 7A, 7B, 8A, 8B, (9), (10)

HEIGHT: 50 to 60 feet

WIDTH: 40 to 50 feet

FRUIT: elongated; oval; 1 to 1.5"; dry and hard; brown; attracts mammals; causes significant litter

GROWTH RATE: moderate; long-lived

HABIT: oval becoming round; moderate density; symmetrical; coarse texture

FAMILY: *Fagaceae*

LEAF COLOR: dark green; copper fall color

LEAF DESCRIPTION: alternate; simple; lobed; oblong to obovate; deciduous; 2 to 5" long

FLOWER DESCRIPTION: brown; not showy; spring

EXPOSURE/CULTURE:

Light Requirement: grows best in full sun

Soil Tolerances: all textures; alkaline to acidic; salt and drought

Aerosol Salt Tolerance: good

PEST PROBLEMS: resistant

BARK/STEMS:

Branches droop slightly as tree grows and will require some pruning for vehicular or pedestrian clearance beneath canopy; should be grown with a single leader; no thorns

Pruning Requirement: needs little pruning to develop strong structure; central leader often dominates young tree; eliminate competing leaders

Limb Breakage: resistant

Current Year Twig: reddish-brown; medium thickness

CULTIVARS AND SPECIES:

There are a number of cultivars for leaf color, crown form, and disease resistance. The most popular is 'Fastigiata', which is distinctly upright or columnar, but varies in spread from 10 to 18 feet due to seedling variation. Leaves turn brown in the fall and persist into the winter. The tree is very tolerant of urban conditions and should be grown and used more often. Powdery mildew is often seen on much of the tree. The cultivar 'Rosehill' is resistant to powdery mildew.

LANDSCAPE NOTES:

Origin: Europe and the Mediterranean region

Usage: parking lot island; buffer strip; highway; shade tree; street tree; urban tolerant

A stately tree, English Oak is hardy in zone 5 and will tolerate a range of soil pH and moisture conditions, including wet soil and dry clay. The crown often appears open, with large branches dominating the round crown. It is low-branching and pyramidal (when young) with a short, gray trunk. The main trunk is normally straight up or slightly bent up through the center of the crown. Branches develop nicely and the tree typically requires little, if any, pruning to create good form and strong structure. English Oak is an underutilized street and shade tree, especially in drier climates.

Planted on 30-foot centers, it makes a nice canopy over the street for summer shade. Roots usually do not lift sidewalks if planted in a strip of soil at least 10 feet wide. Pruning requirements should be minimal once a good central leader is established.

Although not widely used in the United States, the tree appears to be very drought-tolerant and could be used more extensively, particularly in dry climates where powdery mildew may not be a big problem. It also grows well in a wide range of soil from acid to alkaline.

Generally pest-free although mites can be seen on some trees. Powdery mildew can be a serious problem on English Oak in humid areas. Also susceptible to oak wilt, anthracnose, cankers, and gypsy moth. Large acorns can be a nuisance. Seedlings can appear in nearby landscapes.

Quercus robur 'Fastigiata' 'Fastigiata' English Oak

Pronunciation: KWER-cuss ROW-burr

ZONES: 5A, 5B, 6A, 6B, 7A, 7B, 8A, 8B

HEIGHT: 50 to 60 feet

WIDTH: 8 to 15 feet

FRUIT: elongated; oval; 1"; dry and hard; brown; attracts mammals; causes significant litter; persistent

GROWTH RATE: moderate; long-lived

HABIT: oval to round; moderate density; symmetrical; coarse texture

FAMILY: *Fagaceae*

LEAF COLOR: dark green; dull yellow to brown fall color

LEAF DESCRIPTION: alternate; simple; lobed; oblong to obovate; deciduous; leaves persistent into winter; 2 to 5" long

FLOWER DESCRIPTION: brown; not showy; spring

EXPOSURE/CULTURE:

Light Requirement: grows best in full sun

Soil Tolerances: all textures; alkaline to acidic; salt and drought

Aerosol Salt Tolerance: good

PEST PROBLEMS: highly susceptible to mildew

BARK/STEMS:

Branches grow upright and will not interfere with vehicles or pedestrian traffic beneath canopy; should be grown with a single leader; no thorns

Pruning Requirement: needs little pruning to develop strong structure

Limb Breakage: resistant

Current Year Twig: reddish-brown; medium thickness

CULTIVARS AND SPECIES:

'Attention' (mildew resistant) and 'Skyrocket' are very similar, except that canopies may stay even tighter than 'Fastigiata'. 'Skymaster' is slightly wider (15 to 25 feet). 'Pyramich' (50 by 25 feet) is narrowly pyramidal with dark green foliage.

LANDSCAPE NOTES:

Origin: Europe and the Mediterranean region

Usage: parking lot island; buffer strip; highway; street tree; windbreak; urban tolerant

'Fastigiata' or Upright English Oak is an upright, columnar, deciduous tree that eventually matures into a dense, elongated oval shape with a short trunk. It makes a striking landscape specimen. Growing moderately fast, Upright English Oak was first discovered growing wild in a forest in Germany and was propagated by grafting in 1783. The 2.5- to 5-inch-long by 1- to 2.5-inch-wide leaves maintain their dark green color throughout the growing season until they turn brown in autumn. They often remain on the tree for some time before dropping. The attractive, dark brown bark is deeply ridged and furrowed, and the one-inch acorns persist on the tree throughout the winter. Some branches may bend away from the canopy as the tree matures.

Upright English Oak is useful in areas where there is not much room for lateral branch growth. Some people object to the persistent brown leaves in the fall, but others like the tree for this trait. They have been used successfully for planting on 15- to 25-foot centers along an entrance road to a commercial landscape and for downtown tree planting projects where soil and overhead space is extremely limited.

The tree should be grown in full sun on well-drained, acid or slightly alkaline soil. The tree is very tolerant of urban conditions, is adaptable, and should be grown and used more. Few trees in the east are seen without powdery mildew. Locate in full-day sun to help reduce this problem. Trees in drier climates in the midwest and west probably are less affected by powdery mildew.

Powdery mildew is the most common malady. Oak wilt, anthracnose, cankers, gypsy moth, and borers occasionally bother the tree. There are many other potential problems.

Quercus rubra Northern Red Oak

Pronunciation: KWER-cuss ROO-brah

ZONES: 5A, 5B, 6A, 6B, 7A, 7B, 8A

HEIGHT: 60 to 70 feet

WIDTH: 50 to 60 feet

FRUIT: round; 0.75 to 1"; dry and hard; brown; attracts mammals; causes significant litter, but not every year

GROWTH RATE: rapid; long-lived

HABIT: round; dense; symmetrical; coarse texture

FAMILY: *Fagaceae*

LEAF COLOR: dark green; showy, red fall color

LEAF DESCRIPTION: alternate; simple; lobed; obovate; deciduous; 4 to 8" long

FLOWER DESCRIPTION: brown; not showy; spring

EXPOSURE/CULTURE:

Light Requirement: grows best in full sun

Soil Tolerances: all textures; slightly alkaline to acidic; salt; drought; compacted

Aerosol Salt Tolerance: moderate to good

PEST PROBLEMS: susceptible to oak wilt and leaf scorch

BARK/STEMS:

Branches grow mostly upright or horizontal and will not droop; should be grown with a single leader; no thorns

Pruning Requirement: needs little pruning to develop strong structure; prune dormant season only to avoid spreading oak wilt disease

Limb Breakage: resistant

Current Year Twig: brown; medium thickness

LANDSCAPE NOTES:

Origin: Nova Scotia to Minnesota, south to Alabama

Usage: parking lot island; buffer strip; highway; shade tree; street tree; urban tolerant

An adaptable, widely planted Oak with a rapid growth rate, Red Oak is native to rich woodland areas where it can grow to 90 feet tall. Branches and upper trunk are marked with long, light gray longitudinal lines that remind some people of ski trails. Open-grown trees form a rounded crown, which makes a nice shade, park, or street tree. Many communities use Red Oak as a street tree where overhead space is not limiting. Acorns are small and easily cleaned up, broken into small pieces by pedestrians, or eaten by squirrels. The foliage turns a wonderful dark red in the fall.

Trees should be trained to a central leader with major lateral branches spaced along the trunk. Untrained trees can develop many untapered branches growing from the same spot on the trunk, which could become crowded as the tree grows older. However, the tree has been rated as a "low failure rate" tree, indicating that it stays together well in the urban environment.

The tree withstands city conditions well but not high-pH soils. However, it is better in alkaline soils than Pin or Willow Oak. Moderately drought-tolerant in most soils, Red Oak is well suited as a street tree. Best planted in the spring from the field (spring, summer, or fall from containers) in full sun or part shade, Red Oak does best in well-drained soil with a pH less than 7.5. It tolerates compacted soil, air pollution, and moderate salt spray.

Like most other large trees, roots can heave sidewalks. Keep trees at least six feet from pavement if at all possible.

Very susceptible to oak wilt, which is lethal. Chlorosis due to iron deficiency occurs on high pH-soil. Gypsy moth enjoys the foliage. Bacterial leaf scorch can be devastating. Two-lined chestnut borers infest newly planted trees that are not irrigated regularly. Susceptible to orange-striped oakworm.

Quercus shumardii Shumard Oak

Pronunciation: KWER-cuss shoo-MAR-dee-eye

ZONES: 5B, 6A, 6B, 7A, 7B, 8A, 8B, 9A

HEIGHT: 60 to 80 feet

WIDTH: 40 to 50 feet

FRUIT: round; 0.5"; dry and hard; brown; attracts mammals; causes some litter

GROWTH RATE: rapid; long-lived

HABIT: pyramidal when young, oval later; open; irregular silhouette when young; coarse texture

FAMILY: *Fagaceae*

LEAF COLOR: showy, orange to red fall color

LEAF DESCRIPTION: alternate; simple; lobed; oval to obovate; deciduous; 4 to 8" long

FLOWER DESCRIPTION: brown; not showy; spring

EXPOSURE/CULTURE:

Light Requirement: full sun

Soil Tolerances: all textures; alkaline to acidic (provenance dependant); occasionally wet soil; drought

Aerosol Salt Tolerance: moderate

PEST PROBLEMS: resistant

BARK/STEMS:

Branches grow mostly upright or horizontal and will not droop; should be grown with a single leader; no thorns

Pruning Requirement: needs early pruning to develop strong structure; remove upright branches that compete with leader; prune regularly to avoid large cuts—decay from large cuts spreads quickly into trunk

Limb Breakage: resistant

Current Year Twig: brown; medium thickness

LANDSCAPE NOTES:

Origin: North Carolina to Florida, west to Texas and Missouri

Usage: parking lot island; buffer strip; highway; reclamation; shade tree; specimen; street tree; urban tolerant

Shumard Oak forms a large, stately tree with a narrow, rather open, rounded canopy, somewhat reminiscent of Red Oak. The crown spreads with age, becoming round at maturity. The four- to eight-inch-long deciduous leaves are deeply lobed (more so than *Quercus rubra*) and have bristles on the tips of some lobes. Shumard Oak puts on a vivid display of brilliant red to red-orange fall and winter foliage, providing a dramatic landscape statement. *Q. texana* is native to Texas and is closely related.

Shumard Oak has become popular in some areas but is used sparingly in others. It is deserving of wider use in most parts of its range due to its urban adaptability. Planted on 30- to 40-foot-centers, it will form a closed canopy over a two-lane street in 20 to 25 years with good growing conditions. Plant it six feet or more from the curb or other hard surfaces so the large trunk can develop properly. It makes a good street tree if purchased with one central leader.

Several leaders often develop in the nursery. When they are removed to leave one leader, the tree often looks very open and bare. Although this may be somewhat undesirable from an aesthetic standpoint, it creates a stronger tree, which will have a much longer service life than a multiple-leadered tree. The tree fills in as it grows older, forming a coarsely branched, open interior.

A native of the bottomlands of the southeastern United States, Shumard Oak grows well in full sun on a wide variety of soils, wet or dry. Although it prefers moist, rich soil where it will grow rapidly, it will tolerate drier locations. *Q. nuttallii* grows even faster and is well suited for wet sites.

Mites, and occasionally root rot on wet soil, are the major pests. Scale insects, borers, and brown felt fungus can also be a problem. Borers can be especially troublesome in nurseries. Oak wilt will kill Shumard Oak. Acorns cause significant litter in some years, but they provide ample food for wildlife.

Quercus suber Cork Oak

Pronunciation: KWER-cuss SUE-burr

ZONES: 8A, 8B, 9A, 9B, 10A, 10B, 11

HEIGHT: 60 to 80 feet

WIDTH: 60 feet

FRUIT: elongated; oval; 1"; dry and hard; brown; causes some litter

GROWTH RATE: moderately rapid; long-lived

HABIT: weeping oval; dense; nearly symmetrical; fine texture

FAMILY: *Fagaceae*

LEAF COLOR: green; glossy green above, pubescent beneath

LEAF DESCRIPTION: alternate; simple; entire or spiny margins; semi-evergreen, sheds leaves in late spring; 1 to 3" long

FLOWER DESCRIPTION: brown; not showy; spring

EXPOSURE/CULTURE:

Light Requirement: part shade to full sun

Soil Tolerances: all textures; slightly alkaline to acidic; salt and drought

PEST PROBLEMS: resistant

BARK/STEMS:

Fissured, corky bark is a source for commercial cork; in commercial groves, trunks are stripped of corky bark every six or seven years; branches droop as tree grows and will require pruning for clearance beneath canopy; should be grown with a single leader as high up into canopy as possible; no thorns

Pruning Requirement: requires some early pruning to develop strong structure; slow growth rate on aggressive branches to prevent formation of included bark, to generate a dominant trunk; thin canopy to increase wind firmness

Limb Breakage: resistant

Current Year Twig: gray; slender

LANDSCAPE NOTES:

Origin: Mediterranean region

Usage: large parking lot island; buffer strip; highway; shade tree; street tree; urban tolerant

A moderately large tree, Cork Oak is more upright than the Live Oak. This makes it well suited for planting along streets and in parking lot islands where clearance is needed for vehicle passage beneath the canopy. Main branches can be trained to grow up and out from the main trunk, but small-diameter twigs and branches tend to weep toward the ground. The tree is large enough to provide shade for large areas of the landscape. An amazingly durable tree, it is well suited for life in the desert. Give it plenty of room, as the trunk can grow to more than four feet in diameter.

Once established, Cork Oak will thrive in almost any location, but should be saved for sites with plenty of soil for root expansion. The canopy becomes dense and will benefit from occasional thinning. It is a tough, enduring tree that will respond with vigorous growth to plentiful moisture on well-drained soil. It tolerates heat well, growing nicely in the southwestern United States.

Provide plenty of irrigation until the tree is established. Once it is well established, it will survive and grow with only occasional irrigation in the summer. Like other Oaks, care must be taken to develop a strong branch structure early in the life of the tree. Be sure to eliminate multiple trunks and branches that form a narrow angle with the trunk, as these can split from the tree as it grows older. Cut back young competing stems and upright branches to form a dominant trunk.

Be sure that adequate soil space is given to Cork Oak. Although roots will grow under curbs and sidewalks in well-drained soil when planted in confined soil spaces, allowing the tree to thrive in urban sites, in time they lift sidewalks, curbs, and driveways if planted too close. Trees confined to a small root area can become top-heavy unless the canopy is regularly thinned.

Powdery mildew deforms foliage on rapid-growing trees.

Quercus virginiana Southern Live Oak

Pronunciation: KWER-cuss vir-gin-ee-A-nah

ZONES: (7B), 8A, 8B, 9A, 9B, 10A, 10B, 11

HEIGHT: 60 to 80 feet

WIDTH: 60 to 120 feet

FRUIT: elongated; oval; 0.5"; dry and hard; brown; causes some litter

GROWTH RATE: moderately rapid; long-lived

HABIT: spreading and rounded; dense; nearly symmetrical; fine texture

FAMILY: *Fagaceae*

LEAF COLOR: dark green; evergreen to semi-evergreen

LEAF DESCRIPTION: alternate; simple; entire margin; oval; 2 to 5" long

FLOWER DESCRIPTION: brown; not showy; spring

EXPOSURE/CULTURE:

Light Requirement: part shade to full sun

Soil Tolerances: all textures; alkaline to acidic; occasionally wet soil; salt and drought

Aerosol Salt Tolerance: moderate to high

PEST PROBLEMS: resistant, except in parts of Texas

BARK/STEMS:

Branches droop as tree grows and will require pruning for vehicular or pedestrian clearance beneath canopy; should be grown with a single leader as high up into canopy as possible; no thorns

Pruning Requirement: requires pruning to develop strong structure; space branches along trunk; in the spring, pruning paint applied to recent pruning cuts helps prevent the spread of oak wilt

Limb Breakage: resistant

Current Year Twig: gray; slender

CULTIVARS AND SPECIES:

Sand Live Oak, var. *geminata* (*Q. geminata*), grows on sandy soil and is more upright and open-crowned in habit. It may be more suited for street tree planting due to its smaller size. The fast-growing 'Heritage' is recommended for desert areas, and is most common in the southwestern United States. The variety *fusiformis* (*Q. fusiformis*) is native to central and southern Texas; it is susceptible to oak wilt but resistant to root rot. Perhaps more adapted to Texas than *Q. virginiana,* but many nursery operators do not normally differentiate among the Live Oaks.

LANDSCAPE NOTES:

Origin: Virginia along the coast to Texas, including all of Florida

Usage: large parking lot island; buffer strip; highway; reclamation; shade tree; street tree; urban tolerant

A large, sprawling, picturesque tree, usually graced with Spanish moss and strongly reminiscent of the Old South, Live Oak is one of the broadest-spreading of the Oaks, providing large areas of deep, inviting shade. An amazingly durable American native, it can measure its lifetime in centuries if properly located and cared for in the landscape. Unfortunately, Oak wilt has devastated the tree in parts of central Texas. Give it plenty of room, as the trunk can grow to more than six feet in diameter. Construction-impacted trees take a long time to die, giving Live Oak a reputation for being a tough tree. It is usually the last tree to die around a newly constructed building.

Once established, Live Oak will thrive in almost any location and has very good wind resistance. Live Oak is a tough, enduring tree that will respond with vigorous growth to plentiful moisture on well-drained soil. Like other Oaks, care must be taken to develop a strong branch structure early in the life of the tree. Be sure to eliminate young multiple trunks and branches that form a narrow angle with the trunk or included bark, as these can split from the tree as it grows older.

Be sure that adequate soil space is given to Live Oak. Although roots will grow under curbs and sidewalks in sandy soil when planted in confined soil spaces, allowing the tree to thrive in urban sites, in time they lift sidewalks, curbs, and driveways if planted too close. This may be a small price to pay for the bountiful shade cast by a row of healthy trees planted along the street.

There are few major pests, except oak wilt in parts of Texas.

Ravenala madagascariensis Traveler's-Tree

Pronunciation: rav-eh-NAY-lah mad-ah-gas-kar-ee-EN-siss

ZONES: 10A, 10B, 11

HEIGHT: 25 to 30 feet

WIDTH: 18 feet

FRUIT: 0.5"; dry; brown; not showy; no significant litter

GROWTH RATE: moderate; short-lived

HABIT: palm-like; upright; open; irregular silhouette; extremely coarse texture

FAMILY: *Strelitziaceae*

LEAF COLOR: broadleaf evergreen; green

LEAF DESCRIPTION: simple; entire margin; oblong; 10' long; often torn along veins

FLOWER DESCRIPTION: white; showy; fall, spring, summer

EXPOSURE/CULTURE:

Light Requirement: part shade to full sun

Soil Tolerances: all textures; slightly alkaline to acidic; drought-sensitive

Aerosol Salt Tolerance: low

PEST PROBLEMS: somewhat susceptible

BARK/STEMS:

Can be trained to one or many trunks; trunks grow mostly upright and will not droop; showy trunk; no thorns

Pruning Requirement: little required except to remove old leaves; suckers can be removed to create a single-trunked tree or to prevent formation of additional trunks

Limb Breakage: somewhat susceptible

LANDSCAPE NOTES:

Origin: Madagascar

Usage: container or above-ground planter; near a deck or patio; specimen

Traveler's-Tree is ideal for creating an exotic, tropical effect, with its very large, banana-like leaves, each up to 10 feet long, held in fan-shaped formation. The unusual, small, white flowers are held erect in canoe-shaped bracts that contrast nicely with the green foliage. Leaves are usually seen tattered and torn from exposure to wind. Traveler's-Tree will reach a height of 30 feet and a spread of 18 feet, growing at a moderate rate. The common name is derived from the fact that weary travelers would quench their thirst with the rainwater collected in the enlarged sheaths at the base of the leaves. Poke a hole through the base of the thick petiole to release the stored water.

Traveler's-Tree makes a nice tropical accent in a large landscape, but grows too large for most modest-sized yards. Specimens in formal landscapes are often lighted from below at night to display the showy trunk and special form. They are usually planted with a low groundcover or shrub bed around the base of the tree to set it off from the rest of the landscape.

Traveler's-Tree will produce best growth in full sun, though small potted plants may be grown in shade for a period of time. Plants should be grown on fertile soils, high in organic matter, and routinely cared for. Plants are sensitive to frost. Propagation is by division of basal suckers or by seed, which are slow to germinate.

Cercospora leaf spot is a very serious disease problem.

Rhamnus caroliniana Carolina Buckthorn

Pronunciation: RAM-nuss care-o-lin-ee-A-nah

ZONES: 6A, 6B, 7A, 7B, 8A, 8B, 9A, 9B

HEIGHT: 12 to 15 feet

WIDTH: 10 to 15 feet

FRUIT: round; 0.33"; fleshy; red to black; attracts birds; edible; no significant litter; persistent; showy

GROWTH RATE: moderate; short-lived

HABIT: oval; upright; open; irregular silhouette; medium texture

FAMILY: *Rhamnaceae*

LEAF COLOR: bright green; showy, orange fall color
LEAF DESCRIPTION: alternate; simple; entire margin; oval to oblong; deciduous; 3 to 6"
FLOWER DESCRIPTION: yellow; not showy; spring
EXPOSURE/CULTURE:

Light Requirement: mostly shaded to full sun

Soil Tolerances: all textures; alkaline to acidic; drought

PEST PROBLEMS: resistant
BARK/STEMS:

Bark is thin and easily damaged by mechanical impact; branches droop as tree grows and will require pruning for vehicular or pedestrian clearance beneath canopy; routinely grown with multiple trunks, but can be trained to a single trunk; no thorns

Pruning Requirement: requires pruning to develop uniform structure

Limb Breakage: resistant

Current Year Twig: brown; slender

CULTIVARS AND SPECIES:

R. cathartica and *R. davurica* have an interesting, multi-trunked habit and form a nice small canopy, but can become invasive and weedy. Would make a quick cover on a reclamation site.

LANDSCAPE NOTES:

Origin: New York to Florida, west to Nebraska and Texas

Usage: container or above-ground planter; hedge; buffer strip; highway; near a deck or patio; reclamation; standard; specimen

Carolina Buckthorn develops an open crown of many slender branches and is usually seen at 12 to 15 feet in height, although it is capable of reaching 40 feet in a partially shaded location. The bright green, deciduous leaves change to a gorgeous orange-yellow or red in autumn before dropping. The fairly inconspicuous, early summer flowers are greenish-white and are followed by small, showy red fruits, which ripen to black in the fall when their flesh becomes sweet and edible. Birds find the fruits irresistible and they almost inhale them in the fall. The thin, smooth bark is gray with dark markings that are actually quite striking. Carolina Buckthorn is quite attractive in the landscape and is one of the first fruiting plants to show color.

Use this small tree or large shrub in a shrub border to attract birds and for a late summer accent provided by the fruit. It can be planted in mass to form a thicket, which should provide food and cover for a variety of wildlife. It might also be tried as a street or median tree where overhead space is restricted by power lines. It would have to be trained to an upright growth habit by staking and selective thinning to form a nice crown, but the nice foliage and attractive fruit might make it worth a try.

Carolina Buckthorn should be grown in full sun on well-drained soil, acid or alkaline. It is moderately drought-tolerant.

Susceptible to crown rust. A leaf spot will occasionally infect the tree but is of no consequence. Invasive and weedy.

Rhus chinensis Chinese Sumac

Pronunciation: ruse chi-NEN-sis

ZONES: 5B, 6A, 6B, 7A, 7B, 8A, 8B

HEIGHT: 15 to 20 feet

WIDTH: 15 to 20 feet

FRUIT: round; less than 0.5"; fleshy; orange-red; no significant litter; persistent; showy

GROWTH RATE: moderate; short-lived

HABIT: oval; round; moderate density; irregular silhouette; medium texture

FAMILY: *Anacardiaceae*

LEAF COLOR: light green; striking orange to red fall color

LEAF DESCRIPTION: alternate; odd pinnately compound; crenate to serrate margin; oblong to ovate; deciduous; 2 to 5" long

FLOWER DESCRIPTION: white; showy; summer

EXPOSURE/CULTURE:

Light Requirement: grows best in full sun

Soil Tolerances: all textures; alkaline to acidic; occasionally wet soil; drought

PEST PROBLEMS: mostly resistant

BARK/STEMS:

Bark is thin and easily damaged by mechanical impact; branches droop as tree grows and will require pruning for vehicular or pedestrian clearance beneath canopy; routinely grown with multiple trunks, but can be trained to a short, single trunk; no thorns

Pruning Requirement: requires pruning to develop nice tree form

Limb Breakage: susceptible

Current Year Twig: reddish-brown; stout

CULTIVARS AND SPECIES:

'September Beauty' has fantastic fall color. *R. lanceolata* is a single, sometimes multi-trunked small tree, 15 feet tall, with staggeringly beautiful red fall foliage, and should be planted more often.

LANDSCAPE NOTES:

Origin: China, Japan

Usage: container or above-ground planter; buffer strip; highway; near a deck or patio; reclamation; specimen

Chinese Sumac forms a loose, spreading small tree, reaching up to 25 feet in height though most specimens only grow to about 12 to 15 feet tall. The shiny, pinnately compound, five-inch-long leaves change to a brilliant orange, red, or yellow in the fall before dropping. The yellowish-white, summertime flowers appear in 6- to 10-inch-long and wide, terminal panicles and are quite showy. The hairy fruits that follow are orange-red and mature in October.

Training is required to make this large shrub into a tree. Begin by staking the main stem in the upright position for a year or two and develop branches beginning at two to four feet from the ground. Space branches 8 to 12 inches apart and be sure they form a wide angle with the trunk. This will help ensure that they are well attached to the tree. Occasional pinching or heading back of the terminal shoot will increase branching.

Place Chinese Sumac in a prominent location in the landscape in the full sun. It is a nice tree for planting in a low groundcover to display the interesting trunk and branch arrangement. The medium-textured foliage, showy flower display, and bright fall color combine to make this small tree suitable for increased usage. It looks great placed in a container on the deck or patio.

Chinese Sumac should be grown in full sun on well-drained soil, acid or alkaline. It would be well suited for inclusion in a low-maintenance landscape where plants receive little, if any, irrigation. Suckers from the base of the trunk may have to be removed periodically.

Potential problems include aphids, scales, fusarium wilt, leaf spots, and powdery mildew. Verticillium wilt causes wilting of individual stems, followed by death of the foliage. Eventually the entire plant can die. It suckers from the base, but not as much as native Sumacs. Unfortunately, it is difficult to find in nurseries.

Rhus glabra Smooth Sumac

Pronunciation: ruse GLAY-bra

ZONES: 2A, 2B, 3A, 3B, 4A, 4B, 5A, 5B, 6A, 6B, 7A, 7B, 8A, 8B, 9A, 9B

HEIGHT: 10 to 15 feet

WIDTH: 10 to 15 feet

FRUIT: round; less than 0.25"; fleshy, and large, showy clusters; deep red; no significant litter; persistent

GROWTH RATE: moderate; short-lived

HABIT: upright oval; round; open; irregular silhouette; coarse texture

FAMILY: *Anacardiaceae*

LEAF COLOR: dark green; striking orange to red fall color

LEAF DESCRIPTION: alternate; odd pinnately compound; serrate margin; oblong to ovate; deciduous; leaflets 2 to 5" long

FLOWER DESCRIPTION: greenish; summer

EXPOSURE/CULTURE:

Light Requirement: grows best in full sun

Soil Tolerances: all textures; alkaline to acidic; extreme drought

PEST PROBLEMS: mostly resistant

BARK/STEMS:

Bark is thin and easily damaged by mechanical impact; branches mostly upright; routinely grown with multiple trunks, but can be trained to a short, single trunk; suckers readily develop from base of trunk and roots; no thorns

Pruning Requirement: requires pruning to develop nice tree form and to remove suckers

Limb Breakage: susceptible

Current Year Twig: reddish-brown; stout

LANDSCAPE NOTES:

Origin: much of North America

Usage: container or above-ground planter; buffer strip; highway; near a deck or patio; reclamation; specimen

Smooth Sumac forms a loose, spreading small tree, reaching up to 15 feet in height. The shiny, pinnately compound, five-inch-long leaflets change to a brilliant orange or red in the fall before dropping. The yellowish-green, summertime flowers appear in 6- to 10-inch-long and wide, terminal panicles, but are not as showy as those of Chinese Sumac. The hairy fruits that follow are scarlet-red and mature in the fall.

Training is required to make this large shrub into a tree. Begin by staking the main stem in the upright position for a year or two and develop branches beginning at two to four feet from the ground. Space branches six to eight inches apart and be sure they form a wide angle with the trunk. This will help ensure that they are well attached to the tree. Occasional pinching or heading back of the terminal shoot will increase branching and canopy density.

Place Smooth Sumac in a prominent location in the landscape in the full sun. It is a nice tree for planting in a low groundcover to display the interesting trunk and branch arrangement. The coarse-textured foliage, showy flower display, and bright fall color combine to make this small tree suitable for increased usage. It looks great placed in a container on the deck or patio. The fall color displays nicely against an evergreen background.

Sumac should be grown in full sun on well-drained soil, acid or alkaline. It would be well suited for inclusion in a low-maintenance landscape where plants receive little, if any, irrigation. Suckers from the base of the trunk and from the roots may have to be removed periodically to maintain a tree form. They can be left on to allow the plant to form a dense clump.

Potential problems include aphids, scales, fusarium wilt, leaf spots, and powdery mildew. Verticillium wilt causes wilting of individual stems, followed by death of the foliage. Eventually the entire plant can die. Suckers and root sprouts can be a nuisance in a maintained landscape.

Rhus lancea African Sumac

Pronunciation: ruse LAN-see-ah

ZONES: 8B, 9A, 9B, 10A, 10B, 11

HEIGHT: 15 to 20 feet

WIDTH: 15 to 20 feet

FRUIT: round; 0.25"; fleshy; yellow or red in clusters; causes significant litter; showy

GROWTH RATE: moderate; short to moderate life span

HABIT: round and spreading; moderate density; nearly symmetrical; fine texture

FAMILY: *Anacardiaceae*

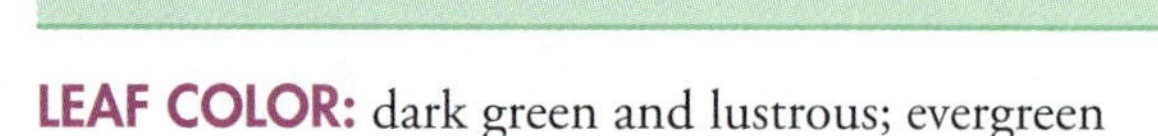

LEAF COLOR: dark green and lustrous; evergreen

LEAF DESCRIPTION: alternate; odd pinnately compound; entire margin; strap-like; 4 to 5" long

FLOWER DESCRIPTION: greenish-yellow; somewhat showy

EXPOSURE/CULTURE:

Light Requirement: grows best in full sun

Soil Tolerances: all textures; alkaline to acidic; drought

PEST PROBLEMS: mostly resistant

BARK/STEMS:

Showy, rough bark; bark is thin and easily damaged by mechanical impact; branches droop as tree grows and will require pruning for vehicular or pedestrian clearance beneath canopy; routinely grown with multiple trunks, but can be trained to a short, single trunk; no thorns

Pruning Requirement: requires pruning to develop nice tree form; thin canopy to enhance wind firmness

Limb Breakage: susceptible

Current Year Twig: reddish-brown; stout

LANDSCAPE NOTES:

Origin: South Africa

Usage: container or above-ground planter; parking lot island; buffer strip; screen or tall hedge; highway; near a deck or patio; specimen; standard

African Sumac forms a light-textured, small tree, reaching up to 20 feet in height. Most specimens only grow to about 12 to 15 feet tall but spread to 20 to 25 feet wide. Unlike many Sumacs, the shiny, pinnately compound, four- to five-inch-long leaves remain on the tree all year. The fruits are yellowish or red and mature in late fall. Trees stay small, making them well suited for planting in many landscape situations.

Training is required to make this large shrub into a tree. Begin by staking the main stem in the upright position for a couple of years and develop branches beginning at two to four feet from the ground. Space branches 8 to 12 inches apart and be sure they form a wide angle with the trunk. This will help ensure that they are well attached to the tree. Occasional pinching or heading back of the terminal shoot when the tree is young will increase branching and form a fuller canopy.

Place African Sumac in a prominent location in the landscape in the full sun. It is a nice tree for planting in a low groundcover to display the interesting trunk and branch arrangement. The fine-textured foliage, small size, and showy flower display combine to make this small tree suitable for increased usage in residential and commercial landscapes. It looks great placed in a container on the deck or patio, or planted in a raised bed.

African Sumac should be grown in full sun on well-drained soil, acid or alkaline. It would be well suited for inclusion in a low-maintenance landscape where plants receive little irrigation. It is highly recommended for the driest climates in southwestern United States.

Potential problems include aphids, scales, fusarium wilt, leaf spots, and powdery mildew. Verticillium wilt may also be a problem.

Robinia pseudoacacia Black Locust

Pronunciation: row-bin-EE-ah sue-doe-ah-CASE-ee-ah

'Frisia'

'Tortuosa'

ZONES: 4A, 4B, 5A, 5B, 6A, 6B, 7A, 7B, 8A, 8B

HEIGHT: 40 to 70 feet

WIDTH: 25 to 35 feet

FRUIT: pod; 3 to 6"; dry and hard; black; attracts wildlife; causes a slight litter problem; persistent

GROWTH RATE: rapid; short to moderate life span

HABIT: oval; upright; open; irregular silhouette; fine texture

FAMILY: *Leguminosae*

LEAF COLOR: blue-green; yellow fall color

LEAF DESCRIPTION: alternate; odd pinnately compound; entire margin; oval to ovate; deciduous, dropping early; 1 to 2" long

FLOWER DESCRIPTION: white; pleasant fragrance; extremely showy; spring

EXPOSURE/CULTURE:

Light Requirement: part shade to full sun

Soil Tolerances: all textures; alkaline to acidic; salt and drought

Aerosol Salt Tolerance: high

PEST PROBLEMS: susceptible

BARK/STEMS:

Branches grow mostly upright and will not droop; showy trunk; should be grown with a single leader; thorns are present on trunk or branches

Pruning Requirement: requires pruning to develop strong structure; prevent formation of included bark by thinning lateral branches

Limb Breakage: branches regularly snap in two with little provocation

Current Year Twig: brown; stout; slender

CULTIVARS AND SPECIES:

'Erecta'—upright form; 'Frisia'—yellowish leaves in early summer, produces many root suckers, and is popular in western United States; 'Purple Robe'—lavender flowers; 'Tortuosa'—beautiful flowers borne on interesting, twisted branches. The bare branches form a sculptured silhouette without pruning.

LANDSCAPE NOTES:

Origin: Appalachian and Ozark Mountains

Usage: reclamation; shade tree

Continued

Robinia pseudoacacia *Continued*

The upright growth and short, irregular branches form an open canopy and cast light shade below the tree, allowing a lawn to thrive. For approximately a 10-day period in late spring, the trees are festooned with 4- to 8-inch-long, dense clusters of extremely fragrant, 1-inch white blossoms (similar to sweet-peas), which are literally alive with the bustling activity of visiting bees. The honey that is produced is quite delicious and much sought-after.

These seeds are widely dispersed by birds and other wildlife, so (along with the root suckering and invasive root system) Black Locust can spread into surrounding landscapes. This feature, along with the thorns found along the branches, should be considered when placing Black Locust in an ornamental garden. It is probably best saved for the reclamation project or roadside planting where trees receive no maintenance. If left to its own devices, it will form dense thickets, even on the poorest soils, a fact that makes it quite useful in reclamation sites.

Although the wood of Black Locust is reputed to be extremely strong and durable (pioneers used it to fashion nails for building ships and houses), the branches are brittle and subject to damage in strong winds. This can be partially corrected by thinning the major branches so they grow to no more than about half the size of the trunk.

Locust borer is a serious pest, although borer-resistant clones are reportedly being developed. Leaf miner is a universal problem, and the trees along the highways in the south can be seen riddled with damage from this pest in summer. Canker, leaf spot, and powdery mildew also affect Black Locust.

Roystonea spp. Royal Palm

Pronunciation: roy-STOW-nee-ah species

ZONES: 10A, 10B, 11

HEIGHT: 50 to 80 feet

WIDTH: 15 to 25 feet

FRUIT: oval; round; less than 0.5"; dry; black to purple; causes some litter; showy

GROWTH RATE: moderate; long-lived

HABIT: palm; open; symmetrical; coarse texture

FAMILY: *Arecaceae*

LEAF COLOR: green

LEAF DESCRIPTION: spiral; odd pinnately compound; entire margin; lanceolate leaflets; 18 to 36" long leaflets

FLOWER DESCRIPTION: yellow; pleasant fragrance; showy; periodic year-round flowering

EXPOSURE/CULTURE:

Light Requirement: part shade to full sun

Soil Tolerances: all textures; slightly alkaline to acidic; wet soil; moderate drought

Aerosol Salt Tolerance: moderate

PEST PROBLEMS: mostly resistant

BARK/STEMS:

Single, 30" diameter trunk; ringed with leaf scars; bright green crown shaft; no thorns

Pruning Requirement: old fronds drop; although not normally practiced, early removal can prevent this

Limb Breakage: resistant

CULTIVARS AND SPECIES:

Roystonea elata is native to Florida and *Roystonea regia* to Cuba. Cuban Royal Palm has less prominent secondary leaf veins and nearly globose fruit. Good luck distinguishing the two! In fact, many nurseries do not distinguish between these two species, as they are very similar.

LANDSCAPE NOTES:

Origin: Florida and Cuba

Usage: parking lot island; buffer strip; highway; specimen; sidewalk cutout; street tree; urban tolerant

Notably popular as street or specimen trees, Royal Palms make a neat, tidy, yet stately landscape element for large landscapes, often reaching 50 to 100 feet in height in almost as many years. The tall, smooth, cement-gray trunks are capped with a glossy, green crown shaft several feet high and a beautiful, broad, dense crown of soft, gently drooping, feathery fronds. Flowers are incredibly fragrant, even from 50 feet away, and are produced periodically throughout the year (but mostly in spring and summer). The old fronds should be removed before they drop, as they can cause injury or damage to plants or property when allowed to fall. About one frond will fall every month or two from a healthy palm.

Royal Palms grow quite rapidly when given an abundance of water and fertilizer in full sun or dappled shade. They withstand strong winds and salt spray very well, but some foliage injury will be evident on Royal Palm located next to the ocean. Not really suited for beachside planting as Sabal Palm is. The young, developing fronds grow in a distorted, frizzled manner in alkaline soil. This is usually attributed to manganese or iron deficiency in the leaves. It can be prevented with regular applications of a special palm fertilizer, or by planting in acid to neutral pH soil.

Palm leaf skeletonizer, giant palm weevil, and scales may attack when the palm is young or recently transplanted. Royal palm bug damages leaves in the spring, but new ones emerge just fine after the insect departs. Any of these can be troublesome in localized areas. Ganoderma butt rot is the most serious problem on Royal Palms. It kills infected trees. The disease often enters the trunk through injuries on the lower trunk and roots. Prevent injury to this area by mulching around the trunk to keep mowers and weed whackers away. Manganese or iron deficiency in alkaline soil reduces the size of the emerging fronds. Severe deficiency causes necrosis. Potassium deficiency causes chlorosis in older leaves. Irrigation is needed in dry climates.

Sabal palmetto Cabbage Palm

Pronunciation: SAY-ball pall-MET-toe

ZONES: 8B, 9A, 9B, 10A, 10B, 11

HEIGHT: 40 to 50 feet

WIDTH: 10 to 12 feet

FRUIT: round; 0.33"; dry; green turning black; attracts wildlife; causes some litter

GROWTH RATE: slow-growing; long-lived

HABIT: palm; open; symmetrical; very coarse

FAMILY: *Arecaceae*

LEAF COLOR: blue-green to green

LEAF DESCRIPTION: spiral; costapalmate; 6 to 8' leaves

FLOWER DESCRIPTION: white; showy; summer

EXPOSURE/CULTURE:

Light Requirement: mostly shaded to full sun

Soil Tolerances: all textures; alkaline to acidic; wet soil; salt and drought

Aerosol Salt Tolerance: high

PEST PROBLEMS: except for recently transplanted palms, nearly pest-free

BARK/STEMS:

Single, 18" thick trunk; some with persistent leaf bases, some without; no thorns; no crown shaft

Pruning Requirement: old fronds drop from palm; prune only drooping lower fronds, not horizontal or upright ones

Limb Breakage: does not break, even in strong hurricane

CULTIVARS AND SPECIES:

Sabal peregrina, planted in Key West, grows to about 25 feet high. *Sabal mexicana (texana)* grows in Texas and looks similar to *Sabal palmetto.*

LANDSCAPE NOTES:

Origin: coastal South Carolina through Florida

Usage: parking lot island; buffer strip; highway; reclamation; specimen; sidewalk cutout; street tree; urban tolerant

Cabbage Palm is capable of reaching 70 feet or more in the woods, but is usually seen at 40 to 50 feet in height. Cabbage Palm is topped with a very dense, 10- to 15-foot-diameter, rounded crown of deeply cut, curved, costapalmate leaves. This is South Carolina's and Florida's state tree, and is well suited as a street planting or clustered in informal groupings of varying size. The four- to six-foot-long, creamy white, showy flower stalks in the summer are followed by small, shiny, green to black fruits, relished by squirrels, raccoons, and other wildlife. Fruits create a short-term mess on nearby sidewalks. Although one of the hardier palms, 11°F killed about 25 percent of the Cabbage Palms in 1983 in Baton Rouge, LA.

Cabbage Palm is about as hurricane-proof as a tree can be, still standing after the Oaks have blown over and Pines have snapped in two. It adapts well to small cutouts in the sidewalk, and can even create shade if planted on 6- to 10-foot centers. Clean the trunk of leaf bases to eliminate a habitat for roaches, but do not injure it by cutting too deeply.

Cabbage Palm is exceptionally easy to transplant and will thrive in full sun or partial shade. It will adapt to slightly brackish water as well as dry, sandy locations and requires no special care once established. It needs to be watered regularly until established, as all cut roots die back to the trunk after transplanting. Leaving fronds tied up for more than a week can encourage fungi infection in the bud. Removing all fronds appears to increase transplant survival if palms cannot be irrigated regularly after planting. A thinner trunk develops for a short period if all fronds are removed.

New transplants (especially those receiving too little water) are particularly susceptible to the giant palm weevil, which kills the palm. Ganoderma butt rot can kill palms. Avoid irrigating or injuring the trunk. Seeds germinate readily in the landscape, generating a proliferation of seedlings. Pulling the seedlings up from beneath the canopy can be a regular nuisance.

Salix alba Weeping Willow

Pronunciation: SAY-licks AL-bah

ZONES: 6A, 6B, 7A, 7B, 8A, 8B, 9A

HEIGHT: 45 to 70 feet

WIDTH: 45 to 70 feet

FRUIT: 0.25"; dry; brown; no litter

GROWTH RATE: rapid; short to moderate life span

HABIT: oval or round and weeping; dense; nearly symmetrical; fine texture

FAMILY: *Salicaceae*

LEAF COLOR: light green; yellow fall color

LEAF DESCRIPTION: alternate; simple; serrate to serrulate margin; lanceolate or linear; deciduous; 1.5 to 4" long

FLOWER DESCRIPTION: yellow; not showy; early spring

EXPOSURE/CULTURE:

Light Requirement: part shade to full sun

Soil Tolerances: all textures; alkaline to acidic; wet soil; salt and drought

Aerosol Salt Tolerance: good

PEST PROBLEMS: sensitive

BARK/STEMS:

Branches droop as tree grows and will require pruning for vehicular or pedestrian clearance beneath canopy; should be grown with a single leader as far into canopy as possible; no thorns

Pruning Requirement: requires pruning to develop good structure; thin main branches to help prevent formation of included bark

Limb Breakage: susceptible to breakage at crotch, due to poor collar formation; wood itself is weak and tends to break; breaks up in ice storms

Current Year Twig: brown; slender

CULTIVARS AND SPECIES:

'Tristis' is probably the most common Weeping Willow in the trade now, surpassing *S. babylonica.*

LANDSCAPE NOTES:

Origin: Asia

Usage: specimen, screen

Often when one envisions a quiet body of water, the graceful, elegant form of a Weeping Willow is seen at the water's edge, the long, light green, pendulous branches reflected in the water, gently swaying with each little breeze. Though it does well in very moist soils, Weeping Willow may also be successfully used as a fast-growing specimen or screen in drier, more open areas, where it should receive regular watering to prevent leaf drop in drought. It will survive drought but loses leaves and is generally a mess without irrigation especially in the southern part of its range. Ultimately reaching a height of 50 feet or more, Weeping Willow should be given plenty of room to develop its broad, rounded crown.

Care should be taken not to locate Weeping Willows near underground water or sewer lines or close to septic tank drain fields, where the roots could cause significant damage. Roots are notoriously aggressive, will spread about three times the distance from the trunk to the edge of the canopy, and often grow on the soil surface. Not for residential lots, this tree is best located near water where soil will be undisturbed. Often planted near retention ponds and lakes for a dramatic softening effect.

Weeping Willow should be grown in full sun or very light shade and will tolerate a wide range of soil conditions, including alkaline pH. All Willows will need initial pruning and training when young to develop a strong, central trunk, with branch crotches as wide as possible. This will increase the longevity of the tree and help overcome the problem of branch breakage at the crotch.

Aggressive roots and messy habit. Willows are favored by a variety of pests too numerous to discuss.

Salix matsudana 'Tortuosa' Corkscrew Willow

Pronunciation: SAY-licks mat-sue-DAY-nah

ZONES: 6A, 6B, 7A, 7B, 8A, (9A), (9B)

HEIGHT: 25 to 35 feet

WIDTH: 15 to 20 feet

FRUIT: less than 0.5"; dry; not showy; no litter

GROWTH RATE: rapid; short to moderate life span

HABIT: oval; upright; moderate density; symmetrical; fine texture

FAMILY: *Salicaceae*

LEAF COLOR: light green; yellow fall color

LEAF DESCRIPTION: alternate; simple; serrate margin; lanceolate to linear; deciduous; 2 to 4" long

FLOWER DESCRIPTION: not showy

EXPOSURE/CULTURE:

Light Requirement: part shade to full sun

Soil Tolerances: all textures; alkaline to acidic; wet soil; salt; drought-sensitive

Aerosol Salt Tolerance: good

PEST PROBLEMS: sensitive, unless on a moist site

BARK/STEMS:

Bark is thin and easily damaged by mechanical impact; branches grow mostly upright and will not droop; should be grown with a single leader; no thorns

Pruning Requirement: requires pruning to develop strong structure; space major branches along trunk and keep them less than half the trunk diameter with pruning to encourage better branch attachment

Limb Breakage: susceptible to breakage at crotch, due to poor collar formation; wood itself is weak and tends to break; breaks in ice storms

Current Year Twig: brown; gray; slender

CULTIVARS AND SPECIES:

'Pendula' has a fine-textured, weeping form reminiscent of a water garden. Plant it near water for a serene effect.

LANDSCAPE NOTES:

Origin: Asia

Usage: specimen; cut branches for florist trade

A small to medium-sized, upright spreading tree, the main ornamental feature of Corkscrew Willow is the contorted and twisted branches and twigs. Branches arise from the trunk at an acute angle and grow up almost parallel to the trunk before they recurve back to the horizontal. The winter branch pattern is most interesting and probably accounts for the popularity of the tree.

The tree makes a nice accent or specimen in a residential or commercial landscape, but does not have the long life needed for a permanent street tree. The tree has the same undesirable features as other Willows, including weak wood, included bark, and insect problems. Weak branch crotches can make the tree fairly prone to breakage as it grows older. Also very prone to trunk and branch decay initiated by mechanical injuries.

Like many Willows, this one does best in moist areas but will tolerate some drought for a short period. Surface roots can be a problem as the tree grows older. Locate carefully, as it is usually short-lived with aggressive roots and breaks apart as it gets older. Not for general planting, but can be used for an occasional specimen. It actually makes quite a nice tree when located on a moist site and pruned to create good structure by eliminating or thinning branches with included bark. Regular pruning is recommended because large pruning wounds can initiate decay, which spreads rapidly inside the tree. The canopy should be thinned to help wind pass through.

Potential problems include aphids, lace bugs, leaf beetles, and borers. Crown gall, leaf spots, canker, and tar spot diseases can be a problem, especially on drier sites. The tree is considered messy because something always seems to be dropping from it.

Sambucus canadensis American Elder

Pronunciation: sam-BEW-kuss kan-ah-DEN-sis

ZONES: 4A, 4B, 5A, 5B, 6A, 6B, 7A, 7B, 8A, 8B, 9A, 9B, 10A, 10B

HEIGHT: 8 to 12 feet

WIDTH: 6 to 10 feet

FRUIT: round; 0.25"; fleshy; purple; attracts wildlife; edible; no significant litter

GROWTH RATE: rapid; short-lived

HABIT: rounded, upright vase; open; irregular silhouette; fine texture

FAMILY: *Caprifoliaceae*

LEAF COLOR: bright green; greenish-yellow fall color

LEAF DESCRIPTION: opposite; odd pinnately compound; serrate margin; oval to lanceolate; deciduous leaflets; 2 to 4" long

FLOWER DESCRIPTION: white; showy; summer

EXPOSURE/CULTURE:

Light Requirement: part shade to full sun

Soil Tolerances: all textures; alkaline to acidic; wet soil; some salt and drought

Aerosol Salt Tolerance: low

PEST PROBLEMS: mostly resistant

BARK/STEMS:

Bark is thin and easily damaged by mechanical impact; branches droop as tree grows; trainable to be grown with a single or multiple trunk; suckers regularly form at the base of the trunk; no thorns

Pruning Requirement: requires pruning to develop upright tree form; head back and remove branches to increase strength and build stronger structure

Limb Breakage: susceptible to breakage at crotch, due to poor collar formation; wood itself is weak and tends to break

Current Year Twig: gray; thick; warty

CULTIVARS AND SPECIES:

'Acutiloba'—leaflets very deeply divided, nice fine-textured plant; 'Aurea'—bright red fruit, yellow leaves; 'Adams'—fruits in dense, large clusters, excellent for baking. There are a variety of other, very attractive species.

LANDSCAPE NOTES:

Origin: eastern North America

Usage: container or above-ground planter; near a deck or patio; wetland reclamation; trainable as a standard; specimen

A fast-growing deciduous shrub, American Elder suckers quite easily and is often seen as a broad, spreading, multi-stemmed plant with long leaves arranged along arching branches. It can be effectively pruned into a nice, single- or multi-stemmed, small, flowering tree, but needs regular pruning to remove suckers growing from the base of the plant. In early summer (northern part of its range) or sporadically all year long (in zones 9 and 10), American Elder is smothered with 6- to 10-inch-wide clusters of yellowish-white blooms. These are followed by a multitude of small, dark purple berries that are quite popular with birds, and can be used in pies and jellies, or fermented to make a wine. The southern version of this plant is sometimes referred to as *S. simpsonii.*

Ideal for use in naturalized landscapes where it will tolerate acid or alkaline soil and even some drought, American Elder performs best in full sun on moist to wet, fertile soils. Plant it in a shrub border or locate it next to the patio for a wonderful flower display. It makes a nice small tree for planting in a container. The plant is often overlooked by the trade, perhaps because it is so commonly found in and along the woods, but it has a place in the garden, although its rather random habit may not make it popular in the commercial landscape. Requires regular pruning to create a neat small tree.

Borers and occasional leaf-chewing insects infest Elder. Cankers, leaf spots, and powdery mildew can also occur.

Sambucus mexicana Mexican Elderberry

Pronunciation: sam-BEW-kuss mex-eh-KAY-nah

ZONES: 7B, 8A, 8B, 9A, 9B, 10A, 10B

HEIGHT: 15 to 25 feet

WIDTH: 25 to 35 feet

FRUIT: round; 0.33"; fleshy; black with white coating that rubs off; attracts birds; edible; causes some litter; persistent until eaten by birds; showy

GROWTH RATE: rapid; moderate life span

HABIT: spreading and round, almost weeping; moderate density; symmetrical; fine texture

FAMILY: *Caprifoliaceae*

LEAF COLOR: light green; no fall color change

LEAF DESCRIPTION: opposite; odd pinnately compound; serrate margin; oval to oblong or ovate; semi-evergreen; sheds when soil dries; 2 to 4" long leaflets

FLOWER DESCRIPTION: white; showy; late spring and fall

EXPOSURE/CULTURE:

Light Requirement: tree grows in full sun

Soil Tolerances: all textures; alkaline to acidic; drought

PEST PROBLEMS: resistant

BARK/STEMS:

Showy, peeling bark; branches droop as tree grows and will require pruning for vehicular or pedestrian clearance beneath canopy; routinely grown, or trainable to be grown, with multiple trunks; can be trained to a short, single trunk; no thorns

Pruning Requirement: requires pruning to develop strong structure and a neat habit, but worth the effort

Limb Breakage: brittle branches

Current Year Twig: brown; green; thick

LANDSCAPE NOTES:

Origin: California to Texas and Mexico

Usage: buffer strip; highway; near a deck or patio; reclamation; specimen

Mexican Elder is a semi-evergreen small tree, reaching up to 30 feet in height. Most open-grown specimens do not reach this height, as they grow wider than tall. The thick, leathery, pinnately compound leaves are often browsed by deer and livestock. The four- to eight-inch-wide, flat cymes of yellow-white blooms appear mainly from April to June, but in its native habitat may also occur at various times throughout the year after heavy rains. The small, blue-black fruits that follow the blooms are quite popular with birds and can be used to make wine or pies, and were reportedly even dried by Native Americans for later use. Fruits appear white because they are covered with a thick, waxy coating. A dye can be made from the stems and has been used to color baskets.

Trees are best located in the open where they can develop their low, wide-spreading, uniform crown. This tree might be planted near a patio, or as a small shade tree in a backyard garden, but the dropping fruit can be messy on brick, concrete, and other hard surfaces. Young trees often require training and pruning to speed formation of a uniform crown. Save the more upright branches and remove the drooping ones to create greater clearance beneath the canopy. It is well suited for planting in a highway median or other low-maintenance area where little, if any, care is available after planting.

Mexican Elder should be grown in full sun on well-drained soil. The tree apparently avoids drought by dropping leaves in dry soil. Irrigation in dry weather prevents this.

The wood is brittle. Remove the root suckers as they form from the base of the trunk and roots for a neater habit.

Sapindus drummondii Western Soapberry

Pronunciation: SAP-in-duss drum-MON-dee-eye

ZONES: 6A, 6B, 7A, 7B, 8A, 8B, 9A, 9B

HEIGHT: 40 to 50 feet

WIDTH: 40 to 50 feet

FRUIT: round; 0.5"; hard; orange-yellow turning black; attracts birds; causes significant litter; persistent; very showy; used for soap; poisonous

GROWTH RATE: moderate; moderate life span

HABIT: rounded vase, nearly weeping; open; symmetrical; medium to fine texture

FAMILY: *Sapindaceae*

LEAF COLOR: green; showy, yellow fall color

LEAF DESCRIPTION: alternate; even pinnately compound; entire margin; lanceolate to oblong leaflets; deciduous; 2 to 4" long leaflets

FLOWER DESCRIPTION: whitish-yellow; showy; spring

EXPOSURE/CULTURE:

Light Requirement: part shade to full sun

Soil Tolerances: all textures; alkaline to acidic; drought

Aerosol Salt Tolerance: none

PEST PROBLEMS: resistant

BARK/STEMS:

Branches on young trees grow mostly upright and will not droop until older; should be grown with a single leader; no thorns

Pruning Requirement: needs some pruning to develop strong structure; space branches along a dominant trunk

Limb Breakage: resistant

Current Year Twig: brown; gray; medium thickness

LANDSCAPE NOTES:

Origin: south central North America into Mexico

Usage: buffer strip; highway; reclamation; shade tree; specimen; sidewalk cutout; street tree; urban tolerant

Western Soapberry, a native of central and western Texas, is an excellent shade or ornamental tree, reaching mature height with an equal spread and forming a billowy, deciduous crown. The small, springtime blooms appear in 6- to 10-inch terminal panicles and are followed on female plants by translucent, yellow-orange, half-inch, grape-like, clustered fruits, which persist through the fall, eventually ripening to black. The common name is derived from the fact that the fruits, when crushed in water, create great quantities of suds; they were used by American and Mexican native peoples as laundry soap, floor wax, and varnish. The fruits are poisonous to humans.

Fruit drops while it is firm and does not rot to create a mess, but people could slip on it and fall. The abundant fruits may create an unwelcome invasion of seedling volunteers, which germinate rapidly in the landscape. Plant it in a lawn area where regular mowings prevent seedlings from developing. Due to the risk of dermatitis and possible poisoning from the fruit, use Western Soapberry as an ornamental or median-strip street tree away from where children could contact the fruit.

Western Soapberry is particularly well suited to urban conditions, tolerating wind, drought, and infertile soils with ease. It transplants easily and establishes with only minimal irrigation. The close-grained, strong wood makes this tree very resistant to wind damage and adaptable to urban landscapes provided the tree is pruned to prevent formation of included bark. Excellent for poor, compacted, or alkaline soil. Root suckers can be a problem in sandy soil but apparently not in clay. There may be a fruitless cultivar, which would make Soapberry suitable for much broader usage, including downtown streets and patios.

Dwarf mistletoe can be quite a problem. Fruit attracts the boxelder bug and is a maintenance nuisance near walks and patios.

Sapindus saponaria Soapberry

Pronunciation: SAP-in-duss sah-poe-NAIR-ee-ah

ZONES: 10A, 10B, 11

HEIGHT: 30 to 40 feet

WIDTH: 25 to 35 feet

FRUIT: round; 0.5"; fleshy; brown to orange; causes some litter; showy

GROWTH RATE: moderate, moderate life span

HABIT: oval to rounded; dense; symmetrical; moderate texture

FAMILY: *Sapindaceae*

LEAF COLOR: dark green; evergreen

LEAF DESCRIPTION: alternate; odd pinnately compound; 2 to 4" long leaflets

FLOWER DESCRIPTION: white; not showy; spring

EXPOSURE/CULTURE:

Light Requirement: part shade to full sun

Soil Tolerances: all textures; alkaline to acidic; salt and drought

Aerosol Salt Tolerance: high

PEST PROBLEMS: pest-free

BARK/STEMS:

Branches droop as tree grows and will require early pruning for vehicular or pedestrian clearance beneath canopy; should be grown with a single leader; no thorns

Pruning Requirement: requires pruning to develop strong structure; space branches along trunk

Limb Breakage: resistant

Current Year Twig: brown; gray; medium thickness

LANDSCAPE NOTES:

Origin: tropical America

Usage: buffer strip; highway; reclamation; shade tree; street tree

Florida Soapberry grows at a moderate rate to mature height. The pinnately compound, evergreen leaves are 12 inches long, with each leaflet 4 inches long. Ten-inch-long panicles of small, white flowers appear during fall, winter, and spring but are fairly inconspicuous. The fleshy fruits that follow are less than an inch long, shiny, and orange-brown. The seeds inside are poisonous, a fact to be considered in the tree's placement in the landscape, especially if children will be present. The bark is rough and gray, forming a picturesque silhouette. The common name of Soapberry comes from the soap-like material made from the berries in tropical countries.

Soapberry can be planted in low-maintenance landscapes for the unusually prominent, orange-colored berries. They require little care other than some initial pruning to direct growth and to develop strong branch structure by spacing branches 6 to 12 inches apart. One central leader can be created by heading back or drop-crotching young lateral branches to reduce their growth rate. This forces growth into the upper canopy and forms a taller tree more suited for urban landscapes, especially if clearance is required beneath the tree for vehicles or pedestrians. The bushy growth habit of this medium-sized tree combines nicely with shrubs and groundcovers planted beneath and around the tree.

Florida Soapberry should be grown in full sun and will tolerate almost any soil. It is highly drought- and salt-tolerant. Rainfall alone can usually maintain a healthy tree after it is well established. No limitations of major concern.

Sapium sebiferum Chinese Tallowtree

Pronunciation: SAY-pee-um seh-BIFF-err-um

ZONES: 8A, 8B, 9A, 9B, 10A, 10B

HEIGHT: 30 to 35 feet

WIDTH: 25 to 30 feet

FRUIT: round; 0.5"; dry and hard; brown, opening to white; causes some litter; persistent; showy; germinates in the landscape

GROWTH RATE: rapid-growing; short-lived

HABIT: oval; round; moderate density; irregular silhouette; medium texture

FAMILY: *Euphorbiaceae*

LEAF COLOR: light green; very striking and dependable orange to red or yellow fall color

LEAF DESCRIPTION: alternate; simple; entire margin; deltoid to ovate or rhomboid; deciduous; 2 to 4" long

FLOWER DESCRIPTION: yellow; not especially showy; spring; attracts bees

EXPOSURE/CULTURE:

Light Requirement: grows best in full sun

Soil Tolerances: all textures; alkaline to acidic; wet soil; moderate drought

Aerosol Salt Tolerance: low

PEST PROBLEMS: nearly pest-free

BARK/STEMS:

Some branches droop slightly as tree grows and will require some pruning for vehicular or pedestrian clearance beneath canopy; should be grown with a single leader; no thorns

Pruning Requirement: needs little pruning to develop interesting branch structure

Limb Breakage: susceptible to breakage at crotch; wood itself is weak and tends to break

Current Year Twig: green; slender

LANDSCAPE NOTES:

Origin: China, Japan

Usage: banned in Florida and in other communities due to invasive habit; specimen; buffer strip; highway median; urban tolerant

With oval, pointed, deciduous leaves and an oval, open canopy, Chinese Tallowtree creates soft, dappled shade, making it ideal for use as a shade tree or lawn specimen. The trunk normally dominates, snaking up through a crown sporting major limbs well spaced along the trunk. Enough light will penetrate to allow lawn grasses to thrive beneath. Brown fruit capsules burst and fall off, leaving behind wax-coated, white, berrylike seeds (hence the common name, Popcorn tree). These berries persist throughout the winter, even after the fluttering, heart-shaped leaves have turned gorgeous autumn shades of red, yellow, or orange and have fallen. Tallowtree is one of the only reliable fall-coloring trees for zones 8B and 9A. The waxy coating on the seeds is extracted by the Chinese for use in candles and soap. The milky sap inside the twigs is poisonous. New growth in spring is red-tinged.

Chinese Tallowtree is easily grown in full sun on a wide range of soils and is particularly drought-resistant and tolerant of compacted and wet soil. The abundant seeds usually create a multitude of unwanted volunteer seedlings. Roots tend to grow quite large near the soil surface and can be a nuisance in the lawn. There are places in Florida and in coastal Texas where the tree has escaped cultivation and is invading native woodlands and the edge of wetlands, so use it with caution, if at all. The wood is brittle, and small to medium-sized branches often split from the tree as it grows to 15 years old. But it is an urban-tough tree with a proven ability to thrive in confined soil spaces such as sidewalk cutouts and highway medians. If you plant this tree, do so in areas with little exposed soil close by to reduce the chance of the tree escaping cultivation. Seeds are carried by birds and can be seen growing quite a distance from the original tree. Caution should be used when considering planting this invasive tree.

Sassafras albidum Sassafras

Pronunciation: SASS-ah-frass AL-beh-dum

ZONES: 5A, 5B, 6A, 6B, 7A, 7B, 8A, 8B, 9A

HEIGHT: 30 to 60 feet

WIDTH: 25 to 40 feet

FRUIT: round; 0.5"; fleshy; blue; attracts wildlife; causes some litter; showy

GROWTH RATE: moderate; long-lived

HABIT: pyramidal when young, oval later; dense; irregular silhouette; medium texture

FAMILY: *Lauraceae*

LEAF COLOR: bright green; stunning orange, red, or yellow fall color, sometimes on same tree

LEAF DESCRIPTION: alternate; simple; lobed; entire margin; oval to ovate; deciduous; 2 to 4" long

FLOWER DESCRIPTION: yellow; pleasant fragrance; showy; spring

EXPOSURE/CULTURE:

Light Requirement: part shade to full sun

Soil Tolerances: all textures; acidic; wet soil; moderate drought

Aerosol Salt Tolerance: some

PEST PROBLEMS: nearly pest-free

BARK/STEMS:

Routinely grown, or trainable to be grown, with multiple trunks; branches grow mostly upright and will not droop; can be trained to grow with a single trunk; no thorns

Pruning Requirement: requires some pruning to develop straight trunk; develop multiple trunks for picturesque growth habit

Limb Breakage: resistant

Current Year Twig: green; medium thickness

LANDSCAPE NOTES:

Origin: Maine to Michigan, to Texas and north Florida

Usage: parking lot island; buffer strip; highway; near a deck or patio; shade tree; specimen; street tree

This lovely, deciduous, native North American tree is pyramidal when young, but later develops into a rounded canopy composed of many short, horizontal branches that give the tree a layered effect. For years, Sassafras was grown for the supposedly medicinal properties of the fragrant roots and bark, but it is the outstanding fall display of foliage that should bring it into the garden today. The wood was used extensively for building boats during the 1920s and 1930s. The large, multi-formed, five-inch leaves, fragrant when crushed, are bright green throughout the summer but are transformed into magical shades of orange-pink, yellow-red, and even scarlet-purple in the cooler months of autumn, brightening the landscape wherever they are found. In spring, before the leaves appear, the yellow, lightly fragrant flowers of Sassafras appear in one- to two-inch-long terminal panicles and are followed by extremely attractive, blue fruits.

Sassafras frequently develops a multiple trunk due to sprouting at the base. Sprouts appear to originate from the root system, forming a cluster of showy, gray, fissured trunks growing from the soil. This characteristic has helped it invade and colonize old fields and other disturbed sites. Prune early in the life of the tree to form a single trunk suitable for urban landscape planting, or grow with multiple trunks for a dramatic specimen. Single-trunked trees are best suited for street tree planting and other urban and suburban areas, and they usually maintain this good form without pruning.

Richer fall colors are displayed on trees grown in full sun. Recovers slowly from transplanting due to coarse root system, making this a candidate for production in copper-treated containers.

Can be bothered by Japanese beetle, promethea moth, sassafras weevil, and scales. Cankers, leaf spots, mildew, wilt, and root rot also have been reported.

Schefflera actinophylla Schefflera

Pronunciation: sheff-LAIR-ah ack-tin-oh-FYE-lah

ZONES: 10A, 10B, 11

HEIGHT: 30 feet

WIDTH: 10 to 15 feet

FRUIT: round; 0.5"; fleshy; purple or red; no significant litter; persistent; showy; sprouts in the landscape

GROWTH RATE: rapid; short-lived

HABIT: upright; moderate density; irregular silhouette; coarse texture

FAMILY: *Araliaceae*

LEAF COLOR: broadleaf evergreen; dark green

LEAF DESCRIPTION: spiral arrangement; palmate; entire or undulate margin; oblong to ovate; 6 to 10" long leaflets; litter ground and decompose slowly

FLOWER DESCRIPTION: red; showy; summer

EXPOSURE/CULTURE:

Light Requirement: part shade to full sun

Soil Tolerances: all textures; alkaline to acidic; wet soil; some salt; moderate drought

Aerosol Salt Tolerance: low

PEST PROBLEMS: somewhat sensitive

BARK/STEMS:

Bark is thin and easily damaged by mechanical impact; routinely grown with multiple trunks; branches grow mostly upright and will not droop; no thorns

Pruning Requirement: requires pruning to develop strong structure

Limb Breakage: susceptible to breakage at crotch, due to poor collar formation; wood itself is weak and tends to break

Current Year Twig: green; stout

CULTIVARS AND SPECIES:

This tree was classified as *Brassaia* until recently. *S. arboricola* grows to about 12 feet, has smaller leaves, and is often used as a hedge or interior container plant.

LANDSCAPE NOTES:

Origin: Australia

Usage: container or above-ground planter; near a deck or patio; specimen; sidewalk cutout; interiorscape; house plant

The large, palmately compound, shiny leaves sit atop the multiple, thin, bare trunks of Schefflera, creating the impression of an exotic, 25-foot-tall plant-umbrella. Schefflera lends a tropical effect to any landscape use, from patio containers to interiorscapes to protected outdoor locations. Capable of reaching 40 feet in height, Schefflera will grow rapidly to create a dense windbreak or screen for property lines. When grown in full sun, trees will produce flowers during the summer, an unusual arrangement of small blooms on three-foot-diameter, stiff terminal clusters. These clusters are held above the foliage and are arranged like the ribs of an inverted umbrella, or like the tentacles of an octopus. The red blooms are followed by reddish-purple, half-inch fruits.

Schefflera will grow in full sun or partial shade on a wide variety of well-drained soils but requires full sun to flower. Trees will display their best growth on rich, moist soil in a full-sun location. There is significant leaf drop on this easily grown tree, creating quite a raking job, but plants will require very little pruning if given enough overhead space to develop. Trees may be topped as desired to create multi-level masses of foliage. This may be desirable, as the lower portions of the trunks lose all their foliage over time, but initiates decay in the trunk. Sometimes the tree is used as a house plant, but it is too often misused by planting it outside too close to a building.

It has naturalized in southern Florida and Hawaii and has been placed on a list of exotic pest plants. The roots are superficial and invasive into gardens and lawns.

No limitations of major concern. Scales and sooty mold are a minor problem. Trees used indoors are susceptible to infestations of spider mites. Alternaria leaf spot can be serious.

Schinus molle California Pepper Tree

Pronunciation: SHY-nuss MOLL-ee

ZONES: 8B, 9A, 9B, 10A, 10B, 11

HEIGHT: 30 to 45 feet

WIDTH: 30 to 45 feet

FRUIT: round in drooping clusters; 0.25"; fleshy; bright red; causes significant litter

GROWTH RATE: moderately rapid; long-lived

HABIT: weeping and spreading, upright oval; moderate density; nearly symmetrical; very fine texture

FAMILY: *Anacardiaceae*

LEAF COLOR: evergreen; unusually light green and delicate

LEAF DESCRIPTION: pinnately compound; 1.5 to 2" long leaflets; entire margin

FLOWER DESCRIPTION: yellowish-white; somewhat showy; spring

EXPOSURE/CULTURE:

Light Requirement: part shade to full sun

Soil Tolerances: all textures; slightly alkaline to acidic; drought

PEST PROBLEMS: resistant

BARK/STEMS:

Gnarled trunk with natural burls; main branches are upright, but twigs droop as tree grows and will require some pruning for clearance beneath canopy; should be grown with a single leader as high up into canopy as possible; no thorns

Pruning Requirement: requires some early pruning to develop strong structure; slow growth rate on aggressive branches to prevent formation of included bark to generate a dominant trunk; thin canopy to increase wind firmness

Limb Breakage: resistant

Current Year Twig: gray; very slender

LANDSCAPE NOTES:

Origin: Peru

Usage: in California and Arizona, buffer strip; highway; shade tree

A moderately large tree, California Pepper is more upright than its close relative, the Brazilian Pepper. Main branches can be trained to grow up and out from the main trunk, but small-diameter twigs and branches weep toward the ground. This gives the tree a decided weeping habit. Combined with rapid growth, this makes the tree popular in many landscapes throughout California. The tree is large enough to provide shade for large areas of the landscape. An amazingly durable tree, it is well suited for life in the desert. Give it plenty of room, as the trunk can grow to more than four feet in diameter.

Once established, this tree will thrive in almost any location, but should be saved for sites with plenty of soil for root expansion. Planted too close, the roots and trunk can eventually engulf and ruin a sidewalk or curb. Surface roots are large and aggressive. Keep it at least 20 feet away from the house or any other structure that you would like to remain intact. The canopy becomes dense and will benefit from occasional thinning. It is a tough, enduring tree that will respond with vigorous growth to plentiful moisture on well-drained soil.

Provide plenty of irrigation until the tree is established. Once it is well established, it will survive and grow with only occasional irrigation in the summer. Develop a strong branch structure early in the life of the tree by pruning to prevent included bark in the main branch crotches. Be sure to eliminate multiple trunks and branches that form a narrow angle with the trunk, as these can split from the tree as it grows older. Cut back to lateral branches competing trunks and upright branches to form a dominant trunk.

Powdery mildew deforms foliage on rapid-growing trees.

Schinus terebinthifolius Brazilian Pepper

Pronunciation: SHY-nuss ter-ah-bin-tha-FOLL-ee-us

ZONES: 9B, 10A, 10B, 11

HEIGHT: 12 to 15 feet

WIDTH: 20 to 30 feet

FRUIT: round in stiff clusters; 0.25"; fleshy; very showy, bright red; causes significant litter; causes allergy in some people

GROWTH RATE: rapid; moderate life span

HABIT: rounded and spreading; moderate density; nearly symmetrical; medium texture

FAMILY: *Anacardiaceae*

LEAF COLOR: evergreen; glossy, dark green

LEAF DESCRIPTION: pinnately compound; 2 to 3" long leaflets; entire or serrate margin

FLOWER DESCRIPTION: yellowish-white; somewhat showy; spring

EXPOSURE/CULTURE:

Light Requirement: full sun

Soil Tolerances: all textures; slightly alkaline to acidic; salt; wet soil; moderate drought

PEST PROBLEMS: resistant

BARK/STEMS:

Branches droop as tree grows and will require pruning when tree is young for clearance beneath canopy; can be trained to several trunks or one, short trunk; no thorns

Pruning Requirement: requires pruning to develop one trunk and upright structure; thin canopy to increase wind firmness and reduce branchiness

Limb Breakage: susceptible

Current Year Twig: gray; stout

LANDSCAPE NOTES:

Origin: not native

Usage: in southwestern United States, buffer strip; highway; near a deck or patio

Brazilian Pepper is used as an ornamental in California, Arizona, and Hawaii, yet is banned in Florida due to its invasive habit. It makes a nice ornamental tree when the canopy is thinned to display the attractive bark and picturesque form. Left to its own, the tree becomes a sprawling mess, drooping to the ground like a large shrub. When it is staked and trained to a single trunk six to eight feet tall, the tree creates enough of a canopy to be used along walks or near a deck or patio. Leaves reportedly contain allelopathic substances that inhibit growth of nearby plants.

The canopy becomes dense and will benefit from occasional thinning. Train main branches to grow up and away from the canopy. Select branches with a wide angle to the trunk for training into main limbs. If one becomes too aggressive, drop-crotch it back to a lateral branch to slow the growth rate. This will help ensure that it remains well attached to the trunk.

An amazingly adaptable tree, it is well suited for life in the desert. Provide plenty of irrigation until the tree is established. Once it is well established, it will survive in almost any location and grow with only occasional irrigation in the summer. Pepper will respond with vigorous growth to plentiful moisture on well-drained soil. It also tolerates wet soil fairly well. It seeds itself into the landscape and can become a pesty weed. Surface roots often form in the lawn and garden.

Powdery mildew deforms foliage on rapid-growing trees. The tree is banned in Florida because it has become a weed, but is used in desert southwest United States landscapes.

Sciadopitys verticillata Japanese Umbrella-Pine

Pronunciation: sigh-ah-DOP-ih-tiss vurr-tih-sill-LAY-tah

ZONES: 5A, 5B, 6A, 6B, 7A, 7B, 8A

HEIGHT: 25 to 40 feet

WIDTH: 15 to 20 feet

FRUIT: oval; 2 to 4"; dry; brown; not showy; not common; no significant litter

GROWTH RATE: slow-growing; long-lived

HABIT: pyramidal; moderate density; symmetrical; fine texture

FAMILY: *Taxodiaceae*

LEAF COLOR: needle-leaf evergreen; dark green

LEAF DESCRIPTION: whorled; simple; entire margin; linear; 2 to 5" long

FLOWER DESCRIPTION: not showy

EXPOSURE/CULTURE:

Light Requirement: shaded to full sun

Soil Tolerances: loam; sand; acidic; requires good drainage; moderate drought

PEST PROBLEMS: currently pest-free

BARK/STEMS:

Trunk is showy, eventually; branches droop as tree grows and will require pruning for vehicular or pedestrian clearance beneath canopy; normally grows with a single leader; no thorns

Pruning Requirement: needs little or no pruning to develop strong structure

Limb Breakage: resistant

Current Year Twig: green; thick

CULTIVARS AND SPECIES:

'Pendula' is listed as a cultivar with pendulous branches. Very rare, even in the best arboreta. 'Wintergreen' may retain greener foliage in winter and could be hardier than the species.

LANDSCAPE NOTES:

Origin: Hondo

Usage: container or above-ground planter; specimen; screen

This small tree is used for its unusual texture and growth habit. The tree will grow very slowly to as much as 40 feet tall (in about 60 years) and spreads about 15 feet. On young plants the branches stick straight out from the single, straight trunk, but become more pendulous with age. The tree grows in a tight pyramid that can be utilized in a rock garden or other small residential landscape site. The foliage is borne on the twigs much like the ribs on an umbrella. Due to the unusual nature of the plant, people either love this tree or will not even look at it.

A bit of shade during the hot part of the day will produce the best plants in the southern part of the range. The orange, peeling bark on older trees is quite attractive but is usually hidden by the foliage. Provide a moist, acidic soil and protection from the wind. The tree is probably best used as a specimen, occasionally for private, protected gardens for a very special effect. A truly unique tree, it is rare in cultivation. Only the most patient nursery operators grow this plant. No serious limitations.

Senna spectabilis (excelsa) Cassia

Pronunciation: SEN-nah speck-TAB-ih-liss

ZONES: 10B, 11

HEIGHT: 15 to 20 feet

WIDTH: 15 to 20 feet

FRUIT: elongated; pod; 6 to 12"; dry; brown; not showy; no significant litter; persistent

GROWTH RATE: moderately rapid; moderate life span

HABIT: rounded vase; dense; symmetrical; fine texture

FAMILY: *Leguminosae*

LEAF COLOR: evergreen; medium green

LEAF DESCRIPTION: alternate; odd pinnately compound; entire margin; oblong to obovate leaflets; 2 to 3" long leaflets

FLOWER DESCRIPTION: yellow; summer and fall; very showy

EXPOSURE/CULTURE:

Light Requirement: flowers best in full sun

Soil Tolerances: all textures; slightly alkaline to acidic; moderate drought

Aerosol Salt Tolerance: low

PEST PROBLEMS: resistant

BARK/STEMS:

Bark is thin and easily damaged by mechanical impact; branches droop as tree grows, and will require pruning for vehicular or pedestrian clearance beneath canopy; routinely grown, or trainable to be grown, with multiple trunks; can be trained to grow with a single trunk; no thorns

Pruning Requirement: requires pruning to develop strong structure and a single leader

Limb Breakage: susceptible to breakage at crotch, due to poor collar formation; wood itself is weak and tends to break

Current Year Twig: green; medium thickness

LANDSCAPE NOTES:

Origin: Brazil

Usage: container or above-ground planter; parking lot island; buffer strip; highway; near a deck or patio; standard; specimen

Cassia, a tree from tropical America, can reach 60 feet in height in its native habitat, but is often much smaller. The pinnately compound leaflets have fuzzy undersides and are three inches long. The bright yellow flowers are 1.5 inches wide but appear in dense racemes up to 2 feet long. The cylindrical seedpods that follow are 12 inches long. The tree is a sight for sore eyes when it is in flower.

This tree is best used in an open, sunny, park-like setting where the bright flowers can be displayed and enjoyed. It is best viewed from a distance to see the flowers, which are displayed at the outer portion of the canopy. It is not quite as dramatic when viewed from under the canopy. The tree forms a large mass of delicate foliage covered with yellow flowers for about two months each year.

Lower branches often reach the ground as they droop under the weight of the flowers. These can be removed to create clearance beneath the tree for pedestrians and vehicles. This would make the tree suited for planting along streets provided major branches were developed with good attachments to the trunk. Unpruned trees branch poorly, and large-diameter limbs often develop, forming a coarse-branching structure. Regularly heading back young lateral branches as they develop from the trunk on small trees helps to create more branching and a more uniformly shaped crown. This also helps to build the diameter of the larger, drooping branches, which will help keep them erect. Occasional pruning during the life of the tree will help maintain a regular shape.

Cassia should be grown in full sun on well-drained soil. It appears to adapt to alkaline soil. No limitations of major concern.

Sequoia sempervirens Coast Redwood

Pronunciation: see-KWOY-yah sem-per-VEE-rens

ZONES: 7A, 7B, (8A), (8B), (9A), (9B), (10A), (10B)

HEIGHT: 60 to 120 feet

WIDTH: 25 to 35 feet

FRUIT: oval to round; 0.75 to 1"; dry and hard; brown; not showy; no significant litter

GROWTH RATE: moderate to rapid; long-lived

HABIT: variable, depending on seed source; pyramidal; moderate density; symmetrical; fine texture

FAMILY: *Taxodiaceae*

LEAF COLOR: evergreen; dark green

LEAF DESCRIPTION: spiral; simple; entire; needle-like (filiform); 0.5 to 1" long

FLOWER DESCRIPTION: not showy

EXPOSURE/CULTURE:

Light Requirement: part shade to full sun

Soil Tolerances: all textures; slightly alkaline to acidic; wet soil; drought-sensitive

PEST PROBLEMS: nearly pest-free (accounts for its long life)

BARK/STEMS:

Branches droop; showy reddish-brown trunk; usually grows with a single leader; no thorns

Pruning Requirement: needs little or no pruning to develop strong trunk and branch structure

Limb Breakage: resistant

Current Year Twig: brown; green; medium to slender

CULTIVARS AND SPECIES:

A number of cultivars have been introduced that are more tolerant of drier conditions. Some, including 'Santa Cruz', are remarkably tolerant of desert conditions, provided they are regularly irrigated. 'Monty' is densely branched with blue-green foliage; 'Sequel' is tightly symmetrical with horizontal branches turned up slightly at the tip. *Sequiodendron giganteum*, Giant Sequoia, grows at high elevations in the Sierra Nevada Mountains and tolerates drier conditions.

LANDSCAPE NOTES:

Origin: northwestern California into Oregon

Usage: specimen; buffer strip; screen; windbreak

Sequoia sempervirens, the Coast Redwoods of California, are among the tallest trees in the world. Redwood is the state tree of California. They can vary greatly when grown from seed, but cultivated varieties are available now which have been vegetatively propagated, and they retain true characteristics. Redwoods grow three to five feet per year and are remarkably pest-free. They live to be many hundreds of years old; some live to several thousand years in their native habitat. Bark is particularly beautiful, turning a bright orange on older trees. Redwood maintains a pyramidal form and dark green foliage throughout the year.

Redwood is tolerant of flooding, growing best along stream banks and flood plains. When a layer of soil is deposited over the root system, the tree simply regenerates a new one from the trunk. Irrigation helps maintain a vigorous tree in other sites. Allow plenty of soil space for proper development. Planted in a row 15 to 20 feet apart, they make a nice screen. In areas outside California and the Northwest, Redwood is probably best used occasionally as a novelty specimen, and may grow poorly in the southeast.

A twig canker can become damaging in nurseries. Fasciation, or flattening of the stem, is occasionally seen, but the cause is unknown. *Sequoia sempervirens* is resistant to oak root fungus. Needles and branchlets often turn brown and fall off; this is not a disease but is natural. Irrigate the tree to help prevent this, if you wish. Galls occasionally form but are of no concern.

Sequoiadendron giganteum Giant Sequoia

Pronunciation: see-KWOY-yah-den-dron jie-GAN-tea-um

ZONES: 6A, 6B, 7A, 7B, (8A), (8B), (9A), (9B), (10A), (10B)

HEIGHT: 60 to 120 feet

WIDTH: 25 to 30 feet

FRUIT: oval to round; 1.5 to 3"; dry and hard; brown; not showy; no significant litter

GROWTH RATE: moderate; long-lived

HABIT: pyramidal or columnar; dense; symmetrical; fine texture

FAMILY: *Taxodiaceae*

LEAF COLOR: bluish-green

LEAF DESCRIPTION: spiral; awl-shaped; less than 0.5" long

FLOWER DESCRIPTION: not showy

EXPOSURE/CULTURE:

Light Requirement: full sun

Soil Tolerances: all textures; slightly alkaline to acidic; drought

PEST PROBLEMS: resistant within its natural range, sensitive outside

BARK/STEMS:

Very showy, orange-brown furrowed bark; branches stay more upright than Coast Redwood; usually grows with a single leader; no thorns

Pruning Requirement: needs little or no pruning to develop strong structure

Limb Breakage: resistant

Current Year Twig: brown; green; medium to slender

CULTIVARS AND SPECIES:

'Pendulum' is the only listed cultivar. It must be staked to the desired height; then it is allowed to droop in a random fashion. Its different appearance provides an exotic feel to any landscape.

LANDSCAPE NOTES:

Origin: Sierra Nevada Mountains in California, 4,500 to 8,500 feet elevation

Usage: specimen; buffer strip; screen; windbreak; possibly a Christmas tree

Arguably, the Giant Sequoias are the largest living organisms in the world. The largest of these giants grows nearly 300 feet tall in Sequoia National Park in California—its trunk is more than 30 feet in diameter! Giant Sequoia grows scattered about in the woods of the native range, not in pure stands, although there are several small groves comprised mostly of Sequoia. The bright orange-brown trunks stand out in the woods among neighboring conifers. Sequoias grow two to three feet per year and are usually pest-free. They live to be many hundreds of years old; some live to several thousand years in their native habitat. They are considered tolerant of drought, receiving little rain between June and October in their native habitat. Most of the 35 to 55 inches of precipitation comes in the form of snow.

Although the tree will not grow more than 60 to 100 feet tall in most landscape sites, trees do get large. Allow plenty of soil space for proper development of the trunk and root system. Planted in a row 15 to 20 feet apart, they make a nice screen or windbreak, even in the eastern United States. Sequoia has good potential use as a lumber tree around the world.

Whereas Redwood is tolerant of flooding, making best growth along stream banks, Giant Sequoia grows on mountain slopes and protected coves. It responds well to irrigation, quickly developing into a nice landscape tree. There are no significant problems within its natural range, but in other places, some insects and diseases can kill large numbers of trees.

Simarouba glauca Paradise-Tree

Pronunciation: sim-mah-ROO-bah GLAW-cah

ZONES: 10B, 11

HEIGHT: 30 to 50 feet

WIDTH: 30 feet

FRUIT: elongated; oval; 1"; fleshy; purple; edible; attracts birds; not showy; causes significant litter

GROWTH RATE: moderate; moderate life span

HABIT: upright, rounded; moderate density; irregular silhouette; coarse texture

FAMILY: *Simaroubaceae*

LEAF COLOR: evergreen; new foliage reddish, older light green

LEAF DESCRIPTION: alternate; even pinnately compound; entire margin; oblong to obovate; 2 to 4" long leaflets

FLOWER DESCRIPTION: yellow and orange; very showy; spring

EXPOSURE/CULTURE:

Light Requirement: part shade to full sun

Soil Tolerances: all textures; alkaline to acidic; wet soil; moderate drought

Aerosol Salt Tolerance: high

PEST PROBLEMS: pest-free

BARK/STEMS:

Trunk is straight in its native habitat, but divides into several trunks in open landscape; should be trained with a single leader; no thorns

Pruning Requirement: requires pruning to develop strong structure; space limbs apart along central trunk and prevent formation of included bark by pruning

Limb Breakage: susceptible, if not properly pruned

Current Year Twig: brown; thick

LANDSCAPE NOTES:

Origin: south Florida into the West Indies

Usage: grouped together in a buffer strip; highway; shade tree; specimen; street tree (best in a protected location)

The pinnately compound, 16-inch leaves of Paradise-Tree have multiple, shiny, leathery, oblong leaflets. An upright tree when young, Paradise-Tree ultimately reaches mature height and spread with a dense, rounded crown. The tiny, inconspicuous, yellowish, springtime blooms on this frost-sensitive tree are followed by small clusters of dark purple, one-inch-long, edible fruits. Although Paradise-Tree produces desirable shade, the seeds and fruits from female trees are messy and will stain hard surfaces. Shallow surface roots are troublesome to sidewalks and driveways and make it difficult to operate a lawn mower beneath the canopy.

The coarse leaf texture and compound foliage allows this tree to stand out. The emerging, reddish foliage turns light green and is displayed very handsomely on a textured canopy, making this a good tree for many landscape sites. It could be used as a boulevard or median street tree for a road where cars do not park regularly. Planting in a residential area could be objectionable only because of the fruit drop. Plant on 25- to 30-foot centers to form a solid canopy over the street.

Paradise-Tree grows in full sun or partial shade on almost any well-drained soil. A native of south Florida, it will grow quickly on rich soils high in organic matter and should be protected from frost. Large trees are reportedly difficult to establish from containers, so consider planting small trees from containers or transplant from a field nursery. Be sure to slice the root ball, or tease roots from the edge of the root ball, if the tree is overgrown in the container.

Sophora affinis Eve's-Necklace, Texas Sophora

Pronunciation: so-FOR-ah ah-FIN-iss

ZONES: 7A, 7B, 8A, 8B, 9A, 9B

HEIGHT: 30 to 35 feet

WIDTH: 20 feet

FRUIT: elongated; pod; 2 to 4"; dry; black; no significant litter; persistent; showy

GROWTH RATE: moderate; moderate life span

HABIT: upright vase; very open; irregular silhouette; fine texture

FAMILY: *Leguminosae*

Photo Courtesy of B. Simpson

LEAF COLOR: dark green; no fall color change

LEAF DESCRIPTION: alternate; odd pinnately compound; entire margin; oval to ovate; deciduous; 0.5" long leaflets

FLOWER DESCRIPTION: pink or white; showy; summer

EXPOSURE/CULTURE:

Light Requirement: mostly shaded to full sun

Soil Tolerances: all textures; alkaline to acidic; good drainage is essential; drought

PEST PROBLEMS: resistant

BARK/STEMS:

Bark is thin and easily damaged by mechanical impact; routinely grown with multiple trunks; branches grow mostly upright and will not droop; showy trunk; no thorns

Pruning Requirement: needs pruning to develop strong structure; stake trunk and grow in full sun for best tree form; tree tends to be viney and requires training in shade

Limb Breakage: resistant

Current Year Twig: brown; green; slender

LANDSCAPE NOTES:

Origin: Oklahoma, Arkansas, Louisiana

Usage: container or above-ground planter; parking lot island; buffer strip; highway; near a deck or patio; reclamation; specimen; possible street tree

Eve's-Necklace, or Texas Sophora, is a native North American, deciduous tree that grows moderately fast to mature height and width. Trees grown in the sun are often shorter. It has an upright silhouette and pinnately compound leaves. Branches often weep slightly, lending a delicate texture to the tree in the sun or shade. The fragrant, white-with-pink blossoms appear in June in dense, two- to six-inch-long racemes, somewhat like wisteria. The black seedpods that follow are up to four inches long; the pod is tightly pinched around each encased seed, giving it almost the appearance of a string of beads (hence its common name).

Texas Sophora often grows like a woody vine in the wild. It can be used as an understory small tree in a partially shaded location, but the crown will not be as dense as when it is grown in full sun, and flowering will be sparse. It makes a nice small tree for planting next to the deck or patio, where it casts light shade. The tree would become more popular with some training and pruning in the nursery to create a more uniform growth habit.

Texas Sophora should be grown in full sun or partial shade on any well-drained soil. Growth is poor and foliage chlorotic in poorly drained soil. Trees are moderately drought-tolerant and will flower most heavily if located in full sun. No limitations of major concern.

Sophora japonica Scholar Tree, Japanese Pagodatree

Pronunciation: so-FOR-ah ja-PON-ee-kah

ZONES: 5A, 5B, 6A, 6B, 7A, 7B, 8A

HEIGHT: 40 to 70 feet

WIDTH: 40 to 70 feet

FRUIT: elongated; pod; 3 to 8"; dry; yellow, turning brown; causes significant litter; persistent; showy

GROWTH RATE: moderate; long-lived

HABIT: oval to round, eventually forming a vase; moderate density; symmetrical; fine texture

FAMILY: *Leguminosae*

LEAF COLOR: medium green to blue-green; muted yellow fall color

LEAF DESCRIPTION: alternate; odd pinnately compound; entire margin; ovate; deciduous; 2" long leaflets

FLOWER DESCRIPTION: white or light yellow; showy; summer; stains pavement

EXPOSURE/CULTURE:

Light Requirement: part shade to full sun

Soil Tolerances: all textures; alkaline to acidic; occasionally wet; drought

Aerosol Salt Tolerance: moderate

PEST PROBLEMS: resistant

BARK/STEMS:

Branches droop as tree grows and will require pruning for vehicular or pedestrian clearance beneath canopy; should be grown with a single leader; no thorns

Pruning Requirement: requires pruning to develop strong structure; head back low, aggressive branches when young to force growth in upper canopy; thin branches to slow growth and keep branches less than half diameter of trunk

Limb Breakage: usually resistant; can break during ice storms

Current Year Twig: green; medium thickness

LANDSCAPE NOTES:

Origin: China, Korea

Usage: parking lot island; buffer strip; highway; near a deck or patio; shade tree; specimen; sidewalk cutout; street tree; urban tolerant

Pagodatree tolerates polluted city conditions, heat, and drought. The very showy, greenish-white to yellow flowers are produced in mid to late summer and provide an airy feel to the tree for several weeks. The species must be at least 10 years old to bloom, but the cultivar 'Regent' reportedly blooms at 6 to 8 years old.

The tree drops flower petals, creating a creamy white carpet for several weeks on the ground, but the petals can temporarily stain sidewalks. The yellow fruit pods form in late summer and are quite showy. They could be a nuisance to some people when they drop later in the winter, but they are small and fairly easily washed away. The leaflets are small, creating light shade beneath the tree, and are mostly washed away with rain or fall into shrub beds or between the grass blades.

Some trees come from the nursery with multiple trunks or branches clustered together at one spot on the trunk. Buy those with one central trunk growing up the center of the tree, or prune the tree to a central leader to create a strong, durable structure. 'Regent' supposedly has a straighter trunk. Space branches along the central leader to ensure good branch attachment. It may take several prunings to train the tree to the proper form.

This urban-tough tree is highly recommended for urban street tree or parking lot planting. It also makes a nice medium-sized patio tree for the residential landscape.

Pruning to a good form can be a challange. Potato leafhopper kills young stems, causing profuse branching or witches' broom on small branches. It usually is not a problem on larger trees.

Sophora secundiflora Texas-Mountain-Laurel

Pronunciation: so-FOR-ah seh-cun-dih-FLOOR-ah

ZONES: 7B, 8A, 8B, 9A, 9B, 10A

HEIGHT: 15 to 25 feet

WIDTH: 10 to 10 feet

FRUIT: elongated pod; 8"; dry and hard; red seeds in tan pod; no significant litter; persistent; showy; seeds are toxic

GROWTH RATE: slow-growing; long-lived

HABIT: upright rounded; open; symmetrical; fine texture

FAMILY: *Leguminosae*

LEAF COLOR: evergreen; dark green

LEAF DESCRIPTION: alternate; pinnately compound; entire margin; oval to oblong; 1" long leaflets

FLOWER DESCRIPTION: purple; pleasant fragrance; showy; spring

EXPOSURE/CULTURE:

Light Requirement: part shade to full sun

Soil Tolerances: all textures; alkaline to acidic; salt; drought

PEST PROBLEMS: somewhat sensitive

BARK/STEMS:

Routinely grown with multiple trunks; could be trained to one, short trunk; branches grow mostly upright; no thorns; showy

Pruning Requirement: needs little pruning to develop strong structure; low branches and foliage are often removed from lower stems to create tree form

Limb Breakage: resistant

Current Year Twig: brown; green; medium thickness

LANDSCAPE NOTES:

Origin: Texas and New Mexico to Mexico

Usage: bonsai; espalier; container or above-ground planter; hedge; parking lot island; buffer strip; highway; near a deck or patio; reclamation; specimen; possible street tree; sidewalk cutout

Texas-Mountain-Laurel is a small North American native evergreen that grows 15 to 25 feet tall with a 10-foot-spread, but is capable of reaching up to 50 feet tall in its native habitat. It has a narrow upright silhouette and dense foliage, which lends itself well to being pruned into a tree form. The four-inch-long leaves are glossy, thick, and leathery. In spring, Texas-Mountain-Laurel is a beautiful sight as it displays its dense, two- to five-inch-long, pendulous clusters of purple-blue, extremely fragrant flowers. Every so often, a white-flowered form can be found. The hairy seedpods that follow are eight inches long and ripen to reveal the inner, bright red seeds. These seeds are quite decorative and have been used to make necklaces, but they are also poisonous. The fissured bark is dark gray to black.

Usually found as a multi-trunked small tree, Texas-Mountain-Laurel can be trained to a single trunk in the nursery. Single-trunked nursery stock would make nice street trees for planting in small soil spaces and where overhead space is limited by wires or other structures. Plant a row of Texas-Mountain-Laurel on 15- or 20-foot centers to form a nice canopy over a walk, or locate it close to a patio or deck. The bark on multi-trunked specimens shows off nicely when lit up at night from beneath the canopy.

Texas-Mountain-Laurel should be grown in full sun or partial shade on well-drained soil. This tough plant will tolerate hot, windy conditions and alkaline or wet soils. It might suffer in compacted soil. Young trees may benefit from afternoon shading from intense summer sun until they become established.

No limitations of major concern except for *Uresiphita reveralis*, which consumes tender foliage on vigorous trees.

Sorbus alnifolia Korean Mountain-Ash

Pronunciation: SORE-buss all-neh-FOE-lee-ah

ZONES: 4A, 4B, 5A, 5B, 6A, 6B, 7A

HEIGHT: 30 to 40 feet

WIDTH: 20 to 25 feet

FRUIT: round; 0.5"; fleshy; green turning red; no significant litter; persistent; attracts wildlife; extremely showy

GROWTH RATE: rapid; moderate life span

HABIT: oval to rounded or upright; dense; symmetrical; medium texture

FAMILY: *Rosaceae*

LEAF COLOR: bright green; showy copper, orange, red, or yellow fall color

LEAF DESCRIPTION: alternate; simple; double serrate or serrate margin; oval to ovate; deciduous; 2 to 4" long

FLOWER DESCRIPTION: white; showy; spring

EXPOSURE/CULTURE:

Light Requirement: grows best in full sun

Soil Tolerances: all textures; slightly alkaline to acidic; occasionally wet; some salt; moderate drought

Aerosol Salt Tolerance: moderate

PEST PROBLEMS: somewhat sensitive

BARK/STEMS:

Bark is thin and easily damaged by mechanical impact; can be trained to a single or multi-trunked tree; branches grow mostly upright and will not droop; no thorns

Pruning Requirement: needs some pruning to develop strong structure; keep branches less than half trunk diameter with regular pruning

Limb Breakage: susceptible

Current Year Twig: brown; medium thickness

LANDSCAPE NOTES:

Origin: China, Japan, Korea

Usage: reclamation; specimen; possible street tree where there is plenty of soil space

Korean Mountain-Ash has two- to four-inch-long, glossy, simple leaves that change to attractive hues of yellow, orange, and rust in fall before dropping. In late spring the trees are decorated with two- to three-inch-diameter, flat-topped clusters of tiny white blossoms, which may appear in great number one year yet sparsely the next. These blooms are followed by the production of small but very showy, red fruits that pendulously hang in 5-inch clusters of 2 to 10 berries. Fruit clusters are not as dense as those of *Sorbus aucuparia*. Fruits are displayed nicely after leaves fall in autumn.

Partially due to the upright habit, weak crotches can develop on branches that are allowed to grow to larger than about half the diameter of the trunk. Pruning early in the life of the tree to prevent main branches from growing too large will improve the structure of the tree and help increase longevity in the landscape.

It is tempting to plant this tree, due to the nice flowers and fruit display, but it may be short-lived and is susceptible to fire blight. However, it is the most resistant of the Mountain-Ashes to borers, which can be devastating to this genus. It is probably best to locate the tree where there is adequate open soil space for root development so the tree will not be subjected to much stress. Provided with some irrigation in a dry summer, this tree will serve the landscape well for a number of years.

Korean Mountain-Ash should be grown in full sun on any well-drained soil, including slightly alkaline. Any required pruning should be done in winter or early spring to avoid exposing the wood to the fire blight disease, which can cause serious damage.

Limitations include aphids, pear leaf blister mite, Japanese leafhopper, mountain-ash sawfly, and scales. Korean Mountain-Ash is somewhat resistant to the borers that affect other *Sorbus* species, and probably should be planted instead of European Mountain-Ash. Crown gall, leaf rusts, and scab have also been reported.

Sorbus aucuparia European Mountain-Ash

Pronunciation: SORE-buss awk-you-PAY-ree-ah

ZONES: 3B, 4A, 4B, 5A, 5B, 6A, 6B

HEIGHT: 30 feet

WIDTH: 15 to 25 feet

GROWTH RATE: moderate; frequently short-lived

FRUIT: round; 0.33"; fleshy; orange or red; attracts birds; causes mushy litter; showy

HABIT: oval, upright vase; moderate density; symmetrical; fine texture

FAMILY: *Rosaceae*

LEAF COLOR: green; showy red or yellow fall color

LEAF DESCRIPTION: alternate; odd pinnately compound; serrate margin; lanceolate to oblong leaflets; deciduous; 1 to 3" long leaflets

FLOWER DESCRIPTION: white; very showy; spring

EXPOSURE/CULTURE:

Light Requirement: full sun

Soil Tolerances: all textures; acidic; salt-sensitive; moderate drought

Aerosol Salt Tolerance: low

PEST PROBLEMS: sensitive especially in eastern United States

BARK/STEMS:

Branches grow mostly upright and will not droop; should be grown with a single leader; no thorns; occasionally splits on southwest side in winter

Pruning Requirement: requires pruning to develop strong structure; thin main branches that have or are developing included bark to slow growth rate; large pruning cuts close slowly; prune regulary to avoid cutting large branches

Limb Breakage: susceptible to breakage at crotch, due to included bark

Current Year Twig: brown; medium thickness

CULTIVARS AND SPECIES:

'Asplenifolia'—leaflets doubly serrated; 'Apricot Queen'—apricot-colored fruits; 'Beissneri'—deeply cut leaflets, red branchlets; 'Brilliant Pink'—pink fruits; 'Michred' and 'Red Strain'—clear red fruits; 'Cole's Columnar'—upright growth habit; 'Fastigiata'—upright growth habit; 'Kirsten Pink' and 'Maidenblush'—fruits pinkish; 'Pendula'—weeping growth habit; 'Scarlet King'—fruits scarlet; 'Wilson's Columnar'—upright growth; 'Xanthocarpa'—yellow fruit, not attractive to birds. Some might make good street trees.

LANDSCAPE NOTES:

Origin: Europe and Asia

Usage: specimen; possible street tree; near a deck or patio

European Mountain-Ash has a rapid growth rate at first but slows down with age, ultimately forming a 30-foot-tall, dense oval. The white flowers are showy, appearing in the spring after the leaves. The tree shows its true colors when the fruit is set in mid-summer. The wonderfully showy, orange-red fruit is borne in heavy clusters of about 40 berries, although cultivars are available with pink, yellow, and red fruits. Fruits are usually eaten by birds and often do not persist on the tree into the winter. Seeds germinate quite readily in some landscapes. The fall color is red to yellow, although sometimes leaves simply drop green.

This tree may be best for specimen planting in a residential or commercial landscape. Extensive use as a street tree is probably not wise, due to the potential pest problems. The trees are sensitive to a variety of insect and disease problems that shorten their life, including borers, aphids, fire blight, and canker. Its outstanding ornamental traits make it worth planting in parks and other open areas where they can be enjoyed.

Vertically oriented branches and multiple trunks make this deciduous tree particularly attractive during the winter, but this same characteristic can also cause branches to break from the trunk, due to poor connections with the trunk (included bark). The tree has no particular soil preference, but is restricted to northern areas due to lack of heat tolerance. It is a tough, urban-tolerant tree, but is susceptible to several pests, including fire blight, which can disfigure or kill the tree.

Spathodea campanulata African Tulip-Tree

Pronunciation: spah-THO-dee-ah cam-pan-you-LAY-tah

ZONES: 10B, 11

HEIGHT: 50 to 60 feet

WIDTH: 35 to 50 feet

FRUIT: elongated pod; 6 to 12"; dry; brown; not showy; causes significant litter

GROWTH RATE: rapid-growing; moderate to long life span

HABIT: upright and oval at first, rounded later; moderate density; nearly symmetrical; coarse texture

FAMILY: *Bignoniaceae*

LEAF COLOR: broadleaf evergreen; medium green

LEAF DESCRIPTION: opposite; odd pinnately compound; entire margin; oval to oblong; 2 to 4" long leaflets

FLOWER DESCRIPTION: orange or yellow; very showy; spring and winter; causes mushy litter

EXPOSURE/CULTURE:

Light Requirement: tree grows and flowers best in full sun

Soil Tolerances: all textures; acidic; moderate salt and drought

Aerosol Salt Tolerance: low to moderate

PEST PROBLEMS: resistant

BARK/STEMS:

Branches eventually droop as tree grows and will require pruning for vehicular or pedestrian clearance beneath canopy; should be grown with a single leader; no thorns

Pruning Requirement: requires pruning to develop strong structure; slow growth rate on aggressive lower branches by thinning to prevent formation of included bark

Limb Breakage: susceptible to breakage at crotch, due to poor collar formation; wood itself is weak and tends to break

Current Year Twig: brown; medium thickness

LANDSCAPE NOTES:

Origin: Africa

Usage: shade tree; specimen

A native of tropical Africa, this large, upright tree has a dense, wide crown and 1.5-foot-long, pinnately-compound, evergreen leaves composed of 4-inch leaflets. Due to its size, it is best located in large, open landscapes and is generally not suited for small residences unless the objective is deep shade. During winter and late spring, African Tulip-Tree produces terminal clusters of beautiful blooms held above the foliage—a profusion of upwardly-facing, orange and yellow flowers, which open several at a time from curved, two-inch-long, fuzzy brown flower buds.

African Tulip-Tree is quite spectacular when in bloom. It is often used as a framing, shade, or specimen tree, but must be planted only in frost-free areas. Also, its soft, brittle wood is easily broken by strong winds, and trees should be located either in sheltered locations or where falling branches will do no damage.

Eliminate major branches that will form included bark as early as possible. Save those that are oriented more horizontally, with stronger attachments to the trunk. Keep them from growing larger than about half the trunk diameter by periodic thinning.

African Tulip-Trees will grow rapidly in full sun on any soil of reasonable drainage and fertility. Plants should be regularly watered until well established and will then require little irrigation.

The fruit and flowers create a mess on pavement and walks. Pnamatiphores are sent up from the roots and can penetrate through pavement. Seeds germinate readily and plants have invaded areas in Hawaii, Florida, and Puerto Rico. Surface roots can become large and invade everything.

Stewartia koreana Korean Stewartia

Pronunciation: stew-ARE-tee-ah core-ee-A-nah

ZONES: 5B, 6A, 6B, 7A, 7B

HEIGHT: 20 to 25 feet

WIDTH: 15 to 25 feet

FRUIT: oval; 1"; dry; red; attracts birds; no significant litter

GROWTH RATE: slow-growing; long-lived

HABIT: upright, pyramidal when young; dense; symmetrical; medium texture

FAMILY: *Theaceae*

LEAF COLOR: dark green; showy, orange, purple, or red fall color

LEAF DESCRIPTION: alternate; simple; serrate margin; oval; deciduous; 1 to 4" long

FLOWER DESCRIPTION: white; showy; summer; attracts bees

EXPOSURE/CULTURE:

Light Requirement: part shade to full sun

Soil Tolerances: all textures; acidic; moderate drought

PEST PROBLEMS: resistant

BARK/STEMS:

Beautiful, exfoliating bark; bark is thin and easily damaged by mechanical impact; routinely grown, or trainable to be grown, with multiple trunks; can be trained to a single trunk for street tree use; branches grow mostly upright and will not droop; no thorns

Pruning Requirement: needs little pruning to develop strong structure; thin canopy to display outstanding bark

Limb Breakage: resistant

Current Year Twig: green; gray; medium thickness

LANDSCAPE NOTES:

Origin: Korea

Usage: container or above-ground planter; espalier; buffer strip; near a deck or patio; specimen; sidewalk cutout; possible street tree

This tree has also been classified as a variety of *S. pteropetiolata* and cultivar or variety of *S. pseudocamellia*. Capable of reaching 50 feet in height, Korean Stewartia is most often seen at 20 to 25 feet. Its short, interwoven branchlets form a dense, pyramidal canopy that casts deep shade below this deciduous tree. Older trees open up to form nearly a vase shape. The one- to four-inch-long by one- to three-inch-wide leaves often turn to lovely shades of orange, red, or purple in the fall, but do not do so reliably.

Over a several-week period in June to July, Korean Stewartia is decorated with lovely, pure white, yellow-centered blossoms, three inches across and flattened, appearing much like a single camellia flower. Each flower is open for about 24 hours and they attract bees. Most other trees have finished flowering by the time these flowers emerge. The bark of Korean Stewartia is probably its most outstanding characteristic, with creamy white, pink, or orange-brown patches showing through the flaking, gray bark. The bark must be seen to understand the excitement generated in gardeners who have grown the tree.

This tree can be used in much the same way as Japanese Stewartia. It is unsurpassed as a specimen, but can also be included in a shrub border, planted near a patio or deck to show off the wonderful bark, or used as a single- or multi-trunked street tree planted on 20-foot centers. Set it off as a specimen by placing it in a bed of low, evergreen groundcover.

Korean Stewartia should be grown in full sun but reportedly looks best where it can receive some shade during the hottest part of the day. However, there are fine-looking specimens growing in poor clay soil in zones 6B and 7B, with no irrigation, which are located in full-day sun. The soil should preferably be moist, acid, and supplemented with organic matter. Pruning is seldom required, as growth rate is slow and branches normally keep in bounds, staying close to the tight canopy.

No limitations of major concern, but it will be hard to find in nurseries.

Stewartia monadelpha Tall Stewartia

Pronunciation: stew-ARE-tee-ah mon-a-DELL-fah

ZONES: 6B, 7A, 7B, 8A, 8B

HEIGHT: 25 to 35 feet

WIDTH: 15 to 25 feet

FRUIT: oval or round; 0.5"; dry; not showy; no significant litter

GROWTH RATE: slow-growing; moderate to long life span

HABIT: upright pyramidal when young, vase later; open; symmetrical; medium texture

FAMILY: *Theaceae*

LEAF COLOR: dark green; showy, red fall color

LEAF DESCRIPTION: alternate; simple; serrate margin; oval to oblong; deciduous; 1.5 to 2.5" long

FLOWER DESCRIPTION: white; showy; summer

EXPOSURE/CULTURE:

Light Requirement: part shade to full sun

Soil Tolerances: all textures; acidic; drought

PEST PROBLEMS: resistant

BARK/STEMS:

Orange-brown, flaking bark; bark is thin and easily damaged by mechanical impact; routinely grown, or trainable to be grown, with multiple trunks; can be trained to a single trunk; branches grow mostly upright and will not droop; no thorns

Pruning Requirement: needs little pruning to develop strong structure; stake trunk (if needed) to form a single leader and space branches along trunk

Limb Breakage: resistant

Current Year Twig: green; gray; slender

LANDSCAPE NOTES:

Origin: Japan

Usage: container or above-ground planter; espalier; parking lot island; buffer strip; near a deck or patio; specimen; sidewalk cutout; possible street tree if irrigated in drought

Tall Stewartia grows 35 feet in height in as many years in the landscape, though it has been known to reach 80 feet in its native habitat. Its young, pyramidal crown matures into an open, multi-trunked form with nearly horizontal branches. The 1.5- to 2.5-inch-long leaves cast light shade below this deciduous tree, and cling well into the fall after changing to an attractive deep red. In June the small, white, cupped flowers appear, opening over a four-week period. Although noticeable, they are not particularly striking. The smooth cinnamon-brown bark is outstanding and helps to make Tall Stewartia quite attractive in the winter landscape.

This tree can be used in much the same way as Japanese Stewartia. It is unsurpassed as a specimen, but can also be included in a shrub border, planted near a patio or deck to show off the wonderful bark, or used as a single- or multi-trunked street tree planted on 20-foot centers. Set it off as a specimen by placing it in a bed of low, evergreen groundcover. Plant it in a large container and move it to where it is needed in the landscape.

Space the thin branches along a central trunk to develop the most durable structure. This technique also allows the tree to develop its most striking form.

Best when grown in partial shade in zone 8B, Tall Stewartia can tolerate a full-sun position if its roots can be shaded by groundcover, mulch, or shrubbery. This, along with the native *S. malocodendron*, may be the best Stewartia species for the Deep South (zone 8). Well-drained, acid to neutral soil is best. No limitations of major concern, and should be grown and planted more often.

Stewartia pseudocamellia Japanese Stewartia

Pronunciation: stew-ARE-tee-ah sue-dough-cah-MEAL-lee-ah

ZONES: 5B, 6A, 6B, 7A, 7B, (8A), (8B)

HEIGHT: 30 to 40 feet

WIDTH: 20 to 30 feet

FRUIT: round; green; 1"; dry; no significant litter

GROWTH RATE: slow-growing; moderate to long life span

HABIT: narrow, upright pyramidal when young, open vase later; dense; symmetrical; medium texture

FAMILY: *Theaceae*

LEAF COLOR: dark green; showy purple, red, or yellow fall color

LEAF DESCRIPTION: alternate; simple; serrulate margin; oval to obovate; deciduous; 2 to 4" long

FLOWER DESCRIPTION: white; showy; summer

EXPOSURE/CULTURE:

Light Requirement: mostly shaded to part sun, full sun if irrigated in the south

Soil Tolerances: all textures; acidic; occasionally wet soil; moderate drought

PEST PROBLEMS: resistant

BARK/STEMS:

Wonderful, exfoliating bark; bark is thin and easily damaged by mechanical impact; can be trained to a single or multi-trunked habit; unpruned trees often develop one central leader; branches grow mostly upright and horizontal; no thorns

Pruning Requirement: needs little pruning to develop strong structure; bark is showy whether trained to single or multiple trunks

Limb Breakage: resistant

Current Year Twig: green; slender

LANDSCAPE NOTES:

Origin: Japan

Usage: container or above-ground planter; espalier; parking lot island; buffer strip; near a deck or patio; specimen; sidewalk cutout; possible street tree if irrigated

An excellent, small to medium-sized, deciduous garden tree, Japanese Stewartia is an all-season performer, exhibiting a distinctive branching pattern in winter, camellia-like flowers in summer, and bright yellow and red foliage in autumn. Most other trees have finished flowering by the time these flowers emerge. Once you have seen it, the fall color is unforgettable. The bark is spectacular, peeling off and exposing contrasting colors. It could be grown for this characteristic alone. Trunks on young trees often grow straight up into the canopy, but trees can be trained to a more open, spreading habit.

The tree branches close to the ground, forming a sinewy pattern on older trees not unlike Crape Myrtle. It would make a nice patio tree, could accent an entry way, or could be grown as a canopy tree over a sidewalk. A row of them on either side of a sidewalk spaced 15 feet apart makes an outstanding "covered walkway." It could be planted as a slow-growing street tree beneath tall power lines due to its small stature.

Japanese Stewartia is a slow grower, reportedly best in acid soil (pH 4.5 to 6.5) with ample moisture and high organic matter content. But trees are also found growing very well without irrigation in poor-quality, compacted clay soil. Some leaf burn may be evident in drier summers in full sun, but this does not appear to affect the tree permanently. Stewartia may prefer some shade in warm climates, where it develops a more open habit, but it does quite well in full-day sun, forming a dense, dark green head of foliage. Transplant as a small tree from a field nursery in early spring or from a container of any size at any time. Irrigation helps prevent scorch in dry weather.

Strelitzia nicolai White Bird-of-Paradise

Pronunciation: streh-LIT-zee-ah NIH-keh-lie

ZONES: 9B, 10A, 10B, 11

HEIGHT: 20 to 30 feet

WIDTH: 10 feet

FRUIT: dry; brown; no litter

GROWTH RATE: moderate; moderate life span

HABIT: palm-like; upright; open; irregular silhouette; very coarse texture

FAMILY: *Strelitziaceae*

LEAF COLOR: broadleaf evergreen; medium green

LEAF DESCRIPTION: spiral; simple; entire margin, but often tattered along veins; oblong; 5 to 8' long; can be messy because leaves decompose slowly

FLOWER DESCRIPTION: white; somewhat showy; year-round

EXPOSURE/CULTURE:

Light Requirement: part shade to full sun

Soil Tolerances: all textures; acidic; moderate salt and drought

Aerosol Salt Tolerance: low to moderate

PEST PROBLEMS: resistant

BARK/STEMS:

Routinely grown with multiple trunks, but can be trained to a single trunk by removing suckers; trunks grow mostly upright and will not droop; showy trunks; no thorns

Pruning Requirement: remove sprouts from base of trunk to form a more open canopy; old, dead leaves should be removed occasionally

Limb Breakage: somewhat susceptible

LANDSCAPE NOTES:

Origin: Africa

Usage: container or above-ground planter; near a deck or patio; specimen

White Bird-of-Paradise is most often planted for its large, banana-like leaves and upright, clumping stalks, which give an exotic feel to the landscape. Plants can reach 20 to 30 feet in height with a spread of 10 feet, though they are often smaller. The five- to eight-foot-long, cold-tender leaves are arranged in a fanlike display from the erect trunks and appear much like Traveler's-Tree. The lower trunk becomes clear of leaves and exposed as the older leaves drop off. Leaves rip along the veins as they blow in strong winds.

The interesting flowers are white and appear to have a dark blue tongue. White Bird-of-Paradise is ideal for entranceways for a dramatic effect or for use at poolside. Plants are not messy, but ragged leaves can be periodically removed for a more tidy appearance. This is a large plant and should be situated accordingly. It is suited for planting close to buildings, due to the narrow canopy.

White Bird-of-Paradise grows well in full sun to light shade on moist, well-drained soil. Plants should be protected from high winds to minimize torn, ragged leaves. It will survive periods of 28°F. with minimal leaf burn and will quickly recover. Prune to remove dead leaves and thin out surplus growth sprouting from the base of the trunk, if you wish.

Except for scales, no limitations of major concern.

Styrax japonicus Japanese Snowbell

Pronunciation: STY-rax jah-PON-ee-kuss

ZONES: 6A, 6B, 7A, 7B, 8A

HEIGHT: 20 to 30 feet

WIDTH: 15 to 25 feet

FRUIT: oval; 0.5"; fleshy; green; no significant litter; persistent

GROWTH RATE: moderate; moderate to long life span

HABIT: rounded vase; moderate density; symmetrical; medium texture

FAMILY: *Styracaceae*

LEAF COLOR: dark green; fairly showy, red or yellow fall color

LEAF DESCRIPTION: alternate; simple; entire margin; oval to oblong; deciduous; 2 to 4" long

FLOWER DESCRIPTION: white; showy; early summer; attracts bees

EXPOSURE/CULTURE:

Light Requirement: part shade to full sun

Soil Tolerances: loam; sand; slightly alkaline to acidic; moderate drought

Aerosol Salt Tolerance: good

PEST PROBLEMS: nearly pest-free

BARK/STEMS:

Branches droop as tree grows and will require pruning for vehicular or pedestrian clearance beneath canopy; routinely grown with multiple trunks; can be trained to a short, single trunk; no thorns

Pruning Requirement: needs little pruning to develop strong structure; pruning required to develop upright growth habit

Limb Breakage: resistant

Current Year Twig: green; slender

CULTIVARS AND SPECIES:

'Carillon'—grows about one foot per year, with a weeping habit, and needs to be trained to tree form; 'Crystal' (zone 5)—upright to fastigiate habit, black-green foliage, crisp white flowers with purple pedicels; 'Emerald Issai'—grows faster than the species and roots easily; 'Pink Chimes'—has pink flowers and supposedly holds up in heat. *S. obassia* has much larger leaves and is native to the eastern United States.

LANDSCAPE NOTES:

Origin: China, Japan

Usage: container or above-ground planter; parking lot island; buffer strip; highway; near a deck or patio; standard; specimen; sidewalk cutout; street tree

Japanese Snowbell is a small, deciduous tree that slowly grows to form a rounded canopy with a horizontal branching pattern. With lower branches removed, it forms a more vase-shaped patio-sized shade tree. The smooth, attractive bark has orange-brown interlacing fissures, adding winter interest to any landscape. The white, bell-shaped, drooping flower clusters of Japanese Snowbell are quite showy in May to June.

Styrax species make excellent small patio trees where the flowers and interesting bark can be viewed up close. Japanese Snowbell also makes a wonderful addition to a mixed shrubbery border. Due to its small stature and vase shape, it can make a nice street tree where overhead space is limited.

Snowbell prefers a peaty, acid soil that is moist but not waterlogged. In colder areas (zone 5), locate this tree in an area protected from winter winds. Plants grow better with a couple of hours of shade in zones 7 and 8, but full sun is fine in the North.

No limitations of major concern. Ambrosia beetle can attack and lead to further decline of stressed plants. Seedlings often emerge from under the canopy.

Swietenia mahagoni Mahogany

Pronunciation: swee-TEE-nee-ah mah-HA-go-nye

ZONES: 10B, 11

HEIGHT: 50 feet

WIDTH: 50 feet

FRUIT: oval; 3 to 6"; dry; brown; causes some litter; persistent; showy

GROWTH RATE: rapid-growing; long-lived

HABIT: oval to rounded vase; moderate density; symmetrical; medium texture

FAMILY: *Meliaceae*

LEAF COLOR: semi-evergreen; medium green

LEAF DESCRIPTION: alternate; even pinnately compound; entire margin; lanceolate to ovate; 2 to 3" long leaflets

FLOWER DESCRIPTION: green; not showy; spring

EXPOSURE/CULTURE:

Light Requirement: part shade to full sun

Soil Tolerances: all textures; alkaline to acidic; occasionally wet; drought

Aerosol Salt Tolerance: high

PEST PROBLEMS: resistant to most lethal pests

BARK/STEMS:

Branches droop somewhat as tree grows and will require pruning for vehicular or pedestrian clearance beneath canopy; should be grown with a single leader; frequently develops included bark; no thorns

Pruning Requirement: requires pruning to develop strong structure

Limb Breakage: resistant until included bark forms in branch crotches

Current Year Twig: brownish-green; medium thickness

CULTIVARS AND SPECIES:

Swietenia macrophylla is a taller tree with a dominant, straight trunk, which could be grown in Florida and used along streets and in other areas. The trunk buttresses, so allow plenty of soil space. Before Hurricane Andrew in 1992, there were several of these trees three feet in diameter at the USDA research station near Miami.

LANDSCAPE NOTES:

Origin: Caribbean, including south Florida

Usage: large parking lot island; buffer strip; reclamation; shade tree; street tree; urban tolerant

This large, semi-evergreen tree forms a loose, rounded canopy and casts light, dappled shade, suitable for maintaining a lawn beneath. It is a popular landscape and street tree in south Florida and other tropical landscapes. The dense, strong wood of Mahogany is quite resistant to wind damage on properly trained trees, but unfortunately many trees are not properly trained. Too often, major limbs originate from the same position on the trunk, creating a multi-trunked, weak-structured tree. Branches sometimes split at the crotch, especially those with included bark. Major limbs should be spaced along the trunk. Trees defoliate briefly in dry winters.

A native of south Florida, Mahogany will grow in full sun or partial shade on a wide range of soil types, and is quite resistant to salt spray. Rapid growth occurs on rich, well-drained soil. Roots can raise sidewalks and curbs when planted too close. It is best to plant Mahogany more than six feet from curbs and sidewalks.

Prune and train the tree while it is young to develop several major limbs spaced several feet apart along a central trunk. Do not allow branches to grow larger than about half the diameter of the trunk. This will improve the tree structure and increase longevity.

Tent caterpillars, tip moth, webworm, scale, leaf notcher, and leaf miner can infest foliage. Borers infest stressed trees. Some of these insects can cause significant problems. Nectria canker can occur but is not abundant. Central leader is difficult to maintain due to tip moth infestation.

Syagrus romanzoffianum Queen Palm

Pronunciation: sigh-A-gruss row-man-zoff-ee-A-num

ZONES: 9B, 10A, 10B, 11

HEIGHT: 25 to 50 feet

WIDTH: 15 to 25 feet

FRUIT: round; 0.5"; fleshy; orange; causes some messy, slimy litter; showy

GROWTH RATE: rapid-growing; moderate life span

HABIT: palm; open; irregular silhouette; fine texture

FAMILY: *Arecaceae*

LEAF COLOR: bright green

LEAF DESCRIPTION: spiral; odd pinnately compound; entire margin; lanceolate; 12 to 18" long leaflets

FLOWER DESCRIPTION: whitish-yellow; very showy; year-round

EXPOSURE/CULTURE:

Light Requirement: full sun

Soil Tolerances: all textures; acid only; occasionally wet; moderate drought

Aerosol Salt Tolerance: moderate

PEST PROBLEMS: resistant

BARK/STEMS:

Single trunk, about 18" thick; ringed with leaf scars; no crown shaft; no thorns

Pruning Requirement: remove dead fronds

Limb Breakage: mostly resistant, although trunks are susceptible to decay

LANDSCAPE NOTES:

Origin: Brazil, Argentina

Usage: parking lot island; buffer strip; highway; near a deck or patio; specimen; sidewalk cutout

This stately, single-trunked palm is crowned by a beautiful head of glossy, soft, pinnate leaves that form a graceful, drooping canopy. The ornamental, bright orange dates are produced in hanging clusters and ripen during the winter months. They fall to create a mushy mess on sidewalks and pavement. The dead fronds are persistent and will require pruning to remove. Queen Palm is popular in commercial or home landscapes planted in rows on 15-foot centers to line a street or walk, in clusters, or occasionally as a specimen. The gray trunk is ringed with old leaf scars.

Growing best in full sun, Queen Palm needs acidic, well-drained soil; it shows severe mineral deficiencies on alkaline soil. This disfigures the palm by stunting the young leaves and can kill it. Unfortunately, Queen Palm is frequently planted in alkaline soil and requires regular preventive applications of manganese and iron to help keep the fronds green. Quick-growing Queen Palm responds well to ample moisture and fertilizer and is slightly salt-tolerant. After planting in the landscape, growth is rapid. This palm is not affected by lethal yellowing disease.

Trunk injury initiates decay, so locate it where it will not be injured by mowers and weed whackers. Maintain mulch around the base of the palm to keep equipment away but do not let it touch the trunk or decay can begin. Overpruning can lead to a lack of vigor and plant decline. Only remove declining fronds that hang below the horizontal.

Queen Palms often exhibit potassium deficiency on the lower foliage. Emerging foliage often develops a frizzled appearance because of manganese deficiency. This is due to improperly locating the palm in an alkaline soil. A regular fertilization program for the life of the palm can help keep the fronds green. Palm leaf skeletonizer and scale are problems for Queen Palm, and Ganoderma butt rot can kill Queen Palm. It probably enters the trunk most often through wounds in the lower trunk and roots. There is no control for butt rot, only prevention. Palms planted too deep often grow slowly and decline.

Syringa pekinensis Peking Lilac

Pronunciation: sigh-RING-gah pea-kin-EN-siss

ZONES: 4A, 4B, 5A, 5B, 6A, 6B, 7A

HEIGHT: 25 to 40 feet

WIDTH: 25 to 30 feet

FRUIT: elongated and flattened; 0.5"; dry; green; no significant litter; persistent; somewhat showy

GROWTH RATE: moderate; moderate life span

HABIT: rounded vase; open; somewhat irregular silhouette; medium texture

FAMILY: *Oleaceae*

LEAF COLOR: green; no fall color change

LEAF DESCRIPTION: opposite; simple; entire margin; ovate; deciduous; 2 to 4" long

FLOWER DESCRIPTION: white; showy; early summer

EXPOSURE/CULTURE:

Light Requirement: full sun to part shade

Soil Tolerances: all textures; slightly alkaline to acidic; salt; moderate drought

Aerosol Salt Tolerance: good

PEST PROBLEMS: resistant, for a lilac

BARK/STEMS:

Branches are thin and droop, requiring pruning for vehicle or pedestrian clearance beneath canopy; routinely grown with multiple trunks; can be trained to grow with a single trunk; no thorns; can be showy

Pruning Requirement: needs little pruning to develop strong structure; pruning is needed to develop a central trunk with an upright habit

Limb Breakage: resistant

Current Year Twig: brown; slender and droopy

CULTIVARS AND SPECIES:

None are readily available, but 'Pendula' and 'Summer Charm' are listed. 'Summer Charm' may have a more uniform canopy outline than many other lilacs.

LANDSCAPE NOTES:

Origin: China

Usage: container or above-ground planter; parking lot island; buffer strip; highway; near a deck or patio; standard; specimen; sidewalk cutout; street tree; urban tolerant

Although a Lilac, this member of the species is quite different in appearance from those with which gardeners are familiar. Its upright habit varies from symmetrical to irregular. This is a medium-sized tree, reaching a height and spread of about 35 feet. The clusters of creamy white flowers are smaller than those of the Japanese Tree Lilac, borne in early summer for about two weeks. They are the most noticed ornamental feature of the tree, but lack the fragrance of the spring-blooming Lilacs—this Lilac's fragrance is more suggestive of privet. The brownish bark rolls up and persists as small patches of shiny, flaky plates. Rapid-growing, upright branches often display a smooth, shiny, tan trunk with a few lenticels circling the stem. The bark on older trees is darkened with prominent lenticels, reminiscent of Black Cherry.

The tree is usually sold as a multi-stemmed specimen, but can be trained as a single-trunked tree suited for streets. It could be used as a street tree, but regular pruning would be required in the first 10 years to remove drooping lower branches. The tree should maintain the upright habit once the main structure of the tree has been established. It will also be popular as a garden specimen or as an accent in a shrub border. It deserves to be in any landscape. It provides shade and a colorful spring show for a deck or patio area. Green fruit clusters are somewhat showy when viewed at close range.

It is perhaps the most pest-resistant Lilac, but that does not mean it is pest-free. Borer and powdery mildew may affect the tree. Occasional irrigation during dry spells helps make it more pest-resistant. Thin, drooping branches can be headed back to help develop an more upright form. This will provide clearance beneath the canopy for pedestrians.

Syringa reticulata Japanese Tree Lilac

Pronunciation: sigh-RING-gah reh-tick-you-LAY-tah

ZONES: 4A, 4B, 5A, 5B, 6A, 6B, 7A

HEIGHT: 20 to 25 feet

WIDTH: 15 to 20 feet

FRUIT: elongated and flattened; 0.5"; dry; green; no significant litter; persistent; showy

GROWTH RATE: slow; moderate life span

HABIT: spreading and rounded vase; dense; irregular silhouette; medium texture

FAMILY: *Oleaceae*

LEAF COLOR: medium green; no fall color change

LEAF DESCRIPTION: opposite; simple; entire margin; ovate; deciduous; 3 to 5" long

FLOWER DESCRIPTION: white; showy; early summer

EXPOSURE/CULTURE:

Light Requirement: full sun

Soil Tolerances: all textures; slightly alkaline to acidic; salt; moderate drought

Aerosol Salt Tolerance: good

PEST PROBLEMS: resistant, for a lilac

BARK/STEMS:

Branches droop; routinely grown with multiple trunks; can be trained to grow with a single trunk; cultivars are best suited for training to an upright habit; no thorns

Pruning Requirement: needs little pruning to develop strong structure; pruning is needed to develop a central trunk with an upright habit

Limb Breakage: resistant

Current Year Twig: brown; thick

CULTIVARS AND SPECIES:

'Ivory Silk'—zones 3 to 6, upright oval, nice flowers borne in alternate years; 'Summer Snow'—zones 3 to 6, upright, round shape, persistent seed pods. Both are dense and make nice street trees.

LANDSCAPE NOTES:

Origin: Japan

Usage: container or above-ground planter; parking lot island; buffer strip; highway; near a deck or patio; standard; specimen; sidewalk cutout; street tree; urban tolerant

Although a Lilac, this member of the species is quite different in appearance from those with which gardeners are familiar. Its upright habit varies from symmetrical to irregular. Cultivars, including 'Ivory Silk' and 'Summer Snow', have a more consistent habit and more flowers. This is a very large shrub or small tree. The huge clusters of creamy white flowers, borne in early summer for about two weeks, are the main ornamental feature, but lack the fragrance of the spring-blooming Lilacs—this Lilac's fragrance is more suggestive of privet. It is being used as a street tree in some parts of the country, particularly in sites with overhead power lines. Japanese Tree Lilac is also popular as a garden specimen or as an accent in a shrub border. It deserves to be in any landscape. It provides shade and a colorful spring show for a deck or patio area. Green fruit clusters are somewhat showy when viewed at close range.

The tree is sold as a multi-stemmed specimen or as a single-trunked street tree. The trunk on the cultivars often grows fairly straight to 10 feet and then branches into a stiff, upright, rounded head of foliage. The bark is somewhat showy with prominent lenticels, reminiscent of Black Cherry. As with other Lilacs, the plant as a shrub may need rejuvenation by pruning every few years as it becomes too big for the site. It is perhaps the most pest-resistant Lilac, but that does not mean it is pest-free. Borers and powdery mildew may affect the tree. Regular irrigation during dry spells helps maintain pest-resistance.

Tabebuia caraiba Trumpet Tree

Pronunciation: tab-uh-BOO-yah kah-RAY-bah

ZONES: 10A, 10B, 11

HEIGHT: 15 to 25 feet

WIDTH: 15 feet

FRUIT: typically absent; elongated pod; 3 to 6"; dry; brown; no significant litter

GROWTH RATE: moderate; moderate life span

HABIT: oval; moderately dense; irregular silhouette; medium texture

FAMILY: *Bignoniaceae*

LEAF COLOR: semi-evergreen; silverish green

LEAF DESCRIPTION: opposite; palmate; entire or undulate margin; oval to oblong; 2 to 4" long leaflets

FLOWER DESCRIPTION: yellow; very showy; winter and spring; bee-pollinated

EXPOSURE/CULTURE:

Light Requirement: part shade to full sun; full sun for best flowering

Soil Tolerances: all textures; alkaline to acidic; drought

Aerosol Salt Tolerance: moderate to good

PEST PROBLEMS: resistant

BARK/STEMS:

Older trees have showy, corky bark; branches droop as tree grows and will require some pruning for vehicular or pedestrian clearance beneath canopy; should be grown with a single leader; no thorns

Pruning Requirement: needs pruning to develop good structure; eliminate or thin branches forming included bark in crotches

Limb Breakage: susceptible to breakage at crotch, due to poor collar formation; wood itself is weak and tends to break

Current Year Twig: brown; medium thickness

CULTIVARS AND SPECIES:

The pink Trumpet Tree (*T. heterophylla*) is the one most suited for street tree planting, as it is reportedly more sturdy and durable than *T. caraiba*. *T. impetigenosa* and *T. umbellata* are hardy to zone 9B, with pink flowers borne on bare branches.

LANDSCAPE NOTES:

Origin: Paraguay

Usage: container or above-ground planter; parking lot island; buffer strip; highway; near a deck or patio; specimen; street tree; urban tolerant

An ideal patio, specimen, or lawn tree, the *Tabebuias* are small, semi-evergreen trees with silvery foliage and deeply furrowed, silvery bark on picturesque, contorted branches and trunk. The crown is usually asymmetrical, with two or three major trunks or branches dominating the canopy. During late winter and sporadically throughout the year, they put on a spectacular flower display of a multitude of two- to three-inch-long, golden-yellow, trumpet-shaped blooms borne in terminal flower clusters. The leaves often drop just before the flowers appear. The tree is awe-inspiring in full bloom.

A native of tropical America, Trumpet Tree can be grown in full sun or partial shade on any reasonably fertile soil with moderate moisture. Trees should be protected from frost. Although some will leaf out again following a light freeze, the tree is often weakened and grows poorly. The wood becomes brittle with age and can break easily in strong winds, but this is not usually a problem because trees are small, with an open canopy. This should not be cause to eliminate this beautiful tree from your tree palette. To the contrary, it is one of the most beautiful trees in flower and has a place in most landscapes.

Thin the canopy periodically to keep it open so the wind will not send branches to the ground. Young trees often need staking for support. This will probably increase the longevity of the tree in the landscape. Some nursery operators strip leaves from transplanted trees. Be sure to cut circling roots at the edge of the rootball.

No limitations of major concern. Plants are prone to circling roots which can prevent successful establishment.

Tabebuia chrysotricha Golden Trumpet Tree

Pronunciation: tab-uh-BOO-yah kry-so-TRICK-ah

ZONES: 10A, 10B, 11

HEIGHT: 25 to 35 feet

WIDTH: 25 to 35 feet

FRUIT: elongated pod; 4 to 8"; dry; brown; no significant litter; persistent; showy

GROWTH RATE: rapid; moderate life span

HABIT: oval; moderate density; irregular silhouette; medium texture

FAMILY: *Bignoniaceae*

LEAF COLOR: silvery above, tan and fuzzy beneath

LEAF DESCRIPTION: opposite; palmate; entire or undulate margin; oval to oblong; deciduous or semi-evergreen; 2 to 4" long leaflets

FLOWER DESCRIPTION: yellow; spectacular in bloom; spring; bee-pollinated

EXPOSURE/CULTURE:

Light Requirement: full sun for best flowering

Soil Tolerances: all textures; alkaline to acidic; moderate drought

Aerosol Salt Tolerance: moderate

PEST PROBLEMS: resistant

BARK/STEMS:

Branches droop as tree grows and will require some pruning for vehicular or pedestrian clearance beneath canopy; should be grown with a single leader; no thorns

Pruning Requirement: needs pruning to develop good structure; thin main branches to slow growth and help to develop strong branch crotches

Limb Breakage: susceptible to breakage at crotch, due to poor collar formation; wood itself is weak and tends to break; trees can fall over after established in the landscape due to circling roots in the nursery container

Current Year Twig: brownish-green; medium thickness

LANDSCAPE NOTES:

Origin: South America

Usage: container or above-ground planter; parking lot island; buffer strip; highway; near a deck or patio; specimen; street tree; urban tolerant

An ideal patio, specimen, or lawn tree, Golden Trumpet Tree is often seen as a small, 15- to 25-foot-tall tree (though it can grow taller), with a rounded, spreading canopy in a wind-protected area. Sometimes evergreen but most often deciduous, Golden Trumpet Tree has four-inch-long silvery leaves with tan, fuzzy undersides. These leaves drop for a short period in April to May. This is when the trees put on their heaviest flowering display, with trumpet-shaped, bright yellow blossoms appearing in dense, 2.5- to 8-inch-long terminal clusters. Some trees produce a small number of flowers sporadically throughout the warm season. The eight-inch-long seed capsules that follow are brown, hairy, and persist on the tree through the winter.

Golden Trumpet Tree is very useful as a median street tree for its vivid flower display, asymmetrical habit, and drought tolerance. Once established in humid climates, it can survive on rainfall alone and produces an excellent flower display each year. It also makes a nice tree for planting close to the patio or deck, where it will cast a light to medium shade below the canopy.

A native of tropical America, Golden Trumpet Tree can be grown best in full sun on any reasonably fertile soil with moderate moisture. Trees should be planted only in frost-free locations. Although some will leaf out again following a light freeze, the tree is often weakened and grows poorly. The wood becomes brittle with age and can break easily in strong winds, so it is not often seen larger than about 30 feet tall. But this should not dampen your desire to plant this wonderful tree, because it provides such enjoyment in the mean time.

Trees planted with circling roots often fall over as they mature. Be sure to slice the root ball on container-grown trees. Young trees often need staking for support.

Tabebuia heptaphylla Pink Trumpet Tree

Pronunciation: tab-uh-BOO-yah hep-tah-FYE-lah

ZONES: 9B, 10A, 10B, 11

HEIGHT: 40 to 50 feet

WIDTH: 35 to 50 feet

FRUIT: elongated pod; 3 to 6"; dry; no significant litter

GROWTH RATE: rapid-growing; moderate to long life span

HABIT: rounded vase; moderate density; symmetrical; medium texture

FAMILY: *Bignoniaceae*

Photo Courtesy of W. Hoyt

LEAF COLOR: silvery green

LEAF DESCRIPTION: opposite; palmate; entire or undulate margin; oval to oblong; deciduous to semi-evergreen; 2 to 4" long leaflets

FLOWER DESCRIPTION: pink; spectacular, second to none among flowering trees; spring; bee-pollinated

EXPOSURE/CULTURE:

Light Requirement: full sun for best flowering

Soil Tolerances: all textures; slightly alkaline to acidic; drought

PEST PROBLEMS: probably resistant

BARK/STEMS:

Branches grow mostly upright and will not droop; should be grown with a single leader; no thorns

Pruning Requirement: requires pruning to develop strong structure; space branches along trunk and keep major limbs less than half diameter of trunk

Limb Breakage: unknown

Current Year Twig: brown; medium thickness

LANDSCAPE NOTES:

Origin: Brazil

Usage: parking lot island; buffer strip; highway; shade tree; specimen; street tree

Pink Trumpet Tree is a wonderful specimen tree, reaching a height of 50 feet. It is covered with terminal panicles of pink to rose-purple, two-inch-wide, showy blossoms in spring. Few, if any, other flowering trees can match the beauty of this tree in bloom! Flowers stand out because there are no leaves on the tree during flowering. They contrast nicely with the light gray bark. The palmately compound leaves bear five leaflets, each about 2.5 inches long.

Pink Trumpet Tree would make a nice tree for planting along a boulevard or residential street where there is plenty of soil space for root development. Prune major limbs so they remain about one-half the diameter of the trunk and become well secured to the trunk. This is a tree you will want to keep around once you see it in flower.

Pink Trumpet Tree should be grown in full sun or partial shade on rich, well-drained soil. Other than availability, there are no limitations of major concern. Young trees often need staking for support.

Tabebuia heterophylla

Pink Trumpet Tree, Roble Blanco

Pronunciation: tab-uh-BOO-yah heh-ter-o-FYE-lah

ZONES: 10A, 10B, 11

HEIGHT: 20 to 30 feet

WIDTH: 15 to 25 feet

FRUIT: elongated pod; 4 to 8"; dry; no significant litter; showy

GROWTH RATE: moderate; moderate life span

HABIT: upright oval; open; irregular silhouette; medium texture

FAMILY: *Bignoniaceae*

LEAF COLOR: evergreen; green

LEAF DESCRIPTION: opposite; palmate; entire or undulate margin; oval to oblong; 2 to 4" long leaflets

FLOWER DESCRIPTION: pink; showy; spring and summer

EXPOSURE/CULTURE:

Light Requirement: full sun for best flowering

Soil Tolerances: all textures; alkaline to acidic; drought

Aerosol Salt Tolerance: moderate

PEST PROBLEMS: mostly resistant

BARK/STEMS:

Branches grow mostly upright and will not droop; should be grown with a single leader; no thorns

Pruning Requirement: requires pruning to develop strong structure; reduce length of aggressive branches to develop a more upright tree for planting along streets

Limb Breakage: susceptible to breakage at crotch, due to poor collar formation

Current Year Twig: brown; medium thickness

CULTIVARS AND SPECIES:

T. rosea and *T. glomerata* are not as susceptible to witches' broom disease, which has limited the usefulness of this tree in Puerto Rico. *T. pallida* and *T. impetiginosa* are similar to *heterophylla*.

LANDSCAPE NOTES:

Origin: Puerto Rico and throughout the Caribbean

Usage: parking lot island; buffer strip; highway; near a deck or patio; specimen; street tree; urban tolerant

Pink Trumpet Tree grows at a moderate rate from a slim pyramid when young to a broad silhouette. The palmately compound leaves are evergreen throughout most of its range, but may be briefly deciduous as the new leaves emerge. The showy display of pink or white, bell-shaped blooms appears throughout the spring and summer and is followed by the production of long, slender seedpods.

Pink Trumpet Tree is well suited for use as a street tree or for other areas, such as parking lot islands and buffer strips, where temperatures are high and soil space is limited. It is one of the most urban-tolerant *Tabebuia.* Although it will not create a canopy over a residential street, because of the upright habit, it is well adapted for street tree planting. Develop high, arching branches several years after planting by removing the lower, drooping branches. This branching habit may take several prunings to accomplish. Pink Trumpet Tree can also be used as a shade tree for a residential property near the patio or deck, or can be planted to provide shade to the driveway.

Pink Trumpet Tree should be grown in full sun on almost any well-drained soil, wet or dry. Established trees are moderately salt-tolerant and highly drought-tolerant. This tree is reported to be more tolerant of urban conditions than the Yellow Trumpet Trees.

Witches' broom disease can defoliate trees in Puerto Rico. Leaf hopper causes early defoliation in drought. Young trees often need staking for support.

Tamarindus indica Tamarind

Pronunciation: tam-ah-RINN-duss INN-dih-kah

ZONES: 10A, 10B, 11

HEIGHT: 40 to 50 feet

WIDTH: 40 to 50 feet

FRUIT: elongated pod; 3 to 6"; dry and hard; brown; edible; causes significant litter; persistent; showy

GROWTH RATE: moderate; long-lived

HABIT: rounded vase; dense; slightly irregular silhouette; very fine texture

FAMILY: *Leguminosae*

LEAF COLOR: evergreen; pale green

LEAF DESCRIPTION: alternate; even pinnately compound; entire margin; oval to oblong; 0.25" long leaflets

FLOWER DESCRIPTION: reddish-yellow; not showy; spring

EXPOSURE/CULTURE:

Light Requirement: part shade to full sun

Soil Tolerances: all textures; alkaline to acidic; occasionally wet; drought

Aerosol Salt Tolerance: moderate

PEST PROBLEMS: pest-free

BARK/STEMS:

Branches droop as tree grows and will require some early pruning for vehicular or pedestrian clearance beneath canopy; should be grown with a single leader; no thorns

Pruning Requirement: requires early pruning to develop strong structure; thin aggressive branches to help develop strong branch crotches without included bark

Limb Breakage: resistant

Current Year Twig: greenish-gray; slender

LANDSCAPE NOTES:

Origin: East Indies

Usage: parking lot island; buffer strip; highway; shade tree; specimen; possible street tree if planted more than 8 feet away; urban tolerant

A frost-tender, tropical, evergreen tree, Tamarind is densely foliated with compound, feathery leaflets that give the broad, spreading crown a light, airy effect. Tamarind may reach a height of 65 feet and a spread of 50 feet, but is more often seen smaller because it has not been widely planted. The delicate leaflets on young trees cast a diffuse, dappled shade which will allow enough sunlight to penetrate for a lawn to thrive beneath this upright, dome-shaped tree. The canopy thickens with age creating dense shade.

The twigs and branches of Tamarind without included bark are very resistant to wind, making it especially useful as a shade or street tree for breezy locations. But Tamarind has only low to moderate salt tolerance, so do not locate it close to the beach. In spring, small red and yellow flowers appear on short racemes and are followed by the production of brittle, brown, six-inch-long, velvety pods. These sticky pods are filled with a sweet-sour, dark brown paste that surrounds two or three seeds. They normally dry up and do not become mushy, but some people will undoubtedly object to the hard fruit falling on sidewalks or streets. Tamarind is grown commercially in the tropics for production of this edible paste, which is used as an ingredient for steak sauce, soft drinks, chutneys, and curries.

Tamarind should be grown only in frost-free regions in full sun on moist, fertile, sandy soil. Care should be taken in the placement of Tamarind, as the seed pods may be messy for a short period if they drop on hard surfaces. Also be sure to maintain a strong tree structure with major branches well-spaced along one central trunk.

No limitations of major concern, though surface roots grow large on old trees and can raise pavement.

Taxodium ascendens Pondcypress

Pronunciation: tax-O-dee-um ah-SENN-dens

ZONES: 5B, 6A, 6B, 7A, 7B, 8A, 8B, 9A, 9B

HEIGHT: 50 to 60 feet

WIDTH: 12 to 20 feet

FRUIT: round; 1"; dry; green turning brown; attracts wildlife; no significant litter

GROWTH RATE: rapid; long-lived

HABIT: columnar; pyramidal; open; symmetrical; fine texture

FAMILY: *Taxodiaceae*

LEAF COLOR: bright green; showy, copper fall color

LEAF DESCRIPTION: ranked; simple; scale-like; deciduous

FLOWER DESCRIPTION: not showy; spring

EXPOSURE/CULTURE:

Light Requirement: part shade to full sun

Soil Tolerances: all textures; slightly alkaline to acidic; flooded soil; moderate salt and drought

Aerosol Salt Tolerance: good

PEST PROBLEMS: resistant

BARK/STEMS:

Branches grow mostly horizontally and will not droop; normally grows with a single leader; no thorns

Pruning Requirement: needs little or no pruning to develop strong structure

Limb Breakage: resistant

Current Year Twig: green; slender

CULTIVARS AND SPECIES:

'Prairie Sentinel' is slightly more narrow than the species.

LANDSCAPE NOTES:

Origin: Virginia to Louisiana, including all of Florida

Usage: reclamation; specimen; street tree; parking lot island; screen

Similar to Baldcypress in that the trunk is perfectly straight 50 to 60 feet tall, Pondcypress has a narrower crown, is smaller, and has a more open habit. It is found along the edges of streams and around the edge of swampy ground where water is standing (Baldcypress is usually found along stream banks). The scale-shaped leaves are arranged in an upright row formation along the branches when young, giving a somewhat stiffer and more upright appearance than Baldcypress. The leaves turn an attractive light brown to copper in fall before dropping, but the bare branches and light brown, ridged bark provide much landscape interest during the winter. The trunk grows unusually thick toward the base, even on young trees. This is thought to provide support for the tree in its wet habitat. The small seeds are used by some birds and squirrels.

Although often seen at water's edge, where it will develop fewer "knees" than Baldcypress, Pondcypress can also be grown in dry locations and could make an attractive street tree for a very narrow space. Cypress knees do not generally form at all on these drier sites. It provides a good vertical accent to the landscape and should be used more often in urban areas. The roots do not appear to lift sidewalks and curbs as readily as some other species. Its delicate foliage affords light, dappled shade, and the heartwood is quite strong and resistant to rot. However, most lumber available at lumber yards today is sapwood and is not resistant to rot.

Pondcypress is ideal for wet locations, such as its native habitat of stream banks and mucky soils, but the trees will also grow quite well on almost any soil, including clay, silt, and sand, except alkaline soils with a pH above 7.5. Its drought-avoidance mechanism allows it to drop leaves in extended dry periods, but little harm appears to come to the tree. Pondcypress is relatively maintenance-free, requiring pruning only to remove dead wood and unwanted lower branches that persist on the tree. It maintains a desirably straight trunk without pruning and does not form double or multiple leaders as do many large trees.

For limitations see *T. distichum.*

Taxodium distichum Baldcypress

Pronunciation: tax-O-dee-um diss-TISH-um

ZONES: 5A, 5B, 6A, 6B, 7A, 7B, 8A, 8B, 9A, 9B, 10A, 10B

HEIGHT: 60 to 80 feet

WIDTH: 25 to 35 feet

FRUIT: oval; 1"; dry; green turning brown; attracts wildlife; no significant litter

GROWTH RATE: rapid; long-lived

HABIT: pyramidal when young, oval later; dense; symmetrical; fine texture

FAMILY: *Taxodiaceae*

LEAF COLOR: light green; showy copper or yellow fall color

LEAF DESCRIPTION: alternate; simple; lanceolate or linear; deciduous; 0.33 to 0.75" long

FLOWER DESCRIPTION: brown; not showy; spring

EXPOSURE/CULTURE:

Light Requirement: grows best in full sun; tolerates partial shade

Soil Tolerances: all textures; slightly alkaline to acidic; flooded soil; moderate salt; drought; compaction

Aerosol Salt Tolerance: moderate

PEST PROBLEMS: resistant

BARK/STEMS:

Branches droop as tree grows and will require pruning for vehicular or pedestrian clearance beneath canopy; normally grown with a single leader; no thorns

Pruning Requirement: needs little or no pruning to develop strong structure

Limb Breakage: resistant

Current Year Twig: greenish turning tan; slender

CULTIVARS AND SPECIES:

'Monarch of Illinois' has a very wide-spreading form and 'Shawnee Brave' has a narrow, pyramidal form, 15 to 20 feet wide. 'Pendens' has drooping branchlets and large cones. *T. distichum* var. *nutans* (*Taxodium ascendens*) is native to wet, boggy areas with standing water, whereas *T. distichum* is more common along streams.

LANDSCAPE NOTES:

Origin: Maryland along the coast to Texas, and up the Mississippi valley

Usage: parking lot island; buffer strip; highway; reclamation; shade tree; specimen; clipped hedge or screen; sidewalk cutout; street tree; urban tolerant

Baldcypress, the state tree of Louisiana, eventually develops into a broad-topped, spreading, open specimen when mature. Trees grow at a moderately fast rate, reaching 40 to 50 feet in about 15 to 25 years. Although it is native to wetlands, growth is often faster on moist, well-drained soil. The trunk is unusually flared at the base.

Although often seen at water's edge, where it will develop "knees" or root projections that extend above the water, Baldcypress can also be grown in dry locations and makes an attractive lawn, street, or shade tree. Cypress knees do not generally form on these drier sites. Cities from Charlotte, North Carolina, Dallas, Texas, to Tampa, Florida, currently use it as a street tree, and it should be used more extensively throughout its range in urban landscapes. It provides a good vertical accent to the landscape and should be used more often in urban areas. Surprisingly, roots do not appear to lift sidewalks and curbs as readily as some other species.

Baldcypress is relatively maintenance-free, requiring pruning only to remove dead wood and unwanted lower branches that persist on the tree. It maintains a straight trunk and a moderately dense canopy and does not form double or multiple leaders as do many other large trees. Cypress makes a very beautiful, fine-textured, tall clipped hedge or screen.

Bagworms, twig blight, canker, and mites can be minor problems. Mites can be particularly troublesome in dry summers without irrigation, causing early leaf browning and defoliation in mid to late summer. Dry soil causes premature leaf fall, but trees suffer no long-term damage.

Taxodium mucronatum Montezuma Baldcypress

Pronunciation: Tax-O-dee-um mew-crow-NAY-tum

ZONES: 8A, 8B, 9A, 9B, 10A, 10B, 11

HEIGHT: 50 to 80 feet

WIDTH: 25 to 35 feet

FRUIT: oval; 1"; dry; green to tan or brown; no significant litter

GROWTH RATE: moderate; long-lived

HABIT: pyramidal; dense; symmetrical; fine texture

FAMILY: *Taxodiaceae*

LEAF COLOR: pale green

LEAF DESCRIPTION: alternate and two-ranked; simple; lanceolate to linear; semi-evergreen

FLOWER DESCRIPTION: not showy; spring

EXPOSURE/CULTURE:

Light Requirement: full sun

Soil Tolerances: all textures; acidic; wet soil; moderate drought

PEST PROBLEMS: resistant

BARK/STEMS:

Branches droop as tree grows and will require pruning for vehicular or pedestrian clearance beneath canopy; should be grown with a single leader; no thorns

Pruning Requirement: needs little pruning to develop strong structure

Limb Breakage: resistant

Current Year Twig: green; slender

LANDSCAPE NOTES:

Origin: southern Texas into Mexico

Usage: specimen

Montezuma Baldcypress, Mexico's national tree, is a huge tree in its native habitat and is broadly pyramidal when young, with a dense crown. Like Baldcypress, it eventually develops into a broad-topped, spreading, open specimen when mature, with pendulous branches. Though it is capable of reaching 100 to 150 feet in height on a single stem, most landscape specimens will not reach this height. The needle-like leaves are deciduous only in the colder sections of its range, remaining evergreen elsewhere. The trunk grows unusually thick toward the base, even on young trees. One of the most vivid examples of this trait is the Tule Cypress found growing in Oaxaca, Mexico, which sports a trunk diameter of 35 to 40 feet, or a circumference of 115 feet—the tree itself is only 140 feet tall.

It provides a good specimen in the landscape and could be tried in urban areas. Its wide-spreading habit in the eastern United States suits it only for open areas such as parks and golf courses. Trees in the western United States are more upright-pyramidal. The roots probably will not lift sidewalks and curbs as readily as some other species. The delicate, feathery foliage affords light, dappled shade. The heartwood of Montezuma Baldcypress is quite strong and resistant to rot. Trees tolerate snow loads well.

Although not truly a swamp tree, as are other *Taxodium,* Montezuma Baldcypress still grows well in moist soil. Its seeds will only germinate in either water or wet soil. Montezuma Baldcypress is relatively maintenance-free, requiring pruning only to remove dead wood and unwanted lower branches that persist on the tree. It maintains a desirably straight trunk and does not form double or multiple leaders as do many other large trees. Growth was stunted in the Auburn University (zone 8) test plots where no irrigation was provided. This tree is not extensively used in the east, but testing should be encouraged.

The tree is not readily available and needs room to expand its canopy. Limitations are probably similar to those of *T. distichum.*

Taxus baccata English Yew

Pronunciation: TAX-us bah-KAY-tah

ZONES: 6A, 6B, 7A, 7B

HEIGHT: 25 to 30 feet

WIDTH: 15 to 30 feet

FRUIT: round; 0.25"; fleshy, red aril; not especially showy; no significant litter; persistent; poisonous

GROWTH RATE: slow-growing; long-lived in well-drained soil

HABIT: pyramidal when young, rounded later; dense; symmetrical; medium texture

FAMILY: *Taxaceae*

LEAF COLOR: evergreen; very dark green

LEAF DESCRIPTION: spiral; simple; linear; needle-leaf evergreen; 0.5 to 1" long

FLOWER DESCRIPTION: greenish-yellow; not showy; spring

EXPOSURE/CULTURE:

Light Requirement: full sun for best growth

Soil Tolerances: all textures; slightly alkaline to acidic; good drainage is essential; drought-sensitive

Aerosol Salt Tolerance: moderate

PEST PROBLEMS: mostly resistant, except for canker

BARK/STEMS:

Extremely showy, exfoliating bark and trunk form; branches droop on older trees, but many cultivars are upright; usually grown with several trunks; no thorns

Pruning Requirement: needs little pruning to develop strong trunk and branch structure

Limb Breakage: resistant, even to snow

Current Year Twig: green; slender

CULTIVARS AND SPECIES:

'Repandens' and 'Amersfoort' are hardier (zone 5); *T. chinensis* is heat tolerant; *T.* x *media* (zone 4) has cultivars with a variety of forms from ground covers to upright columnar trees; *T. caspidata* 'Capital' and 'Dark Green Pyramidal' have a nice form (zone 4B). *Cephalotaxus harringtonia* (zone 5B), *C. fortunei* (zone 6—needs shade), and *C. koreana* (zone 5B) are heat tolerant and thrive on clay, loam, or sand.

LANDSCAPE NOTES:

Origin: Europe and the Mediterranean

Usage: bonsai; hedge or screen; trainable as a standard; specimen; Christmas tree

English Yew is most easily recognized in its trimmed form as a dense hedge or screen, or shaped into topiary, but this dark evergreen makes an outstanding specimen. It is nearly indescribable beautiful when grown in the open as a specimen tree, and should be more commonly grown in this fashion.

The attractive, reddish-brown trunk is often fluted and can become quite large. The inconspicuous flowers appear in spring and are followed by the production of small, showy, red, fleshy fruits, which contain one of the most poisonous seeds known, capable of poisoning both man and livestock. Taxine, the toxic chemical, is found in the leaves, bark, and hard part of the seed. Taxol, derived from the bark, has cancer-fighting properties.

Yew is used in the tree form primarily to create a screen, to develop into a topiary, or to plant as a large clipped specimen. The dense crown makes it especially suited for a screen, provided plants are located in full-day sun and they are given plenty of room to spread. If lateral space is limited, select one of the narrow, upright cultivars.

Trained in a small tree, it is most striking in any well-drained landscape. It is well suited for planting in small landscapes due to its small size and slow growth rate. It is essential that the soil be well-drained; if water stands around the potential planting site after rain, plant another tree, or find a spot with better drainage.

A stem canker that kills trees is beginning to show up on major branches. Taxus mealybug, black vine weevil, taxus scale, and yew-gall midge can be found on Yews, but they rarely limit growth. Plant only in extremely well-drained soil for best growth and survival.

Tecoma stans Yellow-Elder, Yellowbells

Pronunciation: teh-KOE-mah stans

ZONES: 10B, 11

HEIGHT: 20 to 30 feet

WIDTH: 20 to 30 feet

FRUIT: elongated; 4 to 7"; dry; brown; causes some litter; showy

GROWTH RATE: rapid; short-lived

HABIT: oval to round; moderate to dense; somewhat irregular when young, symmetrical later; medium texture

FAMILY: *Bignoniaceae*

LEAF COLOR: light green

LEAF DESCRIPTION: opposite; odd pinnately compound; serrate margin; lanceolate to ovate; semi-evergreen; 3 to 5" long leaflets

FLOWER DESCRIPTION: yellow; showy; year-round

EXPOSURE/CULTURE:

Light Requirement: full sun for best flowering

Soil Tolerances: all textures; alkaline to acidic; drought

Aerosol Salt Tolerance: moderate

PEST PROBLEMS: pest-free

BARK/STEMS:

Branches droop as tree grows and will require pruning for vehicular or pedestrian clearance beneath canopy; can be trained to grow with multiple trunks or a single trunk; no thorns

Pruning Requirement: requires pruning to develop strong, upright structure and to control weedy habit; space branches along a short, central trunk

Limb Breakage: susceptible

Current Year Twig: green; slender

LANDSCAPE NOTES:

Origin: Caribbean region

Usage: container or above-ground planter; espalier; parking lot island; buffer strip; highway; near a deck or patio; specimen; possible street tree

This spreading, fast-growing evergreen shrub or small tree is noted for its brilliant, bell-shaped, fragrant yellow flowers. It is the official flower of the Virgin Islands. Reaching full bloom in fall, Yellow-Elder produces some flowers with each flush of new growth and therefore has some color most of the year. In full bloom, the tree is literally covered with yellow flowers.

Though sometimes trained to a single trunk, Yellow-Elder is most often used as a specimen or mixed into a shrub border. The somewhat weedy growth requires pruning to control shape, but it is worth the effort due to the brilliant flowers. The tree may even need staking for a while to hold it erect. It can be pruned after the flowers have passed to stimulate a new growth flush and additional flowers. Some gardeners maintain the tree in this fashion for many years, much like a Crape Myrtle. Its small stature allows it to be used beneath power lines as a residential street tree in some situations. It would also grow well in a sidewalk cutout.

Growing in full sun on any well-drained soil, Yellow-Elder survives on rain alone in humid climates, making it well suited to naturalized and low-maintenance gardens. It would also make a nice patio tree and is suited for planting in parking lot islands and medians. Normal life expectancy may only be 10 to 15 years, however.

Plants grow easily from seed or from cuttings. Seedlings are easily transplanted and will bloom within two years. Yellow-Elder is relatively pest-free, with chewing insects and scale being only minor problems.

Terminalia catappa Tropical-Almond, Indian-Almond

Pronunciation: tur-mih-NAY-lee-ah kah-TAH-pah

ZONES: 10B, 11

HEIGHT: 30 to 50 feet

WIDTH: 30 to 50 feet

FRUIT: elongated; oval; 2.5"; fleshy; green turning tan; somewhat showy; causes significant litter

GROWTH RATE: moderate; moderate life span

HABIT: variable, pyramidal to vase; moderately dense to open; nearly symmetrical; coarse texture

FAMILY: *Combretaceae*

LEAF COLOR: green; showy, red fall (winter) color

LEAF DESCRIPTION: alternate; simple; entire margin; obovate; deciduous; 8 to 12" long

FLOWER DESCRIPTION: green; not showy; spring

EXPOSURE/CULTURE:

Light Requirement: full sun for best form

Soil Tolerances: all textures; alkaline to acidic; drought

Aerosol Salt Tolerance: high

PEST PROBLEMS: resistant

BARK/STEMS:

Buttressed and wide at base; branches droop as tree grows and will require pruning for vehicular or pedestrian clearance beneath canopy; young trees grow with a central leader and whorled or tiered branches; should be maintained with a single leader; no thorns

Pruning Requirement: vase-shaped trees require pruning to develop strong structure

Limb Breakage: susceptible to breakage at crotch, due to poor collar formation on vase-shaped trees

Current Year Twig: brownish-green; stout

LANDSCAPE NOTES:

Origin: South Africa, East Indies

Usage: parking lot island; buffer strip; highway; shade tree; specimen; street tree; urban tolerant

Tropical-Almond, a deciduous tree, forms a symmetrical, upright silhouette in youth with horizontal branches reaching 35 feet or more in width. The branches are arranged in obvious tiers, giving the tree a pagoda-like shape. As the tree grows older, the crown spreads and flattens on the top to form a wide-spreading vase shape. The large, 12-inch-long and 6-inch-wide, glossy green, leathery leaves change to beautiful shades of red, yellow, and purple before dropping in winter. Due to their large size, the old leaves may be considered a nuisance by some people. The leaves are quickly replaced by new growth, so the tree is bare for only a short period of time.

The inconspicuous blossoms are followed by the edible fruits. These drupes are 2.5 inches long and mature from green to yellow or red during the summer. The outside husk is corky and fibrous, with a thin, green, inner flesh. The inside holds the edible, almond-like kernel. The fruit is high in tannic acid and this could stain cars, pavement, and sidewalks. Aggressive surface roots often form.

The tree may be best suited for planting along the coast as a park or shade tree to provide dense shade. People may object to the large leaves and the fruit that falls from the tree if it is used as a street tree, and the tannic acid may be a problem near parked cars. Branches droop and require regular maintenance pruning to allow for vehicle clearance beneath the canopy. However, it would make a nice tree for a median or along a boulevard where this would cause less of a nuisance.

Tropical-Almond should be grown in full sun on any well-drained soil. Plants are quite tolerant of wind, salt, and drought, but do need protection from freezing temperatures. Trees perform best if mulched and regularly fertilized. Leaf spot disease and thrips can be troublesome.

Thrinax morrisii Key Thatch Palm

Pronunciation: THRY-nacks more-ISS-ee-eye

ZONES: 10B, 11

HEIGHT: 20 feet

WIDTH: 6 to 8 feet

FRUIT: round; less than 0.5" fleshy; white; showy; no significant litter; persistent

GROWTH RATE: slow-growing; moderate life span

HABIT: palm; open; symmetrical; coarse texture

FAMILY: *Arecaceae*

LEAF COLOR: silvery beneath, green above

LEAF DESCRIPTION: spiral; simple; palmate; 36" across

FLOWER DESCRIPTION: white; somewhat showy; spring

EXPOSURE/CULTURE:

Light Requirement: part shade to full sun

Soil Tolerances: all textures; alkaline to acidic; occasionally wet; drought

Aerosol Salt Tolerance: high

PEST PROBLEMS: pest-resistant

BARK/STEMS:

Single, thin trunk; no thorns; no crown shaft

Pruning Requirement: remove dead fronds only

Limb Breakage: resistant

CULTIVARS AND SPECIES:

T. radiata is native to south Florida and the West Indies and has similar culture requirements. The foliage lacks the silvery undersides. *T. parviflora* is native to Jamaica and is not common in the nursery trade.

LANDSCAPE NOTES:

Origin: Florida Keys

Usage: container or above-ground planter; near a deck or patio; specimen

This native North American palm slowly grows to mature height. Its smooth, slender trunk is topped with 3.5-foot-wide, beautiful fronds that are a shimmering silver-white underneath and are a source for thatch. The small but showy, white spring flowers are followed by small, round, fleshy, white fruits. The showy fruits are very prominently held out from the trunk just below the lowest frond.

This palm is small enough to be popular in residential landscapes. It is often planted as a single specimen or in groups of three to accent an area. Due to the coarse texture, it makes a nice entryway palm planted to attract attention to the front door of a building. It often looks best planted in a mulched area or in a bed with a low-growing groundcover.

Key Thatch palm should be grown in full sun or partial shade and is highly drought- and salt-tolerant, making it ideal for seaside applications. No limitations of major concern.

Thuja occidentalis Arborvitae, Northern White-Cedar

Pronunciation: THEW-jah ox-see-den-TAY-liss

ZONES: 2A, 2B, 3A, 3B, 4A, 4B, 5A, 5B, 6A, 6B, 7A, 7B

HEIGHT: 25 to 40 feet

WIDTH: 10 to 12 feet

FRUIT: elongated; oval; 0.5"; dry; brown; no significant litter; persistent

GROWTH RATE: slow-growing; long-lived

HABIT: upright oval or pyramidal; dense; symmetrical; fine texture

FAMILY: *Cupressaceae*

LEAF COLOR: evergreen; green

LEAF DESCRIPTION: alternate; scale-like

FLOWER DESCRIPTION: yellow; not showy; spring

EXPOSURE/CULTURE:

Light Requirement: part shade to full sun

Soil Tolerances: all textures; slightly alkaline to acidic; wet soil; moderate drought

Aerosol Salt Tolerance: low to moderate

PEST PROBLEMS: somewhat sensitive

BARK/STEMS:

Usually grows with a single leader; no thorns

Pruning Requirement: needs little pruning to develop good structure

Limb Breakage: wood is soft, resistant, or susceptible to ice and snow damage

Current Year Twig: brown; green; slender

CULTIVARS AND SPECIES:

White-Cedar has given rise to many cultivars, many of which are shrubs. 'Booth Globe'—low, rounded with a flat top; 'Compacta'—dense and compact; 'Compacta Erecta'—semi-dwarf, pyramidal; 'Douglasi Pyramidalis'—dense, columnar; 'Emerald Green'—good winter color; 'Ericoides'—dwarf, brownish foliage in winter; 'Fastigiata'—narrow, columnar; 'Globosa'—dense, rounded; 'Hetz Junior'—dwarf, wider than it is tall; 'Hetz Midget'—slow grower, quite dwarf, rounded; 'Hovey'—low and rounded; 'Little Champion'—globe shaped; 'Lutea'—yellow foliage; 'Nigra'—dark green foliage in winter, pyramidal; 'Pumila' (Little Gem)—rounded, dwarf; 'Pyramidalis'—narrow pyramidal form; 'Rheingold'—rounded form with yellow to bronze new growth; 'Rosenthalli'—dense, pyramidal; 'Techny'—pyramidal, dark green, hardy; 'Umbraculifera'—flat-topped; 'Wareana'—low and dense, pyramidal; 'Woodwardii'—rounded and spreading.

LANDSCAPE NOTES:

Origin: Manitoba to Nova Scotia, south to North Carolina

Usage: hedge; buffer strip; highway; wetland reclamation; specimen

This slow-growing tree prefers a wet or moist, rich soil. Transplanting is moderately easy if plants are root-pruned and either balled-in-burlap or potted. White-Cedar likes high humidity and tolerates wet soils, though some forms tolerate drought. The foliage turns brownish in winter, especially on cultivars with colored foliage and on exposed sites open to the wind. The superior cultivars have been selected to retain green foliage color throughout the winter.

White-Cedar is best used as a screen or hedge planted on 8- to 10-foot-centers. There are better specimen plants than White-Cedar, but it can be placed at the corner of a building or other area to soften or frame a view. Many of the natural stands in the United States have been cut for timber. Some remain along lakes and rivers throughout the eastern United States.

Bagworms can devour large quantities of foliage very quickly. Mites can discolor foliage quickly in the summer. Leaf blight causes brown spots on the leaves in late spring. The affected foliage appears scorched, then drops.

Thuja plicata Giant Arborvitae, Western Redcedar

Pronunciation: THEW-jah ply-KAY-tah

ZONES: 5B, 6A, 6B, 7A, 7B, 8A, (8B)

HEIGHT: 50 to 70 feet

WIDTH: 15 to 25 feet

FRUIT: elongated; 0.5"; dry; brown; no significant litter; persistent

GROWTH RATE: moderate; long-lived

HABIT: columnar, oval, narrow pyramid; dense; symmetrical; fine texture

FAMILY: *Cupressaceae*

LEAF COLOR: evergreen; dark green

LEAF DESCRIPTION: simple; scale-like

FLOWER DESCRIPTION: yellow; not showy; spring

EXPOSURE/CULTURE:

Light Requirement: mostly shaded to full sun

Soil Tolerances: all textures; alkaline to acidic; occasionally wet soil; drought-sensitive

Aerosol Salt Tolerance: moderate

PEST PROBLEMS: somewhat susceptible

BARK/STEMS:

Showy bark; branches grow mostly upright and horizontal and will not droop; should be grown with a single leader; resists deer browsing; no thorns

Pruning Requirement: needs little or no pruning to develop strong structure

Limb Breakage: susceptible, especially in ice storms

Current Year Twig: brownish green; slender

CULTIVARS AND SPECIES:

'Atrovirens'—excellent glossy green foliage; 'Canadian Gold'—golden foliage; 'Fastigiata' ('Hogan')—well adapted to zone 7, dense columnar silhouette. 'Fastigiata' is very resistant to bagworms; they do not appear to infest this cultivar as much as the species.

LANDSCAPE NOTES:

Origin: south Alaska to Oregon and Idaho

Usage: hedge; screen; buffer strip; highway; specimen; Leyland Cypress replacement

A native western North American tree, Giant-Cedar can reach 180 to 200 feet in height in some areas of the Northwest, but is more often seen at 50 to 70 feet in height in landscapes. Forming an upright pyramidal silhouette with strongly horizontal branches, Giant-Cedar is an evergreen with fragrant, delicate needles that generously clothe the branches, casting dense shade beneath the tree. The insignificant yellow flowers are followed by small, half-inch cones.

Tolerating shearing quite well, Giant-Cedar is ideal for use as a hedge or screen, or as a specimen for a large landscape. The wood of this tree is commercially used in North America for the manufacture of roof shingles, deck boards, and siding. The split trunks were often used by Native Americans for making totem poles and canoes. Due to its narrow crown, it works well close to buildings where soil is frequently alkaline and drainage is poor.

Giant-Cedar naturally occurs on river banks, swamps, and even bogs. It grows best in full sun or partial shade on moist, well-drained, fertile soil, and prefers a moist atmosphere. It grows naturally in alkaline soil, but could be stunted on dry sites. Provide irrigation during the summer or plant in an area with moist soil.

Trees can be defoliated by bagworm, but apparently not as extensively as *Thuja occidentalis*.

Tilia americana American Linden, American Basswood

Pronunciation: TILL-ee-ah ah-mer-eh-KAY-nah

ZONES: 3A, 3B, 4A, 4B, 5A, 5B, 6A, 6B, 7A, 7B, 8A, 8B

HEIGHT: 50 to 80 feet

WIDTH: 35 to 50 feet

FRUIT: round; 0.33"; dry; green turning tan; somewhat showy; no significant litter; persistent

GROWTH RATE: moderate; long-lived

HABIT: pyramidal or oval; dense; coarse texture

FAMILY: *Tiliaceae*

LEAF COLOR: dark green; showy, yellow fall color

LEAF DESCRIPTION: alternate; simple; serrate margin; ovate; deciduous; 4 to 8" long

FLOWER DESCRIPTION: greenish-yellow; pleasant fragrance; showy; summer; attracts bees

EXPOSURE/CULTURE:

Light Requirement: unusually shade tolerant, but grows well in full sun

Soil Tolerances: all textures; alkaline to acidic; salt-sensitive; moderate drought

Aerosol Salt Tolerance: low

PEST PROBLEMS: mostly resistant

BARK/STEMS:

Branches droop as tree grows and will require pruning for vehicular or pedestrian clearance beneath canopy; should be grown with a single leader; no thorns

Pruning Requirement: requires some pruning to develop strong structure; remove upright branches that may compete with leader

Limb Breakage: susceptible in ice storms

Current Year Twig: brown; green; medium thickness

CULTIVARS AND SPECIES:

'Fastigiata' is narrowly pyramidal, with fragrant yellow flowers; 'Legend' is resistant to leaf rust, has a pyramidal shape, and grows with a single, straight trunk and upright, well-spaced branches. Current thinking lumps *T. carolina*, *T. floridana*, *T. georgiana*, and *T. heterophylla* with *T. americana*. These can all be used in wide tree lawns along streets.

LANDSCAPE NOTES:

Origin: Minnesota to Maine, south to Tennessee (*Tilia* genus extends south into Florida)

Usage: shade tree; some cultivars are suited as street trees

The tree is pyramidal when young but develops into a striking specimen, with an upright, oval canopy atop a tall, straight trunk. The lower branches remain on the tree and gently drape toward the ground before sweeping up in a gentle curve. The four- to eight-inch-long, heart-shaped leaves are dark green throughout the year, fading only to pale green or yellow before dropping in autumn. In June, the trees produce abundant, two- to three-inch-wide clusters of very fragrant, light yellow blooms which are extremely attractive to bees, which make a delicious honey from their harvests. The trunk can grow to six feet or more across on mature specimens.

This tree is large and needs plenty of room to develop. Branches should be well spaced along a central trunk to allow development of a durable structure. Left unpruned, weak crotches with included bark can develop, but the wood is flexible so branches usually do not break from the tree unless laden with ice. Plant it as a specimen or shade tree on a commercial property where there is plenty of soil space available for root expansion. Be prepared to remove sprouts periodically from the base of the trunk.

No limitations are usually serious. Some scorch may occur in dry years. The main pests are aphids, although Japanese beetle, borers, lace bug, caterpillars, and others can all be troublesome. Aphids will secrete a honeydew that results in a dark soot over objects below the tree, such as parked cars or lawn furniture. Anthracnose, leaf blight, canker, leaf spots, powdery mildew, and verticillium wilt can also infect the tree. A sudden cold snap can split trunks open.

Tilia x 'Redmond' 'Redmond' Linden, Basswood

Pronunciation: TILL-ee-ah

ZONES: 3A, 3B, 4A, 4B, 5A, 5B, 6A, 6B, 7A, 7B, 8A, 8B

HEIGHT: 50 to 70 feet

WIDTH: 30 to 40 feet

FRUIT: round; 0.33"; dry; green turning tan; somewhat showy; no significant litter on young trees; persistent

GROWTH RATE: moderate to rapid; long-lived

HABIT: tight pyramid; dense; moderately coarse texture

FAMILY: *Tiliaceae*

LEAF COLOR: medium green; showy, yellow fall color

LEAF DESCRIPTION: glossy; alternate; simple; serrate margin; ovate; deciduous; 3 to 6" long

FLOWER DESCRIPTION: greenish-yellow; pleasant fragrance; showy; summer; attracts bees

EXPOSURE/CULTURE:

Light Requirement: unusually shade tolerant, but grows well in full sun

Soil Tolerances: all textures; alkaline to acidic; salt-sensitive; drought-sensitive

Aerosol Salt Tolerance: low

PEST PROBLEMS: mostly resistant

BARK/STEMS:

Branches grow mostly upright and do not droop; should be grown with a single leader; no thorns

Pruning Requirement: requires little pruning to develop strong structure; basal sprouts need occasional removal

Limb Breakage: resistant, but break during ice storms

Current Year Twig: brown; green; medium thickness

CULTIVARS AND SPECIES:

'Fastigiata' is narrowly pyramidal, with fragrant yellow flowers; 'Legend' is resistant to leaf rust, has a pyramidal shape, and grows with a single, straight trunk and upright, well-spaced branches. These can all be used in large tree lawns along streets.

LANDSCAPE NOTES:

Origin: Nebraska

Usage: shade tree; street tree; parking lot island; urban tolerant

The tree is pyramidal when young but develops into a striking specimen, with an upright, oval canopy atop a tall, straight trunk. The lower branches remain on the tree and gently drape toward the ground before sweeping up in a gentle curve. In June, the trees produce abundant, two- to three-inch-wide clusters of very fragrant, light yellow blooms that are extremely attractive to bees, which make a delicious honey from their harvests. The trunk can probably grow to four feet or more across on mature specimens. 'Redmond' compartmentalizes decay very well and is more vigorous than *Tilia cordata*.

This tree is large and needs plenty of room to develop. Branches should be well spaced along a central trunk to allow development of a durable structure. Plant it as a specimen or shade tree on a commercial property where there is plenty of soil space available for root expansion. It is also well suited as a street tree where there is at least a six-foot-wide tree lawn. Be prepared to remove sprouts periodically from the base of the trunk.

No limitations are usually serious. Some scorch may occur in dry years. The main pests are aphids, although Japanese beetle, borers, lace bug, caterpillars, and others can all be troublesome problems. Aphids will secrete a honeydew that results in a dark soot over objects below the tree, such as parked cars or lawn furniture.

Tilia cordata Littleleaf Linden

Pronunciation: TILL-ee-ah core-DAY-tah

ZONES: 4A, 4B, 5A, 5B, 6A, 6B, 7A

HEIGHT: 60 to 70 feet

WIDTH: 35 to 50 feet

FRUIT: round; 0.33"; dry; brown or tan; no significant litter; showy

GROWTH RATE: moderate; long-lived

HABIT: oval or pyramidal; dense; symmetrical; medium texture

FAMILY: *Tiliaceae*

LEAF COLOR: light green; showy yellow fall color

LEAF DESCRIPTION: alternate; simple; serrate margin; cordate to orbiculate; deciduous; 2 to 4" long

FLOWER DESCRIPTION: yellow; pleasant fragrance; showy; summer; attract bees

EXPOSURE/CULTURE:

Light Requirement: part shade to full sun

Soil Tolerances: all textures; alkaline to acidic; salt-sensitive; occasionally wet soil; moderate drought; compaction

Aerosol Salt Tolerance: low

PEST PROBLEMS: mostly tolerant

BARK/STEMS:

Branches droop as tree grows and will require pruning for vehicular or pedestrian clearance beneath canopy; should be grown with a single leader; no thorns

Pruning Requirement: requires some early pruning to develop strong structure; eliminate double leaders; basal sprouts need periodic removal

Limb Breakage: resistant

Current Year Twig: brown; green; slender

CULTIVARS AND SPECIES:

'Chancellor'—upright, becoming pyramidal, straight trunk; 'Corinthian'—narrow pyramid; 'Glenleven'—conical, straight trunk, more open canopy; 'Greenspire'—straight trunk, radially produced branches (may be the lowest maintenance cultivar); 'June Bride'—pyramidal, glossy leaves, profuse flowering, very slow-growing; 'Pyramidalis'—widely pyramidal; 'Rancho'—narrow, upright growth habit, and small leaf, a beautiful tree.

LANDSCAPE NOTES:

Origin: Europe

Usage: above-ground planter; screen; parking lot island; buffer strip; highway; shade tree; specimen; sidewalk cutout; street tree; urban tolerant

Architects enjoy using this tree for its predictably symmetrical shape. Littleleaf Linden is a prolific bloomer, with small, fragrant flowers appearing in late June into July. Many bees are attracted to the flowers, and the dried flowers persist on the tree for some time. Japanese beetles often skeletonize Linden foliage in certain areas in the northern part of its range. Defoliation can be nearly total and mature trees can be killed by repeated, severe infestations. Planting Linden in areas with severe infestations of this pest may not be wise. However, at least one reference reports that although defoliation by Japanese beetles is common, control is seldom needed.

Multiple trunks and upright, equally dominant branches are often crowded on the trunk, and they may be poorly attached to the tree. The cultivars 'Glenleven', 'Greenspire', and 'June Bride' have been developed for their single, straight trunks. Linden often sprouts from the base of the trunk, and these sprouts will have to be removed from time to time.

The tree grows in sun or partial shade, will tolerate alkaline soil if it is moist, and transplants well. It is not particularly tolerant of drought, scorching at the leaf margins in summer drought, but this apparently does little long-term harm. It is more tolerant of heat and compact soil than American Basswood.

Aphids exude a sticky liquid, which can be a nuisance beneath the canopy. Trunks occasionally split when cold weather occurs suddenly. Winter trunk protection is essential until thicker bark develops. Grafted cultivars can form many suckers.

Tilia tomentosa Silver Linden

Pronunciation: TILL-ee-ah toe-men-TOE-sah

ZONES: 4B, 5A, 5B, 6A, 6B, 7A, 7B, 8A, 8B, (9A), (9B), (10A), (10B)

HEIGHT: 50 to 70 feet

WIDTH: 40 to 60 feet

FRUIT: oval; 0.33"; dry; tan; no significant litter; persists and dries on tree

GROWTH RATE: moderate to fast; long-lived

HABIT: oval; pyramidal; dense; symmetrical; coarse texture

FAMILY: *Tiliaceae*

LEAF COLOR: dark green above, silver undersides; showy yellow fall color

LEAF DESCRIPTION: alternate; simple; double serrate to serrate margin; cordate; deciduous; 2 to 5" long

FLOWER DESCRIPTION: yellowish; pleasant fragrance; showy; summer; attracts bees

EXPOSURE/CULTURE:

Light Requirement: part shade to full sun

Soil Tolerances: all textures; alkaline to acidic; occasionally wet; salt; moderate drought

Aerosol Salt Tolerance: moderate

PEST PROBLEMS: mostly resistant

BARK/STEMS:

Branches droop as tree grows and will require pruning for vehicular or pedestrian clearance beneath canopy; should be grown with a single leader; no thorns

Pruning Requirement: requires pruning to develop strong structure; eliminate upright, aggressive branches and double leaders; basal sprouts need periodic removal

Limb Breakage: resistant

Current Year Twig: pubescent; medium thickness

CULTIVARS AND SPECIES:

'Green Mountain', a rapid-growing form with a dense canopy, is reportedly resistant to Japanese beetle and gypsy moth; 'Princeton' compartmentalizes decay well; 'Sterling' is reportedly resistant to gypsy moth and Japanese beetle; 'Wandell', a broadly pyramidal form, has leaves that reportedly are resistant to Japanese beetle.

LANDSCAPE NOTES:

Origin: southern Europe and Asia

Usage: parking lot island; buffer strip; highway; shade tree; specimen; street tree; urban tolerant

Rapidly growing to mature height and spread, Silver Linden could be quite popular for use as a shade, specimen, or street tree. A deciduous tree, Silver Linden has a pyramidal form when young but develops into a rounded silhouette, and often has multiple trunks. Casting dense shade below the tree, the four- to five-inch-long dark green leaves are bright silver and fuzzy below, causing the trees to appear as if they are shimmering with each little breeze. The leaves turn yellow before dropping in autumn. In early summer, the trees are perfumed with extremely fragrant clusters of small, yellow-white blossoms, but the flowers are difficult to see because of the dense cover of the large leaves.

This tree is large and needs plenty of room to develop. Plant it as a specimen or shade tree on a commercial property where there is plenty of soil space for root expansion. Well-suited for large tree lawns along streets and for large parking lot islands. Be prepared to remove sprouts periodically from the base of the trunk on young trees.

Silver Linden should be grown in full sun on moist, well-drained soil, acid or slightly alkaline. This tree is moderately tolerant of pollution, soil compaction, heat, and drought, making it an ideal street or shade tree. It appears to tolerate drought, better than other Lindens. Some report that it is risky to transplant in the fall.

Reportedly less susceptible to Japanese beetles than other Lindens. Perhaps due to the leaf pubescence, other pest susceptibility is probably similar to that of other Lindens.

Tipuana tipu Tipu Tree, Pride-of-Bolivia

Pronunciation: tip-oo-AH-nah TI-pooh

ZONES: (9B), 10A, 10B, 11

HEIGHT: 30 to 50 feet

WIDTH: 50 to 80 feet

FRUIT: pod; about 2.5" long; dry; green turning brownish; causes some litter; persistent

GROWTH RATE: rapid; moderate longevity

HABIT: rounded dome; open; nearly symmetrical; fine texture

FAMILY: *Leguminosae*

LEAF COLOR: semi-evergreen; light green

LEAF DESCRIPTION: alternate; pinnately compound; leaflets up to 1.5" long

FLOWER DESCRIPTION: yellow; very showy; pleasant fragrance; spring and summer; causes some litter

EXPOSURE/CULTURE:

Light Requirement: full sun for best flowering

Soil Tolerances: all textures; slightly alkaline to acidic; moderate drought

Aerosol Salt Tolerance: poor

PEST PROBLEMS: resistant

BARK/STEMS:

Branches grow mostly upright, eventually drooping on old trees; should be grown with a single leader; no thorns

Pruning Requirement: space limbs along trunk and eliminate crossed and touching branches; remove or thin aggressive, upright branches

Limb Breakage: resistant on properly trained trees

Current Year Twig: moderate thickness

LANDSCAPE NOTES:

Origin: Bolivia, Brazil

Usage: parking lot island; buffer strip; highway; light shade; sidewalk cutout; street tree; urban tolerant

Each summer, the tree is entirely covered with apricot-yellow blooms that engulf the landscape in color. Silhouetted against a clear blue sky, there is nothing quite like a Tipu Tree in full flower. Flowers drop to cover the ground for several weeks, adding additional beauty to the yard. However, they can become slippery as they decompose on a walk, street, or driveway. The tree casts light shade through its compound leaves, making it easy to maintain a lawn beneath the canopy. The wood of Tipu is used as a substitute for real rosewood (*Dalbergia*).

Strive to develop one central trunk as far up into the canopy as possible, with upright, spreading branches spaced several feet apart along the trunk. Slow the growth on aggressive, upright branches by thinning them. This will help ensure that included bark will not form in the crotches of major limbs. Purchase good-quality trees with one leader to reduce the pruning requirement in the landscape.

Tipu Tree adapts to the rigors of street tree planting, provided it is not mechanically injured at the base of the trunk. Keep mowers and other equipment away from the tree. Tipu grows well in sidewalk cutouts, but, like other medium and large-sized trees, will eventually lift the walk. Allow plenty of soil space for proper root and lower trunk development, as trunks tend to flare out at the base. They are popular in California, less so in Florida.

Limitations include falling seed pods; cankers, borers, pod gall midge, and other potential problems may occur on tough sites.

Trachycarpus fortunei Windmill Palm

Pronunciation: tray-kee-KAR-puss for-TUNE-ee-eye

ZONES: 8A, 8B, 9A, 9B, 10A, 10B

HEIGHT: 20 to 40 feet

WIDTH: 6 to 10 feet

FRUIT: round; 0.5"; fleshy; blue-black; no significant litter

GROWTH RATE: slow-growing; moderate longevity

HABIT: palm; open; symmetrical; medium texture

FAMILY: *Arecaceae*

LEAF COLOR: dark green above, blue-green below

LEAF DESCRIPTION: spiral; simple; palmate; 3 to 4' across

FLOWER DESCRIPTION: whitish-yellow; pleasant fragrance; not showy; summer; males and females on separate plants

EXPOSURE/CULTURE:

Light Requirement: mostly shaded to full sun

Soil Tolerances: all textures; alkaline to acidic; drought

Aerosol Salt Tolerance: moderate

PEST PROBLEMS: mostly resistant, except for lethal yellowing in Florida

BARK/STEMS:

Showy, like burlap; single, 8" thick trunk; no thorns; no crown shaft; persistent fronds

Pruning Requirement: remove old fronds as they die

Limb Breakage: resistant

LANDSCAPE NOTES:

Origin: China

Usage: container or above-ground planter; near a deck or patio; specimen

The erect, single trunk of Windmill Palm is covered with dense, brown, hairlike fibers, and the 3-foot-wide, fan-shaped fronds extend from 1.5-foot-long, rough-edged petioles. The trunk appears to be wrapped in burlap. A very slow-growing palm, Windmill Palm can reach 40 feet in height, but is often seen much smaller at about 10 to 20 feet tall. Windmill Palm works well as a framing tree, accent, specimen, patio, or urn subject. It is ideal for use as an accent in a shady shrub border or by a front entryway. It does well in confined areas and is hardy to 10°F or lower.

There are fine examples of mass plantings where palms are spaced 6 to 10 feet apart around a patio or sitting area. They have also been used very successfully to line an entry walk to a large building, where the slightly fragrant flowers add another dimension to the landscape. This adds formal elegance to any structure, especially one with a glass facade.

Windmill Palm should be grown in shade or partial shade on fertile soil to look its best, but it is also tolerant of full sun on well-drained soils when given ample moisture. Plants should be watered faithfully. Protection from harsh winds will minimize leaf tearing, but plants can be used successfully close to the shore, as they are quite tolerant of salt and wind. This is one of only a few palms that performs poorly in very hot climates.

Limitations include scales and palm aphids, root rot, moderate susceptibility to lethal yellowing disease, and leaf spots.

Tristania conferta Brisbane Box

Pronunciation: triss-TANE-ee-ah kon-FUR-tah

ZONES: 9B, 10A, 10B, 11

HEIGHT: 30 to 50 feet

WIDTH: 30 to 40 feet

FRUIT: round; up to 1"; dry; brown and woody; no significant litter; persistent

GROWTH RATE: moderate; moderate life span

HABIT: upright oval; moderate density; somewhat irregular; medium texture

FAMILY: *Myrtaceae*

LEAF COLOR: evergreen

LEAF DESCRIPTION: alternate; simple; entire, undulating margin; oval to oblong; 4 to 6" long

FLOWER DESCRIPTION: white, each like a pinwheel; in showy clusters partially hidden by foliage; summer

EXPOSURE/CULTURE:

Light Requirement: full sun

Soil Tolerances: all textures; slightly alkaline to acidic; drought

PEST PROBLEMS: resistant

BARK/STEMS:

Showy bark, exfoliating in reddish-brown and gray patches; bark is thin and easily damaged by mechanical impact; branches grow mostly upright and do not droop; can be trained to grow with a single trunk or multiple trunks; no thorns

Pruning Requirement: except to eliminate double leaders and upright branches with included bark, little pruning is needed to develop strong structure

Limb Breakage: resistant

Current Year Twig: brownish-green; slender

CULTIVARS AND SPECIES:

T. laurina is smaller, with several trunks and a tight canopy. It is often used as a small garden accent or clipped into a shrub.

LANDSCAPE NOTES:

Origin: Malaya and Australia

Usage: parking lot island; buffer strip; shade tree; street tree; sidewalk cutout; highway

Growing to 150 feet tall in its native habitat, Brisbane Box can reach 40 to 50 feet in height and spreads of 30 to 40 feet in the landscape. The tree is often planted for its showy, smooth, mottled, and peeling bark. Even the bark on young branches exfoliates in a showy pattern. The coarse-textured foliage is clustered at the tips of the branches and hides most of the delicately frilled flowers. A peek under the canopy in the summer will reveal the flowers borne profusely along the branches.

Where there are no overhead restrictions, this makes a nice medium-sized street tree. Its moderate size makes it suited for planting in sidewalk cutouts and in other areas with limited soil space. To train for street tree use, cut back horizontal stems and branches for the first several years when they begin to droop or grow nearly upright. This will help the tree maintain the upright habit and central trunk needed for planting in parking lot islands, buffer strips, and along streets. The first permanent lateral branch can be selected at about eight feet from the ground. Space major branches 8 to 12 inches apart on the central trunk.

Multiple-trunked specimens are best suited for planting in gardens and lawns as accent or shade trees. The trunks are remarkably showy and will be the source of comments from neighbors and friends. This wonderful tree is suited for many uses in any landscape.

Plants should be watered faithfully and protected from heavy frost. Once established, trees require only occasional irrigation in the summer. No major limitations.

Tsuga canadensis Canadian Hemlock

Pronunciation: SUE-gah can-ah-DEN-siss

ZONES: 4A, 4B, 5A, 5B, 6A, 6B, 7A

HEIGHT: 50 to 70 feet

WIDTH: 30 to 35 feet

FRUIT: oval; 0.5 to 1" long; dry; brown; no significant litter; persistent

GROWTH RATE: slow-growing; long-lived

HABIT: pyramidal; moderate density; symmetrical; fine texture

FAMILY: *Pinaceae*

LEAF COLOR: soft green

LEAF DESCRIPTION: two-ranked; linear; needle-leaf evergreen; 0.25 to 0.66" long

FLOWER DESCRIPTION: green; yellow; not showy; spring

EXPOSURE/CULTURE:

Light Requirement: mostly shaded to part shade; tolerates full sun

Soil Tolerances: all textures; acidic; salt- and drought-sensitive

Aerosol Salt Tolerance: low

PEST PROBLEMS: sensitive

BARK/STEMS:

Branches droop as tree grows and will require pruning for vehicular or pedestrian clearance beneath canopy; normally grows with a single leader; no thorns

Pruning Requirement: needs little pruning to develop strong structure; eliminate double leaders

Limb Breakage: resistant

Current Year Twig: brown; gray; slender

CULTIVARS AND SPECIES:

'Bennett'—globose habit, branches weeping at the tips, dark green, grows in partial shade, 3-foot height and spread; 'Sargentii'—dwarf, weeping mound form commonly available that will grow 6 to 12 feet tall (in about 80 years) and 10 to 20 feet wide; 'White Gentsch'—rounded, flattened dwarf, 4 feet high and wide, foliage white at the tips.

T. heterophylla and *T. mertensiana* are similar trees native to the northwestern United States *T. caroliniana* is native to the southeastern United States but is not well known.

LANDSCAPE NOTES:

Origin: north Georgia to Minnesota and Nova Scotia

Usage: bonsai; clipped hedge; screen; specimen

Canadian Hemlock may be one of the most beautiful conifers, with its soft green needles attached to gently arching branches that often reach the ground. Specimen trees look their best with lower branches left on and touching the ground. The strong, conical form is supported by one, straight central trunk. Canadian Hemlock is tolerant of shearing and can be maintained at any height by shearing provided it is grown in the sun. Unfortunately, infestations of woolly adelgid have been severe in Virginia, Maryland, Pennsylvania, and other areas. This limits the tree's usefulness in these areas.

Hemlock transplants well balled-in-burlap, and will grow in moderately dry or moist, acidic soil, though it prefers moist, well-drained sites. Root rot and bark splitting are common on sites that are constantly wet. Some die-back can be expected after transplanting if irrigation is not managed just right. Best growth is in partial shade in mountainous valleys, but specimens grow well in full sun in the middle and northern part of their range. Perfect beneath other trees or in a large, mulched bed where soil is cool and some sun sneaks through during portions of the day. Not for planting in a lawn in competition with turf unless a wide area of mulch is maintained around the tree. Plants grown in full shade will be thin and will not produce the screening effect so characteristic of Hemlock in full sun or partial shade. This is a picky plant but well worth the extra effort to cultivate it.

Woolly adelgid and scale infestations can devastate trees in the mid-Atlantic states and other areas of the northeast. Needle rust, cankers, and nonparasitic bark splitting can occur on heavy, poorly drained soil. Weevils, bagworm, mites, and sapsucker woodpecker can also be troublesome. Hemlock loopers defoliate plants in part of New England. Trees sometimes blow over on exposed sites due to shallow roots.

Tsuga canadensis 'Lewisii' 'Lewisii' Hemlock

HEIGHT: probably to about 20 feet

WIDTH: 10 feet

GROWTH RATE: very slow-growing; long-lived

HABIT: tight pyramid; dense; perfectly symmetrical; fine texture

LANDSCAPE NOTES:

Usage: bonsai; clipped hedge; screen; specimen in a dwarf garden

The tight canopy and slow growth rate make this a special, small tree. In some ways it could be thought of as a larger, faster growing version of Dwarf Alberta Spruce. It is best used as a specimen plant in a rock garden or dwarf conifer collection. Hemlock is also adapted to clipping into a hedge or screen, a use that should be compatible with this cultivar.

Tsuga canadensis 'Sargentii' Weeping Hemlock

HEIGHT: probably to about 10 feet

WIDTH: 25 feet

GROWTH RATE: slow-growing; long-lived

HABIT: weeping; dense; irregular; fine texture

LANDSCAPE NOTES:

Usage: specimen

The weeping, spreading habit of this special tree is quite remarkable. It is featured in many older gardens where special care and some pruning and training over the decades have helped it develop a pleasing silhouette. Young trees should be staked to form an upright trunk capable of supporting the canopy off the ground. Stake the tree as tall as possible, because most subsequent growth will weep toward the ground. Left to its own devices, the tree sprawls across the earth. Give the tree plenty of room to develop, as it can grow very wide. It looks best when the sides of the tree do not have to be trimmed to provide clearance for a walk or patio.

Ulmus alata Winged Elm

Pronunciation: ULL-muss ah-LAY-tah

ZONES: 6A, 6B, 7A, 7B, 8A, 8B, 9A, 9B

HEIGHT: 45 to 70 feet

WIDTH: 30 to 40 feet

FRUIT: oval; 0.5"; dry; brown; no significant litter

GROWTH RATE: rapid-growing; long-lived

HABIT: variable; pyramidal, upright, rounded vase; moderate density; irregular silhouette; fine texture

FAMILY: *Ulmaceae*

LEAF COLOR: dark green; showy yellow fall color

LEAF DESCRIPTION: alternate; simple; double serrate margin; oblong to ovate; deciduous; 2 to 3" long

FLOWER DESCRIPTION: greenish-red; fall; not showy

EXPOSURE/CULTURE:

Light Requirement: part shade to full sun

Soil Tolerances: all textures; alkaline to acidic; wet soil; drought

Aerosol Salt Tolerance: moderate

PEST PROBLEMS: mostly resistant

BARK/STEMS:

Trunk buttresses on older trees; branches grow mostly upright and will not droop; should be grown with a single leader; no thorns

Pruning Requirement: requires regular pruning to develop strong structure; space branches along a central trunk; eliminate competing upright trunks and branches

Limb Breakage: resistant if no included bark develops

Current Year Twig: brown or gray; stout with corky wings

LANDSCAPE NOTES:

Origin: Missouri to Virginia, Texas, and Florida

Usage: parking lot island; buffer strip; highway; reclamation; shade tree; specimen; sidewalk cutout; street tree; urban tolerant

Usually seen at 40 to 50 feet high, Winged Elm can reach 90 feet in height in the woods, with a 30- to 40-foot spread and a 30-inch trunk. Canopy form is variable from pyramidal to vase or rounded. A North American native, this fast-growing, deciduous tree is quickly identified by the corky, winglike projections that appear on opposite sides of twigs and branches. The size of the wings varies greatly from one tree to another. Branches rise through the crown, then bend outward in a sweeping manner. The tree is found growing in wet sites as well as dry, rocky ridges, making it a very adaptable tree for urban planting.

Winged Elm will thrive in full sun or partial shade, growing relatively quickly on any soil. It is an extremely sturdy (following proper pruning) and adaptable tree and is well suited as a shade or street tree. It is suited to parking lot islands and other confined soil spaces, because roots do not appear to lift sidewalks and curbs readily.

It must be pruned regularly at an early age to eliminate double and multiple trunks. Select branches that form a wide angle with the trunk, eliminating those with narrow crotches. Strive to produce a central trunk with major lateral limbs spaced along it. This trunk will usually not be straight (unless it is staked), but this is fine. Trees look open and lanky following proper pruning and this may be one reason the tree has not been very popular. After this initial training period, though, the tree fills in nicely to make a well-adapted, beautiful shade tree. Purchase trees with good form in the nursery and be selective, as form varies greatly from one tree to the next.

The biggest problems are Dutch elm disease and phloem necrosis, which can kill trees. To protect a community from widespread tree loss, do not plant a large number of these trees. Some trees are susceptible to powdery mildew, which causes varying degrees of leaf color changes in early fall, right before leaves drop. Mites can yellow the foliage but usually cause no permanent damage. Scale insects can infest Winged Elm along branches. They blend in with the corky wings and are hard to see.

Ulmus americana American Elm

Pronunciation: ULL-muss ah-mare-eh-CAY-nah

ZONES: 2A, 2B, 3A, 3B, 4A, 4B, 5A, 5B, 6A, 6B, 7A, 7B, 8A, 8B, 9A, 9B

HEIGHT: 80 to 120 feet

WIDTH: 60 to 120 feet

FRUIT: round; 0.5"; dry; green; attracts birds; no significant litter

GROWTH RATE: rapid; moderate life span

HABIT: upright vase; moderate density; irregular silhouette; medium texture

FAMILY: *Ulmaceae*

LEAF COLOR: dark green; yellow fall color

LEAF DESCRIPTION: alternate; simple; double serrate margin; oblong to ovate; deciduous; 3 to 6" long

FLOWER DESCRIPTION: green; not showy; spring

EXPOSURE/CULTURE:

Light Requirement: part shade to full sun

Soil Tolerances: all textures; alkaline to acidic; flooded soil; salt; drought

Aerosol Salt Tolerance: moderate

PEST PROBLEMS: very sensitive

BARK/STEMS:

Lower trunk flares, becoming very wide; main branches grow mostly upright and will not droop, though small branches droop; should be grown with a single leader; no thorns

Pruning Requirement: requires pruning to develop strong structure; space major branches along trunk and thin so they remain less than half diameter of trunk

Limb Breakage: susceptible to breakage at crotch on older trees, due to included bark

Current Year Twig: brown; slender

CULTIVARS AND SPECIES:

Consider using 'Liberty' (some DED), 'Delaware #2', 'Valley Forge', 'Princeton', and 'Washington', or the hybrid elms.

LANDSCAPE NOTES:

Origin: Nova Scotia to Florida, Texas, and North Dakota

Usage: specimen in a landscape with a high maintenance budget or a commitment to elms

This native North American tree grows quickly when young, forming a broad or upright, vase-shaped silhouette. Trunks on older trees could grow to seven feet across. In early spring, before the new leaves unfold, the rather inconspicuous, small, green flowers appear on pendulous stalks. These blooms are followed by green, wafer-like seedpods that mature soon after flowering is finished. The seeds are quite popular with both birds and wildlife. American Elm must be at least 15 years old before it will bear seed. The copious amount of seeds can create a mess on hard surfaces for a period of time. Trees have an extensive but shallow root system.

Once a very popular and long-lived (300-plus years) shade and street tree, American Elm suffered a dramatic decline with the introduction of Dutch elm disease, a fungus spread by a bark beetle. The wood of American Elm is very hard and it was a valuable timber tree used for lumber, furniture, and veneer. Native Americans once made canoes out of American Elm trunks, and early settlers steamed the wood so it could be bent to make barrels and wheel hoops. It was also used for the rockers on rocking chairs.

If you plant American Elm, select a disease-resistant one, such as 'New Harmony', and plan on implementing a monitoring program to watch for symptoms of Dutch elm disease. It is vital to the health of existing trees that a program be in place to administer special, immediate, remedial care to these sensitive trees when disease is detected.

Dutch elm disease (DED) has killed many trees, but a religous maintenance program can save many remaining trees. Other disease problems include phloem necrosis (now called yellows), especially in the eastern United States, leaf spot diseases, bacterial leaf scorch, and cankers. Insect pests include leaf beetles, bark beetles, elm borer, gypsy moth, mites, and scales.

Ulmus crassifolia Cedar Elm

Pronunciation: ULL-muss crass-eh-FOE-lee-ah

ZONES: 6A, 6B, 7A, 7B, 8A, 8B, 9A, 9B

HEIGHT: 50 to 90 feet

WIDTH: 40 to 60 feet

FRUIT: oval; 0.75 to 1.5"; dry; green; no significant litter

GROWTH RATE: moderate; moderate to long life span

HABIT: rounded or upright vase; moderate density; irregular silhouette; fine texture

FAMILY: *Ulmaceae*

LEAF COLOR: dark green; very showy, yellow fall color

LEAF DESCRIPTION: alternate; simple; double serrate to serrate margin; oval to oblong; deciduous; 1 to 2.5" long

FLOWER DESCRIPTION: green; summer; not showy

EXPOSURE/CULTURE:

Light Requirement: part shade to full sun

Soil Tolerances: all textures; alkaline to acidic; wet soil; drought

Aerosol Salt Tolerance: moderate

PEST PROBLEMS: somewhat susceptible

BARK/STEMS:

Branches droop as tree grows and will require early pruning for vehicular or pedestrian clearance beneath canopy; should be grown with a single leader as far up into canopy as possible; no thorns

Pruning Requirement: requires pruning to develop strong structure; space main branches along trunk and prevent included bark by thinning major branches

Limb Breakage: susceptible to breakage at crotch, due to included bark; wood itself tends to break

Current Year Twig: brown or gray; slender

LANDSCAPE NOTES:

Origin: Texas to Mississippi

Usage: parking lot island; buffer strip; highway; reclamation; shade tree; sidewalk cutout; street tree; urban tolerant

Cedar Elm is a native North American, deciduous tree that can reach 90 feet in height with a spread of 60 feet and forms a rounded silhouette. The stiff and rough-textured leaves fade to bright yellow or red-brown before dropping in fall. The inconspicuous, green, summertime flowers are followed by the production of winged seeds in late summer or early fall.

This would be a low-maintenance shade and street tree except for its somewhat drooping branches, which are susceptible to breakage at the crotches of major limbs. Some of this could be avoided by maintaining a regular pruning and training program in the early years after transplanting. Keep branches less than about one-half the diameter of the trunk to prevent formation of included bark. Cedar Elm has been used extensively (almost exclusively in some local areas) in Texas as a street tree for many years due to its adaptability to wet, poor soil conditions. However, it is always best to diversify the tree species in an area so that if a major problem arises on one species, it will only affect a portion of the tree population in the community.

Cedar Elm should be grown in full sun on any soil, acid or alkaline. It is very drought tolerant once established and tolerates wet soil well. Its popularity is due to the ease of transplanting and adaptability to urban soil conditions.

Elm leaf beetles can feed on foliage. Aphids can also drop copious amounts of honeydew beneath the canopy. Dutch elm disease kills trees. Powdery mildew can be a problem in some years. Mistletoe can engulf the tree.

Ulmus hybrids hybrid Elm

Pronunciation: ULL-muss

ZONES: (4A), (4B), 5A, 5B, 6A, 6B, 7A, 7B, (8A), (8B)

HEIGHT: 50 to 75 feet

WIDTH: 30 to 70 feet

FRUIT: oval; 0.5"; dry; brown; somewhat showy; some cause significant litter

GROWTH RATE: rapid; longevity unknown

HABIT: pyramidal to vase, depending on cultivar; moderate to dense; irregular when young, nearly symmetrical later; medium texture

FAMILY: *Ulmaceae*

LEAF COLOR: dark green; showy, yellow fall color

LEAF DESCRIPTION: alternate; simple; serrate to serrulate margin; oval to obovate or ovate; deciduous; 2 to 5" long

FLOWER DESCRIPTION: green; inconspicuous

EXPOSURE/CULTURE:

Light Requirement: part shade to full sun

Soil Tolerances: all textures; alkaline to acidic; occasionally wet soil, once established; drought

Aerosol Salt Tolerance: good

PEST PROBLEMS: somewhat sensitive

BARK/STEMS:

Branches grow mostly upright and spreading and do not droop; should be grown with a single leader as far up into canopy as possible; no thorns

Pruning Requirement: requires pruning to develop strong structure; space branches along trunk and thin aggressive main branches to prevent formation of included bark

Limb Breakage: resistant if properly pruned

Current Year Twig: brownish-gray; slender

CULTIVARS AND SPECIES:

'Accolade' has a nice vase shape; 'Cathedral' is vase-shaped and spreading at maturity and resists Dutch elm disease (DED); 'Frontier'—young trees are oval or pyramidal, one of the most extensively tested, shows moderate resistance to elm leaf beetle and DED; 'Homestead'—good resistance to Dutch elm disease but sensitive to elm leaf beetle; 'Patriot' has good resistance to DED (as good as 'Homestead'), tolerates elm leaf beetle even more so than 'Urban'; 'Pioneer'—canopy is more rounded than others, good resistance to Dutch elm disease; 'Regal'—grows three to four feet each year with a central leader and wide branch angles; 'Sapporo Autumn Gold' needs strict attention to pruning to eliminate multiple leaders; 'Urban'—zone 4, nearly pyramidal with a more-or-less dominant central trunk, good resistance to Dutch elm disease but is sensitive to elm leaf beetle and cankers. *U. davidiana* has a vase shape and appears resistant to DED.

LANDSCAPE NOTES:

Origin: hybrid

Usage: parking lot island; buffer strip; highway; reclamation; shade tree; specimen; street tree; urban tolerant

The hybrid Elms form a graceful, upright, rounded or oval canopy of long, arching branches. Some specimens grow in the typical vase-shaped elm form; others are more upright. Some even display a central leader.

Elms often develop several dominant trunks or codominant stems. As trees grow older, these potentially weak points can split, shortening the life of the tree. Space major branches along a central trunk and keep major limbs less than half the diameter of the trunk to help prevent this.

Elms grow in full sun on a wide range of soils, adapting easily to extremes in pH (including alkaline) or moisture and tolerating cold, urban heat, and wind. Aggressive roots can probably break sidewalks and raise pavement if trees are improperly located.

None of these trees has been extensively tested. They have been around for only about 20 years, so their true resistance to Dutch elm disease is not certain. Use them with caution by planting small numbers. This will help evaluate these new elms, which have a high potential for becoming important trees in the urban landscape.

Ulmus parvifolia Chinese Elm, Lacebark Elm

Pronunciation: ULL-muss par-veh-FOE-lee-ah

ZONES: (5A), 5B, 6A, 6B, 7A, 7B, 8A, 8B, 9A, 9B, 10A, 10B

HEIGHT: 40 to 50 feet

WIDTH: 35 to 50 feet

FRUIT: oval; 0.5"; dry; brown; somewhate showy; no significant litter

GROWTH RATE: rapid; moderate to long life span

HABIT: rounded or weeping vase; moderate density; irregular to nearly symmetrical; fine texture

FAMILY: *Ulmaceae*

LEAF COLOR: dark green; purple, red, or yellow fall color

LEAF DESCRIPTION: alternate; simple; serrate to serrulate margin; oval to obovate or ovate; deciduous; 2 to 4" long

FLOWER DESCRIPTION: green; fall; inconspicuous

EXPOSURE/CULTURE:

Light Requirement: part shade to full sun

Soil Tolerances: all textures; alkaline to acidic; occasionally wet soil, once established; drought

Aerosol Salt Tolerance: moderate

PEST PROBLEMS: resistant

BARK/STEMS:

Beautiful, exfoliating bark, variable; bark is thin and easily damaged by mechanical impact; branches droop as tree grows and will require pruning for vehicular or pedestrian clearance beneath canopy; should be grown with a single leader; no thorns

Pruning Requirement: requires pruning to develop strong structure; space branches along trunk and thin aggressive main branches to prevent formation of included bark

Limb Breakage: breaks in ice storms; otherwise resistant if properly pruned

Current Year Twig: brownish-gray; slender

CULTIVARS AND SPECIES:

Allee ('Emer II')—70 by 50 feet with a vase habit, zone 5B; Athena ('Emer I')—nice, rounded, uniform canopy; 'Dynasty'—vase-shaped, susceptible to breakage, red fall color; 'Frosty'—small, white-margined leaf; 'Pathfinder'—extensively tested, broad, upright branches; 'True Green'—evergreen.

LANDSCAPE NOTES:

Origin: China, Japan, and Korea

Usage: bonsai; parking lot island; buffer strip; highway; reclamation; shade tree; specimen; street tree; urban tolerant

Chinese Elm forms a graceful, upright, rounded canopy of long, arching branches. Some specimens grow in the typical vase-shaped elm form. In colder regions of the country in fall, leaves are transformed into various shades of red, purple, or yellow. The tree is nearly evergreen in the southern extent of its range. The showy, exfoliating bark reveals random, mottled patterns of gray, green, orange, and brown, adding great textural and visual interest especially to its winter silhouette.

Chinese Elm is sometimes topped in the nursery to create a full head of foliage, causing branches to originate from one point on the trunk. Some of these branches may split out from the tree as it ages. Space major branches along a central trunk to prevent this.

Chinese Elm will grow in full sun on a wide range of soils, and it adapts easily to extremes in pH (including alkaline) and moisture, and tolerates cold, urban heat, and wind. Very suitable for street tree pits, parking lot islands, and other confined soil spaces. Trees will look their best when grown in moist, well-drained, fertile soil, but they adapt to drought and the extremes of urban sites.

Shows considerable resistance to elm leaf beetle, Japanese beetle, Dutch elm disease, and phloem necrosis (yellows). Cankers may develop on young trunks in excessively wet soil. Twig blight and canker can be an occasional problem. Trees may sprout from the trunk base or from roots close to the trunk. Avoid injury to surface roots and lower trunk. Mistletoe can be troublesome, especially in Texas.

Ulmus parvifolia 'Drake' 'Drake' Chinese Elm

Pronunciation: ULL-muss par-veh-FOE-lee-ah

ZONES: 7A, 7B, 8A, 8B, 9A, 9B, 10A, 10B

HEIGHT: 40 feet

WIDTH: 50 feet

FRUIT: oval; 0.5"; dry; brown; somewhat showy; no significant litter

GROWTH RATE: rapid; moderate life span

HABIT: rounded or weeping; moderate density; irregular to nearly symmetrical; fine texture

FAMILY: *Ulmaceae*

LEAF COLOR: green

LEAF DESCRIPTION: alternate; simple; serrate to serrulate margin; oval to obovate or ovate; semi-deciduous; 2 to 4" long

FLOWER DESCRIPTION: green; fall; not showy

EXPOSURE/CULTURE:

Light Requirement: part shade to full sun

Soil Tolerances: all textures; alkaline to acidic; occasionally wet soil, once established; drought

Aerosol Salt Tolerance: moderate

PEST PROBLEMS: resistant

BARK/STEMS:

Beautiful, exfoliating bark; bark is thin and easily damaged by mechanical impact; branches droop as tree grows and will require pruning for vehicular or pedestrian clearance beneath canopy; should be grown with a single leader; no thorns

Pruning Requirement: requires pruning to develop strong structure; space branches along trunk and thin aggressive main branches to prevent formation of included bark

Limb Breakage: resistant if properly pruned

Current Year Twig: brownish-gray; slender

LANDSCAPE NOTES:

Origin: species is from China, Japan, and Korea

Usage: bonsai; parking lot island; buffer strip; highway; reclamation; shade tree; specimen; street tree; urban tolerant

Drake Elm forms a graceful, rounded canopy of long, arching branches. Some specimens grow in the typical vase-shaped elm form, but most are wider than tall. In colder regions of the country in fall, leaves are transformed into various shades of red, purple, or yellow. The tree is evergreen in the southern extent of its range. The showy, exfoliating bark reveals random, mottled patterns of gray, green, orange, and brown, adding great textural and visual interest especially to its winter silhouette.

Chinese Elm will grow in full sun on a wide range of soils. It adapts easily to extremes in pH (including alkaline) and moisture and tolerates cold, urban heat, and wind. Very suitable for street tree pits, parking lot islands, and other confined soil spaces. Trees will look their best when grown in moist, well-drained, fertile soil, but they adapt to drought and the extremes of urban sites.

Select trees with branches spaced along one trunk. It is not essential that this trunk be straight. Buy from nurseries that understand how to train and prune this tree for street and parking lot use; otherwise you may be trimming and pruning low drooping branches on a regular basis. Trees with a trunk less than about two inches in diameter often require staking and some early pruning to prevent leaning and blowover, due to a heavy crown and unstable root system. Nursery operators often train trees to a single, straight trunk by staking at an early age. Leave branches on the lower trunk during this training period to develop a thicker trunk.

Shows considerable resistance to elm leaf beetle, Japanese beetle, Dutch elm disease, and phloem necrosis. Cankers may develop on young trunks in excessively wet soil. Twig blight can be an occasional problem. Trees may sprout from the trunk base or from roots close to the trunk. Avoid injury to surface roots and lower trunk.

Ulmus pumila Siberian Elm

Pronunciation: ULL-muss poo-MILL-ah

ZONES: 4A, 4B, 5A, 5B, 6A, 6B, 7A, 7B, 8A, 8B, 9A, 9B

HEIGHT: 50 to 70 feet

WIDTH: 35 to 50 feet

FRUIT: round; 0.5"; dry; brown; causes significant litter

GROWTH RATE: rapid; moderate life span

HABIT: vase; moderate density; irregular silhouette; medium texture

FAMILY: *Ulmaceae*

LEAF COLOR: green; showy yellow fall color

LEAF DESCRIPTION: alternate; simple; serrate margin; oval to lanceolate; deciduous; 1.5 to 3" long

FLOWER DESCRIPTION: green; not showy; spring

EXPOSURE/CULTURE:

Light Requirement: full sun

Soil Tolerances: all textures; alkaline to acidic; wet soil; salt and drought

Aerosol Salt Tolerance: moderate to high

PEST PROBLEMS: somewhat sensitive

BARK/STEMS:

Small branches droop as tree grows and will require pruning for vehicular or pedestrian clearance beneath canopy; should be grown with a single leader; small branches randomly die and persist in canopy; no thorns

Pruning Requirement: requires pruning to prevent formation of included bark

Limb Breakage: susceptible to breakage at crotch, due to poor collar formation; wood itself is weak and tends to break, especially in ice storms

Current Year Twig: green or gray; slender

LANDSCAPE NOTES:

Origin: China, Korea, Manchuria, and Siberia

Usage: windbreak

This rapid-growing deciduous tree has a rounded canopy with somewhat drooping branches by the time it reaches mature height and spread. The glossy green, 1.5- to 3-inch-long by 0.5- to 1-inch-wide leaves turn pale yellow in fall before dropping. The inconspicuous, green, springtime flowers are produced in small clusters among the leaves and are followed by half-inch-long, flat, winged seedpods that mature during early summer. The Chinese Elm is easily distinguished from Siberian Elm because Chinese Elm flowers in the fall.

The wood of Siberian Elm is fairly brittle and subject to damage during storms, which creates a lot of twig litter on the lawn afterward. Because major limbs split from the crotches on older trees, this is considered by many people a tree to avoid. Most urban tree managers and horticulturists do not recommend planting this tree. If you are managing existing Siberian Elms along streets or in yards, be sure to have them checked for structural integrity. Bark often becomes included between major limbs and the trunk, resulting in crotch splitting and possible branch failure.

Siberian Elm grows well in full sun on almost any soil. This tree is easily grown and will tolerate a variety of adverse conditions, such as poor soil, drought, and salt. It is probably best saved for reclamation sites, shelterbelt windbreaks, or other out-of-the-way locations. It has been commonly planted in the drier parts of the high plain states.

Elm leaf beetles shred the foliage of this tree, one of the most preferred hosts. The tree is considered weak-wooded and has invasive surface roots. Seedlings germinate in the landscape. Although wetwood disease is found in this and many other elms, many trees appear to grow fine. Resists Dutch elm disease and yellows disease.

Veitchia merrillii Christmas Palm, Manila Palm

Pronunciation: VEE-chee-ah meh-RILL-ee-eye

ZONES: 10B, 11

HEIGHT: 25 feet

WIDTH: 8 feet

FRUIT: oval; 0.5 to 1" long; fleshy; red; causes some messy litter; persistent; showy

GROWTH RATE: moderate; moderate life span

HABIT: palm; open; symmetrical; medium texture

FAMILY: *Arecaceae*

LEAF COLOR: bright green

LEAF DESCRIPTION: spiral; odd pinnately compound; entire margin; lanceolate; 10 to 18" long leaflets; 3 to 4' long fronds

FLOWER DESCRIPTION: whitish-yellow; not showy; summer

EXPOSURE/CULTURE:

Light Requirement: part shade to full sun

Soil Tolerances: all textures; alkaline to acidic; moderate drought

Aerosol Salt Tolerance: moderate

PEST PROBLEMS: moderately sensitive

BARK/STEMS:

Single, 6 to 8" thick trunk; bright green crown shaft; showy and ringed with leaf scars; no thorns

Pruning Requirement: remove dead fronds

Limb Breakage: resistant

LANDSCAPE NOTES:

Origin: Philippines

Usage: container or above-ground planter; near a deck or patio; specimen; sidewalk cutout

This stocky, single-trunked palm, with stiffly arched, six-foot-long fronds is noted for the fall and winter appearance of the very showy clusters of glossy, bright red fruits, which hang below the leaves at the base of the crown shaft. Reaching 25 feet in height, though often much smaller, Christmas Palm has a very neat appearance and is well suited to use as a patio, terrace, specimen, or framing tree. Unfortunately, the palm is very susceptible to lethal yellowing disease and probably should not be extensively planted. Fortunately, there are other *Veitchia* resistant to the disease, including *V. macdanielsii* and *V. montgomeryana*, but these are much taller palms with thicker trunks.

Christmas Palm should receive some shade while young, but will grow in full sun when older on a wide variety of well-drained soils, including limestone soils. Trees should be used only in frost-free areas. Consider substituting one of the other *Veitchia* mentioned here, the Carpentaria Palm, *Ptychosperma macarthurii*, *Wodyetia bifurcata,* or a variety of other palms resistant to lethal yellowing disease.

Viburnum odoratissimum Sweet Viburnum

Pronunciation: vie-BURR-num o-door-ah-TISS-eh-mum

ZONES: 8B, 9A, 9B, 10A, 10B

HEIGHT: 25 feet

WIDTH: 15 to 25 feet

FRUIT: round; less than 0.5"; fleshy; red to black; no significant litter

GROWTH RATE: slow to moderate; moderate life span

HABIT: rounded vase; dense; symmetrical; medium texture

FAMILY: *Caprifoliaceae*

LEAF COLOR: dull dark green

LEAF DESCRIPTION: opposite; simple; serrate margin; oval; broadleaf evergreen; 3 to 5" long

FLOWER DESCRIPTION: white; pleasant fragrance; showy; spring

EXPOSURE/CULTURE:

Light Requirement: part shade to full sun

Soil Tolerances: all textures; alkaline to acidic; some salt; drought

Aerosol Salt Tolerance: low to moderate

PEST PROBLEMS: resistant

BARK/STEMS:

Bark is thin and easily damaged by mechanical impact; routinely grown with multiple trunks, but can be trained to a single trunk; branches grow mostly upright once tree is large; no thorns

Pruning Requirement: requires pruning to develop tree form and remove low branches

Limb Breakage: resistant

Current Year Twig: green; thick

CULTIVARS AND SPECIES:

'Emerald Lustre' has larger leaves; 'Nanum' is a dwarf form. *V. odoratissimum* var. *awabuki* has large glossy leaves and flower panicles, and beautiful red berries.

LANDSCAPE NOTES:

Origin: India to Japan

Usage: container or above-ground planter; hedge or screen; parking lot island; buffer strip; highway; near a deck or patio; shade tree; specimen; sidewalk cutout; street tree; urban tolerant

Large, leathery leaves and clusters of extremely fragrant, small, white flowers (which completely cover the plant in springtime) make Sweet Viburnum a longtime landscape favorite. The large, dull green leaves form a moderately coarse texture. For some reason, it has fallen out of favor as a small tree in recent years. This adaptable tree should be rediscovered by landscape architects and others specifying small trees for urban and suburban landscapes. Often used as a screen or clipped hedge, its dense, spreading, evergreen habit makes Sweet Viburnum suitable for use as a small tree. It reaches 25 to 30 feet tall and wide at maturity, with a dense, multi-branched, rounded canopy. The flowers are often followed by small, red berries, which turn black when ripe and are moderately showy.

Sweet Viburnum grows quickly in full sun or partial shade on a wide variety of soils. It is tolerant to drought, holding up very well in unirrigated landscapes in humid climates once established. Relatively maintenance-free, Sweet Viburnum grown as a tree will require pruning to direct growth and to form an upright, tree form. Suckers need to be periodically removed. A nice, neat, compact canopy develops without any training once the main structure of the tree is established. Some interior sprouts and branches should be removed to make the trunk show and to create a neater appearance. This would be a good tree for planting along a street where power lines or other obstructions limit overhead space. It can be trained to one trunk or grown as a multi-stemmed specimen.

No serious limitations. Occasionally aphids, bagworms, scales thrips, mites, and canker may occur.

Viburnum odoratissimum var. *awabuki*
Awabuki Viburnum
Pronunciation: vie-BURR-num o-door-ah-TISS-eh-mum

ZONES: (8B), 9A, 9B, 10A, 10B

HEIGHT: 15 to 20 feet

WIDTH: 15 to 20 feet

FRUIT: round; less than 0.5"; fleshy; red to black; no significant litter

GROWTH RATE: slow to moderate; moderate life span

HABIT: rounded vase; dense; symmetrical; medium texture

FAMILY: *Caprifoliaceae*

LEAF COLOR: glossy; medium green

LEAF DESCRIPTION: opposite; simple; serrate margin; oval; broadleaf evergreen; 3 to 5" long

FLOWER DESCRIPTION: white; pleasant fragrance; showy; spring

EXPOSURE/CULTURE:

Light Requirement: part shade to full sun

Soil Tolerances: all textures; alkaline to acidic; some salt; drought

Aerosol Salt Tolerance: low

PEST PROBLEMS: resistant

BARK/STEMS:

Bark is thin and easily damaged by mechanical impact; routinely grown with multiple trunks; can be trained to a single trunk by removing suckers and sprouts; branches droop; no thorns

Pruning Requirement: requires pruning to develop tree form; thin the many sprouts that develop inside canopy to develop tree form

Limb Breakage: somewhat susceptible

Current Year Twig: green; thick

LANDSCAPE NOTES:

Origin: India to Japan

Usage: container or above-ground planter; hedge or screen; parking lot island; buffer strip; highway; near a deck or patio; shade tree; specimen; sidewalk cutout; street tree; urban tolerant

Large, leathery, glossy leaves and clusters of fragrant, small, white flowers (which completely cover the plant in springtime) make this Viburnum a good, small tree for many landscapes. This adaptable cultivar should be rediscovered by landscape architects and others specifying small trees for urban and suburban landscapes. Often used as a screen or clipped hedge, its dense, spreading, evergreen habit makes Awabuki Viburnum suitable for use as a small tree. It reaches 20 feet tall and wide at maturity, with a dense, multi-branched, rounded canopy. The flowers are often followed by small, red berries, which turn black when ripe and are very showy.

Awabuki Viburnum grows quickly in full sun or partial shade on a wide variety of soils. It is tolerant to drought, holding up very well in unirrigated landscapes once established. Sweet Viburnum grown as a tree will require pruning to direct growth and to form an upright, tree form. A nice, neat, compact canopy develops without any training once the main structure of the tree is established. Interior sprouts and branches should be removed to make the trunk show and to create a neater, more open appearance. Trees tend to grow all winter, making them sensitive to freezes. New growth will be killed back depending on the maturity of the leaves and the severity of the cold.

Once trained into a tree form, this would be a good tree for planting along a street where power lines or other obstructions limit overhead space. It can be trained to one trunk or grown as a multi-stemmed specimen. Growth is rapid in the first few years, making it well suited for nursery production.

No serious limitations. Occasionally aphids, bagworms, scales, thrips, mites, and canker may occur.

Viburnum plicatum var. *tomentosum*
Doublefile Viburnum
Pronunciation: vie-BUR-num ple-KA-tum

ZONES: (5A), 5B, 6A, 6B, 7A, 7B, 8A

HEIGHT: 10 to 15 feet

WIDTH: 15 to 20 feet

FRUIT: oval; 0.33"; fleshy; red, turning black; attracts birds; no significant litter; persistent; showy

GROWTH RATE: moderate; moderate life span

HABIT: round and spreading with horizontal branching; dense; nearly symmetrical; medium texture

FAMILY: *Cornaceae*

LEAF COLOR: dark green; reddish purple fall color

LEAF DESCRIPTION: opposite; simple; serrate margin; ovate; deciduous; 2 to 5" long

FLOWER DESCRIPTION: white; very showy, held above foliage; spring

EXPOSURE/CULTURE:

Light Requirement: mostly shaded to partial sun; full sun with irrigation

Soil Tolerances: well-drained clay; loam; sand; slightly alkaline to acidic; moderate drought in shade, drought-sensitive in the south

PEST PROBLEMS: somewhat sensitive

BARK/STEMS:

Showy on older trees; branches droop as tree grows and will require pruning for vehicular or pedestrian clearance beneath canopy; normally grown with multiple trunks, but can be trained to grow with a single trunk; no thorns

Pruning Requirement: needs pruning to develop upright structure; remove lower branches to form a tree

Limb Breakage: resistant

Current Year Twig: grayish-brown; medium thickness

CULTIVARS AND SPECIES:

There are many cultivars with various sizes, forms, and flower color.

LANDSCAPE NOTES:

Origin: China, Japan

Usage: near a deck or patio; specimen

Doublefile Viburnum grows 10 to 15 feet tall, spreading even wider if not pruned. It can be trained with one central trunk or as a picturesque multi-trunked tree. Most plants are grown as multi-trunked shrubs. The flowers consist of four or five lobes that are white as snow, and they stand up on the twig above the new foliage. The plant is a joy to behold during the two-week flowering period. The bright red fruits are devoured by birds.

Branches on the lower half of the canopy grow horizontally and droop to the ground, creating a wonderful landscape feature. In time, this lends a strikingly horizontal impact to the landscape, particularly if some branches are thinned to open up the crown. Old plants can be cut back to rejuvenate them.

The tree is not suited for parking lot planting, but can be grown in a wide street median, if provided with less than full-day sun and some irrigation. It can be used in a shrub border to add spring and fall color, or as a specimen in a lawn or groundcover bed. It can be grown in sun or shade, but very shaded trees will be less dense, grow more quickly and taller, and have poor fall color and fewer flowers. Trees prefer part shade (preferably in the afternoon), especially in the southern end of its range.

Viburnum prefers a deep, rich, well-drained, sandy or loamy soil. It is not recommended for heavy, wet soils unless it is grown on a raised bed to keep roots on the dry side. The roots rot in soils without adequate drainage.

There are reports of twig die-back, but otherwise no serious limitations.

Viburnum rufidulum Rusty Blackhaw

Pronunciation: vie-BURR-num roo-FID-you-lum

ZONES: 5A, 5B, 6A, 6B, 7A, 7B, 8A, 8B, 9A, 9B

HEIGHT: 15 to 25 feet

WIDTH: 15 to 20 feet

FRUIT: round; 0.33"; fleshy; red turning blue; scarce; no significant litter

GROWTH RATE: slow-growing; moderate longevity

HABIT: rounded vase; moderate density; irregular silhouette; medium texture

FAMILY: *Caprifoliaceae*

LEAF COLOR: dark green; purplish to red, showy, long-lasting fall color

LEAF DESCRIPTION: opposite; simple; serrulate margin; obovate to ovate; deciduous; 2 to 3" long

FLOWER DESCRIPTION: white; showy; spring

EXPOSURE/CULTURE:

Light Requirement: mostly shaded to full sun

Soil Tolerances: all textures; alkaline to acidic; occasionally wet soil; drought

Aerosol Salt Tolerance: none

PEST PROBLEMS: pest-free

BARK/STEMS:

Grows to about 6 inches thick; routinely grown with multiple trunks; branches droop; no thorns

Pruning Requirement: needs pruning to develop a neat, well-structured canopy; remove some inner sprouts to clean canopy

Limb Breakage: resistant

Current Year Twig: brown; slender to medium thickness

LANDSCAPE NOTES:

Origin: southeastern United States

Usage: container or above-ground planter; parking lot island; buffer strip; highway; near a deck or patio; reclamation; specimen; sidewalk cutout; street tree

A native of the well-drained, upland woods of southeastern North America, Rusty Blackhaw forms a multiple or (occasionally) single-trunked small tree or large shrub, reaching up to 25 feet in height with an equal spread. The dark bark is blocky, resembling older Flowering Dogwood bark. Trunks usually grow no thicker than six inches and arch away from the tree, forming a pleasing, vase-shaped crown. Leaves are three inches long, leathery, and extremely glossy. The tree is covered in springtime with striking five-inch-wide clusters of small, white blooms. These flowers are followed by clusters of dark blue, waxy, one-half-inch-long fruits that are extremely popular with wildlife and will occasionally persist on the plant from September throughout the autumn, if not eaten. In fall, Rusty Blackhaw puts on a brilliant display of scarlet red to purple foliage.

This would be a good tree for planting beneath power lines and in other limited-space areas. Useful as a hedge, specimen, or border tree, this deciduous tree adapts well to urban areas. Shoots arise from the root system, sometimes as far out as the dripline. This could be a maintenance problem when planted in a bed of shrubs, groundcover, or mulch, but sprouts would be routinely cut with regular mowing when planted in a lawn. Pests are usually not a major problem.

Rusty Blackhaw will grow and look nice in full sun or partial shade on any reasonably fertile, well-drained soil. The tree grows in a shady spot but forms a more open habit. Flowering is significantly reduced in the shade. Although tolerant of drought, it will not thrive in compacted soil.

Usually pest-free. Viburnum aphid is gray to dark green and feeds in clusters at the tips of the branches, causing leaf curl. Viburnums have a variety of other problems that could affect this plant. Fruit is scarce, so propagation is challenging.

Viburnum sieboldii Siebold Viburnum

Pronunciation: vie-BURR-num see-BOWL-dee-eye

ZONES: 4A, 4B, 5A, 5B, 6A, 6B, 7A, 7B

HEIGHT: 20 to 30 feet

WIDTH: 10 to 12 feet

FRUIT: oval; 0.33"; fleshy; red, turning black; attracts birds; no significant litter; very showy

GROWTH RATE: moderate; moderate life span

HABIT: upright vase; moderate density; irregular silhouette; coarse texture

FAMILY: *Caprifoliaceae*

LEAF COLOR: dark green

LEAF DESCRIPTION: opposite; simple; crenate to serrate margin; oval to obovate; broadleaf evergreen; 3 to 6" long

FLOWER DESCRIPTION: white; showy; spring

EXPOSURE/CULTURE:

Light Requirement: part shade to full sun

Soil Tolerances: all textures; occasionally wet; slightly alkaline to acidic; moderate drought

Aerosol Salt Tolerance: moderate

PEST PROBLEMS: resistant

BARK/STEMS:

Routinely grown with multiple trunks; branches grow mostly upright and will not droop; no thorns

Pruning Requirement: requires little pruning to develop strong structure

Limb Breakage: resistant

Current Year Twig: gray; thick

CULTIVARS AND SPECIES:

'Seneca' has very firm fruits that are not readily eaten by birds and provide a much longer fruit effect than the species. It is preferred over the species because of the extended fruiting effect.

LANDSCAPE NOTES:

Origin: Japan

Usage: hedge; buffer strip; highway; near a deck or patio; specimen; sidewalk cutout

Most often grown as a large multi-trunked shrub or small tree, Siebold Viburnum only reaches about 20 feet high (occasionally to 30 feet) and 10 to 12 feet wide, creating an upright silhouette with short, firm branches. The 2- to 6-inch-long by 1.5- to 3-inch-wide leaves give off a pungent odor when crushed, and have been known to give an occasional autumn display of red or purple color, although typically there is no fall color. In late May, the plant is covered with three- to six-inch-diameter clusters of off-white, tiny blossoms. These blooms are followed by small, half-inch, pinkish-red fruits that ripen to dark blue or black from August to October, and are held above the foliage, making them an easy target for the birds who find them a delectable treat. They are extremely showy for a period of about two weeks.

It is best used as a specimen or patio tree where there is plenty of soil space for root expansion. It provides nice scale to a small landscape, but is also suited for planting in mass, or on 15- to 25-foot centers along a boulevard or entrance road to a commercial landscape. The multiple trunks twist as they grow, giving a formal, Oriental effect.

Siebold Viburnum should be grown in full sun or partial shade on moist, well-drained soils, acid or alkaline. Leaves will show evidence of scorch if soil dries out, but this appears to be a drought-avoidance mechanism that causes little permanent damage. No limitations of major concern.

Vitex agnus-castus Chastetree

Pronunciation: VEE-tex AG-nuss KASS-tuss

ZONES: (5), (6), 7B, 8A, 8B, 9A, 9B, 10A, 10B

HEIGHT: 10 to 15 feet

WIDTH: 15 to 20 feet

FRUIT: fleshy; black; no significant litter

GROWTH RATE: rapid when young; moderate life span

HABIT: rounded, upright-spreading vase; moderate density; nearly symmetrical; fine texture

FAMILY: *Verbenaceae*

50 to 60 years old

'Alba'

LEAF COLOR: blue-green

LEAF DESCRIPTION: opposite; palmate; entire margin; lanceolate; deciduous; 2 to 6" long leaflets

FLOWER DESCRIPTION: lavender; pleasant fragrance; showy; spring to early summer; attracts bees

EXPOSURE/CULTURE:

Light Requirement: part shade to full sun

Soil Tolerances: all textures; slightly alkaline to acidic; drought

Aerosol Salt Tolerance: moderate

PEST PROBLEMS: resistant

BARK/STEMS:

Branches droop as tree grows and will require pruning for vehicular or pedestrian clearance beneath canopy; can be trained to a single or multiple trunk; no thorns

Pruning Requirement: requires pruning to develop strong structure and tree form

Limb Breakage: resistant

Current Year Twig: green; slender to medium thickness

LANDSCAPE NOTES:

Origin: Europe and Asia

Usage: container or above-ground planter; buffer strip; highway; near a deck or patio; trainable as a standard; specimen

Chastetree can be grown as a large, deciduous, multi-stemmed shrub or small tree, and is noteworthy for its showy, summer display (late springtime in warm climates) of fragrant, upward-pointing, terminal panicles of lavender blooms, which are quite attractive to butterflies and bees. The tree is often planted where honey is marketed to promote excellent honey production. The trunk is gray and blocky and somewhat ornamental. The sage-scented leaves of Chastetree are shaped like a hand (palmate) and were once believed to have sedative effects. *Vitex* seeds itself into landscaped beds and can become somewhat weedy.

Chastetree is used effectively in a mixed shrubbery border or as a specimen. It is usually seen with a multiple trunk but can be trained in the nursery into a tree with one or several trunks, if so desired. Occasionally used as a highway median tree, because it will not grow up and into power lines, but branches tend to droop and would hinder traffic visibility if planted too close to the street. Highway median planting would be fine if there is adequate horizontal space for the crown to develop and spread. The flowers attract bees, so locate it accordingly.

Chastetree prefers a loose, well-drained soil that is moist or on the dry side (not wet), but will tolerate drained clay or sandy soils. The tree often suffers from die-back in organic, mucky, or other soil that is kept too moist, such as in the New Orleans or Dallas areas. Chastetree should be planted in full sun or light shade, and will tolerate hot weather extremely well, moderate salt-air exposure, and alkaline soil. *Vitex negundo* is more cold-hardy.

Leaf spot can almost defoliate the tree. Root rot can cause decline in soils that are too moist.

Vitex negundo 'Heterophylla' Cut-Leaf Chastetree

Pronunciation: VEE-tecks neh-GUN-doe 'het-er-uh-FYE-lah'

ZONES: 6A, 6B, 7A, 7B, 8A, 8B, 9A, 9B

HEIGHT: 10 to 15 feet

WIDTH: 10 to 15 feet

FRUIT: fleshy; no significant litter

GROWTH RATE: rapid when young; moderate life span

HABIT: rounded vase; open; irregular silhouette; symmetrical; fine texture

FAMILY: *Verbenaceae*

LEAF COLOR: blue-green

LEAF DESCRIPTION: opposite; palmate; serrate margin; lanceolate; deciduous; 2 to 4" long, delicate leaflets

FLOWER DESCRIPTION: lavender; pleasant fragrance; showy; summer

EXPOSURE/CULTURE:

Light Requirement: part shade to full sun for best flowering

Soil Tolerances: all textures; slightly alkaline to acidic; moderate drought

PEST PROBLEMS: resistant

BARK/STEMS:

Bark is thin and easily damaged by mechanical impact; branches droop as tree grows and will require pruning for vehicular or pedestrian clearance beneath canopy; routinely grown with multiple trunks; can be trained to grow with a short, single trunk; no thorns

Pruning Requirement: requires pruning to develop good structure and tree form; canopy needs trimming to keep tree looking neat

Limb Breakage: resistant

Current Year Twig: green; slender

LANDSCAPE NOTES:

Origin: Africa, Asia, Philippines

Usage: buffer strip; highway; specimen; near a deck or patio; sidewalk cutout

Cut-Leaf Chastetree can be grown as a large, deciduous, multi-stemmed shrub or small tree, and is noteworthy for its showy, summer display (late springtime in warm climates) of fragrant, upward-pointing, terminal panicles of lavender blooms, which are quite attractive to butterflies and bees. The tree is often planted where honey is marketed to promote excellent honey production. The trunk is gray and blocky and somewhat ornamental. The highly dissected leaves of Cut-leaf Chastetree are shaped like Cut-leaf Japanese Maple and were once believed to have sedative effects. *Vitex* seeds itself into landscaped beds and can become somewhat weedy.

Chastetree is used effectively in a mixed shrubbery border or as a specimen. It is usually seen as a shrub with a multiple trunk but can be trained in the nursery into a tree with one or several trunks, if so desired. Occasionally used as a highway median tree, because it will not grow up and into power lines, but branches tend to droop and would hinder traffic visibility if planted too close to the street. Highway median planting would be fine if there is adequate horizontal space for the crown to develop and spread. The flowers attract bees, so locate it accordingly.

Chastetree prefers a loose, well-drained soil that is moist or on the dry side (not wet), but will tolerate drained clay or sandy soils. The tree often suffers from die-back in organic, mucky, or other soil that is kept too moist, such as in the New Orleans or Dallas areas. Chastetree should be planted in full sun or light shade, and will tolerate hot weather extremely well, some salty-air exposure, and alkaline soil.

Leaf spot can almost defoliate the tree. Root rot can cause decline in soils that are too moist.

Washingtonia filifera Desert Fan Palm

Pronunciation: wash-ing-TONE-ee-ah fuh-LIFF-ur-ah

ZONES: 9A, 9B, 10A, 10B, 11

HEIGHT: 40 to 60 feet

WIDTH: 10 to 15 feet

FRUIT: round; 0.33"; fleshy; black; no significant litter

GROWTH RATE: moderate; long-lived

HABIT: palm; symmetrical; coarse texture

FAMILY: *Arecaceae*

LEAF COLOR: gray-green

LEAF DESCRIPTION: spiral; costapalmate; 4 to 6' long fronds; toothed petioles

FLOWER DESCRIPTION: whitish-yellow; showy, extending well beyond foliage; spring

EXPOSURE/CULTURE:

Light Requirement: part sun to full sun

Soil Tolerances: well-drained clay, loam, and sand; alkaline to acidic; drought

Aerosol Salt Tolerance: moderate

PEST PROBLEMS: resistant

BARK/STEMS:

Single, 3 to 4' thick trunk; no crown shaft

Pruning Requirement: old fronds often hang on trunk and are usually removed to provide neater appearance

Limb Breakage: resistant

LANDSCAPE NOTES:

Origin: California desert

Usage: street tree; specimen

Commonly seen at 40 to 50 feet, but capable of soaring to 80 feet in height, Desert Fan Palm is quickly recognized as related to the straight, single-trunked palm, *Washingtonia robusta*. However, Desert Fan Palm is better suited to the home landscape in dry climates because it grows more slowly and is shorter. This also allows it to be used in more garden applications, such as large containers or group as a mass planting. It does not grow well when overirrigated because it will develop trunk or root rot.

The lower leaves usually persist on the tree after they die, forming a dense, brown, shaggy covering below the living, broad, fan-shaped leaves that give it the common name of Petticoat Palm. These dead fronds are known to be a fire hazard and a popular bedding roost for rodents; because of this, the law in some areas requires that they be removed.

Plant this palm only on extremely well-drained soil, to prevent trunk or root rot. Moderate salt tolerance allows it to be used close to the coast in several areas of the country. This palm could be tried more in well-drained sites as a replacement for *W. robusta*, which grows very tall with a skinny trunk, but overirrigation and rainy weather could initiate root rot. *W. filifera* is shorter, has a thicker trunk, and is better suited for planting in dry landscapes.

Limitations include scales while young, palm weevil in old age, palm leaf skeletonizer and a variety of scales at any time. Seedlings germinate freely in the landscape near Washington Palms and the tree can invade surrounding land.

Washingtonia robusta Washington Palm

Pronunciation: wash-ing-TONE-ee-ah row-BUSS-tah

ZONES: (8B), 9A, 9B, 10A, 10B, 11

HEIGHT: 60 to 120 feet

WIDTH: 10 to 15 feet

FRUIT: round; 0.33"; fleshy; black; no significant litter

GROWTH RATE: rapid; long-lived

HABIT: palm; open; symmetrical; coarse texture

FAMILY: *Arecaceae*

LEAF COLOR: bright green

LEAF DESCRIPTION: spiral; costapalmate; 5 to 6' across; toothed petioles

FLOWER DESCRIPTION: white; very showy; summer

EXPOSURE/CULTURE:

Light Requirement: part shade to full sun

Soil Tolerances: all textures; alkaline to acidic; wet soil; drought

Aerosol Salt Tolerance: moderate

PEST PROBLEMS: resistant

BARK/STEMS:

Single, 16 to 24" thick trunk; no crown shaft

Pruning Requirement: old fronds are persistent and must be removed in many situations

Limb Breakage: resistant

CULTIVARS AND SPECIES:

W. filifera is shorter, has a thicker trunk, and is best suited for planting in dry climates. It suffers and often dies from root rot when irrigated. Select *W. robusta* in an irrigated landscape and for the eastern United States.

LANDSCAPE NOTES:

Origin: Mexico

Usage: specimen; sidewalk cutout; street tree

Commonly seen at 50 to 80 feet, but capable of soaring to 120 feet in height, Washington Palm is quickly recognized as the much-used, straight, single-trunked street palm of years past. The lower leaves persist on the tree after they die, forming a dense, brown, shaggy covering below the living, broad, fan-shaped leaves that give it the common name of Petticoat Palm. These dead fronds are known to be a fire hazard and a popular bedding roost for rodents; because of this, the law in some areas requires that they be removed. The sharply barbed leaf petioles and tall, thin trunks make frond removal a rather unpleasant task, but some people think the rapid growth rate and statuesque appearance more than make up for this trouble.

Washington Palm makes a dramatic statement in the large landscape and creates a striking accent for multi-storied homes, but often grows out of scale in most landscapes with one-story buildings because all of the fronds are at the top of the palm. It eventually looks like a telephone pole with a green hat.

Washington Palm needs full sun for best growth but will endure some shade while young. It will tolerate poor soil and drought, and is hardy to about 20°F. Transplant with a large root ball to ensure survival. Removing fronds at transplantation helps survival if irrigation is in short supply.

Coconut mealybug, palm leaf skeletonizer, palm platid planthopper, and a variety of scales infest this palm. Roots rot in wet soil.

Wodyetia bifurcata Foxtail Palm

Pronunciation: wood-EE-tee-ah bye-fur-KAY-tah

ZONES: 10A, 10B, 11

HEIGHT: 25 feet

WIDTH: 8 feet

FRUIT: oval; 2" long; fleshy; green turning red; no significant litter; persistent; showy

GROWTH RATE: rapid; moderate life span

HABIT: palm; open; symmetrical; fine texture

FAMILY: *Arecaceae*

LEAF COLOR: bright green

LEAF DESCRIPTION: spiral; pinnately compound; 10 to 18" long leaflets arranged entirely around frond; 5 to 8' long fronds

FLOWER DESCRIPTION: white; not especially showy

EXPOSURE/CULTURE:

Light Requirement: part shade to full sun

Soil Tolerances: all textures; alkaline to acidic; moderate drought

Aerosol Salt Tolerance: moderate

PEST PROBLEMS: resistant

BARK/STEMS:

Single, 8 to 10" thick trunk; long crown shaft; showy and ringed with widely spaced leaf scars; no thorns

Pruning Requirement: remove dead fronds

Limb Breakage: resistant

LANDSCAPE NOTES:

Origin: Northeastern Australia

Usage: container or above-ground planter; near a deck or patio; specimen; sidewalk cutout; street tree

This stocky, single-trunked palm has delicately arched, long fronds spaced out along a single trunk. The crown shaft is longer than on most other small palms, adding to its ornamental characteristics. Reaching at most 30 feet in height, Foxtail Palm has a very neat appearance and is well suited to use as a patio, terrace, specimen, or framing tree. Despite its thicker trunk, it could help replace the disease-sensitive Christmas Palm in south Florida.

A grouping of Foxtail Palms brings new life to any area of the landscape. The frond segments are held out from the frond like a bottlebrush or foxtail and swing in the lightest breeze, creating a quieting shuffle. Their fine texture complements water features. The small size make them suited for planting alone in a small residential lot as a specimen, or in a container or above-ground planter.

Foxtail Palm grows best in full sun, but tolerates a wide variety of well-drained soils, including limestone soils. It is considered tolerant of drought, but responds to irrigation and good fertility with rapid growth. Trees tolerate only a light frost.

No limitations of major concern. Keep palms green with regular applications of a palm fertilizer, especially where soil pH is greater than 7.0.

Xanthoceras sorbifolium Yellowhorn

Pronunciation: zan-tho-SEAR-us sore-bih-FOE-lee-um

ZONES: 4A, 4B, 5A, 5B, 6A, 6B, 7A

HEIGHT: 30 to 40 feet

WIDTH: 30 to 40 feet

FRUIT: capsule; 2"; dry; no significant litter

GROWTH RATE: moderate; moderate life span

HABIT: rounded; open; nearly symmetrical; medium texture

FAMILY: *Sapindaceae*

LEAF COLOR: medium green

LEAF DESCRIPTION: alternate; pinnately compound; incised to serrate margin; deciduous; 2" long leaflets

FLOWER DESCRIPTION: snow-white; very showy; spring

EXPOSURE/CULTURE:

Light Requirement: flowers best in full sun, will grow in part sun

Soil Tolerances: all textures; alkaline to acidic; salt and drought

Aerosol Salt Tolerance: moderate

PEST PROBLEMS: resistant

BARK/STEMS:

Branches droop as tree grows and will require pruning for vehicular or pedestrian clearance beneath canopy; usually seen with several trunks, but can be trained to a short, single trunk; no thorns

Pruning Requirement: space branches along a short, central trunk; prune in dormant season

Limb Breakage: probably somewhat susceptible at branches with included bark

Current Year Twig: brown; medium thickness

LANDSCAPE NOTES:

Origin: China

Usage: above-ground planter; buffer strip; highway; specimen; near a deck or patio

This little-known, small tree deserves to be placed into production in the nursery trade. Once nursery operators and buyers see the tree in flower, they will want it. The tree simply explodes into a white mass in the spring, along with the Redbuds and Crabapples. It appears to grow with several twisted trunks, not unlike its close relative, the Goldenraintree. The ultimate form of the tree is akin to a small version of the Raintree.

Its small stature makes it very suitable for placing anywhere a bright-white, springtime accent is needed in a residential or commercial landscape. As with other white-flowering trees, it is best to locate the tree in front of a darker background to properly display the flowers. Nursery operators should seek out this little-known tree and make it available in the trade.

Limitations are unknown.

Yucca elephantipes Spineless Yucca

Pronunciation: YOU-cah eh-leh-fan-TIE-pees

ZONES: 9B, 10A, 10B, 11

HEIGHT: 30 feet

WIDTH: 15 feet

FRUIT: oval; 0.5"; fleshy; brown; no significant litter

GROWTH RATE: moderately rapid; long-lived

HABIT: upright; open; irregular silhouette; coarse texture

FAMILY: *Agavaceae*

LEAF COLOR: dull green

LEAF DESCRIPTION: spiral; simple; entire margin; lanceolate to linear; evergreen; 30 to 40" long; no sharp, terminal spine

FLOWER DESCRIPTION: white; very showy; summer

EXPOSURE/CULTURE:

Light Requirement: part shade to full sun

Soil Tolerances: all textures; alkaline to acidic; drought

Aerosol Salt Tolerance: moderate

PEST PROBLEMS: resistant

BARK/STEMS:

Routinely grown with multiple trunks; branches grow mostly upright and will not droop; showy trunk; no thorns

Pruning Requirement: remove sprouts at base of trunk for a more open canopy

Limb Breakage: resistant

CULTIVARS AND SPECIES:

A Spineless Yucca cultivar with striped foliage, 'Variegata', may be found in some nurseries, and may be gaining in popularity.

LANDSCAPE NOTES:

Origin: Mexico and Central America

Usage: container or above-ground planter; buffer strip; highway; near a deck or patio; specimen; sidewalk cutout; food

A dramatic landscape element, Spineless Yucca is the tallest of the Yuccas, reaching 30 feet in height with a single, thick, rough trunk topped with straplike, 4-foot-long leaves. The trunk can grow to four feet thick. Sprouts often grow from the base of the trunk, forming a multitrunked tree. Spineless Yucca grows fairly rapidly but usually stays under 20 feet in height, and is ideal for use in succulent gardens or large planters. Unlike its close relative, Spanish Bayonet, Spineless Yucca can be used within range of people, as it lacks the formidable, terminal spine and has harmless leaves. It was introduced into Florida in 1956 as a substitute for the spiny Spanish Bayonet, but grows much larger.

The 2- to 3-foot-tall bloom is produced on top of the stalks once the plant is 8 to 10 feet tall. Blooms are edible, high in calcium and potassium, and can be used in salads. Leaves contain large amounts of ascorbic acid. This is a very nutritious plant!

Spineless Yucca grows easily in full sun or partial shade on any well-drained soil. Do not plant Yucca unless drainage is superior. Yucca grows well as a houseplant in a well-lighted area.

Propagation is by seed or by cuttings of any size. Suckers at the base of the plant root quite easily.

Pests include yucca moth borers, scale, and black weevil, which bore into roots and stems. Roots rot in soils kept too moist. Do not irrigate Yucca once it is established. Leaf spots sometimes infect Yucca but do no real harm to the plant.

Zelkova serrata Japanese Zelkova

Pronunciation: zell-COE-vah sir-RAY-tah

ZONES: 5A, 5B, 6A, 6B, 7A, 7B, 8A, 8B, 9A

HEIGHT: 55 to 80 feet

WIDTH: 50 to 75 feet

FRUIT: oval; less than 0.25"; fleshy; no significant litter

GROWTH RATE: moderate; long-lived

HABIT: vase; moderate density; symmetrical; fine texture

FAMILY: *Ulmaceae*

LEAF COLOR: medium green; showy copper, orange, red, or yellow fall color

LEAF DESCRIPTION: alternate; simple; serrate margin; oblong to ovate; deciduous; 1.25 to 2" long

FLOWER DESCRIPTION: not showy; spring

EXPOSURE/CULTURE:

Light Requirement: full sun

Soil Tolerances: all textures; slightly alkaline to acidic; occasionally wet soil; drought; compacted

Aerosol Salt Tolerance: moderate

PEST PROBLEMS: resistant

BARK/STEMS:

Showy trunk with exfoliating bark; branches grow mostly upright and will not droop; should be grown with a single leader; no thorns

Pruning Requirement: requires pruning to develop strong structure; space main branches along a single trunk; help prevent formation of included bark by thinning aggressive branches to slow their growth

Limb Breakage: susceptible to breakage at crotch, due to poor collar formation

Current Year Twig: brown; slender

CULTIVARS AND SPECIES:

'Green Vase' resembles the shape of American Elms, makes a great city street tree, and produces a taller and narrower tree than 'Village Green'; 'Village Green' is more winter-hardy than 'Green Vase' and may have a straighter trunk.

LANDSCAPE NOTES:

Origin: Japan, Korea

Usage: bonsai; parking lot island; buffer strip; highway; shade tree; sidewalk cutout; street tree; urban tolerant

Zelkova is often listed as a replacement for American Elm, because it has roughly the same vase shape, but the canopy is shorter and more rounded. (No tree will truly match the grace and elegance of the American Elm.) Zelkova is a large tree, with a trunk capable of growing to four feet or more in diameter. Branches are more numerous and smaller in diameter than American Elm. Leaves turn a brilliant yellow, orange, or burnt umber in the fall.

Unfortunately, the branches on Zelkova are often clumped together on the trunk at one point, forming multiple trunks with included bark. This is not a desirable form for planting in urban areas and should be prevented or corrected on existing trees. Purchase trees with branches spaced along the trunk so they can develop a secure hold on the trunk. These will be hard to find, but insist on it! Be sure that branches remain less than about half the diameter of the trunk to maintain a strong, durable form.

The tree will tolerate most soil types, including those with a pH to about 7.5. It is reportedly risky to transplant in the fall. It makes a wonderful street tree, even in restricted-soil tree pits, and is almost pest-free. The crowns will eventually grow together if trees are planted on 30-foot centers, forming a wonderful shaded street corridor. This tough, urban tree is often planted along streets in downtown and residential areas.

Normally disease-free, as it resists Dutch elm disease and elm leaf beetle. Zelkova is subject to canker diseases, particularly if the trunk is repeatedly wounded. Avoid wounding and maintain tree health. Japanese beetles devour foliage.

Ziziphus jujuba Chinese Date or Jujube

Pronunciation: ZIZ-ih-fuss joo-JOO-bah

ZONES: 6A, 6B, 7A, 7B, 8A, 8B, 9A, 9B, 10A, 10B

HEIGHT: 25 to 35 feet

WIDTH: up to 30 feet

FRUIT: oval 0.5 to 1"; fleshy; red turning black; edible; attracts mammals; causes significant mushy litter; showy

GROWTH RATE: moderate; moderate life span

HABIT: spreading, rounded vase; open; irregular silhouette; fine texture

FAMILY: *Rhamnaceae*

LEAF COLOR: dark green; yellow fall color

LEAF DESCRIPTION: alternate; simple; crenate to serrulate margin; glossy; lanceolate to ovate; deciduous; 2 to 4" long

FLOWER DESCRIPTION: yellow; not showy; spring

EXPOSURE/CULTURE:

Light Requirement: part shade to full sun

Soil Tolerances: all textures; alkaline to acidic; wet soil; extremely drought tolerance

Aerosol Salt Tolerance: low

PEST PROBLEMS: resistant

BARK/STEMS:

Branches grow mostly horizontal and droop on older trees; should be grown with a short, single leader; thorns are present on trunk or branches; suckers form freely from trunk base

Pruning Requirement: requires pruning to develop strong structure

Limb Breakage: resistant

Current Year Twig: brown; medium thickness

LANDSCAPE NOTES:

Origin: India

Usage: fruit tree

Chinese Jujube is an interesting deciduous tree with spiny, gnarled branches and an open, irregular form. Growing at a moderate pace, Chinese Jujube reaches anywhere from 15 to 35 feet in height with a spread of 10 to 30 feet and can be trained to a short, single trunk. Most unpruned plants grow with several trunks with mottled, gray-black, rough, shaggy bark. The one- to two-inch-long leaves have a paler underside and sharp spines at the base of each leaf. Fall color is often a showy yellow, but not consistently. In spring, small clusters of yellow or white, fragrant blossoms appear, hidden in foliage between the leaf and stem.

The one-inch-long green fruits ripen to dark red and finally black. Eaten either fresh, candied, canned, or dried like dates, these fruits are quite sweet. Even young, two-year-old trees are able to produce these delectable treats, but be forewarned that these fruits can create quite a litter problem. Locate the tree so the fruit drops in a mulch bed or on the lawn, not on a sidewalk, patio, or driveway.

Jujube has an upright habit when young, developing a number of aggressive, upright branches clustered on the trunk. These should be spaced more than 12 inches along the trunk and headed back or thinned regularly to prevent included bark from forming in the crotches. This training technique will encourage good development of strong structure and increase longevity of the tree.

Chinese Jujube should be grown in full sun or partial shade on any well-drained soil, acid or alkaline. Plants do not grow well on heavy clay or swampy soils. Fruits rarely form in dry years in dry climates. No limitations of major concern. Roots sprout new stems.

Using Appendices

Several of the Appendices are cross-referenced. For example, Appendix 19 indicates which of the trees listed with showy flowers also have showy fall foliage color. Appendices 7, 8, and 9 list small-, medium-, and large-maturing trees, respectively, but also indicate the drought tolerance of each of the trees.

- Appendix 1. Site Analysis Form for Selecting the Right Tree for a Planting Site
- Appendix 2. Drought- and Wet-Site-Tolerant Trees
- Appendix 3. Trees Tolerant of Moderate or High Soil or Aerosol Salt
- Appendix 4. Trees Tolerant of Alkaline Soil pH
- Appendix 5. Trees Tolerating Poor Drainage
- Appendix 6. Trees Tolerating Shaded or Mostly Shaded Sites
- Appendix 7. Trees That Grow Less Than 30 Feet Tall
- Appendix 8. Trees That Grow 30 to 50 Feet Tall
- Appendix 9. Trees Capable of Growing More Than 50 Feet Tall
- Appendix 10. Trees with a Columnar or Narrow Canopy Usually Not More Than About 25 Feet Wide
- Appendix 11. Trees That Can Develop Large-Diameter Surface Roots or Numerous Smaller Surface Roots
- Appendix 12. Trees That Are Sensitive to Pests or Diseases That Can Cause Noticeable Damage or Tree Death
- Appendix 13. Trees That Are Susceptible to Breakage
- Appendix 14. Nonconiferous Trees Requiring Only a Moderate Amount of Pruning to Develop Good Structure
- Appendix 15. Trees Bearing Little or No Fruit
- Appendix 16. Trees with Messy Fruit or Leaves
- Appendix 17. Trees Commonly Grown with a Multi-trunk Habit
- Appendix 18. Trees with Outstanding Bark or Trunk Structure
- Appendix 19. Trees with Showy Flowers
- Appendix 20. Trees with Good to Exceptional Fall Foliage Color
- Appendix 21. Trees Native to North America
- Appendix 22. State Trees of the United States
- Appendix 23. Approximate Order of Flowering from Spring Through Summer

Appendix 1

Site Analysis Form for Selecting the Right Tree for a Planting Site

What USDA hardiness zone is the planting site located in?

____

____

What is the average annual rainfall in the area?

☐ Less than 20 inches ☐ More than 20 inches

Will the tree be planted:

☐ In the ground
☐ In containers or in above-ground planters
☐ Near the coast

What is the light exposure at the site?

☐ More than six hours of full sun
☐ Between two and five hours of direct sun
☐ Two to five hours of sun with significant reflected light
☐ Less than two hours of full sun, or filtered shade most of the day

What is the soil pH at the planting site? (Have it tested; do not guess.)

____

____

How fast does water drain through the soil at the planting site? Test this by digging a hole 18 inches deep and filling it up with water. If the water drains away in an hour or two, the drainage is fast. If it takes up to a day for the water to drain away, drainage is moderate. If it takes longer than a day, the drainage is slow.

☐ Slow ☐ Moderate ☐ Fast

What is the distance between the top of the water table and the soil surface? To test this, dig several holes on the site about two feet deep and wait two or three hours. If any water appears in the hole, the site probably has a high water table.

____

____

How will the site be irrigated?

☐ Hardly at all
☐ During the establishment period only
☐ During establishment and then only during extended drought
☐ The trees will be regularly irrigated

(Note: Establishment period is about 6 to 12 months per inch of trunk diameter. For example, a 2-inch-diameter tree takes about 12 to 24 months to establish.)

What is the soil texture?

☐ Clay ☐ Loam ☐ Sand

What is the soil density?

☐ The soil is compacted and hard
☐ The soil is loose

Will the tree be planted in a tree lawn or streetscape (the grassy strip between the curb and the sidewalk)? ____________________

If so, how wide is the tree lawn?

☐ Three to four feet
☐ Four to six feet
☐ Six to eight feet
☐ More than eight feet wide

Will the tree be planted along a street without a sidewalk?

☐ Yes ☐ No

If so, how far from the edge of the road will the tree be planted?

____

____

Will the tree be planted in a sidewalk cutout?

☐ Yes ☐ No

Will the tree be planted in a parking lot?

☐ Yes ☐ No

If so, will it be planted in a sidewalk cutout, parking lot island, buffer strip, or narrow linear strip of soil?

____

____

For a parking lot island:
What is the square footage of the parking lot island?

____

For a buffer strip or linear strip:
What is the width of the buffer strip or linear planting strip?

____

Will the tree be planted in an open lawn area or in a shrub bed?
☐ Yes
☐ No

What is the approximate size of this area?

____

Will the tree be planted within eight feet of a sidewalk, driveway, or other hard surface?

____

____

Will an adjacent sidewalk or roadway receive de-icing salts?

____

____

Is there a swimming pool, vegetable garden, masonry wall, septic tank, or drain field within 50 feet of the planting site?
☐ Yes
☐ No

If so, how far away is it?

____

Are overhead wires within 30 feet of the planting site?
☐ Yes
☐ No

If so:

What is the horizontal distance between the planting hole and the wire?

____

What is the distance between the ground and the lowest wire?

____

Is there a street light or security-type light within 35 feet of the planting hole?
☐ Yes
☐ No

If so:

Do you want the tree branches to stay clear of the light so they will not have to be pruned?

____

Are you willing to provide some pruning to train the branches to grow over the light?

____

What is the horizontal distance between the light and the planting hole?

____

How tall is the light?

____

Is the planting site within 35 feet of a building?
☐ Yes
☐ No

If so:

What is the horizontal distance between the planting hole and the building?

____

Approximately how tall is the building?

____

Do you want to eliminate trees that could drop messy fruit, large leaves, or twigs during an extended period?
☐ Yes
☐ No

Would you like to eliminate trees that are known to be susceptible to breakage?
☐ Yes
☐ No

What is your budget for pruning trees?

____

Do you want to plant only native trees?
☐ Yes
☐ No

List any other attributes that you would like your trees to have:

____

____

____

____

____

____

Appendix 2

Drought- and Wet-Site Tolerant Trees

Abbreviation Key:

f. = Forma
spp. = Species
var. = Variety

Note: When a species is listed, its cultivars are also considered tolerant. While these plants tolerate sites that are periodically wet, most grow best without periodic flooding.

Acacia auriculiformis—Earleaf Acacia
Acer negundo—Boxelder
Acer saccharinum—Silver Maple
Ailanthus altissima—Tree-of-Heaven

Betula populifolia—Gray Birch

Carpinus betulus 'Fastigiata'—'Fastigiata' European Hornbeam
Carya illinoensis—Pecan
Catalpa spp.—Catalpa
Celtis laevigata—Sugarberry, Sugar Hackberry
Celtis occidentalis—Common Hackberry
Chorisia speciosa—Floss-Silk Tree
Conocarpus erectus—Buttonwood
Conocarpus erectus var. *sericeus*—Silver Buttonwood
Crataegus viridis 'Winter King'—'Winter King' Hawthorn
Cupaniopsis anacardiopsis—Carrotwood

Ficus aurea—Strangler Fig, Golden Fig
Ficus benjamina—Weeping Fig
Ficus lyrata—Fiddleleaf Fig
Fraxinus pennsylvanica—Green Ash
Fraxinus uhdei—Shamel Ash
Fraxinus velutina—Velvet Ash, Modesto Ash, Arizona Ash

Ginkgo biloba—Maidenhair Tree, Ginkgo
Gleditsia triacanthos var. *inermis*—Thornless Honeylocust
Guaiacum sanctum—Lignumvitae

Ilex decidua—Possumhaw
Ilex opaca—American Holly
Ilex vomitoria—Yaupon Holly

Koelreuteria spp.—Goldenraintree

Maclura pomifera—Osage-Orange, Bois-D'Arc
Magnolia virginiana—Sweetbay Magnolia, Swamp Magnolia
Melaleuca quinquinervia—Punk Tree, Melaleuca
Morus alba—White Mulberry
Myrica cerifera—Southern Waxmyrtle, Southern Bayberry

Nyssa sylvatica—Blackgum, Sourgum, Black Tupelo

Persea borbonia—Redbay
Persea palustris—Swamp Bay
Phellodendron amurense—Amur Corktree, Chinese Corktree
Pinus elliottii—Slash Pine
Pinus mugo—Mugo Pine, Swiss Mountain Pine
Platanus occidentalis—Sycamore, American Planetree
Platanus orientalis—Oriental Planetree
Platanus x acerifolia—London Planetree
Plumeria alba—White Frangipani
Plumeria rubra—Frangipani
Podocarpus falcatus—Podocarpus
Podocarpus gracilior—Weeping Podocarpus, Fern Podocarpus
Podocarpus latifolius—Podocarpus
Podocarpus macrophyllus—Podocarpus, Yew-Pine, Japanese Yew
Podocarpus macrophyllus var. *angustifolius*—Podocarpus
Podocarpus nagi—Nagi Podocarpus, Broadleaf Podocarpus
Pongamia pinnata—Pongam, Karum Tree, Poonga-Oil Tree
Populus deltoides—Eastern Cottonwood
Prunus serotina—Black Cherry
Psidium littorale—Cattley Guava, Strawberry Guava
Pyrus calleryana cultivars—Callery Pear

Quercus acutissima—Sawtooth Oak
Quercus bicolor—Swamp White Oak
Quercus glauca—Blue Japanese Oak, Ring-Cupped Oak
Quercus macrocarpa—Bur Oak
Quercus michauxii—Swamp Chestnut Oak
Quercus nigra—Water Oak
Quercus nuttallii—Nuttall Oak
Quercus palustris—Pin Oak
Quercus phellos—Willow Oak

Rhus chinensis—Chinese Sumac
Rhus glabra—Smooth Sumac
Rhus lancea—African Sumac
Roystonea spp.—Royal Palm

Sabal palmetto—Cabbage Palm
Salix spp.—Weeping Willow
Sapium sebiferum—Chinese Tallowtree, Popcorn Tree, Tallowtree
Swietenia mahagoni—Mahogany, West Indies Mahogany

Tamarindus indica—Tamarind
Taxodium ascendens—Pondcypress
Taxodium distichum—Baldcypress
Thrinax morrisii—Key Thatchpalm

Ulmus alata—Winged Elm
Ulmus americana—American Elm
Ulmus crassifolia—Cedar Elm
Ulmus hybrids—Hybrid Elm
Ulmus parvifolia—Chinese Elm, Lacebark Elm
Ulmus pumila—Siberian Elm

Washingtonia robusta—Washington Palm, Mexican Washington Palm

Zelkova serrata—Japanese Zelkova, Saw-Leaf Zelkova

Appendix 3

Trees Tolerant of Moderate or High Soil or Aerosol Salt

Legend:
* = Very good salt tolerance
S = Some authors list as salt-sensitive
V = Varies with cultivar

Note: When a species is listed, its cultivars are also considered tolerant.

Acacia auriculiformis—Earleaf Acacia
Acacia farnesiana—Sweet Acacia, Huisache
Acer campestre—Hedge Maple (*)
Acer cissifolium—Ivy-Leaf Maple
Acer ginnala—Amur Maple
Acer griseum—Paperbark Maple
Acer negundo—Boxelder
Acer rubrum var. *drummondii*—Drummond Red Maple
Acer platanoides—Norway Maple (*)
Acer pseudoplatanus—Sycamore Maple, Planetree Maple (*)
Acer saccharinum—Silver Maple (*)
Acer tataricum—Tataricum Maple
Acer triflorum—Three-Flower Maple
Acer truncatum—Purpleblow Maple
Acoelorrhaphe wrightii—Paurotis Palm
Aesculus hippocastanum—Horsechestnut, European Horsechestnut (*)
Aesculus x carnea—Red Horsechestnut
Ailanthus altissima—Tree-of-Heaven (*)
Albizia julibrissin—Mimosa, Silktree
Alnus glutinosa—Common Alder, Black Alder, European Alder (S)
Amelanchier spp.—Serviceberry (S)
Amphitecna latifolia—Black-Calabash (*)
Araucaria araucana—Monkey-Puzzletree
Araucaria bidwillii—False Monkey-Puzzletree
Araucaria heterophylla—Norfolk-Island-Pine

Bauhinia spp.—Orchid-Tree
Beaucarnea recurvata—Ponytail
Betula platyphylla 'Whitespire'—Japanese Whitespire Birch
Betula nigra—River Birch
Betula papyrifera—Paper Birch, Canoe Birch (*)
Betula populifolia—Gray Birch (*)
Bischofia javanica—Toog Tree, Bischofia
Bismarckia nobilis—Bismarck Palm
Bucida buceras—Black Olive, Oxhorn Bucida (*)
Bursera simaruba—Gumbo-Limbo (*)
Butia capitata—Pindo Palm

Callistemon citrinus—Red Bottlebrush, Lemon Bottlebrush
Callistemon viminalis—Weeping Bottlebrush
Calophyllum brasiliense—Santa Maria (*)
Carya illinoensis—Pecan
Carya ovata—Shagbark Hickory
Cassia fistula—Golden-Shower
Cassia javonica—Pink-and-White-Shower
Catalpa spp.—Catalpa
Celtis australis—Mediterranean Hackberry, European Hackberry
Celtis laevigata—Sugarberry, Sugar Hackberry (S)
Celtis occidentalis—Common Hackberry (S)
Chrysalidocarpus lutescens—Yellow Butterfly Palm, Bamboo Palm, Areca Palm
Chrysophyllum oliviforme—Satinleaf
Clusia rosea—Pitch-Apple, Florida Clusia (*)
Coccoloba diversifolia—Pigeon-Plum (*)
Coccoloba uvifera—Seagrape (*)
Coccothrinax argentata—Silverpalm, Thatchpalm (*)
Cocos nucifera 'Malayan Dwarf'—'Malayan Dwarf' Coconut Palm (*)
Conocarpus erectus—Buttonwood (*)
Conocarpus erectus var. *sericeus*—Silver Buttonwood (*)
Cordia boissieri—Wild-Olive, Anacahuita
Cordia sebestena—Geiger-Tree (*)
Cotinus coggygria—Smoketree, Wig-Tree, Smokebush
Cupaniopsis anacardiopsis—Carrotwood
x Cupressocyparis leylandii—Leyland Cypress
Cupressus macrocarpa—Monterey Cypress
Cupressus sempervirens—Italian Cypress

Diospyros virginiana—Common Persimmon

Elaeagnus angustifolia—Russian-Olive, Oleaster (*)
Eriobotrya deflexa—Bronze Loquat
Eriobotrya japonica—Loquat
Erythrina crista-galli—Cockspur Coral Tree
Erythrina variegata var. *orientalis*—Coral Tree
Eucommia ulmoides—Hardy Rubber Tree
Eugenia spp.—Stopper, Eugenia

Feijoa sellowiana—Guava, Feijoa, Pineapple Guava (*)
Ficus aurea—Strangler Fig, Golden Fig
Ficus benjamina—Weeping Fig
Ficus elastica—Rubber Tree, India-Rubber Fig
Ficus lyrata—Fiddleleaf Fig
Ficus retusa—Cuban-Laurel
Ficus rubiginosa—Rusty Fig
Fraxinus americana—White Ash (*)
Fraxinus excelsior—Common Ash, European Ash (*)
Fraxinus pennsylvanica—Green Ash
Fraxinus uhdei—Shamel Ash
Fraxinus velutina—Velvet Ash, Modesto Ash, Arizona Ash

Geijera parviflora—Australian-Willow
Ginkgo biloba—Maidenhair Tree, Ginkgo
Gleditsia triacanthos var. *inermis*—Thornless Honeylocust (*)
Guaiacum sanctum—Lignumvitae (*)
Gymnocladus dioicus—Kentucky Coffeetree

Hibiscus syriacus—Rose-of-Sharon, Shrub-Althea

Ilex cassine—Dahoon Holly
Ilex cornuta 'Burfordii'—Burford Holly
Ilex opaca—American Holly (*)

Ilex vomitoria—Yaupon Holly (*)
Ilex x attenuata 'East Palatka'—'East Palatka' Holly

Jatropha integerrima—Peregrina, Fire-Cracker
Juglans cinerea—White Walnut (*)
Juglans nigra—Black Walnut (*)
Juniperus chinensis 'Torulosa'—'Torulosa' Juniper (*)
Juniperus scopulorum—Rocky Mountain Juniper, Colorado Redcedar
Juniperus silicicola—Southern Redcedar (*)
Juniperus virginiana—Eastern Redcedar (*)

Koelreuteria bipinnata—Chinese Flame-Tree, Bougainvillea Goldenraintree
Koelreuteria elegans—Flamegold
Koelreuteria paniculata—Goldenraintree, Varnish-Tree

Lagerstroemia fauriei—Japanese Crape-Myrtle
Lagerstroemia indica—Crape-Myrtle
Lagerstroemia speciosa—Queens Crape-Myrtle
Latania spp.—Latan Palm
Larix decidua—European Larch
Leptospermum laevigatum—Australian Tea Tree
Ligustrum japonicum—Japanese Privet, Wax-Leaf Privet
Ligustrum lucidum—Glossy Privet, Tree Ligustrum
Livistona chinensis—Chinese Fan Palm
Lysiloma bahamensis—Wild-Tamarind, Bahama Lysiloma (*)

Maclura pomifera—Osage-Orange, Bois-D'Arc
Magnolia grandiflora—Southern Magnolia
Malpighia glabra—Barbados Cherry
Malus spp.—Crabapple (V)
Mangifera indica—Mango
Manilkara zapota—Sapodilla (*)
Melaleuca quinquinervia—Punk Tree, Melaleuca
Melia azedarach—Chinaberry
Morus alba (fruitless cultivars)—White Mulberry (*)
Murraya paniculata—Orange Jessamine
Myrica cerifera—Southern Waxmyrtle, Southern Bayberry (*)

Nerium oleander—Oleander
Noronhia emarginata—Madagascar Olive (*)
Nyssa sylvatica—Blackgum, Sourgum, Black Tupelo

Ochrosia elliptica—Ochrosia (*)
Olea europaea—European Olive
Ostrya virginiana—American Hophornbeam, Eastern Hophornbeam (S)

Pandanus utilis—Screw-pine (*)
Parkinsonia aculeata—Jerusalem-Thorn (*)
Paulownia tomentosa—Princess-Tree, Empress-Tree, Paulownia
Persea borbonia—Redbay (*)
Phellodendron amurense—Amur Corktree, Chinese Corktree
Phoenix canariensis—Canary Island Date Palm
Phoenix dactylifera—Date Palm (*)
Phoenix reclinata—Senegal Date Palm
Picea glauca—White Spruce
Picea pungens—Colorado Spruce, Colorado Blue Spruce, Blue Spruce (*)
Pinus clausa—Sand Pine
Pinus elliottii—Slash Pine
Pinus mugo—Mugo Pine, Swiss Mountain Pine
Pinus nigra—Austrian Pine, Black Pine (*)
Pinus parviflora—Japanese White Pine
Pinus pinea—Stone Pine, Italian Stone Pine, Umbrella Pine (*)
Pinus ponderosa—Ponderosa Pine
Pinus radiata—Monterey Pine (*)
Pinus sylvestris—Scotch Pine
Pinus thunbergiana—Japanese Black Pine (*)
Pithecellobium flexicaule—Ebony Blackbead, Texas-Ebony
Pittosporum spp.—Pittosporum (*)
Platanus occidentalis—Sycamore, American Planetree (S)
Platanus orientalis—Oriental Planetree
Platanus x acerifolia—London Planetree
Plumeria alba—White Frangipani (*)
Plumeria rubra—Frangipani (*)
Podocarpus macrophyllus—Podocarpus, Yew-Pine, Japanese Yew
Podocarpus nagi—Nagi Podocarpus, Broadleaf Podocarpus
Pongamia pinnata—Pongam, Karum Tree, Poonga-Oil Tree
Populus alba—White Poplar (*)
Populus deltoides—Eastern Cottonwood (*)
Populus freemontii—Fremont Poplar
Populus nigra 'Italica'—Lombardy Poplar (*)
Prosopis glandulosa—Mesquite, Honey Mesquite
Prunus caroliniana—Cherry-Laurel, Carolina Laurelcherry
Prunus cerasifera—Purple Leaf Plum
Prunus maackii—Amur Chokecherry, Manchurian Cherry
Prunus sargentii—Sargent Cherry
Prunus serrulata 'Kwanzan'—Kwanzan Cherry
Pyrus calleryana cultivars—Callery Pear cultivars

Quercus acutissima—Sawtooth Oak
Quercus agrifolia—Coastal Live Oak
Quercus alba—White Oak (*)
Quercus bicolor—Swamp White Oak
Quercus cerris—Turkey Oak, Moss-Cupped Oak
Quercus coccinea—Scarlet Oak
Quercus falcata—Southern Red Oak, Spanish Oak
Quercus ilex—Holly Oak
Quercus imbricaria—Shingle Oak, Northern Laurel Oak
Quercus macrocarpa—Bur Oak
Quercus muehlenbergii—Chinkapin Oak, Chestnut Oak
Quercus phellos—Willow Oak
Quercus robur—English Oak (*)
Quercus rubra—Northern Red Oak (S)
Quercus shumardii—Shumard Oak
Quercus virginiana—Southern Live Oak, Live Oak

Robinia pseudoacacia—Black Locust, Common Locust (*)
Roystonea spp.—Royal Palm

Sabal palmetto—Cabbage Palm (*)
Salix alba—Weeping Willow (*)
Salix matsudana 'Tortuosa'—Corkscrew Willow, Pekin Willow, Hankow Willow (*)
Sapindus saponaria—Florida Soapberry, Wingleaf Soapberry (*)
Schefflera actinophylla—Schefflera, Queensland Umbrella-Tree
Schefflera arboricola—Dwarf Schefflera
Schinus terebinthifolius—Brazilian Pepper Tree (*)
Simarouba glauca—Paradise-Tree (*)
Sophora japonica—Scholar Tree, Japanese Pagoda Tree
Styrax japonicus—Japanese Snowbell
Swietenia mahagoni—Mahogany, West Indies Mahogany (*)
Syagrus romanzoffianum—Queen Palm
Syringa pekinensis—Peking Lilac (*)
Syringa reticulata—Japanese Tree Lilac (*)

Tabebuia caraiba—Trumpet Tree
Tabebuia chrysotricha—Golden Trumpet Tree
Tabebuia heterophylla—Pink Trumpet Tree
Tabebuia impetiginosa—Purple Tabebuia

Tamarindus indica—Tamarind
Taxodium ascendens—Pondcypress
Taxodium distichum—Baldcypress
Tecoma stans—Yellow-Elder, Yellow Trumpet-Flower
Terminalia catappa—Tropical-Almond, India-Almond (*)
Terminalia muelleri—Muellers Terminalia
Thrinax morrisii—Key Thatchpalm (*)
Tilia tomentosa—Silver Linden
Trachycarpus fortunei—Windmill Palm

Ulmus alata—Winged Elm
Ulmus americana—American Elm
Ulmus crassifolia—Cedar Elm
Ulmus hybrids—Hybrid Elm
Ulmus parvifolia—Chinese Elm, Lacebark Elm
Ulmus pumila—Siberian Elm (*)

Veitchia merrillii—Christmas Palm, Manila Palm
Viburnum odoratissimum—Sweet Viburnum
Vitex agnus-castus—Chastetree, Vitex

Washingtonia filifera—Desert Palm, California Washingtonia Palm
Washingtonia robusta—Washington Palm, Mexican Washington Palm
Wodyetia bifurcata—Foxtail Palm

Yucca elephantipes—Spineless Yucca, Soft-Tip Yucca

Zelkova serrata—Japanese Zelkova, Saw-Leaf Zelkova

Appendix 4

Trees Tolerant of Alkaline Soil pH

Legend:
* = Only plant in soil with a pH less than 7.7
FT = Tolerates wet soil

Abies cilicica—Cilician Fir (*)
Abies concolor—White Fir, Colorado Fir (*)
Abies nordmanniana—Nordmann Fir (*)
Acacia auriculiformis—Earleaf Acacia (FT)
Acacia farnesiana—Sweet Acacia, Huisache (FT)
Acacia wrightii—Wright Acacia, Wright Catclaw
Acer barbatum var. *caddo*—Caddo Florida Maple, Southern Sugar Maple
Acer buergeranum—Trident Maple (*)
Acer campestre—Hedge Maple (*)
Acer ginnala—Amur Maple
Acer grandidentatum—Bigtooth Maple, Rocky Mountain Sugar Maple, Canyon Maple
Acer griseum—Paperbark Maple (*)
Acer japonicum—Fullmoon Maple (*)
Acer leucoderme—Chalk Maple (*)
Acer macrophyllum—Bigleaf Maple (*, FT)
Acer miyabei—Miyabe Maple
Acer negundo—Boxelder (FT)
Acer palmatum—Japanese Maple (*)
Acer platanoides—Norway Maple (FT)
Acer pseudoplatanus—Sycamore Maple, Planetree Maple (FT)
Acer saccharinum—Silver Maple (*, FT)
Acer saccharum—Sugar Maple
Acer tataricum—Tataricum Maple (FT)
Acer triflorum—Three-Flowered Maple (*)
Acer truncatum—Purpleblow Maple (*)
Acoelorrhaphe wrightii—Paurotis Palm (*, FT)
Aesculus flava—Yellow Buckeye, Sweet Buckeye (*, FT)
Aesculus glabra—Ohio Buckeye, Fetid Buckeye (*, FT)
Aesculus hippocastanum—Horsechestnut, European Horsechestnut (*, FT)
Aesculus indica—Indian Horsechestnut (*)
Aesculus pavia—Red Buckeye (*, FT)
Aesculus x carnea—Red Horsechestnut (*, FT)
Ailanthus altissima—Tree-of-Heaven (FT)
Albizia julibrissin—Mimosa, Silktree (*, FT)
Alnus glutinosa—Common Alder, Black Alder, European Alder (FT)
Alnus rhombifolia—White Alder (FT)
Amelanchier spp.—Serviceberry (*)
Amphitecna latifolia—Black-Calabash
Aralia spinosa—Devils-Walkingstick, Hercules-Club (FT)
Araucaria araucana—Monkey-Puzzletree
Araucaria bidwillii—False Monkey-Puzzletree
Araucaria heterophylla—Norfolk-Island-Pine
Arbutus texana—Texas Madrone
Arbutus unedo—Strawberry-Tree
Asimina triloba—Pawpaw (*, FT)

Bauhinia spp.—Orchid-Tree (*)
Beaucarnea recurvata—Ponytail
Betula utilis var. *jacquemontii*—Jacquemontii Birch (*)
Betula platyphylla 'Whitespire'—Japanese Whitespire Birch (*)
Betula papyrifera—Paper Birch, Canoe Birch (*, FT)
Betula pendula—European Birch (*, FT)
Betula populifolia—Gray Birch (*, FT)
Bischofia javanica—Toog Tree, Bischofia (FT)
Bismarckia nobilis—Bismarck Palm (*, FT)
Brahea armata—Mexican Blue Palm
Bucida buceras—Black Olive, Oxhorn Bucida (*)
Bulnesia arborea—Bulnesia (*)
Bursera simaruba—Gumbo-Limbo
Butia capitata—Pindo Palm (*)

Caesalpinia granadillo—Bridalveil Tree
Caesalpinia pulcherrima—Dwarf Poinciana, Barbados Flowerfence
Calliandra haematocephala—Powderpuff (*)
Callistemon citrinus—Red Bottlebrush, Lemon Bottlebrush (*)
Callistemon viminalis—Weeping Bottlebrush (*)
Calocedrus decurrens—California Incense-Cedar (*)
Calodendron capense—Cape Chestnut (*)
Calophyllum brasiliense—Santa Maria
Camellia oleifera—Tea-Oil Camellia (*)
Carpentaria acuminata—Carpentaria Palm (*, FT)
Carpinus betulus 'Fastigiata'—'Fastigiata' European Hornbeam (FT)
Carpinus caroliniana—American Hornbeam, Blue-Beech, Ironwood (*, FT)
Carya glabra—Pignut Hickory (*, FT)
Carya illinoensis—Pecan (FT)
Carya ovata—Shagbark Hickory (*, FT)
Caryota mitis—Fishtail Palm
Cassia alata—Candlebrush (*)
Cassia fistula—Golden-Shower
Cassia javonica—Pink-and-White-Shower
Cassia leptophylla—Gold Medallion Tree
Cassia suratensis—Glaucous Cassia
Castanea mollissima—Chinese Chestnut (*)
Catalpa spp.—Catalpa (FT)
Cedrus atlantica—Atlas Cedar (*)
Cedrus deodara—Deodar Cedar (*)
Cedrus libani—Cedar-of-Lebanon (*)
Celtis australis—Mediterranean Hackberry, European Hackberry (*, FT)
Celtis laevigata—Sugarberry, Sugar Hackberry (*, FT)
Celtis occidentalis—Common Hackberry (FT)
Celtis sinensis—Japanese Hackberry, Chinese Hackberry (*, FT)
Cercidiphyllum japonicum—Katsuratree (*)
Cercis canadensis—Eastern Redbud
Cercis canadensis var. *texensis*—Texas Redbud
Cercis mexicana—Mexican Redbud
Cercis reniformis 'Oklahoma'—Oklahoma Redbud
Chamaecyparis lawsoniana—Lawson Falsecypress, Port-Orford-Cedar (*)
Chilopsis linearis—Desert-Willow

Chionanthus retusus—Chinese Fringetree (*, FT)
Chitalpa tashkentensis—Chitalpa (*)
Chorisia speciosa—Floss-Silk Tree (FT)
Chrysalidocarpus lutescens—Yellow Butterfly Palm, Bamboo Palm, Areca Palm (*, FT)
Chrysophyllum oliviforme—Satinleaf (FT)
Cinnamomum camphora—Camphor-Tree (*)
Citrus spp.—Citrus (*)
Cladrastis kentukea—American Yellowwood, Virgilia (FT)
Clusia rosea—Pitch-Apple, Florida Clusia (FT)
Coccoloba diversifolia—Pigeon-Plum
Coccoloba uvifera—Seagrape
Coccothrinax argentata—Silverpalm, Thatchpalm
Cocos nucifera 'Malayan Dwarf'—'Malayan Dwarf' Coconut Palm
Conocarpus erectus—Buttonwood (FT)
Conocarpus erectus var. *sericeus*—Silver Buttonwood (FT)
Cordia boissieri—Wild-Olive, Anacahuita (FT)
Cordia sebestena—Geiger-Tree
Cornus alternifolia—Pagoda Dogwood (*)
Cornus controversa—Giant Dogwood (*)
Cornus drummondii—Roughleaf Dogwood (*)
Cornus florida—Flowering Dogwood (*)
Cornus kousa—Kousa Dogwood, Chinese Dogwood (*)
Cornus mas—Cornelian-Cherry
Cornus walteri—Walter Dogwood (*)
Corylus colurna—Turkish Filbert, Turkish Hazel (FT)
Cotinus coggygria—Smoketree, Wig-Tree, Smokebush
Cotinus obovatus—American Smoketree, Chittamwood
Crataegus spp.—Hawthorn (FT)
Crescentia cujete—Calabash Tree
Cupaniopsis anacardiopsis—Carrotwood (*, FT)
x Cupressocyparis leylandii—Leyland Cypress
Cupressus arizonica var. *arizonica*—Arizona Cypress
Cupressus glabra 'Carolina Safire'—Smooth-Barked Arizona Cypress
Cupressus sempervirens—Italian Cypress
Cydonia sinensis—Chinese Quince (*)

Delonix regia—Royal Poinciana
Diospyros kaki—Japanese Persimmon
Diospyros texana—Texas Persimmon
Diospyros virginiana—Common Persimmon (*, FT)

Elaeagnus angustifolia—Russian-Olive, Oleaster (FT)
Eriobotrya deflexa—Bronze Loquat
Eriobotrya japonica—Loquat
Erythrina crista-galli—Cockspur Coral Tree
Erythrina variegata var. *orientalis*—Coral Tree
Eucalyptus citriodora—Lemon-Scented Gum
Eucalyptus ficifolia—Red-Flowering Gum
Eucommia ulmoides—Hardy Rubber Tree
Eugenia spp.—Stopper, Eugenia (FT)
Evodia danielii—Korean Evodia, Bebe Tree

Fagus sylvatica—European Beech (*)
Feijoa sellowiana—Guava, Feijoa, Pineapple Guava (*)
Ficus aurea—Strangler Fig, Golden Fig (FT)
Ficus benjamina—Weeping Fig (FT)
Ficus elastica—Rubber Tree, India-Rubber Fig
Ficus lyrata—Fiddleleaf Fig (FT)
Ficus retusa—Cuban-Laurel (FT)
Ficus rubiginosa—Rusty Fig
Firmiana simplex—Chinese Parasoltree (*)
Fraxinus americana—White Ash (FT)
Fraxinus excelsior—Common Ash, European Ash
Fraxinus ornus—Flowering Ash (*)
Fraxinus pennsylvanica—Green Ash (FT)
Fraxinus texensis—Texas Ash
Fraxinus uhdei—Shamel Ash (FT)
Fraxinus velutina—Velvet Ash, Modesto Ash, Arizona Ash (FT)

Geijera parviflora—Australian-Willow (*)
Ginkgo biloba—Maidenhair Tree, Ginkgo (FT)
Gleditsia triacanthos var. *inermis*—Thornless Honeylocust
Grevillea robusta—Silk-Oak (*)
Guaiacum sanctum—Lignumvitae
Gymnocladus dioicus—Kentucky Coffeetree (FT)

Hovenia dulcis—Japanese Raisintwree (FT)
Hydrangea paniculata—Panicle Hydrangea

Ilex cassine—Dahoon Holly (*, FT)
Ilex cornuta 'Burfordii'—Burford Holly
Ilex decidua—Possumhaw (FT)
Ilex latifolia—Lusterleaf Holly (*)
Ilex opaca—American Holly (*, FT)
Ilex rotunda—Round-Leaf Holly (*, FT)
Ilex vomitoria—Yaupon Holly (FT)
Ilex vomitoria 'Pendula'—Weeping Yaupon Holly (FT)
Ilex x 'Nellie R. Stevens'—'Nellie R. Stevens' Holly (*)

Jacaranda mimosifolia—Jacaranda (*)
Jatropha integerrima—Peregrina, Fire-Cracker
Juglans spp.—Walnut (*, FT)
Juniperus ashei—Ashe Juniper, Mountain-Cedar
Juniperus chinensis 'Torulosa'—'Torulosa' Juniper
Juniperus deppeana 'Mcfetter'—McFetter Alligator Juniper
Juniperus scopulorum—Rocky Mountain Juniper, Colorado Redcedar
Juniperus silicicola—Southern Redcedar
Juniperus virginiana—Eastern Redcedar

Kalopanax pictus—Castor-Aralia, Prickly Castor-Oil Tree
Koelreuteria spp.—Goldenraintree (FT)

Laburnum x alpinum—Goldenchain Tree
Lagerstroemia fauriei—Japanese Crape-Myrtle
Lagerstroemia indica—Crape-Myrtle (*)
Lagerstroemia speciosa—Queens Crape-Myrtle (*)
Latania spp.—Latan Palm (*)
Larix decidua—European Larch (*)
Leucaena retusa—Goldenball Leadtree, Littleleaf Leucaena, Littleleaf Leadtree
Ligustrum japonicum—Japanese Privet, Wax-Leaf Privet (*)
Ligustrum lucidum—Glossy Privet, Tree Ligustrum
Litchi chinensis—Lychee (*)
Livistona chinensis—Chinese Fan Palm
Lysiloma bahamensis—Wild-Tamarind, Bahama Lysiloma

Maackia amurensis—Amur Maackia
Maclura pomifera—Osage-Orange, Bois-D'Arc
Magnolia acuminata—Cucumbertree, Cucumber Magnolia (FT)
Magnolia acuminata var. *cordata*—Yellow Magnolia (*)
Magnolia denudata—Yulan Magnolia (*)
Magnolia grandiflora—Southern Magnolia (*, FT)
Magnolia macrophylla—Bigleaf Magnolia (*)
Magnolia stellata var. *kobus*—Kobus Magnolia, Northern Japanese Magnolia (*)
Malpighia glabra—Barbados Cherry

Malus spp.—Crabapple
Mangifera indica—Mango
Manilkara zapota—Sapodilla
Melaleuca quinquinervia—Punk Tree, Melaleuca (FT)
Melia azedarach—Chinaberry
Morus alba—White Mulberry (FT)
Murraya paniculata—Orange Jessamine (*)
Musa spp.—Banana
Myrica cerifera—Southern Waxmyrtle, Southern Bayberry (FT)

Neodypsis decaryi—Triangle Palm
Nerium oleander—Oleander
Noronhia emarginata—Madagascar Olive

Ochrosia elliptica—Ochrosia
Olea europaea—European Olive
Ostrya virginiana—American Hophornbeam, Eastern Hophornbeam

Pandanus utilis—Screw-pine (FT)
Parkinsonia aculeata—Jerusalem-Thorn
Parrotia persica—Persian Parrotia (*)
Paulownia tomentosa—Princess-Tree, Paulownia (*, FT)
Peltophorum spp.—Yellow Poinciana (FT)
Persea americana—Avocado
Persea borbonia—Redbay (FT)
Phellodendron amurense—Amur Corktree, Chinese Corktree (FT)
Phoenix canariensis—Canary Island Date Palm
Phoenix dactylifera—Date Palm (*)
Phoenix reclinata—Senegal Date Palm
Photinia spp.—Photinia, Red-Top (*)
Picea abies—Norway Spruce (*)
Picea omorika—Serbian Spruce (*)
Picea orientalis—Oriental Spruce (*)
Picea pungens—Colorado Blue Spruce (*)
Pinus bungeana—Lacebark Pine (*)
Pinus cembra—Swiss Stone Pine (*)
Pinus cembra 'Nana'—Dwarf Swiss Stone Pine (*)
Pinus cembroides—Mexican Pinyon, Pinyon Pine
Pinus clausa—Sand Pine (*)
Pinus eldarica—Mondell Pine (*)
Pinus flexilis—Limber Pine (*)
Pinus mugo—Mugo Pine, Swiss Mountain Pine (FT)
Pinus nigra—Austrian Pine (*)
Pinus palustris—Longleaf Pine (*)
Pinus pinea—Stone Pine (*)
Pinus ponderosa—Ponderosa Pine (*)
Pinus sylvestris—Scotch Pine (*)
Pinus thunbergiana—Japanese Black Pine
Pistacia chinensis—Chinese Pistache
Pithecellobium flexicaule—Ebony Blackbead, Texas-Ebony
Pittosporum spp.—Pittosporum (*)
Platanus spp.—Sycamore (FT)
Plumeria spp.—Frangipani
Podocarpus spp.—Podocarpus
Pongamia pinnata—Pongam, Karum Tree, Poonga-Oil Tree (*)
Populus spp.—Poplar (FT)
Prosopis glandulosa—Mesquite, Honey Mesquite
Prunus 'Hally Jolivette'—Hally Jolivette Cherry
Prunus americana—American Plum (*)
Prunus caroliniana—Cherry-Laurel, Carolina Laurelcherry
Prunus cerasifera—Purple Leaf Plum (*)
Prunus maackii—Amur Chokecherry, Manchurian Cherry (*)
Prunus mexicana—Mexican Plum (*)
Prunus sargentii—Sargent Cherry (*)
Prunus serotina—Black Cherry (*, FT)
Prunus serrulata 'Kwanzan'—Kwanzan Cherry (*, FT)
Prunus umbellata—Flatwoods Plum (*)
Prunus x incamp 'Okame'—'Okame' Cherry (*)
Pseudotsuga menziesii—Douglas-fir
Psidium littorale—Cattley Guava, Strawberry Guava (FT)
Pterostyrax hispida—Fragrant Epaulette Tree, Wisteria-Tree (*)
Pyrus calleryana cultivars—Callery Pear (FT)

Quercus acutissima—Sawtooth Oak (*, FT)
Quercus agrifolia—Coastal Live Oak (*)
Quercus austrina—Bluff Oak (*)
Quercus bicolor—Swamp White Oak (*, FT)
Quercus cerris—Turkey Oak, Moss-Cupped Oak (*)
Quercus falcata var. *pagodifolia*—Cherrybark Oak (*)
Quercus ilex—Holly Oak (*)
Quercus laurifolia—Laurel Oak (*, FT)
Quercus lobata—Valley Oak, California White Oak (*)
Quercus macrocarpa—Bur Oak (FT)
Quercus muehlenbergii—Chinkapin Oak, Chestnut Oak (FT)
Quercus nigra—Water Oak (*, FT)
Quercus robur—English Oak
Quercus robur 'Fastigiata'—'Fastigiata' English Oak
Quercus rubra—Northern Red Oak (*)
Quercus shumardii—Shumard Oak
Quercus suber—Cork Oak (*)
Quercus texana—Texas Red Oak, Texas Oak
Quercus virginiana—Southern Live Oak, Live Oak

Ravenala madagascariensis—Travelers-Tree (*)
Rhamnus caroliniana—Carolina Buckthorn
Rhus chinensis—Chinese Sumac (FT)
Rhus glabra—Smooth Sumac (FT)
Rhus lancea—African Sumac (*, FT)
Robinia pseudoacacia—Black Locust, Common Locust
Roystonea spp.—Royal Palm (*, FT)

Sabal palmetto—Cabbage Palm (FT)
Salix spp.—Willow (FT)
Sambucus canadensis—American Elder, Common Elder (FT)
Sambucus mexicana—Mexican Elder
Sapindus drummondii—Western Soapberry
Sapindus saponaria—Florida Soapberry, Wingleaf Soapberry
Sapium sebiferum—Chinese Tallowtree, Popcorn Tree, Tallowtree (FT)
Schefflera actinophylla—Schefflera, Queensland Umbrella-Tree (FT)
Schefflera arboricola—Dwarf Schefflera
Schinus molle—California Pepper Tree (*)
Schinus terebinthifolius—Brazilian Pepper Tree (*, FT)
Senna spectabilis—Cassia (*)
Sequoiadendron giganteum—Giant Sequoia (*)
Sequoia sempervirens—Coast Redwood (*, FT)
Simarouba glauca—Paradise-Tree
Sophora affinis—Eves-Necklace, Texas Sophora
Sophora japonica—Scholar Tree, Japanese Pagoda Tree
Sophora secundiflora—Texas-Mountain-Laurel, Mescalbean
Sorbus alnifolia—Korean Mountain-Ash (*)
Styrax japonicus—Japanese Snowbell (*)
Swietenia mahagoni—Mahogany, West Indies Mahogany (FT)
Syringa pekinensis—Peking Lilac (*)
Syringa reticulata—Japanese Tree Lilac (*)

Tabebuia spp.—Trumpet Tree
Tamarindus indica—Tamarind (FT)
Taxodium ascendens—Pondcypress (*, FT)

Taxodium distichum—Baldcypress (*, FT)
Taxus baccata—English Yew (*)
Tecoma stans—Yellow-Elder, Yellow Trumpet-Flower
Terminalia catappa—Tropical-Almond, India-Almond
Terminalia muelleri—Muellers Terminalia
Thrinax morrisii—Key Thatchpalm (FT)
Thuja occidentalis—White-Cedar, Arborvitae, Northern White-Cedar (*, FT)
Thuja plicata—Giant Arborvitae, Giant-Cedar, Western Redcedar (FT)
Tilia americana—American Linden, Basswood, American Basswood
Tilia cordata—Littleleaf Linden
Tilia tomentosa—Silver Linden
Tipuana tipu—Tipuana (*)
Trachycarpus fortunei—Windmill Palm
Tristania conferta—Brisbane Box (*)

Ulmus alata—Winged Elm (FT)
Ulmus americana—American Elm (FT)
Ulmus crassifolia—Cedar Elm (FT)
Ulmus hybrids—Hybrid Elm (FT)
Ulmus parvifolia—Chinese Elm, Lacebark Elm (FT)
Ulmus pumila—Siberian Elm (FT)

Veitchia merrillii—Christmas Palm, Manila Palm
Viburnum odoratissimum—Sweet Viburnum (FT)
Viburnum odoratissimum var. *awabuki*—Awabuki Sweet Viburnum (FT)
Viburnum plicatum f. 'Tomentosa'—Doublefile Viburnum (*)
Viburnum rufidulum—Rusty Blackhaw, Southern Blackhaw
Viburnum sieboldii—Siebold Viburnum (*, FT)
Vitex spp.—Chastetree, Vitex (*)

Washingtonia filifera—Desert Palm, California Washingtonia Palm
Washingtonia robusta—Washington Palm, Mexican Washington Palm (FT)
Wodyetia bifurcata—Foxtail Palm

Xanthoceras sorbifolium—Yellowhorn

Yucca elephantipes—Spineless Yucca, Soft-Tip Yucca

Zelkova serrata—Japanese Zelkova, Saw-Leaf Zelkova (*, FT)
Ziziphus jujuba—Chinese Date, Common Jujube, Chinese Jujube (FT)

Appendix 5

Trees Tolerating Poor Drainage

Legend:
* = Withstands prolonged flooding

Acacia auriculiformis—Earleaf Acacia
Acacia farnesiana—Sweet Acacia, Huisache
Acer macrophyllum—Bigleaf Maple
Acer negundo—Boxelder (*)
Acer platanoides—Norway Maple
Acer pseudoplatanus—Sycamore Maple
Acer rubrum—Red Maple, Swamp Maple (*)
Acer saccharinum—Silver Maple (*)
Acer tataricum—Tatarian Maple
Acer x freemanii—Freeman Maple
Acoelorrhaphe wrightii—Paurotis Palm (*)
Aesculus flava—Yellow Buckeye, Sweet Buckeye
Aesculus glabra—Ohio Buckeye (*)
Aesculus hippocastanum—Horsechestnut, European Horsechestnut
Aesculus pavia—Red Buckeye (*)
Aesculus x carnea—Red Horsechestnut
Ailanthus altissima—Tree-of-Heaven
Albizia julibrissin—Mimosa, Silktree
Alnus glutinosa—Common Alder, Black Alder, European Alder (*)
Alnus rhombifolia—White Alder
Aralia spinosa—Devils-Walkingstick, Hercules-Club (*)
Asimina triloba—Pawpaw

Betula nigra—River Birch (*)
Betula papyrifera—Paper Birch
Betula pendula—European Birch
Betula populifolia—Gray Birch
Bischofia javanica—Toog Tree, Bischofia (*)
Bismarckia nobilis—Bismarck Palm

Carpentaria acuminata—Carpentaria Palm
Carpinus betulus 'Fastigiata'—'Fastigiata' European Hornbeam
Carpinus caroliniana—American Hornbeam, Blue-Beech, Ironwood
Carya aquatica—Water Hickory
Carya glabra—Pignut Hickory
Carya illinoensis—Pecan
Carya ovata—Shagbark Hickory
Catalpa spp.—Catalpa
Celtis australis—Mediterranean Hackberry, European Hackberry
Celtis laevigata—Sugarberry, Sugar Hackberry (*)
Celtis occidentalis—Common Hackberry (*)
Celtis sinensis—Japanese Hackberry, Chinese Hackberry
Chionanthus retusus—Chinese Fringetree
Chionanthus virginicus—Fringetree, Old-Mans-Beard
Chorisia speciosa—Floss-Silk Tree
Chrysalidocarpus lutescens—Yellow Butterfly Palm, Bamboo Palm, Areca Palm
Chrysophyllum oliviforme—Satinleaf
Cladrastis kentukea—American Yellowwood, Virgilia
Clusia rosea—Pitch-Apple, Florida Clusia
Conocarpus erectus—Buttonwood (*)
Conocarpus erectus var. *sericeus*—Silver Buttonwood (*)
Cordia boissieri—Wild-Olive, Anacahuita
Corylus colurna—Turkish Filbert, Turkish Hazel
Crataegus laevigata—English Hawthorn
Crataegus viridis 'Winter King'—'Winter King' Hawthorn
Cupaniopsis anacardiopsis—Carrotwood (*)

Dalbergia sissoo—Indian Rosewood
Diospyros virginiana—Common Persimmon

Elaeagnus angustifolia—Russian-Olive, Oleaster
Eugenia spp.—Stopper, Eugenia

Ficus aurea—Strangler Fig, Golden Fig
Ficus benjamina—Weeping Fig
Ficus lyrata—Fiddleleaf Fig
Ficus retusa—Cuban-Laurel
Fraxinus americana—White Ash
Fraxinus pennsylvanica—Green Ash (*)
Fraxinus uhdei—Shamel Ash (*)
Fraxinus velutina—Velvet Ash, Modesto Ash, Arizona Ash (*)

Ginkgo biloba—Maidenhair Tree, Ginkgo
Gleditsia triacanthos var. *inermis*—Thornless Honeylocust (*)
Gordonia lasianthus—Loblolly-Bay, Sweet-Bay (*)
Guaiacum sanctum—Lignumvitae (*)
Gymnocladus dioicus—Kentucky Coffeetree

Halesia carolina—Carolina Silverbell
Hibiscus syriacus—Rose-of-Sharon, Shrub-Althea
Hovenia dulcis—Japanese Raisintree

Ilex cassine—Dahoon Holly (*)
Ilex decidua—Possumhaw (*)
Ilex opaca—American Holly
Ilex rotunda—Round-Leaf Holly (*)
Ilex verticillata—Winterberry
Ilex vomitoria—Yaupon Holly

Juglans nigra—Black Walnut

Koelreuteria spp.—Goldenraintree

Liquidambar formosana—Formosa Sweetgum, Chinese Sweetgum
Liquidambar styraciflua—Sweetgum
Liriodendron tulipifera—Tuliptree, Tulip-Poplar, Yellow-Poplar

Magnolia acuminata—Cucumbertree, Cucumber Magnolia
Magnolia grandiflora—Southern Magnolia
Magnolia virginiana—Sweetbay Magnolia, Swamp Magnolia (*)
Melaleuca quinquinervia—Punk Tree, Melaleuca (*)
Metasequoia glyptostroboides—Dawn Redwood (*)

Morus alba—White Mulberry
Musa spp.—Banana
Myrica cerifera—Southern Waxmyrtle, Southern Bayberry (*)

Nyssa ogeche—Ogeechee Tupelo, Ogeechee-Lime (*)
Nyssa sinensis—Chinese Tupelo
Nyssa sylvatica—Blackgum, Sourgum, Black Tupelo (*)

Osmanthus americanus—Devilwood, Wild-Olive

Pandanus utilis—Screw-pine (*)
Paulownia tomentosa—Princess-Tree, Empress-Tree, Paulownia (*)
Peltophorum pterocarpum—Yellow Poinciana
Persea borbonia—Redbay (*)
Persea palustris—Swamp Bay
Phellodendron amurense—Amur Corktree, Chinese Corktree
Pinckneya pubens—Pinckneya, Fevertree
Pinus elliottii—Slash Pine
Pinus glabra—Spruce Pine (*)
Pinus radiata—Monterey Pine
Pinus taeda—Loblolly Pine
Platanus occidentalis—Sycamore, American Planetree (*)
Platanus orientalis—Oriental Planetree
Platanus racemosa—Western Sycamore (*)
Platanus x acerifolia—London Planetree (*)
Populus deltoides—Eastern Cottonwood (*)
Populus freemanii—Fremont Poplar
Populus nigra 'Italica'—Lombardy Poplar
Prunus serotina—Black Cherry
Prunus serrulata 'Kwanzan'—Kwanzan Cherry
Psidium littorale—Cattley Guava, Strawberry Guava
Pyrus calleryana cultivars—Callery Pear

Quercus acutissima—Sawtooth Oak
Quercus bicolor—Swamp White Oak (*)
Quercus glauca—Blue Japanese Oak, Ring-Cupped Oak
Quercus imbricaria—Shingle Oak, Northern Laurel Oak
Quercus laurifolia—Laurel Oak
Quercus lyrata—Overcup Oak (*)
Quercus macrocarpa—Bur Oak (*)
Quercus michauxii—Swamp Chestnut Oak
Quercus muehlenbergii—Chinkapin Oak, Chestnut Oak
Quercus nigra—Water Oak (*)
Quercus nuttallii—Nuttall Oak (*)
Quercus palustris—Pin Oak (*)
Quercus phellos—Willow Oak

Rhus chinensis—Chinese Sumac
Rhus glabra—Smooth Sumac
Rhus lancea—African Sumac
Roystonea spp.—Royal Palm

Sabal palmetto—Cabbage Palm (*)
Salix spp.—Weeping Willow (*)
Sambucus canadensis—American Elder, Common Elder (*)
Sapium sebiferum—Chinese Tallowtree, Popcorn Tree, Tallowtree
Schefflera actinophylla—Schefflera, Queensland Umbrella-Tree
Schinus terebinthifolius—Brazilian Pepper Tree
Sequoia sempervirens—Coast Redwood (*)
Stewartia pseudocamellia—Japanese Stewartia
Swietenia mahagoni—Mahogany, West Indies Mahogany

Tamarindus indica—Tamarind
Taxodium ascendens—Pondcypress (*)
Taxodium distichum—Baldcypress (*)
Taxodium mucronatum—Montezuma Baldcypress, Mexican Cypress (*)
Thrinax morrisii—Key Thatchpalm
Thuja occidentalis—White-Cedar, Arborvitae, Northern White-Cedar (*)
Thuja plicata—Giant Arborvitae, Giant-Cedar, Western Redcedar

Ulmus alata—Winged Elm (*)
Ulmus americana—American Elm (*)
Ulmus crassifolia—Cedar Elm
Ulmus hybrids—Hybrid Elm
Ulmus parvifolia—Chinese Elm, Lacebark Elm
Ulmus pumila—Siberian Elm

Viburnum rufidulum—Rusty Blackhaw, Southern Blackhaw
Viburnum sieboldii—Siebold Viburnum

Washingtonia robusta—Washington Palm, Mexican Washington Palm (*)

Zelkova serrata—Japanese Zelkova, Saw-Leaf Zelkova
Ziziphus jujuba—Chinese Date, Common Jujube, Chinese Jujube

Appendix 6

Trees Tolerating Shaded to Mostly Shaded Sites

Legend:
F = Has good fall foliage color
S = Has showy flowers
P = Has persistent, showy fruit

Note: Cultivars of the species mentioned also generally tolerate shade.

Acer buergeranum—Trident Maple (F, S)
Acer campestre—Hedge Maple
Acer cissifolium—Ivy-Leaf Maple
Acer ginnala—Amur Maple (F, S)
Acer griseum—Paperbark Maple (F)
Acer japonicum—Fullmoon Maple, Fernleaf Maple (F, S)
Acer leucoderme—Chalk Maple, Whitebark Maple (F)
Acer palmatum—Japanese Maple (F)
Acer saccharum—Sugar Maple (F)
Acer triflorum—Three-Flowered Maple (F)
Aesculus pavia—Red Buckeye (S)
Amelanchier spp.—Serviceberry (F, S)
Asimina triloba—Pawpaw (F, S)

Butia capitata—Pindo Palm (S)

Carpinus caroliniana—American Hornbeam, Blue-Beech, Ironwood (F)
Caryota mitis—Fishtail Palm
Cercis canadensis—Eastern Redbud (F, S)
Chamaecyparis nootkatensis 'Pendula'—Nootka Falsecypress
Chionanthus retusus—Chinese Fringetree (F, S)
Chionanthus virginicus—Fringetree, Old-Mans-Beard (F, S)
Clusia rosea—Pitch-Apple, Florida Clusia
Coccothrinax argentata—Silverpalm, Thatchpalm
Cornus alternifolia—Pagoda Dogwood (F, S)
Cornus florida—Flowering Dogwood (F, S)
Cornus mas—Cornelian-Cherry (F, S)
Cornus walteri—Walter Dogwood (F, S)

Fagus grandifolia—American Beech (F)

Gordonia lasianthus—Loblolly-Bay, Sweet-Bay (S)

Halesia spp.—Silverbell (S)
Hamamelis mollis—Chinese Witch-Hazel
Hamamelis virginiana—Witch-Hazel (F)
Howea forsterana—Sentry Palm, Kentia Palm

Ilex cassine—Dahoon Holly (P)
Ilex decidua—Possumhaw (P)
Ilex latifolia—Lusterleaf Holly
Ilex opaca—American Holly (P)
Ilex verticillata—Winterberry (P)
Ilex vomitoria—Yaupon Holly (P)

Myrica cerifera—Southern Waxmyrtle, Southern Bayberry

Podocarpus macrophyllus—Podocarpus, Yew-Pine, Japanese Yew (S)
Podocarpus nagi—Nagi Podocarpus, Broadleaf Podocarpus
Prunus caroliniana—Cherry-Laurel, Carolina Laurelcherry (S)
Prunus x incamp 'Okame'—'Okame' Cherry (S)

Rhamnus caroliniana—Carolina Buckthorn

Sambucus canadensis—American Elder, Common Elder (S)
Sciadopitys verticillata—Japanese Umbrella-Tree
Schefflera arboricola—Dwarf Schefflera
Sophora affinis—Eves-Necklace, Texas Sophora (P)
Stewartia spp.—Stewartia (F, S)

Taxodium spp.—Baldcypress, Pondcypress
Taxus baccata—English Yew
Thuja plicata—Giant Arborvitae, Giant-Cedar, Western Redcedar
Trachiocarpus fortunei—Windmill Palm
Tsuga canadensis—Canadian Hemlock, Eastern Hemlock

Viburnum odoratissimum—Sweet Viburnum
Viburnum rufidulum—Rusty Blackhaw, Southern Blackhaw (F, S)
Viburnum sieboldii—Siebold Viburnum (P, S)

Appendix 7

Trees That Grow Less Than 30 Feet Tall

Legend:
* = May eventually grow slightly taller
M = Some tolerance to drought
P = Poor tolerance to drought
S = Tolerates shady sites

Neither P nor M indicates good drought tolerance.

Note: Most trees in this list are suited for planting in above-ground containers. In dry, desert climates, most trees are less drought-tolerant than indicated here.

Acacia farnesiana—Sweet Acacia
Acacia wrightii—Wright Acacia
Acer buegeranum—Trident Maple (M, S)
Acer campestre—Hedge Maple (*)
Acer cissifolium—Ivy-Leaf Maple (M, S)
Acer ginnala—Amur Maple (S)
Acer griseum—Paperbark Maple (P, S)
Acer japonicum—Fullmoon Maple (P, S)
Acer palmatum—Japanese Maple (P, S)
Acer platanoides 'Globosum'—'Globosum' Norway Maple (M)
Acer tataricum—Tatarian Maple (M)
Acer triflorum—Three-Flowered Maple (P, S)
Acer truncatum—Purpleblow Maple (S)
Acoelorrhaphe wrightii—Paurotis Palm (M)
Aesculus californica—California Buckeye
Aesculus pavia—Red Buckeye (M, S)
Albizia julibrissin—Mimosa, Silktree
Amelanchier spp.—Serviceberry (M, S)
Aralia spinosa—Devils-Walkingstick (M)
Arbutus unedo—Strawberry-Tree (M)
Asimina triloba—Pawpaw (P, S)

Bauhinia spp.—Orchid-Tree (*)
Beaucarnea recurvata—Ponytail
Betula pendula 'Youngii'—'Youngii' European Birch (P)
Bulnesia arborea—Bulnesia (*)
Bursera simaruba—Gumbo-Limbo (*)
Butia capitata—Pindo Palm (S)

Caesalpinia pulcherrima—Dwarf Poinciana
Calliandra haematocephala—Powderpuff
Calliandra surinamensis—Pink Powderpuff
Callistemon citrinus—Red Bottlebrush
Callistemon viminalis—Weeping Bottlebrush (M)
Camellia oleifera—Tea-Oil Camellia (M, S)
Carpinus caroliniana—American Hornbeam (*, S)
Caryota mitis—Fishtail Palm (S)
Cassia alata—Candlebrush
Cassia leptophylla—Gold Medallion Tree
Cassia surattensis—Glaucous Cassia (M)
Cedrus atlantica 'Glauca Pendula'—Weeping Atlas Cedar
Celtis reticulata—Western Hackberry
Cercidiphyllum japonicum 'Pendula'—'Pendula' Katsuratree (M)
Cercidium floridum—Palo Verde
Cercis canadensis—Eastern Redbud (M, S)
Cercis canadensis var. *texensis*—Texas Redbud (S)
Cercis mexicana—Mexican Redbud (S)
Cercis occidentalis—Western Redbud (S)
Cercis reniformis 'Oklahoma'—Oklahoma Redbud (S)
Chilopsis linearis—Desert-Willow
Chionanthus retusus—Chinese Fringetree (M, S)
Chionanthus virginicus—Fringetree (M, S)
x Chitalpa tashkentensis—Chitalpa
Chrysalidocarpus lutescens—Yellow Butterfly Palm (M, S)
Citrus spp.—Citrus (M)
Clusia rosea—Pitch-Apple (S)
Coccoloba diversifolia—Pigeon-Plum (*)
Coccoloba uvifera—Seagrape (*)
Coccothrinax argentata—Silverpalm (S)
Conocarpus erectus—Buttonwood (*)
Conocarpus erectus var. *sericeus*—Silver Buttonwood
Cordia boissieri—Wild-Olive
Cordia sebestena—Geiger-Tree
Cornus alternifolia—Pagoda Dogwood (M, S)
Cornus controversa—Giant Dogwood (M, S)
Cornus drummondii—Roughleaf Dogwood (M, S)
Cornus florida—Flowering Dogwood (M, S)
Cornus kousa—Kousa Dogwood, Chinese Dogwood (M, S)
Cornus mas—Cornelian-Cherry (M, S)
Cornus nuttallii—Western Dogwood (M, S)
Cornus walteri—Walter Dogwood (M, S)
Cotinus coggygria—Smoketree
Cotinus obovatus—American Smoketree
Crataegus aestivalis—May Hawthorn
Crataegus laevigata—English Hawthorn
Crataegus phaenopyrum—Washington Hawthorn
Crataegus viridis 'Winter King'—'Winter King' Green Hawthorn
Crataegus x lavallei—Lavalle Hawthorn
Crescentia cujete—Calabash Tree (M)
Cryptomeria japonica 'Elegans'—'Elegans' Japanese-Cedar

Diospyros kaki—Japanese Persimmon (M)
Diospyros texana—Texas Persimmon

Elaeagnus angustifolia—Russian-Olive
Eriobotrya deflexa—Bronze Loquat
Eriobotrya japonica—Loquat
Erythrina caffre—Kaffir Bloom Coral Tree
Erythrina crista-galli—Cockspur Coral Tree
Eucalyptus ficifolia—Red-Flowering Gum (M)
Eugenia spp.—Stopper, Eugenia
Evodia danielii—Korean Evodia

Fagus sylvatica 'Purpurea Pendula'—'Purpurea Pendula' European Beech (M, S)
Feijoa sellowiana—Guava, Feijoa
Franklinia alatamaha—Franklin-Tree (M)
Fraxinus ornus—Flowering Ash (*)

Guaiacum sanctum—Lignumvitae

Halesia carolina—Carolina Silverbell (P, S)
Halesia diptera—Two-Winged Silverbell (P, S)
Hamamelis mollis—Chinese Witch-Hazel (M, S)
Hamamelis virginiana—Witch-Hazel (S)
Hibiscus syriacus—Rose-of-Sharon (P)
Howea forsterana—Sentry Palm, Kentia Palm (S)
Hydrangea paniculata—Panicle Hydrangea

Ilex cassine—Dahoon Holly (M, S)
Ilex cornuta 'Burfordii'—Burford Holly
Ilex decidua—Possumhaw (M, S)
Ilex latifolia—Lusterleaf Holly (S)
Ilex rotunda—Round-Leaf Holly (M, S)
Ilex serrata—Fine-Toothed Holly
Ilex verticillata—Winterberry (P, S)
Ilex vomitoria—Yaupon Holly (S)
Ilex vomitoria 'Pendula'—Weeping Yaupon Holly (S)
Ilex x attenuata 'Fosteri'—Fosters Holly
Ilex x 'Nellie R. Stevens'—'Nellie R. Stevens' Holly

Jatropha integerrima—Peregrina, Fire-Cracker (M)

Koelreuteria bipinnata—Bougainvillea Goldenraintree (*)
Koelreuteria paniculata—Goldenraintree (*)

Laburnum x alpinum—Goldenchain Tree (P)
Lagerstroemia indica—Crape-Myrtle
Latania spp.—Latan Palm
Leptospermum laevigatum—Australian Tea Tree
Leucaena retusa—Goldenball Leadtree
Ligustrum japonicum—Japanese Privet (M)
Ligustrum lucidum—Tree Ligustrum (*)
Litchi chinensis—Lychee
Lithocarpus henryi—Henry Tanbark Oak

Maackia amurensis—Amur Maackia
Magnolia acuminata var. *cordata*—Yellow Magnolia (*, M)
Magnolia denudata—Yulan Magnolia (*, M)
Magnolia grandiflora 'Glen St. Mary'—'Glen St. Mary' Southern Magnolia (M)
Magnolia grandiflora 'Little Gem'—'Little Gem' Southern Magnolia (*, M)
Magnolia stellata—Star Magnolia (M)
Magnolia stellata var. *kobus*—Kobus Magnolia, Northern Japanese Magnolia (*, M)
Magnolia x soulangiana—Saucer Magnolia (M)
Malpighia glabra—Barbados Cherry
Malus spp.—Crabapple
Melia azedarach—Chinaberry
Morus alba—White Mulberry
Murraya paniculata—Orange Jessamine
Musa spp.—Banana (P)
Myrica californica—California Waxmyrtle (M, S)
Myrica cerifera—Southern Waxmyrtle (S)

Neodypsis decaryi—Triangle Palm
Nerium oleander—Oleander
Noronhia emarginata—Madagascar Olive

Ochrosia elliptica—Ochrosia
Osmanthus americanus—Devilwood
Osmanthus fragrans—Sweet Osmanthus
Osmanthus x fortunei—Fortunes Osmanthus
Ostrya virginiana—American Hophornbeam (S)

Pandanus utilis—Screw-pine
Parkinsonia aculeata—Jerusalem-Thorn
Phoenix roebelenii—Pygmy Date Palm (M)
Photinia glabra—Red-Leaf Photinia, Red-Top
Photinia serrulata—Chinese Photinia
Photinia villosa—Oriental Photinia
Photinia x fraseri—Fraser Photinia
Pinckneya pubens—Pinckneya, Fevertree (M, P)
Pinus bungeana—Lacebark Pine (*, M)
Pinus cembra 'Nana'—Dwarf Swiss Stone Pine
Pinus densiflora 'Pendula'—Weeping Japanese Red Pine (M)
Pinus densiflora 'Umbraculifera'—'Umbraculifera' Japanese Red Pine (M)
Pinus edulis—Pinyon Pine
Pinus mugo—Mugo Pine, Swiss Mountain Pine (M)
Pinus strobus 'Nana'—'Nana' Eastern White Pine (P)
Pinus strobus 'Pendula'—'Pendula' Eastern White Pine (P)
Pinus sylvestris 'Watereri'—'Watereri' Scotch Pine
Pinus taeda 'Nana'—'Nana' Loblolly Pine
Pinus thunbergiana—Japanese Black Pine
Pithecellobium flexicaule—Ebony Blackbead, Texas-Ebony
Pittosporum spp.—Pittosporum (M)
Plumeria alba—White Frangipani
Plumeria rubra—Frangipani
Prosopis spp.—Mesquite
Prunus americana—American Plum (M)
Prunus angustifolia—Chickasaw Plum
Prunus cerasifera—Purple Leaf Plum (M)
Prunus 'Hally Jolivette'—Hally Jolivette Cherry (M)
Prunus maackii—Amur Chokecherry (*, M)
Prunus mexicana—Mexican Plum (M)
Prunus mume—Japanese Apricot (M, P)
Prunus persica—Peach (M, P)
Prunus sargentii—Sargent Cherry (*, M)
Prunus serrulata 'Kwanzan'—Kwanzan Cherry (M, P)
Prunus subhirtella var. *'autumnalis'*—Autumnalis Higan Cherry (M)
Prunus subhirtella 'Pendula'—Weeping Higan Cherry (M)
Prunus triloba var. *multiplex*—Flowering-Almond (M, P)
Prunus umbellata—Flatwoods Plum (M)
Prunus x incamp 'Okame'—'Okame' Cherry (M, P)
Prunus x yedoensis—Yoshino Cherry (M, P)
Pseudocydonia sinensis—Chinese Quince (M)
Psidium littorale—Cattley Guava, Strawberry Guava
Pterostyrax hispida—Fragrant Epaulettetree (M)

Quercus glauca—Blue Japanese Oak (*)

Rhamnus caroliniana—Carolina Buckthorn (S)
Rhus chinensis—Chinese Sumac
Rhus glabra—Smooth Sumac
Rhus lancea—African Sumac
Robinia pseudoacacia 'Tortuosa'—'Tortuosa' Black Locust
Robinia pseudoacacia 'Umbraculifera'—Umbrella Black Locust

Sambucus canadensis—American Elder, Common Elder (M, P, S)
Sambucus mexicana—Mexican Elder
Sapium sebiferum—Chinese Tallowtree (M)
Schefflera actinophylla—Schefflera (*, M)
Schefflera arboricola—Dwarf Schefflera (S)
Schinus terebinthifolius—Brazilian Pepper Tree (M)
Sciadopitys verticillata—Japanese Umbrella-Pine (M, S)
Senna spectabilis—Cassia (M)
Sophora affinis—Eves-Necklace, Texas Sophora (S)
Sophora secundiflora—Texas-Mountain-Laurel
Sorbus alnifolia—Korean Mountain-Ash (M)

Sorbus aucuparia—European Mountain-Ash (M)
Stewartia koreana—Korean Stewartia (*, M, S)
Stewartia monadelpha—Tall Stewartia (*, S)
Stewartia pseudocamellia—Japanese Stewartia (*, M, S)
Strelitzia nicolai—White Bird-of-Paradise (M)
Styrax japonicus—Japanese Snowbell (M)
Syringa pekinensis—Peking Lilac (M)
Syringa reticulata—Japanese Tree Lilac (M)

Tabebuia spp.—Trumpet Tree
Taxus baccata—English Yew (M, P, S)
Tecoma stans—Yellow-Elder
Terminalia muelleri—Muellers Terminalia
Thrinax morrisii—Key Thatchpalm
Trachycarpus fortunei—Windmill Palm (S)
Tsuga canadensis 'Lewisii'—Lewis Hemlock (P, S)
Tsuga canadensis 'Sargentii'—Weeping Canadian Hemlock (M, S)

Veitchia merrillii—Christmas Palm, Manila Palm (M)
Viburnum odoratissimum—Sweet Viburnum (S)
Viburnum odoratissimum var. *awabuki*—Awabuki Sweet Viburnum (S)
Viburnum plicatum var. *tomentosum*—Doublefile Viburnum (M, P)
Viburnum rufidulum—Rusty Blackhaw, Southern Blackhaw (S)
Viburnum sieboldii—Siebold Viburnum (M, S)
Vitex agnus-castus—Chastetree, Vitex
Vitex agnus-castus 'Alba'—'Alba' Chastetree, 'Alba' Vitex
Vitex negundo 'Heterophylla'—Cutleaf Chastetree (M)

Wodyetia bifurcata—Foxtail Palm

Xanthoceras sorbifolium—Yellowhorn (*)

Yucca elephantipes—Spineless Yucca, Soft-Tip Yucca

Ziziphus jujuba—Chinese Jujube

Appendix 8

Trees That Grow 30 to 50 Feet Tall

Legend:
* = Capable of growing taller
M = Some tolerance to drought
P = Poor tolerance to drought

Neither letter indicates good tolerance to drought.

Abies firma—Japanese Fir
Acacia auriculiformis—Earleaf Acacia
Acer barbatum—Florida Maple, Southern Sugar Maple (M)
Acer barbatum var. *caddo*—Caddo Florida Maple, Southern Sugar Maple
Acer grandidentatum—Bigtooth Maple, Rocky Mountain Sugar Maple, Canyon Maple (M)
Acer leucoderme—Chalk Maple, Whitebark Maple
Acer miyabei—Miyabe Maple (M)
Acer negundo—Boxelder
Acer platanoides 'Columnare'—'Columnare' Norway Maple (M)
Acer platanoides 'Crimson King'—'Crimson King' Norway Maple (M)
Acer platanoides 'Olmsted'—'Olmsted' Norway Maple (M)
Acer rubrum 'Autumn Flame'—'Autumn Flame' Red Maple (M)
Acer rubrum 'Bowhall'—'Bowhall' Red Maple (M)
Acer rubrum 'Gerling'—'Gerling' Red Maple (M)
Acer rubrum 'October Glory'—'October Glory' Red Maple (M)
Acer rubrum 'Red Sunset'—'Red Sunset' Red Maple (M)
Acer saccharum 'Endowment'—'Endowment' Sugar Maple (M)
Acer saccharum 'Goldspire'—'Goldspire' Sugar Maple (M)
Acer saccharum 'Newton Sentry'—'Newton Sentry' Sugar Maple (M)
Acer saccharum 'Temple's Upright'—'Temple's Upright' Sugar Maple (M)
Aesculus glabra—Ohio Buckeye, Fetid Buckeye (*)
Aesculus indica—Indian Horsechestnut (M)
Aesculus x carnea—Red Horsechestnut (M)
Alnus glutinosa—Common Alder, Black Alder, European Alder (M)
Arbutus menzesii—Pacific Madrone (*)
Arbutus texana—Texas Madrone

Betula ermanii—Erman Maple (P)
Betula nigra—River Birch (M)
Betula papyrifera—Paper Birch, Canoe Birch (P)
Betula pendula—European Birch (P)
Betula platyphylla 'Whitespire'—Japanese Whitespire Birch (P)
Betula populifolia—Gray Birch (M)
Betula utilis var. *jacquemontii*—Jacquemontii Birch (M)
Bischofia javanica—Toog Tree, Bischofia (M)
Bismarckia nobilis—Bismarck Palm
Brahea armata—Mexican Blue Palm
Brahea edulis—Guadelupe Palm
Bucida buceras—Black Olive, Oxhorn Bucida
Bumelia lanuginosa—Chittamwood, Gum Bumelia
Bursera simaruba—Gumbo-Limbo
Caesalpinia granadillo—Bridalveil Tree (M)
Calodendron capense—Cape Chestnut (M)
Calophyllum brasiliense—Santa Maria
Carpentaria acuminata—Carpentaria Palm (P)
Carpinus betulus 'Fastigiata'—'Fastigiata' European Hornbeam (M)
Cassia fistula—Golden-Shower
Cassia javanica—Pink-and-White-Shower (M)
Castanea mollissima—Chinese Chestnut (M)
Catalpa spp.—Catalpa
Cedrus atlantica—Atlas Cedar
Cedrus deodara—Deodar Cedar
Cedrus libani—Cedar-of-Lebanon
Cercidiphyllum japonicum—Katsuratree (M)
Chamaecyparis nootkatensis 'Pendula'—Nootka Falsecypress, Alaska-Cedar
Chamaecyparis obtusa—Hinoki Falsecypress (M)
Chamaecyparis pisifera 'Filifera'—Sawara Falsecypress (M)
Chorisia speciosa—Floss-Silk Tree
Chrysophyllum oliviforme—Satinleaf
Cinnamomum camphora—Camphor-Tree
Cladrastis kentukea—American Yellowwood, Virgilia (M)
Corylus colurna—Turkish Filbert, Turkish Hazel
Cryptomeria japonica—Japanese-Cedar (*)
Cupaniopsis anacardiopsis—Carrotwood
x Cupressocyparis leylandii—Leyland Cypress
Cupressus arizonica var. *arizonica*—Arizona Cypress
Cupressus macrocarpa—Monterey Cypress (M)
Cupressus sempervirens—Italian Cypress

Dalbergia sissoo—Indian Rosewood (M)
Delonix regia—Royal Poinciana
Diospyros virginiana—Common Persimmon

Erythrina variegata var. *orientalis*—Coral Tree
Eucalyptus ficifolia—Red-Flowering Gum
Eucommia ulmoides—Hardy Rubber Tree

Fagus sylvatica 'Pendula'—Weeping European Beech (M)
Ficus elastica—Rubbertree
Ficus lyrata—Fiddleleaf Fig
Firmiana simplex—Chinese Parasoltree
Fraxinus texensis—Texas Ash
Fraxinus velutina—Velvet Ash, Modesto Ash, Arizona Ash (*, M)

Geijera parviflora—Australian-Willow (M)
Ginkgo biloba 'Autumn Gold'—'Autumn Gold' Maidenhair Tree, 'Autumn Gold' Ginkgo
Gleditsia triacanthos var. *inermis* 'Imperial'—'Imperial' Thornless Honeylocust
Gordonia lasianthus—Loblolly-Bay, Sweet-Bay (P)

Halesia monticola—Mountain Silverbell
Hovenia dulcis—Japanese Raisintree (M)

Ilex opaca—American Holly
Ilex x attenuata 'East Palatka'—'East Palatka' Holly
Ilex x attenuata 'Savannah'—Savannah Holly

Jacaranda mimosifolia—Jacaranda (M)
Juglans californica—California Black Walnut
Juglans regia—English Walnut (M, P)
Juniperus ashei—Ashe Juniper, Mountain-Cedar
Juniperus deppeana 'Mcfetter'—McFetter Alligator Juniper
Juniperus scopulorum—Rocky Mountain Juniper, Colorado Redcedar
Juniperus scopulorum 'Tolleson's Green Weeping'—'Tolleson's Green Weeping' Rocky Mountain Juniper
Juniperus silicicola—Southern Redcedar
Juniperus virginiana—Eastern Redcedar

Kalopanax pictus—Castor-Aralia, Prickly Castor-Oil Tree
Koelreuteria elegans—Flamegold

Lagerstroemia fauriei—Japanese Crape-Myrtle
Lagerstroemia speciosa—Queens Crape-Myrtle
Litchi chinensis—Lychee (M)
Lithocarpus densiflora—Tanbark Oak
Livistona chinensis—Chinese Fan Palm (M)
Lysiloma spp.—Wild-Tamarind

Maclura pomifera—Osage-Orange, Bois-D'Arc
Magnolia grandiflora 'Bracken's Brown Beauty'—'Bracken's Brown Beauty' Southern Magnolia (M)
Magnolia grandiflora 'Hasse'—'Hasse' Southern Magnolia (M)
Magnolia grandiflora 'Majestic Beauty'—'Majestic Beauty' Southern Magnolia (M)
Magnolia grandiflora 'Samuel Sommer'—'Samuel Sommer' Southern Magnolia (M)
Magnolia macrophylla—Bigleaf Magnolia (M)
Magnolia virginiana—Sweetbay Magnolia, Swamp Magnolia (M)
Mangifera indica—Mango (M)
Manilkara zapota—Sapodilla
Melaleuca linarifolia—Flax Leaf Paper Bark
Melaleuca quinquinervia—Punk Tree, Melaleuca

Nyssa ogeche—Ogeechee Tupelo, Ogeechee-Lime (M)
Nyssa sinensis—Chinese Tupelo
Nyssa sylvatica—Blackgum, Sourgum, Black Tupelo (*)

Olea europaea—European Olive
Ostrya virginiana—American Hophornbeam, Eastern Hophornbeam
Oxydendrum arboreum—Sourwood, Sorrel-Tree (M)

Parrotia persica—Persian Parrotia
Paulownia tomentosa—Princess-Tree, Paulownia
Peltophorum dubium—Copperpod
Peltophorum pterocarpum—Yellow Poinciana
Persea americana—Avocado (M)
Persea borbonia—Redbay
Phellodendron amurense—Amur Corktree, Chinese Corktree
Phellodendron amurense 'Macho'—'Macho' Amur Corktree, 'Macho' Chinese Corktree
Phoenix canariensis—Canary Island Date Palm
Phoenix dactylifera—Date Palm
Phoenix reclinata—Senegal Date Palm (M)
Picea abies 'Inversa'—'Inversa' Norway Spruce (M)
Picea glauca—White Spruce
Picea omorika—Serbian Spruce (*, M)
Picea orientalis—Oriental Spruce
Picea pungens—Colorado Blue Spruce (*, M)
Pinus cembra—Swiss Stone Pine
Pinus cembroides—Mexican Pinyon, Pinyon Pine
Pinus clausa—Sand Pine
Pinus densiflora—Japanese Red Pine (M)
Pinus echinata—Shortleaf Pine (M)
Pinus eldarica—Mondell Pine
Pinus flexilis—Limber Pine (*)
Pinus glabra—Spruce Pine (*, M)
Pinus halepensis—Aleppo Pine
Pinus nigra—Austrian Pine
Pinus parviflora—Japanese White Pine
Pinus pinea—Stone Pine
Pinus radiata—Monterey Pine (M, P)
Pinus sylvestris—Scotch Pine (M)
Pinus virginiana—Virginia Pine, Scrub Pine
Pistacia chinensis—Chinese Pistache
Podocarpus falcatus—Podocarpus (M)
Podocarpus gracilior—Weeping Podocarpus, Fern Podocarpus (M)
Podocarpus latifolius—Podocarpus (M)
Podocarpus macrophyllus—Podocarpus, Yew-Pine, Japanese Yew
Podocarpus nagi—Nagi Podocarpus, Broadleaf Podocarpus
Pongamia pinnata—Pongam, Karum Tree, Poonga-Oil Tree
Populus nigra 'Italica'—Lombardy Poplar
Prunus caroliniana—Cherry-Laurel, Carolina Laurelcherry (M)
Prunus serotina—Black Cherry
Pyrus calleryana cultivars—Callery Pear

Quercus acutissima—Sawtooth Oak
Quercus agrifolia—Coastal Live Oak
Quercus austrina—Bluff Oak (*)
Quercus cerris—Turkey Oak, Moss-Cupped Oak (*)
Quercus ilex—Holly Oak
Quercus kelloggii—California Black Oak
Quercus lobata—Valley Oak, California White Oak
Quercus lyrata—Overcup Oak
Quercus nuttallii—Nuttall Oak
Quercus suber—Cork Oak (*)
Quercus texana—Texas Red Oak
Quercus virginiana—Live Oak (*)

Ravenala madagascariensis—Travelers-Tree (P)
Robinia pseudoacacia 'Frisia'—'Frisia' Black Locust, 'Frisia' Common Locust
Roystonea spp.—Royal Palm (*)

Sabal spp.—Cabbage Palm (*)
Salix alba—Weeping Willow (M, P)
Salix matsudana 'Tortuosa'—Corkscrew Willow, Pekin Willow, Hankow Willow (P)
Sapindus drummondii—Western Soapberry
Sapindus saponaria—Florida Soapberry, Wingleaf Soapberry
Sassafras albidum—Sassafras (M)
Schinus molle—California Pepper Tree (*)
Simarouba glauca—Paradise-Tree (M)
Sophora japonica 'Princeton Upright'—'Princeton Upright' Japanese Pagoda Tree
Styrax obassia—Fragrant Snowbell
Swietenia mahagoni—Mahogany, West Indies Mahogany (*)
Syagrus romanzoffianum—Queen Palm (M)

Terminalia catappa—Tropical-Almond, India-Almond
Thuja occidentalis—Northern White-Cedar (M)

Thuja plicata—Giant Arborvitae, Western Redcedar (*, M)
Tilia cordata 'Greenspire'—'Greenspire' Littleleaf Linden (M)
Tilia cordata 'June Bride'—'June Bride' Littleleaf Linden (M)
Tilia cordata 'Rancho'—'Rancho' Littleleaf Linden (M)
Tipuana tipu—Tipuana (M)
Tristania conferta—Brisbane Box

Ulmus alata—Winged Elm (*)
Ulmus crassifolia—Cedar Elm
Ulmus parvifolia—Chinese Elm, Lacebark Elm
Ulmus parvifolia 'Drake'—'Drake' Chinese Elm, 'Drake' Lacebark Elm
Ulmus parvifolia 'Dynasty'—'Dynasty' Chinese Elm, 'Dynasty' Lacebark Elm
Ulmus parvifolia 'Sempervirens'—Weeping Chinese Elm, Weeping Lacebark Elm

Appendix 9

Trees Capable of Growing More Than 50 Feet Tall

Legend:
* = Can become quite large fairly quickly
M = Some tolerance to drought
P = Poor tolerance to drought

Neither letter indicates good drought tolerance.

Abies cilicica—Cilician Fir
Abies concolor—White Fir, Colorado Fir (*)
Abies nordmanniana—Nordmann Fir (*)
Acer macrophyllum—Bigleaf Maple (*, M)
Acer platanoides—Norway Maple (M)
Acer platanoides 'Cleveland'—'Cleveland' Norway Maple (M)
Acer platanoides 'Emerald Queen'—'Emerald Queen' Norway Maple (M)
Acer platanoides 'Erectum'—'Erectum' Norway Maple (M)
Acer platanoides 'Schwedleri'—'Schwedleri' Norway Maple (M)
Acer platanoides 'Summershade'—'Summershade' Norway Maple (M)
Acer platanoides 'Superform'—'Superform' Norway Maple (M)
Acer pseudoplatanus—Sycamore Maple, Planetree Maple (*)
Acer rubrum—Red Maple, Swamp Maple (M, P)
Acer saccharinum—Silver Maple
Acer saccharum—Sugar Maple (M)
Acer saccharum 'Green Mountain'—'Green Mountain' Sugar Maple (M)
Acer x freemanii—Freeman Maple (M)
Aesculus flava—Yellow Buckeye, Sweet Buckeye (M)
Aesculus glabra—Ohio Buckeye, Fetid Buckeye (M)
Aesculus hippocastanum—Horsechestnut, European Horsechestnut (M)
Ailanthus altissima—Tree-of-Heaven
Alnus rhombifolia—White Alder (P)
Araucaria araucana—Monkey-Puzzletree
Araucaria bidwillii—False Monkey-Puzzletree (M)
Araucaria heterophylla—Norfolk-Island-Pine

Calocedrus decurrens—California Incense-Cedar
Carya glabra—Pignut Hickory
Carya illinoensis—Pecan (*)
Carya ovata—Shagbark Hickory
Celtis spp.—Hackberry
Chamaecyparis lawsoniana—Lawson Falsecypress, Port-Orford-Cedar (M)
Cocos nucifera 'Malayan Dwarf'—'Malayan Dwarf' Coconut Palm
Cunninghamia lanceolata—China-Fir
Cupressus glabra 'Carolina Safire'—Smooth-Barked Arizona Cypress

Eucalyptus citriodora—Lemon-Scented Gum (*, M)
Fagus grandifolia—American Beech (*, P)
Fagus sylvatica—European Beech (*, M)
Ficus aurea—Strangler Fig, Golden Fig (*)
Ficus benjamina—Weeping Fig (*)
Ficus retusa—Cuban-Laurel (*)
Ficus rubiginosa—Rusty Fig (*)
Fraxinus americana—White Ash (*, M)
Fraxinus excelsior—Common Ash, European Ash
Fraxinus pennsylvanica—Green Ash (*)
Fraxinus uhdei—Shamel Ash (M)

Ginkgo biloba—Maidenhair Tree, Ginkgo
Gleditsia triacanthos var. *inermis*—Thornless Honeylocust
Grevillea robusta—Silk-Oak (*)
Gymnocladus dioicus—Kentucky Coffeetree

Juglans nigra—Black Walnut (*)

Larix decidua—European Larch (M)
Liquidambar spp.—Sweetgum (M)
Liriodendron tulipifera—Tuliptree, Tulip-Poplar, Yellow-Poplar (*, M)

Magnolia acuminata—Cucumbertree, Cucumber Magnolia (M)
Magnolia grandiflora—Southern Magnolia (M)
Metasequoia glyptostroboides—Dawn Redwood (*, M)

Picea abies—Norway Spruce (M)
Pinus canariensis—Canary Island Pine
Pinus coulteri—Coulters Pine
Pinus elliottii—Slash Pine
Pinus palustris—Longleaf Pine
Pinus ponderosa—Ponderosa Pine
Pinus strobus—Eastern White Pine (M)
Pinus taeda—Loblolly Pine
Platanus spp.—Sycamore (*, M)
Populus alba—White Poplar (*)
Populus deltoides—Eastern Cottonwood (*)
Populus fremontii—Fremont Poplar
Pseudolarix kaempferi—Golden Larch (M)
Pseudotsuga menziesii—Douglas Fir (P)
Pterocarya stenoptera—Chinese Wingnut (*)

Quercus alba—White Oak (M)
Quercus bicolor—Swamp White Oak (M)
Quercus coccinea—Scarlet Oak (M)
Quercus falcata—Southern Red Oak, Spanish Oak
Quercus imbricaria—Shingle Oak, Northern Laurel Oak
Quercus laurifolia—Laurel Oak
Quercus macrocarpa—Bur Oak
Quercus michauxii—Swamp White Oak
Quercus muehlenbergii—Chinkapin Oak, Chestnut Oak
Quercus nigra—Water Oak
Quercus palustris—Pin Oak (M)
Quercus phellos—Willow Oak
Quercus prinus—Chestnut Oak, Rock Oak, Basket Oak
Quercus robur—English Oak
Quercus robur 'Fastigiata'—'Fastigiata' English Oak

Quercus rubra—Northern Red Oak
Quercus shumardii—Shumard Oak

Robinia pseudoacacia—Black Locust, Common Locust
Roystonea spp.—Royal Palm

Sequoia sempervirens—Coast Redwood (*, P)
Sequoiadendron giganteum—Giant Sequoia (*, M)
Sophora japonica—Scholar Tree, Japanese Pagoda Tree
Spathodea campanulata—African Tulip-Tree (M)

Tamarindus indica—Tamarind
Taxodium ascendens—Pondcypress (M)
Taxodium distichum—Baldcypress
Taxodium mucronatum—Montezuma Baldcypress, Mexican-Cypress (M)

Tilia americana—American Linden, Basswood, American Basswood (M)
Tilia cordata—Littleleaf Linden
Tilia tomentosa—Silver Linden (M)
Tilia x 'Redmond'—'Redmond' Basswood (M, P)
Tsuga canadensis—Canadian Hemlock, Eastern Hemlock (P)

Ulmus americana—American Elm
Ulmus hybrids—Hybrid Elm
Ulmus pumila—Siberian Elm

Washingtonia filifera—Desert Palm, California Washingtonia Palm
Washingtonia robusta—Washington Palm, Mexican Washington Palm

Zelkova serrata—Japanese Zelkova, Saw-Leaf Zelkova

Appendix 10

Trees with a Columnar or Narrow Canopy Usually Not More Than About 25 Feet Wide

Legend:
* = May grow wider than 25 feet

Abies concolor—White Fir, Colorado Fir
Abies cilicica—Cilician Fir
Abies nordmanniana—Nordmann Fir
Abies firma—Japanese Fir
Acer barbatum var. *caddo*—Caddo Florida Maple, Southern Sugar Maple
Acer campestre 'Queen Elizabeth'—'Queen Elizabeth' Hedge Maple
Acer griseum—Paperbark Maple
Acer nigrum 'Greencolumn'—'Greencolumn' Black Maple
Acer platanoides 'Columnare'—'Columnare' Norway Maple
Acer platanoides 'Columnarbroad'—Parkway Norway Maple
Acer platanoides 'Crimson King'—'Crimson King' Norway Maple
Acer platanoides 'Crimson Sentry'—'Crimson Sentry' Norway Maple
Acer platanoides 'Drummondii'—Drummond Norway Maple
Acer platanoides 'Erectum'—'Erectum' Norway Maple
Acer platanoides 'Olmsted'—'Olmsted' Norway Maple (*)
Acer rubrum 'Bowhall'—'Bowhall' Red Maple
Acer saccharinum 'Pyramidale'—'Pyramidale' Silver Maple
Acer saccharum 'Endowment Columnar'—'Endowment Columnar' Sugar Maple
Acer saccharum 'Goldspire'—'Goldspire' Sugar Maple
Acer saccharum 'Newton Sentry'—'Newton Sentry' Sugar Maple
Acer saccharum 'Seneca Chief'—'Seneca Chief' Sugar Maple
Acer saccharum 'Temple's Upright'—'Temple's Upright' Sugar Maple
Acer x freemanii 'Armstrong'—Armstrong Freeman Maple
Acer x freemanii 'Celzam'—Celebration Freeman Maple
Acer x freemanii 'Scarlet Sentinal'—'Scarlet Sentinal' Freeman Maple
Acoelorrhaphe wrightii—Paurotis Palm
Aesculus flava—Yellow Buckeye, Sweet Buckeye
Aesculus x carnea—Red Horsechestnut (*)
Alnus glutinosa 'Pyramidalis'—'Pyramidalis' Black Alder
Alnus rhombifolia—White Alder
Amelanchier spp.—Serviceberry
Araucaria araucana—Monkey-Puzzletree
Araucaria bidwillii—False Monkey-Puzzletree
Araucaria heterophylla—Norfolk-Island-Pine
Arbutus texana—Texas Madrone

Betula ermanii—Erman Maple
Betula pendula—European Birch
Betula platyphylla 'Whitespire'—Japanese Whitespire Birch
Betula populifolia—Gray Birch
Betula utilis var. *jacquemontii*—Jacquemontii Birch
Bismarckia nobilis—Bismarck Palm
Brahea armata—Mexican Blue Palm
Butia capitata—Pindo Palm

Callistemon viminalis—Weeping Bottlebrush
Calocedrus decurrens—California Incense-Cedar
Carpentaria acuminata—Carpentaria Palm
Carpinus betulus 'Fastigiata'—'Fastigiata' European Hornbeam
Caryota mitis—Fishtail Palm
Catalpa spp.—Catalpa (*)
Chamaecyparis lawsoniana—Lawson Falsecypress, Port-Orford-Cedar
Chamaecyparis nootkatensis 'Pendula'—Nootka Falsecypress, Alaska-Cedar
Chamaecyparis obtusa—Hinoki Falsecypress
Chrysalidocarpus lutescens—Yellow Butterfly Palm, Bamboo Palm, Areca Palm
Chrysophyllum oliviforme—Satinleaf
Coccoloba diversifolia—Pigeon-Plum
Coccothrinax argentata—Silverpalm, Thatchpalm
Cocos nucifera 'Malayan Dwarf'—'Malayan Dwarf' Coconut Palm
Cornus mas 'Golden Glory'—'Golden Glory' Cornelian Dogwood
Crataegus phaenopyrum 'Fastigiata'—'Fastigiata' Washington Hawthorn
Cryptomeria japonica—Japanese-Cedar
Cunninghamia lanceolata—China-Fir
x Cupressocyparis leylandii—Leyland Cypress
Cupressus arizonica var. *arizonica*—Arizona Cypress
Cupressus glabra 'Carolina Safire'—Smooth-Barked Arizona Cypress
Cupressus sempervirens—Italian Cypress

Diospyros virginiana—Common Persimmon (*)

Eucalyptus citriodora—Lemon-Scented Gum (*)
Eugenia spp.—Stopper, Eugenia

Firmiana simplex—Chinese Parasoltree
Fraxinus americana 'Skyline'—'Skyline' White Ash
Fraxinus pennsylvanica 'Skyward'—Skyward' Green Ash
Fraxinus uhdei 'Majestic Beauty'—'Majestic Beauty' Shamel Ash

Geijera parviflora—Australian-Willow
Ginkgo biloba (males)—Maidenhair Tree, Ginkgo
Ginkgo biloba 'Fairmont'—'Fairmont' Ginkgo
Ginkgo biloba 'Fastigiata'—'Fastigiata' Ginkgo
Ginkgo biloba 'Lakeview'—'Lakeview' Ginkgo
Ginkgo biloba 'Princeton Sentry'—'Princeton Sentry' Ginkgo
Gordonia lasianthus—Loblolly-Bay, Sweet-Bay
Grevillea robusta—Silk-Oak (*)

Hibiscus syriacus—Rose-of-Sharon, Shrub-Althea
Hovenia dulcis—Japanese Raisintree
Howea forsterana—Sentry Palm, Kentia Palm

Ilex cassine—Dahoon Holly
Ilex opaca—American Holly
Ilex vomitoria 'Pendula'—Weeping Yaupon Holly
Ilex x attenuata 'East Palatka'—'East Palatka' Holly
Ilex x attenuata 'Fosteri'—Fosters Holly
Ilex x attenuata 'Savannah'—Savannah Holly
Ilex x 'Nellie R. Stevens'—'Nellie R. Stevens' Holly

Juniperus ashei—Ashe Juniper, Mountain-Cedar
Juniperus chinensis 'Torulosa'—'Torulosa' Juniper
Juniperus deppeana 'Mcfetter'—McFetter Alligator Juniper
Juniperus scopulorum—Rocky Mountain Juniper, Colorado Redcedar
Juniperus silicicola—Southern Redcedar (*)
Juniperus virginiana 'Canaertii'—'Canaertii' Eastern Redcedar
Juniperus virginiana 'Keteleeri'—'Keteleeri' Eastern Redcedar
Juniperus virginiana 'Pendula'—'Pendula' Eastern Redcedar
Juniperus virginiana 'Skyrocket'—'Skyrocket' Eastern Redcedar

Koelreuteria paniculata 'Fastigiata'—'Fastigiata' Goldenraintree

Laburnum x alpinum—Goldenchain Tree
Latania spp.—Latan Palm
Larix decidua—European Larch (*)
Liquidambar styraciflua 'Festival'—'Festival' Sweetgum
Livistona chinensis—Chinese Fan Palm

Maackia amurense 'Macho'—'Macho' Amur Corktree
Magnolia acuminata—Cucumbertree, Cucumber Magnolia (*)
Magnolia 'Galaxy'—Hybrid Magnolia
Magnolia grandiflora 'Bracken's Brown Beauty'—'Bracken's Brown Beauty' Magnolia
Magnolia grandiflora 'Glen St. Mary'—'Glen St. Mary' Southern Magnolia
Magnolia grandiflora 'Hasse'—'Hasse' Southern Magnolia
Magnolia grandiflora 'Little Gem'—'Little Gem' Southern Magnolia
Magnolia grandiflora 'Majestic Beauty'—'Majestic Beauty' Southern Magnolia
Magnolia stellata var. *kobus* 'Wadas Memory'—'Wadas Memory' Magnolia
Magnolia virginiana—Sweetbay Magnolia, Swamp Magnolia
Malus baccata 'Columnaris'—Columnar Siberian Crabapple
Malus 'Beauty'—'Beauty' Crabapple
Malus 'Centurion'—'Centurion' Crabapple
Malus 'Evelyn'—'Evelyn' Crabapple
Malus 'Liset'—'Liset' Crabapple
Malus 'Madonna'—'Madonna' Crabapple
Malus 'Marshell Oyama'—'Marshell Oyama' Crabapple
Malus 'Pink Spires'—'Pink Spires' Crabapple
Malus 'Red Barron'—'Red Barron' Crabapple
Malus 'Sentinel'—Sentinel' Crabapple
Malus 'Silver Moon'—'Silver Moon' Crabapple
Malus 'Strathmore'—'Strathmore' Crabapple
Malus 'Tschonoskii'—'Tschonoskii' Crabapple
Malus 'Velvet Pillar'—'Velvet Pillar' Crabapple
Malus 'White Candle'—'White Candle' Crabapple
Malus 'Yunnanensis'—'Yunnanensis' Crabapple
Melaleuca quinquinervia—Punk Tree, Melaleuca
Metasequoia glyptostroboides—Dawn Redwood (*)
Musa spp.—Banana

Neodypsis decaryi—Triangle Palm

Osmanthus fragrans—Sweet Osmanthus
Oxydendrum arboreum—Sourwood, Sorrel-Tree (*)

Phoenix canariensis—Canary Island Date Palm
Phoenix dactylifera—Date Palm
Phoenix reclinata—Senegal Date Palm
Picea omorika—Serbian Spruce
Picea orientalis—Oriental Spruce
Picea pungens—Colorado Spruce, Colorado Blue Spruce, Blue Spruce
Pinckneya pubens—Pinckneya, Fevertree
Pinus bungeana—Lacebark Pine
Pinus canariensis—Canary Island Pine
Pinus cembra—Swiss Stone Pine
Pinus cembra 'Nana'—Dwarf Swiss Stone Pine
Pinus cembroides—Mexican Pinyon, Pinyon Pine
Pinus densiflora 'Umbraculifera'—'Umbraculifera' Japanese Red Pine
Pinus eldarica—Mondell Pine
Pinus flexilis 'Columnaris'—Columnar Limber Pine
Pinus strobus 'Fastigiata'—'Fastigiata' Eastern White Pine
Pinus strobus 'Nana'—'Nana' Eastern White Pine
Pinus sylvestris 'Watereri'—'Watereri' Scotch Pine
Podocarpus macrophyllus var. *angustifolius*—Podocarpus
Podocarpus nagi—Nagi Podocarpus, Broadleaf Podocarpus (*)
Populus alba 'Pyramidalis'—'Pyramidalis' White Poplar
Populus nigra 'Italica'—Lombardy Poplar
Prunus sargentii 'Columnaris'—Columnar Sargent Cherry
Prunus serrulata 'Amanogawa'—'Amanogawa' Oriental Cherry
Pseudolarix kaempferi—Golden Larch (*)
Pseudotsuga menziesii 'Fastigiata'—'Fastigiata' Douglas Fir
Pseudocydonia sinensis—Chinese Quince
Pyrus calleryana 'Autumn Blaze'—'Autumn Blaze' Callery Pear
Pyrus calleryana 'Capital'—'Capital' Callery Pear
Pyrus calleryana 'Chanticleer'—'Chanticleer' Callery Pear
Pyrus calleryana 'Redspire'—'Redspire' Callery Pear
Pyrus calleryana 'Whitehouse'—'Whitehouse' Callery Pear

Quercus robur 'Fastigiata'—'Fastigiata' English Oak
Quercus robur 'Pyramich'—'Pyramich' English Oak
Quercus robur 'Skyrocket'—'Skyrocket' English Oak

Ravenala madagascariensis—Travelers-Tree
Robinia pseudoacacia 'Tortuosa'—'Tortuosa' Black Locust
Robinia pseudoacacia 'Umbraculifera'—Umbrella Black Locust
Roystonea spp.—Royal Palm

Sabal spp.—Cabbage Palm
Salix matsudana 'Tortuosa'—Corkscrew Willow, Pekin Willow, Hankow Willow
Schefflera actinophylla—Schefflera, Queensland Umbrella-Tree
Sciadopitys verticillata—Japanese Umbrella-Pine
Sequoia sempervirens—Redwood
Sequoiadendron giganteum—Giant Sequoia (*)
Sophora japonica 'Columnaris'—Columnar Japanese Pagoda Tree
Sorbus aucuparia—European Mountain-Ash
Stewartia spp.—Stewartia
Strelitzia nicolai—White Bird-of-Paradise, Giant Bird-of-Paradise
Syagrus romanzoffianum—Queen Palm
Syringa reticulata 'Ivory Silk'—'Ivory Silk' Japanese Tree Lilac

Taxodium ascendens—Pondcypress
Taxodium distichum—Baldcypress (*)

Thrinax morrisii—Key Thatchpalm
Thuja occidentalis—White-Cedar, Arborvitae, Northern White-Cedar
Thuja plicata—Giant Arborvitae, Giant-Cedar, Western Redcedar
Tilia americana 'Fastigiata'—'Fastigiata' American Basswood (*)
Tilia americana 'Redmond'—'Redmond' American Basswood (*)
Tilia cordata 'Corzam'—'Corzam' Littleleaf Linden
Tilia cordata 'June Bride'—'June Bride' Littleleaf Linden
Trachycarpus fortunei—Windmill Palm

Veitchia merrillii—Christmas Palm, Manila Palm
Viburnum sieboldii—Siebold Viburnum

Washingtonia filifera—Desert Palm, California Washingtonia Palm
Washingtonia robusta—Washington Palm, Mexican Washington Palm
Wodyetia bifurcata—Foxtail Palm

Yucca elephantipes—Spineless Yucca, Soft-Tip Yucca

Appendix 11

Trees That Can Develop Large-Diameter Surface Roots or Numerous Smaller Surface Roots

Note: Surface roots are especially prone to develop in compact, clay soil.

Acer platanoides—Norway Maple
Acer rubrum—Red Maple, Swamp Maple
Acer saccharinum—Silver Maple
Acer x freemanii 'Armstrong'—Armstrong Freeman Maple
Ailanthus altissima—Tree-of-Heaven

Bischofia javanica—Toog Tree, Bischofia
Bursera simaruba—Gumbo-Limbo

Carya illinoensis—Pecan
Celtis spp.—Hackberry
Cercidiphyllum japonicum—Katsuratree
Chorisia speciosa—Floss-Silk Tree
Cinnamomum camphora—Camphor-Tree
Coccoloba uvifera—Seagrape

Dalbergia sissoo—Indian Rosewood
Delonix regia—Royal Poinciana

Erythrina variegata var. *orientalis*—Coral Tree

Eucalyptus citriodora—Lemon-Scented Gum

Fagus spp.—Beech
Ficus spp.—Fig
Fraxinus spp.—Ash

Gleditsia triacanthos var. *inermis*—Thornless Honeylocust
Grevillea robusta—Silk-Oak

Jacaranda mimosifolia—Jacaranda
Juglans spp.—Walnut

Liquidambar spp.—Sweetgum
Lysiloma bahamensis—Wild-Tamarind, Bahama Lysiloma

Maclura pomifera—Osage-Orange, Bois-D'Arc
Manilkara zapota—Sapodilla
Melaleuca quinquinervia—Punk Tree, Melaleuca

Paulownia tomentosa—Princess-Tree
Peltophorum pterocarpum—Yellow Poinciana
Phellodendron amurense—Amur Corktree, Chinese Corktree
Pinus elliottii—Slash Pine
Platanus spp.—Sycamore
Populus spp.—Poplar
Pterocarya stenoptera—Chinese Wingnut

Quercus spp.—Oak

Robinia pseudoacacia—Black Locust

Salix alba—Weeping Willow
Sapium sebiferum—Chinese Tallowtree, Popcorn Tree, Tallowtree
Schefflera actinophylla—Schefflera, Queensland Umbrella-Tree
Spathodea campanulata—African Tulip-Tree
Swietenia mahagoni—Mahogany, West Indies Mahogany

Ulmus americana—American Elm
Ulmus crassifolia—Cedar Elm
Ulmus hybrids—Hybrid Elm
Ulmus pumila—Siberian Elm

Appendix 12

Trees That Are Sensitive to Pests or Diseases That Can Cause Noticeable Damage or Tree Death

Legend:
T = Only a problem in Texas
* = Especially sensitive to one or more problems

Note: Many trees on this list are only affected in some parts of the planting range. Check with local specialists about the appropriateness of planting these trees. They might suggest that you consider planting a different tree.

Acer negundo—Boxelder
Acer pseudoplatanus—Sycamore Maple
Acer saccharinum—Silver Maple
Acer saccharum—Sugar Maple
Aesculus flava (octandra)—Yellow Buckeye
Aesculus glabra—Ohio Buckeye
Aesculus hippocastanum—Horsechestnut
Aesculus indica—Indian Horsechestnut
Albizia julibrissin—Mimosa (*)
Amelanchier spp.—Serviceberry

Betula ermanii—Erman Birch (*)
Betula papyrifera—Paper Birch (*)
Betula pendula—European Birch (*)
Betula populifolia—Gray Birch (*)
Betula utilis var. *jacquemontii*—Jacquemontii Birch (*)
Bischofia javanica—Bischofia

Callistemon viminalis—Bottlebrush
Carya illinoensis—Pecan
Caryota mitis—Fishtail Palm
Catalpa spp.—Catalpa
Citrus spp.—Citrus
Cocos nucifera—Coconut Palm
Cornus florida—Flowering Dogwood
Crataegus laevigata—English Hawthorn
Cryptomeria japonica—Japanese-Cedar
x Cupressocyparis leyandii—Leyland Cypress
Cupressus macrocarpa—Monterey Cypress (*)
Cupressus sempervirens—Italian Cypress

Diospyros virginiana—Common Persimmon

Elaeagnus angustifolia—Russian-Olive
Erythrina variegata var. *orientalis*—Coral Tree

Fraxinus spp.—Ash

Howea forsterana—Kentia Palm

Juglans spp.—Walnut

Latania spp.—Latan Palm
Livistona chinensis—Chinese Fan Palm

Malus spp.—Crabapple (some sensitive, others not)
Mangifera indica—Mango

Nerium oleander—Oleander

Photinia spp.—Photinia, Red-Top
Picea pungens—Colorado Blue Spruce
Pinckneya pubens—Pinckneya, Fevertree
Pinus elliottii—Slash Pine
Pinus nigra—Austrian Pine (*)
Pinus parviflora—Japanese White Pine
Pinus pinea—Stone Pine
Pinus strobus—Eastern White Pine
Pinus sylvestris—Scotch Pine (*)
Pinus thunbergiana—Japanese Black Pine (*)
Platanus spp.—Sycamore (*)
Populus nigra 'Italica'—Lombardy Poplar (*)
Prunus cerasifera—Purple Leaf Plum
Prunus mume—Japanese Apricot
Prunus persica—Peach
Prunus serrulata 'Kwanzan'—Kwanzan Cherry
Prunus subhirtella 'Autumnalis'—'Autumnalis' Higan Cherry
Prunus subhirtella 'Pendula'—Weeping Higan Cherry
Prunus x incamp 'Okame'—'Okame' Cherry
Prunus x yedoensis—Yoshino Cherry

Quercus fusiformis—Texas Live Oak (*)
Quercus rubra—Northern Red Oak
Quercus shumardii—Shumard Oak
Quercus virginiana—Live Oak (*, T)

Ravenala madagascariensis—Travelers-Tree
Robinia pseudoacacia—Black Locust, Common Locust

Salix matsudana 'Tortuosa'—Corkscrew Willow
Sorbus aucuparia—European Mountain-Ash (*)

Tsuga canadensis—Eastern Hemlock (*)

Ulmus alata—Winged Elm (*)
Ulmus americana—American Elm (*)
Ulmus crassifolia—Cedar Elm (*)

Veitchia merrillii—Christmas Palm, Manila Palm

Appendix 13

Trees That Are Susceptible to Breakage

> Note: Trees that develop included bark in the branch crotches are most likely to break apart. The likelihood of breaking can be minimized with appropriate pruning.

Acacia auriculiformis—Earleaf Acacia
Acer negundo—Boxelder
Acer saccharinum—Silver Maple
Ailanthus altissima—Tree-of-Heaven
Albizia julibrissin—Mimosa, Silktree
Amphitecna latifolia—Black-Calabash

Bauhinia spp.—Orchid-Tree

Caesalpinia pulcherrima—Dwarf Poinciana
Carya illinoensis—Pecan
Celtis spp.—Hackberry
Cercidiphyllum japonicum—Katsuratree
Cercis canadensis—Eastern Redbud
Citrus spp.—Citrus
Cladrastis kentukea—American Yellowwood
Coccoloba uvifera—Seagrape

Dalbergia sissoo—Indian Rosewood
Delonix regia—Royal Poinciana

Evodia danielii—Korean Evodia

Ficus elastica—Fig
Firmiana simplex—Chinese Parasoltree
Fraxinus spp.—Ash

Gleditsia triacanthos var. inermis—Thornless Honeylocust
Grevillea robusta—Silk-Oak

Hibiscus syriacus—Rose-of-Sharon
Hydrangea paniculata—Panicle Hydrangea

Jacaranda mimosifolia—Jacaranda

Laburnum hybrids—Goldenchain Tree

Mangifera indica—Mango
Melia azedarach—Chinaberry
Morus alba—White Mulberry
Myrica cerifera—Southern Waxmyrtle

Nerium oleander—Oleander

Peltophorum spp.—Yellow Poinciana
Persea americana—Avocado
Persea borbonia—Redbay
Pinus bungeana—Lacebark Pine
Pinus densiflora 'Umbraculifera'—'Umbraculifera' Japanese Red Pine
Pinus elliottii—Slash Pine
Pinus halepensis—Aleppo Pine
Pinus taeda—Loblolly Pine
Plumeria spp.—Frangipani
Populus spp.—Poplar
Prunus serotina—Black Cherry
Pyrus calleryana 'Bradford'—'Bradford' Callery Pear

Quercus macrocarpa—Bur Oak
Quercus nigra—Water Oak
Quercus palustris—Pin Oak

Rhus spp.—Sumac
Robinia pseudoacacia—Black Locust

Salix spp.—Willow
Sambucus canadensis—American Elder
Sapium sebiferum—Chinese Tallowtree
Schefflera actinophylla—Schefflera
Schinus terebinthifolius—Brazilian Pepper
Sorbus alnifolia—Korean Mountain-Ash
Sorbus aucuparia—European Mountain-Ash
Spathodea campanulata—African Tulip-Tree
Swietenia mahagoni—Mahogany

Tabebuia spp.—Trumpet Tree
Terminalia catappa—Tropical-Almond

Ulmus crassifolia—Cedar Elm
Ulmus pumila—Siberian Elm

Zelkova serrata—Japanese Zelkova

Appendix 14

Nonconiferous Trees Requiring Only a Moderate Amount of Pruning to Develop Good Structure

Legend:
* = May need pruning to balance the canopy

Note: Conifers typically require little pruning to develop good structure.
When only the genus or species of tree is listed, most cultivars also require only moderate pruning.

Acer barbatum—Florida Maple, Southern Sugar Maple
Acer barbatum var. *caddo*—Caddo Florida Maple, Southern Sugar Maple
Acer buergeranum—Trident Maple
Acer griseum—Paperbark Maple
Acer japonicum 'Acontifolium'—'Acontifolium' Fullmoon Maple, Fernleaf Maple
Acer miyabei—Miyabe Maple
Acer platanoides 'Emerald Queen'—'Emerald Queen' Norway Maple
Acer platanoides 'Superform'—'Superform' Norway Maple
Acer pseudoplatanus—Sycamore Maple, Planetree Maple
Acer rubrum 'Autumn Flame'—'Autumn Flame' Red Maple
Acer rubrum 'Red Sunset'—'Red Sunset' Red Maple
Acer saccharum and cultivars—Sugar Maple
Acer triflorum—Three-Flower Maple
Acer truncatum—Purpleblow Maple
Acer x freemanii 'Armstrong'—Armstrong Freeman Maple
Aesculus spp.—Horsechestnut, Buckeye
Alnus glutinosa 'Pyramidalis'—'Pyramidalis' Black Alder, 'Pyramidalis' European Alder
Alnus rhombifolia—White Alder
Amelanchier spp.—Serviceberry
Araucaria bidwillii—False Monkey-Puzzletree
Araucaria heterophylla—Norfolk-Island-Pine
Arbutus menzesii—Pacific Madrone
Arbutus texana—Texas Madrone
Asimina triloba—Pawpaw

Betula spp.—Birch
Bumelia lanuginosa—Chittamwood, Gum Bumelia, Gum Elastic Buckthorn

Caesalpinia granadillo—Bridalveil Tree
Callistemon citrinus—Red Bottlebrush, Lemon Bottlebrush
Callistemon viminalis—Weeping Bottlebrush
Calocedrus decurrens—California Incense-Cedar
Calodendron capense—Cape Chestnut
Calophyllum brasiliense—Santa Maria
Camellia oleifera—Tea-Oil Camellia
Carpentaria acuminata—Carpentaria Palm
Carpinus betulus 'Fastigiata'—'Fastigiata' European Hornbeam
Carpinus caroliniana—American Hornbeam, Blue-Beech, Ironwood
Carya spp.—Hickory
Catalpa spp.—Catalpa
Cercidiphyllum japonicum—Katsuratree
Chionanthus spp.—Fringetree
Chrysophyllum oliviforme—Satinleaf
Coccoloba diversifolia—Pigeon-Plum
Cornus alternifolia—Pagoda Dogwood
Cornus controversa—Giant Dogwood
Cornus drummondii—Roughleaf Dogwood
Cornus florida—Flowering Dogwood
Cornus kousa—Kousa Dogwood, Chinese Dogwood, Japanese Dogwood
Cornus mas—Cornelian-Cherry
Cornus walteri—Walter Dogwood
Corylus colurna—Turkish Filbert, Turkish Hazel
Cotinus coggygria—Smoketree, Wig-Tree, Smokebush
Cotinus obovatus—American Smoketree, Chittamwood
Crataegus spp.—Hawthorn

Diospyros texana—Texas Persimmon
Diospyros virginiana—Common Persimmon

Eucalyptus citriodora—Lemon-Scented Gum
Eucalyptus ficifolia—Red-Flowering Gum
Eucommia ulmoides—Hardy Rubber Tree
Eugenia spp.—Stopper, Eugenia

Fagus spp.—Beech
Firmiana simplex—Chinese Parasoltree
Franklinia alatamaha—Franklin-Tree, Franklinia
Fraxinus americana 'Autumn Applause'—'Autumn Applause' White Ash
Fraxinus americana 'Autumn Purple'—'Autumn Purple' White Ash
Fraxinus ornus—Flowering Ash
Fraxinus oxycarpa 'Raywood'—Raywood Ash, Claret Ash
Fraxinus pennsylvanica 'Summit'—'Summit' Green Ash
Fraxinus texensis—Texas Ash

Ginkgo biloba—Maidenhair Tree, Ginkgo (*)
Gordonia lasianthus—Loblolly-Bay, Sweet-Bay
Guaiacum sanctum—Lignumvitae
Gymnocladus dioicus—Kentucky Coffeetree (*)

Halesia spp.—Silverbell
Hamamelis mollis—Chinese Witch-Hazel (*)
Hibiscus syriacus—Rose-of-Sharon, Shrub-Althea
Hovenia dulcis—Japanese Raisintree

Ilex latifolia—Lusterleaf Holly
Ilex opaca—American Holly
Ilex x attenuata 'East Palatka'—'East Palatka' Holly
Ilex x attenuata 'Fosteri'—Fosters Holly
Ilex x attenuata 'Savannah'—Savannah Holly

Juglans nigra—Black Walnut (*)

Kalopanax pictus—Castor-Aralia, Prickly Castor-Oil Tree (*)

Lagerstroemia fauriei—Japanese Crape-Myrtle
Lagerstroemia indica—Crape-Myrtle
Ligustrum lucidum—Glossy Privet, Tree Ligustrum
Liquidambar spp.—Sweetgum
Liriodendron tulipifera—Tuliptree, Tulip-Poplar, Yellow-Poplar
Lithocarpus densiflora—Tanbark Oak

Magnolia acuminata—Cucumbertree, Cucumber Magnolia
Magnolia grandiflora—Southern Magnolia
Magnolia macrophylla—Bigleaf Magnolia (*)
Magnolia stellata—Star Magnolia
Magnolia stellata var. *kobus*—Kobus Magnolia, Northern Japanese Magnolia (*)
Magnolia virginiana—Sweetbay Magnolia, Swamp Magnolia
Magnolia x soulangiana—Saucer Magnolia (*)
Malpighia glabra—Barbados Cherry
Manilkara zapota—Sapodilla
Melaleuca quinquinervia—Punk Tree, Melaleuca
Morus alba—White Mulberry (*)

Nyssa ogeche—Ogeechee Tupelo, Ogeechee-Lime
Nyssa sinensis—Chinese Tupelo
Nyssa sylvatica—Blackgum, Sourgum, Black Tupelo

Osmanthus spp.—Osmanthus
Ostrya virginiana—American Hophornbeam, Eastern Hophornbeam
Oxydendrum arboreum—Sourwood, Sorrel-Tree

Pandanus utilis—Screw-pine
Parrotia persica—Persian Parrotia
Photinia serrulata—Chinese Photinia
Photinia villosa—Oriental Photinia
Pinckneya pubens—Pinckneya, Fevertree (*)
Platanus spp.—Sycamore, Planetree
Platanus orientalis—Oriental Planetree
Platanus racemosa—Western Sycamore
Platanus x acerifolia—London Planetree
Podocarpus falcatus—Podocarpus
Podocarpus gracilior—Weeping Podocarpus, Fern Podocarpus
Podocarpus latifolius—Podocarpus
Podocarpus macrophyllus—Podocarpus, Yew-Pine, Japanese Yew
Podocarpus macrophyllus var. *angustifolius*—Podocarpus
Podocarpus nagi—Nagi Podocarpus, Broadleaf Podocarpus
Populus alba—White Poplar (*)
Populus deltoides—Eastern Cottonwood (*)
Populus freemanii—Fremont Poplar
Populus nigra 'Italica'—Lombardy Poplar
Prunus caroliniana—Cherry-Laurel, Carolina Laurelcherry
Prunus cerasifera—Purple Leaf Plum
Prunus maackii—Amur Chokecherry, Manchurian Cherry
Prunus mexicana—Mexican Plum
Prunus sargentii—Sargent Cherry
Prunus serotina—Black Cherry
Prunus serrulata 'Kwanzan'—Kwanzan Cherry
Prunus subhirtella 'Autumnalis'—'Autumnalis' Higan Cherry
Prunus subhirtella 'Pendula'—Weeping Higan Cherry (*)
Prunus umbellata—Flatwoods Plum
Prunus x incamp 'Okame'—'Okame' Cherry
Prunus x yedoensis—Yoshino Cherry
Pseudocydonia sinensis—Chinese Quince
Psidium littorale—Cattley Guava, Strawberry Guava

Quercus acutissima—Sawtooth Oak
Quercus alba—White Oak
Quercus austrina—Bluff Oak
Quercus bicolor—Swamp White Oak
Quercus cerris—Turkey Oak, Moss-Cupped Oak
Quercus coccinea—Scarlet Oak
Quercus falcata—Southern Red Oak, Spanish Oak
Quercus glauca—Blue Japanese Oak, Ring-Cupped Oak
Quercus imbricaria—Shingle Oak, Northern Laurel Oak
Quercus lobata—Valley Oak, California White Oak
Quercus lyrata—Overcup Oak
Quercus macrocarpa—Bur Oak
Quercus michauxii—Swamp Chestnut Oak
Quercus muehlenbergii—Chinkapin Oak, Chestnut Oak
Quercus nuttallii—Nuttall Oak
Quercus palustris—Pin Oak
Quercus robur—English Oak
Quercus rubra—Northern Red Oak
Quercus texana—Texas Red Oak, Texas Oak

Robinia pseudoacacia 'Tortuosa'—'Tortuosa' Black Locust
Robinia pseudoacacia 'Umbraculifera'—Umbrella Black Locust

Sambucus mexicana—Mexican Elder
Sapindus drummondii—Western Soapberry
Sapindus saponaria—Florida Soapberry, Wingleaf Soapberry
Sassafras albidum—Sassafras
Sophora secundiflora—Texas-Mountain-Laurel, Mescalbean
Sorbus alnifolia—Korean Mountain-Ash
Stewartia koreana—Korean Stewartia
Stewartia monadelpha—Tall Stewartia
Stewartia pseudocamellia—Japanese Stewartia
Styrax japonicus—Japanese Snowbell
Styrax obassia—Fragrant Snowbell
Syringa pekinensis—Peking Lilac
Syringa reticulata—Japanese Tree Lilac

Tabebuia spp.—Trumpet Tree (*)
Terminalia catappa—Tropical-Almond, India-Almond
Terminalia muelleri—Muellers Terminalia
Tilia americana 'Fastigiata'—'Fastigiata' American Basswood
Tilia americana 'Redmond'—'Redmond' American Basswood
Tilia cordata—Littleleaf Linden
Tristania conferta—Brisbane Box (*)
Tsuga canadensis—Canadian Hemlock, Eastern Hemlock
Tsuga canadensis 'Lewisii'—Lewis Hemlock

Ulmus x 'Urban'—'Urban' Elm
Ulmus x 'Regal'—'Regal' Elm

Viburnum rufidulum—Rusty Blackhaw, Southern Blackhaw
Viburnum sieboldii—Siebold Viburnum

Xanthoceras sorbifolium—Yellowhorn (*)

Yucca elephantipes—Spineless Yucca, Soft-Tip Yucca

Appendix 15

Trees Bearing Little or No Fruit

Legend:
* = Fruit usually not produced in North America
O = Occasionally produces fruit
R = Fruit usually remains on the tree

Acer rubrum 'Autumn Flame'—'Autumn Flame' Red Maple
Acer rubrum 'Northwood'—'Northwood' Red Maple
Acer x freemanii 'Autumn Blaze'—'Autumn Blaze' Freeman Maple (O)
Acer x freemanii 'Clezam'—'Clezam' Freeman Maple (O)
Aesculus hippocastanum 'Baumannii'—'Baumannii' Horsechestnut
Aesculus x carnea 'Briotii'—Ruby Red Horsechestnut (O)
Araucaria spp.—Monkey-Puzzletree, Norfolk-Island-Pine (*)

Bauhinia blakeana—Hong Kong Orchid-Tree

Cercis canadensis 'Flame'—'Flame' Eastern Redbud
x Chitalpa tashkentensis—Chitalpa
x Cupressocyparis leylandii—Leyland Cypress
Cupressus arizonica var. *arizonica*—Arizona Cypress
Cupressus glabra 'Carolina Safire'—Smooth-Barked Arizona Cypress
Cupressus sempervirens—Italian Cypress

Fraxinus americana 'Autumn Applause'—'Autumn Applause' White Ash
Fraxinus americana 'Autumn Purple'—'Autumn Purple' White Ash
Fraxinus oxycarpa 'Raywood'—Raywood Ash, Claret Ash
Fraxinus pennsylvanica 'Marshall's Seedless'—'Marshall's Seedless' Green Ash (O)
Fraxinus pennsylvanica 'Newport'—'Newport' Green Ash
Fraxinus pennsylvanica 'Patmore'—'Patmore' Green Ash
Fraxinus pennsylvanica 'Summit'—'Summit' Green Ash (O)

Geijera parviflora—Australian-Willow
Ginkgo biloba 'Autumn Gold'—'Autumn Gold' Ginkgo
Ginkgo biloba 'Fairmont'—'Fairmont' Ginkgo
Ginkgo biloba 'Fastigiata'—'Fastigiata' Ginkgo
Ginkgo biloba 'Lakeview'—'Lakeview' Ginkgo
Ginkgo biloba 'Princeton Sentry'—'Princeton Sentry' Ginkgo
Gleditsia triacanthos var. *inermis* 'Imperial'—'Imperial' Thornless Honeylocust
Gleditsia triacanthos var. *inermis* 'Impcole'—'Impcole' Thornless Honeylocust (O)
Gleditsia triacanthos var. *inermis* 'Moraine'—'Moraine' Thornless Honeylocust (O)
Gleditsia triacanthos var. *inermis* 'Shademaster'—Shademaster Thornless Honeylocust (O)
Gleditsia triacanthos var. *inermis* 'Skyline'—'Skyline' Thornless Honeylocust (O)
Gymnocladus dioicus (male trees)—Kentucky Coffeetree

Liquidambar styraciflua 'Rotundiloba'—'Rotundiloba' Sweetgum

Maclura pomifera 'Witchita', 'White Shield'—Fruitless Osage-Orange
Magnolia x soulangiana—Saucer Magnolia
Malus 'Spring Snow'—'Spring Snow' Crabapple
Morus alba (fruitless cultivars)—Fruitless White Mulberry

Osmanthus fragrans—Sweet Osmanthus

Phellodendron amurense 'Macho'—'Macho' Amur Corktree
Phellodendron amurense 'Shademaster'—'Shademaster' Amur Corktree
Prunus 'Accolade'—'Accolade' Flowering Cherry
Prunus cerasifera 'Krauter Vesuvius'—'Krauter Vesuvius' Cherry Plum
Prunus serrulata 'Amanogawa'—'Amanogawa' Oriental Cherry (O)
Prunus serrulata 'Kwanzan'—'Kwanzan' Oriental Cherry
Prunus x blireiana—Blireiana Plum
Pyrus calleryana 'Autumn Blaze'—'Autumn Blaze' Callery Pear

Robinia pseudoacacia 'Umbraculifera'—Umbrella Black Locust

Tilia tomentosa and cultivars—Silver Linden (R)

Ulmus x 'Frontier'—'Frontier' Elm

Appendix 16

Trees with Messy Fruit or Leaves

Legend:
C = Cultivars, varieties, or species without fruit are available
H = Fruit is hard
L = Leaves often litter the ground in summer
M = Fruit is mushy
P = Fruit is papery and not mushy
S = Fruit is soft and not mushy
T = Twigs drop and litter the ground

Acer negundo—Boxelder (P)
Acer saccharinum—Silver Maple (P)
Aesculus californica—California Buckeye (H, L)
Aesculus flava—Yellow Buckeye, Sweet Buckeye (H)
Aesculus glabra—Ohio Buckeye, Fetid Buckeye (H, L)
Aesculus hippocastanum—Horsechestnut, European Horsechestnut (C, H, L)
Aesculus indica—Indian Horsechestnut (H)
Aesculus x carnea—Red Horsechestnut (C, H, L)
Ailanthus altissima—Tree-of-Heaven (P)
Albizia julibrissin—Mimosa, Silktree (L, P)
Arbutus menzesii—Pacific Madrone (M)
Arbutus texana—Texas Madrone (M)
Arbutus unedo—Strawberry-Tree (M)
Asimina triloba—Pawpaw (M)

Bauhinia spp.—Orchid-Tree (C, L, P, T)
Bischofia javanica—Toog Tree, Bischofia (M, T)
Butia capitata—Pindo Palm (M)

Calophyllum brasiliense—Santa Maria (M)
Carya glabra—Pignut Hickory (H)
Carya illinoensis—Pecan (H, T)
Carya ovata—Shagbark Hickory (H)
Cassia fistula—Golden-Shower (P)
Cassia javanica—Pink-and-White-Shower (P)
Castanea mollissima—Chinese Chestnut (H)
Catalpa spp.—Catalpa (P)
Celtis spp.—Hackberry (M, T)
Cinnamomum camphora—Camphor-Tree (M)
Citrus spp.—Citrus (M)
Clusia rosea—Pitch-Apple, Florida Clusia (H)
Coccoloba diversifolia—Pigeon-Plum (M)
Coccoloba uvifera—Seagrape (L, M, T)
Cocos nucifera 'Malayan Dwarf'—'Malayan Dwarf' Coconut Palm (H)
Cordia sebestena—Geiger-Tree (H)
Cornus mas—Cornelian-Cherry (M)
Corylus colurna—Turkish Filbert, Turkish Hazel (H)
Cupaniopsis anacardiopsis—Carrotwood (M)

Delonix regia—Royal Poinciana (P)
Diospyros kaki—Japanese Persimmon (M)
Diospyros virginiana—Common Persimmon (M)

Eriobotrya deflexa—Bronze Loquat (M)
Eriobotrya japonica—Loquat (M)
Eugenia spp.—Stopper, Eugenia (M)

Feijoa sellowiana—Guava, Feijoa, Pineapple Guava (M)
Ficus spp.—Fig (L, M)
Firmiana simplex—Chinese Parasoltree (P)
Fraxinus americana—White Ash (P, C)
Fraxinus excelsior—Common Ash, European Ash (P)
Fraxinus pennsylvanica—Green Ash (C, P)
Fraxinus uhdei—Shamel Ash (P)
Fraxinus velutina—Velvet Ash, Modesto Ash, Arizona Ash (C, P)

Ginkgo biloba—Maidenhair Tree, Ginkgo (M)
Gleditsia triacanthos var. *inermis*—Thornless Honeylocust (C, H)
Gymnocladus dioicus—Kentucky Coffeetree (C, H)

Hovenia dulcis—Japanese Raisintree (M)

Jacaranda mimosifolia—Jacaranda (P)
Juglans spp.—Walnut (H, L)

Leucaena retusa—Goldenball Leadtree, Littleleaf Leucaena, Littleleaf Leadtree (M)
Ligustrum lucidum—Glossy Privet, Tree Ligustrum (M)
Liquidambar formosana—Formosa Sweetgum, Chinese Sweetgum (H)
Liquidambar styraciflua—Sweetgum (C, H)
Liriodendron tulipifera—Tuliptree, Tulip-Poplar, Yellow-Poplar (L, P)
Litchi chinensis—Lychee (M)

Maackia amurensis—Amur Maackia (C, H)
Maclura pomifera—Osage-Orange, Bois-D'Arc (C, M)
Magnolia grandiflora—Southern Magnolia (H, L)
Magnolia stellata var. *kobus*—Kobus Magnolia, Northern Japanese Magnolia (H)
Malpighia glabra—Barbados Cherry (M)
Malus spp.—Crabapple (H, M)
Mangifera indica—Mango (M, T)
Manilkara zapota—Sapodilla (M)
Melia azedarach—Chinaberry (H)
Morus alba—White Mulberry (C, M)
Musa spp.—Banana (M)

Nyssa spp.—Tupelo (M)

Ochrosia elliptica—Ochrosia (H)
Olea europaea—European Olive (C, M)

Pandanus utilis—Screw-pine (H, L)
Paulownia tomentosa—Princess-Tree, Empress-Tree, Paulownia (H)
Persea americana—Avocado (M, T)
Persea borbonia—Redbay (M)

Phellodendron amurense—Amur Corktree, Chinese Corktree (C, M)
Phoenix canariensis—Canary Island Date Palm (H, M)
Phoenix dactylifera—Date Palm (H, M)
Pinus spp.—Pines (H, L, T)
Pistacia chinensis—Chinese Pistache (M)
Pithecellobium flexicaule—Ebony Blackbead, Texas-Ebony (P)
Platanus spp.—Sycamore, Planetree (S)
Pongamia pinnata—Pongam, Karum Tree, Poonga-Oil Tree (M)
Populus alba—White Poplar (L, S)
Populus deltoides—Eastern Cottonwood (L, S)
Populus freemanii—Fremont Poplar (L, S)
Prosopis glandulosa—Mesquite, Honey Mesquite (M)
Prunus americana—American Plum (M)
Prunus angustifolia—Chickasaw Plum (M)
Prunus caroliniana—Cherry-Laurel, Carolina Laurelcherry (M)
Prunus cerasifera—Purple Leaf Plum (M, C)
Prunus mexicana—Mexican Plum (M)
Prunus mume—Japanese Apricot (M)
Prunus persica—Peach (M)
Prunus sargentii—Sargent Cherry (M)
Prunus serotina—Black Cherry (M)
Pseudocydonia sinensis—Chinese Quince (M)
Psidium littorale—Cattley Guava, Strawberry Guava (M)
Pterocarya stenoptera—Chinese Wingnut (P)

Quercus spp.—Oaks (H)

Roystonea spp.—Royal Palm (L)

Sabal palmetto—Cabbage Palm (M)
Sambucus mexicana—Mexican Elder (M)
Sapindus drummondii—Western Soapberry (H)
Sapindus saponaria—Florida Soapberry, Wingleaf Soapberry (H)
Schinus terebinthifolius—Brazilian Pepper Tree (M)
Simarouba glauca—Paradise-Tree (M)
Sophora japonica—Scholar Tree, Japanese Pagoda Tree (P)
Sorbus aucuparia—European Mountain-Ash (M)
Spathodea campanulata—African Tulip-Tree (L)
Swietenia mahagoni—Mahogany, West Indies Mahogany (S)
Syagrus romanzoffianum—Queen Palm (M)

Tamarindus indica—Tamarind (H)
Terminalia catappa—Tropical-Almond, India-Almond (H, L)
Terminalia muelleri—Muellers Terminalia (H, L)

Ulmus hybrids—Hybrid Elm (T)
Ulmus parvifolia—Chinese Elm, Lacebark Elm (T)
Ulmus pumila—Siberian Elm (T)

Ziziphus jujuba—Chinese Date, Common Jujube, Chinese Jujube (M)

Appendix 17

Trees Commonly Grown with a Multitrunk Habit

Legend:
* = Would be very difficult to train to one trunk
B = Bigger species are not suited for multitrunk habit

Note: Trees with multitrunk habit are strongest if they do not develop included bark in the crotches.

Acacia farnesiana—Sweet Acacia, Huisache (*)
Acacia wrightii—Wright Acacia, Wright Catclaw
Acer buergeranum—Trident Maple
Acer campestre—Hedge Maple
Acer cissifolium—Ivy-Leaf Maple (*)
Acer ginnala—Amur Maple
Acer griseum—Paperbark Maple
Acer japonicum 'Acontifolium'—'Acontifolium' Fullmoon Maple, Fernleaf Maple (*)
Acer palmatum—Japanese Maple (*)
Acer tataricum—Tatarian Maple
Acer triflorum—Three-Flower Maple (*)
Acer truncatum—Purpleblow Maple
Acoelorrhaphe wrightii—Paurotis Palm (*)
Aesculus californica—California Buckeye (*)
Aesculus pavia—Red Buckeye
Albizia julibrissin—Mimosa, Silktree (*)
Amelanchier spp.—Serviceberry
Aralia spinosa—Devils-Walkingstick, Hercules-Club
Arbutus texana—Texas Madrone
Arbutus unedo—Strawberry-Tree
Asimina triloba—Pawpaw

Bauhinia spp.—Orchid-Tree (*, B)
Beaucarnea recurvata—Ponytail (*)
Bumelia lanuginosa—Chittamwood, Gum Bumelia, Gum Elastic Buckthorn

Caesalpinia pulcherrima—Dwarf Poinciana, Barbados Flowerfence (*)
Calliandra haematocephala—Powderpuff (*)
Callistemon citrinus—Red Bottlebrush, Lemon Bottlebrush
Callistemon viminalis—Weeping Bottlebrush
Camellia oleifera—Tea-Oil Camellia (*)
Carpinus caroliniana—American Hornbeam, Blue-Beech, Ironwood
Caryota mitis—Fishtail Palm (*)
Cassia suratensis—Glaucous Cassia (*)
Castanea mollissima—Chinese Chestnut (*)
Cercidiphyllum japonicum—Katsuratree
Cercis canadensis—Eastern Redbud
Cercis mexicana—Mexican Redbud
Cercis occidentalis—Western Redbud, California Redbud
Cercis reniformis 'Oklahoma'—Oklahoma Redbud
Chilopsis linearis—Desert-Willow (*)
Chionanthus retusus—Chinese Fringetree
Chionanthus virginicus—Fringetree, Old-Mans-Beard (*)
x Chitalpa tashkentensis—Chitalpa (*)
Chrysalidocarpus lutescens—Yellow Butterfly Palm, Bamboo Palm, Areca Palm (*)
Conocarpus erectus—Buttonwood
Conocarpus erectus var. *sericeus*—Silver Buttonwood
Cordia boissieri—Wild-Olive, Anacahuita (*)
Cordia sebestena—Geiger-Tree
Cornus alternifolia—Pagoda Dogwood
Cornus controversa—Giant Dogwood
Cornus drummondii—Roughleaf Dogwood
Cornus kousa—Kousa Dogwood, Chinese Dogwood, Japanese Dogwood (*)
Cornus mas—Cornelian-Cherry (*)
Cotinus coggygria—Smoketree, Wig-Tree, Smokebush (*)
Cotinus obovatus—American Smoketree, Chittamwood
Crataegus spp.—Hawthorn, Apple Hawthorn
Crescentia cujete—Calabash Tree (*)

Diospyros texana—Texas Persimmon

Elaeagnus angustifolia—Russian-Olive, Oleaster (*)
Erythrina crista-galli—Cockspur Coral Tree (*)
Erythrina variegata var. *orientalis*—Coral Tree (*)
Eucalyptus ficifolia—Red-Flowering Gum
Eugenia spp.—Stopper, Eugenia
Evodia danielii—Korean Evodia, Bebe Tree (*)

Feijoa sellowiana—Guava, Feijoa, Pineapple Guava (*)
Franklinia alatamaha—Franklin-Tree, Franklinia

Guaiacum sanctum—Lignumvitae (*)

Halesia spp.—Silverbell
Hamamelis mollis—Chinese Witch-Hazel
Hamamelis virginiana—Witch-Hazel (*)
Hibiscus syriacus—Rose-of-Sharon, Shrub-Althea (*)
Hydrangea paniculata—Panicle Hydrangea (*)

Ilex cornuta 'Burfordii'—Burford Holly (*)
Ilex decidua—Possumhaw (*)
Ilex latifolia—Lusterleaf Holly
Ilex rotunda—Round-Leaf Holly
Ilex serrata—Fine-Toothed Holly (*)
Ilex verticillata—Winterberry (*)
Ilex vomitoria—Yaupon Holly (*)
Ilex vomitoria 'Pendula'—Weeping Yaupon Holly

Jatropha integerrima—Peregrina, Fire-Cracker (*)
Juniperus ashei—Ashe Juniper, Mountain-Cedar
Juniperus chinensis 'Torulosa'—'Torulosa' Juniper
Juniperus deppeana 'Mcfetter'—McFetter Alligator Juniper
Juniperus scopulorum—Rocky Mountain Juniper, Colorado Redcedar
Juniperus scopulorum 'Tolleson's Green Weeping'—'Tolleson's Green Weeping' Rocky Mountain Juniper (*)

Lagerstroemia fauriei—Japanese Crape-Myrtle
Lagerstroemia indica—Crape-Myrtle
Leptospermum laevigatum—Australian Tea Tree (*)
Leucaena retusa—Goldenball Leadtree, Littleleaf Leucaena, Littleleaf Leadtree (*)
Ligustrum japonicum—Japanese Privet, Wax-Leaf Privet (*)
Ligustrum lucidum—Glossy Privet, Tree Ligustrum
Lithocarpus henryi—Henry Tanbark Oak

Maackia amurensis—Amur Maackia
Maclura pomifera—Osage-Orange, Bois-D'Arc
Magnolia denudata—Yulan Magnolia
Magnolia stellata—Star Magnolia (*)
Magnolia stellata var. *kobus*—Kobus Magnolia
Magnolia x soulangiana—Saucer Magnolia (*)
Malpighia glabra—Barbados Cherry (*)
Malus spp.—Crabapple (*)
Melaleuca quinquinervia—Punk Tree, Melaleuca
Murraya paniculata—Orange Jessamine (*)
Musa spp.—Banana (*)
Myrica cerifera—Southern Waxmyrtle, Southern Bayberry (*)

Noronhia emarginata—Madagascar Olive

Olea europaea—European Olive
Osmanthus americanus—Devilwood, Wild-Olive
Osmanthus fragrans—Sweet Osmanthus
Osmanthus x fortunei—Fortunes Osmanthus (*)
Ostrya virginiana—American Hophornbeam, Eastern Hophornbeam

Parkinsonia aculeata—Jerusalem-Thorn (*)
Parrotia persica—Persian Parrotia
Phoenix reclinata—Senegal Date Palm
Photinia glabra—Red-Leaf Photinia, Red-Top
Photinia serrulata—Chinese Photinia (*)
Photinia villosa—Oriental Photinia (*)
Photinia x fraseri—Fraser Photinia
Pinus bungeana—Lacebark Pine
Pinus cembra 'Nana'—Dwarf Swiss Stone Pine
Pinus cembroides—Mexican Pinyon, Pinyon Pine
Pinus densiflora—Japanese Red Pine
Pinus densiflora 'Umbraculifera'—'Umbraculifera' Japanese Red Pine (*)
Pinus mugo—Mugo Pine, Swiss Mountain Pine
Pinus strobus 'Nana'—'Nana' Eastern White Pine
Pinus strobus 'Pendula'—'Pendula' Eastern White Pine
Pinus sylvestris 'Watereri'—'Watereri' Scotch Pine (*)
Pinus thunbergiana—Japanese Black Pine (*)
Pithecellobium flexicaule—Ebony Blackbead, Texas-Ebony
Pittosporum spp.—Pittosporum
Plumeria alba—White Frangipani
Plumeria rubra—Frangipani
Podocarpus falcatus—Podocarpus
Podocarpus gracilior—Weeping Podocarpus, Fern Podocarpus
Podocarpus latifolius—Podocarpus
Podocarpus macrophyllus—Podocarpus, Yew-Pine, Japanese Yew
Podocarpus nagi—Nagi Podocarpus, Broadleaf Podocarpus
Prosopis glandulosa—Mesquite, Honey Mesquite
Prunus spp.—Cherry
Pseudocydonia sinensis—Chinese Quince
Psidium littorale—Cattley Guava, Strawberry Guava

Quercus glauca—Blue Japanese Oak, Ring-Cupped Oak

Ravenala madagascariensis—Travelers-Tree (*)
Rhamnus caroliniana—Carolina Buckthorn (*)
Rhus chinensis—Chinese Sumac
Rhus glabra—Smooth Sumac
Rhus lancea—African Sumac
Robinia pseudoacacia 'Tortuosa'—'Tortuosa' Black Locust
Robinia pseudoacacia 'Umbraculifera'—Umbrella Black Locust

Sambucus canadensis—American Elder, Common Elder (*)
Sambucus mexicana—Mexican Elder
Sassafras albidum—Sassafras
Schefflera arboricola—Dwarf Schefflera
Sophora secundiflora—Texas-Mountain-Laurel, Mescalbean
Stewartia koreana—Korean Stewartia
Stewartia monadelpha—Tall Stewartia
Stewartia pseudocamellia—Japanese Stewartia
Strelitzia nicolai—White Bird-of-Paradise, Giant Bird-of-Paradise
Styrax japonicus—Japanese Snowbell
Styrax obassia—Fragrant Snowbell
Syringa pekinensis—Peking Lilac
Syringa reticulata—Japanese Tree Lilac

Taxus baccata—English Yew (*)
Tecoma stans—Yellow-Elder, Yellow Trumpet-Flower (*)
Tristania conferta—Brisbane Box
Tsuga canadensis 'Sargentii'—Weeping Canadian Hemlock (*)
Tsuga canadensis 'Lewisii'—Lewis Hemlock (*)

Viburnum odoratissimum—Sweet Viburnum (*)
Viburnum odoratissimum var. *awabuki*—Awabuki Sweet Viburnum (*)
Viburnum plicatum f. 'Tomentosa'—Doublefile Viburnum (*)
Viburnum rufidulum—Rusty Blackhaw, Southern Blackhaw (*)
Viburnum sieboldii—Siebold Viburnum
Vitex agnus-castus—Chastetree, Vitex (*)
Vitex agnus-castus 'Alba'—'Alba' Chastetree, 'Alba' Vitex (*)

Xanthoceras sorbifolium—Yellowhorn (*)

Yucca elephantipes—Spineless Yucca, Soft-Tip Yucca (*)

Appendix 18

Trees with Outstanding Bark or Trunk Structure

Legend:
* = One of the showiest trunks
** = The best of the best

Acer buergeranum—Trident Maple
Acer cissifolium—Ivy-Leaf Maple
Acer griseum—Paperbark Maple (**)
Acer miyabei—Miyabe Maple (*)
Acer palmatum—Japanese Maple
Acer tataricum—Tatarian Maple
Acer triflorum—Three-Flower Maple (*)
Acer truncatum—Purpleblow Maple
Acoelorrhaphe wrightii—Paurotis Palm
Amelanchier spp.—Serviceberry
Arbutus menzesii—Pacific Madrone (**)
Arbutus texana—Texas Madrone

Beaucarnea recurvata—Ponytail
Betula spp—Birch (**)
Bursera simaruba—Gumbo-Limbo (**)

Caesalpinia granadillo—Bridalveil Tree (*)
Callophyllum brasiliense—Santa-Maria
Camellia oleifera—Tea-Oil Camellia
Carpinus caroliniana—American Hornbeam, Blue-Beech, Ironwood
Carya ovata—Shagbark Hickory
Cercidiphyllum japonicum—Katsuratree
Chionanthus retusus—Chinese Fringetree
Chorisia speciosa—Floss-Silk Tree (*)
Chrysalidocarpus lutescens—Yellow Butterfly Palm, Bamboo Palm, Areca Palm
Coccoloba diversifolia—Pigeon-Plum (*)
Conocarpus erectus—Buttonwood
Conocarpus erectus var. *sericeus*—Silver Buttonwood
Cornus kousa—Kousa Dogwood, Chinese Dogwood, Japanese Dogwood
Cornus mas—Cornelian-Cherry
Corylus colorna—Turkish Filbert
Cryptomeria japonica—Japanese-Cedar (*)

Diospyros texana—Texas Persimmon
Diospyros virginiana—Common Persimmon

Erythrina crista-galli—Cockspur Coral Tree
Erythrina variegata var. *orientalis*—Coral Tree
Eucalyptus citriodora—Lemon-scented Gum
Eugenia spp.—Stopper, Eugenia

Fagus spp.—Beech (*)
Ficus spp.—Fig
Firmiana simplex—Chinese Parasoltree
Franklinia alatamaha—Franklin-Tree, Franklinia

Guaiacum sanctum—Lignumvitae (*)
Gymnocladus dioicus—Kentucky Coffeetree

Halesia spp.—Silverbell
Hovenia dulcis—Japanese Raisintree

Ilex rotunda—Round Leaf Holly

Juniperus spp.—Juniper

Kalopanix pictus—Castor-Aralia

Lagerstroemia spp.—Crape-Myrtle
Leptospermum laevigatum—Australian Tea Tree

Maackia amurensis—Amur Maackia (*)
Manilkara zapota—Sapodilla
Melaleuca quinquinervia—Punk Tree, Melaleuca
Metasequoia glyptostroboides—Dawn Redwood (**)
Murraya paniculata—Orange Jessamine

Olea europaea—European Olive
Ostrya virginiana—American Hophornbeam, Eastern Hophornbeam

Parkinsonia aculeata—Jerusalem-Thorn
Parrotia persica—Persian Parrotia (*)
Persea borbonia—Redbay
Phellodendron amurense—Amur Corktree, Chinese Corktree
Phoenix spp.—Date Palms
Pinus bungeana—Lacebark Pine (*)
Pinus densiflora—Japanese Red Pine (*)
Pinus densiflora 'Umbraculifera'—'Umbraculifera' Japanese Red Pine (*)
Pinus palustris—Longleaf Pine
Pinus sylvestris 'Watereri'—'Watereri' Scotch Pine
Prunus maackii—Amur Chokecherry (**)
Prunus x yedoensis—Yoshino Cherry
Pseudocydonia sinensis—Chinese Quince (**)
Psidium littorale—Cattley Guava, Strawberry Guava

Quercus agriflia—Coast Live Oak
Quercus austrina—Bluff Oak
Quercus bicolor—Swamp White Oak
Quercus macrocarpa—Bur Oak
Quercus michauxii—Swamp Chestnut Oak
Quercus suber—Cork Oak
Quercus stellata—Post Oak
Quercus virginiana—Live Oak

Ravenala madagascariensis—Travelers-Tree
Roystonea spp.—Royal Palm

Sambucus mexicana—Mexican Elder
Sequoia sempervirens—Redwood (*)
Sequoiadendron giganteum—Giant Sequoia (*)
Stewartia koreana—Korean Stewartia (**)
Stewartia monadelpha—Tall Stewartia (**)

Stewartia pseudocamellia—Japanese Stewartia (**)
Syringa pekinensis—Peking Lilac
Syringa reticulata—Japanese Tree Lilac

Tabebuia spp.—Trumpet Tree
Tamarindus indica—Tamarind
Taxodium distichum—Baldcypress
Taxus baccata—English Yew (*)
Thuja occidentalis—Northern White-Cedar
Tristania conferta—Brisbane Box (**)

Ulmus parvifolia—Chinese Elm

Veitchia merrillii—Christmas Palm

Yucca elephantipes—Spineless Yucca, Soft-Tip Yucca

Zelkova serrata—Japanese Zelkova

Appendix 19

Trees with Showy Flowers

Legend:
* = Exceptionally nice flower display in most years
** = In a class by themselves
*** = One of the showiest trees on the planet
F = Also have good fall foliage color

Acacia auriculiformis—Earleaf Acacia
Acacia farnesiana—Sweet Acacia (*)
Acacia wrightii—Wright Acacia (*)
Acer buergeranum—Trident Maple (F)
Acer ginnala—Amur Maple (F)
Acer japonicum—Fullmoon Maple (F)
Acer platanoides—Norway Maple (F)
Acer rubrum—Red Maple, Swamp Maple (F)
Acer tataricum—Tatarian Maple (F)
Acer truncatum—Purpleblow Maple (F)
Acer x freemanii 'Armstrong'—Armstrong Freeman Maple (F)
Acoelorrhaphe wrightii—Paurotis Palm
Aesculus flava—Yellow Buckeye (**, F)
Aesculus glabra—Ohio Buckeye (**, F)
Aesculus hippocastanum—Horsechestnut (**)
Aesculus indica—Indian Horsechestnut (**)
Aesculus pavia—Red Buckeye (**)
Aesculus x carnea—Red Horsechestnut (**)
Ailanthus altissima—Tree-of-Heaven
Albizia julibrissin—Mimosa, Silktree
Amelanchier spp.—Serviceberry (*, F)
Aralia spinosa—Devils-Walkingstick (F)
Arbutus menzesii—Pacific Madrone
Arbutus texana—Texas Madrone
Arbutus unedo—Strawberry-Tree
Asimina triloba—Pawpaw (F)

Bauhinia spp.—Orchid-Tree (*)
Beaucarnea recurvata—Ponytail (*)
Brahea armata—Mexican Blue Palm (**)
Bulnesia arborea—Bulnesia (**)
Butia capitata—Pindo Palm

Caesalpinia granadillo—Bridalveil Tree
Caesalpinia pulcherrima—Dwarf Poinciana (*)
Calliandra spp.—Powderpuff
Callistemon spp.—Bottlebrush
Calodendron capense—Cape Chestnut (**)
Camellia oleifera—Tea-Oil Camellia
Cassia spp.—Cassia (**)
Castanea mollissima—Chinese Chestnut (F)
Catalpa spp.—Catalpa (*, F)
Cercis spp.—Redbud (*, F)
Chilopsis linearis—Desert-Willow
Chionanthus retusus—Chinese Fringetree (**, F)
Chionanthus virginicus—Fringetree (**, F)
Chitalpa tashkentensis—Chitalpa (*)
Chorisia speciosa—Floss-Silk Tree (**)
Citrus spp.—Citrus
Cladrastis kentukea—American Yellowwood (**, F)
Cordia boissieri—Wild-Olive (*)
Cordia sebestena—Geiger-Tree (**)
Cornus alternifolia—Pagoda Dogwood (F)
Cornus coreana—Dogwood (F)
Cornus drummondii—Roughleaf Dogwood (F)
Cornus florida—Flowering Dogwood (F)
Cornus kousa—Kousa Dogwood, Chinese Dogwood, Japanese Dogwood (*, F)
Cornus mas—Cornelian-Cherry (F)
Cornus walteri—Walter Dogwood (F)
Cotinus coggygria—Smoketree (F)
Cotinus obovatus—American Smoketree (F)
Crataegus spp.—Hawthorn

Delonix regia—Royal Poinciana (***)

Eriobotrya deflexa—Bronze Loquat
Eriobotrya japonica—Loquat
Erythrina crista-galli—Cockspur Coral Tree (*)
Erythrina variegata var. *orientalis*—Coral Tree
Evodia danielii—Korean Evodia (*)
Evodia hupehensis—Hupeh Evodia (*)

Firmiana simplex—Chinese Parasoltree (F)
Franklinia alatamaha—Franklin-Tree (**, F)
Fraxinus ornus—Flowering Ash (F)

Gordonia lasianthus—Loblolly-Bay (*)
Guaiacum augustifolium—Guayacan
Guaiacum officinale—Lignumvitae (*)
Guaiacum sanctum—Lignumvitae (*)

Halesia carolina—Carolina Silverbell (**)
Halesia monticola—Mountain Silverbell (**)
Hamamelis mollis—Chinese Witch-Hazel
Hibiscus syriacus—Rose-of-Sharon (*)
Hovenia dulcis—Japanese Raisintree
Hydrangea paniculata—Panicle Hydrangea (*)

Jacaranda mimosifolia—Jacaranda (**)
Jatropha integerrima—Fire-Cracker (*)

Kalopanax pictus—Castor-Aralia
Koelreuteria spp.—Goldenraintree (**, F)

Laburnum alpinum—Scotch Laburnum, Goldenchain Tree (***)
Laburnum x watereri—Goldenchain Tree (***)
Lagerstroemia fauriei—Japanese Crape-Myrtle (**, F)
Lagerstroemia indica—Crape-Myrtle (*, F)
Lagerstroemia speciosa—Queens Crape-Myrtle (*)
Leucaena retusa—Goldenball Leadtree
Ligustrum japonicum—Japanese Privet
Ligustrum lucidum—Tree Ligustrum (*)
Liriodendron tulipifera—Tuliptree (F)
Livistona chinensis—Chinese Fan Palm

Magnolia acuminata—Cucumbertree (F)
Magnolia denudata—Yulan Magnolia (*, F)
Magnolia 'Elizabeth'—Elizabeth Magnolia (*)
Magnolia grandiflora—Southern Magnolia (*)
Magnolia stellata var. *kobus*—Kobus Magnolia (F)
Magnolia 'Little Girl Hybrids'—Hybrid Magnolia (*)
Magnolia macrophylla—Bigleaf Magnolia (F)
Magnolia stellata—Star Magnolia (*, F)
Magnolia virginiana—Sweetbay Magnolia
Magnolia 'Wada's Memory'—Wada's Magnolia
Magnolia x soulangiana—Saucer Magnolia (**)
Malus spp.—Crabapple (**, F)
Mangifera indica—Mango
Melaleuca quinquinervia—Punk Tree, Melaleuca
Murraya paniculata—Orange Jessamine
Musa spp.—Banana

Nerium oleander—Oleander (*)

Oxydendrum arboreum—Sourwood (F)

Pandanus spp.—Screw-Pine
Parkinsonia aculeata—Jerusalem-Thorn (*)
Paulownia tomentosa—Princess-Tree (**)
Peltophorum dubium—Copperpod (**)
Peltophorum pterocarpum—Yellow Poinciana (**)
Photinia spp.—Red-Leaf Photinia, Red-Top (*)
Pinckneya pubens—Pinckneya, Fevertree (**)
Pistacia chinensis—Chinese Pistache (F)
Pithecellobium flexicaule—Texas-Ebony
Pithecellobium pallens—Tenaxa, Ape's Earring
Pittosporum spp.—Pittosporum
Plumeria alba—White Frangipani (*)
Plumeria obtusa—Singapore Plumeria (*)
Plumeria rubra—Frangipani (*)
Podocarpus macrophyllus—Podocarpus
Pongamia pinnata—Pongam
Prosopis glandulosa—Mesquite
Prosopis pubescens—Tornillo
Prunus americana—American Plum (*, F)
Prunus angustifolia—Chickasaw Plum
Prunus caroliniana—Cherry-Laurel
Prunus cerasifera—Purple Leaf Plum
Prunus 'Hally Jolivette'—Hally Jolivette Cherry
Prunus maackii—Amur Chokecherry (F)
Prunus mexicana—Mexican Plum
Prunus mume—Japanese Apricot (*, F)
Prunus persica—Peach (F)
Prunus sargentii—Sargent Cherry (**, F)
Prunus serotina—Black Cherry (F)
Prunus serrulata 'Kwanzan'—Kwanzan Cherry (**, F)
Prunus subhirtella 'Autumnalis'—'Autumnalis' Higan Cherry (*)
Prunus subhirtella 'Pendula'—Weeping Higan Cherry
Prunus triloba var. *multiplex*—Flowering-Almond (*)
Prunus umbellata—Flatwoods Plum
Prunus x incamp 'Okame'—'Okame' Cherry
Prunus x yedoensis—Yoshino Cherry (*)

Pseudocydonia oblonga—Quince
Pseudocydonia sinensis—Chinese Quince
Pyrus calleryana cultivars—Callery Pear (*, F)

Rhus chinensis—Chinese Sumac (F)
Rhus glabra—Smooth Sumac (F)
Rhus lancea—African Sumac
Robinia pseudoacacia—Black Locust (*)
Robinia pseudoacacia 'Tortuosa'—'Tortuosa' Black Locust (**)
Roystonea spp.—Royal Palm

Sabal mexicana (texana)—Texas Palm
Sabal palmetto—Cabbage Palm
Sambucus canadensis—American Elder
Sambucus mexicana—Mexican Elder
Sapindus drummondii—Western Soapberry (F)
Sassafras albidum—Sassafras (F)
Schefflera actinophylla—Schefflera
Senna spectabilis—Cassia (**)
Sophora japonica—Scholar Tree, Japanese Pagoda Tree (*)
Sophora secundiflora—Texas-Mountain-Laurel (*)
Sorbus alnifolia—Korean Mountain-Ash (F)
Sorbus aucuparia—European Mountain-Ash (F)
Spathodea campanulata—African Tulip-Tree (**)
Stewartia koreana—Korean Stewartia (F)
Stewartia monadelpha—Tall Stewartia (F)
Stewartia pseudocamellia—Japanese Stewartia (F)
Strelitzia nicolai—White Bird-of-Paradise
Styrax japonicus—Japanese Snowbell (F)
Syagrus romanzoffianum—Queen Palm
Syringa pekinensis—Peking Lilac
Syringa reticulata—Japanese Tree Lilac (*, F)

Tabebuia spp.—Trumpet Tree (**)
Tecoma stans—Yellow-Elder (**)
Thrinax morrisii—Key Thatchpalm
Tipuana tipu—Tipuana (**)

Veitchia merrillii—Christmas Palm
Veitchia montgomeryana—Montgomery Palm
Viburnum odoratissimum—Sweet Viburnum
Viburnum odoratissimum var. *awabuki*—Awabuki Sweet Viburnum
Viburnum plicatum var. *tomentosum*—Doublefile Viburnum (*)
Viburnum rufidulum—Rusty Blackhaw (F)
Viburnum sieboldii—Siebold Viburnum
Vitex agnus-castus—Chastetree (*)
Vitex agnus-castus 'Alba'—'Alba' Chastetree (*)
Vitex negundo 'Heterophylla'—Cut-Leaf Chastetree

Washingtonia filifera—Desert Palm
Washingtonia robusta—Washington Palm
Wodyetia bifurcata—Foxtail Palm

Xanthoceras sorbifolium—Yellowhorn (*)

Yucca elephantipes—Spineless Yucca

Appendix 20

Trees with Good to Exceptional Fall Foliage Color

Legend:
* = Exceptional fall foliage color
** = In a class by themselves
S = Also have showy flowers

Acer barbatum—Florida Maple (*)
Acer buergeranum—Trident Maple (S)
Acer campestre—Hedge Maple
Acer cissifolium—Ivy-Leaf Maple
Acer ginnala—Amur Maple (*, S)
Acer glabrum—Rocky Mountain Maple
Acer grandidentatum—Bigtooth Maple (*)
Acer griseum—Paperbark Maple (*)
Acer japonicum—Fullmoon Maple (*, S)
Acer leucoderme—Chalk Maple, Whitebark Maple
Acer macrophyllum—Bigleaf Maple (*)
Acer miyabei—Miyabe Maple
Acer negundo—Boxelder, Ash-leafed Maple
Acer nigrum—Black Maple (*)
Acer palmatum—Japanese Maple (*)
Acer platanoides—Norway Maple (*, S)
Acer rubrum—Red Maple, Swamp Maple (**, S)
Acer saccharinum—Silver Maple
Acer saccharum—Sugar Maple (**)
Acer tataricum—Tatarian Maple (S)
Acer triflorum—Three-Flowered Maple (*)
Acer truncatum—Purpleblow Maple (S)
Acer x freemanii 'Armstrong'—Armstrong Freeman Maple (**, S)
Aesculus flava—Yellow Buckeye, Sweet Buckeye (*, S)
Aesculus glabra—Ohio Buckeye, Fetid Buckeye (*, S)
Aesculus hippocastanum—Horsechestnut, European Horsechestnut (S)
Amelanchier spp.—Downy Serviceberry, Juneberry (*, S)
Aralia spinosa—Devils-Walkingstick, Hercules-Club (S)
Asimina triloba—Pawpaw (*, S)

Betula ermanii—Erman Maple
Betula jacquemontii—Jacquemontii Birch
Betula papyrifera—Paper Birch, Canoe Birch
Betula pendula—European Birch
Betula platyphylla 'Whitespire'—Japanese Whitespire Birch
Betula populifolia—Gray Birch

Carpinus betulus 'Fastigiata'—'Fastigiata' European Hornbeam
Carpinus caroliniana—American Hornbeam (*)
Carya glabra—Pignut Hickory (*)
Carya illinoensis—Pecan
Carya ovata—Shagbark Hickory (*)
Carya tomentosa—Mockernut Hickory (*)
Castanea mollissima—Chinese Chestnut (S)
Catalpa spp.—Catalpa (S)
Celtis spp.—Hackberry
Cercidiphyllum japonicum—Katsuratree (**)
Cercis spp.—Redbud (S)
Chionanthus retusus—Chinese Fringetree (S)
Chionanthus virginicus—Fringetree (S)
Cladrastis kentukea—American Yellowwood (S)
Cornus alternifolia—Pagoda Dogwood (S)
Cornus coreana—Dogwood (S)
Cornus drummondii—Roughleaf Dogwood (S)
Cornus florida—Flowering Dogwood (*, S)
Cornus kousa—Kousa Dogwood (S)
Cornus mas—Cornelian-Cherry (S)
Cornus walteri—Walter Dogwood (S)
Cotinus coggygria—Smoketree, Wig-Tree, Smokebush (S)
Cotinus obovatus—American Smoketree (**, S)

Diospyros virginiana—Common Persimmon (*)

Fagus grandifolia—American Beech
Fagus sylvatica—European Beech
Firmiana simplex—Chinese Parasoltree
Franklinia alatamaha—Franklin-Tree (S)
Fraxinus americana—White Ash
Fraxinus americana 'Autumn Purple'—'Autumn Purple' White Ash (**)
Fraxinus excelsior—Common Ash, European Ash
Fraxinus ornus—Flowering Ash (S)
Fraxinus pennsylvanica—Green Ash (*)
Fraxinus texensis—Texas Ash
Fraxinus uhdei—Shamel Ash (*)
Fraxinus velutina—Velvet Ash, Modesto Ash

Ginkgo biloba—Maidenhair Tree, Ginkgo (*)
Gleditsia triacanthos var. *inermis*—Thornless Honeylocust (*)
Gymnocladus dioicus—Kentucky Coffeetree (*)

Koelreuteria spp.—Goldenraintree (S)

Lagerstroemia indica—Crape-Myrtle (*, S)
Larix spp.—Larch
Liquidambar spp.—Sweetgum
Liriodendron tulipifera—Tuliptree

Maclura pomifera—Osage-Orange, Bois-D'Arc (*)
Magnolia acuminata—Cucumbertree, Cucumber Magnolia (*, S)
Magnolia denudata—Yulan Magnolia (S)
Magnolia macrophylla—Bigleaf Magnolia (S)
Magnolia stellata—Star Magnolia (*, S)
Magnolia stellata var. *kobus*—Kobus Magnolia (*, S)
Malus spp.—Crabapple (S)
Metasequoia glyptostroboides—Dawn Redwood
Morus alba—White Mulberry

Nyssa ogeche—Ogeechee Tupelo
Nyssa sinensis—Chinese Tupelo
Nyssa sylvatica—Blackgum (**)

Ostrya virginiana—American Hophornbeam, Eastern Hophornbeam (*)
Oxydendrum arboreum—Sourwood, Sorrel-Tree (**, S)

Parrotia persica—Persian Parrotia (*)
Phellodendron amurense—Amur Corktree (**)
Pistacia chinensis—Chinese Pistache (**, S)
Platanus spp.—Sycamore
Populus deltoides—Eastern Cottonwood (*)
Populus nigra 'Italica'—Lombardy Poplar
Prunus americana—American Plum (S)
Prunus maackii—Amur Chokecherry, Manchurian Cherry (S)
Prunus mume—Japanese Apricot (S)
Prunus persica—Peach (S)
Prunus sargentii—Sargent Cherry (*, S)
Prunus serotina—Black Cherry (S)
Prunus serrulata 'Kwanzan'—Kwanzan Cherry (S)
Pseudolarix kaempferi—Golden Larch (*)
Pyrus calleryana cultivars—Callery Pear (*, S)
Pyrus calleryana 'Autumn Blaze'—'Autumn Blaze' Callery Pear (**, S)

Quercus acutissima—Sawtooth Oak (*)
Quercus alba—White Oak
Quercus austrina—Bluff Oak
Quercus bicolor—Swamp White Oak
Quercus coccinea—Scarlet Oak
Quercus imbricaria—Shingle Oak
Quercus laevis—Turkey Oak
Quercus lyrata—Overcup Oak
Quercus macrocarpa—Bur Oak
Quercus michauxii—Swamp White Oak
Quercus muehlenbergii—Chinkapin Oak
Quercus palustris—Pin Oak (*)
Quercus rubra—Northern Red Oak
Quercus shumardii—Shumard Oak (*)
Quercus texana—Texas Red Oak

Rhamnus caroliniana—Carolina Buckthorn
Rhus chinensis—Chinese Sumac (**, S)
Rhus glabra—Smooth Sumac (**, S)
Robinia pseudoacacia—Black Locust (S)

Sapindus drummondii—Western Soapberry (S)
Sapium sebiferum—Chinese Tallowtree (**, S)
Sassafras albidum—Sassafras (**, S)
Sorbus alnifolia—Korean Mountain-Ash (*, S)
Sorbus aucuparia—European Mountain-Ash
Stewartia koreana—Korean Stewartia (S)
Stewartia monadelpha—Tall Stewartia (S)
Stewartia pseudocamellia—Japanese Stewartia (S)
Styrax japonicus—Japanese Snowbell (S)
Syringa reticulata—Japanese Tree Lilac

Taxodium ascendens—Pondcypress
Taxodium distichum—Baldcypress
Tilia americana—American Linden
Tilia cordata—Littleleaf Linden

Ulmus alata—Winged Elm
Ulmus americana—American Elm
Ulmus crassifolia—Cedar Elm (*)
Ulmus hybrids—Hybrid Elm

Viburnum rufidulum—Rusty Blackhaw, Southern Blackhaw (*, S)

Zelkova serrata—Japanese Zelkova, Saw-Leaf Zelkova (*)

Appendix 21

Trees Native to North America

Abies amabilis—Pacific Silver Fir
Abies balsamea—Balsam Fir
Abies concolor—White Fir, Colorado Fir
Abies fraseri—Fraser Fir
Abies lasiocarpa—Alpine Fir
Abies procera—Noble Fir
Acacia farnesiana—Sweet Acacia, Huisache
Acacia wrightii—Wright Acacia, Wright Catclaw
Acer barbatum—Florida Maple, Southern Sugar Maple
Acer glabrum—Rocky Mountain Maple
Acer grandidentatum—Bigtooth Maple, Rocky Mountain Sugar Maple, Canyon Maple
Acer leucoderme—Chalk Maple, Whitebark Maple
Acer macrophyllum—Bigleaf Maple
Acer negundo—Boxelder, Ash-leafed Maple
Acer nigrum—Black Maple
Acer rubrum—Red Maple, Swamp Maple
Acer saccharinum—Silver Maple
Acer saccharum—Sugar Maple
Acer x freemanii—Freeman Maple
Acoelorrhaphe wrightii—Paurotis Palm
Aesculus flava—Yellow Buckeye, Sweet Buckeye
Aesculus glabra—Ohio Buckeye, Fetid Buckeye
Aesculus pavia—Red Buckeye
Alnus rhombifolia—White Alder
Amelanchier arborea—Downy Serviceberry, Juneberry
Amelanchier canadensis—Shadblow Serviceberry, Downy Serviceberry
Amelanchier laevis—Allegheny Serviceberry
Amelanchier x grandiflora—Apple Serviceberry
Aralia spinosa—Devils-Walkingstick, Hercules-Club
Arbutus menzesii—Pacific Madrone
Arbutus texana—Texas Madrone
Asimina triloba—Pawpaw

Betula lenta—Yellow Birch
Betula nigra—River Birch
Betula papyrifera—Paper Birch, Canoe Birch
Betula populifolia—Gray Birch
Brahea armata—Mexican Blue Palm
Brahea brandegeei—San Jose Hesper Palm
Brahea edulis—Guadaloupe Palm
Brahea elegans—Franceschi Palm
Bumelia lanuginosa—Chittamwood, Gum Bumelia, Gum Elastic Buckthorn
Bursera simaruba—Gumbo-Limbo

Calocedrus decurrens—California Incense-Cedar
Carpinus caroliniana—American Hornbeam, Blue-Beech, Ironwood
Carya aquatica—Water Hickory
Carya glabra—Pignut Hickory
Carya illinoensis—Pecan
Carya ovata—Shagbark Hickory
Carya tomentosa—Mockernut Hickory
Catalpa bignonioides—Southern Catalpa
Catalpa speciosa—Northern Catalpa
Celtis laevigata—Sugarberry, Sugar Hackberry
Celtis occidentalis—Common Hackberry
Celtis reticulata—Western Hackberry
Cercidium floridum—Blue Paloverde
Cercidium torreyanum—Paloverde
Cercis canadensis—Eastern Redbud
Cercis canadensis var. *texensis*—Texas Redbud
Cercis mexicana—Mexican Redbud
Cercis occidentalis—Western Redbud, California Redbud
Chamaecyparis lawsoniana—Lawson Falsecypress, Port-Orford-Cedar
Chamaecyparis nootkatensis 'Pendula'—Nootka Falsecypress, Alaska-Cedar
Chilopsis linearis—Desert-Willow
Chionanthus virginicus—Fringetree, Old-Mans-Beard
Chrysophyllum oliviforme—Satinleaf
Cladrastis kentukea—American Yellowwood, Virgilia
Clusia rosea—Pitch-Apple, Florida Clusia
Coccoloba diversifolia—Pigeon-Plum
Coccoloba uvifera—Seagrape
Coccothrinax argentata—Silverpalm, Thatchpalm
Conocarpus erectus—Buttonwood
Conocarpus erectus var. *sericeus*—Silver Buttonwood
Cordia boissieri—Wild-Olive, Anacahuita
Cordia sebestena—Geiger-Tree
Cornus alternifolia—Pagoda Dogwood
Cornus florida—Flowering Dogwood
Cornus nuttallii—Western Dogwood
Cotinus obovatus—American Smoketree, Chittamwood
Crataegus marshalli—Marshalls Hawthorn
Crataegus phaenopyrum—Washington Hawthorn
Crataegus viridis 'Winter King'—'Winter King' Green Hawthorn
Cupressus arizonica—Arizona Cypress
Cupressus glabra 'Carolina Safire'—Smooth-Barked Arizona Cypress

Diospyros texana—Texas Persimmon
Diospyros virginiana—Common Persimmon

Eugenia confusa—Red Stopper, Ironwood
Eugenia foetida—Spanish Stopper
Eugenia spp.—Stopper, Eugenia

Fagus grandifolia—American Beech
Ficus aurea—Strangler Fig, Golden Fig
Franklinia alatamaha—Franklin-Tree, Franklinia
Fraxinus americana—White Ash
Fraxinus cuspidata—Fragrant Ash
Fraxinus pennsylvanica—Green Ash
Fraxinus quadrangulata—Blue Ash
Fraxinus texensis—Texas Ash
Fraxinus velutina—Velvet Ash, Modesto Ash, Arizona Ash

Gleditsia triacanthos var. *inermis*—Thornless Honeylocust
Gordonia lasianthus—Loblolly-Bay, Sweet-Bay

Guaiacum augustifolium—Guayacan
Guaiacum officinale—Lignumvitae
Guaiacum sanctum—Lignumvitae
Gymnocladus dioicus—Kentucky Coffeetree

Halesia carolina—Carolina Silverbell
Halesia monticola—Mountain Silverbell
Hamamelis virginiana—Witch-Hazel

Ilex cassine—Dahoon Holly
Ilex decidua—Possumhaw
Ilex opaca—American Holly
Ilex verticillata—Winterberry
Ilex vomitoria—Yaupon Holly
Ilex vomitoria 'Pendula'—Weeping Yaupon Holly
Ilex x attenuata 'East Palatka'—'East Palatka' Holly
Ilex x attenuata 'Savannah'—Savannah Holly

Juglans californica—Southern California Black Walnut
Juglans cinerea—Butternut, White Walnut
Juglans nigra—Black Walnut
Juniperus ashei—Ashe Juniper, Mountain-Cedar
Juniperus deppeana 'Mcfetter'—McFetter Alligator Juniper
Juniperus scopulorum—Rocky Mountain Juniper, Colorado Redcedar
Juniperus silicicola—Southern Redcedar
Juniperus virginiana—Eastern Redcedar

Larix larcina—Tamarack, Eastern Larch
Leucaena retusa—Goldenball Leadtree, Littleleaf Leucaena
Liquidambar styraciflua—Sweetgum
Liriodendron tulipifera—Tuliptree, Tulip-Poplar, Yellow-Poplar
Lithocarpus densiflora—Tanbark Oak
Lysiloma bahamensis—Cuban Tamarind, Bahama Lysiloma
Lysiloma latisiliqua—Wild-Tamarind

Maclura pomifera—Osage-Orange, Bois-D'Arc
Magnolia acuminata—Cucumbertree, Cucumber Magnolia
Magnolia grandiflora—Southern Magnolia
Magnolia macrophylla—Bigleaf Magnolia
Magnolia virginiana—Sweetbay Magnolia, Swamp Magnolia
Myrica californica—California Waxmyrtle
Myrica cerifera—Southern Waxmyrtle, Southern Bayberry

Nyssa aquatica—Water Tupelo
Nyssa ogeche—Ogeechee Tupelo, Ogeechee-Lime
Nyssa sylvatica—Blackgum, Sourgum, Black Tupelo

Ostrya virginiana—American Hophornbeam, Eastern Hophornbeam
Oxydendrum arboreum—Sourwood, Sorrel-Tree

Parkinsonia texana—Paloverde
Persea borbonia—Redbay
Picea glauca—White Spruce
Picea pungens—Colorado Spruce, Colorado Blue Spruce
Picea sitchensis—Sitka Spruce
Pinckneya pubens—Pinckneya, Fevertree
Pinus cembroides—Mexican Pinyon, Pinyon Pine
Pinus clausa—Sand Pine
Pinus coulteri—Coulters Pine
Pinus echinata—Shortleaf Pine
Pinus edulis—Pinyon, Pinyon Pine
Pinus elliottii—Slash Pine
Pinus flexilis—Limber Pine
Pinus glabra—Spruce Pine
Pinus monophylla—Single-leaf Pinyon Pine
Pinus monticola—Western White Pine
Pinus palustris—Longleaf Pine
Pinus ponderosa—Ponderosa Pine
Pinus radiata—Monterey Pine
Pinus resinosa—Red Pine
Pinus strobus—Eastern White Pine
Pinus taeda—Loblolly Pine
Pinus virginiana—Virginia Pine, Scrub Pine
Pistacia texana—Texas Pistachio
Pithecellobium flexicaule—Ebony Blackbead, Texas-Ebony
Pithecellobium pallens—Tenaxa, Ape's Earring
Platanus occidentalis—Sycamore, American Planetree
Platanus racemosa—Western Sycamore
Populus deltoides—Eastern Cottonwood
Populus fremontii—Fremont Poplar
Prosopis glandulosa—Mesquite, Honey Mesquite
Prosopis pubescens—Tornillo
Prunus americana—American Plum
Prunus angustifolia—Chickasaw Plum
Prunus caroliniana—Cherry-Laurel, Carolina Laurelcherry
Prunus mexicana—Mexican Plum
Prunus serotina—Black Cherry
Prunus umbellata—Flatwoods Plum
Pseudotsuga menziesii—Douglas Fir

Quercus agrifolia—Coastal Live Oak
Quercus alba—White Oak
Quercus austrina—Bluff Oak
Quercus bicolor—Swamp White Oak
Quercus coccinea—Scarlet Oak
Quercus falcata—Southern Red Oak, Spanish Oak
Quercus fusiformis—Texas Live Oak
Quercus garryana—Oregon White Oak, Garry Oak
Quercus geminata—Sand Live Oak
Quercus kelloggii—California Black Oak
Quercus hemisphaerica—Laurel Oak
Quercus imbricaria—Shingle Oak, Northern Laurel Oak
Quercus laurifolia—Laurel Oak
Quercus laevis—Turkey Oak
Quercus lobata—Valley Oak, California White Oak
Quercus lyrata—Overcup Oak
Quercus macrocarpa—Bur Oak
Quercus michauxii—Swamp White Oak
Quercus muehlenbergii—Chinkapin Oak, Chestnut Oak
Quercus nigra—Water Oak
Quercus nuttallii—Nuttall Oak
Quercus pagodifolia—Cherrybark Oak
Quercus palustris—Pin Oak
Quercus phellos—Willow Oak
Quercus prinus—Chestnut Oak, Rock Oak, Basket Oak
Quercus rubra—Northern Red Oak
Quercus shumardii—Shumard Oak
Quercus texana—Texas Red Oak, Texas Oak
Quercus virginiana—Southern Live Oak, Live Oak

Rhamnus caroliniana—Carolina Buckthorn
Rhus glabra—Smooth Sumac
Rhus typhina—Staghorn Sumac
Robinia pseudoacacia—Black Locust, Common Locust
Roystonea elata—Royal Palm

Sabal mexicana (texana)—Texas Palm, Mexican Palm
Sabal palmetto—Cabbage Palm
Sambucus canadensis—American Elder, Common Elder
Sambucus mexicana—Mexican Elder

Sapindus drummondii—Western Soapberry
Sapindus saponaria—Florida Soapberry, Wingleaf Soapberry
Sassafras albidum—Sassafras
Sequoia sempervirens—Coast Redwood
Sequoiadendron giganteum—Giant Sequoia
Simarouba glauca—Paradise-Tree
Sophora affinis—Eves-Necklace, Texas Sophora
Sophora secundiflora—Texas-Mountain-Laurel, Mescalbean
Swietenia mahagoni—Mahogany, West Indies Mahogany

Taxodium ascendens—Pondcypress
Taxodium distichum—Baldcypress
Taxodium mucronatum—Montezuma Baldcypress, Mexican-Cypress
Thrinax morrisii—Key Thatchpalm
Thrinax radiata—Florida Thatchpalm
Thuja occidentalis—White-Cedar, Arborvitae, Northern White-Cedar
Thuja plicata—Giant Arborvitae, Western Redcedar
Tilia americana—American Linden, American Basswood
Tsuga canadensis—Canadian Hemlock, Eastern Hemlock
Tsuga heterophylla—Western Hemlock
Tsuga mertensiana—Mountain Hemlock

Ulmus alata—Winged Elm
Ulmus americana—American Elm
Ulmus crassifolia—Cedar Elm

Viburnum rufidulum—Rusty Blackhaw, Southern Blackhaw

Washingtonia filifera—Desert Palm, California Washingtonia Palm
Washingtonia robusta—Washington Palm, Mexican Washington Palm

Yucca elephantipes—Spineless Yucca, Soft-Tip Yucca

Appendix 22

State Trees of the United States

Alabama—*Pinus palustris*, Southern Pine
Alaska—*Picea sitchensis*, Sitka Spruce
Arizona—*Cercidium torreyanum*, Paloverde
Arkansas—*Pinus echinata*, Shortleaf Pine
California—*Sequioa sempervirens* and *Sequioadendron giganteum*, Redwood
Colorado—*Picea pungens*, Colorado Blue Spruce
Connecticut—*Quercus alba*, White Oak
Delaware—*Ilex opaca*, American Holly
D.C.—*Quercus coccinia*, Scarlet Oak
Florida—*Sabal palmetto*, Cabbage Palm
Georgia—*Quercus virginiana*, Live Oak
Hawaii—*Aleurites moluccane*, Kukui
Idaho—*Pinus monticola*, Western White Pine
Illinois—*Quercus alba*, White Oak
Indiana—*Liriodendron tulipifera*, Yellow-Poplar
Iowa—*Quercus spp.*, Oak
Kansas—*Populus deltoides*, Cottonwood
Kentucky—*Gymnocladus dioicus*, Kentucky Coffeetree
Loiusiana—*Taxodium distichum*, Baldcypress
Maine—*Pinus strobus*, White Pine
Maryland—*Quercus alba*, White Oak
Massachusetts—*Ulmus americana*, American Elm
Michigan—*Pinus strobus*, White Pine
Minnesota—*Pinus resinosa*, Red Pine
Mississippi—*Magnolia grandiflora*, Southern Magnolia
Missouri—*Cornus florida*, Flowering Dogwood
Montana—*Pinus ponderosa*, Ponderosa Pine
Nebraska—*Populus deltoides*, Cottonwood
Nevada—*Pinus monophylla*, Single-leaf Pinyon Pine
New Hampshire—*Betula papyrifera*, Paper Birch
New Jersey—*Quercus rubra*, Northern Red Oak
New Mexico—*Pinus edulis*, Pinyon
New York—*Acer saccharum*, Sugar Maple
North Carolina—*Pinus echinata*, Shortleaf Pine
North Dakota—*Ulmus americana*, American Elm
Ohio—*Aesculus glabra*, Ohio Buckeye
Oklahoma—*Cercis canadensis*, Eastern Redbud
Oregon—*Pseudotsuga menziesii*, Douglas Fir
Pennsylvania—*Tsuga canadensis*, Eastern Hemlock
Rhode Island—*Acer rubrum*, Red Maple
South Carolina—*Sabal palmetto*, Cabbage Palm
South Dakota—*Picea glauca* var. *densata*, White Spruce
Tennessee—*Liriodendron tulipifera*, Yellow-Poplar
Texas—*Carya illinoensis*, Pecan
Utah—*Picea pungens*, Colorado Blue Spruce
Vermont—*Acer saccharum*, Sugar Maple
Virginia—*Cornus florida*, Flowering Dogwood
Washington—*Tsuga heterophylla*, Western Hemlock
West Virginia—*Acer saccharum*, Sugar Maple
Wisconsin—*Acer saccharum*, Sugar Maple
Wyoming—*Populus deltoides*, Cottonwood

Appendix 23

Approximate Order of Flowering from Spring through Summer

Imagine a landscape with many of these trees planted to create nearly six months of continuous color.

Cornus mas—Cornelian-Cherry
Bauhinia spp.—Orchid Tree
Prunus spp.—Cherry
Magnolia stellata—Star Magnolia
Magnolia x soulangiana—Japanese Magnolia
Amelanchier spp.—Serviceberry
Malus spp.—Crabapple
Magnolia spp.—Magnolia
Cercis spp.—Redbud
Pyrus spp.—Callery Pear
Caragana spp.—Peashrub
Aesculus spp.—Horsechestnut
Crateagus—Hawthorn
Cornus florida—Flowering Dogwood
Cornus alternifolia—Pagoda Dogwood
Sorbus spp.—Mountain-Ash
Viburnum spp.—Viburnum
Syringa vulgaris—Lilac
Chionanthus spp.—Fringetree
Cornus kousa—Kousa Dogwood
Fraxinus ornus—Flowering Ash
Calodendron capense—Cape Chestnut
Crateagus lavellei—Hawthorn
Tipuana tipu—Tipuana
Robinia pseudoacacia—Black Locust
Lirodendron tulipifera—Tuliptree (flowers along with azaleas)
Cladrastis kentukea—Yellowwood
Castanea spp.—Chinese Chestnut
Syringa reticulata—Tree Lilac
Syringa pekinensis—Peking Lilac
Ligustrum spp.—Privet
Lagerstroemia spp.—Crape Myrtle
Cotinus spp.—Smoketree
Peltophorum spp.—Yellow Poinciana, Copperpod
Koelreuteria paniculata—Goldenraintree
Evodia—Korean Evodia
Sophora japonica—Japanese Pagodatree
Koelreuteria elegans—Goldenraintree

Bibliography

Abbott, J. D., and R. E. Gough. 1987. Seasonal development of highbush blueberry roots under sawdust mulch. *J. Amer. Soc'y Horticultural Sci.* 112(1):60–62.

Adams, W. D. 1976. *Trees for Southern Landscapes.* Houston, TX: Gulf Publishing. 86pp.

Adriance, G. W., and H. E. Hampton. 1949. Root distribution in citrus, as influenced by environment. *J. Amer. Soc'y Horticultural Sci.* 53:103–8.

Arnold, H. F. 1993. *Trees in urban design.* 2d ed. New York: Van Nostrand Reinhold. 197pp.

Arnold, M. A., and D. K. Struve. 1989. Green ash establishment following transplant. *J. Amer. Soc'y Horticultural Sci.* 114(4):591–95.

Atkinson, D. 1973. Seasonal changes in the length of white unsuberized roots on raspberry plants under irrigated conditions. *J. Horticultural Sci.* 48:413–19.

Atkinson, D., and G. C. White. 1976. Soil management with herbicides: The response of soils and plants. *Proceedings British Crop Protection Conf.—Weeds* 3:873–84.

Bailey, L. H., and E. Z. Bailey. 1976. *Hortus Third.* New York: MacMillan. 1290pp.

Ball, J. 1992. Response of the bronze birch borer to pruning wounds on paper birch. *J. Arboriculture* 18:294–97.

Bassuk, N. L. 1988. *Recommended urban trees.* Ithaca, NY: Cornell University Press. 11pp.

Bolgaino, C. July/August 1993. The aliens. *American Forests* 17–21, 53–54.

Brickell, C. D., et al., eds. 1980. *International Code of Nomenclature for Cultivated Plants—1980.* The Hague: Dr. W. Junk, Publishers. 32pp.

Brockman, C. F. 1968. *Trees of North America.* New York: Golden Press. 280pp.

Broschot, T. K., and A. W. Meerow. 1991. *Betrock's Reference Guide to Florida Landscape Plants.* Cooper City, FL: Betrock Informations Systems. 428pp.

Brown, C. L., and L. K. Kirkman. 1990. *Trees of Georgia and Adjacent States.* Portland, OR: Timber Press. 292pp.

Bunger, M. T., and H. J. Thomson. 1938. Root development as a factor in the success of windbreak trees in the southern high plains. *J. Forestry* 36:790–803.

Burn, B. 1984. *North American Trees.* New York: Bonanza Books. 96pp.

Bush, C. S., and J. F. Morton. *Native Trees and Plants for Florida Landscaping.* Bulletin No. 193. Gainesville, FL: Florida Department of Agriculture. 144pp.

Bushey, D. J. 1937. Root extension of shade trees. *Proceedings International Shade Tree Conf.* 13:22–30.

Byrnes, R. L. 1976. Effects of soil amendments in variable ratios and irrigation levels on soil conditions and the establishment and growth of *Pittosporum tobira.* M.S. Thesis, University of Florida, Gainesville, FL.

Calloway, D. J. 1994. *The world of magnolias.* Portland, OR: Timber Press. 260pp.

Chong, C., G. P. Lumis, R. A. Cline, and H. J. Reissmann. 1987. Growth and chemical composition of *Populus deltoides* x *nigra* grown in Field-Grow fabric containers. *J. Envtl. Horticulture* 5(2):45–48.

Cockrell, R. A. 1976. *Trees of the Berkeley Campus.* Berkeley, CA: Division of Agricultural Sciences, University of California. 97pp.

Coffelt, M. A., and P. B. Schultz. 1993. Host suitability of the orangestriped oakworm (*Lepidoptera:Saturniidae*), *J. Envtl. Horticulture* 11:182–86.

Coker, E. G. 1958. Root studies. XII. Root systems of apple on Malling rootstocks on five soil series. *J. Horticultural Sci.* 33:71–79.

Conover, C. A., and E. W. McElwee. 1966. *Selected Trees for Florida Homes.* Bulletin 182. Gainesville, FL: Agricultural Extension Service, IFAS. 77pp.

Coombes, A. J. 1992. *Trees.* New York: Dorling Kindersley. 320pp.

Costello, L. R., and K. S. Jones. 1992. *WUCOLS Project: Water use classification of landscape species.* University of California, Cooperative Extension Service, San Mateo and San Francisco Counties, California.

Couenberg, E. 1994. Amsterdam tree soil. In *The landscape below ground,* G. W. Watson and D. Neely, eds. Proceedings of the International Workshop on Tree Root Development in Urban Soils. Savoy, IL: International Society of Arboriculture.

Courtright, G. 1979. *Trees and Shrubs for Temperate Climates.* Beaverton, OR: Timber Press. 239pp.

Coutts, M. P. 1986. Components of tree stability. *Forestry* 59(2):171–97.

Davis, R. G. 1994. Ornamental pears. *Amer. Nursery* 179:30–45.

Davis, W. B. 1981. Landscape trees for the great central valley of California. Leaflet 2580. Berkeley, CA: Division of Agriculture Services, University of California. 9pp.

Day, R. J., and J. R. Cary. 1974. Differences in post-planting soil-moisture relations of container-grown tube and plug stock affect the field survival and growth of black spruce. In *Proceedings of the North American Containerized Forest Tree Seedling Symposium.* Great Plains Agricultural Council Publication. 68pp.

DeGraaf, R. M., and G. M. Witman. 1979. *Trees, shrubs and vines for attracting birds.* Amherst, MA: University of Massachusetts Press. 194pp.

Delahaut, K. A., and E. R. Hasselkus. 1994. 1993 apple scab rating of the crabapple collection at Longnecker Gardens. *Amer. Nursery* 179:89–90.

Dirr, M. 1984. *All about evergreens.* San Francisco: Ortho Books. 96pp.

Dirr, M. A. 1990. *Manual of woody landscape plants.* 4th ed. Champaign, IL: Stipes Publishing. 1007pp.

Dobson, M. C. 1991. De-icing salt damage to trees and shrubs. Forestry Bulletin 101. Farnham, Surrey, UK: Forestry Commission. 64pp.

Doichev, K. 1977. Root distribution of M.7 clonal apple rootstocks grafted with Golden Delicious as affected by different methods and rates of irrigation. *Grad. Lozar. Nauka* 14:19–24.

Dorris, L. June 1993. *Chitalpa tashkentensis,* Chitalpa. *Arbor Age* 30–31.

Doxon, E. L., and R. Kirksey. 1993. Trees for the southern great plains. *HortTechnology* 3(4):440–41.

Edlin, H. L., and M. Nimmo. 1974. *The world of trees.* New York: Crown Publishers. 128pp.

Eis, S. 1974. Root system morphology of western hemlock, western red cedar, and Douglas fir. *Canadian J. Forest Research* 4:28–38.

Fare, D. C., C. H. Gilliam, and H. G. Ponder. 1985. Root distribution of two field-grown *Ilex. HortScience* 10:1129–30.

Ferguson, B., ed. 1982. *All about trees.* San Francisco: Ortho Books. 112pp.

Florida Department of Education. 1977. *Dig manual—A guide to the identification and selection of Florida ornamental plants.* Tallahassee, FL: Author. 224pp.

Ford, H. W., W. Reuther, and P. F. Smith. 1956. Effect of nitrogen on root development of Valencia orange trees. *J. Amer. Soc'y Horticultural Sci.* 70:237–44.

Fraser, A. I. 1962. The soil and roots as factors in tree stability. *Forestry* 35:117–27.

Fuller, D. L., and W. A. Meadows. 1987. Influence of production systems on root regeneration following transplanting of five woody ornamental species. *Proceedings Southern Nursery Association Research Conf.* 33:120–25.

Gilman, E. F. 1988. Root and top growth response of five woody ornamental species to Fabric Field-Grow containers, bed height and trickle irrigation. *Proceedings Southern Nursery Association Research Conf.* 34:148–51.

Funk, S. A. 1990. *Urban forestry notebook.* Seattle, WA: Center for Urban Horticulture, University of Washington. 89pp.

Garrett, J. H. 1991. *Plants of the metroplex III.* Dallas, TX: Lantana Press. 90pp.

Gates, G. July 1993. *Thuja plicata. American Nurseryman* 186.

Geer, G. B. 1980. *Big trees of southeastern Pennsylvania.* Philadelphia, PA: The Morris Arboretum of the University of Pennsylvania. 20pp.

Gerhold, H. D., W. N. Wandell, N. L. Lacasse, and R. D. Schein, eds. 1993. *Street tree fact sheets.* University Park, PA: Municipal Tree Restoration Program, The Pennsylvania State University. 245pp.

Gibbons, M. 1993. *Palms.* Secaucus, NJ: Chartwell Books. 80pp.

Gilman, E. F. 1988. Field grown hibiscus response to nitrogen rate. *Proceedings Florida State Horticultural Soc'y* 101:99–101.

Gilman, E. F., H. W. Beck, D. G. Watson, P. Fowler, and N. R. Morgan. 1996. Southern trees: An expert system for selecting trees. University of Florida, Cooperative Extension Service Monograph 3, CD-ROM.

Gilman, E. F., and S. P. Brown. 1992. Florida guide to environmetal landscapes. University of Florida. Circular 941. 32pp.

Gilman, E. F, G. W. Knox, C. Neal, and U. Yadav. 1994. Microirrigation affects growth and root distribution of trees in fabric containers. *HortTechnology* 4:43–45.

Gilman, E. F., I. A. Leone, and F. B. Flower. 1981. The adaptability of 19 woody species in vegetating a former sanitary landfill. *Forest Sci.* 27(1):13–18.

———. 1981. Influence of soil gas contamination on tree root growth. *Plant and Soil* 65:3–10.

———. 1981. Vertical root distribution of American Basswood in sanitary landfill soil. *Forest Sci.* 27(4):725–29.

Gilman, E. F., T. H. Yeager, R. Newton, and S. Davis. 1988. Response of field-grown trees to root pruning. *Proceedings Southern Nursery Association Research Conf.* 33:104–5.

Goode, J. E., and K. J. Hyrycz. 1964. The response of Laxton's Superb apple trees to different soil moisture conditions. *J. Horticultural Sci.* 39:254–76.

Grace, J. 1983. *Ornamental conifers.* Portland, OR: Timber Press. 220pp.

Green, T. L. 1992. Crabapples recommended for the midwest. Lisle, IL: The Morton Arboretum. 1p.

———. 1986. How to choose a crabapple for the home landscape. The Morton Arboretum Plant Information Bulletin No. 30–31. Lisle, IL: The Morton Arboretum. 8pp.

———. 1992. Outstanding crabapples. *Fine Gardening* 24:26–30.

Greene, W. December 1993. *Robinia psuedo-acacia,* Black Locust. *Arbor Age* 45.

Hackett, C. 1972. A method of applying nutrients locally to roots under controlled conditions, and some morphological effects of locally applied nitrate on the branching of wheat roots. *Australian J. Biological Sci.* 25:1169–80.

Halcomb, M. A. November 1993. More borers with plastic tree guards. *Landscape and Nursery Digest* 15.

Halfacre, R. G., and A. R. Shawcroft. 1989. *Landscape plants of the southeast.* Raleigh, NC: Sparks Press. 426pp.

Haller, J. M. January 1994. *Liquidambar styraciflua. Arbor Age* 27.

Harlow, W. M. 1957. *Trees of the eastern and central United States and Canada.* New York: Dover Publications. 288pp.

Harlow, W. M., and E. S. Harrar. 1969. *Textbook of dendrology.* New York: McGraw-Hill. 512pp.

Harrington, C. A., J. C. Brissette, and W. C. Carlson. 1989. Root system structure in planted and seeded loblolly and shortleaf pine. *Forest Sci.* 35(2):469–80.

Harris, J. R., N. L. Bassuk, and S. D. Day. 1994. Plant establishment and root growth research at the Urban Horticulture Institute. In *The landscape below ground,* G. W. Watson & D. Neely, eds. Proceedings of the International Workshop on Tree Root Development in Urban Soils. Savoy, IL: International Society of Arboriculture.

Hay, R. L., and F. W. Woods. 1968. Distribution of available carbohydrates in planted loblolly pine root systems. *Forest Sci.* 14(3):301–3.

Hayward, P. 1993. Conifers for connoisseurs. *Amer. Nursery* 178:82–87.

Head, G. C. 1967. Effects in seasonal changes in shoot growth on the amount of unsuberized root on apple and plum trees. *J. Horticultural Sci.* 42:169–80.

Heriteau, J. 1990. *The National Arboretum book of outstanding garden plants.* New York: Simon and Schuster. 292pp.

Herman, D. July 1993. *Quercus macrocarpa,* Bur Oak. *Arbor Age* 48–49.

Hodel, D. R. 1988. *Exceptional trees for Los Angeles.* Los Angeles, CA: California Arboretum Foundation. 80pp.

Hodgkins, E. J., and N. G. Nichols. 1977. Extent of main lateral roots in natural longleaf pine as related to position and age of the tree. *Forest Sci.* 23:161–66.

Hook, D., and C. Brown, 1973. Root adaptation and relative flood tolerance of five hardwood species. *Forest Sci.* 19:225–29.

Hopkins, R. M., and W. H. Patrick. 1964. Combined effect of oxygen content and soil compaction on root development. *Soil Sci.* 108:408–13.

Ingram, D. L., U. Yadav, and C. A. Neal. 1987. Production system comparisons for selected woody plants in Florida. *HortScience* 22(6):1285–87.

ISA pruning guidelines. 1995. Savoy, IL: International Society of Arboriculture.

Kansas Urban Forestry Council. 1992. *Preferred tree species for south central Kansas.* Manhatten, KS: State and Extension Forestry. 4pp.

Koslowski, T. T. 1986. Soil aeration and growth of forest trees (review article). *Scandinavian J. Forest Research* 1:113–23.

———. 1994. *Bleeding from branches of pruned trees.* HortScript No. 5. Berkeley, CA: University of California Cooperative Extension Service.

Laiche, A. J., W. W. Kilby, and J. P. Overcash. 1983. Root and shoot growth of field- and container-grown pecan nursery trees five years after transplanting. *HortScience* 18:328–29.

Leibee, G. L. June 15, 1993. Goldenrain trees beget bugs. *Amer. Nurseryman* 95–96.

Levin, I., R. Assaf, and B. Bravdo. 1980. Irrigation water status and nutrient uptake in an apple orchard. In *The mineral nutrition of fruit trees* 255–64, D. Atkinson, J. E. Jackson, R. O. Sharples, and W. M. Waller, eds. Borough Green, UK: Butterworths.

Lindgren, O., and G. Orlander. 1978. A study on root development and stability of 6- to 7-year old container plants. In *Proceedings of symposium on the root form of planted trees* 147–54, E. Van Eerden and J. M. Kinghorn, eds. British Columbia Ministry of Forests/Canadian Forest Service Report 8. Victoria, BC: Province of British Columbia, Ministry of Forests. 357pp.

Little, E. L. 1979. *Checklist of United States trees (native and naturalized).* Agricultural Handbook No. 541. Washington, DC: United States Department of Agriculture, Forest Service. 375pp.

Little, E. L., R. O. Woodbury, and F. H. Wadsworth. 1974. *Trees of Puerto Rico and the Virgin Islands.* Agricultural Handbook No. 449. Washington, DC: United States Department of Agriculture, Forest Service. 1024pp.

Little, S., and H. A Somes. 1964. *Root systems of direct-seeded and various planted loblolly, shortleaf and pitch pines.* United States Forest Service Research Paper NE-26. 13pp.

Long, D. 1961. Developing and maintaining street trees. *Proceedings International Shade Tree Conf.* 37:172.

Longwood Gardens. 1989. *Deciduous trees.* Kennett Square, PA: Author. 154pp.

Lyford, W. H. 1975. Rhizography of non-woody roots of trees in the forest floor. In *The development and function of roots* 179–96, J. G. Torrey and D. T. Clarkson, eds. New York: Academic Press.

Martin, E. C. 1983. *Landscape plants in design.* Westport, CT: AVI Publishing. 496pp.

Mason, G. F., D. S. Bhar, and R. J. Hilton. 1970. Root growth studies on mugho pine. *Canadian J. Botany* 48:43–47.

Maxwell, L. S., and B. M. Maxwell. 1986. *Florida trees and palms.* Tampa, FL: Lewis S. Maxwell, Publisher. 119pp.

McCain A. H. 1988. *Foliage and branch diseases of landscape trees.* Leaflet 2616. Oakland, CA: Division of Agriculture and Natural Resources. 4pp.

McLean, D. May 1989. *Coccoloba diversifolia. Nursery Digest* 57.

McPherson, E. G., and G. H. Graves. 1984. *Ornamental and shade trees for Utah.* Logan, UT: Cooperative Extension Service, Utah State University. 144pp.

Menninger, E. A. 1964. *Seaside plants of the world.* New York: Hearthside Press. 303pp.

Monselise, S. P. 1947. The growth of citrus roots and shoots under different cultural conditions. Palist. *J. Botany* 6:43–54.

Morgan, N. R. 1985. *Trees for Eugene.* Eugene, OR: City of Eugene, Oregon. 12pp.

Morris, L. A., and R. F. Lowery. 1988. Influence of site preparation on soil conditions affecting stand establishment and tree growth. *Southern J. Applied Forestry* 12(3):170–78.

Morton, J. F. 1977. *Exotic plants for house and garden.* New York: Golden Press. 160pp.

Muller, K. K., R. E. Broder, and W. Beittel. 1974. *Trees for Santa Barbara.* Santa Barbara, CA: Santa Barbara Botanic Garden. 248pp.

New Jersey Shade Tree Federation. 1990. *Trees for New Jersey streets.* New Brunswick, NJ: Author. 29pp.

New Pronouncing Dictionary of Plant Names. 1990. Chicago: Florists' Publishing. 64pp.

Nichols, T. J., and A. A. Alm. 1983. Root development of container-reared, nursery-grown, and naturally regenerated pine seedlings. *Canadian J. Forest Research* 13:239–45.

Phillips, R. 1978. *Trees of North America and Europe.* New York: Random House. 224pp.

Plant Selection Committee of the Ohio Nurseryman's Association. May 15, 1992. Made for the midwest. *Amer. Nursery* 39–49.

Platt, Rutherford. 1952. *American trees.* New York: Dodd, Mead. 256pp.

Poor, J. M., ed. 1984. *Plants that merit attention. Vol. 1—Trees.* Portland, OR: Timber Press.

Ranney, T. G., and K. A. Powell. 1991. *Recommended trees for urban landscapes: Proven performers for difficult sites.* North Carolina State University, Horticultural Information Leaflet 617. 6pp.

Ranney, T. G., and J. F. Walgenbach. 1993. Feeding preference of Japanese beetles for Taxa of birch, cherry and crabapple. *J. Envtl. Horticulture* 10(3):177–80.

Reich, P. B., R. O. Teskey, P. S. Johnson, and T. M. Hinckley. 1980. Periodic root and shoot growth in oak. *Forest Sci.* 26(4):590–98.

Reuther, W. 1973. Climate and citrus behavior. In *The citrus industry* 280–337, W. Reuther, ed. vol. 3. Berkeley, CA: University of California.

Roberts, J. 1976. A study of root distribution and growth in a *Pinus sylvestris* L. (Scots pine) plantation in East Anglia. *Plant and Soil* 44:607–21.

Rogers, D. J., and C. Rogers. 1991. *Woody ornamentals for Deep South gardens.* Pensacola, FL: University of West Florida Press. 296pp.

Rogers, W. S. 1934. Root studies III. Pears, gooseberry and black currant root systems under different soil fertility conditions with some observation on root stock and scion effect in pears. *J. Pomology & Horticultural Sci.* 11:1–18.

Rogers, W. S., and G. C. Head. 1969. Factors affecting the distribution and growth of roots of perennial woody species. In *Root growth,* W. J. Whittington, ed. London: Butterworths.

Russell, M. B. 1952. Soil aeration and plant growth. In *Soil aeration and plant growth.* New York: Academic Press.

Russell, R. S. 1977. Plant root systems: Their function and interaction with the soil. London: McGraw-Hill, Ltd. 298 pp.

Santamour, F. S., Jr. 1986. Wound compartmentalization in tree cultivars: Addendum. *J. Arboriculture* 12:227–32.

Savill, P. S. 1983. The effects of drainage and ploughing of surface water gleys on rooting and windthrow of Sitka spruce in Northern Ireland. *Forestry* 49:133–41.

Schnelle, M. A. September 15, 1992. *Evodia. Amer. Nursery* 130.

———. 1993. *Quercus muehlenbergii. Amer. Nursery* 178:198.

Schubert, T. H. 1979. *Trees for urban use in Puerto Rico and the Virgin Islands.* Washington, DC: USDA Forest Service. 91pp.

Settergren, C., and R. E. McDermott. 1979. *Trees of Missouri.* Columbia, MO: University of Missouri-Columbia Agricultural Experiment Station. 123pp.

Shigo, A. L. 1990. *Modern arboriculture: A systems approach to the care of trees and their associates.* Durham, NH: Shigo and Trees Associates. 424pp.

Siebenthaler, J. February 1994. *Tabebuia* spp. *Arbor Age* 34.

Simpson, B. J. 1988. *A field guide to Texas trees.* Austin, TX: Texas Monthly Press. 372pp.

Simpson, B. J., and B. W. Hipp. March 1, 1993. Maples of the southwest. *Amer. Nursery* 26–35.

Smiley, T. E., and B. R. Fraedrich. 1994. Treatments for compacted soil. Paper presented at the International Society of Arboriculture's Research Conference, Halifax, CA.

Smith-Fiola, D. March 15, 1993. Maskell Scale: A new threat to Japanese black pine. *Amer. Nurseryman* 95–96.

Storer, A. J. 1994. *Pitch canker disease.* HortScript No. 6. Berkeley, CA: University of California Cooperative Extension Service.

Stresau, F. B. 1986. *Florida, my Eden.* Fort Salerno, FL: Florida Classics Library. 299pp.

Struve, D. K., and B. C. Moser. 1984. Root system and root regeneration characteristics of pin oak and scarlet oak. *HortScience* 19:123–25.

Struve, D. K., T. D. Sydnor, and R. Rideout. 1989. Root system configuration affects transplanting of honeylocust and English oak. *J. Arboriculture* 15(8):129–34.

Sutton, R. F., and R. W. Tinus. 1983. Root and root system terminology. Monograph 24. *Forest Sci.* 29(4):1–137.

Svihra, P., and A. H. McCain. 1994. *Ash anthracnose.* HortScript No. 2. Berkeley, CA: University of California Cooperative Extension Service.

Swanson, T. C. November 1993. Free soil surveys. *Arbor Age* 46–47.

Taylor, A. 1974. Trickle irrigation experiments in the Goulburn Valley. *Victoria Horticultural Digest* 61:4–8.

University of Florida. 1987. *Ornamental horticulture plant identification manual.* Vol. 2. Gainesville, FL: Author. 210pp.

———. 1989. *Ornamental horticulture plant identification manual.* Vol. 1. Gainesville, FL: Author. 234pp.

Van Eerden and Kinghorn, eds. 1978. *Proceedings: Root form of planted trees.* British Columbia Minestry of Forests/Canadian Forest Service Report 8. Victoria, BC: Province of British Columbia, Ministry of Forests. 357pp.

Vines, Robert A. 1982. *Trees of North Texas.* Austin, TX: University of Texas Press. 466pp.

Voss, E. G., et al., eds. 1983. *International code of botanical nomenclature.* The Hague/Boston: Dr. W. Junk, Publishers. 472pp.

Wade, G. L., and G. E. Smith. 1985. Effect of root disturbance on establishment of container grown *Ilex crenata* 'Compacta' in the landscape. *Proceedings Southern Nurserymen's Association Research Conf.* 30:110–11.

Wandell, W. N. 1989. *Handbook of landscape tree cultivars.* Gladstone, IL: East Prairie Publishing. 313pp.

Wasowski, S. 1991. *Native Texas plants.* 2d ed. Houston, TX: Gulf Publishing. 406pp.

Watkins, J. V., and T. J. Sheehan. 1975. *Florida landscape plants.* Gainesville, FL: The University Presses of Florida. 420pp.

Watkins, J. V., and H. S. Wolfe. 1961. *Your Florida garden.* Gainesville, FL: University of Florida Press. 392pp.

Watson, G. W., ed. 1991. *Selecting and planting trees.* Lisle, IL: The Morton Arboretum. 25pp.

Watson, G. W., and S. Clark. 1993. Regeneration of girdling roots after removal. *J. Arboriculture* 19:278–80.

Watson, G. W., and G. Kupkowski. 1991. Soil moisture uptake by green ash trees after transplanting. *J. Envtl. Horticulture* 9:227–30.

Watson, G. W., and D. Neely. 1994. *The landscape below ground.* Proceedings of the International Workshop on Tree Root Development in Urban Soils. Savoy, IL: International Society of Arboriculture.

Weller, F. 1966. Horizontal distribution of absorbing roots and the utilization of fertilizers in apple orchards. *Erwobstbsobstbau* 8:181–84 [in German].

White, G. C., and R. I. C. Halloway. 1967. The influence of Simizine on a straw mulch on the establishment of apple trees in grassed down on cultivated soil. *J. Horticultural Sci.* 42:377–89.

Widrlechner, M. P., E. R. Hasselkus, D. E. Herman, J. K. Iles, J. C. Pair, E. T. Paparozzi, R. E. Schutzki, and D. K. Wildung. 1992. Performance of landscape plants from Yugoslavia in the north central United States. *J. Envtl. Horticulture* 10(4):192–98.

Williamson, J. F., ed. 1990. *Sunset western gardening book.* Menlo Park, CA: Lane Publishing. 592pp.

Wilson, B. 1967. Root growth around barriers. *Botany Gazette* 128(2):79–82.

Ziza, R. P., H. G. Halverson, and B. B. Stout. 1980. *Establishment and early growth of conifers on compacted soils in urban areas.* Forest Service Research Paper NE 451. Broomal, PA: United States Forest Service.

References

Abod, S. A., and A. D. Webster. 1990. Shoot and root pruning effects on the growth and water relations of young *Malus, Tilia,* and *Betula* transplants. *J. Amer. Soc'y Horticultural Sci.* 65:451–59.

American National Standards Institute. 1990. *American standard for nursery stock (ANSI Z60.1).* American Association of Nurserymen. Washington, D.C.

Appleton, B. L. December 1993. Less may be better. *Arbor Age* 8–12.

———. 1994. Use and misuse of tree trunk protective wraps, paints and guards. Technical Bulletin No. 1, Forestry Report R8–FR 44, Urban and Community Forestry Assistance Program, Southern Region, USDA Forest Service.

Arnold, M. A. 1996. Mechanical correction and chemical avoidance of circling roots differentially affect post-transplant root regeneration and field establishment of container-grown shumard oak. *J. Amer. Soc'y Horticultural Sci.* 121:258–63.

———. March 1995. Enthusiasm for excellence. *Amer. Nursery* 12–13.

Arnold, M. A., and D. K. Struve. 1993. Root distribution and mineral uptake of coarse-textured trees grown in cupric hydroxide-treated containers. *HortScience* 28(10):988–92.

Atkinson, D. 1980. The distribution and effectiveness of the roots of tree crops. *Horticultural Reviews* 2:424–90.

Atkinson, D., M. G. Johnson, D. Mattam, and E. R. Mercer. 1979. Effect of orchard soil management on the uptake of nitrogen by established apple trees. *J. Sci. Food & Agriculture* 27:253–57.

Bacon, G. J., and E. P. Bachelard. 1978. The influence of nursery conditioning treatments on some physiological responses of recently transplanted seedlings of *Pinus caribaea* Mor. var. *hondurensis* B&G. *Australian Forest Research* 8:171–83.

Barker, P. 1993. Root barriers for controlling damage to sidewalks. In *The landscape below ground,* G. W. Watson and D. Neely, eds. Proceedings of the International Workshop on Tree Root Development in Urban Soils. Savoy, IL: International Society of Arboriculture.

Bates, R. M., and A. X. Niemiera. 1996. Effect of transplanting on shoot water potential of bare-root Washington hawthorn and Norway Maple trees. *J. Envtl. Horticulture* 14:1–4.

Beard, J. B., and W. H. Danial. 1965. Effect of temperature and cutting on the growth of creeping bentgrass roots. *Agronomy J.* 57:249–50.

Beeson, R. C., Jr. 1995. Relationship of root regeneration potential of live oak to season and physiological state. Abstract presented at Int'l Soc'y Arboriculture Annual Meeting, Hilton Head, SC.

Beeson, R. C., Jr., and E. F. Gilman. 1992a. Diurnal water stress during landscape establishment of slash pine differs among three production methods. *J. Arboriculture* 18:281–87.

———. 1992b. Water stress and osmotic adjustment during post-digging acclimatization of *Quercus virginiana* produced in fabric containers. *J. Envtl. Horticulture* 10:208–14.

———. 1995. Irrigation and fertilizer placement affect root and canopy growth of trees produced in in-ground fabric containers. *J. Envtl. Horticulture* 13:133–36.

Benson, A. D., and K. P. Shephard. 1977. Effects of nursery practice on *Pinus radiata* seedling characteristics and field performance: II. Nursery root wrenching. *New Zealand J. Forest Research* 7:68–76.

Bernhardt, E., and T. J. Swiecki. 1993. The state of urban forestry in California–1992. California Department of Forestry and Fire Protection 19 pp.

Bevington, K. B., and W. S. Castle. 1985. Annual root growth pattern of young citrus trees in relation to shoot growth, soil temperature, and soil water content. *J. Amer. Soc'y Horticultural Sci.* 110(6):840–45.

Birdel, R., C. E. Whitcomb, and B. L. Appleton. 1983. Planting techniques for tree spade dug trees. *J. Arboriculture.* 9:282–84.

Biswell, H. H., and J. E. Weaver. 1933. Effect of frequent clipping on the development of roots and tops of grasses in prairie soil. *Ecology* 14:368–90.

Blessing, S. C., and M. N. Dana. 1988. Post-transplant root system expansion in *Juniperus chinensis* as influenced by production system, mechanical root disruption, and soil type. *J. Envtl. Horticulture* 5:155–58.

Bridwell, F. M. 1994. *Landscape Plants: Their Identification, Culture, and Use.* Albany, NY: Delmar Publishers.

Broschot, T. K., and H. M. Donselman. 1984a. Regrowth of severed palm roots. *J. Arboriculture* 10:238–40.

———. 1984b. Root regeneration in transplanted palms. *Principles* 28:90–91.

Burger, D. W., P. Svihra, and R. Harris. 1992. Treeshelter use in producing container-grown trees. *HortScience* 27:30–32.

Burns, R. M., and B. H. Honkala, technical coordinators. 1990a. *Silvics of North America: 1. Conifers.* Agriculture Handbook 654. Washington, DC: U.S. Department of Agriculture, Forest Service. vol 1. 675pp.

———. 1990b. *Silvics of North America: 2. Hardwoods.* Agriculture Handbook 654. Washington, DC: U.S. Department of Agriculture, Forest Service. vol. 2. 877pp.

Carlson, W. C., C. L. Preisig, and L. C. Promnitz. 1980. Comparative root system morphologies of seeded-in-place, bareroot and container-cultured seedlings after outplanting. *Canadian J. Forest Research* 10(3):250–56.

Coile, T. S. 1937. Distribution of forest tree roots in North Carolina Piedmont soils. *J. Forestry* 36:247–57.

Coker, E. G. 1959. Root development in grass and clean cultivation. *J. Horticultural Sci.* 34:111–21.

Conover, C. A., and J. N. Joiner. 1974. *Influence of fertilizer source and rate on growth of woody ornamentals transplanted to stress conditions.* Florida Agricultural Experimental Statistics J. Series 5469.

Cool, R. A. 1976. Trees spade vs. bare root planting. *J. Arboriculture.* 2:92–95.

Corley, W. L, L. L. Goodroad, and C. D. Robacker. March 1988. Initial growth response to landscape fertilizers. *Amer. Nursery* 1:117–18.

Costello, L. R., and J. L. Paul. 1975. Moisture relations in transplanting container plants. *HortScience* 10:371–72.

Coutts, M. P. 1983. Development of the structural root system of Sitka spruce. *Forestry* 56(1):1–16.

Coutts, M. P., and J. J. Philipson. 1976. The influence of mineral nutrition on the root development of trees. I. The growth of Sitka spruce with divided root systems. *J. Experimental Botany* 27:1102–11.

Craul, P. J. 1992. *Urban soil in landscape design.* New York: John Wiley & Sons. 396pp.

Cullen, P. W., A. K. Turner, and J. H. Wilson. 1972. The effect of irrigation depth on root growth of some pasture species. *Plant and Soil* 37:345–52.

Cutler, D. F., P. E. Gasson, and M. C. Farmer. 1990. The wind blown tree survey: Analyses of results. *Arboriculture J.* 14:265–86.

Dana, M. N., and S. C. Blessing. 1994. Post-transplant root growth and water relations of *Thuja occidentalis* from field and containers. In *The landscape below ground,* G. W. Watson and D. Neely, eds. Proceedings of the International Workshop on Tree Root Development in Urban Soils. Savoy, IL: International Society of Arboriculture.

Davey, P. H. 1930. *Fertilization of shade trees. Part II. Chemical fertilizers for conifers?* Bulletin 5. Keny, OH: Davey Tree Expert Co.

Day, S. D., and N. L. Bassuk. 1994. Reducing mechanical impedance of roots in compacted soils increases root and shoot growth. *HortScience* 29: 481.

Day, S. D., and N. L. Bassuk. 1994. A review of the effects of soil compaction and amelioration treatments on landscape trees. *J. Arboriculture* 20:9–17.

Day, S. D., N. L. Bassuk, and H. van Es. 1995. Effects of four compaction remediation methods for landscape trees on soil aeration, mechanical impedance, and tree establishment. *J. Envtl. Horticulture* 13:64–71.

Derr, J. F., and B. L. Appleton. 1989. Weed control with landscape fabrics. *J. Envtl. Horticulture* 7:129–33.

Doss, B. D., D. A. Ashley, and O. L. Bennett. 1962. Effect of soil moisture regime on root distribution on warm season forage grasses. *Agronomy J.* 54:569–72.

Edwards, J. H., R. R. Bruce, B. D. Horton, J. L. Chesness, and E. J. Wehunt. 1982. Soil cation and water distribution as affected by NH_4NO_3 applied through a drip irrigation system. *J. Amer. Soc'y Horticultural Sci.* 107(6):1142–48.

Eissenstat, D. M., and M. M. Caldwell. 1988. Seasonal timing of root growth in favorable microsites. *Ecology* 69(3):870–73.

Elam, P., and J. Baker. 1996. Fruit inhibition in *Quercus* species using growth regulators. *J. Arboriculture* 22:109–10.

Englert, J. M., K. Warren, L. H. Fuchigami, and T. H. H. Chen. 1993. Antidesiccant compounds improve the survival of bare-root deciduous nursery trees. *J. Amer. Soc'y Horticulture Sci.* 118:228–35.

Fales, S. L., and R. C. Wakefield. 1981. Effects of turfgrass on establishment of woody plants. *Agronomy J.* 73:605–10.

Florida Department of Agriculture. 1996. *Grades and standards for nursery stock.* Tallahassee, FL.

Ford, E. D., and J. D. Deans. 1977. Growth of a Sitka spruce plantation: Spatial distribution and seasonal fluctuations of lengths, weights and carbohydrate concentrations of fine roots. *Plant and Soil* 47:463–85.

Fordham, R. 1972. Observations on the growth of roots and shoots of tea (*Camellia sinensis* L.) in Southern Malawi. *J. Horticultural Sci.* 47:221–29.

Gale, M. R., and D. F. Grigal. 1987. Vertical root distribution of northern tree species in relation to successional status. *Canadian J. Forest Research* 17(8):829–34.

Gholz, H. L., L. C. Hendry, and W. P. Cropper. 1985. Organic matter dynamics of fine roots in plantations of slash pine (*Pinus elliottii*) in north Florida. *Canadian J. Forest Research* 16:529–38.

Gilliam, C. H., G. S. Cobb, and D. C. Fare. 1986. Effects of pruning on root and shoot growth of *Ilex crenata* 'Compacta'. *J. Envtl. Horticulture* 4:41–43.

Gilman, E. F. 1987. Response of hibiscus to soil applied nitrogen. *Proceedings Florida State Horticultural Soc'y* 100:356–57.

———. 1988a. Predicting root spread from trunk diameter and branch spread. *J. Arboriculture* 14(4):85–89.

———. 1988b. Root initiation in root-pruned hardwoods. *HortScience* 23:775.

———. 1988c. Tree root spread in relation to branch dripline and harvestable root ball. *HortScience* 23(2):351–53.

———. 1989a. Effects of injected and surface fertility on hibiscus growth in bare ground, mulch and turf. *Proceedings Florida State Horticultural Soc'y* 102:144–45.

———. 1989b. Plant form in relation to root spread. *J. Envtl. Horticulture* 7(3):88–90.

———. 1989c. Tree root depth relative to landfill tolerance. *HortScience* 24(5):857.

———. 1990a. Growth dynamics following planting of cultivars of *Juniperus chinensis. J. Amer. Soc'y Horticultural Sci.* 116(4):637–41.

———. 1990b. Tree root growth and development. I. Form, spread, depth and periodicity. *J. Envtl. Horticulture* 8(4):215–20.

———. 1990c. Tree root growth and development. II. Response to culture, management and planting. *J. Envtl. Horticulture* 8(4):220–27.

———. 1992. Effect of root pruning prior to transplanting on establishment of southern magnolia in the landscape. *J. Arboriculture* 18:197–200.

———. 1994. Establishing trees in the landscape. In *The landscape below ground,* G. W. Watson and D. Neely, eds. Proceedings of the International Workshop on Tree Root Development in Urban Soils. Savoy, IL: International Society of Arboriculture.

———. 1995a. Selecting trees from the nursery. In *Your Florida landscape,* R. J. Black and K. Ruppert, eds. University of Florida, Gainesville, FL.

———. 1995b. Root barriers affect root distribution. In *Trees and building sites.* G. W. Watson and D. Neely, eds. Proceedings of an International Conference, Savoy, IL: International Society of Arboriculture.

Gilman, E. F., and R. C. Beeson. 1995. Copper hydroxide affects root distribution of *Ilex cassine* in plastic containers. *Hort Technology* 5:48–49.

———. 1996. Production method affects tree establishment in the landscape. *J. Envtl. Horticulture* 14:81–87.

Gilman, E. F., R. C. Beeson, and R. J. Black. 1992. Comparing root balls on laurel oak transplanted from the wild with those of nursery and container grown trees. *J. Arboriculture* 18:124–29.

Gilman, E. F., and R. J. Black. 1997. Comparing live oak planted from containers with those planted B&B. *J. Envtl. Horticulture* in press.

Gilman, E. F., and M. E. Kane. 1990. Root growth of red maple following planting in two different soils from containers. *HortScience* 25(5):527–28.

Gilman, E. F., I. A. Leone, and F. B. Flower. 1987. Effect of soil compaction and oxygen content on vertical and horizontal root distribution. *J. Envtl. Horticulture* 5(1):33–36.

Gilman, E. F., and T. H. Yeager. 1987. Root pruning *Quercus virginiana* to promote a compact root system. *Proceedings Southern Nursery Association Research Conf.* 32:340–42.

———. 1990. Fertilizer type and nitrogen rate affects field-grown laurel oak and Japanese ligustrum. *Proceedings Florida State Horticultural Soc'y* 103:370–72.

Gilman, E. F, T. H. Yeager, and D. Weigle. 1990. Nitrogen leaching from cypress wood chips. *HortScience* 25(11):1388–91.

———. 1996. Fertilizer, irrigation, and root ball slicing affects burford holly growth after planting. *J. Envtl. Horticulture* 14:105–10.

Good, G. L., and T. E. Corell. 1982. Field trials indicate the benefits and limits of fall planting. *Amer. Nurseryman* 156(8):31–34.

Goode, J. E., K. H. Higgs, and K. J. Hyrycz. 1978. Trickle irrigation of apple trees and the effect of liquid feeding with NO_3^- and K^+ compared with normal manuring. *J. Horticultural Sci.* 53:307–16.

Grabosky, J., and N. Bassuk. 1995. A new urban soil to safely increase rooting volumes under sidewalks. *J. Arboriculture* 21(4):187–201.

Green, T. L., and G. W. Watson. 1989. Effects of turfgrass and mulch on the establishment and growth of bare-root sugar maple. *J. Arboriculture* 15:268–72.

Hamilton, J., and M. Drosdoff. 1946. The effect of cultivation, watering and time of fertilization on growth of transplanted one-year-old tung trees. *Proceedings Amer. Soc'y Horticultural Sci.* 47:161–68.

Harris, J. R., and N. Bassuk. 1994. Seasonal effects on transplantability of scarlet oak, green ash, Turkish hazelnut, and tree lilac. *J. Arboriculture* 20:310–17.

———. 1995. Effects of defoliation and antitranspirants on transplant response of scarlet oak, green ash and turkish filbert. *J. Arboriculture* 21:33–36.

Harris, J. R., and E. F. Gilman. 1991. Production system affects growth and root regeneration of leyland cypress, laurel oak and slash pine. *J. Arboriculture* 17:64–69.

———. 1993. Production method affects growth and post-transplant establishment of 'East Palatka' holly. *J. Amer. Soc'y Horticultural Sci.* 118(2):194–200.

Harris, J. R., P. R. Knight, and J. K. Fanelli. 1997. Fall transplanting improves establishment of balled-and-burlapped fringe tree (*Chionathus virginicus L.*). *J. Envtl. Horticulture* 15:in press.

Harris, R. W. 1992. *Arboriculture: Integrated management of landscape trees, shrubs and vines.* Englewood Cliffs, NJ: Prentice Hall.

Harris, R. W., J. P. Paul, and A. T. Leiser. 1977. Fertilizing woody plants. University of California Agricultural Science Leaflet 2958.

Hauer, R. J., W. Wang, and J. O. Dawson. 1993. Ice storm damage to urban trees. *J. Arboriculture* 19:187–95.

Henderson, J. C., and D. L. Hensley. 1992. Influence of field-grow fabric containers and various soil amendments on the growth of green ash. *J. Envtl. Horticulture* 10(4):218–21.

Hensley, D. L. 1994. Harvest method has no influence on the growth of transplanted green ash. *Amer. Nurseryman* 179(6):87–88.

Hensley, D. L., R. E. McNeil, and R. Sundheim. 1988. Management influences on growth of transplanted *Magnolia grandiflora*. *J. Arboriculture* 14:204–7.

Hermann, R. K. 1977. Growth and production of tree roots: A review. In *The below ground ecosystem: A synthesis of plant associated processes* 7–28, J. K. Marshall, ed. Range Science Service, Colorado State University, Range Science Department, No. 26.

Hild, A. L., and D. L. Morgan. 1993. Mulch effects on crown growth of five southwestern shrub species. *J. Envtl. Horticulture* 11:41–43.

Huguet, J. G. 1976. Influence of a localized irrigation on the rooting of young apple trees. *Annals Agronomy* 27:343–61.

Ingram, D. L. 1981. Characteristics of temperature fluctuations and woody plant growth in white poly bags and conventional containers. *HortScience* 16:762–63.

Ingram, D. L., R. J. Black, and C. R. Johnson. 1981. Effect of backfill composition and fertilization on establishment of container grown plants in the landscape. *Proceedings Florida State Horticultural Soc'y* 94:198–200.

Jensen, P., and S. Petterson. 1977. Effects of some internal and environmental factors on ion uptake efficiency in roots of pine seedlings. *Swedish Coniferous Forest Project Technical Report* 6:1–19.

Kelsey, P. 1995. Mulch induced chlorosis. Paper presented at International Soc'y of Arboriculture Annual Meeting, Hilton Head, SC.

Kirby, W. W., and G. F. Potter. 1956. Research on care of young trees in tung orchards. *Mississippi Farm Research* 19(3):8.

Kjelgren, R. 1994. Growth and water relations of Kentucky coffee tree in protective shelters during establishment. *HortScience* 29(7):777–80.

Kopinga, J. 1993. Aspects of damage to asphalt road pavings caused by tree roots. In *The landscape below ground,* G. W. Watson and D. Neely, eds. Proceedings of the International Workshop on Tree Root Development in Urban Soils. Savoy, IL: International Society of Arboriculture.

Kramer, P. J. 1983. *Water relations of plants.* New York: Academic Press.

Larson, M. M. 1975. Pruning northern red oak nursery seedlings: Effects on root regeneration and early growth. *Canadian J. Forest Research* 5:381–86.

Lichter, J., and P. Lindsey. 1993. Soil compaction and site construction: Assessment and case studies. In *The landscape below ground,* G. W. Watson and D. Neely, eds. Proceedings of the International Workshop on Tree Root Development in Urban Soils. Savoy, IL: International Society of Arboriculture.

Lichter, J. M., and L. R. Costello. 1994. An evaluation of volume excavation and soil core sampling techniques for measuring soil bulk density. *J. Arboriculture* 20(3):160–64.

Lindsey, P. 1994a. *The design of structural soil mixes for trees in urban areas. Part I. Growing points.* vol. 1. University of California at Davis.

———. 1994b. *The design of structural soil mixes for trees in urban areas. Part II. Growing points.* vol. 2. University of California at Davis.

Lindsey, P., and N. Bassuk. 1992. Redesigning the urban forest from the ground below: A new approach to specifying adequate soil volumes for street trees. *J. Arboriculture* 16:25–39.

Litzow, M., and H. Pellett. 1983a. Influence of mulch materials on growth of green ash. *J. Arboriculture* 9:7.

———. 1983b. Materials for potential use in sun-scald prevention. *J. Arboriculture* 9(2):35–38.

Lumis, G. P. August 15, 1990. Wire baskets: A further look. *Amer. Nursery* 128–31.

Lyr, H., and G. Hoffman. 1967. Growth rates and growth periodicity of tree roots. *International Review Forest Research* 2:181–236.

Magley, S. B., and D. K. Struve. 1983. Effects of three transplant methods on survival, growth, and root regeneration of caliper pin oaks. *J. Envtl. Horticulture* 1:59–62.

Marler, T. E., and F. S. Davies. 1987. Growth of bare-root and container grown 'Hamlin' orange trees in the field. *Proceedings Florida State Horticultural Soc'y* 100:89–93.

Mathene, N., and J. Clark. 1995. In *Trees and building sites,* G. W. Watson and D. Neely, eds. Proceedings of an International Conference. Savoy, IL: International Society of Arboriculture.

May, L. H., F. H. Chapman, and D. Aspinall. 1964. Quantitative studies of root development. I. The influence of nutrient concentration. *Australian J. Biological Sci.* 18:25–35.

McMinn, R. G. 1963. Characteristics of Douglas-fir root systems. *Canadian J. Botany* 41:105–22.

Meerow, A. W. 1992. *Betrock's guide to landscape palms.* Cooper City, FL: Betrock Information Systems. 153pp.

Messinger, A. S., and B. A. Hurby. 1990. Response of interveinally chlorotic red maple trees treated with Medicaps or by soil acidification. *J. Envtl. Horticulture* 8:5–9.

Milbocker, D. 1994. Producing trees in low-profile containers. In *The landscape below ground,* G. W. Watson and D. Neely, eds. Proceedings of the International Workshop on Tree Root Development in Urban Soils. Savoy, IL: International Society of Arboriculture.

Mitchell, P. D., and J. D. F. Black. 1968. Distribution of peach roots under pasture and cultivation. *Australian J. Experimental Agriculture & Animal Husbandry* 8:106–11.

Mullen, R. D. 1966. Root pruning of nursery stock. *Forestry Chronicles* 42:256–64.

Neely, D., E. B. Himelick, and W. R. Crowley. 1970. Fertilization of established trees: A report of field studies. *Illinois Natural History Survey Bulletin* 30(4):235–66.

Nelms, L. R., and L. A. Spomer. 1983. Water retention of container soil transplanted into ground beds. *HortScience* 18(6):863–66.

Newman, S. E., and M. W. Follett. 1987. Effects of three container designs on growth of *Quercus laurifolia* michx. *Proceedings Southern Nursery Association Annual Meeting* 32:128–31.

Niemiera, A. X., and R. D. Wright. 1982. Growth of *Ilex crenata* Thunb. 'Helleri' at different substrate nitrogen levels. *HortScience* 17(3):354–55.

Ovington, J. D., and G. Murray. 1968. Seasonal periodicity of root growth of birch trees. In *Methods of productivity studies in root systems and rhizosphere organisms* 146–54, M. S. Ghilarov, V. A. Kovda, L. N. Novicchkova-Juanova, L. E. Rodin, and V. M. Sveshnikova, eds. Leningrad Publishing House "NAUKA": USSR Academy of Sciences.

Owen, N. P., C. S. Sadof, and M. Raupp. 1991. The effect of plastic tree wrap on borer incidence in dogwood. *J. Arboriculture* 17 (2):29–31.

Paine, T. D., C. C. Hanlon, D. R. Pittenger, D. M. Ferrin, and M. K. Malinoski. 1992. Consequences of water and nitrogen management on growth and aesthetic quality of drought-tolerant woody landscape plants. *J. Envtl. Horticulture* 10:94–99.

Pan, E., and N. Bassuk. 1985. Effects of soil type and compaction on growth of *Ailanthus* seedlings. *J. Envtl. Horticulture* 3:158–62.

Patterson, J. C. 1976. Soil compaction and its effects on urban vegetation. In *Better trees for metropolitan landscapes* 91–102, F. Santamour, J. D. Gerhold, and S. Little, eds. Symposium of Proceedings USDA Forest Service General Technical Report NE-22.

Patterson, J. C., and C. J. Bates. 1994. Long-term, light-weight aggregate performance as soil aggregates. In *The landscape below ground,* G. W. Watson, and D. Neely, eds. Proceedings of an International Workshop on tree root development in urban soils. Savoy, IL. International Society of Arboriculture.

Perry, E., and G. W. Hickman. 1992. Growth response of newly planted valley oak trees to supplemental fertilizers. *J. Envtl. Horticulture* 10(4):242–44.

Perry, T. O. 1982. The ecology of tree roots and the practical significance thereof. *J. Arboriculture* 9:197–211.

Ponder, H. G., and A. L. Kenworthy. 1976. Trickle irrigation of shade trees growing in the nursery. *J. Amer. Soc'y Horticultural Sci.* 101:104–07.

Preisig, C. L., W. C. Carlson, and L. C. Promnitz. 1979. Comparative root morphologies of seeded-in-place, bareroot and containerized Douglas-fir seedlings after outplanting. *Canadian J. Forest Research* 9:399–405.

Privett, D. W., and R. L. Hummel. 1992. Root and shoot growth of 'Coral Beauty' Cotoneaster and leyland cypress produced in porous and nonporous containers. *J. Envtl. Horticulture* 10:133–36.

Ranney, T. G., N. L. Bassuk, and T. H. Whitlow. 1989. Effect of transplanting practices on growth and water relations of

'Colt' cherry trees during reestablishment. *J. Envtl. Horticulture* 7(1):41–45.

Reynolds, E. R. C. 1974. The distribution pattern of fine roots of trees. In *International symposium on the ecology and physiology of root growth* 101–12, G. Hoffman, ed. Berlin: Akademie-Verlag GmbH.

Rhichardson, H. H. 1956. Studies of root growth of Acer saccharinum. IV. The effect of different shoot and root temperatures on root growth. *Koninklijke Nederlandse Akadem,* i.e., *Van Wetenschappen, Verhandelingen Afdeling Natuurkunde Tweede Reeks* (abstract) 59:428.

Rogers, W. S., and M. C. Vyvyan. 1934. Root studies V. Root stock and soil effect on apple root systems. *J. Pomology & Horticultural Sci.* 12:110–50.

Ruter, J. M. 1993a. Growth and landscape performance of three landscape plants produced in conventional and pot-in-pot production systems. *J. Envtl. Horticulture* 11(3):124–27.

———. 1993b. Growth of three species produced in a pot-in-pot production system. *Proceedings Southern Nurserymen's Association Research Conf.* 38:100–2.

Schulte, J. R., and C. E. Whitcomb. 1975. Effects of soil amendments and fertilizer levels on the establishment of silver maple. *J. Arboriculture* 1:192–95.

Shoup, S., R. Reavis, and C. E. Whitcomb. 1981. Effects of pruning and fertilizers on establishment of bareroot deciduous trees. *J. Arboriculture* 7(6):155–57.

Singh, K. P., and S. K. Srivastava. 1985. Seasonal variations in the spatial distribution of root tips in teak (*Tectonia grandis* Linn. f.) plantations in the Varanasi Forest Division, India. *Plant and Soil* 84:93–104.

Skroch, W. A., M. A. Powell, T. E. Bilderback, and P. A. Henry. 1992. Mulches: Durability, aesthetic value, weed control and temperature. *J. Envtl. Horticulture* 10:43–45.

Smalley, T., and C. B. Wood. 1995. Effect of backfill amendment on growth of red maple. *J. Arboriculture* 21:247–50.

Smith, P. F. 1965. Effect of nitrogen source and placement on the root development of Valencia orange trees. *Proceedings Florida State Horticultural Soc'y* 78:55–59.

Somerville, A. 1979. Root anchorage and root morphology of *Pinus radiata* on a range of ripping treatments. *New Zealand J. Forestry Sci.* 9(3):294–315.

Stout, B. B. 1956. *Studies of the root systems of deciduous trees.* Black Forest Bulletin 15. Cornwall-on-the-Hudson, NY: Harvard University. 45pp.

Struve, D. K. 1993. Effect of copper-treated containers on transplant survival and regrowth of four tree species. *J. Envtl. Horticulture* 11(4):196–99.

Struve, D. K. 1994. Street tree establishment. In D. Neely and G. Watson (ed). *The landscape below ground,* G. W. Watson and D. Neely, eds. Proceedings of the International Workshop on Tree Root Development in Urban Soils. Savoy, IL: International Society of Arboriculture.

Struve, D. K., M. A. Arnold, R. C. Beeson, Jr., J. M. Ruter, S. Svenson, and W. T. Witte. 1994. The copper connection. *Amer. Nurseryman* 180(4):52–61.

Swanson, B. T. 1977. Transplanting woody plants effectively. *Amer. Nurseryman* 146(8):7–8.

van de Werken, H. 1981. Fertilization and other factors enhancing the growth rate of young shade trees. *J. Arboriculture* 7:33–37.

Wagar, J. A. 1985. Reducing surface routing of trees with control planters and wells. *J. Arboriculture* 11:165–71.

Wagar, J. A., and A. L. Franklin. 1994. Sidewalk effects on soil moisture and temperature. *J. Arboriculture* 20(4):237–38.

Warren. 1993. Growth and nutrient concentration in flowering dogwood after nitrogen fertilization and dormant root pruning. *J. Arboriculture* 19:57–63.

Watson, G. W. 1985. Tree size affects root regeneration and top growth after transplanting. *J. Arboriculture* 11:37–40.

———. 1988. Organic mulch and grass competition influence tree root development. *J. Arboriculture* 14(8):200–3.

———. 1994. Root growth response to fertilizers. *J. Arboriculture* 20:4–8.

———. 1994. Dissecting trees planted in wire baskets. Abstract. Presented at International Society of Arboriculture Annual Conference, Savoy, IL.

Watson, G. W., S. Clark, and K. Johnson. 1990. Formation of girdling roots. *J. Arboriculture* 16(8):197–202.

Watson, G. W., and E. B. Himelick. 1982a. Root distribution of nursery trees and its relationship to transplanting success. *J. Arboriculture* 8:225–29.

———. 1982b. Seasonal variation in root growth regeneration of transplanted trees. *J. Arboriculture* 8:305–10.

Watson, G. W., E. B. Himelick, and T. Smiley. 1986. Twig growth of eight species of shade trees following transplanting. *J. Arboriculture* 12(10): 241–45.

Watson, G. W., G. Kupkowski, and K. G. von der Heide-Spravka. 1992. The effect of backfill soil texture and planting hole shape on root regeneration of transplanted green ash. *J. Arboriculture* 18:124–29.

Watson, G. W., and T. D. Sydnor. 1987. The effect of root pruning on the root system of nursery trees. *J. Arboriculture* 13:126–30.

Whitcomb, C. E. 1985. *Know it and grow it II.* Stillwater, OK: Lacebark Publications. 740pp.

———. 1986a. Fabric Field-Grow containers enhance root growth. *Amer. Nursery* 163:49–52.

———. 1986b. Landscape plant production, establishment and maintenance. Stillwater, OK: Lacebark Publications. 680pp.

Whitcomb, C. E., and J. D. Williams. 1985. Stair-step container for improved root growth. *HortScience* 20:66–67.

Wilson, W. F. 1964. *Structure and growth of woody roots of Acer rubrum L.* Harvard Forest Paper No. 11. Petersham, MA: Harvard University.

Wright, R. D., and E. B. Hale. 1983. Growth of three shade tree genera as influenced by irrigation and nitrogen rates. *J. Envtl. Horticulture* 1:5–6.

Yeager, T. H., R. Wright, D. Fare, C. Gilliam, J. Johnson, T. Bilderback, and R. Zonday. 1993. Six state survey of container nursery nitrate nitrogen runoff. *J. Envtl. Horticulture* 11:206–8.

Yeager, T. H., D. L. Ingram, and C. A. Larsen. 1989. Response of Ligustrum and Azalea to surface and growth medium-incorporated fertilizer applications. *Proceedings Florida State Horticultural Soc'y* 102:269–71.

Yeager, T. H., and R. D. Wright. 1981. Influence of nitrogen and phosphorus on shoot:root ratio of *Ilex crenata* Thunb. 'Helleri'. *HortScience* 16:564–65.

Subject Index

Index to Scientific and Common Names